Engineering Mechanics: DYNAMICS

Engineering Mechanics:

R.C. Hibbeler

DYNAMICS

SECOND EDITION

Macmillan Publishing Co., Inc.
New York
Collier Macmillan Publishers
London

Printed in the United States of America

Macmillan Publishing Co., Inc.
866 Third Avenue, New York, New York 10022

Collier Macmillan Canada, Ltd.

Library of Congress Cataloging in Publication Data

Hibbeler, R C
Engineering mechanics, dynamics.

Includes index.
1. Dynamics. I. Title.
TA352.H5 1978 620.1'04 77-23154
ISBN 0-02-354040-0

Printing: 4 5 6 7 8 Year: 1 2 3 4

Preface

The purpose of this book is to provide the student with a clear and thorough presentation of the theory and application of the principles of engineering mechanics. Emphasis is placed on developing the student's ability to analyze problems—a most important skill for any engineer. Furthermore, the Système International or SI system of units is used for numerical work since this system is intended in time to become the worldwide standard for measurement.

The contents of each chapter are organized into well-defined sections. Selected groups of sections contain the development and explanation of specific topics, illustrative example problems, and a set of problems designed to test the student's ability to apply the theory. Many of the problems depict realistic situations encountered in engineering practice. It is hoped that this realism will both stimulate the student's interest in engineering mechanics and provide a means for developing the skill to reduce any such problem from its physical description to a model or symbolic representation to which the principles of mechanics may be applied. In any set, the problems are arranged in order of increasing difficulty. Furthermore, the answers to all but every fourth problem, which is indicated by an asterisk, are listed in the back of the book. SI units are used in all the numerical examples and problems; however, for the convenience of some instructors, every fifth problem is stated *twice,* once in SI units and again in FPS units.

Besides a change from FPS to SI units and the addition of many new problems, this book differs from the author's first edition: *Engineering Mechanics: Dynamics* in many respects. Most of the text material has been completely rewritten so that topics within each section are categorized into subgroups, defined by bold face titles. The purpose of this is to present a structured method for introducing each new definition or concept and to provide a convenient means for later reference or review of the material.

Another unique feature used throughout this book is the "Procedure for Analysis." This guide to problem solving, which was initially presented in Sec. 9-3 of the first edition of *Engineering Mechanics: Statics*, is essentially a step-by-step set of instructions which provide the student with a logical and orderly method to follow when applying the theory. As in the first edition, the example problems are solved using this outlined method for solution in order to clarify application of the steps.

Since mathematics provides a systematic means of applying the principles of mechanics, the student is expected to have prior knowledge of algebra, geometry, trigonometry, and some calculus. Vector analysis is introduced at points where it is most applicable. Its use often provides a convenient means for presenting concise derivations of the theory, and it makes possible a simple and systematic solution of many complicated three-dimensional problems. Occasionally, the example problems are solved using several different methods of analysis so that the student develops the ability to use mathematics as a tool, whereby the solution of any problem may be carried out in the most direct and effective manner.

The contents of this book are presented in 11 chapters.* In particular, the kinematics of a particle is discussed in Chapter 12,† followed by a discussion of particle kinetics in Chapter 13 (equations of motion), Chapter 14 (work and energy), and Chapter 15 (impulse and momentum). A similar sequence of presentation is given for the planar motion of a rigid body: Chapter 16 (planar kinematics), Chapter 17 (equations of motion), Chapter 18 (work and energy), and Chapter 19 (impulse and momentum). If desired, it is possible to cover Chapters 12 through 19 in the following order with no loss in continuity: Chapters 12 and 16 (kinematics), Chapters 13 and 17 (equations of motion), Chapters 14 and 18 (work and energy), and Chapters 15 and 19 (impulse and momentum).

Time permitting, some of the material involving spatial rigid-body motion may be included in the course. The kinematics and kinetics of this motion are discussed in Chapters 20 and 21, respectively. Chapter 22 (vibrations) may be included if the student has the necessary mathematical background. Sections of the book which are considered to be beyond the scope of the basic dynamics course are indicated by a star and may be omitted. Note, however, that this more advanced material provides a suitable reference for basic principles when it is covered in more advanced courses.

The author has endeavored to write this book so that it will appeal to both the student and the instructor. Many people helped in its development. I wish to acknowledge the valuable suggestions and comments

*A discussion of units and a review of vector analysis is given in Appendixes A and B, respectively.

†The first 11 chapters of this sequence form the contents of *Engineering Mechanics: Statics*.

made by M. H. Clayton, North Carolina State University; D. I. Cook, University of Nebraska; D. Krajcinovic, University of Illinois at Chicago Circle; W. Lee, United States Naval Academy; G. Mavrigian, Youngstown State University; F. Panlilio, Union College; H. A. Scarton, Rensselaer Polytechnic Institute; W. C. Van Buskirk, Tulane University; and P. K. Mallick, Illinois Institute of Technology. Many thanks are also extended to all of the author's students and to the professionals who have provided suggestions and comments. Although the list is too long to mention, I hope that others who have given help will accept this anonymous recognition. Lastly, I should like to acknowledge the able assistance of my wife, Cornelie, who has furnished a great deal of her time and energy in helping to prepare the manuscript for publication.

Russell C. Hibbeler

Contents

12
Kinematics of a Particle

12–1. Introductory Remarks, Kinematics of Particles

Engineering mechanics consists of a study of both statics and dynamics. *Statics* deals with the equilibrium of bodies at rest or moving with constant velocity, whereas *dynamics* deals with bodies having accelerated motion. In general, dynamics is more complicated than statics, since the forces acting on the body must be related to the body's acceleration. The subject of dynamics is usually divided into two parts: (1) *kinematics* is concerned with the geometrical aspects of motion, and (2) *kinetics* is concerned with the analysis of the forces causing the motion. For simplicity in presenting the theory of both kinematics and kinetics, particle dynamics will be discussed first, followed by topics in rigid-body dynamics.

Particle Motion. Recall that a *particle* is defined as a small portion of matter such that its dimension or size is of no consequence in the analysis of a physical problem. In most problems encountered, one is interested in bodies of a finite size, such as rockets, projectiles, or vehicles. Such objects may be considered as particles, provided motion of the body is characterized by motion of its mass center and any rotation of the body can be neglected.

In general, the "kinematics" of a particle is characterized by specifying the particle's displacement, velocity, and acceleration. This chapter begins with the study of the *absolute motion* of a particle, which is motion measured with respect to a *fixed coordinate system.* In this regard, motion along a straight line will be studied before introducing the more general motion along a curved path. Afterwards, the *relative motion* between two particles will be considered, using a translating coordinate system.

12–2. Rectilinear Velocity and Acceleration of a Particle

The simplest motion of a particle is motion occurring along a straight-line path, called *rectilinear motion.*

Position. Consider the particle at point P shown in Fig. 12–1. The coordinate s which is measured from the fixed origin O, is used to define the *position* of the particle at any given instant. If s is positive, the particle is located to the right of the origin; if s is negative, the particle is located to the left. Ordinarily, this position is measured in metres (m).

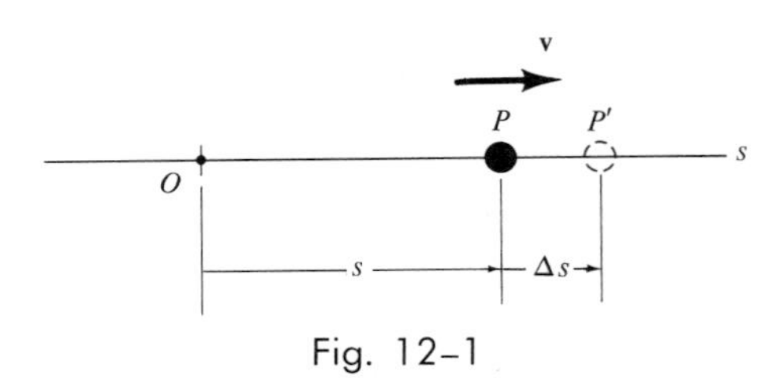

Fig. 12–1

Displacement. The *displacement* of the particle is defined as the *change* in its *position.* This is represented by the symbol Δs. When the particle's final position P' is to the right of its initial position P, Δs is positive, Fig. 12–1; when the displacement is to the left, Δs is negative.

The displacement of a particle must be distinguished from the distance the particle travels. Specifically, the *distance* traveled is defined as the *total length of path* traversed by the particle—*which is always positive.*

Velocity. Consider now that the particle moves through a positive displacement Δs from P to P' during the time interval Δt, Fig. 12–1. The *average velocity* of the particle during this time interval is defined as

$$v_{avg} = \frac{\Delta s}{\Delta t} \tag{12–1}$$

By taking smaller and smaller values of Δt, and consequently smaller and smaller values of Δs, we obtain the *instantaneous velocity,* defined as

$$v = \lim_{\Delta t \to 0} \frac{\Delta s}{\Delta t}$$

or

$$v = \frac{ds}{dt} \tag{12–2}$$

For both the average velocity and instantaneous velocity, the *direction* is either positive or negative depending upon whether the displacement is positive or negative. For example, if the particle is moving to the right as shown in Fig. 12–1, the velocity is positive. The *magnitude* of the velocity is known as the *speed.* If the displacement is expressed in metres (m) and the time in seconds (s), the speed is expressed as m/s.

Occasionally the term "average speed" is used. The *average speed,* $(v_{sp})_{avg}$, is defined as the total distance of the path traveled by a particle, s_T, divided by the elapsed time Δt, i.e.,

$$(v_{sp})_{avg} = \frac{s_T}{\Delta t} \tag{12–3}$$

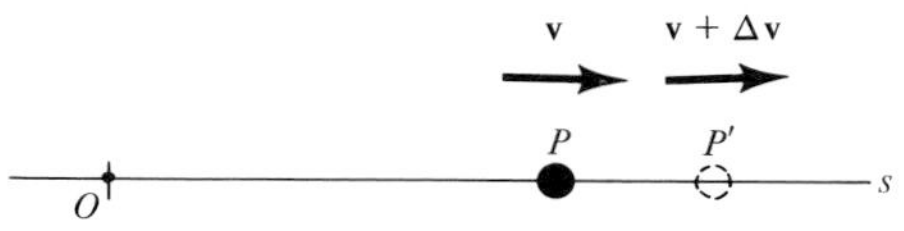

Fig. 12–2

Acceleration. Provided the instantaneous velocities for the particle are known at the two points P and P', the *average acceleration* for the particle during the time interval Δt is defined as

$$a_{avg} = \frac{\Delta v}{\Delta t} \tag{12–4}$$

where Δv represents the difference in the velocities during the time interval Δt, Fig. 12–2.

The *instantaneous acceleration* at time t is found by taking smaller and smaller values of Δt, and corresponding smaller and smaller values of Δv, so that

$$a = \lim_{\Delta t \to 0} \left(\frac{\Delta v}{\Delta t} \right)$$

or

$$a = \frac{dv}{dt} \tag{12–5}$$

Taking the second time derivative of Eq. 12–2, we can also write

$$a = \frac{d^2 s}{dt^2} \tag{12–6}$$

Both the average and instantaneous acceleration can be either positive or negative. In particular, when the particle is *slowing down,* the velocity change is negative and the particle is said to be *decelerating.* Also, note that when *the velocity is constant, the acceleration is zero.* Units commonly used to express the magnitude of acceleration are m/s^2.

A differential relation involving the displacement, velocity, and acceleration along the path may be obtained by solving for the time differential dt in Eqs. 12–2 and 12–5 and equating, i.e.,

$$dt = \frac{ds}{v} = \frac{dv}{a}$$

so that

$$a\,ds = v\,dv \tag{12–7}$$

Constant Acceleration. When the acceleration is constant, $a = a_c$, each of the three kinematic equations $a = dv/dt$, $v = ds/dt$, and $a\,ds = v\,dv$ may be integrated to obtain formulas that relate a_c, v, s, and t.

To determine the *velocity as a function of time,* integrate $a = dv/dt = a_c$, assuming that initially $v = v_1$ at $t = 0$.

$$\int_{v_1}^{v} dv = \int_0^t a_c\,dt$$

$$v - v_1 = a_c(t - 0)$$

$$v = v_1 + a_c t \tag{12–8}$$

To determine the *displacement as a function of time,* integrate $v = ds/dt = v_1 + a_c t$, assuming that initially $s = s_1$ at $t = 0$.

$$\int_{s_1}^{s} ds = \int_0^t (v_1 + a_c t)\,dt$$

$$s - s_1 = v_1(t - 0) + a_c(\tfrac{1}{2}t^2 - 0)$$

$$s = s_1 + v_1 t + \tfrac{1}{2}a_c t^2 \tag{12–9}$$

To determine the *velocity as a function of displacement,* either solve for t in Eq. 12–8 and substitute into Eq. 12–9, or integrate $v\,dv = a_c\,ds$, assuming that initially $v = v_1$ when $s = s_1$.

$$\int_{v_1}^{v} v\,dv = \int_{s_1}^{s} a_c\,ds$$

$$\tfrac{1}{2}v^2 - \tfrac{1}{2}v_1^2 = a_c(s - s_1)$$

$$v^2 = v_1^2 + 2a_c(s - s_1) \tag{12–10}$$

The magnitudes and signs of s_1, v_1, and a_c, used in these equations, are determined from the chosen origin and positive direction of the s axis.

It is important to remember that the above equations are useful *only when the acceleration is constant.* A common example of constant accelerated motion occurs when a body falls freely toward the earth. If air resistance is neglected and the distance of fall is short, then the constant *downward* acceleration of the body is approximately 9.81 m/s^2.*

*The proof is given in Example 13–3.

PROCEDURE FOR ANALYSIS

When a functional relationship between *any two* of the quantities a, v, s, and t is known, the functional relations describing the other kinematic quantities can be obtained by either the proper differentiation or integration* of the equations $a = dv/dt$, $v = ds/dt$ or $a\,ds = v\,dv$. In attempting to solve a problem, it should be realized that each of these equations relates *three quantities.* Hence when a quantity is known as a function of another quantity, the third quantity is obtained *by choosing the kinematic equation which relates all three.* For example, suppose that the acceleration is known as a function of displacement, $a = f(s)$. The velocity can be determined from $a\,ds = v\,dv$ by substituting $f(s)$ for a, since $f(s)\,ds = v\,dv$ may be integrated.† The velocity *cannot* be obtained by using $a = dv/dt$, since a is not a function of time, i.e., $f(s)\,dt = dv$ *cannot* be integrated. Proceeding on this basis, four common types of problems which are often encountered, and their method for solution, are given as follows:

1. *Acceleration given as a function of time,* $a = f(t)$. To find the velocity as a function of time, substitute into $a = dv/dt$, which yields $dv = f(t)\,dt$, and integrate to obtain $v = h(t)$. The displacement as a function of time is obtained by substituting for v into $v = ds/dt$, which gives $ds = h(t)\,dt$. Integration yields $s = g(t)$.
2. *Acceleration given as a function of velocity,* $a = f(v)$. To find the velocity as a function of time, substitute into $a = dv/dt$, which yields $dv = f(v)\,dt$ or $dv/f(v) = dt$, and integrate to obtain $v = h(t)$. The displacement as a function of time is obtained by substituting for v into $v = ds/dt$, which gives $ds = h(t)\,dt$. Integration yields $s = g(t)$.
3. *Acceleration given as a function of displacement,* $a = f(s)$. To find the velocity as a function of displacement, substitute into $a\,ds = v\,dv$, which yields $f(s)\,ds = v\,dv$, and integrate to obtain $v = h(s)$. The displacement as a function of time is obtained by substituting for v into $v = ds/dt$, which gives $h(s) = ds/dt$ or $ds/h(s) = dt$. Integration yields $s = g(t)$.
4. *Acceleration is constant,* $a = a_c$. Rather than integrating, use one of the appropriate derived equations, 12–8, 12–9, or 12–10.

*Some standard differentiation and integration formulas are given in Appendix C.

†The position s_1 and velocity v_1 must be known at a given instant in order to evaluate either the constant of integration if an indefinite integral is used, or the limits of integration if a definite integral is used.

Example 12–1

A small projectile is fired vertically downward into a fluid medium with an initial velocity of 60 m/s. If fluid resistance causes a deceleration of the projectile which is equal to $a = (-0.4v^3)\ \text{m/s}^2$, where v is measured in m/s, determine both the velocity v and position s four seconds after the projectile is fired.

Solution

Since a is given as a function of velocity, $a = (-0.4v^3)\ \text{m/s}^2$, to obtain velocity v as a function of time it is necessary to use $a = dv/dt$, since this equation relates v, a, and t. (Why not use Eq. 12–8, $v = v_1 + a_c t$?) If the downward direction is assumed positive, then integrating, with the initial condition that $v = 60$ m/s at $t = 0$, yields*

$$(+\downarrow) \qquad a = \frac{dv}{dt} = -0.4v^3$$

$$\int_{60}^{v} \frac{dv}{-0.4v^3} = \int_0^t dt$$

$$\frac{1}{0.8}\frac{1}{v^2}\Bigg|_{60}^{v} = t$$

$$\frac{1}{0.8}\left[\frac{1}{v^2} - \frac{1}{(60)^2}\right] = t$$

$$v = \left\{\left[\frac{1}{(60)^2} + 0.8t\right]^{-1/2}\right\}\ \text{m/s}$$

Here the positive root is taken, since the projectile is moving downward. When $t = 4$ s,

$$v = 0.559\ \text{m/s} \qquad \textit{Ans.}$$

Knowing the velocity as a function of time, the position s as a function of time is obtained from $v = ds/dt$, since this equation relates s, v, and t. Using the initial condition $s = 0$ at $t = 0$, we have

$$(+\downarrow) \qquad v = \frac{ds}{dt} = \left[\frac{1}{(60)^2} + 0.8t\right]^{-1/2}$$

$$\int_0^s ds = \int_0^t \left[\frac{1}{(60)^2} + 0.8t\right]^{-1/2} dt$$

$$s = \frac{2}{0.8}\left[\frac{1}{(60)^2} + 0.8t\right]^{1/2}\Bigg|_0^t$$

*The *same result* is obtained by evaluating a constant of integration rather than using definite limits on the integral. For example, integrating $dt = dv/-0.4v^3$ yields $t = 1/0.8(1/v^2) + C$. Using the condition that at $t = 0$, $v = 60$ m/s, the constant of integration is $C = -1/0.8[1/(60)^2]$.

$$s = \frac{1}{0.4}\left\{\left[\frac{1}{(60)^2} + 0.8t\right]^{1/2} - \frac{1}{60}\right\} \text{m}$$

When $t = 4$ s,

$$s = 4.43 \text{ m} \qquad \textit{Ans.}$$

Example 12–2

A boy tosses a ball in the vertical direction off the side of a cliff, as shown in Fig. 12–3. If the initial velocity of the ball is 15 m/s upward, and the ball is released 40 m from the bottom of the cliff, determine (a) the maximum height s_B reached by the ball and (b) the speed of the ball just before it hits the ground. During the entire time the ball is in motion, it is subjected to a constant downward acceleration of 9.81 m/s² due to gravity. Neglect the effect of air resistance.

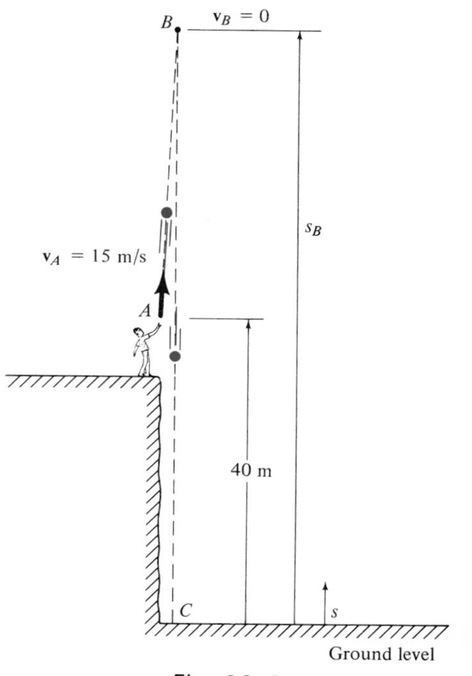

Fig. 12–3

Solution

Part (a). The coordinate axis for position $s = 0$ is taken at the base of the cliff as shown in the figure. At the maximum height s_B, the velocity $v_B = 0$. Furthermore, the ball is thrown from an initial height of $s_A = +40$ m. Since the ball is thrown *upward* at $t = 0$, it is subjected to a velocity of $v_A = +15$ m/s (positive since it is in the same direction as positive displacement). For the entire motion, the acceleration is *constant* such that $a_c = -9.81$ m/s² (negative since it acts in a direction *opposite* to positive velocity or positive displacement). Since a_c is *constant,* throughout the entire motion, the displacement may be related to velocity at points A and B using Eq. 12–10, i.e.,

$$(+\uparrow) \qquad v_B^2 = v_A^2 + 2a_c(s_B - s_A)$$
$$0 = (15)^2 + 2(-9.81)(s_B - 40)$$

so that

$$s_B = 51.5 \text{ m} \qquad \textit{Ans.}$$

Part (b). To obtain the velocity v_C of the ball just before it hits the ground, Eq. 12–10 can be applied between points B and C, Fig. 12–3,

$$(+\uparrow) \qquad v_C^2 = v_B^2 + 2a_c(s_C - s_B)$$
$$= 0 + 2(-9.81)(0 - 51.5)$$
$$v_C = -31.8 \text{ m/s} \qquad \textit{Ans.}$$

The negative root was chosen since the ball is moving *downward.*

Similarly, Eq. 12–10 may also be applied between points A and C, i.e.,

$$(+\uparrow) \qquad v_C^2 = v_A^2 + 2a_c(s_C - s_A)$$
$$= 15^2 + 2(-9.81)(0 - 40)$$
$$v_C = -31.8 \text{ m/s} \qquad \textit{Ans.}$$

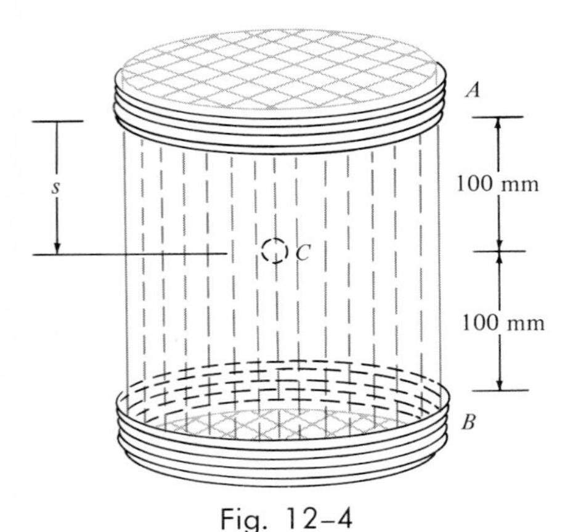

Fig. 12–4

Example 12–3

A metallic particle is subjected to the influence of a magnetic field such that it travels vertically through a fluid, from plate A to plate B, Fig. 12–4. If the particle is released from rest at C, $s = 100$ mm, and the acceleration is measured as $a = (4s)$ m/s^2, where s is in metres, determine (a) the velocity of the particle when it reaches plate B, $s = 200$ mm, and (b) the time it needs to travel from C to B.

Solution

Part (a). Knowing the acceleration as a function of displacement, the velocity as a function of displacement can be obtained by using $v\,dv = a\,ds$. Why? Realizing that $v = 0$ at $s = 100$ mm $= 0.1$ m, we have

$$(+\downarrow) \qquad v\,dv = a\,ds$$

$$\int_0^v v\,dv = \int_{0.1}^s 4s\,ds$$

$$\tfrac{1}{2}v^2\Big|_0^v = \frac{4}{2}s^2\Big|_{0.1}^s$$

$$v = 2(s^2 - 0.01)^{1/2} \qquad (1)$$

When $s = 200$ mm $= 0.2$ m,

$$v_B = 0.3464 \text{ m/s} = 346.4 \text{ mm/s} \qquad \textit{Ans.}$$

The positive root is chosen since the particle is traveling downwards, i.e., in the $+s$ direction.

Part (b). Since the velocity is known as a function of displacement, Eq. (1), the time for the particle to travel from C to B can be obtained using $v = ds/dt$, where $s = 0.1$ m at $t = 0$.

$$(+\downarrow) \qquad ds = v\,dt$$

$$= 2(s^2 - 0.01)^{1/2}\,dt$$

$$\int_{0.1}^s \frac{ds}{(s^2 - 0.01)^{1/2}} = \int_0^t 2dt$$

$$\ln\left(s + \sqrt{s^2 - 0.01}\right)\Big|_{0.1}^s = 2t\Big|_0^t$$

$$\ln\left(s + \sqrt{s^2 - 0.01}\right) + 2.30 = 2t$$

When $s = 200$ mm $= 0.2$ m,

$$t = \frac{\ln\left(0.2 + \sqrt{(0.2)^2 - 0.01}\right) + 2.30}{2} = 0.657 \text{ s} \qquad \textit{Ans.}$$

Example 12–4

A particle moves along a horizontal straight line such that its velocity is given by $v = (3t^2 - 6t)$ m/s, where t is the time in seconds. If the particle is initially located at the origin O, determine (a) the distance traveled during the time interval $t = 0$ to $t = 3.5$ s, (b) the average velocity and the average speed of the particle during this time interval, and (c) the instantaneous acceleration at $t = 3.5$ s.

Solution

Part (a). Since the velocity is related to time, a function that relates displacement to time may be found by integrating $v = ds/dt$ with the condition that at $t = 0$, $s = 0$.

$(\overset{+}{\rightarrow})$

$$ds = v\,dt$$

$$= (3t^2 - 6t)\,dt$$

$$\int_0^s ds = 3\int_0^t t^2\,dt - 6\int_0^t t\,dt$$

$$s = (t^3 - 3t^2)\text{ m} \qquad (1)$$

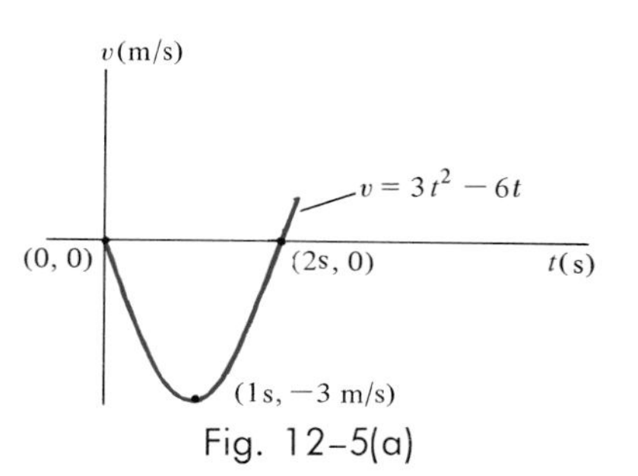

Fig. 12–5(a)

In order to determine the distance traveled in 3.5 s, it is necessary to investigate the path of motion during this time. The graph of the velocity function, Fig. 12–5*a*, reveals that for $0 \leqslant t < 2$ s, the velocity is *negative,* which means the particle is traveling to the *left,* and for $t > 2$ s, the velocity is *positive* and hence the particle is traveling to the *right*. Specifically, since the velocity changes sign when $t = 2$ s, the particle reverses its direction at this time. The particle's position when $t = 0$, $t = 2$ s, and $t = 3.5$ s can be computed from Eq. (1). This yields

$$s|_{t=0} = 0, \qquad s|_{t=2\text{ s}} = -4\text{ m}, \qquad s|_{t=3.5\text{ s}} = 6.12\text{ m}$$

The path is shown in Fig. 12–5*b*. Hence, the distance traveled in 3.5 s is

$$s_T = 4 + 4 + 6.12 = 14.12\text{ m} \qquad \textit{Ans.}$$

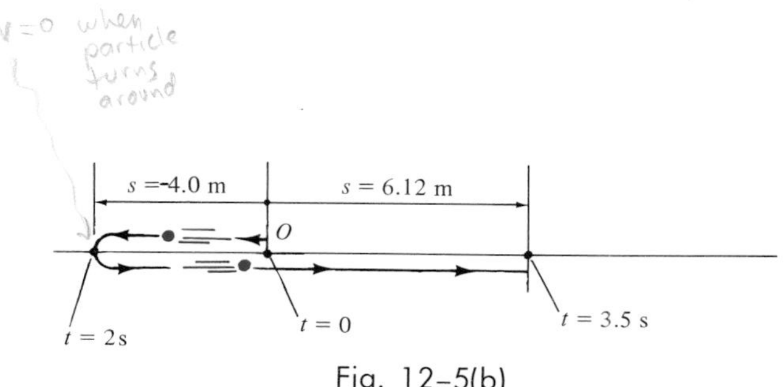

Fig. 12–5(b)

Part (b). The *displacement* from $t = 0$ to $t = 3.5$ s is

$$\Delta s = s|_{t=3.5\text{ s}} - s|_{t=0} = 6.12 - 0 = 6.12 \text{ m}$$

so that the average velocity is

$$v_{\text{avg}} = \frac{\Delta s}{\Delta t} = \frac{6.12}{3.5 - 0} = 1.75 \text{ m/s} \qquad \textit{Ans.}$$

The average speed is defined in terms of the *distance traveled* s_T. Hence,

$$(v_{\text{sp}})_{\text{avg}} = \frac{s_T}{\Delta t} = \frac{14.12}{3.5 - 0} = 4.03 \text{ m/s} \qquad \textit{Ans.}$$

Part (c). Knowing the velocity as a function of time, the acceleration is obtained from

$$(\overset{+}{\rightarrow}) \qquad a = \frac{dv}{dt} = \frac{d}{dt}(3t^2 - 6t)$$
$$= (6t - 6) \text{ m/s}^2$$

Thus, when $t = 3.5$ s,

$$a = 6(3.5) - 6$$
$$= 15 \text{ m/s}^2 \qquad \textit{Ans.}$$

Problems

12–1. After traveling a distance of 100 m, a car reaches a speed of 75 km/h, starting from rest. Determine the car's *constant* acceleration.

12–2. A car is traveling at a speed of 100 km/h when the brakes are suddenly applied, causing a constant deceleration of 4 m/s². Determine the time required to stop the car and the distance traveled before stopping.

12–3. When $t = 0$, a car has a speed of 25 m/s and a constant deceleration of 3 m/s². Determine the velocity of the car when $t = 4$ s. What is the displacement of the car during the 4-s time interval? How much time is needed to stop the car?

* **12–4.** A particle is moving downward along a straight line through a fluid medium such that its speed is measured as $v = (4t)$ m/s, where t is in seconds. If the particle is released from rest at $s = 0$, determine its position when $t = 2$ s.

12–5. A missile is fired vertically such that its altitude is defined by $s = (2t^3 + 2.5t^2 + 14t)$ m, where t is measured in seconds. Determine the missile's position, velocity, and acceleration at $t = 5$ s.

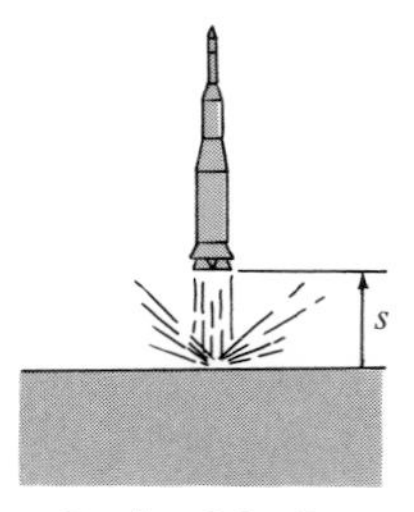

Prob. 12–5

12–5a. Solve Prob. 12–5 if $s = (5t^3 - 2t^2 + 6t)$ ft.

12–6. A car is initially traveling to the right with a speed of 25 m/s. If it is subjected to a constant deceleration of 3 m/s^2 (directed to the left), determine its velocity when $t = 4$ s. What is the displacement during this time interval?

12–7. A particle is moving along a straight-line path such that its position is defined by $s = (10t^2 + 20)$ mm, where t is measured in seconds. Determine (a) the displacement of the particle during the time interval from $t = 1$ s to $t = 5$ s, (b) the average velocity of the particle during this time interval, and (c) the acceleration at $t = 1$ s.

* **12–8.** The cork from a champagne bottle is fired vertically upward. If it takes a *total* of 8 s to rise from the bottle top and then fall to its initial elevation, determine the initial velocity at which it left the bottle. What is the total distance the cork travels during this time? *Note:* In reality, air resistance on a light object, such as a cork, has an appreciable effect on the motion and should be accounted for in the analysis. This is discussed in Chapter 13.

12–9. A particle moves along a straight-line path such that its position is defined by $s = (0.4t^3 - 16t^2 + 3)$ mm, where t is measured in seconds. From time $t = 0$, determine (a) the distance the particle must travel to reduce its velocity to zero, and (b) the time at which the acceleration is zero.

12–10. A particle moves along a straight-line path such that in 2 s it moves from an initial position of $s_A = +0.5$ m to a position $s_B = -1.5$ m. Then in another 4 s, it moves from s_B to $s_C = 2.5$ m. Determine the particle's average velocity and average speed during the 6-s time interval.

12–10a. Solve Prob. 12–10 if $s_A = 2$ ft, $s_B = -4$ ft, and $s_C = 3$ ft.

12–11. A small metal particle passes downward through a fluid medium while being subjected to the attraction of a magnetic field such that its position is observed to be $s = (10t^3 - 2t)$ mm, where t is measured in seconds. Determine (a) the particle's displacement from $t = 1$ s to $t = 3$ s, and (b) the velocity and acceleration of the particle when $t = 5$ s.

* **12–12.** A race car, originally at rest, has a constant acceleration of 4 m/s^2. When it reaches a maximum speed of 200 km/h, it is subjected to a constant deceleration until it stops. Determine the deceleration and the total elapsed time of motion if the distance traveled is 500 m.

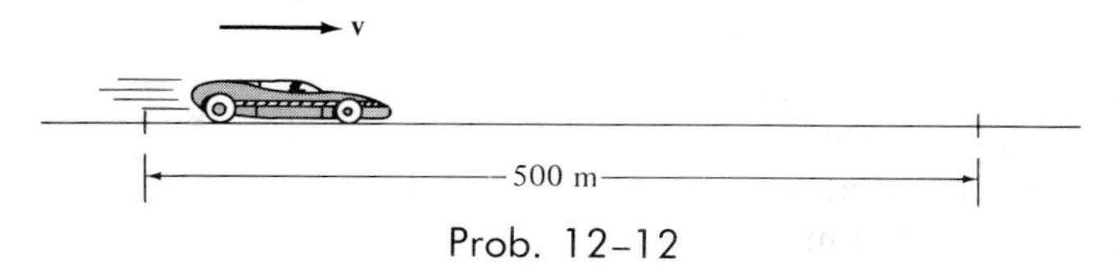

Prob. 12–12

12–13. A train is initially traveling along a straight track at a speed of 40 km/h. For 5 s it is subjected to a constant deceleration of 0.6 m/s^2, and then for the next 8 s it has a constant deceleration of a_c. Determine the magnitude of a_c so that the train stops at the end of the 13-s time period.

12–14. A car, originally at rest, moves along a straight-line path such that it attains a velocity of 15 m/s in 60 m with constant acceleration. Then after being subjected to *another* constant acceleration, it attains a final velocity of 35 m/s when $s = 200$ m. Determine the average velocity and average acceleration of the car for the entire 200-m displacement.

12–15. At the same instant, two cars A and B start from rest at a stop line. Car A has a constant acceleration of $a_A = 8 \text{ m/s}^2$, while car B has an acceleration of $a_B = (2t^{3/2}) \text{ m/s}^2$, where t is measured in seconds. Determine the distance between the cars when A reaches a speed of $v_A = 120$ km/h.

12–15a. Solve Prob. 12–15, provided $a_A = 10 \text{ ft/s}^2$, $a_B = (1.5t^{1/2}) \text{ ft/s}^2$, and $v_A = 70$ mi/h (1 mi = 5280 ft).

* **12–16.** A particle moves along a straight-line path with an acceleration of $a = (3t^2 - 2) \text{ mm/s}^2$, where t is measured in seconds. When $t = 0$, the particle is located 100 mm to the *left* of the origin; and when $t = 2$ s, it is 500 mm to the *left* of the origin. If the positive position is measured to the *right* of the origin, determine the particle's position when $t = 4$ s.

12-17. A projectile, initially at the origin, moves downward through a fluid medium such that its velocity is defined as $v = 2600(1 - e^{-0.3t})$ mm/s, where t is measured in seconds. Determine the displacement of the projectile during the first 2 s.

12-18. A particle moves along a horizontal straight-line path with an acceleration of $a = (kt^3 + 4)$ mm/s^2, where t is measured in seconds. Determine the constant k and compute the particle's velocity when $t = 3$ s, knowing that $v = 120$ mm/s when $t = 1$ s, and that $v = -100$ mm/s when $t = 2$ s. The positive direction is measured to the right.

12-19. A particle moves along a straight-line path with an acceleration of $a = (5/s)$ m/s^2, where s is measured in metres. Determine the particle's velocity when $s = 2$ m if it is initially released from rest when $s = 1$ m.

* **12-20.** The speed of a particle traveling downward within a liquid is measured as a function of its displacement as $v = (125 - s)$ mm/s, where s is given in millimetres. Determine (a) the particle's deceleration when it is located at point A, where $s_A = 100$ mm from its original position, (b) the distance the particle travels before it stops, and (c) the time needed to stop the particle.

* **12-20a.** Solve Prob. 12–20 if $v = (6 - s)$ ft/s, where s is given in ft, and $s_A = 3$ ft.

12-21. As a body is projected to a high altitude above the earth's *surface,* the variation of the acceleration of gravity with respect to altitude y above the *surface* of the earth must be taken into account. Neglecting air resistance, this acceleration is determined from the formula $a = -g_o\,[R^2/(R + y)^2]$, where g_o is the constant gravitational acceleration at sea level, R is the radius of the earth, and the positive direction is measured upwards. If $g_o = 9.81$ m/s^2 and $R = 6356$ km, determine the minimum initial velocity (escape velocity) at which a projectile should be shot vertically from the earth's surface so that it does not fall back to the earth. *Hint:* This requires that $v = 0$ as $y \to \infty$.

12-22. Accounting for the variation of gravitational acceleration a with respect to altitude y (see Prob. 12–21), derive an equation that relates the velocity of a freely falling particle to its altitude y above the surface of the earth. Assume that the particle is released from rest at an altitude of y_o from the earth's surface. With what velocity does the particle strike the earth if it is released from rest at an altitude of $y_o = 500$ km? Use the numerical data in Prob. 12–21.

12-23. If the effects of atmospheric resistance are accounted for, a freely falling body has an acceleration defined by the equation $a = 9.81[1 - v^2(10^{-4})]$ m/s^2, where v is measured in m/s and the positive direction is downward. If the body is released from rest at a *very high altitude,* determine (a) the velocity at time $t = 5$ s, and (b) the body's terminal or maximum attainable velocity (as $t \to \infty$).

12–3. Graphical Solutions

In some cases it is difficult to obtain a continuous mathematical function which describes the position, velocity, or acceleration of a particle during a time interval. When this occurs, a graphical procedure may be used to obtain curves or graphs that can be plotted to describe the motion.

Given the *s-t* Graph, Construct the *v-t* and *a-t* Graphs. Provided the position s of a particle can be *experimentally determined* for several instants of time t, an s-t graph for the particle can be plotted. Such a graph, which consists of a parabolic curve and two straight-line segments, is shown in Fig. 12–6a. The ordinates give the position of the particle from

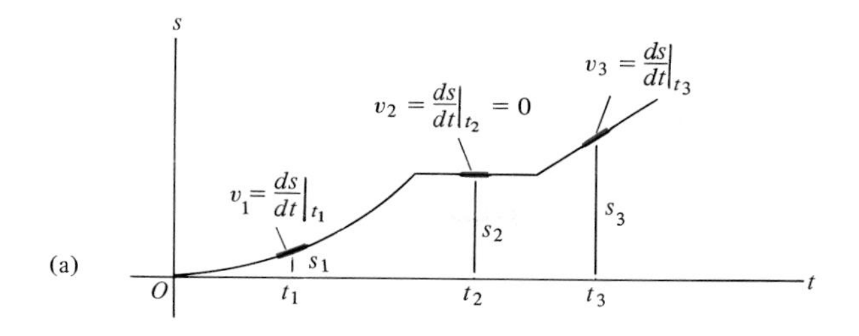

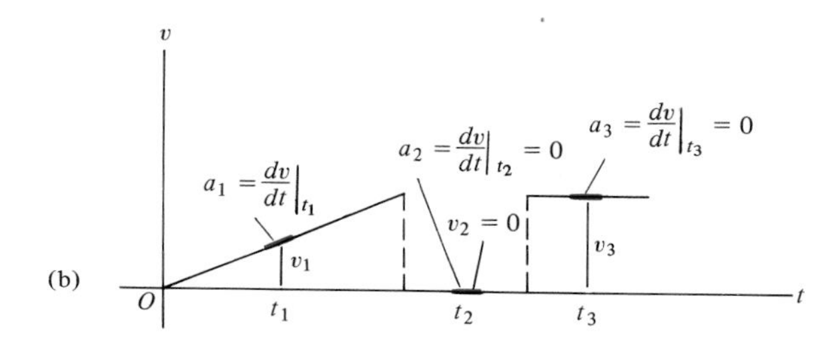

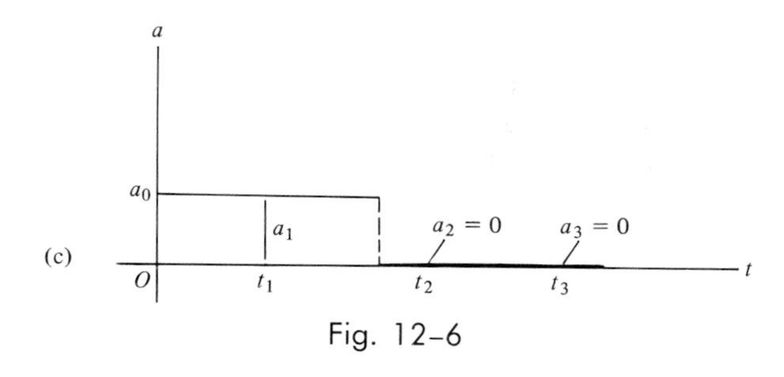

Fig. 12–6

the origin at any instant. Since the velocity $v = ds/dt$ (Eq. 12–2), the v-t graph can be established from measurements of the "slope" at various points along the s-t graph. (In a graphical sense, the slope is measured with a ruler and protractor.) For example, measurements of the slopes v_1, v_2, and v_3 at the intermediate points (t_1, s_1), (t_2, s_2), and (t_3, s_3) on the s-t graph yield the construction shown in Fig. 12–6*b*. In a similar manner, the a-t graph (acceleration versus time), Fig. 12–6*c*, may be constructed given the fact that the particle's acceleration $a = dv/dt$ (Eq. 12–5). In this case, the accelerations a_1, a_2, and a_3 are "slopes" measured from the v-t graph at points (t_1, v_1), (t_2, v_2), and (t_3, v_3), as shown in Fig. 12–6*b*.

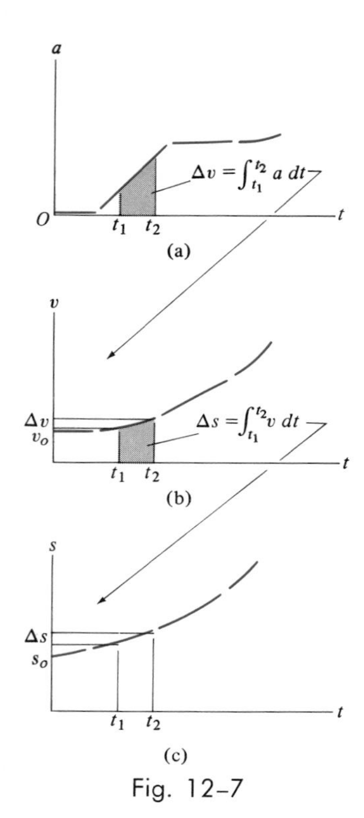

Fig. 12–7

Given the *a-t* Graph, Construct the *v-t* and *s-t* Graphs. If the *a-t* graph is given, Fig. 12–7*a*, using integration the *v-t* and *s-t* graphs may be constructed. For example, since $a = dv/dt$ or $dv = a\,dt$, then between two points on the *v-t* graph, $\Delta v = v_2 - v_1 = \int_{t_1}^{t_2} a\,dt$. Hence, for a finite time $\Delta t = t_2 - t_1$, the change in speed Δv, Fig. 12–7*b*, is equal to the area $\int_{t_1}^{t_2} a\,dt$ shown shaded in Fig. 12–7*a*. (In a graphical sense, any small area may be approximated by a trapezoid or rectangle.) In a similar manner, $v = ds/dt$ or $ds = v\,dt$; then $\Delta s = s_2 - s_1 = \int_{t_1}^{t_2} v\,dt$. Graphically, this equation indicates that the small area under the *v-t* graph during the time interval $\Delta t = t_2 - t_1$, shown shaded in Fig. 12–7*b*, represents the displacement Δs of the particle during the same time interval Δt, Fig. 12–7*c*.

Using this method, one begins with knowing the particle's initial velocity v_o and initial position s_o. Then *adding* (algebraically) to this the small area increments Δv and Δs determined under the *a-t* and *v-t* graphs, one determines successive points, $v_1 = v_o + \Delta v$, $s_1 = s_o + \Delta s$, etc., for the *v-t* and *s-t* graphs. Notice that an algebraic addition of areas is necessary, since areas lying above the *t* axis correspond to an increase in *v* or *s* ("positive" area), whereas those lying below the *t* axis indicate a decrease in *v* or *s* ("negative" area).

Since integration is required to construct the *v-t* and *s-t* graphs, given the *a-t* graph, then "integrating" an *a-t* graph that is constant (zero-degree curve) yields a *v-t* graph that is linear (first-degree curve), and "integrating" a linear *v-t* graph yields an *s-t* graph that is parabolic (second-degree curve), Fig. 12–7. In general then, if the *a-t* graph is a polynomial of degree *n*, the *v-t* and *s-t* graphs are polynomials of degrees $n + 1$ and $n + 2$, respectively.

Given the *a-s* Graph, Construct the *v-s* Graph. In some cases an *a-s* graph is known, so that the *v-s* graph can be constructed at various points by using $v\,dv = a\,ds$ (Eq. 12–7). Integrating this equation between the limits $v = v_1$ at $s = s_1$, and $v = v_2$ at $s = s_2$, we have

$$\int_{v_1}^{v_2} v\,dv = \int_{s_1}^{s_2} a\,ds$$

$$\tfrac{1}{2}(v_2^2 - v_1^2) = \int_{s_1}^{s_2} a\,ds$$

Thus, small segments of area under the *a-s* graph, $\int_{s_1}^{s_2} a\,ds$, shown colored in Fig. 12–8*a*, equal one-half the difference in the squares of the speed, $\tfrac{1}{2}(v_2^2 - v_1^2)$. By approximation of the area, $\int_{s_1}^{s_2} a\,ds$, it is possible to compute the value of v_2 at s_2 if an initial value of v_1 at s_1 is known, i.e., $v_2 = (2\int_{s_1}^{s_2} a\,ds + v_1^2)^{1/2}$, Fig. 12–8*b*. The *v-s* graph is constructed in this manner, starting from the initial velocity v_o.

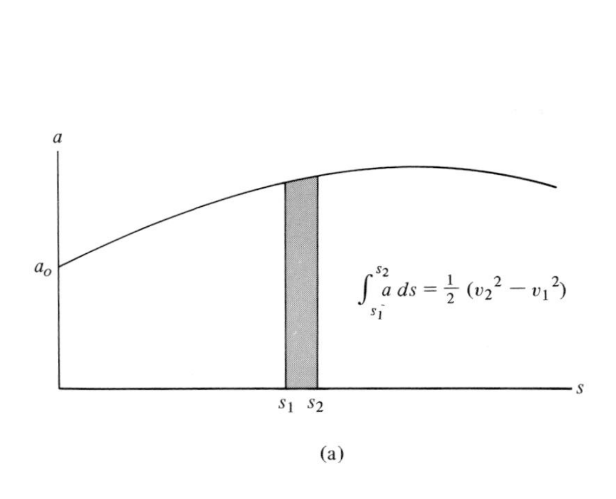

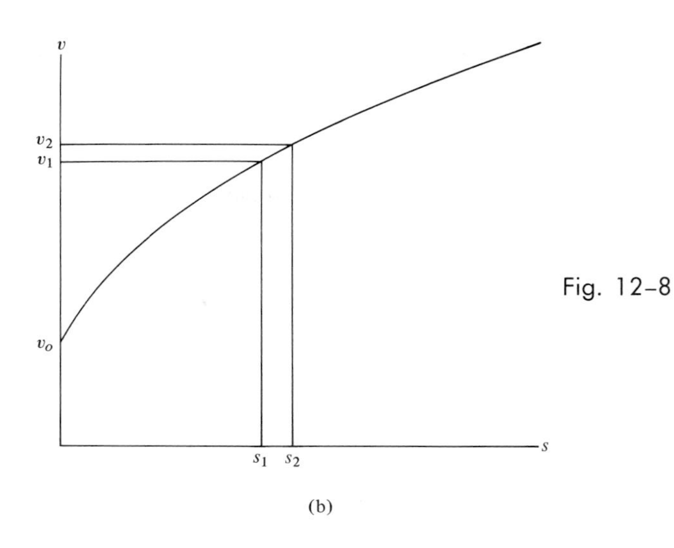

Fig. 12–8

Given the *v*-*s* Graph, Construct the *a*-*s* Graph. If the v-s graph is known, the a-s graph is constructed at various points using the following procedure. At any point $C(s, v)$, Fig. 12–9a, the slope dv/ds of the v-s graph is determined. The perpendicular line drawn from this slope intersects the s axis at point B. The distance AB represents the magnitude of the particle's acceleration a, Fig. 12–9b, since the two colored triangles shown in Fig. 12–9a are similar. (Recall that $a\,ds = v\,dv$, so that $\tan\theta = dv/ds = a/v$.) Of course, one must work with a consistent set of units when using this procedure; for example, if v is measured in m/s and s in metres, then a will be measured in m/s².

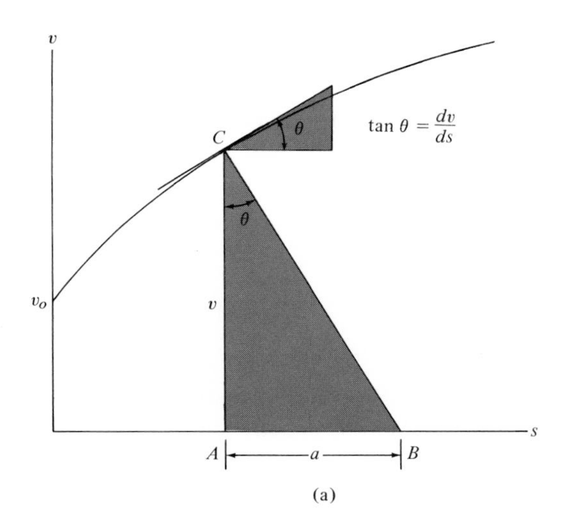

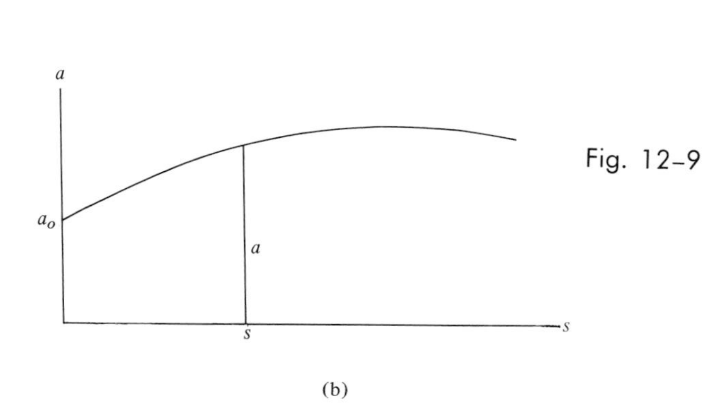

Fig. 12–9

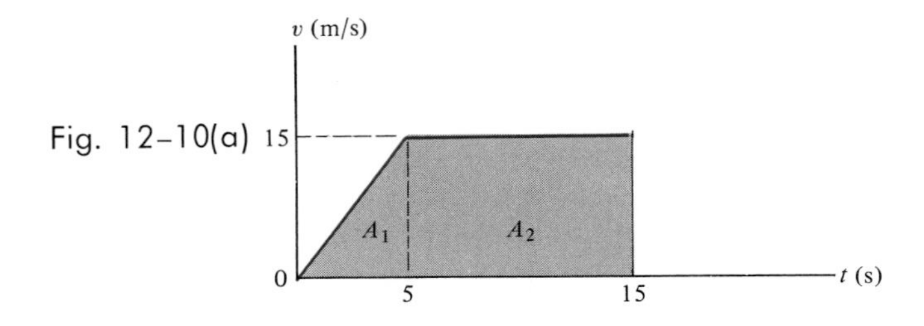

Fig. 12–10(a)

Example 12-5

A car moves along a straight-line road such that its velocity is described by the graph shown in Fig. 12–10*a*. If $s = 20$ m when $t = 0$, construct the *a-t* and *s-t* graphs for $0 \leqslant t \leqslant 15$ s.

Solution

a-t Graph. Since $a = dv/dt$, the *a-t* graph is constructed graphically by measuring the slope of points on the *v-t* graph. In this case the *v-t* graph consists of two line segments, Fig. 12–10*a*, for which the slope of each line segment is

Fig. 12–10(b)

$0 < t < 5$ s; $\qquad a = \dfrac{\Delta v}{\Delta t} = \dfrac{15 - 0}{5 - 0} = 3 \text{ m/s}^2$

$5 \text{ s} < t < 15$ s; $\qquad a = \dfrac{\Delta v}{\Delta t} = \dfrac{15 - 15}{15 - 5} = 0$

The results are shown in Fig. 12–10*b*.

s-t Graph. Since $\Delta s = \int_{t_1}^{t_2} v\, dt$, the *s-t* graph is constructed graphically by measuring segments of area under the *v-t* graph, Fig. 12–10*a*, during various time intervals Δt.

$t_1 = 0;$ $\qquad s_1 = 20$ m

$0 < t < 5$ s; $\qquad \Delta s = s_2 - s_1 = A_1 = \frac{1}{2}(5)(15) = 37.5$ m

$t_2 = 5$ s; $\qquad s_2 = 20 + 37.5 = 57.5$ m

$5 \text{ s} < t < 15$ s; $\qquad \Delta s = s_3 - s_2 = A_2 = (15)(15 - 5) = 150$ m

$t_3 = 15$ s; $\qquad s_3 = 57.5 + 150 = 207.5$ m

These three points are plotted in Fig. 12–10*c*. Since the *v-t* graph is integrated (theoretically) to obtain the *s-t* graph, the sloping line ($0 < t < 5$ s) on the *v-t* graph becomes parabolic on the *s-t* graph, whereas the horizontal line ($5 \text{ s} < t < 15$ s) on the *v-t* graph becomes a sloping line on the *s-t* graph.

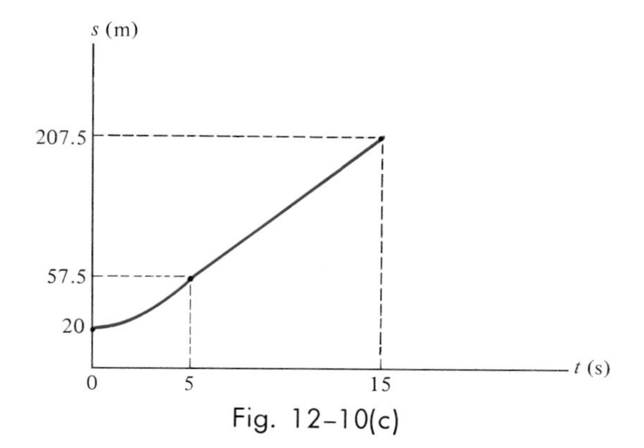

Fig. 12–10(c)

The shape of the s-t graph may be checked by noting that the slope of this graph at any instant t is equivalent to v at that instant $(ds/dt = v)$. For example, since v is linearly *increasing* during the time interval $0 < t < 5$ s, the parabolic line segment of the s-t graph within this region is concave *upward;* that is, the slope measured from one point to the next is increasing.

Since the two line segments on the v-t graph in Fig. 12–10a can be expressed in equation form, it is also possible to establish the a-t and s-t graphs by differentiation and integration, respectively. For example, when $0 \leqslant t < 5$ s, noting that $s_1 = 20$ m at $t_1 = 0$,

$$v = 3t \quad \text{(Fig. 12–10}a\text{)}$$

$$a = \frac{dv}{dt} = 3 \quad \text{(Fig. 12–10}b\text{)}$$

$$ds = v\,dt = 3t\,dt$$

$$\int_{20}^{s} ds = \int_{0}^{t} 3t\,dt$$

$$s = 1.5t^2 + 20 \quad \text{(Fig. 12–10}c\text{)}$$

When $5 \text{ s} < t \leqslant 15$ s, noting that $s_2 = 57.5$ m at $t_2 = 5$ s,

$$v = 15 \quad \text{(Fig. 12–10}a\text{)}$$

$$a = \frac{dv}{dt} = 0 \quad \text{(Fig. 12–10}b\text{)}$$

$$ds = v\,dt = 15\,dt$$

$$\int_{57.5}^{s} ds = \int_{5}^{t} 15\,dt$$

$$s = 15t - 17.5 \quad \text{(Fig. 12–10}c\text{)}$$

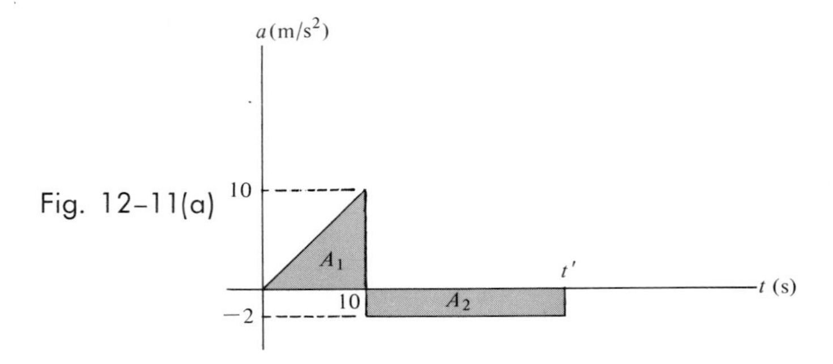

Fig. 12–11(a)

Example 12–6

A rocket sled starting from rest travels along a straight track such that it accelerates in the manner shown on the graph, Fig. 12–11*a*, for $t = 10$ s, and then decelerates at a constant rate. Draw the v-t graph and determine the time t' needed to stop the sled.

Solution

In general, $\Delta v = \int_{t_1}^{t_2} a\,dt$, so that the v-t graph is constructed graphically by measuring segments of area under the a-t graph, Fig. 12–11*a*, during various time intervals Δt. Since the sled is initially at rest,

$t_1 = 0;$ $\quad v_1 = 0$

$0 < t < 10$ s; $\quad \Delta v = v_2 - v_1 = A_1 = \frac{1}{2}(10)(10) = 50$ m/s

$t_2 = 10$ s; $\quad v_2 = 50 + 0 = 50$ m/s

The equation defining the v-t graph for $0 \leqslant t < 10$ s can be determined by direct integration, since $a = (t)$ m/s² for $0 \leqslant t < 10$ s, Fig. 12–11*a*. Hence,

$$dv = a\,dt = t\,dt$$

$$\int_0^v dv = \int_0^t t\,dt$$

$$v = \tfrac{1}{2}t^2$$

As a check, note that when $t = 10$ s, $v = 50$ m/s, as previously calculated. The result is plotted in Fig. 12–11*b*.

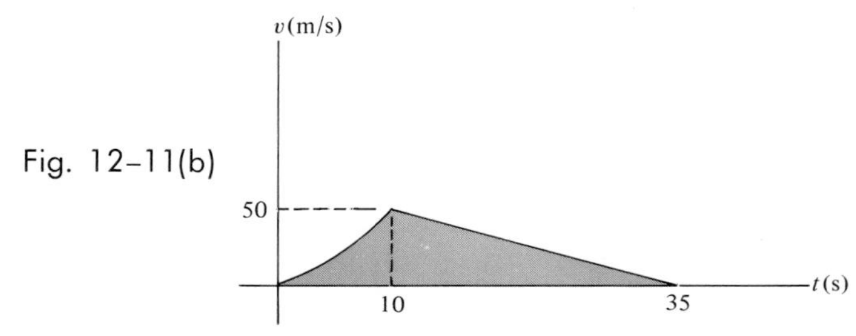

Fig. 12–11(b)

When the sled stops, the change in velocity Δv from $t = 0$ to $t = t'$ must be zero. Hence, the area under the a-t graph from $0 \leqslant t \leqslant t'$ must be zero. This requires that

$$A_1 + A_2 = 0$$
$$50 + (-2)(t' - 10) = 0$$
$$t = 35 \text{ s} \qquad \textit{Ans.}$$

To obtain the equation defining the v-t graph for $10 \text{ s} < t \leqslant 35 \text{ s}$, note that $a = -2 \text{ m/s}^2$, Fig. 12–11a, and $v_2 = 50$ m/s at $t = 10$ s. Thus,

$$dv = a\,dt = -2\,dt$$
$$\int_{50}^{v} dv = \int_{10}^{t} -2\,dt$$
$$v = -2t + 70$$

When $t = 35$ s, $v = 0$ as required, Fig. 12–11b.

Problems

*12–24. The speed of a car traveling up a hill is plotted as shown. Compute the total distance the car moves until it stops ($t = 80$ s). Plot the a-t graph.

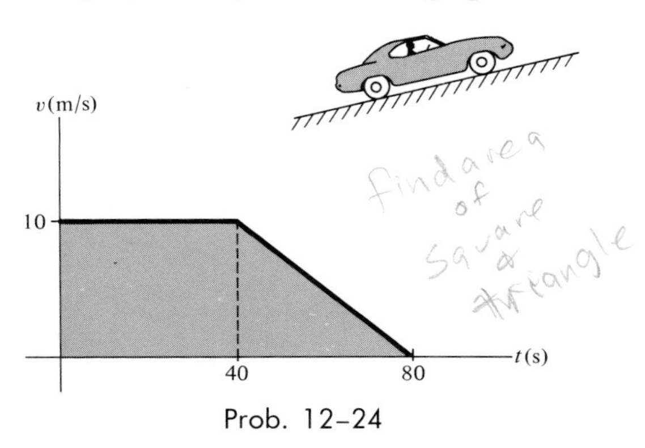

Prob. 12–24

12–25. The s-t graph for a train has been experimentally determined. From the data, construct the v-t and a-t graphs for the motion; $0 \leqslant t \leqslant t_2$. On the graph $t_1 = 30$ s, $s_1 = 360$ m, and $t_2 = 40$ s, $s_2 = 600$ m. Furthermore, for $0 \leqslant t \leqslant 30$ s, the curve is $s = (0.4t^2)$ m, where t is in seconds.

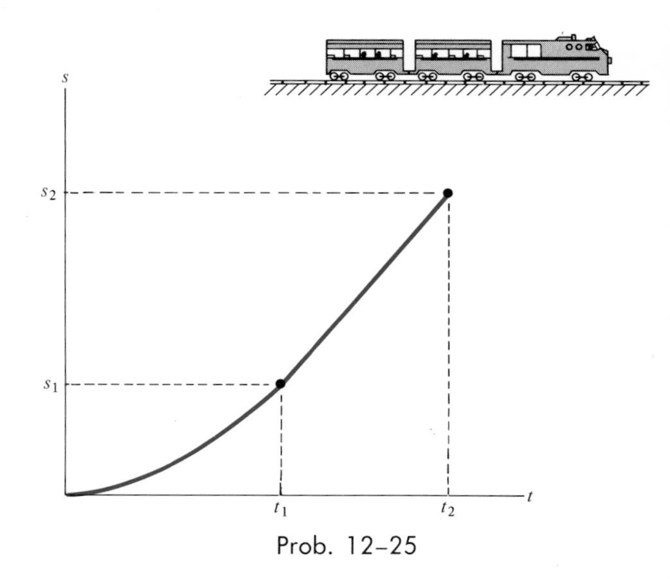

Prob. 12–25

12–25a. Solve Prob. 12–25 if on the graph $s_1 = 1800$ ft and $t_1 = 30$ s, and $s_2 = 3000$ ft at $t_2 = 40$ s. For $0 \leqslant t \leqslant 30$ s, the curve is $s = (2t^2)$ ft.

12–26. A two-stage missile is fired vertically from rest with an acceleration as shown. In 15 s the first stage A burns out and the second stage B ignites. Plot the v-t and s-t graphs which describe the motion of the second stage for $0 \leqslant t \leqslant 20$ s.

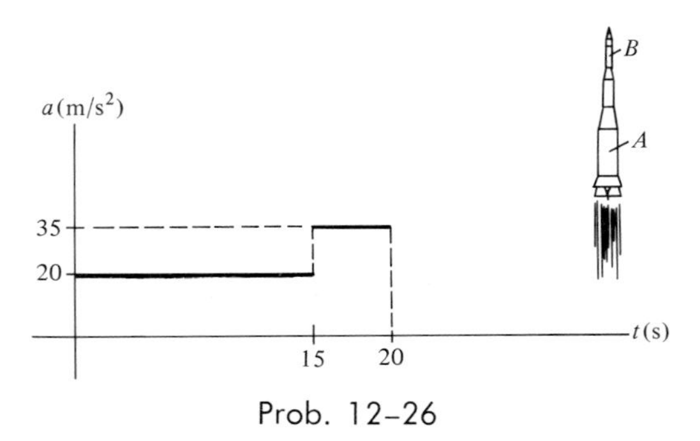

Prob. 12–26

12–27. A race car starting from rest moves along a straight track with an acceleration as shown, where $t \geqslant 10$ s, $a = 10$ m/s^2. Determine the time t for the car to reach a speed of 75 m/s and construct the v-t graph that describes the motion until the time t.

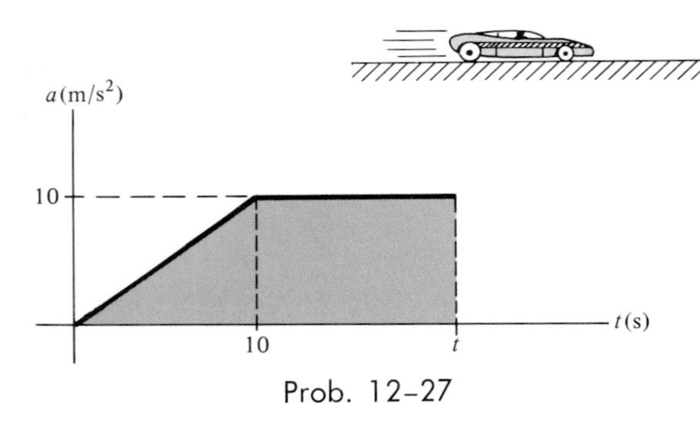

Prob. 12–27

***12–28.** A man riding upward in a freight elevator accidentally drops a package off the elevator when it is 20 m from the ground. If the elevator maintains a constant upward speed of 5 m/s, determine how high it is from the ground the instant the package hits the ground. Draw the v-t curve for the package during the time it is in motion. Assume that the package was released with the same upward speed as the elevator, and that the acceleration of the package is constant, acting downward with a magnitude of 9.81 m/s^2.

12–29. A car starts from rest and travels along a straight road such that its acceleration varies in accordance with the graph. Construct the v-t and s-t graphs for the motion.

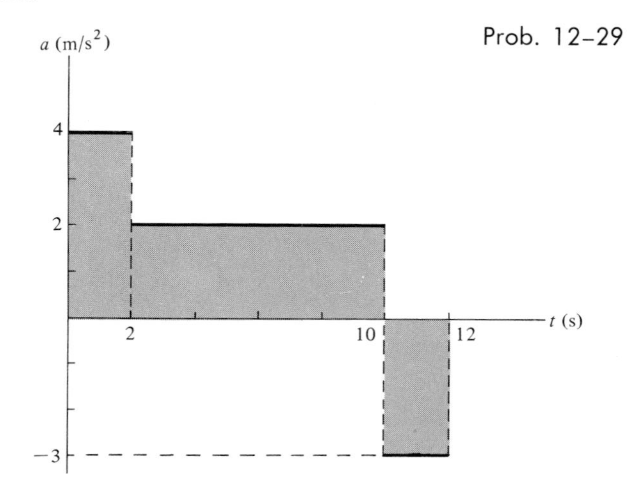

Prob. 12–29

12–30. The a-s graph for a rocket moving along a straight horizontal track has been experimentally determined. If the rocket starts from rest, determine its speed at the instants $s = 20$ m, 50 m, and 70 m, respectively. On the graph, $s_1 = 50$ m, $a_1 = 10$ m/s^2, and $s_2 = 70$ m, $a_2 = 20$ m/s^2.

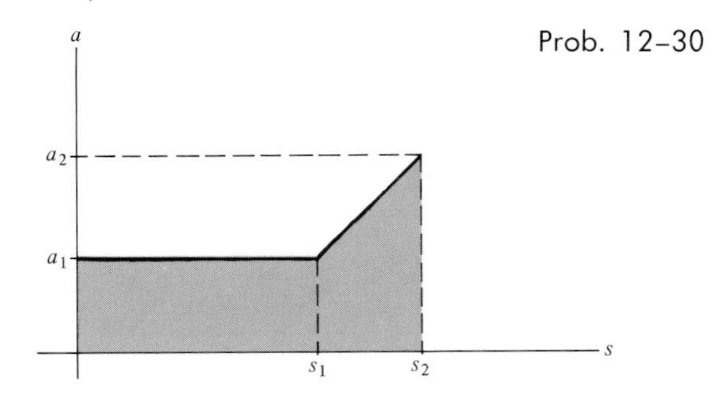

Prob. 12–30

12–30a. Solve Prob. 12–30 when $s = 50$ ft, 150 ft, and 200 ft, if on the graph $s_1 = 150$ ft, $a_1 = 6$ ft/s^2, and $s_2 = 200$ ft, $a_2 = 10$ ft/s^2.

12–31. A rocket is fired vertically from rest with an acceleration that varies linearly with time, as shown. Compute both the height h traveled by the rocket and the rocket's velocity at the instant the fuel burns out

($t = 50$ s). *Suggestion:* Solve the problem by determining the equation of the two lines of the a-t graph, and then integrating.

Prob. 12–31

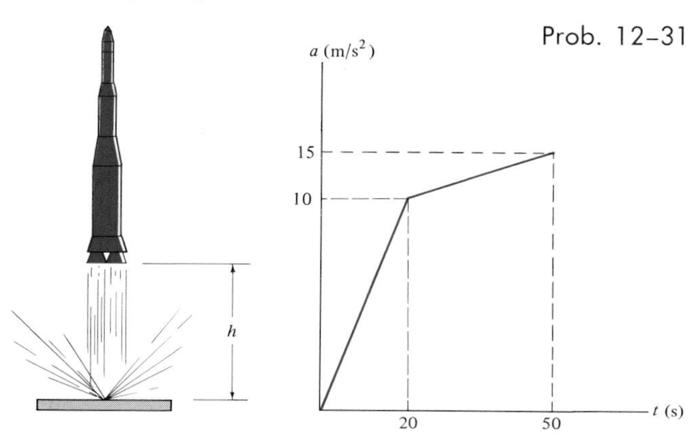

***12–32.** From experimental data, the motion of a jet plane while traveling along a runway is defined by the v-t graph shown. Construct the s-t and a-t graphs for the motion.

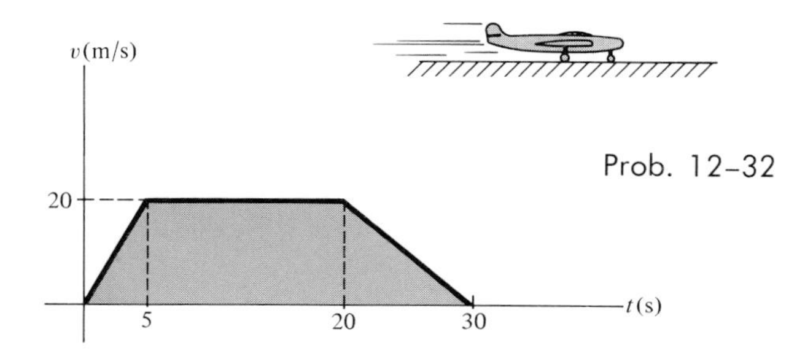

Prob. 12–32

12–33. Two cars start from rest side by side and race along a straight track. Car A accelerates at 4 m/s² for 10 s and then maintains a constant speed. Car B accelerates at 5 m/s² until reaching a constant speed of 25 m/s and then maintains this speed. Construct the a-t, v-t, and s-t graphs for each car until $t = 15$ s. What is the distance between the two cars when $t = 15$ s?

12–34. The v-s graph was determined experimentally to describe the straight-line motion of a rocket sled. Using the data of this graph, determine the acceleration of the sled when $s = 100$ m, and when $s = 200$ m.

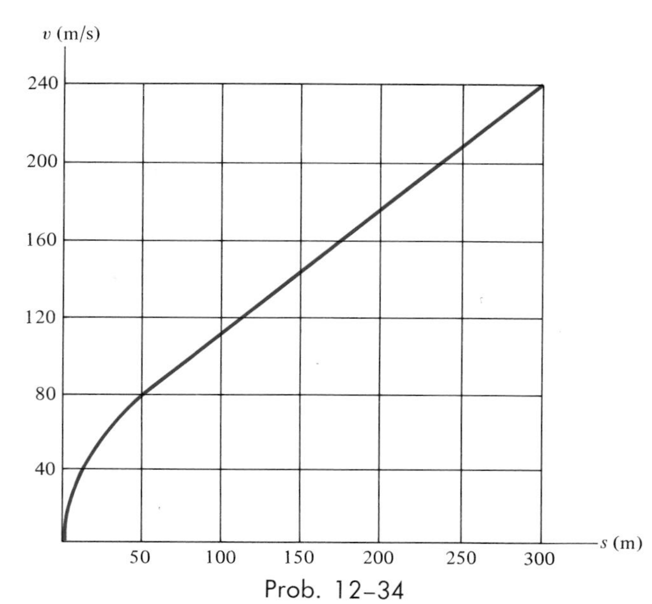

Prob. 12–34

12–35. The rocket car is subjected to a constant acceleration of $a_c = 6$ m/s² until $t_1 = 15$ s. The brakes are then applied, which causes a deceleration at the rate shown until the car stops. Determine the maximum speed of the car and the time t when the car stops.

Prob. 12–35

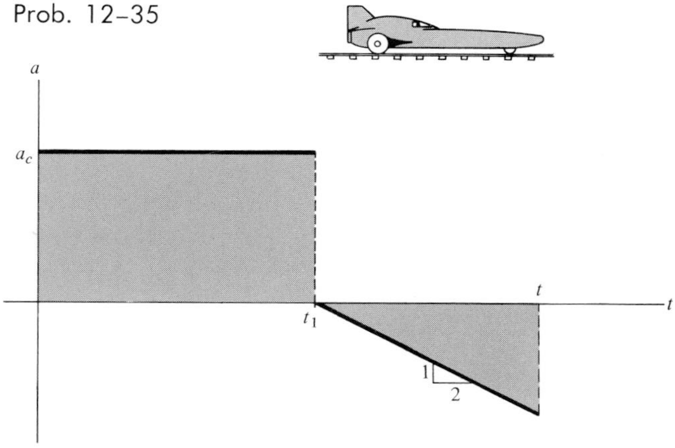

12–35a. Solve Prob. 12–35 if $a_c = 15$ ft/s² until $t_1 = 10$ s, and the graph maintains the constant downward slope of 1 to 2 as shown.

12–4. Curvilinear Velocity and Acceleration of a Particle

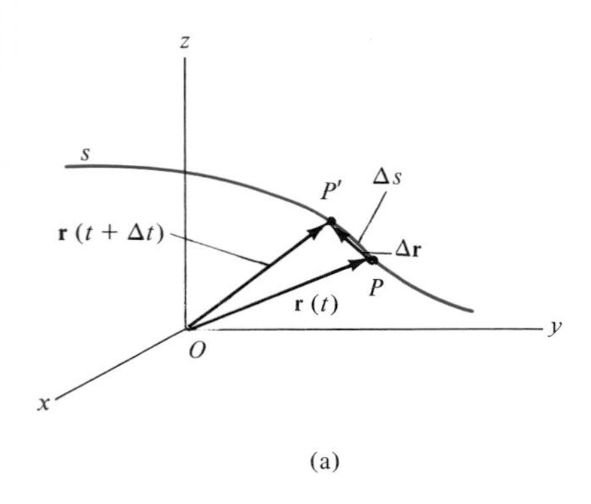

(a)

When a particle moves along a curved path, the motion is called *curvilinear motion.* Because the path is often represented in three dimensions, *vector analysis** will be used to formulate the particle's position, velocity, and acceleration. In this section some general aspects of curvilinear motion will be discussed and in subsequent sections three types of orthogonal coordinate systems commonly used to describe this motion will be introduced.

Position. The position of a particle, located at point $P(x, y, z)$ on a space curve, will be designated by the *position vector* $\mathbf{r} = \mathbf{r}(t)$ shown in Fig. 12–12*a*. This vector is a function of time t since, in general, both its magnitude and direction change as the particle moves along the curve s.

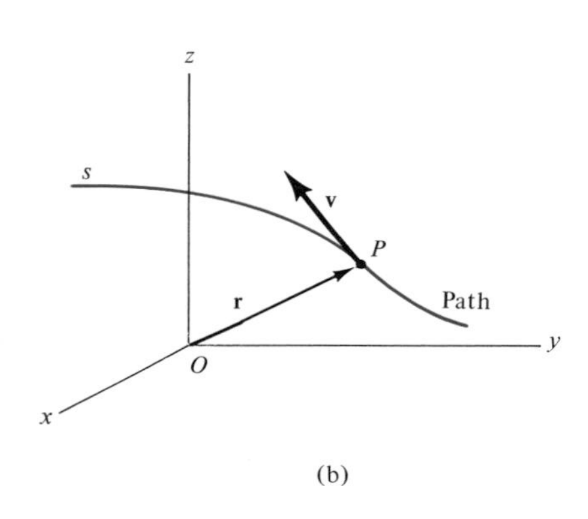

(b)

Displacement. Suppose that during a small time interval Δt, the particle moves a distance Δs along the curve to a new position $P'(x + \Delta x, y + \Delta y, z + \Delta z)$, defined by $\mathbf{r}(t + \Delta t) = \mathbf{r}(t) + \Delta\mathbf{r}$, Fig. 12–12*a*. The *displacement* $\Delta\mathbf{r}$ of the particle is determined by vector subtraction, i.e., $\Delta\mathbf{r} = \mathbf{r}(t + \Delta t) - \mathbf{r}(t)$.

Velocity. During the time Δt, the *average velocity* of the particle is defined as

$$\mathbf{v}_{\text{avg}} = \frac{\Delta\mathbf{r}}{\Delta t} \tag{12–11}$$

The *instantaneous velocity* is determined from this equation by letting $\Delta t \rightarrow 0$, so that $\Delta\mathbf{r}$ approaches the tangent to the curve at point P. Hence,

$$\mathbf{v} = \lim_{\Delta t \to 0} \frac{\Delta\mathbf{r}}{\Delta t}$$

or

$$\mathbf{v} = \frac{d\mathbf{r}}{dt} \tag{12–12}$$

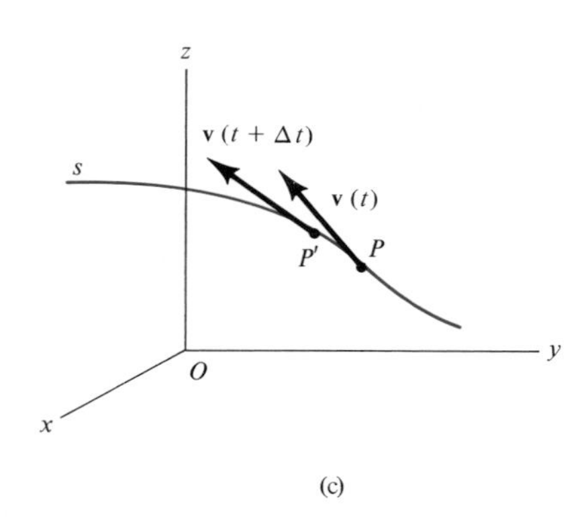

(c)

Fig 12–12(a–c)

As shown in Fig. 12–12*b*, the *direction* of $\mathbf{v}$ is *always tangent to the path of motion*. The *magnitude* of $\mathbf{v}$, which is called the *speed,* may be obtained by noting that the magnitude of the displacement $\Delta\mathbf{r}$ is the length of the

*A summary of some of the important concepts of vector analysis is given in Appendix B.

straight-line segment from P to P', Fig. 12–12a. Realizing that this length, Δr, approaches the arc length Δs as $\Delta t \rightarrow 0$, we have

$$v = \lim_{\Delta t \to 0} \frac{\Delta r}{\Delta t} = \lim_{\Delta t \to 0} \frac{\Delta s}{\Delta t}$$

or

$$v = \frac{ds}{dt} \qquad (12\text{–}13)$$

Thus, the *speed* may be obtained by differentiating the path function s with respect to time.

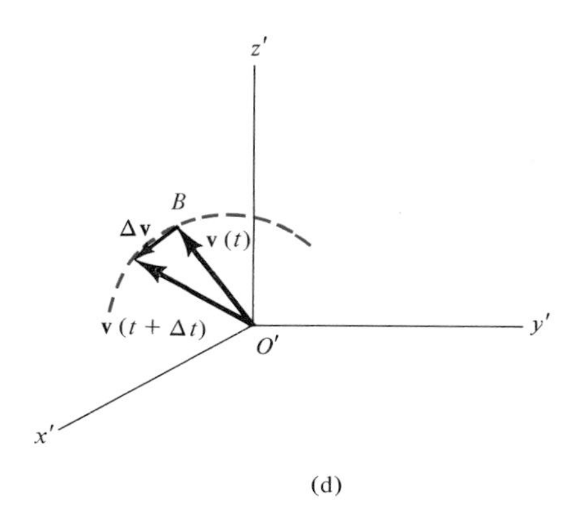

(d)

Acceleration. If the particle has a velocity $\mathbf{v}(t)$ at time t and a velocity $\mathbf{v}(t + \Delta t) = \mathbf{v}(t) + \Delta\mathbf{v}$ at $t + \Delta t$, Fig. 12–12c, then the *average acceleration* of the particle during the time interval Δt is

$$\mathbf{a}_{avg} = \frac{\Delta \mathbf{v}}{\Delta t} \qquad (12\text{–}14)$$

where $\Delta\mathbf{v} = \mathbf{v}(t + \Delta t) - \mathbf{v}(t)$. To study this time rate of change of $\mathbf{v}$, consider a set of x', y', z' axes measured in units of speed, Fig. 12–12d. The two velocity vectors in Fig. 12–12c are plotted in Fig. 12–12d such that their tails are located at the fixed origin O' and their tips reach points on the dashed curve. This curve is called a *hodograph,* and when constructed, it describes the locus of points for the tip of the velocity vector in the same manner that the *path s* describes the locus of points for the tip of the position vector $\mathbf{r}$, Fig. 12–12b.

To obtain the *instantaneous acceleration,* let $\Delta t \rightarrow 0$ so that $\Delta\mathbf{v} \rightarrow 0$, Eq. 12–14. In the limit, $\Delta\mathbf{v}$ becomes *tangent to the hodograph* at B and we have

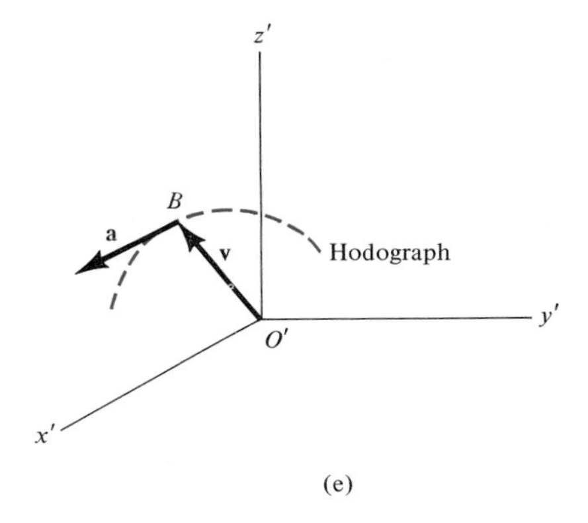

(e)

$$\mathbf{a} = \lim_{\Delta t \to 0} \frac{\Delta \mathbf{v}}{\Delta t}$$

or

$$\mathbf{a} = \frac{d\mathbf{v}}{dt} \qquad (12\text{–}15)$$

Using Eq. 12–12, we can also write

$$\mathbf{a} = \frac{d^2\mathbf{r}}{dt^2} \qquad (12\text{–}16)$$

By definition of the derivative, $\mathbf{a}$ acts *tangent to the hodograph* at B, Fig. 12–12e, and therefore, in general, $\mathbf{a}$ is *not* tangent to the path of motion, Fig. 12–12f.

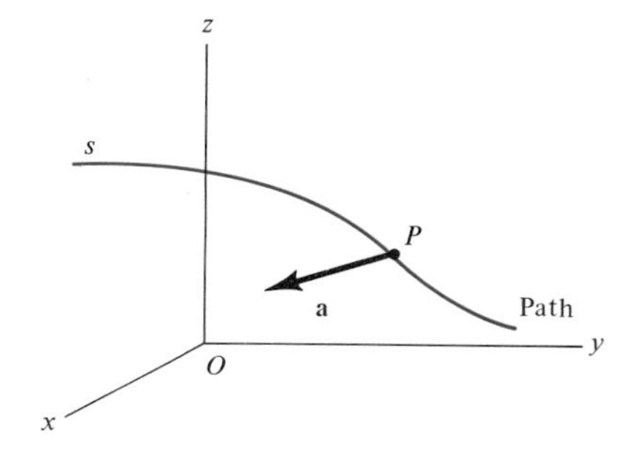

(f)

Fig. 12–12(d–f)

12–5. Curvilinear Motion of a Particle: Rectangular Components

In some cases the motion of a particle can be conveniently expressed in terms of its x, y, and z components. If the path extends into three dimensions, then **i, j,** and **k** unit vectors are used to formulate the kinematic quantities.

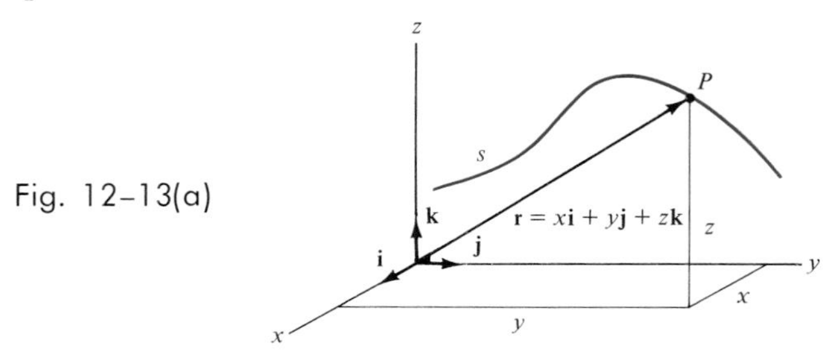

Fig. 12–13(a)

Position. If at a given instant the particle is located at point $P(x, y, z)$ along the fixed curved path s, Fig. 12–13a, its location is defined by the *position vector*

$$\mathbf{r} = x\mathbf{i} + y\mathbf{j} + z\mathbf{k} \tag{12–17}$$

Because of the motion and the geometry of the path, the x, y, z components of $\mathbf{r}$ are generally all functions of time; i.e., $x = x(t)$, $y = y(t)$, and $z = z(t)$, so that $\mathbf{r} = \mathbf{r}(t)$.

In accordance with the discussion in Appendix B, the *magnitude* of $\mathbf{r}$ is defined from Eq. B–3 as

$$r = \sqrt{x^2 + y^2 + z^2} \tag{12–18}$$

and the *direction* is specified by the components of the unit vector $\mathbf{u}_r = \mathbf{r}/r$.

Velocity. The first time derivative of $\mathbf{r}$ yields the velocity $\mathbf{v}$ of the particle. Hence,

$$\mathbf{v} = \frac{d\mathbf{r}}{dt} = \frac{d}{dt}(x\mathbf{i}) + \frac{d}{dt}(y\mathbf{j}) + \frac{d}{dt}(z\mathbf{k})$$

In taking the derivative, it is necessary to account for changes in *both* the magnitude and direction of the vector's components. From Eq. B–23 of Appendix B, $(d/ds\,[f(s)\,\mathbf{A}(s)] = (df/ds)\mathbf{A} + f(s)\,d\mathbf{A}/ds)$ the derivative of the $\mathbf{i}$ component of $\mathbf{v}$ is

$$\frac{d}{dt}(x\mathbf{i}) = \frac{dx}{dt}\mathbf{i} + x\frac{d\mathbf{i}}{dt}$$

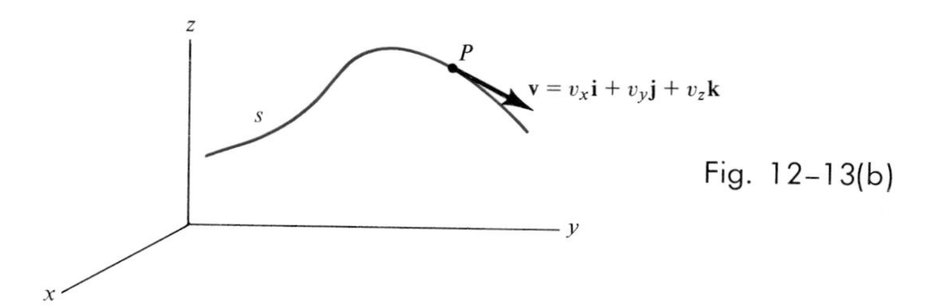

Fig. 12–13(b)

The second term on the right side is zero, since the x, y, z reference frame is *fixed,* and therefore the *direction* (and the *magnitude*) of **i** does not change with time. Differentiation of the **j** and **k** components may be carried out in a similar manner, which yields the final result,

$$\mathbf{v} = \frac{d\mathbf{r}}{dt} = v_x\mathbf{i} + v_y\mathbf{j} + v_z\mathbf{k} \tag{12–19}$$

where

$$\begin{aligned} v_x &= \dot{x} \\ v_y &= \dot{y} \\ v_z &= \dot{z} \end{aligned} \tag{12–20}$$

The "dot" notation $\dot{x}$, $\dot{y}$, $\dot{z}$ represents the first time derivatives of $x = x(t)$, $y = y(t)$, and $z = z(t)$, respectively.

The velocity has a *magnitude* defined by

$$v = \sqrt{v_x^2 + v_y^2 + v_z^2} \tag{12–21}$$

and a *direction* that is specified by the components of the unit vector $\mathbf{u}_v = \mathbf{v}/v$. This direction is *always tangent* to the path, as shown in Fig. 12–13b.

Acceleration. The acceleration of the particle is obtained by taking the first time derivative of Eq. 12–19 (or the second time derivative of Eq. 12–17). Using dots to represent the derivatives of the components, we have,

$$\mathbf{a} = \frac{d\mathbf{v}}{dt} = a_x\mathbf{i} + a_y\mathbf{j} + a_z\mathbf{k} \tag{12–22}$$

where

$$\begin{aligned} a_x &= \dot{v}_x = \ddot{x} \\ a_y &= \dot{v}_y = \ddot{y} \\ a_z &= \dot{v}_z = \ddot{z} \end{aligned} \tag{12–23}$$

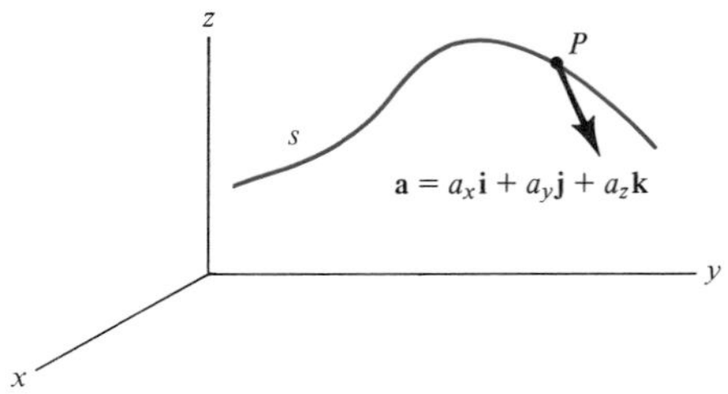

Fig. 12–13(c)

Hence, a_x, a_y, and a_z represent, respectively, the first time derivatives of the time functions v_x, v_y, and v_z; or the second time derivatives of the time functions x, y, and z.

The acceleration has a *magnitude* defined by

$$a = \sqrt{a_x^2 + a_y^2 + a_z^2} \tag{12–24}$$

and a *direction* specified by the components of the unit vector $\mathbf{u}_a = \mathbf{a}/a$. Since $\mathbf{a}$ represents the time rate of *change* in velocity, in general $\mathbf{a}$ will *not* be tangent to the path traveled by the particle, Fig. 12–13c.

Example 12–7

The motion of a bead B sliding down along the spiral path shown in Fig. 12–14 is defined by the position vector $\mathbf{r} = \{0.5 \sin(2t)\mathbf{i} + 0.5 \cos(2t)\mathbf{j} - 0.2t\mathbf{k}\}$ m, where t is given in seconds and the arguments for sine and cosine are given in radians (π rad $= 180°$). Determine the location of the bead when $t = 0.75$ s, and the magnitudes of the bead's velocity and acceleration at this instant.

Solution

Evaluating $\mathbf{r}$ at $t = 0.75$ s yields

$$\begin{aligned}\mathbf{r}\big|_{t=0.75\text{ s}} &= \{0.5 \sin(1.5 \text{ rad})\mathbf{i} + 0.5 \cos(1.5 \text{ rad})\mathbf{j} - 0.2(0.75)\mathbf{k}\} \text{ m} \\ &= \{0.499\mathbf{i} + 0.035\mathbf{j} - 0.150\mathbf{k}\} \text{ m} \qquad \textit{Ans.}\end{aligned}$$

The magnitude of this vector represents the distance of the bead from the origin O, which is

$$r = \sqrt{(0.499)^2 + (0.035)^2 + (-0.150)^2} = 0.522 \text{ m} \qquad \textit{Ans.}$$

The direction of $\mathbf{r}$ is obtained from the components of the unit vector

$$\begin{aligned}\mathbf{u}_r = \frac{\mathbf{r}}{r} &= \frac{0.499}{0.522}\mathbf{i} + \frac{0.035}{0.522}\mathbf{j} - \frac{0.150}{0.522}\mathbf{k} \\ &= 0.956\mathbf{i} + 0.067\mathbf{j} - 0.287\mathbf{k}\end{aligned}$$

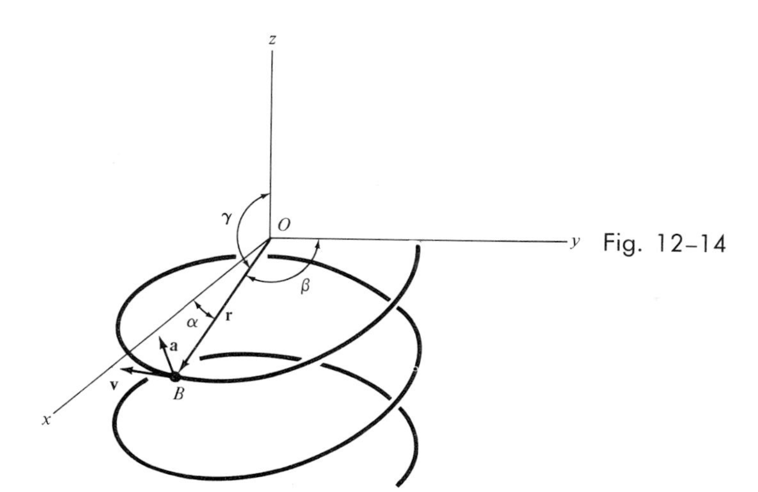

Fig. 12–14

Hence, the direction angles α, β, and γ, shown in Fig. 12–14, are

$$\alpha = \cos^{-1}(0.956) = 17.1° \qquad \textit{Ans.}$$
$$\beta = \cos^{-1}(0.067) = 86.2° \qquad \textit{Ans.}$$
$$\gamma = \cos^{-1}(-0.287) = 106.7° \qquad \textit{Ans.}$$

The velocity is defined by

$$\mathbf{v} = \frac{d\mathbf{r}}{dt} = \{1\cos(2t)\mathbf{i} - 1\sin(2t)\mathbf{j} - 0.2\mathbf{k}\}\ \text{m/s}$$

Hence, in 0.75 s the magnitude of velocity, or speed, is

$$\begin{aligned} v &= \sqrt{v_x^2 + v_y^2 + v_z^2} \\ &= \sqrt{(1\cos(1.5\text{ rad}))^2 + (-1\sin(1.5\text{ rad}))^2 + (-0.2)^2} \\ &= 1.02\text{ m/s} \end{aligned} \qquad \textit{Ans.}$$

The velocity is tangent to the path as shown in Fig. 12–14.

The acceleration is defined by

$$\mathbf{a} = \frac{d\mathbf{v}}{dt} = \{-2\sin(2t)\mathbf{i} - 2\cos(2t)\mathbf{j}\}\ \text{m/s}^2$$

Hence, in 0.75 s the magnitude of acceleration is

$$\begin{aligned} a &= \sqrt{a_x^2 + a_y^2 + a_z^2} \\ &= \sqrt{(-2\sin(1.5\text{ rad}))^2 + (-2\cos(1.5\text{ rad}))^2 + 0} \\ &= 2.00\text{ m/s}^2 \end{aligned} \qquad \textit{Ans.}$$

As shown in Fig. 12–14, the acceleration is *not* tangent to the path; instead, if needed, its coordinate direction angles α, β, and γ are defined from the components of the unit vector $\mathbf{u}_a = \mathbf{a}/a$.

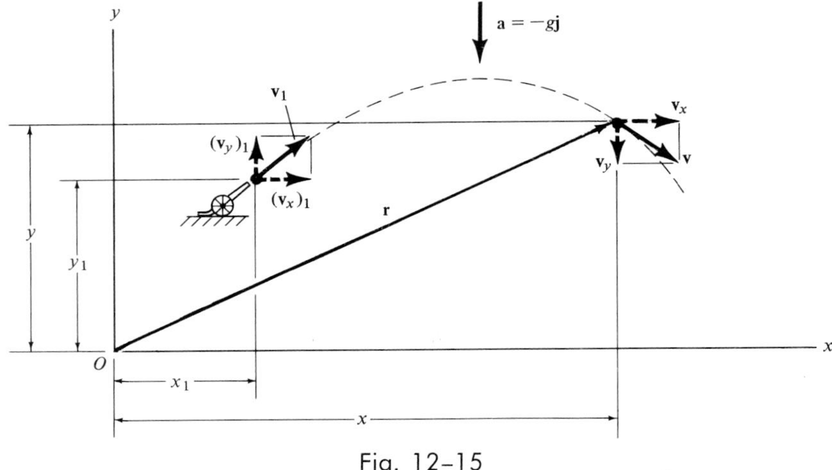

Fig. 12–15

12–6. Motion of a Projectile

Rectangular components are often used to study the free-flight motion of a projectile. To illustrate the concepts involved in the kinematic analysis, consider a projectile fired from a gun located at point (x_1, y_1), as shown in Fig. 12–15. For simplicity, the path is defined in the x-y plane such that the initial velocity is $\mathbf{v}_1$, having components $(\mathbf{v}_x)_1$ and $(\mathbf{v}_y)_1$. Provided air resistance is neglected, the only force acting on the projectile is its weight, which creates a *constant downward acceleration* of approximately $a_c = g = 9.81 \text{ m/s}^2$.*

Horizontal Motion. The component of acceleration in the x direction is $a_x = 0$. Since the positive direction is to the right, application of the constant acceleration equations, 12–8, 12–9, and 12–10, yields

$$(\overset{+}{\rightarrow})v = v_1 + a_c t; \qquad v_x = (v_x)_1$$

$$(\overset{+}{\rightarrow})s = s_1 + v_1 t + \tfrac{1}{2}a_c t^2; \qquad x = x_1 + (v_x)_1 t$$

$$(\overset{+}{\rightarrow})v^2 = v_1^2 + 2a_c(s - s_1); \qquad v_x = (v_x)_1$$

The first and last equations indicate that *the velocity component in the x direction remains constant throughout the motion.*

Vertical Motion. Since the positive y axis is directed upward, the component of acceleration in the y direction is $a_y = -g$. Hence, from Eqs. 12–8, 12–9, and 12–10, we get

*This assumes that the earth's gravitational field does not vary with altitude. See Example 13–3.

$$(+\uparrow)v = v_1 + a_c t; \qquad v_y = (v_y)_1 - gt$$
$$(+\uparrow)s = s_1 + v_1 t + \tfrac{1}{2}a_c t^2; \qquad y = y_1 + (v_y)_1 t - \tfrac{1}{2}gt^2$$
$$(+\uparrow)v^2 = v_1^2 + 2a_c(s - s_1); \qquad v_y^2 = (v_y)_1^2 - 2g(y - y_1)$$

Recall that the last equation can be formulated on the basis of eliminating the time t between the first two equations, and therefore *only two of the above three equations are independent of one another.*

At any instant, the resultant position **r** and velocity **v** are defined as the *vector sum* of their horizontal and vertical components which are determined from the above equations and shown in Fig. 12–15.

PROCEDURE FOR ANALYSIS

The following two-step procedure should be used when solving problems concerned with free-flight projectile motion.

Step 1: Sketch the trajectory of the particle, establish the x,y coordinate axes, and between any *two points* on the path, specify the given problem data and the unknowns. In all cases the acceleration of gravity is $g = 9.81\ \text{m/s}^2$ downward. The particle's initial and final velocities should be represented in terms of their x and y components.

Step 2: Only three independent equations are available for solution, and therefore there can be no more than three unknowns when these equations are applied between *two points* on the path. In particular, motion in the *horizontal or x direction* is described by $x_2 = x_1 + (v_x)_1 t$; also, $(v_x)_2 = (v_x)_1$. In the *vertical or y direction,* only *two* of the three equations $(v_y)_2 = (v_y)_1 + a_c t$, $(v_y)_2^2 = (v_y)_1^2 + 2a_c(y_2 - y_1)$ and $y_2 = y_1 + (v_y)_1 t + \frac{1}{2}a_c t^2$ can be used for the solution. Depending upon the known data in the y direction, a choice should be made as to which two of these three equations give the most direct solution.

The following examples numerically illustrate some applications of projectile motion using the above procedure.

Example 12–8

A cannonball is fired from point A with a muzzle velocity $\mathbf{v}_1$, as shown in Fig. 12–16. If the cannon is located at an elevation h above the ground, determine the equation that describes the trajectory AB.

Solution

Step 1: When the cannonball is at the arbitrary point $P(x, -y)$, the three unknowns between points A and P are the velocity components v_x and v_y at P and the time of flight, t. The initial velocity has components

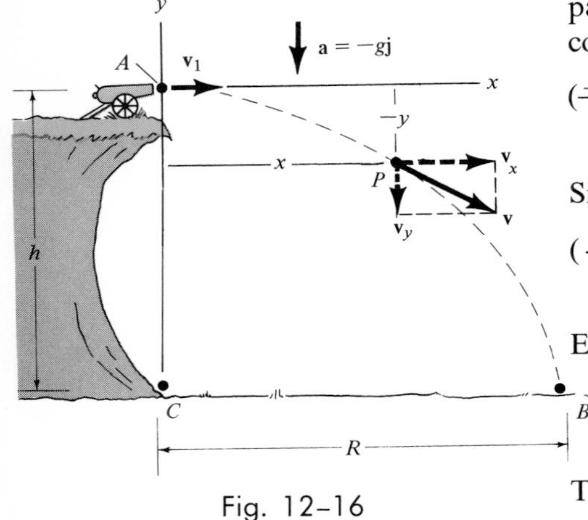

Fig. 12–16

$(v_x)_1 = v_1$ and $(v_y)_1 = 0$, and the acceleration at any point has components $a_x = 0$, $a_y = -g$, Fig. 12–16.
Step 2: To obtain the equation of the path, $y = f(x)$, the position coordinates x and y can each be expressed in terms of the unknown parameter t. Eliminating t yields the locus $y = f(x)$. In particular, the x coordinate is related to t by the equation

$(\overset{+}{\rightarrow})$
$$s_x = (s_x)_1 + (v_x)_1 t$$
$$x = 0 + v_1 t \qquad (1)$$

Similarly, y is related to t by the equation

$(+\uparrow)$
$$s_y = (s_y)_1 + (v_y)_1 t + \tfrac{1}{2}a_c t^2$$
$$y = 0 + 0 - \tfrac{1}{2}gt^2 \qquad (2)$$

Eliminating t between Eqs. (1) and (2) and solving for y yields

$$y = \frac{-g}{2v_1^2}x^2 \qquad \textit{Ans.}$$

This parabolic path AB is shown in Fig. 12–16.

Note that the *range* R, or horizontal distance to where the cannonball strikes the ground, is determined by substituting $y = -h$ into the above equation and solving for $x = R$, i.e., $R = v_1\sqrt{2h/g}$. Furthermore, if an object was released from rest at A at the instant the cannon was fired, it would strike the ground at C at the same instant when the cannonball strikes the ground at B. Why? (Refer to Eq. (2).)

Example 12–9

A ball is thrown from a position 1.5 m above the ground to the roof of a 30-m-high building, as shown in Fig. 12–17. If the initial velocity of the ball is 40 m/s, inclined at an angle of 60° from the horizontal, determine (a) the maximum height h attained, and (b) the range or horizontal distance d from the point where the ball was thrown to where it strikes the roof.

Solution

Part (a). *Step 1:* When the motion is analyzed between points O and A, the three unknowns are represented as the height h, horizontal distance d_{OA}, and time of flight t_{OA}. Here it is necessary only to determine h. Establishing the origin of coordinates at O, Fig. 12–17, the ball's initial velocity has components of

$$(v_x)_O = 40 \cos 60° = 20 \text{ m/s}$$
$$(v_y)_O = 40 \sin 60° = 34.64 \text{ m/s}$$

When the ball is at A, $(v_x)_A = (v_x)_O = 20$ m/s, since the horizontal component of velocity is always constant. Also, $(v_y)_A = 0$. Why?

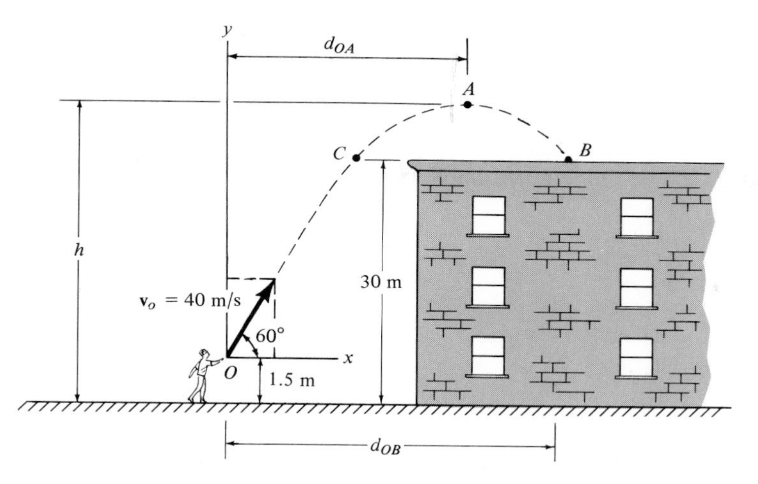

Fig. 12–17

Step 2: Since the ball's vertical components of velocity are known at points O and A, and the position at O is $(s_y)_O = 0$, and at A, $(s_y)_A = (h - 1.5)$; the height h can be computed *directly* without consideration of the (unknown) time t_{OA} by using the equation

$$(+\uparrow) \qquad (v_y)_A^2 = (v_y)_O^2 + 2a_c[(s_y)_A - (s_y)_O]$$

$$0 = (34.64)^2 + 2(-9.81)[(h - 1.5) - 0]$$

$$h = 62.7 \text{ m} \qquad \textit{Ans.}$$

Part (b). *Step 1:* When the motion is analyzed between points O and B, the three unknowns are represented as the distance d_{OB}, time of flight t_{OB}, and vertical component of velocity $(v_y)_B$. (The horizontal component of velocity is $(v_x)_B = (v_x)_O = 20$ m/s.)
Step 2: The distance d_{OB} is related to the time t_{OB} by the equation

$$(\overset{+}{\rightarrow}) \qquad (s_x)_B = (s_x)_O + (v_x)_O t_{OB}$$

$$d_{OB} = 0 + 20t_{OB} \qquad (1)$$

The time t_{OB} can also be related to the *known* initial velocity in the y direction and the initial and final elevations of the ball by the equation

$$(+\uparrow) \qquad (s_y)_B = (s_y)_O + (v_y)_O t_{OB} + \tfrac{1}{2}a_c t_{OB}^2$$

$$(30 - 1.5) = 0 + 34.64t_{OB} + \tfrac{1}{2}(-9.81)t_{OB}^2$$

Solving for the two roots, using the quadratic formula, we have

$$t_{OC} = 0.951 \text{ s} \quad \text{and} \quad t_{OB} = 6.111 \text{ s}$$

The first root (shortest time) is designated as t_{OC}, since it represents the time needed for the ball to reach point C, which has the same elevation as point B, Fig. 12–17. Substituting t_{OB} into Eq. (1) and solving for d_{OB} yields

$$d_{OB} = 20(6.111) = 122.2 \text{ m} \qquad \textit{Ans.}$$

Example 12–10

When a ball is kicked from A as shown in Fig. 12–18, it just clears the top of a wall at B. Knowing that the distance from A to the wall is 20 m and the wall is 4 m high, determine the initial speed at which the ball was kicked. In the calculation, neglect the size of the ball.

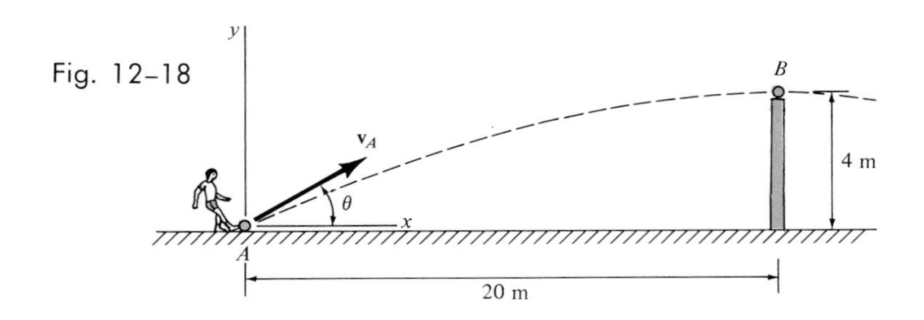

Fig. 12–18

Solution

Step 1: The unknowns are represented by the initial speed v_A, angle of inclination θ, and the time t_{AB} to travel from A to B, Fig. 12–18. At the highest point, B, the velocity $(v_y)_B = 0$, and $(v_x)_B = (v_x)_A = v_A \cos\theta$.

Step 2: Motion in the horizontal direction from A to B requires that

$$(\overset{+}{\rightarrow}) \qquad (s_x)_B = (s_x)_A + (v_x)_A t_{AB}$$

$$20 = 0 + (v_A \cos\theta) t_{AB} \qquad (1)$$

The following two equations apply to motion in the vertical direction:

$$(+\uparrow) \qquad (v_y)_B = (v_y)_A + a_c t_{AB}$$

$$0 = v_A \sin\theta - 9.81 t_{AB} \qquad (2)$$

$$(+\uparrow) \qquad (v_y)_B^2 = (v_y)_A^2 + 2a_c[(s_y)_B - (s_y)_A]$$

$$0 = v_A^2 \sin^2\theta + 2(-9.81)(4 - 0) \qquad (3)$$

To obtain v_A, eliminate t_{AB} from Eqs. (1) and (2), which yields

$$v_A^2 \sin\theta \cos\theta = 196.2 \qquad (4)$$

Solve for v_A^2 in Eq. (4) and substitute into Eq. (3), so that

$$\frac{\sin\theta}{\cos\theta} = \tan\theta = \frac{2(9.81)(4)}{196.2} = 0.4$$

$$\theta = \tan^{-1}(0.4) = 21.8^\circ$$

Then using Eq. (4), the required initial speed is

$$v_A = \sqrt{\frac{196.2}{(\sin 21.8^\circ)(\cos 21.8^\circ)}} = 23.9 \text{ m/s}$$

*12–36. The curvilinear motion of a particle is defined by $x = 3t^2$, $y = 4t + 2$, and $z = 6t^3 - 8$, where the x, y, z position is given in metres and the time in seconds. Determine the magnitude and direction of both the velocity and acceleration of the particle when $t = 2$ s.

12–37. If the velocity of a particle is defined as $\mathbf{v}(t) = \{0.8t^2\mathbf{i} + 12t^{1/2}\mathbf{j} + 5\mathbf{k}\}$ m/s, determine the magnitude and direction of the particle's acceleration when $t = 2$ s.

12–38. The velocity of a particle is given by $\mathbf{v} = \{16t^2\mathbf{i} + 4t^3\mathbf{j} + (5t + 2)\mathbf{k}\}$ m/s, where t is measured in seconds. If the particle is at the origin when $t = 0$, determine the magnitude of the particle's acceleration when $t = 2$ s. Also, what is the x, y, z coordinate position of the particle at this instant?

12–39. A particle moves with curvilinear motion in the x-y plane such that the y component of motion is described by the equation $y = 7t^3$, where y is in metres and t is in seconds. If the particle starts from rest at the origin when $t = 0$, and maintains a *constant* acceleration in the x direction of 12 m/s², determine the particle's speed when $t = 2$ s.

***12–40.** A golf ball lands a horizontal distance of $R = 75$ m away from the point at which it was struck. If it is shot at an angle of $\theta = 30°$ from the horizontal, determine the initial speed at which it left the tee.

***12–40a.** Solve Prob. 12–40 if $R = 150$ ft and $\theta = 15°$. The constant downward acceleration due to gravity is $g = 32.2$ ft/s².

12–41. A bird is flying with a speed of 6 m/s at a constant altitude of 10 m. If a worm accidentally drops from the bird's mouth, determine the speed and compute the angle θ at which the worm strikes the ground. What is the distance between the bird and the worm at this instant? Assume that when the worm is initially dropped it has a horizontal velocity of 6 m/s.

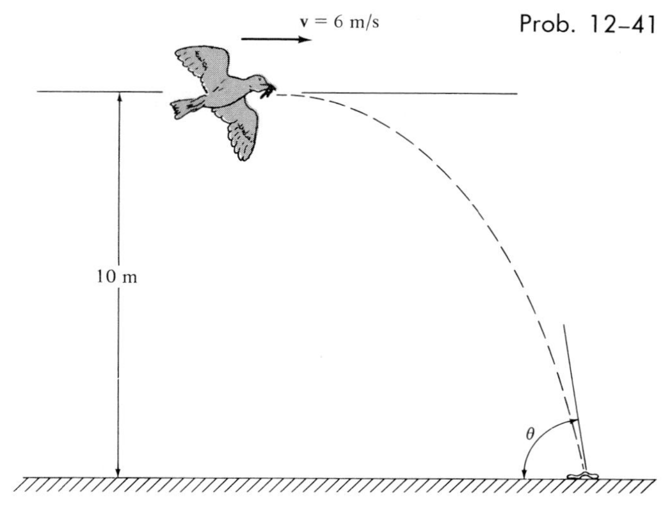

Prob. 12–41

12–42. The nozzle of a garden hose discharges water at the rate of 15 m/s. If the nozzle is held at the ground level and directed $\theta = 30°$ from the ground, determine the maximum height reached by the water and the horizontal distance from the nozzle to where the water strikes the ground.

12–43. Show that if a projectile is fired at an angle θ from the horizontal with an initial velocity v_1, the *maximum* range (farthest horizontal distance) the projectile can travel is given by $R_{max} = v_1^2/g$, where g is the acceleration of gravity. What is the angle θ for this condition?

***12–44.** Determine the minimum speed v_A which the toboggan must have when it approaches the jump at point A so that it reaches the other side of the gorge.

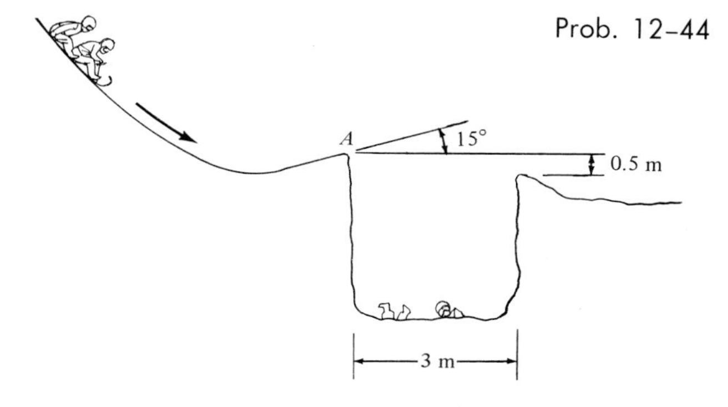

Prob. 12–44

12–45. A basketball is tossed from A at an angle of 30° from the horizontal. Determine the speed v_A at which the ball is released in order to make the basket B. With what speed does the ball pass through the hoop if $R = 10$ m, $h_A = 1.5$ m, and $h_B = 3$ m? Neglect the size of the ball in the calculation.

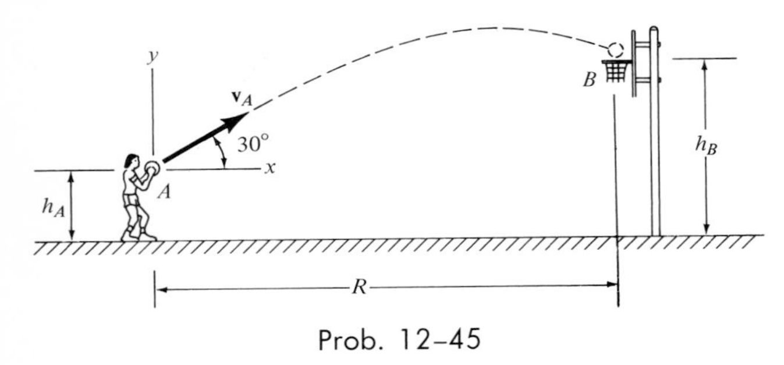

Prob. 12–45

12–45a. Solve Prob. 12–45 if in the diagram $h_A = 6$ ft, $h_B = 10$ ft, and $R = 25$ ft. The constant acceleration due to gravity is $g = 32.2$ ft/s².

12–46. A boy at A throws a ball 45° from the horizontal such that it strikes the slope at B. Determine the speed at which the ball is thrown and the time of flight.

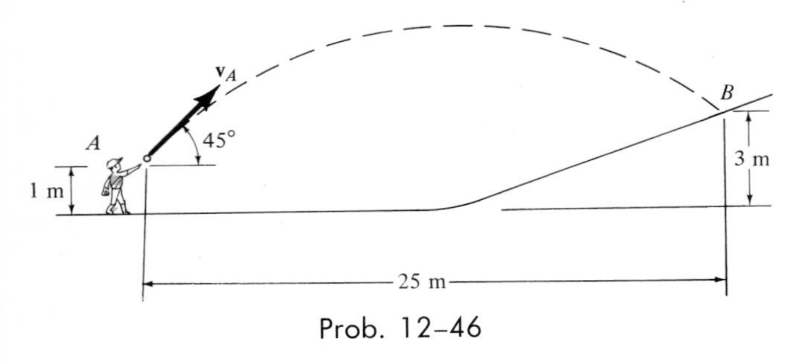

Prob. 12–46

12–47. Small packages traveling on the conveyor belt fall off into a 1-m-long loading car. If the conveyor is running at a constant speed of $v_c = 2$ m/s, determine the smallest and largest distance R at which the end A of the car may be placed from the conveyor so that the packages enter the car.

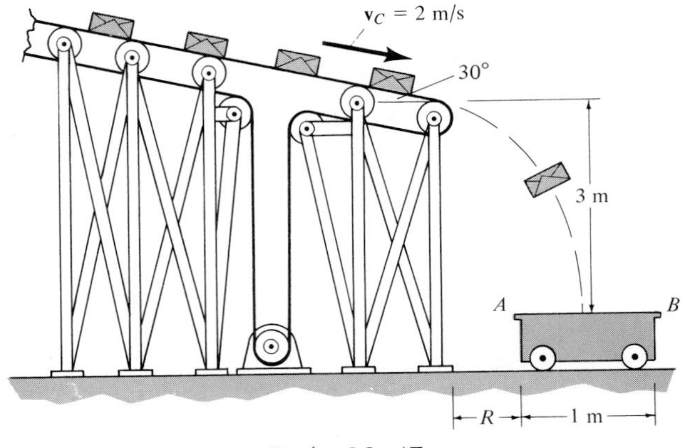

Prob. 12–47

*** 12–48.** A toy missile is fired with an initial velocity of 15 m/s, perpendicular to a 30° slope at A. If the missile travels in free flight, determine the range R of the path of motion.

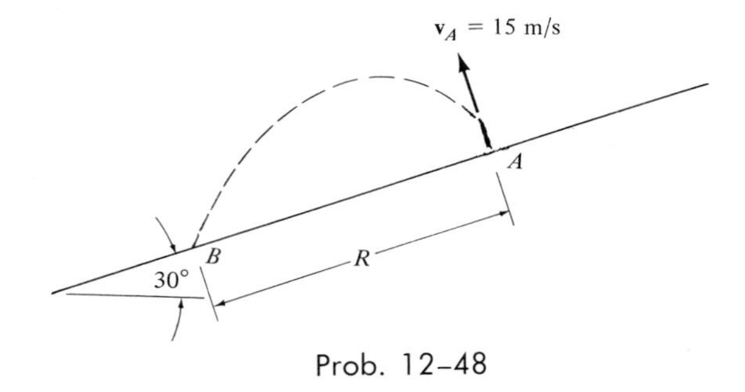

Prob. 12–48

12–49. The man throws a ball with an initial speed of 20 m/s. Determine the two possible angles θ_1 and θ_2 of release so that the ball strikes point B.

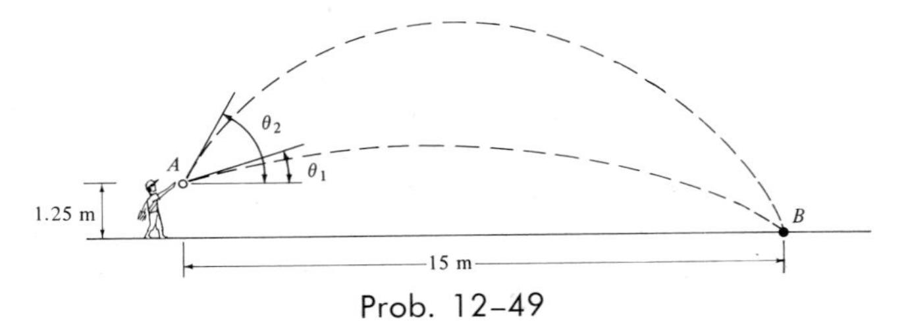

Prob. 12–49

12–50. Two projectiles are to be fired from a cannon, each projectile having a muzzle speed of $v_m = 400$ m/s. If it is required that the target, located $R = 12$ km away, be struck by both projectiles at the *same time,* determine the angles θ_1 and θ_2 at which the cannon barrel must be inclined when each projectile is fired. Also, determine the time between each firing. The cannon is first fired from θ_1 ($> \theta_2$), then it is rotated down to θ_2 and fired again. Neglect the height of the cannon in the calculation.

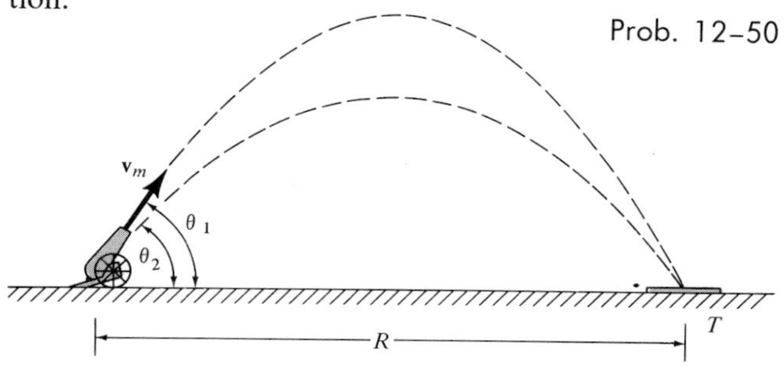

Prob. 12–50

12–50a. Solve Prob. 12–50 if $v_m = 1500$ ft/s and $R = 35(10^3)$ ft. The constant downward acceleration due to gravity is $g = 32.2$ ft/s^2.

12–51. A boy throws a snowball horizontally with a speed of $v_1 = 12$ m/s off a bridge in an effort to hit the top surface AB of a truck traveling directly underneath the boy on the bridge. If the truck maintains a constant speed of $v_t = 15$ m/s, and the snowball is released at the instant point B on the top of the truck appears at point C, determine the position s where the snowball strikes the top of the truck. (Experimental verification is not recommended.)

Prob. 12–51

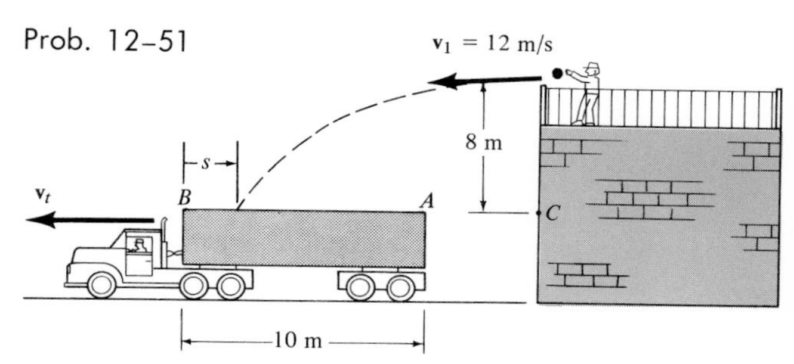

***12–52.** Solve Prob. 12–51 assuming that the truck has a constant acceleration of $a_t = 1.5$ m/s^2 and that its initial velocity when point B is at point C is $v_t = 15$ m/s.

12–7. Curvilinear Motion of a Particle: Cylindrical Components

Cylindrical Components. For some engineering problems it is often convenient to express the path of motion using a system of fixed *cylindrical coordinates.* For example, consider the position $P(r, \theta, z)$ of the particle located on the space curve s shown in Fig. 12–19*a*. The *axial component z* is identical to that used for rectangular coordinates. The *radial component r* is perpendicular to the z axis, with the positive direction extending *away* from the origin O. Finally, the *transverse component* θ is the angle between the positive x and r axes. This angle is generally measured in radians, where 1 rad $= 180°/\pi$. Using the right-hand rule, that is, with the thumb pointing in the positive z direction, the fingers curl in the direction of *positive* θ.

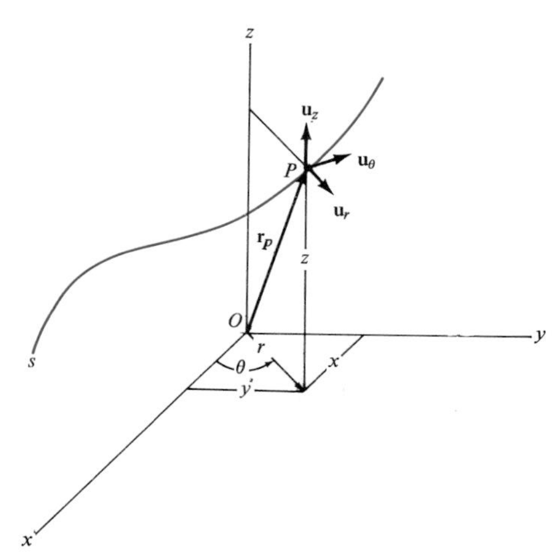

Fig. 12–19(a)

From Fig. 12–19*a*, r, θ, and z are related to x, y, and z by the transformation equations

$$x = r \cos \theta$$
$$y = r \sin \theta \qquad (12\text{–}25)$$
$$z = z$$

or

$$r = \sqrt{x^2 + y^2}$$

$$\theta = \tan^{-1}\frac{y}{x} \qquad (12\text{–}26)$$

$$z = z$$

Unit Vectors. An orthogonal set of unit vectors acting in the positive directions of r, θ, and z will be designated as $\mathbf{u}_r$, $\mathbf{u}_\theta$, and $\mathbf{u}_z$, respectively, Fig. 12–19*a*. As shown in Figs. 12–19*b* and 12–19*c*, these vectors are related to the **i, j, k** Cartesian unit vectors by the transformation equations

$$\mathbf{u}_r = \cos\theta\mathbf{i} + \sin\theta\mathbf{j} \qquad (12\text{–}27)$$

$$\mathbf{u}_\theta = -\sin\theta\mathbf{i} + \cos\theta\mathbf{j} \qquad (12\text{–}28)$$

$$\mathbf{u}_z = \mathbf{k} \qquad (12\text{–}29)$$

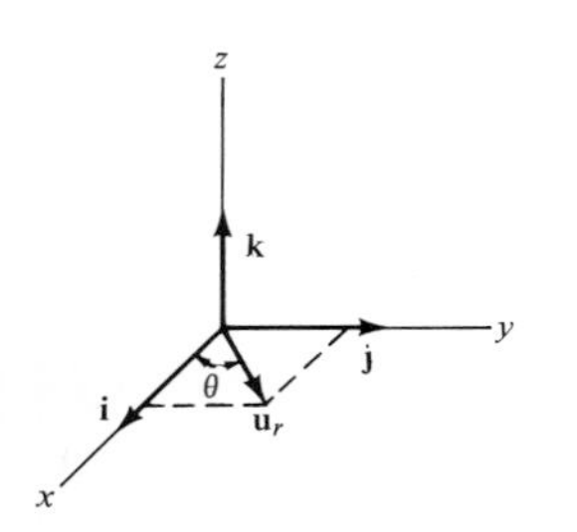

(b)

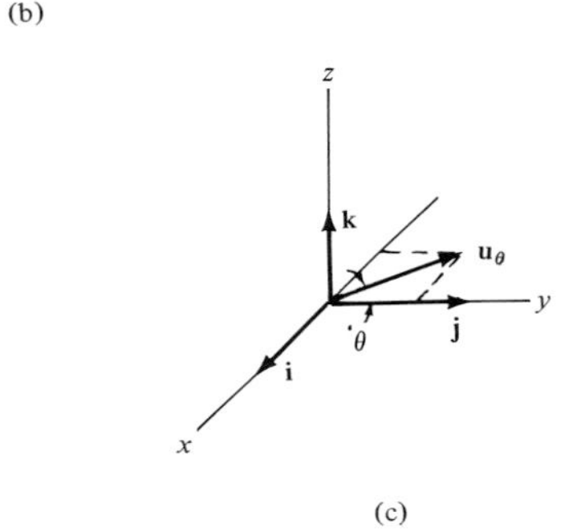

(c)

Fig. 12–19(b,c)

Position. At any instant of time, the position vector $\mathbf{r}_P$, which locates the particle, Fig. 12–19*a*, is expressed only in terms of components in the $\mathbf{u}_r$ and $\mathbf{u}_z$ directions, i.e.,

$$\mathbf{r}_P = r\mathbf{u}_r + z\mathbf{u}_z \qquad (12\text{–}30)$$

Although $\mathbf{u}_\theta$ is not involved in this formulation, notice that $\mathbf{u}_r$ is a function of θ, Eq. 12–27, so that the location of the particle is uniquely specified either from the x, y, z or r, θ, z axes.

The magnitude of $\mathbf{r}_P$ is

$$r_P = \sqrt{r^2 + z^2} \qquad (12\text{–}31)$$

Velocity. The instantaneous velocity $\mathbf{v}$ is obtained by taking the time derivative of Eq. 12–30. Using a dot to represent time differentiation, we have

$$\mathbf{v} = \dot{\mathbf{r}}_P = \dot{r}\mathbf{u}_r + r\dot{\mathbf{u}}_r + \dot{z}\mathbf{u}_z + z\dot{\mathbf{u}}_z \qquad (12\text{–}32)$$

The term $\dot{\mathbf{u}}_r = d\mathbf{u}_r/dt$ is determined from Eq. 12–27, in which case

$$\dot{\mathbf{u}}_r = -(\sin\theta)\dot{\theta}\mathbf{i} + (\cos\theta)\dot{\theta}\mathbf{j} = \dot{\theta}(-\sin\theta\mathbf{i} + \cos\theta\mathbf{j})$$

Using Eq. 12–28, we obtain

$$\dot{\mathbf{u}}_r = \dot{\theta}\mathbf{u}_\theta \qquad (12\text{–}33)$$

Since $\mathbf{u}_z$ does not change with time for a fixed x, y, z reference, the last term in Eq. 12–32 is zero. Hence, Eq. 12–32 can be written in component form as

$$\mathbf{v} = v_r\mathbf{u}_r + v_\theta\mathbf{u}_\theta + v_z\mathbf{u}_z \qquad (12\text{–}34)$$

where

$$v_r = \dot{r}$$
$$v_\theta = r\dot{\theta} \qquad (12\text{–}35)$$
$$v_z = \dot{z}$$

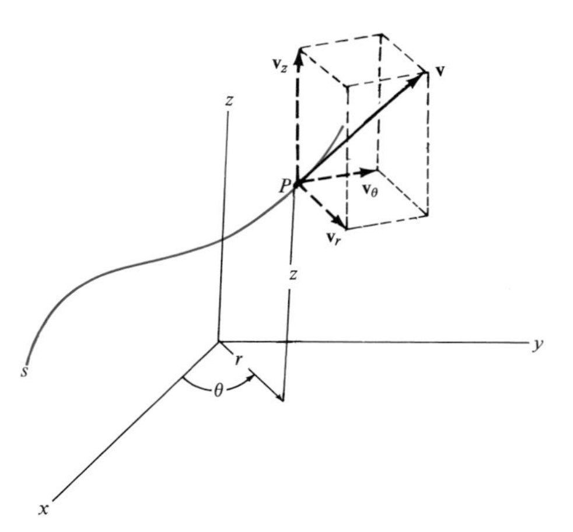

Fig. 12–19(d)

In particular, note that the term $\dot{\theta} = d\theta/dt$ provides a measure of the time rate of change of the angle θ. Common units used for this measurement are rad/s.

The *magnitude* of velocity is simply

$$v = \sqrt{v_r^2 + v_\theta^2 + v_z^2} \qquad (12\text{–}36)$$

and the *direction* of **v** is, of course, tangent to the path at point P, Fig. 12–19*d*.

Acceleration. Taking the time derivative of Eq. 12–34 using Eq. 12–35, and recalling that $\dot{\mathbf{u}}_z = \mathbf{0}$, we obtain the particle's instantaneous acceleration **a**.

$$\mathbf{a} = \dot{\mathbf{v}} = \ddot{r}\mathbf{u}_r + \dot{r}\dot{\mathbf{u}}_r + \dot{r}\dot{\theta}\mathbf{u}_\theta + r\ddot{\theta}\mathbf{u}_\theta + r\dot{\theta}\dot{\mathbf{u}}_\theta + \ddot{z}\mathbf{u}_z \qquad (12\text{–}37)$$

Using Eq. 12–28 to evaluate the term involving $\dot{\mathbf{u}}_\theta$ yields

$$\dot{\mathbf{u}}_\theta = -(\cos\theta)\dot{\theta}\mathbf{i} - (\sin\theta)\dot{\theta}\mathbf{j} = -\dot{\theta}[\cos\theta\mathbf{i} + \sin\theta\mathbf{j}]$$

From Eq. 12–27

$$\dot{\mathbf{u}}_\theta = -\dot{\theta}\mathbf{u}_r \qquad (12\text{–}38)$$

Hence, with Eqs. 12–33 and 12–38, Eq. 12–37 may be written as

$$\mathbf{a} = a_r\mathbf{u}_r + a_\theta\mathbf{u}_\theta + a_z\mathbf{u}_z \qquad (12\text{–}39)$$

where

$$a_r = \ddot{r} - r\dot{\theta}^2$$
$$a_\theta = r\ddot{\theta} + 2\dot{r}\dot{\theta} \qquad (12\text{–}40)$$
$$a_z = \ddot{z}$$

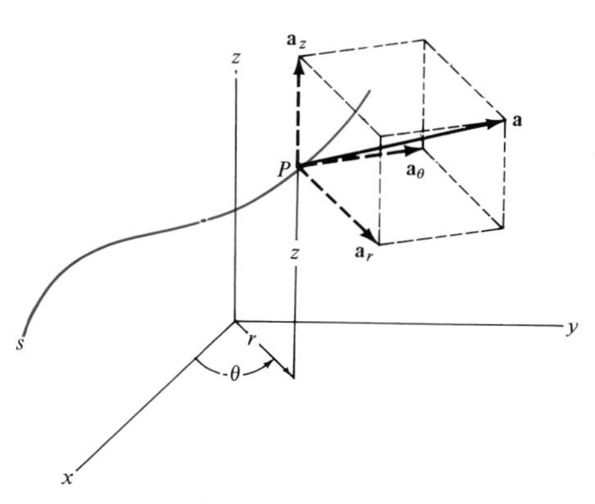

Fig. 12–19(e)

In particular, the term $\ddot{\theta} = d^2\theta/dt^2 = d/dt(d\theta/dt)$ is a measure of the change made in the rate of change of θ during an instant of time. Common units for this measurement are rad/s^2.

The *magnitude* of acceleration is simply

$$a = \sqrt{a_r^2 + a_\theta^2 + a_z^2} \qquad (12\text{–}41)$$

The *direction* is determined from the vector addition of each of its three components. In general, **a** will *not* be tangent to the path, Fig. 12–19*e*.

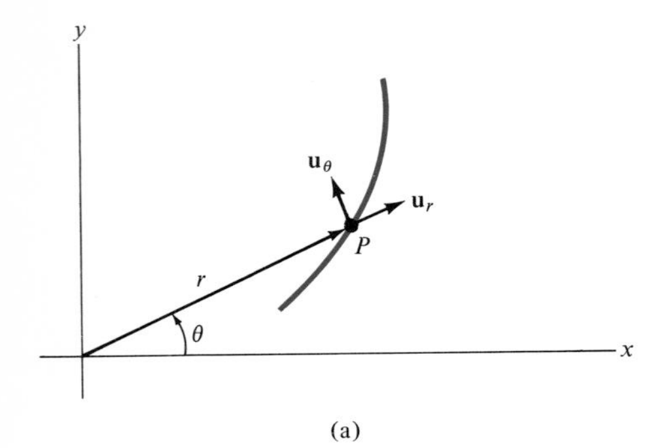

(a)

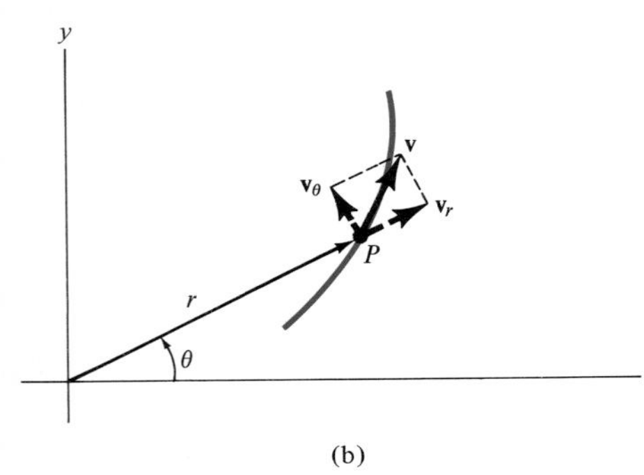

(b)

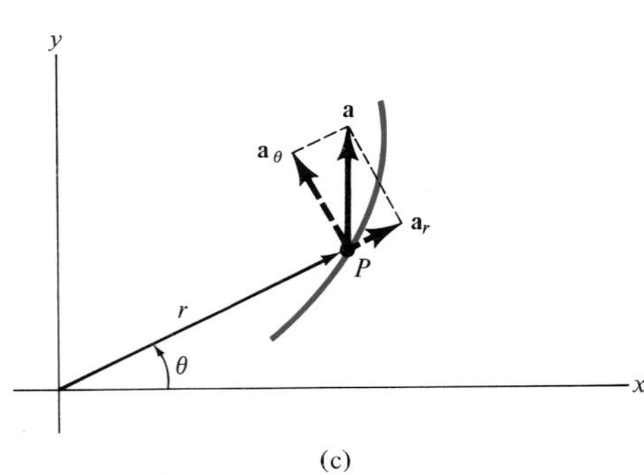

(c)

Fig. 12–20

Polar Coordinates. When motion of the particle is restricted to the x-y plane, $z = 0$ and polar coordinates r and θ can be used to describe the motion. In this case the above equations reduce to a simpler form:

Position. Fig. 12–20*a:*

$$\mathbf{r} = r\mathbf{u}_r \tag{12–42}$$

Velocity. Fig. 12–20*b:*

$$\mathbf{v} = v_r\mathbf{u}_r + v_\theta\mathbf{u}_\theta \tag{12–43}$$

where

$$\begin{aligned} v_r &= \dot{r} \\ v_\theta &= r\dot{\theta} \end{aligned} \tag{12–44}$$

The *magnitude* of **v** is

$$v = \sqrt{(\dot{r})^2 + (r\dot{\theta})^2} \tag{12–45}$$

Acceleration. Fig. 12–20*c:*

$$\mathbf{a} = a_r\mathbf{u}_r + a_\theta\mathbf{u}_\theta \tag{12–46}$$

where

$$\begin{aligned} a_r &= \ddot{r} - r\dot{\theta}^2 \\ a_\theta &= r\ddot{\theta} + 2\dot{r}\dot{\theta} \end{aligned} \tag{12–47}$$

The *magnitude* of **a** is

$$a = \sqrt{(\ddot{r} - r\dot{\theta}^2)^2 + (r\ddot{\theta} + 2\dot{r}\dot{\theta})^2} \tag{12–48}$$

PROCEDURE FOR ANALYSIS

Application of the preceding polar equations for computing the particle's velocity and acceleration requires a *straightforward substitution* once r and the time derivatives $\dot{r}$, $\ddot{r}$, $\dot{\theta}$, and $\ddot{\theta}$ have been computed and evaluated at the instant considered. Two types of problems generally occur:

1) If the coordinates are specified as time-parametric equations, $r = r(t)$ and $\theta = \theta(t)$, the derivatives are simply $\dot{r} = dr(t)/dt$, $\ddot{r} = d^2r(t)/dt^2$, $\dot{\theta} = d\theta(t)/dt$, and $\ddot{\theta} = d^2\theta(t)/dt^2$. (See Examples 12–11 and 12–12.)
2) If the time-parametric equations of motion are not given, it will be necessary to specify the path $r = f(\theta)$ and compute the relationship between the time derivatives using the chain rule of calculus. In this case,

$$\dot{r} = \frac{df(\theta)}{d\theta}\dot{\theta} \quad \text{and,} \quad \ddot{r} = \left(\frac{d^2f(\theta)}{d\theta^2}\dot{\theta}\right)\dot{\theta} + \frac{df(\theta)}{d\theta}\ddot{\theta}$$

Thus, if two of the *four* time derivatives $\dot{r}$, $\ddot{r}$, $\dot{\theta}$, and $\ddot{\theta}$ are *known,* the other two can be obtained from these equations. (See Example 12–13.) In some problems, however, two of these time derivatives may *not* be known; instead, the magnitude of the particle's velocity or acceleration may be specified. If this is the case $v^2 = \dot{r}^2 + (r\dot{\theta})^2$ and $a^2 = (\ddot{r} - r\dot{\theta}^2)^2 + (r\ddot{\theta} + 2\dot{r}\dot{\theta})^2$ may be used to obtain the necessary relationships involving $\dot{r}$, $\ddot{r}$, $\dot{\theta}$, and $\ddot{\theta}$.

Motion in three dimensions requires a simple extension of the above procedure to include $\dot{z}$ and $\ddot{z}$.

Example 12–11

The rod AB shown in Fig. 12–21 has a constant angular rotation of 5 rad/s and is advancing upward along the axis of the bolt at a constant rate of 10 mm/s. Express the velocity and acceleration of point A on the rod in terms of its cylindrical components.

Solution

Since Eqs. 12–35 and 12–40 will be used for the solution, it is necessary to obtain the first and second time derivatives of r, θ, and z. Noting that the radius r, angular rate of rotation $\dot{\theta}$, and velocity along the z axis are all *constant,* we have

$$r = 100 \text{ mm}, \qquad \dot{r} = 0, \qquad \ddot{r} = 0$$
$$\dot{\theta} = 5 \text{ rad/s}, \qquad \ddot{\theta} = 0$$
$$\dot{z} = 10 \text{ mm/s}, \qquad \ddot{z} = 0$$

Both $\dot{\theta}$ and $\dot{z}$ are positive, since they increase θ and z, respectively. Using the data, we therefore have

$$\begin{aligned}\mathbf{v}_A &= \dot{r}\mathbf{u}_r + r\dot{\theta}\mathbf{u}_\theta + \dot{z}\mathbf{u}_z \\ &= 0 + 100(5)\mathbf{u}_\theta + 10\mathbf{u}_z \\ &= \{500\mathbf{u}_\theta + 10\mathbf{u}_z\} \text{ mm/s} \qquad \textit{Ans.}\end{aligned}$$

$$\begin{aligned}\mathbf{a}_A &= (\ddot{r} - r\dot{\theta}^2)\mathbf{u}_r + (r\ddot{\theta} + 2\dot{r}\dot{\theta})\mathbf{u}_\theta + \ddot{z}\mathbf{u}_z \\ &= [0 - 100(5)^2]\mathbf{u}_r + [100(0) + 2(0)(5)]\mathbf{u}_\theta + 0\mathbf{u}_z \\ &= \{-2500\mathbf{u}_r\} \text{ mm/s}^2 \qquad \textit{Ans.}\end{aligned}$$

The path of motion for point A is a spiral. From the results the velocity $\mathbf{v}_A$ is tangent to this path and the acceleration $\mathbf{a}_A$ is directed toward the z axis, Fig. 12–21.

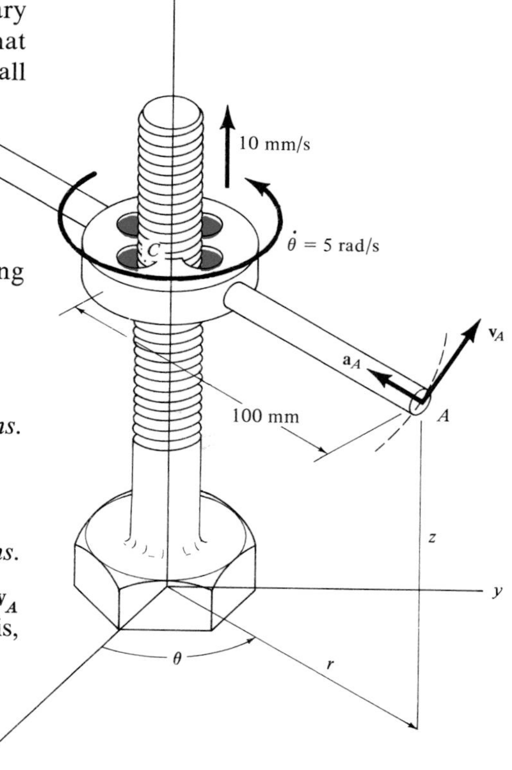

Fig. 12–21

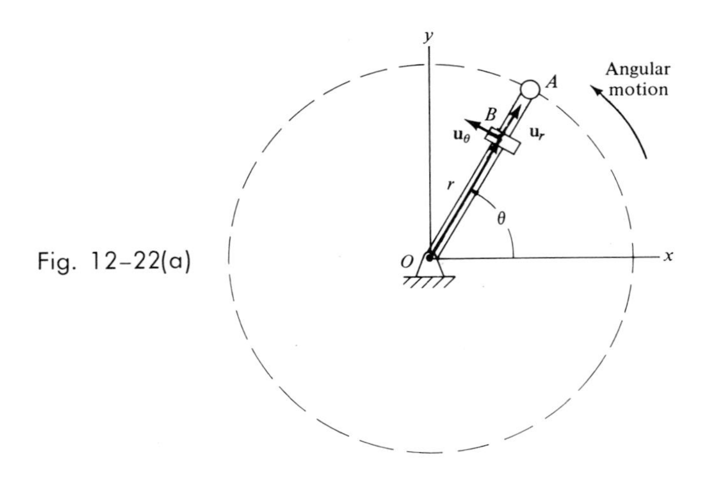

Fig. 12–22(a)

Example 12–12

The rod OA, shown in Fig. 12–22*a*, is rotating in the x-y plane such that at any instant $\theta = (t^{2/3})$ rad. At the same time, the collar B is sliding outward along OA so that $r = (100t^2)$ mm. If in both cases t is measured in seconds, determine the velocity and acceleration of the collar when $t = 1$ s.

Solution

Since motion occurs in the x-y plane (or r-θ plane) Eqs. 12–44 and 12–47 will be used for the solution. Computing the time derivatives, and evaluating each at $t = 1$ s, we have

$$r = 100t^2 \Big|_{t=1\text{ s}} = 100 \text{ mm} \qquad \theta = t^{2/3} \Big|_{t=1\text{ s}} = 1 \text{ rad} = 57.3^\circ$$

$$\dot{r} = 200t \Big|_{t=1\text{ s}} = 200 \text{ mm/s} \qquad \dot{\theta} = \tfrac{2}{3}t^{-1/3} \Big|_{t=1\text{ s}} = 0.667 \text{ rad/s}$$

$$\ddot{r} = 200 \Big|_{t=1\text{ s}} = 200 \text{ mm/s}^2 \qquad \ddot{\theta} = \tfrac{2}{9} - t^{-2/3} \Big|_{t=1\text{ s}} = -\ 0.222 \text{ rad/s}^2$$

The velocity of the collar is

$$\begin{aligned} \mathbf{v} &= \dot{r}\mathbf{u}_r + r\dot{\theta}\mathbf{u}_\theta \\ &= 200\mathbf{u}_r + 100(0.667)\mathbf{u}_\theta \\ &= \{200\mathbf{u}_r + 66.7\mathbf{u}_\theta\} \text{ mm/s} \end{aligned}$$

The magnitude of $\mathbf{v}$ is

$$v = \sqrt{(200)^2 + (66.7)^2} = 210.8 \text{ mm/s} \qquad \textit{Ans.}$$

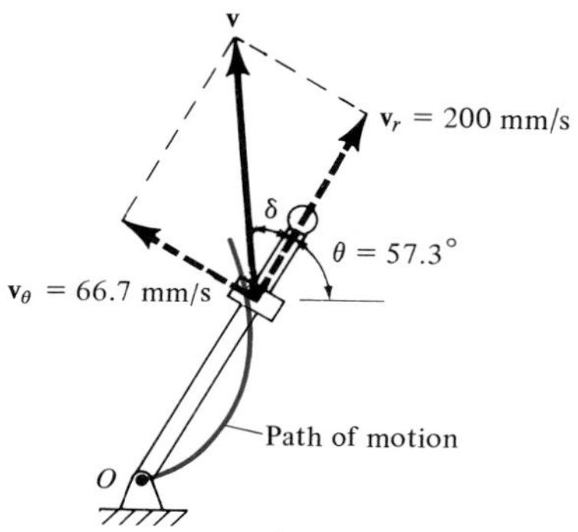

Fig. 12-22(b)

From Fig. 12-22*b*,

$$\delta = \tan^{-1}\left(\frac{66.7}{200}\right) = 18.4^\circ$$

So that the angle which $\mathbf{v}$ makes with the x axis is

$$\delta + 57.3^\circ = 75.7^\circ \quad \overset{\mathbf{v}}{\measuredangle}\,75.7^\circ \qquad \textit{Ans.}$$

The acceleration of the collar is

$$\begin{aligned}\mathbf{a} &= (\ddot{r} - r\dot{\theta}^2)\mathbf{u}_r + (r\ddot{\theta} + 2\dot{r}\dot{\theta})\mathbf{u}_\theta \\ &= [200 - 100(0.667)^2]\mathbf{u}_r + [100(-0.222) + 2(200)0.667]\mathbf{u}_\theta \\ &= \{155.5\mathbf{u}_r + 244.5\mathbf{u}_\theta\} \text{ mm/s}^2\end{aligned}$$

The magnitude of $\mathbf{a}$ is

$$a = \sqrt{(155.5)^2 + (289.0)^2} = 328.2 \text{ mm/s}^2 \qquad \textit{Ans.}$$

From Fig. 12-22*c*,

$$\phi = \tan^{-1}\left(\frac{289.0}{155.5}\right) = 61.7^\circ$$

so that the angle which $\mathbf{a}$ makes with the x axis is

$$\phi + 57.3^\circ = 119.0^\circ \quad \overset{\mathbf{a}}{\measuredangle}\,119.0^\circ \qquad \textit{Ans.}$$

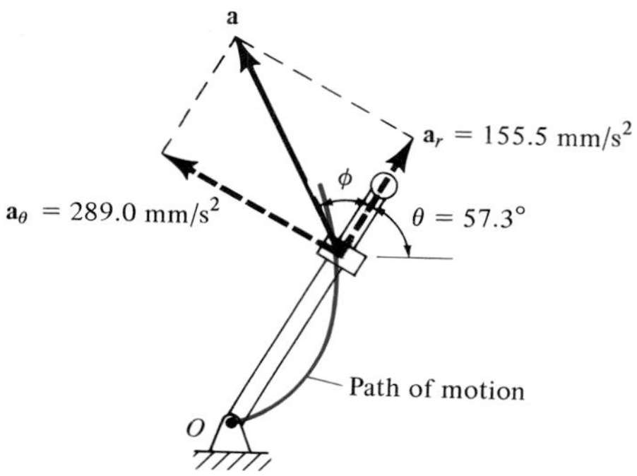

Fig. 12-22(c)

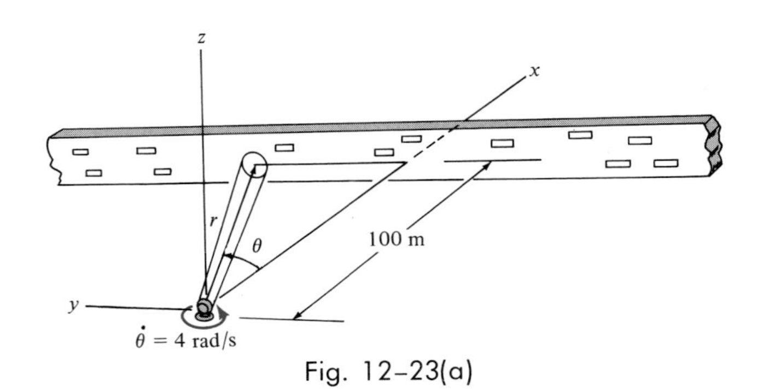

Fig. 12–23(a)

Example 12–13

The searchlight shown in Fig. 12–23*a* casts a spot of light along the face of a vertical wall that is located 100 m from the searchlight. Determine the magnitudes of the velocity and acceleration at which the spot appears to travel across the wall at the instant $\theta = 45°$. The searchlight is rotating about the z axis at a constant rate of $\dot{\theta} = 4$ rad/s.

Solution

Equations 12–44 and 12–47 will be used for the solution. To compute the necessary time derivatives, however, it is first necessary to relate r to θ. From Fig. 12–23*a*, this relation is

$$r = \frac{100}{\cos\theta} = 100 \sec\theta$$

Using the chain rule of calculus, noting that $d(\sec\theta) = \sec\theta \tan\theta \, d\theta$, and $d(\tan\theta) = \sec^2\theta \, d\theta$, we have

$$\begin{aligned}
\dot{r} &= 100 \sec\theta \tan\theta\dot{\theta} \\
\ddot{r} &= 100(\sec\theta \tan\theta\dot{\theta}) \tan\theta\dot{\theta} + 100 \sec\theta(\sec^2\theta\dot{\theta})\dot{\theta} + 100 \sec\theta \tan\theta(\ddot{\theta}) \\
&= 100 \sec\theta \tan^2\theta(\dot{\theta})^2 + 100 \sec^3\theta(\dot{\theta})^2 + 100 \sec\theta \tan\theta\ddot{\theta}
\end{aligned}$$

Since $\dot{\theta} = 4$ rad/s $=$ constant, then $\ddot{\theta} = 0$, and the above equations become

$$\begin{aligned}
\dot{r} &= 400 \sec\theta \tan\theta \\
\ddot{r} &= 1600(\sec\theta \tan^2\theta + \sec^3\theta)
\end{aligned}$$

The velocity of the light spot is thus

$$\begin{aligned}
\mathbf{v} &= \dot{r}\mathbf{u}_r + r\dot{\theta}\mathbf{u}_\theta \\
&= 400 \sec\theta \tan\theta\mathbf{u}_r + 100 \sec\theta\,(4)\mathbf{u}_\theta
\end{aligned}$$

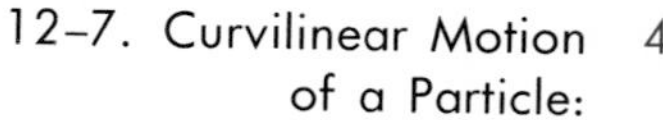

(b)

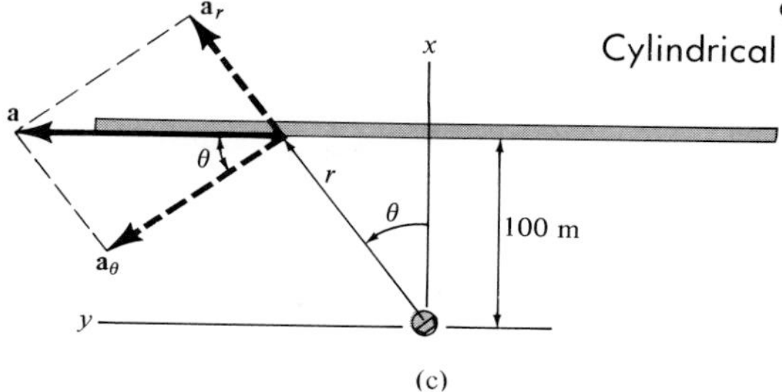

(c)

Fig. 12–23(b,c)

The magnitude of $\mathbf{v}$ at $\theta = 45°$ is, therefore,

$$v = \sqrt{v_r^2 + v_\theta^2} = \sqrt{(400 \sec 45° \tan 45°)^2 + (400 \sec 45°)^2}$$
$$= 800 \text{ m/s} \qquad \textit{Ans.}$$

As shown in Fig. 12-23*b*, the light spot travels with *rectilinear motion* along the wall. This can be seen from the above formulation by noting that

$$\frac{v_r}{v_\theta} = \frac{400 \sec\theta \tan\theta}{400 \sec\theta} = \tan\theta$$

The acceleration of the light spot is

$$\mathbf{a} = (\ddot{r} - r\dot{\theta}^2)\mathbf{u}_r + (r\ddot{\theta} + 2\dot{r}\dot{\theta})\mathbf{u}_\theta$$
$$= [1600(\sec\theta \tan^2\theta + \sec^3\theta) - 100 \sec\theta(4)^2]\mathbf{u}_r$$
$$+ [100 \sec\theta(0) + 2(400 \sec\theta \tan\theta)4]\mathbf{u}_\theta$$

Using the trigonometric identity $\sec^2\theta = 1 + \tan^2\theta$ in order to simplify the result, we have

$$\mathbf{a} = (3200 \sec\theta \tan^2\theta)\mathbf{u}_r + (3200 \sec\theta \tan\theta)\mathbf{u}_\theta$$

The magnitude of acceleration when $\theta = 45°$ is therefore

$$a = \sqrt{a_r^2 + a_\theta^2}$$
$$= \sqrt{(3200 \sec 45° \tan^2 45°)^2 + (3200 \sec 45° \tan 45°)^2}$$
$$= 6400 \text{ m/s}^2 \qquad \textit{Ans.}$$

Since the velocity does not change direction, the acceleration is directed along the wall as shown in Fig. 12–23*c*. This can be shown mathematically since

$$\frac{a_r}{a_\theta} = \frac{3200 \sec\theta \tan^2\theta}{3200 \sec\theta \tan\theta} = \tan\theta$$

12–53. A particle is moving along a circular path of 400-mm radius such that its position as a function of time is given by $\theta = (2t^2)$ rad, where t is in seconds. Determine the magnitude of the particle's acceleration when $\theta = 30°$. The particle starts from rest when $\theta = 0°$.

12–54. For a short time, measured from point A, the position of a roller coaster along its path is defined by the equations $r = 20$ m, $\theta = (0.2t)$ rad, and $z = (-10 \cos \theta)$ m, where t is measured in seconds. Determine the magnitudes of the roller coaster's velocity and acceleration at the instant $t = 3$ s.

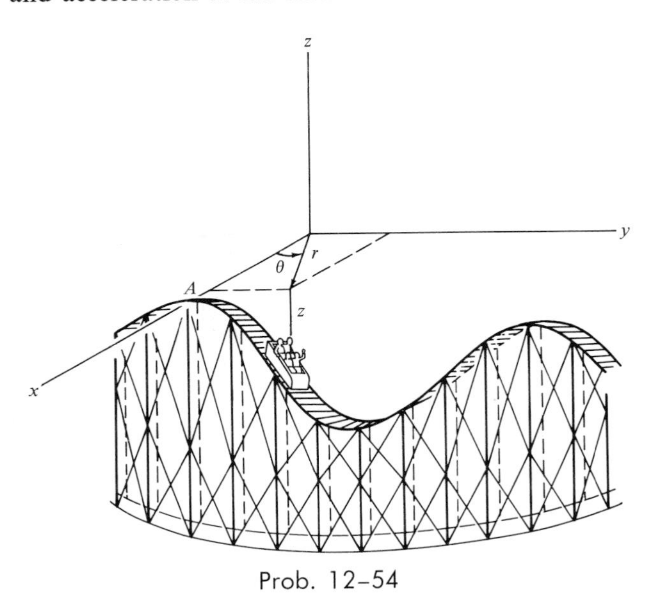

Prob. 12–54

12–55. The position of a particle may be defined in terms of its rectangular components as $x = 3t, y = -4t$, and $z = 2t^2$, where x, y, and z are measured in metres and t is measured in seconds. Express the velocity and acceleration of the particle in terms of its cylindrical components, and determine the magnitudes of the particle's velocity and acceleration when $t = 1.5$ s.

12–55a. Solve Prob. 12–55 if $x = 3t^2$, $y = 4t^2$, and $z = -3t$, where x, y, and z are in feet and t is in seconds. Evaluate the velocity and acceleration at $t = 2$ s.

* **12–56.** The small washer is sliding down the cord OA. When it is halfway down the cord, its speed is 200 mm/s and its acceleration (directed toward point O) has a magnitude of 10 mm/s^2. Express the velocity and acceleration of the washer at this point in terms of its cylindrical vector components.

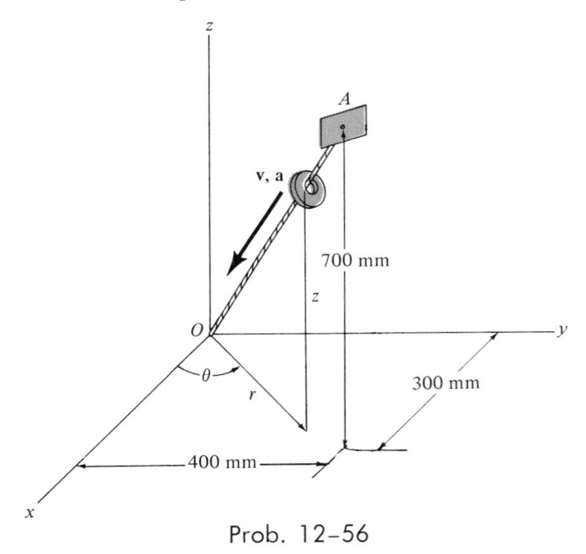

Prob. 12–56

12–57. The time rate of change of acceleration is referred to as the *jerk,* which is often used as a means of measuring passenger discomfort. Calculate this vector, $\dot{\mathbf{a}}$, in terms of its cylindrical components, using Eqs. 12–40.

12–58. A particle P moves along the spiral path $r = (100/\theta)$ mm with a constant speed of $v =$

500 mm/s. Determine the magnitudes v_r and v_θ as functions of θ and evaluate each at $\theta = 1$ rad.

Prob. 12–58

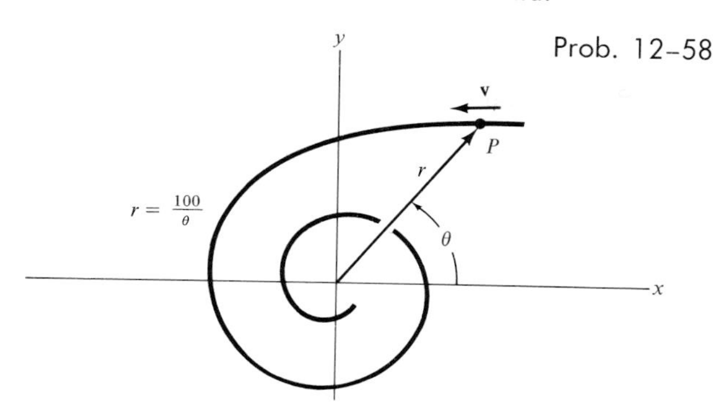

12–59. A cameraman standing at A is following the movement of a race car, B, which is traveling along a straight track at a constant speed of 30 m/s. Determine the angular rate at which he must turn in order to keep the camera directed on the car at the instant $\theta = 60°$.

Prob. 12–59

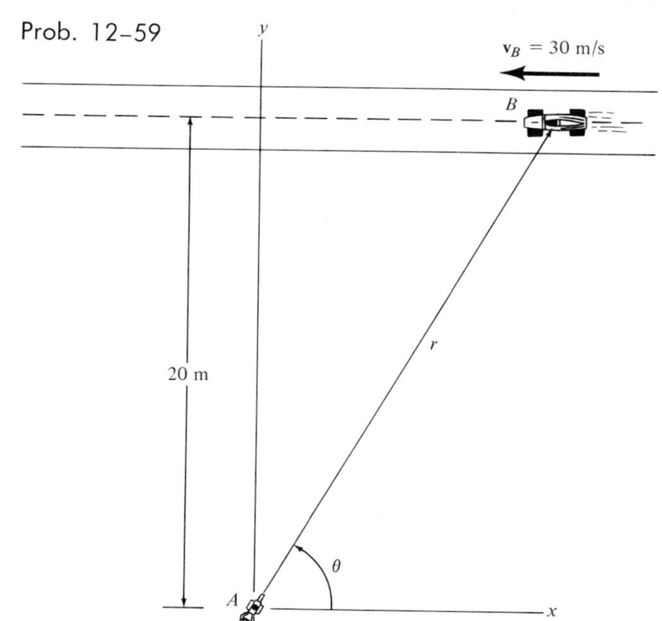

***12–60.** The motion of particle B is controlled by the rotation of the grooved link OA. If the link is rotating at a constant angular rate of $\dot{\theta} = 6$ rad/s, determine the magnitudes of the velocity and acceleration of B at the instant $\theta = \pi/2$ rad. The spiral path is defined by the equation $r = (40\,\theta)$ mm, where θ is measured in radians.

Prob. 12–60

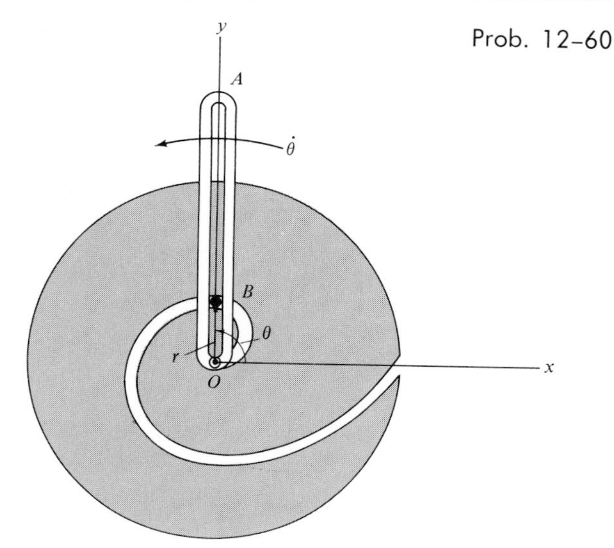

***12–60a.** Solve Prob. 12–60 if $\dot{\theta} = 5$ rad/s and the curve is defined by $r = (0.1\theta)$ ft, where θ is in radians.

12–61. The rod OA rotates counterclockwise with a constant angular rate of $\dot{\theta} = 5$ rad/s. Two pin-connected slider blocks, located at B, move freely on OA and the curved rod whose shape is a limaçon described by the equation $r = 100(2 - \cos\theta)$ mm. Determine the speed of the slider blocks at the instant $\theta = 120°$.

Prob. 12–61

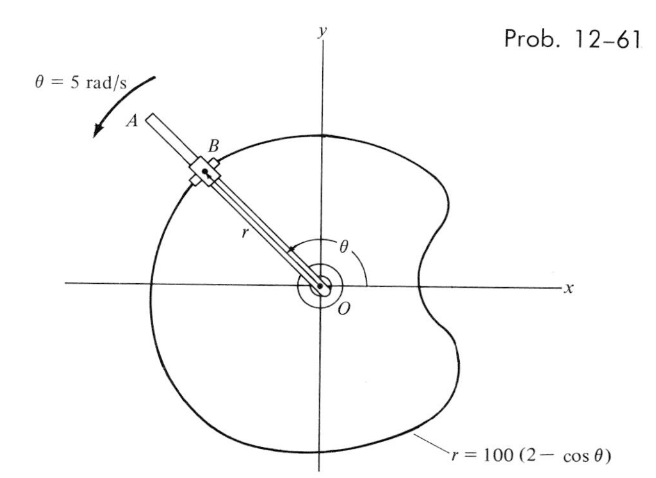

12-62. Determine the magnitude of acceleration of the slider blocks at B in Prob. 12–61 when $\theta = 120°$.

12-63. The slotted link is pinned at O, and as a result of rotation it drives the peg P along the horizontal guide. Compute the magnitudes of the velocity and acceleration of P as a function of θ if $\theta = (2t)$ rad, where t is measured in seconds.

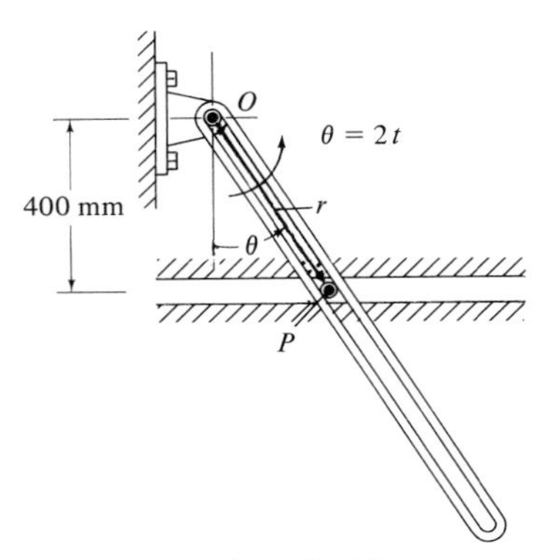

Prob. 12–63

*** 12-64.** A cameraman standing at A is following the movement of a race car B which is traveling around a curved track at a constant speed of 30 m/s. Determine the angular rate $\dot{\theta}$ at which the man must turn in order to keep the camera directed on the car at the instant $\theta = 30°$.

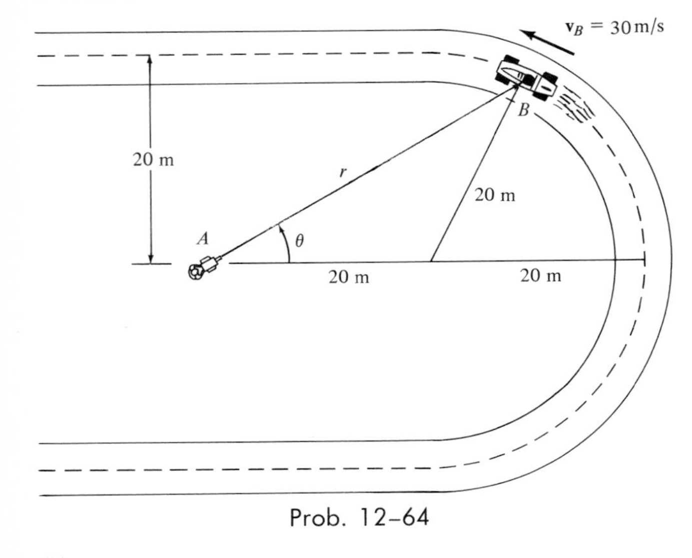

Prob. 12–64

12-65. The cylindrical cam C is held fixed while the rod AB and bearings E and F rotate about the z axis of the cam at a constant rate of $\dot{\theta} = 2$ rad/s. If the rod is free to slide through the bearings, determine the magnitudes of the velocity and acceleration of the guide D on the rod as a function of θ if the guide follows the groove in the cam. The groove is defined by the equations $r = 50$ mm and $z = (50 \cos \theta)$ mm, where $a = 50$ mm on the diagram.

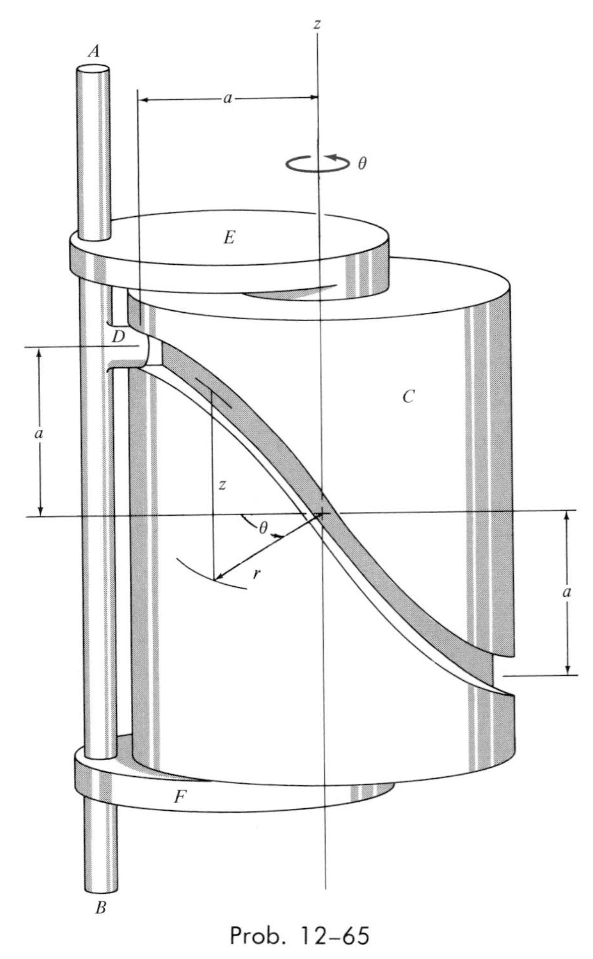

Prob. 12–65

12-65a. Solve Prob. 12–65 if $\dot{\theta} = 3$ rad/s, $r = 0.25$ ft, $z = (0.25 \cos \theta)$ ft, and $a = 0.25$ ft.

12–66. The peg B is confined to move in the circular groove and is forced into motion by the slotted guide A. If the guide is moving upward with a constant velocity of 4 m/s, determine the magnitude of the velocity of B at the instant $\theta = 45°$. *Hint:* For part of the solution, express y as a function of θ; then take the time derivative to determine $\dot{y}$ as a function of $\dot{\theta}$. $\dot{\theta}$ is obtained by noting that $\dot{y} = 4$ m/s.

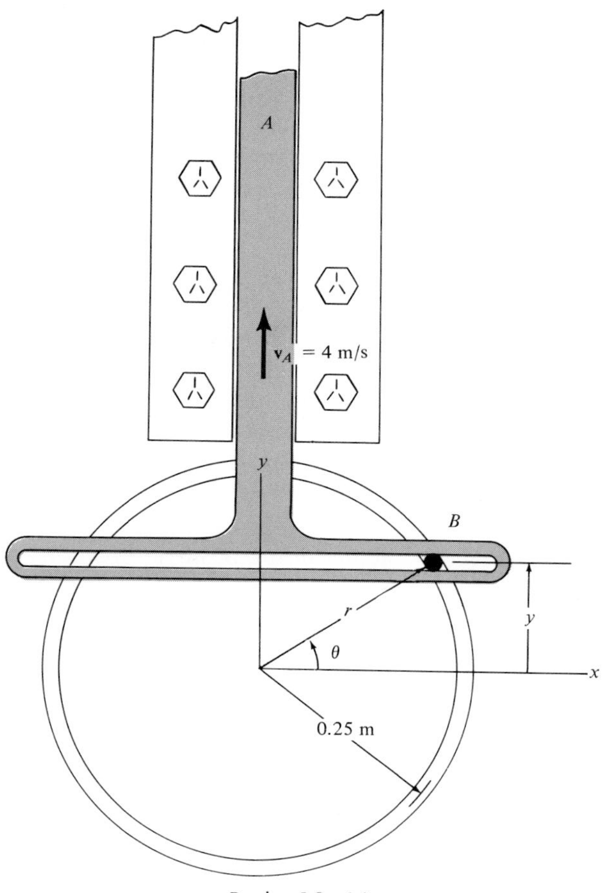

Prob. 12–66

12–67. Determine the acceleration of the peg in Prob. 12–66 at the instant $\theta = 45°$. Use the data in Prob. 12–66.

***12–68.** The automobile is traveling from a parking deck down along a cylindrical spiral ramp at a constant speed of $v = 2$ m/s. If the ramp descends a distance of 10 m for every full revolution, $\theta = 2\pi$ rad, determine the magnitude of the car's acceleration as it moves along the ramp, $r = 8$ m. *Hint:* For part of the solution, note that the tangent to the ramp at any point acts at an angle of $\phi = \tan^{-1}[10/2\pi(8)] = 11.25°$ from the horizontal. Use this to determine the velocity components v_z and v_θ, which in turn are used to determine $\dot{\theta}$ and $\dot{z}$.

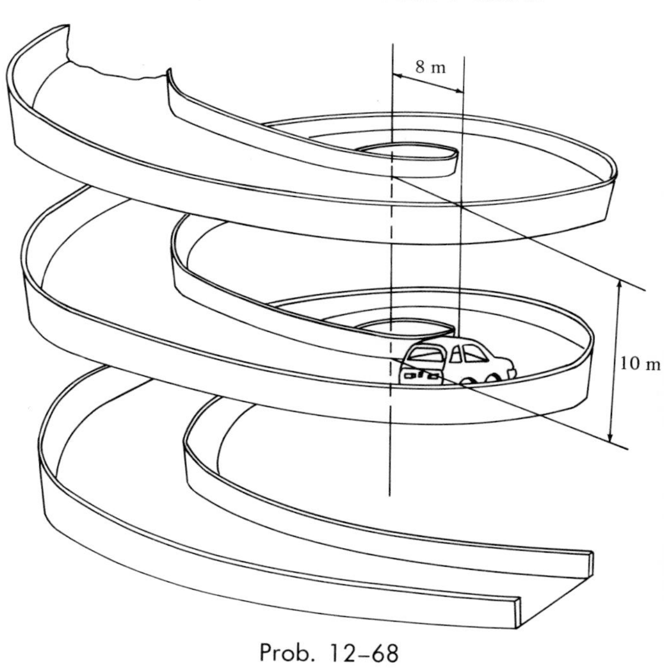

Prob. 12–68

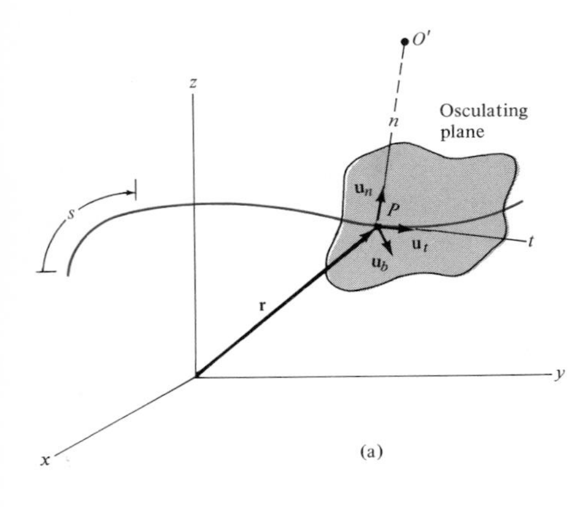

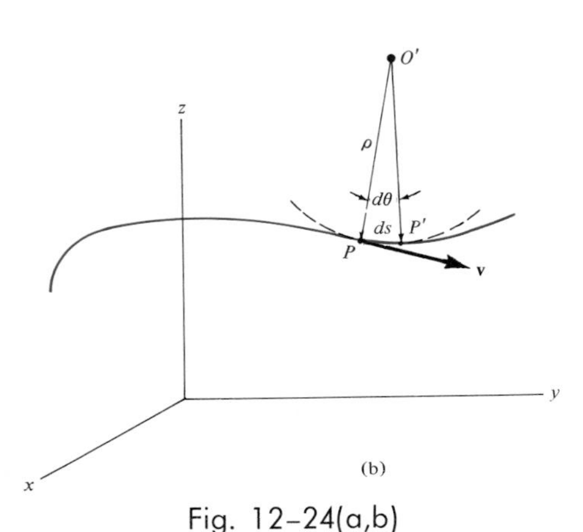

Fig. 12–24(a,b)

12–8. Curvilinear Motion of a Particle: Normal and Tangential Components

Normal and Tangential Components. When the path along which the particle is moving is *known*, it is often convenient to describe the motion using n and t coordinates which act normal and tangent to the path, respectively, and at the instant considered, have their *origin located at the particle P*. A set of such coordinates is shown in Fig. 12–24*a*. The positive tangent axis t is directed along the space curve in the direction of *increasing s*. This direction is always uniquely specified. There is, however, an infinite number of straight lines which can be constructed normal to the tangent axis at P. In order to make a unique choice for the n axis, it is necessary to consider the fact that geometrically the curve consists of a series of *differential arc segments ds*. As shown in Fig. 12–24*b*, each segment ds is constructed from the arc of a *unique circle* having a *radius of curvature* ρ (rho) and *center of curvature O′*. The normal axis n which will be chosen is directed from P *towards* the center of curvature O'. This normal is called the *principal normal* to the curve at P.

The plane containing the t and n axes is referred to as the *osculating plane*, Fig. 12–24*a*. For most applications, the path coordinates n and t are *used for solving planar motion problems,* so that the osculating plane remains fixed in the plane of motion. Although it is possible to obtain the principal normal n and the direction of the osculating plane for space curves*, it is generally easier to define the spatial motion of a particle by using other coordinate systems, e.g., x, y, z, or r, θ, z.

Unit Vectors. A set of unit vectors acting in the positive directions of t and n will be designated as $\mathbf{u}_t$ and $\mathbf{u}_n$, respectively. Since $\mathbf{u}_t$ and $\mathbf{u}_n$ are always perpendicular to one another, and lie in the osculating plane, for spatial motion a third unit vector, $\mathbf{u}_b$, defines a *binormal axis* at P which is perpendicular to $\mathbf{u}_t$ and $\mathbf{u}_n$, Fig. 12–24*a*. Using the vector cross product, these three unit vectors are related by the equation $\mathbf{u}_b = \mathbf{u}_t \times \mathbf{u}_n$.

Velocity. The instantaneous velocity of the particle traveling over the curved path is defined as $\mathbf{v} = d\mathbf{r}/dt$, where $\mathbf{r}$ is a position vector directed from the origin of a fixed coordinate system to the particle, Fig. 12–24*a*. As indicated in Sec. 12–4, $\mathbf{v}$ has a *direction* that is always tangent to the curve, defined by $\mathbf{u}_t$; and a *magnitude* that is determined by taking the time derivative of the path function s, i.e., $v = \dot{s} = ds/dt$ (Eq. 12–13). Hence,

$$\mathbf{v} = v\mathbf{u}_t \tag{12–49}$$

*See Prob. 12–81.

where

$$v = \dot{s} \tag{12-50}$$

Acceleration. The acceleration of the particle is the time rate of change of the velocity vector. Thus, using Eq. 12–49, we have

$$\mathbf{a} = \dot{\mathbf{v}} = \dot{v}\mathbf{u}_t + v\dot{\mathbf{u}}_t \tag{12-51}$$

In order to compute the time derivative $\dot{\mathbf{u}}_t$, note that as the particle moves along the arc ds in time dt, $\mathbf{u}_t$ preserves its magnitude of unity; however, it changes its *direction,* so that it becomes $\mathbf{u}_t'$, Fig. 12–24*c*. The change $d\mathbf{u}_t$ in $\mathbf{u}_t$ is shown in Fig. 12–24*d*. Here $d\mathbf{u}_t$ stretches between two points lying on an infinitesimal arc of radius $u_t = 1$. Hence, $d\mathbf{u}_t$ has a *magnitude* of $du_t = (1)\,d\theta$ and its *direction* is defined by $\mathbf{u}_n$. Consequently, $d\mathbf{u}_t = d\theta\mathbf{u}_n$ or $\dot{\mathbf{u}}_t = \dot{\theta}\mathbf{u}_n$. Since $ds = \rho\, d\theta$ or $\dot{\theta} = \dot{s}/\rho$, then

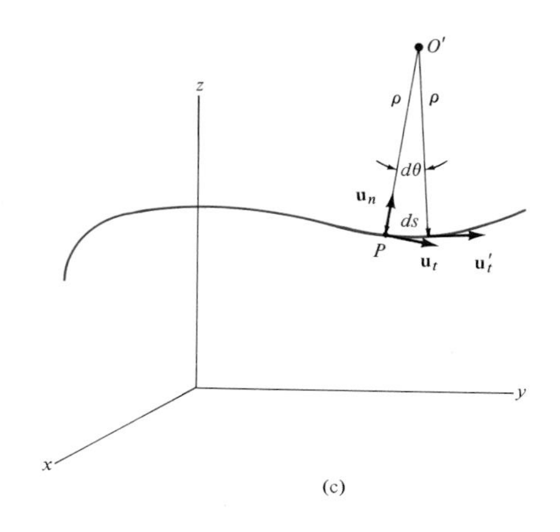

(c)

$$\dot{\mathbf{u}}_t = \frac{\dot{s}}{\rho}\mathbf{u}_n = \frac{v}{\rho}\mathbf{u}_n$$

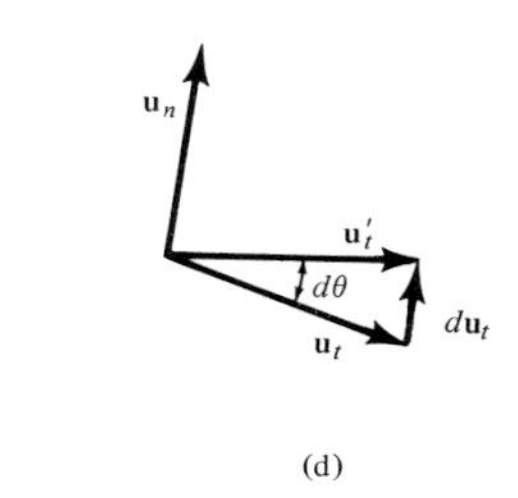

(d)

Hence, Eq. 12–51 can be written as the sum of two acceleration components,

$$\mathbf{a} = a_t\mathbf{u}_t + a_n\mathbf{u}_n \tag{12-52}$$

where

$$a_t = \dot{v} = v\frac{dv}{ds} \tag{12-53}$$

and

$$a_n = \frac{v^2}{\rho} \tag{12-54}$$

These two components are shown in Fig. 12–24*e*, in which case the *magnitude* of acceleration is

$$a = \sqrt{a_t^2 + a_n^2} \tag{12-55}$$

For application of Eq. 12–54, note that if the path s lies in the x-y plane such that $y = f(x)$, the radius of curvature at any point (x, y) can be determined from the equation*

$$\rho = \left| \frac{[1 + (dy/dx)^2]^{3/2}}{d^2y/dx^2} \right| \tag{12-56}$$

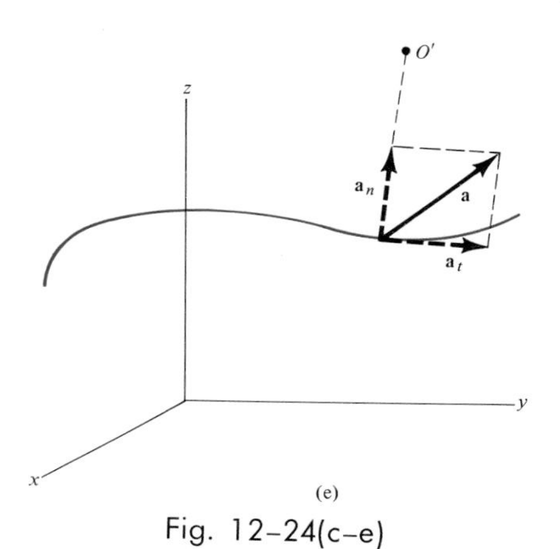

(e)

Fig. 12–24(c–e)

*The derivation of this result is given in any standard calculus text.

PROCEDURE FOR ANALYSIS

When a particle is moving along a *known* curved path, the motion can be expressed in terms of its normal and tangential components using the above equations.

At the instant considered, the origin of coordinates is established at the particle and the positive n axis is directed towards the path's center of curvature; whereas, the positive t axis acts in the direction of motion.

The particle's *velocity* $\mathbf{v}$ always acts in the positive tangential direction. The tangential component of acceleration, $\mathbf{a}_t$, creates a change in magnitude of velocity. This component acts in the direction of motion if the particle's speed is increasing or in the opposite direction if the speed is decreasing. The magnitudes of $\mathbf{v}$ and $\mathbf{a}_t$ are related to the path s and time t by the equations of *rectilinear motion,* namely, $v = \dot{s}$, $a_t = \dot{v}$, and $v\,dv = a_t ds$; or if a_t is *constant,* then $s = s_1 + v_1 t + \frac{1}{2}(a_t)_c t^2$, $v = v_1 + (a_t)_c t$ and $v^2 = v_1^2 + 2(a_t)_c(s - s_1)$ apply.

The normal component of acceleration, $\mathbf{a}_n$, creates a change in the direction of the particle's velocity. This component is *always* directed toward the center of curvature of the path, i.e., along the positive n axis. The magnitude of $a_n = v^2/\rho$, where ρ is computed from Eq. 12–56 if the path is expressed as $y = f(x)$.

The following examples numerically illustrate application of the above equations.

Example 12–14

A skier starting at A travels with a constant speed of 6 m/s along the parabolic path $y = \frac{1}{20}x^2$, shown in Fig. 12–25. Determine his velocity and acceleration at the instant he arrives at B. In the calculation, neglect the size of the skier.

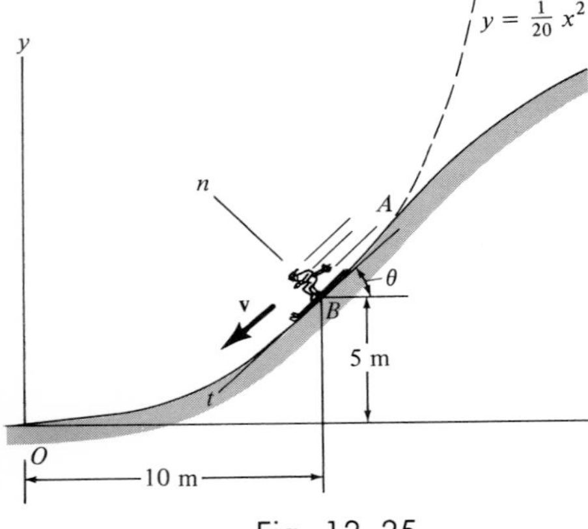

Fig. 12–25

Solution

In this problem the osculating plane, which contains the n and t axes, coincides with the x-y plane of the path. By definition, the velocity is always directed tangent to the path. Since $y = \frac{1}{20}x^2$, $dy/dx = \frac{1}{10}x$, then $dy/dx|_{x=10} = 1$. Hence, at B, $\mathbf{v}$ makes an angle of $\theta = \tan^{-1} 1 = 45°$ with the x axis, Fig. 12–25. Therefore,

$$v = 6 \text{ m/s} \quad 45°\!\searrow \mathbf{v} \qquad \textit{Ans.}$$

The acceleration is computed by using $\mathbf{a} = \dot{v}\mathbf{u}_t + (v^2/\rho)\mathbf{u}_n$. It is first necessary, however, to determine the radius of curvature of the path at B (10 m, 5 m). Since $d^2y/dx^2 = \frac{1}{10}$, then from Eq. 12–56,

$$\rho = \left|\frac{[1 + (dy/dx)^2]^{3/2}}{d^2y/dx^2}\right| = \left|\frac{[1 + (\frac{1}{10}x)^2]^{3/2}}{\frac{1}{10}}\right|_{x=10\text{ m}} = 28.3 \text{ m}$$

The acceleration becomes

$$\mathbf{a}_B = \dot{v}\mathbf{u}_t + \frac{v^2}{\rho}\mathbf{u}_n$$
$$= 0\mathbf{u}_t + \frac{(6)^2}{28.3}\mathbf{u}_n$$
$$= \{1.27\mathbf{u}_n\}\ \text{m/s}^2$$

Since $\mathbf{a}_B$ acts in the direction of the positive n axis, it makes an angle of $\theta + 90° = 135°$ with the positive x axis. Hence,

$$a_B = 1.27\ \text{m/s}^2 \quad \mathbf{a}_B \measuredangle 135° \qquad \textit{Ans.}$$

Example 12–15

A race car C travels around the horizontal circular track that has a radius of 150 m, Fig. 12–26. If the car increases its speed at a constant rate of 7 m/s² starting from rest, determine the time needed for it to reach an acceleration of 80 m/s².

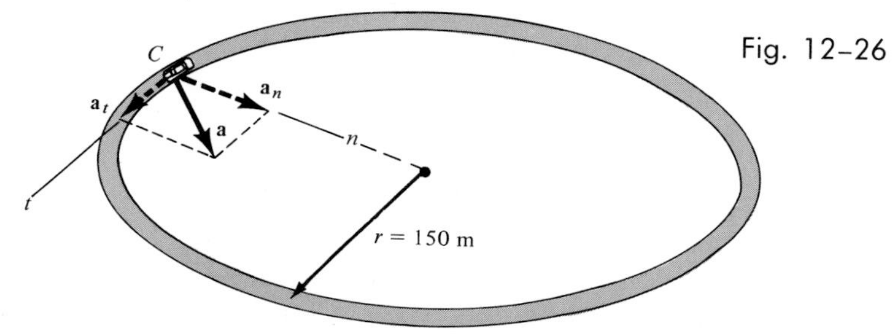

Fig. 12–26

Solution

In general, the acceleration is defined as $\mathbf{a} = \mathbf{a}_t + \mathbf{a}_n$. Since $\mathbf{a}_t$ represents the change in the velocity magnitude, i.e., $\dot{v}$, then $a_t = 7$ m/s². The normal component $\mathbf{a}_n$, representing the change in the velocity direction, is $a_n = v^2/\rho$. The speed expressed as a function of time, can be obtained either by integrating $a_t = \dot{v} = 7$ m/s², or by using the constant-acceleration formula, $v = v_1 + a_c t$. In either case, realizing that $v_1 = 0$ (at $t = 0$), we have $v = 7t$. Since $\rho = 150$ m, then

$$a_n = \frac{v^2}{\rho} = \frac{(7t)^2}{150} = 0.327t^2\ \text{m/s}^2$$

Adding $\mathbf{a}_t$ and $\mathbf{a}_n$ vectorially, Fig. 12–26, the magnitude of acceleration can be expressed as

$$a = \sqrt{a_t^2 + a_n^2}$$
$$80 = \sqrt{(7)^2 + (0.327t^2)^2}$$

Solving for t,

$$0.327t^2 = \sqrt{(80)^2 - (7)^2}$$
$$t = 15.6\ \text{s} \qquad \textit{Ans.}$$

Example 12-16

A car starts from rest at point A and travels along the horizontal track ABC shown in Fig. 12-27a. During the motion, the increase in speed is $a_t = (0.2t)\ \text{m/s}^2$, where t is in seconds. Determine the car's acceleration when it arrives at point B.

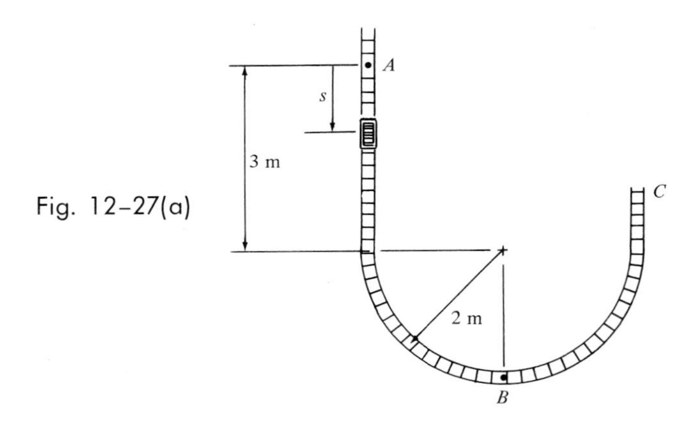

Fig. 12-27(a)

Solution

The acceleration is computed from $\mathbf{a} = \dot{v}\mathbf{u}_t + (v^2/\rho)\mathbf{u}_n$. To use this equation, however, it is first necessary to formulate v and $\dot{v}$ so that they may be evaluated at B. Since $v = 0$ at $t = 0$, then from Eq. 12-53, we have

$$a_t = \dot{v} = \frac{dv}{dt} = 0.2t \tag{1}$$

$$\int_0^v dv = \int_0^t 0.2t\, dt$$

$$v = 0.1t^2 \tag{2}$$

To determine the velocity at point B, Fig. 12-27a, it is first necessary to determine s as a function of time, then obtain the time needed for the car to arrive at B in order to evaluate $\dot{v}$ and v (Eqs. (1) and (2)) at this point. Using Eq. 12-50 and Eq. (2), we have

$$v = \frac{ds}{dt} = 0.1t^2$$

$$\int_0^s ds = \int_0^t 0.1t^2\, dt$$

$$s = 0.0333t^3 \tag{3}$$

From the geometry of Fig. 12–27*a*, $s_B = 3 + 2\pi(2)/4 = 6.14$ m. Hence, from Eq. (3), the time needed to reach *B* is

$$6.14 = 0.0333t_B^3$$
$$t_B = 5.69 \text{ s}$$

Substituting this value into Eqs. (1) and (2) yields the tangential acceleration component and the speed at *B*, namely,

$$(a_t)_B = \dot{v}_B = 0.2(5.69) = 1.14 \text{ m/s}^2$$
$$v_B = 0.1(5.69)^2 = 3.24 \text{ m/s}$$

At *B*, $\rho_B = 2$ m; therefore,

$$\mathbf{a}_B = \dot{v}_B\mathbf{u}_t + \frac{v_B^2}{\rho_B}\mathbf{u}_n$$
$$= 1.14\mathbf{u}_t + \frac{(3.24)^2}{2}\mathbf{u}_n$$
$$= \{1.14\mathbf{u}_t + 5.25\mathbf{u}_n\} \text{ m/s}^2$$

The magnitude of $\mathbf{a}_B$ is

$$a_B = \sqrt{(1.14)^2 + (5.25)^2} = 5.37 \text{ m/s}^2$$

The direction is determined as shown in Fig. 12–27*b*, where

$$\theta = \tan^{-1}\left(\frac{5.25}{1.14}\right) = 77.7°$$

Thus,

$$a_B = 5.37 \text{ m/s}^2 \quad \mathbf{a}_B \measuredangle 77.7° \qquad \textit{Ans.}$$

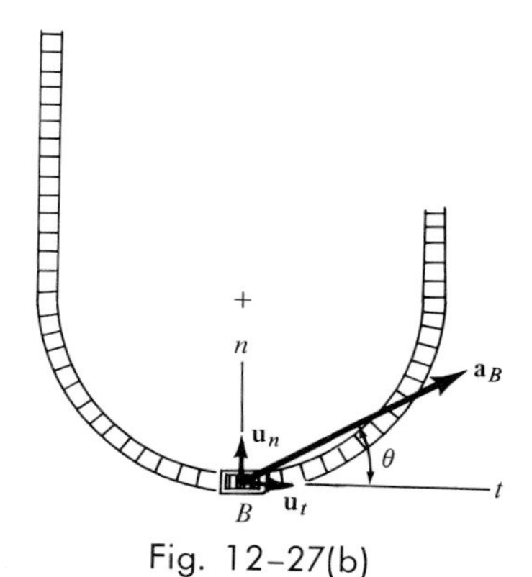

Fig. 12–27(b)

Problems

12–69. A boat is traveling along a circular path having a radius of 30 m. Determine the magnitude of the boat's acceleration if at a given instant the boat's speed is $v = 6$ m/s and the rate of increase in speed is $\dot{v} = 2$ m/s².

12–70. An automobile is traveling with a *constant speed* along a horizontal circular curve that has a radius of $\rho = 250$ m. If the magnitude of acceleration is $a = 1.5$ m/s², determine the speed at which the automobile is traveling.

12–70a. Solve Prob. 12–70 if $\rho = 800$ ft and $a = 6$ ft/s².

12–71. A train travels along a horizontal circular curve that has a radius of 600 m. If the speed of the train is uniformly increased from 40 km/h to 60 km/h in 5 s, determine the magnitude of the acceleration at the instant the speed of the train is 50 km/h.

*12–72. At a given instant, the automobile A has a speed of 25 m/s and an acceleration of 3 m/s² acting in the direction shown. Determine the radius of curvature of the path at point A and the rate of increase of the automobile's speed.

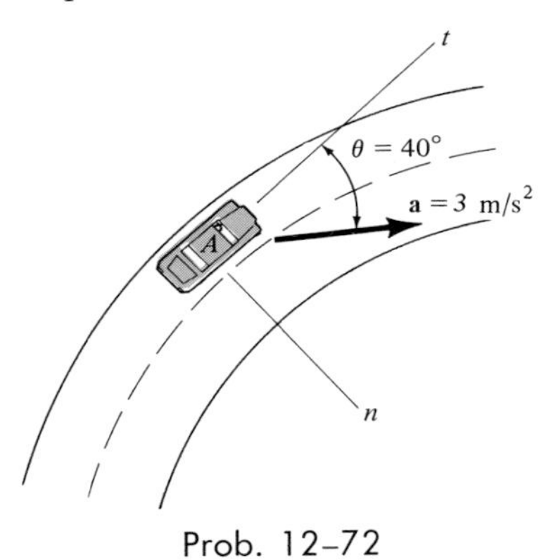

Prob. 12–72

12–73. The satellite S travels around the earth in a circular path with a constant speed of 20 Mm/h. If the magnitude of acceleration is 2.5 m/s², determine the altitude h. Assume the earth's diameter to be 12 713 km.

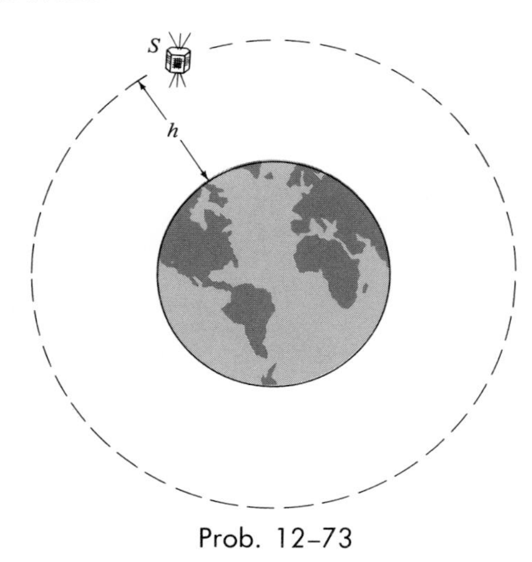

Prob. 12–73

12–74. A train is traveling with a constant speed of 20 m/s along the curved path shown. Determine the magnitude of the acceleration of the front of the train, B, at the instant it reaches the crossing at point A ($y = 0$). The coordinates are measured in metres.

Prob. 12–74

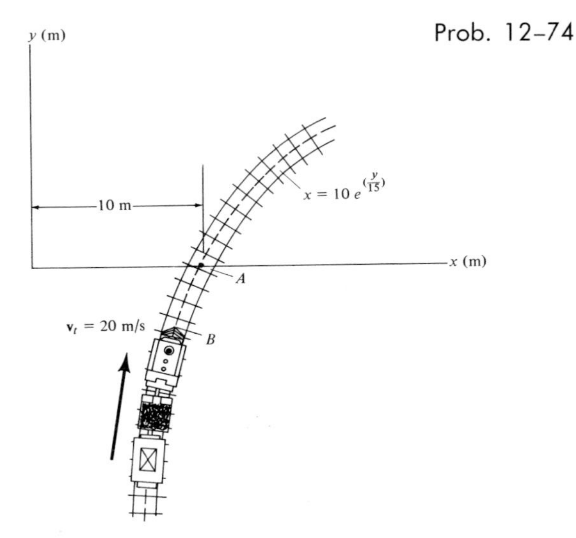

12–75. A sled is traveling down along a curve which can be approximated by the parabola $y = \frac{1}{4}x^2$. When point B on the runner is coincident with point A on the curve ($x_A = 2$ m, $y_A = 1$ m), the speed of B is measured as $v_B = 10$ m/s and the increase in speed is $\dot{v}_B = 3$ m/s². Determine the magnitude of the acceleration of point B at this instant.

Prob. 12–75

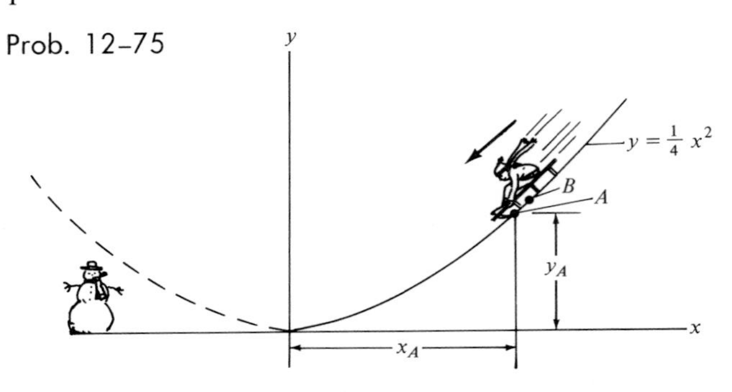

12–75a. Solve Prob. 12–75 if the curve is defined by $y = \frac{1}{4}x^2$, $x_A = 8$ ft, $y_A = 16$ ft, $v_B = 4$ ft/s, and $\dot{v}_B = 0.9$ ft/s².

*12–76. The small bead B travels with a constant speed of 300 mm/s along the curve. Determine the acceleration of the bead when it is located at point (200 mm, 100 mm), and sketch this vector on the curve.

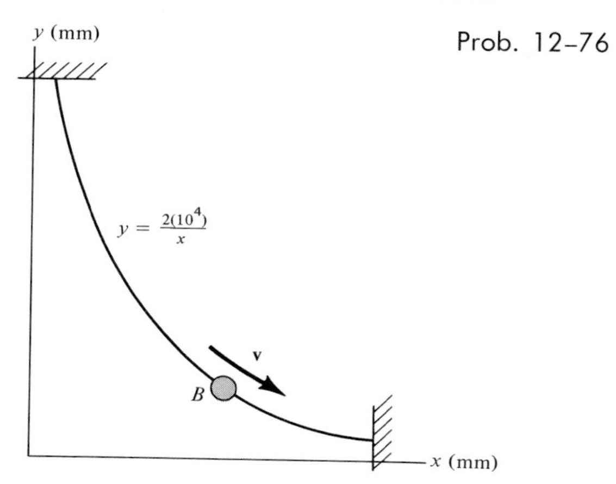

Prob. 12–76

12–77. The particle P moves along the curve $y = (x^2 - 4)$ m with a constant speed of 5 m/s. Determine the point on the curve where the maximum acceleration occurs, and compute its value.

12–78. A race car starting from rest at A increases its speed along the circular track, $\rho = 25$ m, at the rate of $a_t = (0.4s)$ m/s², where s is measured in metres. Determine the distance s the car must travel to attain a total acceleration of 4 m/s².

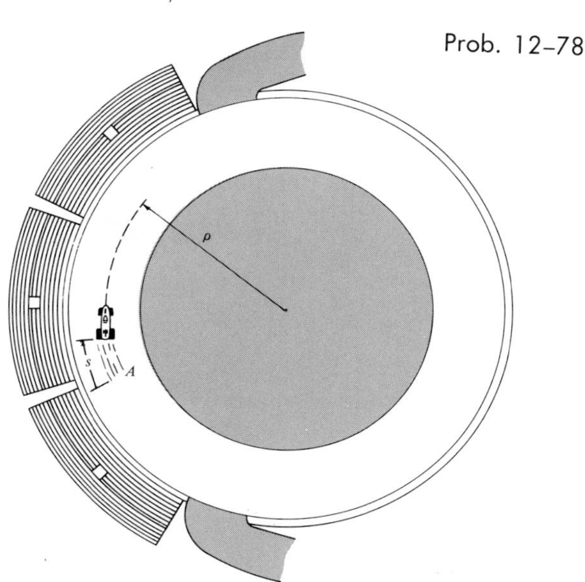

Prob. 12–78

12–79. If the race car in Prob. 12–78 has an initial speed of $v_A = 10$ m/s at A, compute the time needed for the car to travel 25 m. Assume $a_t = (0.4s)$ m/s² and $\rho = 25$ m.

* **12–80.** A package is dropped from the plane, which is flying with a constant horizontal velocity of $v_A = 80$ m/s. Determine the tangential and normal components of acceleration and the radius of curvature of the path of motion at the moment the package is released at A, where it has a horizontal velocity of $v_A = 80$ m/s and $h = 0.5$ km, and *just before* it strikes the ground at B.

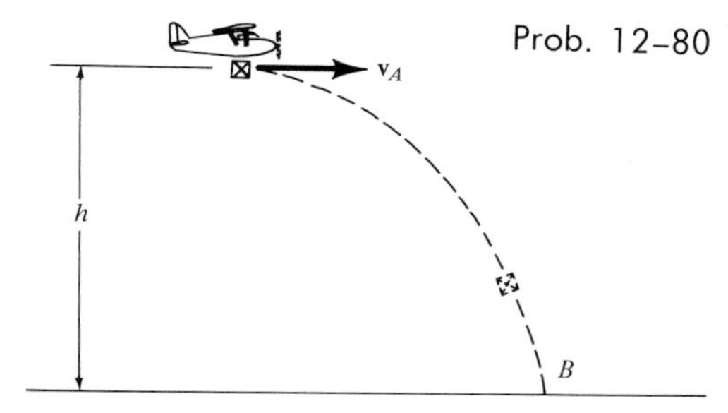

Prob. 12–80

* **12–80a.** Solve Prob. 12–80 if $v_A = 170$ ft/s and $h = 200$ ft. The constant downward acceleration of gravity is $g = 32.2$ ft/s².

12–81. A particle P travels along an elliptical spiral path

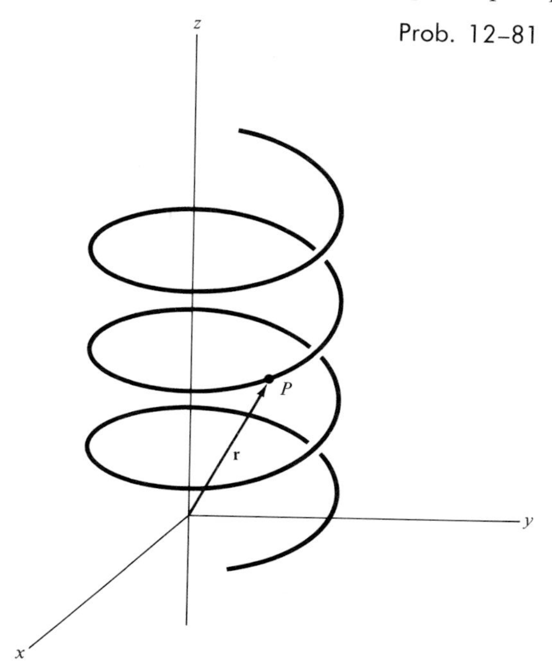

Prob. 12–81

such that its position vector **r** is defined by $\mathbf{r} = \{2\cos(0.1t)\mathbf{i} + 1.5\sin(0.1t)\mathbf{j} + (2t)\mathbf{k}\}$ m, where t is measured in seconds and the arguments for the sine and cosine are given in radians. When $t = 10$ s, determine the coordinate direction angles α, β, and γ which the binormal axis to the osculating plane makes with the x, y, and z axes. *Hint:* Solve for the velocity $\mathbf{v}_P$ and acceleration $\mathbf{a}_P$ of the particle in terms of their **i, j, k** components. The binormal is parallel to $\mathbf{a}_P \times \mathbf{v}_P$. Why?

12–82. The motion of a particle along a fixed path is defined by the parametric equations $r = 1.5$ m, $\theta = 2t$, and $z = t^2$, where r and z are in metres and θ is in radians, when t is measured in seconds. Determine the unit vector that specifies the direction of the binormal axis to the osculating plane with respect to a set of fixed x, y, z coordinate axes when $t = 0.25$ s. *Hint:* Formulate the particle's velocity $\mathbf{v}_P$ and acceleration $\mathbf{a}_P$ in terms of their **i, j, k** components. The binormal is parallel to $\mathbf{a}_P \times \mathbf{v}_P$. Why?

12–9. Absolute-Dependent-Motion Analysis of Two Particles

In some types of problems the motion of one particle will *depend* upon the corresponding motion of another particle. This dependency generally occurs if the particles are interconnected by inextensible cords which are wrapped around pulleys. For example, the movement of block A downward along the inclined plane in Fig. 12–28 will cause a corresponding movement of block B up the other incline. To prove this, the location of the blocks is specified from the *fixed point O* using the coordinates s_A and s_B. Since the cord has a fixed length, these coordinates which extend along the *changing portions* of the cord are related by the equation

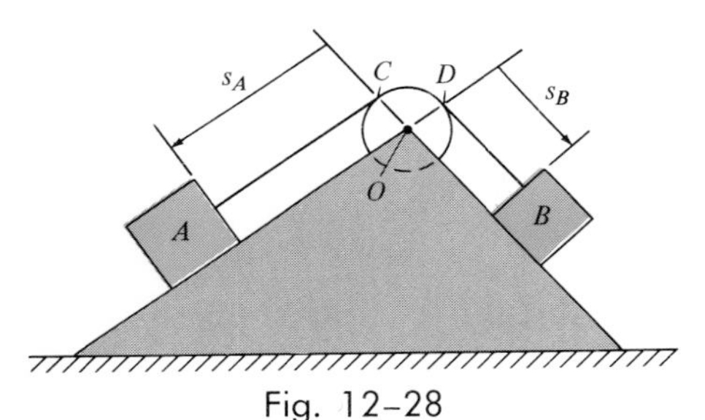

Fig. 12–28

$$s_A + s_B = l$$

where l is a *constant* and represents the length of cord *excluding* the *constant arc CD*. (The cord length over arc CD does not change as the blocks move.) Taking the time derivative of the above expression yields a relation between the velocities of the blocks, i.e.,

$$\frac{ds_B}{dt} + \frac{ds_A}{dt} = 0$$

or

$$v_B = -v_A$$

The negative sign indicates that positive motion of block A (downward in the direction of increasing s_A) causes a corresponding negative motion (upward) of block B.

In a similar manner, time differentiation of the velocities yields the relation between the accelerations of the blocks,

$$a_B = -a_A$$

A more complicated example involving the dependent motions of blocks with pulleys is shown in Fig. 12–29*a*. Since there are two separate cables in this system, the motions of A and B may be related by first separating the cables, as shown in Figs. 12–29*b* and 12–29*c*, and then studying the motions of each cable. From Fig. 12–29*b*, the positions of A and the center of pulley C are defined by s_A and s_C, where both coordinates are measured from a horizontal datum passing through the *fixed point* D. The magnitudes of these coordinates are related by the equation

$$s_A + 2s_C = l_1$$

Here l_1 represents the total vertical cable length which remains *constant* as C and A move. (The circular segments EF and GH and segment IJ are not included in this length, since they are of no consequence to the changing coordinates s_A and s_C.) In a similar manner, the position of C may be related to the position of B from the same fixed datum, Fig. 12–29*c*. This requires

$$s_B + (s_B - s_C) = l_2$$
$$2s_B - s_C = l_2$$

where l_2 represents the constant vertical cable length (excluding segments KL and MN). If s_C is eliminated between the above two equations, the equation defining the position of the two blocks becomes

$$s_A + 4s_B = 2l_2 + l_1$$

Taking successive time derivatives, realizing that l_1 and l_2 are constants, we obtain

$$v_A = -4v_B$$
$$a_A = -4a_B$$

Thus, the velocity and acceleration of block A is four times that of block B, but in the opposite direction.

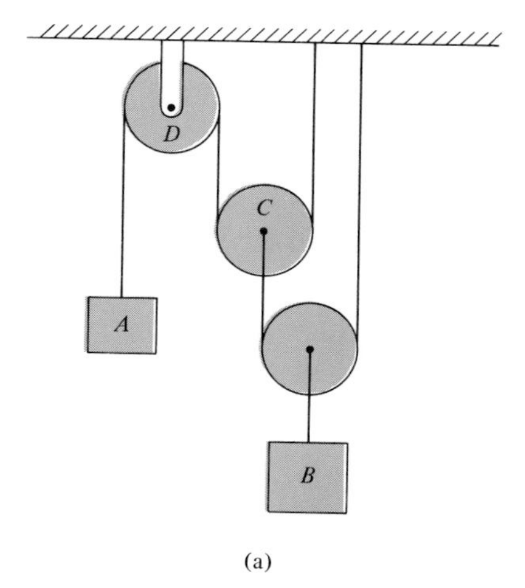

(a)

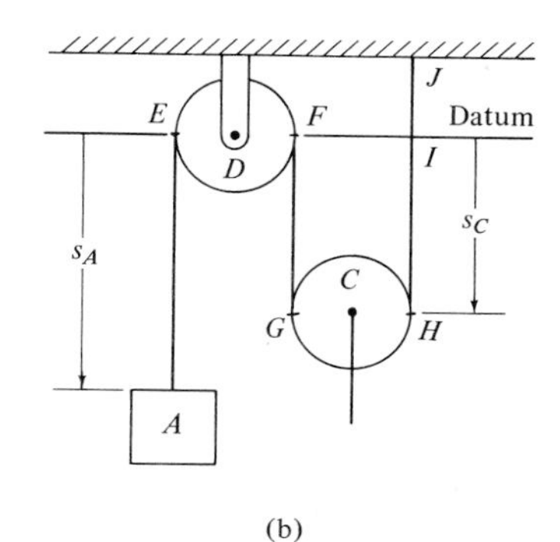

(b)

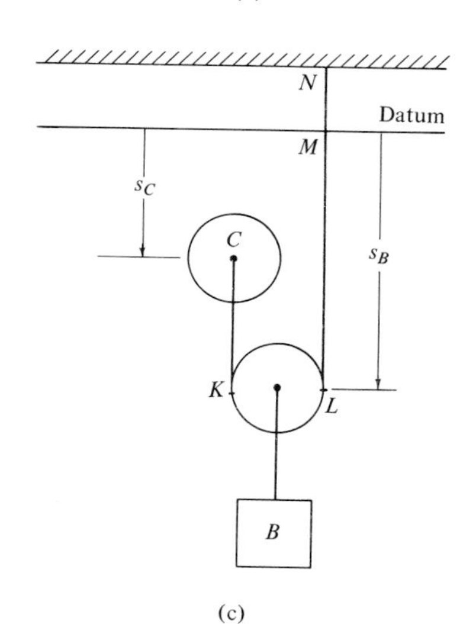

(c)

Fig. 12–29

PROCEDURE FOR ANALYSIS

The dependent motion of one particle located on a cable may be related to that of another by first establishing rectilinear coordinates that define the *position* of each of the particles, measured from a *fixed point* (or datum). It is necessary that each of the coordinates selected has the *same direction* as the path of motion of the particle to which it is directed. Then, using geometry or trigonometry, the coordinates are related to that constant portion of the cable length which *changes its position* as the particles move. By taking two successive time derivatives of this equation, one obtains the required velocity and acceleration equations which relate the motions of the two particles.

Example 12–17

The freight elevator shown in Fig. 12–30 is operated by an electric motor located at A. If the motor winds up its attached cable at a speed of 15 m/s, determine the speed at which the elevator rises.

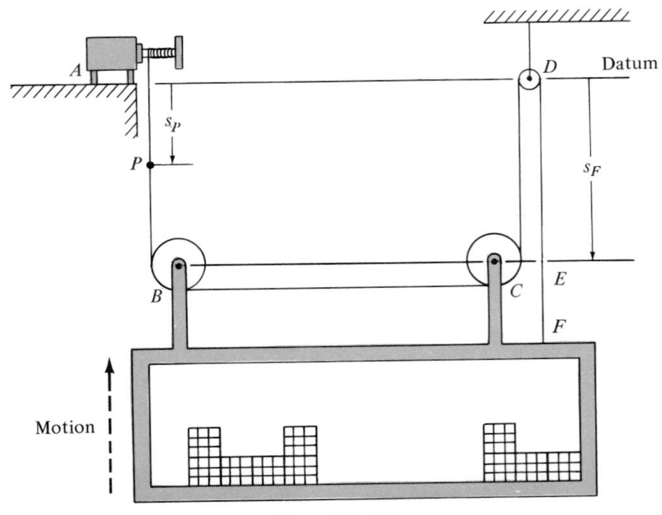

Fig. 12–30

Solution

If a datum is established through the center of pulley D (a fixed point), Fig. 12–30, then the vertical coordinate s_P is used to locate an arbitrary point P on the cable which is moving upward toward the motor with a speed of 15 m/s, whereas s_F determines the vertical position of the elevator, defined by the center of the pulleys at B and C. Excluding segment EF, the constant vertical length l of cable, which is changing in Fig. 12–30, can be related to the coordinates s_F and s_P by the equation

$$2s_F + (s_F - s_P) = l$$

or

$$3s_F - s_P = l$$

Taking the time derivative, we have

$$3v_F - v_P = 0$$

Since point P is moving upward (decreasing s_P), $v_P = -15$ m/s, and therefore the velocity of the elevator is

$$3v_F - (-15 \text{ m/s}) = 0$$

$$v_F = -5 \text{ m/s (upward)} \qquad \textit{Ans.}$$

Example 12-18

A man at A is hoisting a safe S as shown in Fig. 12-31 by walking to the right with a constant velocity of $v_A = 0.5$ m/s. Determine the velocity and acceleration of the safe when it reaches the window elevation at E. The rope is 30 m long and passes over a small pulley D.

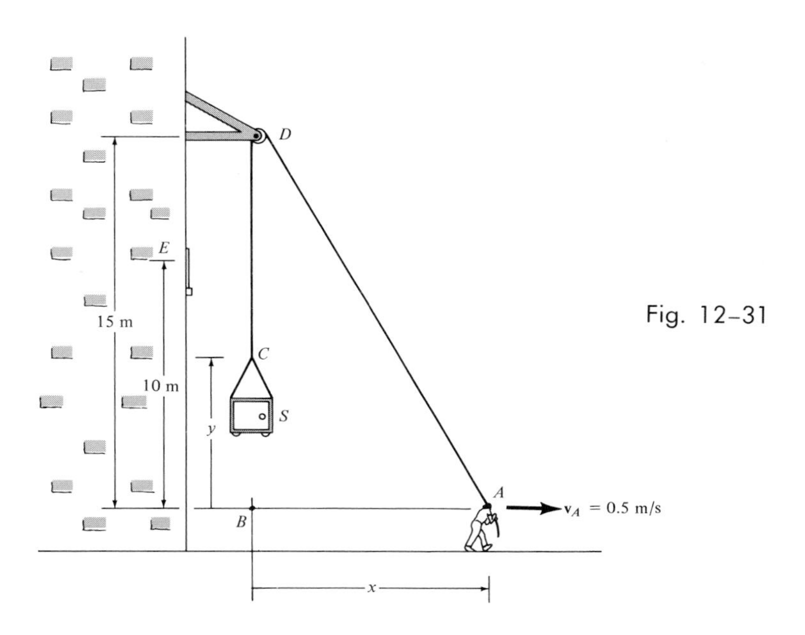

Fig. 12-31

Solution

This problem is unlike the previous example, since rope segment DA changes both direction and magnitude. However, following the Procedure for Analysis, the ends of the rope, which define the positions of S and A, are specified by means of the x and y coordinates measured from the fixed point B. These two coordinates act in the *direction of motion* of the ends of the rope, and may be related since the rope has a fixed length $l = 30$ m, which at all times is equal to the length of segment DA plus CD. Using the Pythagorean theorem to determine l_{DA}, we have

$$l_{DA} = \sqrt{(15)^2 + x^2} \quad \text{also,} \quad l_{CD} = 15 - y$$

Hence,

$$\begin{aligned} l &= l_{DA} + l_{CD} \\ 30 &= \sqrt{(15)^2 + x^2} + (15 - y) \\ y &= \sqrt{225 + x^2} - 15 \end{aligned} \tag{1}$$

Taking the time derivative, where $v_S = dy/dt$ and $v_A = dx/dt$, yields

$$v_S = \frac{dy}{dt} = \frac{1}{2}\frac{2x}{\sqrt{225 + x^2}}\frac{dx}{dt}$$

$$= \frac{x}{\sqrt{225 + x^2}} v_A \qquad (2)$$

When $y = 10$ m, x is determined from Eq. (1), i.e.,

$$(10 + 15)^2 = 225 + x^2, \qquad x = 20 \text{ m}$$

Hence, from Eq. (2) with $v_A = 0.5$ m/s,

$$v_S = \frac{20}{\sqrt{225 + (20)^2}}(0.5) = 0.4 \text{ m/s} = 400 \text{ mm/s} \qquad \textit{Ans.}$$

The acceleration of the safe, a_S, is determined by taking the time derivative of Eq. (2). Since $v_A = 0.5$ m/s, then $a_A = dv_A/dt = 0$, and we have

$$a_S = \frac{d^2y}{dt^2} = \left[\frac{-x^2}{(225 + x^2)^{3/2}}\frac{dx}{dt} + \frac{1}{(225 + x^2)^{1/2}}\frac{dx}{dt}\right] v_A$$

$$= \frac{225 v_A^2}{(225 + x^2)^{3/2}}$$

At $x = 20$ m, with $v_A = 0.5$ m/s, the acceleration becomes

$$a_S = \frac{225(0.5)^2}{[225 + (20)^2]^{3/2}} = 0.00360 \text{ m/s}^2 = 3.60 \text{ mm/s}^2 \qquad \textit{Ans.}$$

12–10. Relative-Motion Analysis of Two Particles Using Translating Axes

Throughout this chapter the absolute motion of a particle has been determined using a single fixed reference frame for measurement. There are many cases, however, where the path of motion for a particle is complicated, so that it may be feasible to analyze the motion in parts by using two or more frames of reference. For example, the motion of a particle located at the tip of an airplane propeller, while the plane is in flight, is more easily described if one observes first the motion of the airplane from a fixed reference and then superimposes (vectorially) the circular motion of the particle measured from a reference attached to the airplane. Any type of coordinates—rectangular, cylindrical, etc.—may be chosen to describe these two different motions.

In this section only *translating frames of reference* will be considered for the analysis. Relative-motion analysis of particles using rotating frames of reference will be treated in Secs. 16–10 and 20–5, since such an analysis depends upon prior knowledge of the kinematics of line segments.

Position. Consider particles A and B, which move along the arbitrary paths aa and bb, respectively, as shown in Fig. 12–32. The *absolute position* of each particle, $\mathbf{r}_A$ and $\mathbf{r}_B$, is measured from the common origin O of the *fixed* x, y, z reference frame. The origin of a second frame of reference x', y', z' is attached to and moves with particle A. The axes of this frame do not rotate; rather they are *only permitted to translate* relative to the fixed frame. The *relative position* of B with respect to A is observed from this moving frame and is designated by $\mathbf{r}_{B/A}$, called a *relative-position vector*. Using vector addition, the three vectors shown in Fig. 12–32 can be related by the equation*

$$\mathbf{r}_B = \mathbf{r}_A + \mathbf{r}_{B/A} \qquad (12\text{–}57)$$

Velocity. An equation that relates the velocities of the particles can be determined by taking the time derivative of Eq. 12–57, i.e.,

$$\frac{d\mathbf{r}_B}{dt} = \frac{d\mathbf{r}_A}{dt} + \frac{d\mathbf{r}_{B/A}}{dt}$$

Here $\mathbf{v}_B = d\mathbf{r}_B/dt$ and $\mathbf{v}_A = d\mathbf{r}_A/dt$ refer to *absolute velocities,* since they are measured from the fixed frame of reference. Designating the *relative velocity* as $\mathbf{v}_{B/A} = d\mathbf{r}_{B/A}/dt$, the above equation can be written as

*An easy way to remember the setup of this equation, and others like it, is to note the "cancellation" of the subscript A between the two terms, i.e., $\mathbf{r}_B = \mathbf{r}_{\not A} + \mathbf{r}_{B/\not A}$.

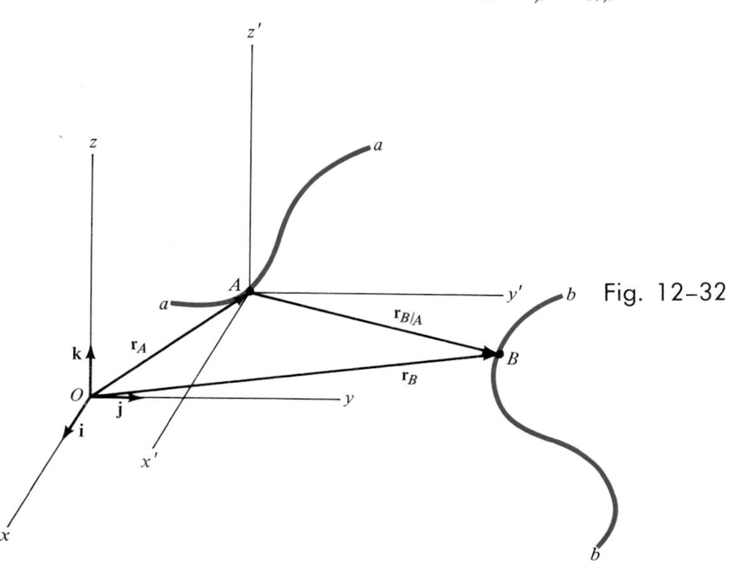

Fig. 12–32

$$\mathbf{v}_B = \mathbf{v}_A + \mathbf{v}_{B/A} \tag{12-58}$$

This equation states that the absolute velocity of B is equal to the absolute velocity of A plus (vectorially) the relative velocity of B with respect to A, as measured by a *translating observer.*

Acceleration. The time derivative of Eq. 12–58 yields a similar vector relationship between the *absolute* and *relative accelerations* of particles A and B,

$$\mathbf{a}_B = \mathbf{a}_A + \mathbf{a}_{B/A} \tag{12-59}$$

Here $\mathbf{a}_{B/A}$ is the acceleration of B as seen by an observer located at A and translating with the x', y', z' reference frame.

PROCEDURE FOR ANALYSIS

When applying the relative-position equation, $\mathbf{r}_B = \mathbf{r}_A + \mathbf{r}_{B/A}$, it is first necessary to specify the location of the fixed x, y, z and translating x', y', z' axes. Usually, the origin A of the translating axes is located at a point having a *known position,* $\mathbf{r}_A$, Fig. 12–32. A graphical representation of the vector addition $\mathbf{r}_B = \mathbf{r}_A + \mathbf{r}_{B/A}$ should be shown, and both the known and unknown quantities labeled on this sketch. Since vector addition forms a triangle there can be at most *two unknowns,* represented by the magnitudes and directions of the vector quantities. These unknowns can be solved for either graphically, using trigonometry (law of sines, law of cosines), or by resolving each of the three vectors $\mathbf{r}_B$, $\mathbf{r}_A$, and $\mathbf{r}_{B/A}$ into rectangular components, thereby generating two scalar equations. The latter method is illustrated in the example problems which follow.

The relative-motion equations, $\mathbf{v}_B = \mathbf{v}_A + \mathbf{v}_{B/A}$ and $\mathbf{a}_B = \mathbf{a}_A + \mathbf{a}_{B/A}$ are applied in the same manner as explained above, except that in this case the origin of the fixed x, y, z axes does not have to be specified.

Example 12–19

Water drips from a faucet at the rate of five drops per second, as shown in Fig. 12–33. Determine the vertical separation between two consecutive drops after the lower drop has attained a velocity of 3 m/s.

Solution

If the first and second drops of water are denoted as B and A, respectively, Fig. 12–33, the separation $s_{B/A}$ (position of B with respect to A) can be determined by the equation

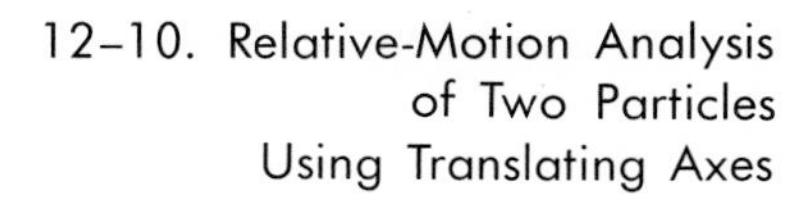

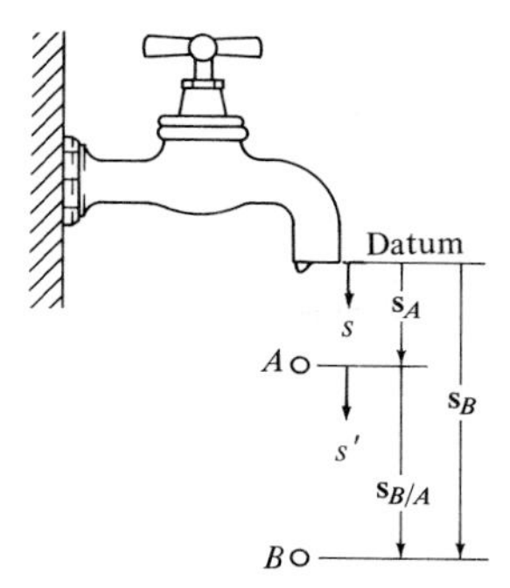

Fig. 12–33

$$s_{B/A} = s_B - s_A \qquad (1)$$

Here the origin of the fixed frame of reference is located at the head of the faucet (datum) with positive s downward. The origin of the moving frame s' is at drop A. Since each drop is subjected to rectilinear motion, having a constant downward acceleration of $a_c = g = 9.81 \text{ m/s}^2$, Eq. 12–8 may be applied with the initial condition $v_1 = 0$ to determine the time needed for drop B to attain a velocity of 3 m/s.

$(+\downarrow)$
$$v = v_1 + a_c t$$
$$3 = 0 + 9.81 t_B$$
$$t_B = 0.306 \text{ s}$$

Since one drop falls every fifth of a second (0.2 s), drop A falls for a time

$$t_A = 0.306 \text{ s} - 0.2 \text{ s} = 0.106 \text{ s}$$

before drop B attains a velocity of 3 m/s.

The position of each drop can be found by means of Eq. 12–9, with the condition $s_1 = 0$, $v_1 = 0$.

$(+\downarrow)$
$$s = s_1 + v_1 t + \tfrac{1}{2} a_c t^2$$
$$s_B = 0 + 0 + \tfrac{1}{2}(9.81)(0.306)^2$$
$$= 0.4593 \text{ m}$$
$$s_A = 0 + 0 + \tfrac{1}{2}(9.81)(0.106)^2$$
$$= 0.0551 \text{ m}$$

Applying Eq. (1) yields

$$s_{B/A} = s_B - s_A = 0.4593 - 0.0551 = 0.4042 \text{ m}$$
$$= 404.2 \text{ mm} \qquad \textit{Ans.}$$

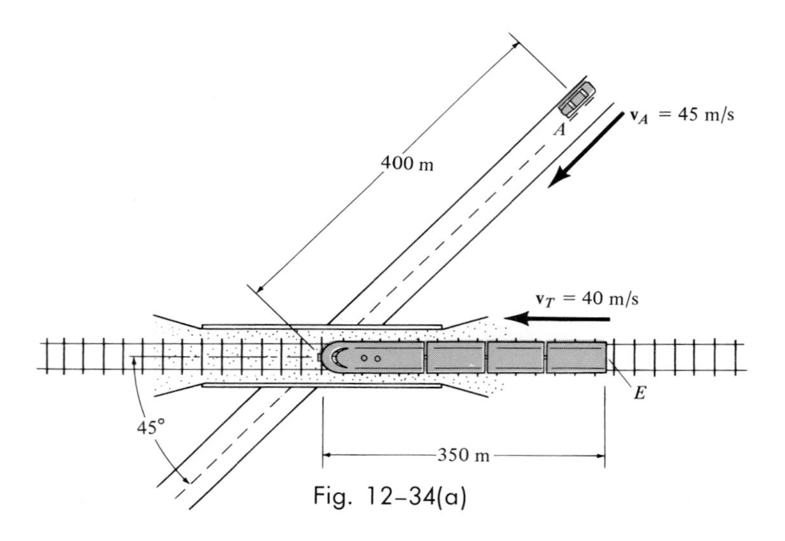

Fig. 12–34(a)

Example 12–20

A 350-m-long train traveling at a constant speed of 40 m/s crosses over a road, as shown in Fig. 12–34*a*. If an automobile A is traveling at 45 m/s and is 400 m from the crossing at the instant the front of the train reaches the crossing, determine (a) the relative velocity of the train with respect to the automobile, and (b) the distance from the automobile to the end of the last car of the train at this instant.

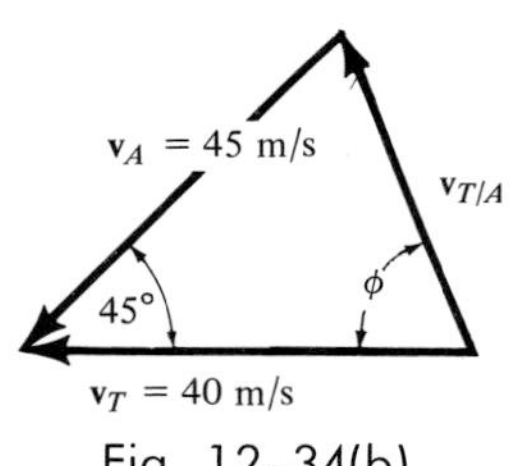

Fig. 12–34(b)

Solution

Part (a). The relative velocity of the train with respect to the automobile, $\mathbf{v}_{T/A}$, is found by using the equation

$$\mathbf{v}_T = \mathbf{v}_A + \mathbf{v}_{T/A} \tag{1}$$

The vector addition is shown in Fig. 12–34*b*. The two unknowns are represented by the magnitude and direction of $\mathbf{v}_{T/A}$, i.e., $v_{T/A}$ and ϕ. For the solution, each of the three vectors in Eq. (1) will be resolved into horizontal and vertical components, thereby generating two scalar equations; i.e.,

$$\mathbf{v}_T = \mathbf{v}_A + \mathbf{v}_{T/A}$$

$$\underset{\leftarrow}{40 \text{ m/s}} = \underset{45°\swarrow}{45 \text{ m/s}} + \underset{\phi\nwarrow}{v_{T/A}}$$

Thus, assuming the positive directions to the left and upward, we have

$$(\overset{+}{\leftarrow}) \qquad 40 = 45 \cos 45° + v_{T/A} \cos \phi \tag{2}$$

$$(+\uparrow) \qquad 0 = -45 \sin 45° + v_{T/A} \sin \phi \tag{3}$$

Rearranging terms and dividing Eq. (3) by Eq. (2),

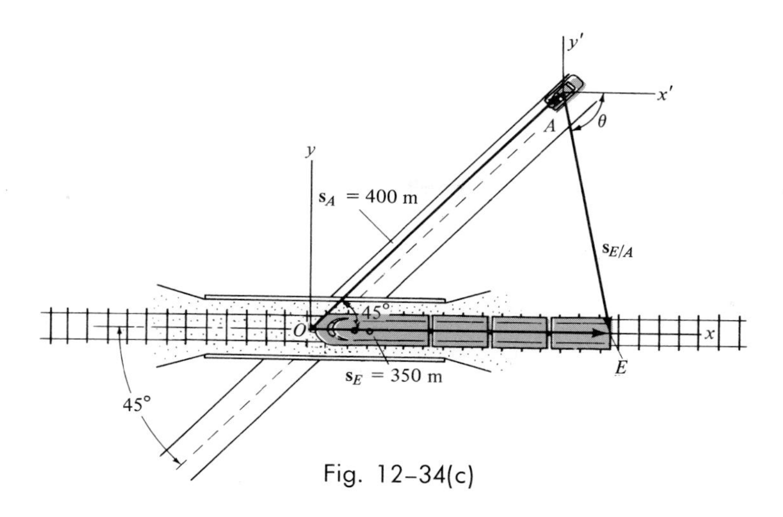

Fig. 12-34(c)

$$\frac{v_{T/A}\sin\phi}{v_{T/A}\cos\phi} = \tan\phi = \frac{45\sin 45°}{40 - 45\cos 45°} = 3.89$$

$$\phi = \tan^{-1}(3.89) = 75.6° \qquad \textit{Ans.}$$

so that, from Eq. (3),

$$v_{T/A} = 45\sin 45°/\sin 75.6°$$
$$= 32.9 \text{ m/s} \qquad \textit{Ans.}$$

Hence, as the motorist approaches the bridge, he sees the train approaching at a speed of 32.9 m/s. Then, once he has crossed under the bridge, the train is seen to move away from him at this speed.

Part (b). With the origin of the fixed x, y coordinate system at the crossing, and the origin of the translating x', y' coordinate system at the automobile, Fig. 12-34*c*, the required distance $\mathbf{s}_{E/A}$ can be found by solving the vector equation $\mathbf{s}_E = \mathbf{s}_A + \mathbf{s}_{E/A}$. The two unknowns are represented by $s_{E/A}$ and θ, Fig. 12-34*c*. Hence,

$$\mathbf{s}_E = \mathbf{s}_A + \mathbf{s}_{E/A}$$
$$\underset{\rightarrow}{350 \text{ m}} = \underset{\nearrow 45°}{400 \text{ m}} + \underset{\searrow \theta}{s_{E/A}}$$

$(\overset{+}{\rightarrow})$ $$350 = 400\cos 45° + s_{E/A}\cos\theta$$

$(+\uparrow)$ $$0 = 400\sin 45° - s_{E/A}\sin\theta$$

The solution follows the same procedure used in solving Eqs. (2) and (3). The results are

$$s_{E/A} = 290.7 \text{ m}, \qquad \theta = 76.6° \qquad \textit{Ans.}$$

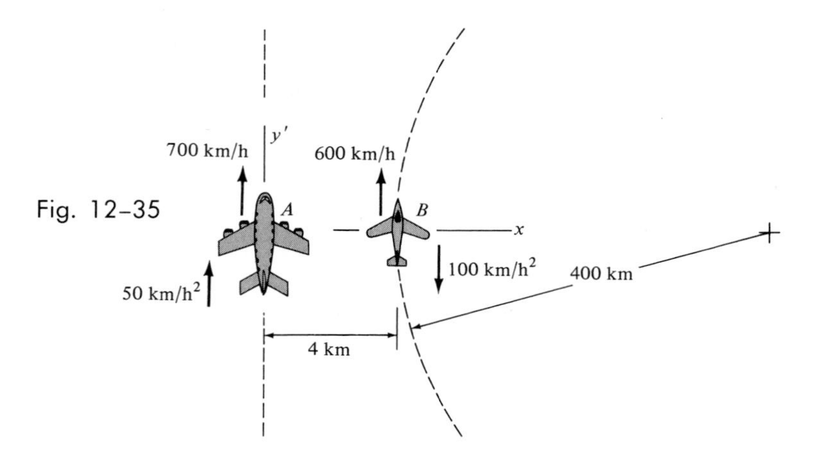

Fig. 12–35

Example 12–21

Two jet planes are flying horizontally at the same elevation, as shown in Fig. 12–35. Plane A is flying along a straight-line path, and at the instant shown it has a speed of 700 km/h and an acceleration of 50 km/h^2. Plane B is flying along a circular path at 600 km/h and decreasing its speed at the rate of 100 km/h^2. Determine the velocity and acceleration of B as measured by the pilot in A.

Solution

Plane A is traveling with rectilinear motion and a *translating frame of reference, x', y', z',* will be attached to it. Applying the relative-velocity equation in scalar form since the velocity vectors of both planes are parallel at the instant shown, we have

$$(+\uparrow) \qquad v_B = v_A + v_{B/A}$$

$$600 = 700 + v_{B/A}$$

$$v_{B/A} = -100 \text{ km/h}\downarrow \qquad \textit{Ans.}$$

Plane B has both tangential and normal components of acceleration, since it is flying along a *curved path.* From Eq. 12–54, the magnitude of the normal component is

$$(a_B)_n = \frac{v_B^2}{\rho} = \frac{(600)^2}{400} = 900 \text{ km/h}^2$$

Applying the relative-acceleration equation, we have

$$\mathbf{a}_B = \mathbf{a}_A + \mathbf{a}_{B/A}$$

$$(\mathbf{a}_B)_t + (\mathbf{a}_B)_n = \mathbf{a}_A + \mathbf{a}_{B/A}$$

$$\underset{\downarrow}{100} + \underset{\rightarrow}{900} = \underset{\uparrow}{50} + \underset{\measuredangle^{\theta}}{a_{B/A}}$$

Resolving these vectors into components acting in the x' and y' directions yields the two scalar equations

$$(\overset{+}{\rightarrow}) \qquad 0 + 900 = 0 + a_{B/A} \cos\theta$$

$$(+\uparrow) \qquad -100 + 0 = 50 - a_{B/A} \sin\theta$$

Solving,

$$a_{B/A} = 912.4 \text{ km/h}^2, \qquad \theta = 9.46^\circ \qquad \textit{Ans.}$$

Notice that the solution to this problem is possible using a translating frame of reference, since the pilot in plane A is "translating." Observation of plane A with respect to plane B, however, must be obtained using a *rotating* set of axes attached to plane B. (This assumes, of course, that the pilot of B does not turn his eyes to follow the motion of A.) The analysis for this case is given in Example 16–18.

Problems

12–83. The mine car is being pulled up the inclined plane using the motor M and the rope-and-pulley arrangement shown. Determine the speed v_P at which a point P on the cable must be traveling toward the motor to move the car up the plane with a constant speed of $v = 2$ m/s.

Prob. 12–83

***12–84.** If block B moves down the inclined plane at a speed of 4 m/s, determine the speed at which block A moves up the other inclined plane.

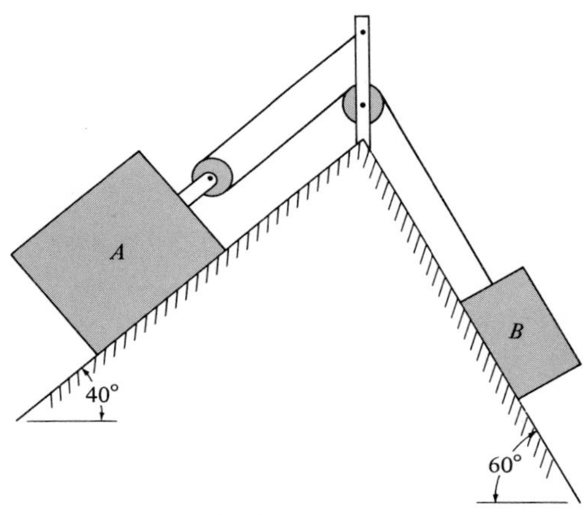

Prob. 12–84

12–85. Determine the speed v_P at which point P on the cable must be traveling toward the motor M to lift platform A at $v_A = 2$ m/s.

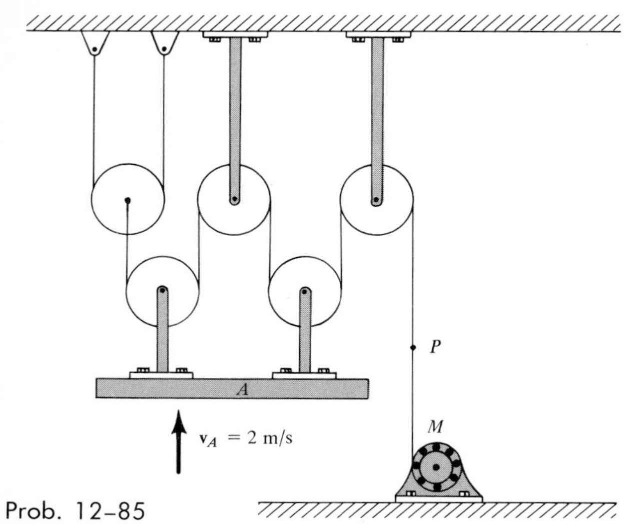

Prob. 12–85

12–85a. Solve Prob. 12–85 if $v_A = 5$ ft/s.

12–86. If platform B is lowered with a speed of $v_B = 1.5$ m/s, determine the speed at which platform A rises. Both platforms A and B remain horizontal.

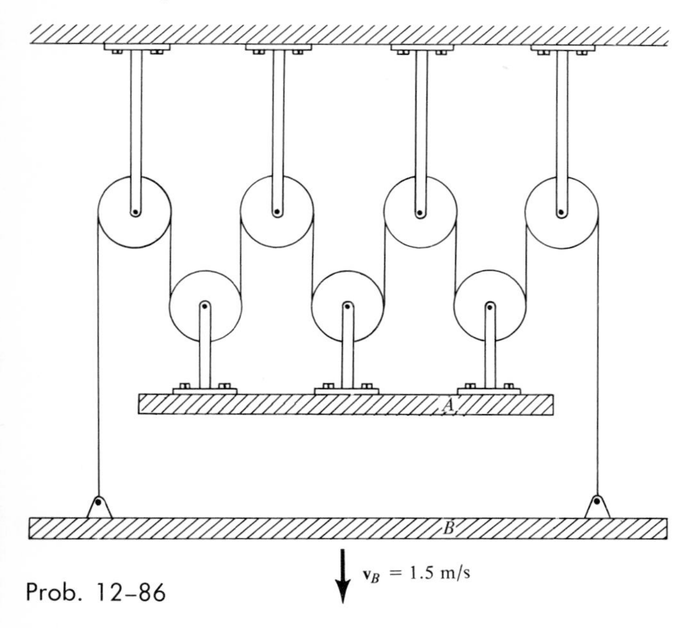

Prob. 12–86

12–87. Determine the speed of the hook at A if the end of the cable at B is pulled downward with a speed of $v_B = 6$ m/s.

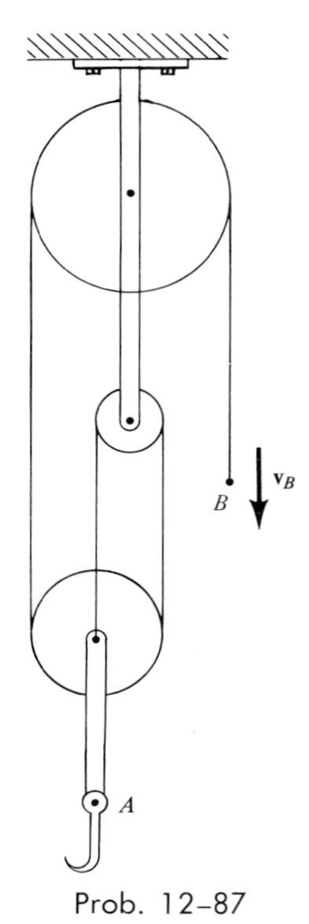

Prob. 12–87

* **12–88.** The device shown is used for catapulting the slider A. If this is done by using the hydraulic cylinder H to draw rod BC in at a rate of $v_{BC} = 5$ m/s, determine the velocity of the slider.

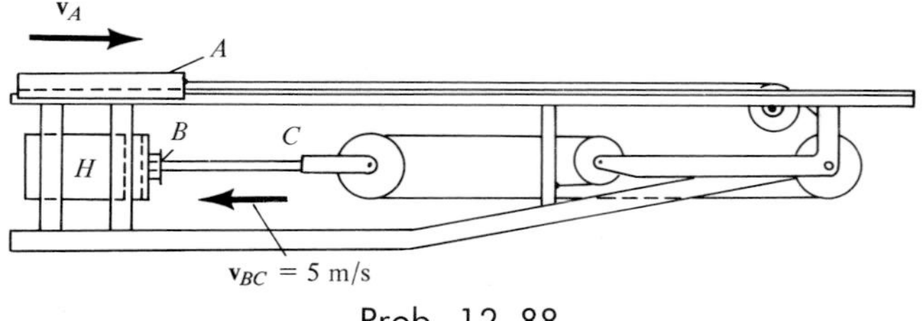

Prob. 12–88

12–89. Three blocks, A, B, and C, move along a straight-line path with constant velocities. If the relative velocity of A with respect to B is 8 m/s (moving to the right), and the relative velocity of B with respect to C is -3 m/s (moving to the left), determine the absolute velocities of A and B. The velocity of C is 6 m/s to the right.

12–90. Two planes, A and B, are flying at the same altitude. If their velocities are $v_A = 600$ km/h and $v_B = 500$ km/h, such that the angle between their straight-line courses is $\theta = 75°$, determine the velocity of plane B with respect to plane A.

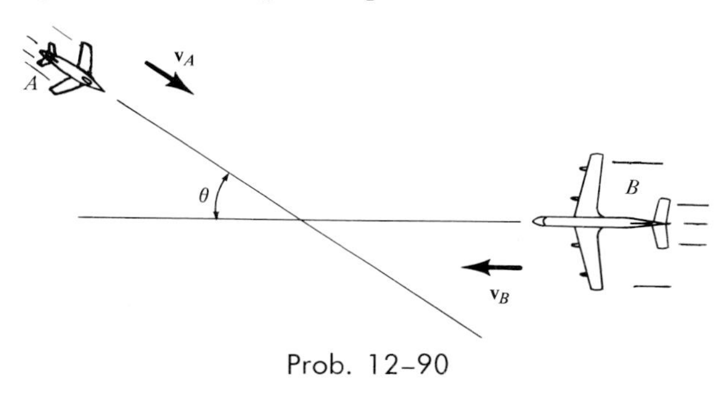

Prob. 12–90

12–90a. Solve Prob. 12–90 if $v_A = 400$ mi/h, $v_B = 450$ mi/h, and $\theta = 60°$.

12–91. Two billiard balls, A and B, are moving with constant velocities of $v_A = 0.5$ m/s and $v_B = 1.5$ m/s, respectively. Determine the relative velocity of A with respect to B at the instant shown ($t = 0$), and the distance between the two balls when $t = 1.25$ s.

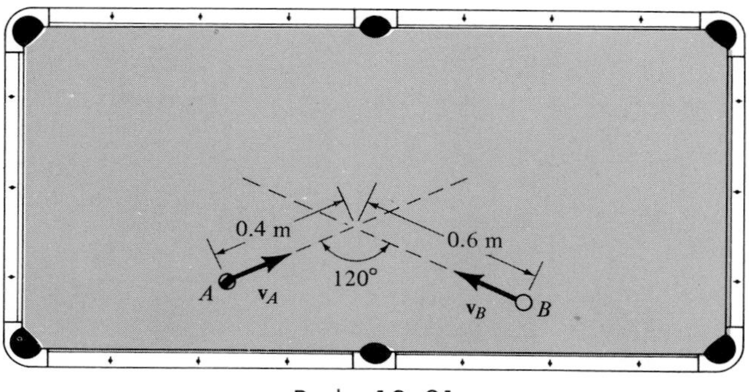

Prob. 12–91

* **12–92.** Sand falls from rest 0.5 m vertically onto a chute, as shown. If the sand is sliding at a velocity of $v_C = 2$ m/s down the chute, determine the relative velocity $\mathbf{v}_{A/C}$ of the sand just falling on the chute at A with respect to the sand sliding down the chute. The chute is inclined at an angle of 40° with the horizontal.

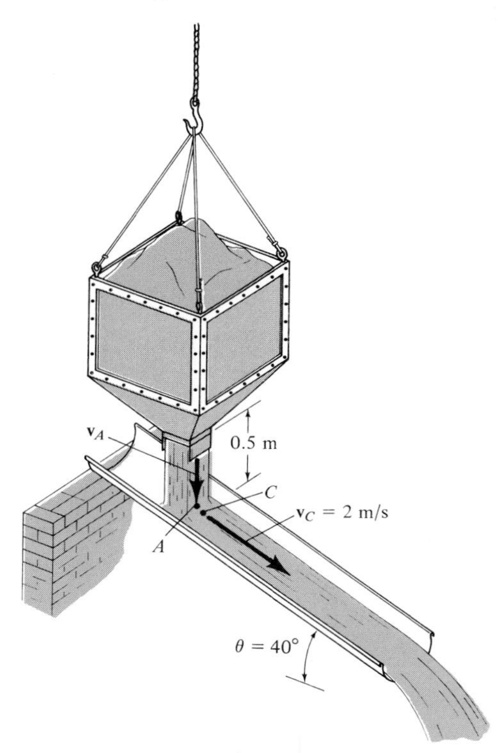

Prob. 12–92

12-93. An aircraft carrier is traveling forward with a velocity of 50 km/h. At the instant shown, the plane at A has just taken off and has attained a forward horizontal airspeed of 200 km/h, measured from still water. If the plane at B is traveling along the runway of the carrier at 175 km/h in the direction shown, determine the velocity of A with respect to B.

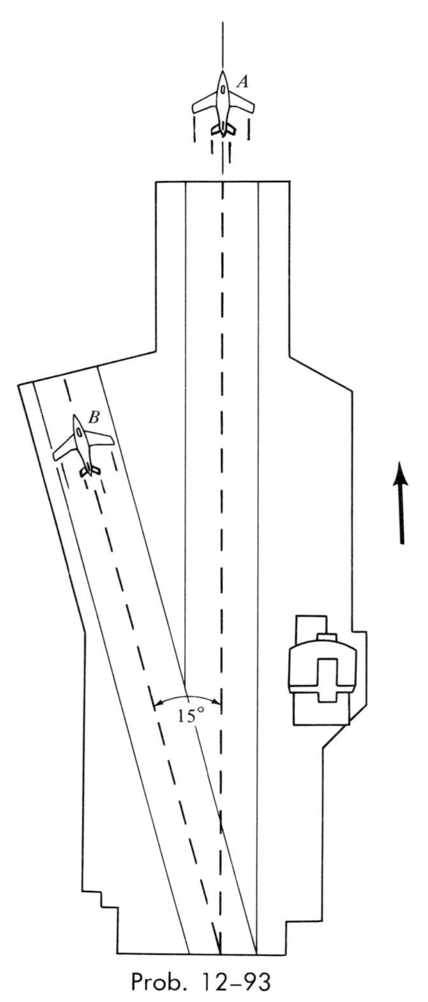

Prob. 12–93

12-94. The pilot of fighter plane F is following 1.5 km behind the pilot of bomber B. Both planes are originally traveling at 120 m/s. In an effort to pass the bomber, the pilot in F gives his plane a constant acceleration of 12 m/s². Determine the speed at which the pilot in the bomber sees the pilot of the fighter plane pass if at the start of the passing operation the bomber is decelerating at 3 m/s². Neglect the effect of any turning.

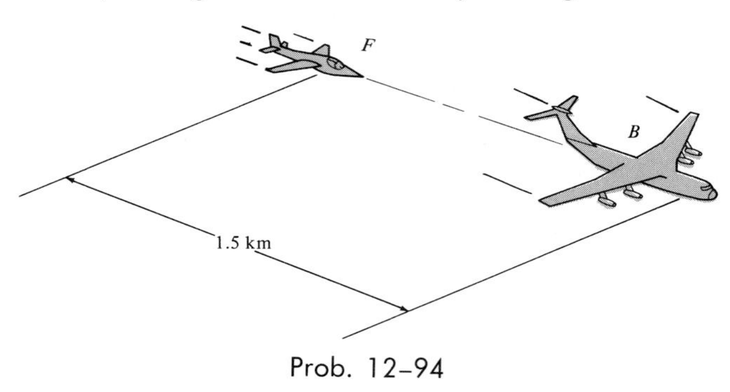

Prob. 12–94

12-95. A toy car C is traveling along the straight path $s_C = 0.4$ m with a constant velocity of $v_C = 0.2$ m/s. If at the same instant a second car B, starting from rest, is directed towards A as shown, determine the constant acceleration it must have to cause a collision when it reaches A; $d_x = 0.6$ m, $d_y = 1.2$ m. What is the relative velocity and relative acceleration of car B with respect to car C just before the collision occurs?

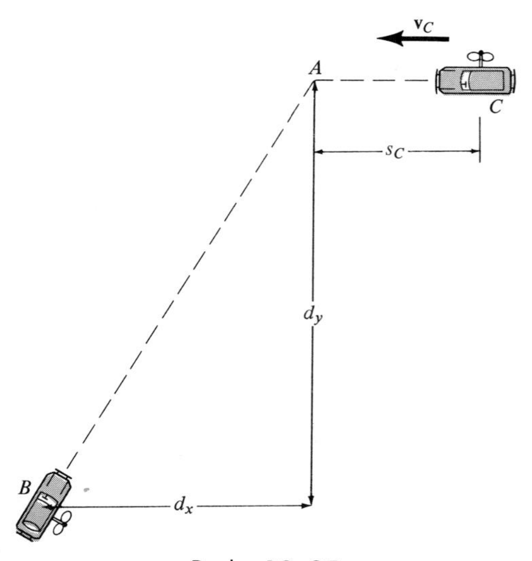

Prob. 12–95

12–95a. Solve Prob. 12–95 if $s_C = 2$ ft, $v_C = 0.25$ ft/s, $d_x = 3$ ft, and $d_y = 5$ ft.

*** 12–96.** The block B is suspended from a cable that is attached to the block at E, wraps around three pulleys, and is tied to the back of a truck. If the truck starts from rest when x_D is zero, and moves forward with a constant acceleration of $a_D = 0.5$ m/s², determine the speed of the block at the instant $x_D = 2$ m. Neglect the size of the pulleys in the calculation. When $x_D = 0$, $x_C = 5$ m, so that points C and D are at the same elevation. *Hint:* Relate the coordinates x_C and x_D using the problem geometry, then take the time derivative.

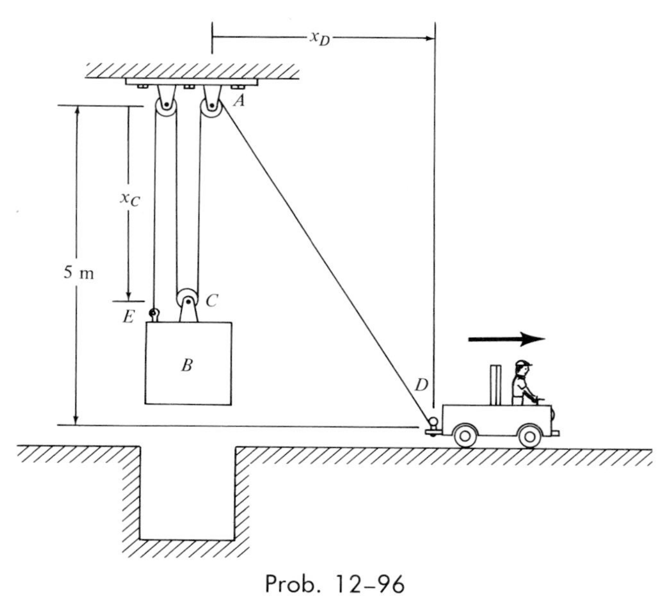

Prob. 12–96

12–97. At the instant shown, car A has a speed of 20 km/h which is being increased at the rate of 300 km/h² as the car enters an expressway. At the same instant, car B is decelerating at 250 km/h² while traveling forward at 100 km/h. Determine the velocity and acceleration of A with respect to the driver of B.

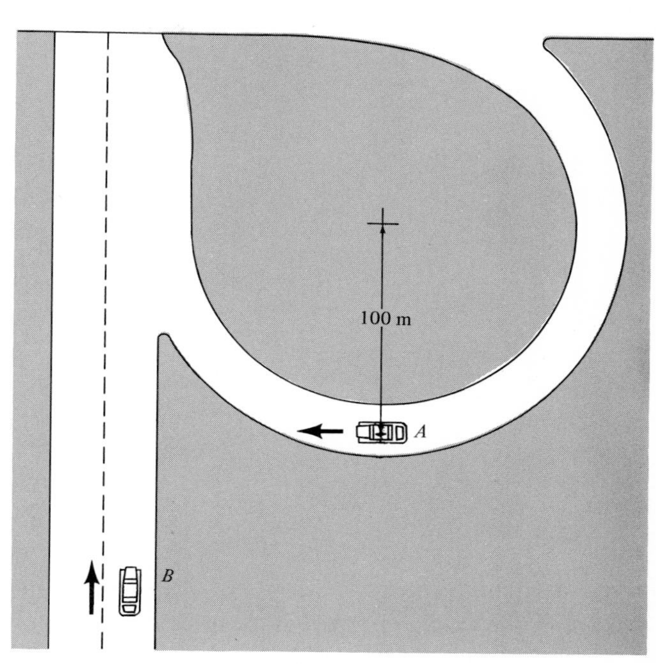

Prob. 12–97

12–98. At the instant shown, cars A and B are traveling at speeds of 20 km/h and 65 km/h, respectively. If B is accelerating at 1200 km/h² while A maintains a constant speed, determine the velocity and acceleration of A with respect to B.

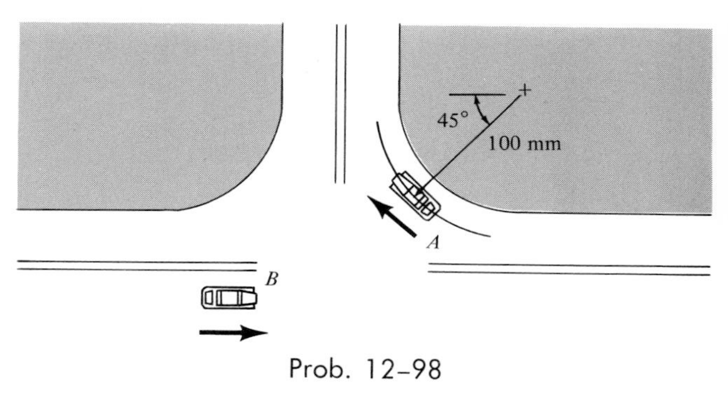

Prob. 12–98

12–99. An automobile traveling with a constant speed of 15 m/s passes directly underneath a streetlight. Determine the speed v_S at which the tip of the automobile's shadow S moves along the ground. *Hint:* Relate the coordinates x_A and x_S using the problem geometry, then take the time derivative.

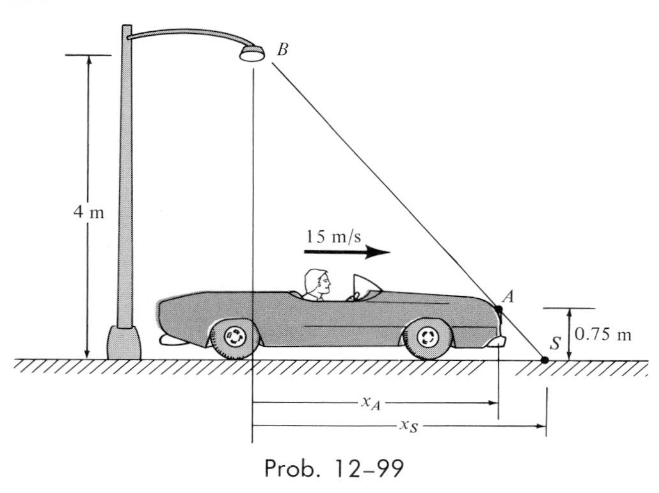

Prob. 12–99

*** 12–100.** The crate C is being dragged across the ground by the hauling truck T. If the truck is traveling at a constant speed of $v_T = 1.5$ m/s, determine the speed of the crate for any angle θ of the rope; $h = 4$ m. The rope has a length of $l = 20$ m and passes over a pulley of negligible size at A. *Hint:* Relate the coordinates x_T and x_C to the length of the rope and take the time derivative. Then substitute the geometric relation between x_C and θ.

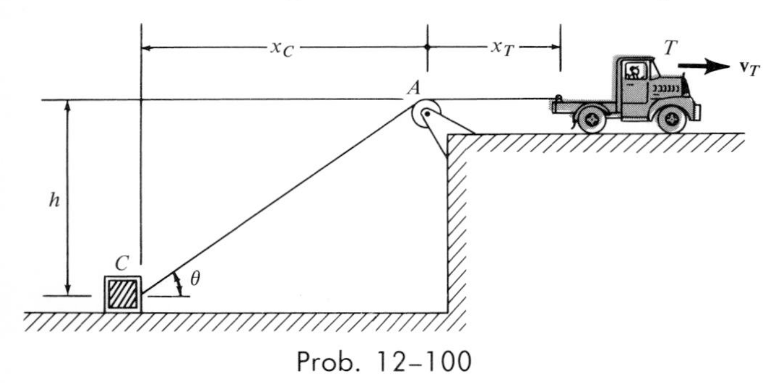

Prob. 12–100

*** 12–100a.** Solve Prob. 12–100 if $v_T = 5$ ft/s, $h = 20$ ft, and $l = 150$ ft.

12–101. The crate C is being lifted by moving the roller at A downward with a constant speed of $v_A = 4$ m/s along the guide. Determine the velocity and acceleration of the crate at the instant $s = 1$ m. When the roller is at B, the crate rests on the ground. Neglect the size of the pulleys in the calculation. *Hint:* Relate the coordinates x_C and x_A using the problem geometry, then take the first and second time derivatives.

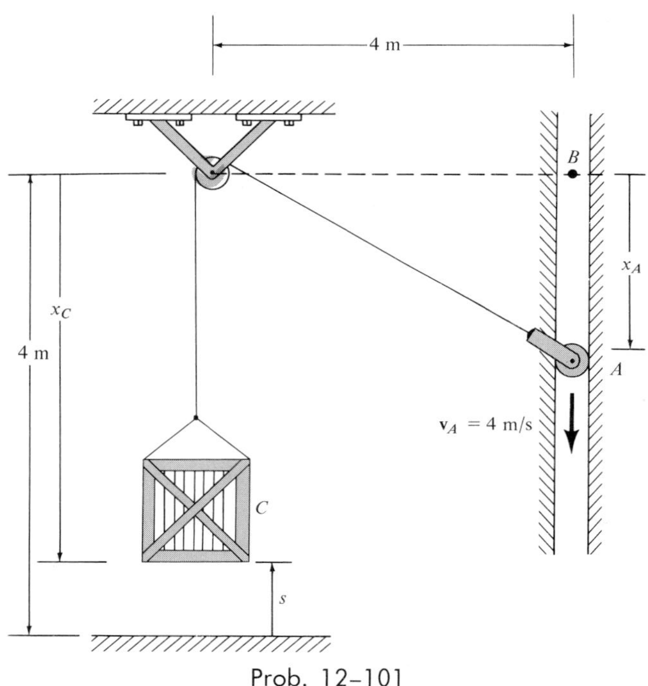

Prob. 12–101

13 Kinetics of a Particle: Forces and Accelerations

13–1. Newton's Laws of Motion

Many of the earlier notions about dynamics were dispelled after 1590 when Galileo performed experiments to study the motions of pendulums and falling bodies. The conclusions drawn from these experiments gave some insight as to the effects of forces acting on bodies in motion. The general motion of a body subjected to forces was not known, however, until 1687, when Isaac Newton first stated three basic laws governing the motion of a particle. In a slightly reworded form, Newton's three laws of motion can be stated as follows:

First Law. A particle originally at rest, or moving in a straight line with a constant velocity, will continue to remain in this state provided the particle is not subjected to an unbalanced force.

Second Law. A particle acted upon by an unbalanced force **F** receives an acceleration **a** that has the same direction as the force and a magnitude that is directly proportional to the force.

Third Law. For every force acting on a particle, the particle exerts an equal, opposite, and collinear reactive force.

The first and third laws were used extensively in developing the concepts of statics. Although these laws are also considered in dynamics, Newton's second law of motion forms the basis for most of this study, since this law relates the accelerated motion of a particle to the forces that act on it. In this regard, measurements of force and acceleration can be recorded in a laboratory so that in accordance with the second law, if a known unbalanced force $\mathbf{F}_1$ is applied to a particle, the acceleration $\mathbf{a}_1$ of the particle may be measured. Since the force and acceleration are directly proportional, the constant of proportionality, m, may be determined from the ratio $m = F_1/a_1$. Provided the units of measurement are consistent, a different unbalanced force $\mathbf{F}_2$ applied to the particle will create an acceleration $\mathbf{a}_2$, such that $F_2/a_2 = m$. In both cases the ratio will be the same and the acceleration and the force, both being vector quantities, will have the same direction. The scalar m is called the *mass* of the particle. Being constant during any acceleration, m provides a quantitative measure of the resistance of the particle to a change in its velocity.

If the mass of the particle is m, Newton's second law of motion may be written in mathematical form as

$$\mathbf{F} = m\mathbf{a} \tag{13–1}$$

This equation, which is referred to as the *equation of motion,* is one of the most important formulations in mechanics. In 1905, however, Albert Einstein placed limitations on its use for describing general particle motion. In developing the theory of relativity, he had discovered that *time* was not an absolute quantity as assumed by Newton; as a result, it has been shown that the equation of motion fails to *accurately* predict the behavior of a particle, especially when the particle's speed approaches the speed of light ($3.0(10^8)$ m/s). Developments of the theory of quantum mechanics by Schrödinger and his colleagues indicate further that conclusions drawn from using this equation are also invalid when particles move within an atomic distance of one another. For the most part, however, these requirements regarding particle speed and size are generally not encountered in engineering problems, so that these effects will not be considered in this book.

Newton's Law of Gravitational Attraction. Shortly after formulating his three laws of motion for a particle, Newton postulated a law governing the mutual attraction between any two particles. In mathematical form this law can be expressed as

$$F = G\frac{m_1 m_2}{r^2} \tag{13–2}$$

where

F = the force of attraction between the two particles

G = the universal constant of gravitation; according to experimental evidence $G = (6.673(10^{-11})\ \text{m}^3)/(\text{kg} \cdot \text{s}^2)$

m_1, m_2 = the mass of each of the two particles

r = the distance between the centers of both particles

Any two particles or bodies have a mutually attractive (gravitational) force acting between them. In the case of a particle located at or near the surface of the earth, however, the only attractive force having any sizable magnitude is that of the earth's gravitation. This force is termed the "weight" and, for our purpose, it will be the only gravitational force considered.

Mass and Weight. Mass and weight are two terms which are often confused when using either SI or FPS units.* Specifically, the *mass* of a body may be regarded as a quantitative measure of the resistance of matter to a *change* in its velocity. This property is more fundamental (or absolute) than specifying the weight of a body, because the *weight,* or gravitational force which one body exerts upon another, changes in magnitude depending upon the location where the measurement for weight is made.

SI System of Units. Since the SI system of units is an absolute system, the mass of the body is specified and the weight must be calculated. Hence, if a body has a mass of m (kg) and is located at a point where the acceleration due to gravity is g (m/s^2), then since $F = ma$, the weight W is expressed in *newtons* as $W = mg$ (N), Fig. 13-1*a*. In particular, if the body is located on the earth at sea level and at a latitude of 45° (considered the "standard location"), the acceleration due to gravity is $g = 9.806\ 65$ m/s^2. For calculations, the value $g = 9.81$ m/s^2 will be used, so that

$$W = mg\ (\text{N}) \quad (g = 9.81\ \text{m/s}^2) \tag{13–3}$$

Hence, from this equation, a body of mass 1 kg has a weight of 9.81 N; a 2-kg body weighs 19.62 N; and so on.

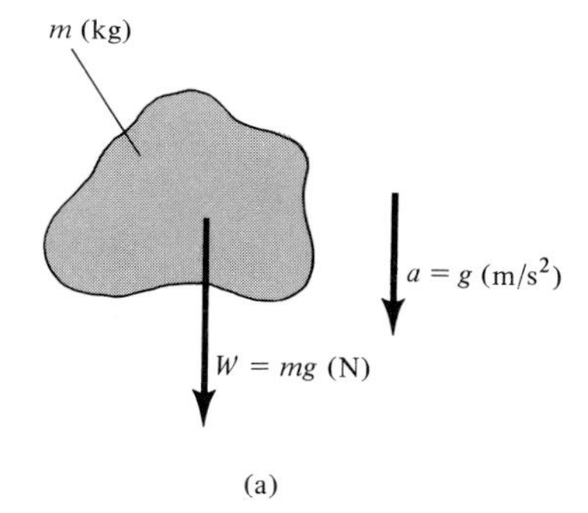

(a)

FPS System of Units. Since the FPS system of units is a *gravitational system,* the weight of the body is specified and the mass must be calculated. Hence, if a body has a weight of W (lb) and is located at a point where the acceleration due to gravity is g (ft/s^2), then from $F = ma$, the mass is expressed in *slugs* as $m = W/g$ (slug), Fig. 13–1*b*. Since the acceleration of gravity at the standard location is approximately 32.2 ft/s^2 ($= 9.81$ m/s^2), the mass of the body measured in slugs is

$$m = \frac{W}{g}\ (\text{slug}) \quad (g = 32.2\ \text{ft/s}^2) \tag{13–4}$$

Therefore, a body weighing 32.2 lb has a mass of 1 slug; a 64.4-lb body has a mass of 2 slugs; and so on.

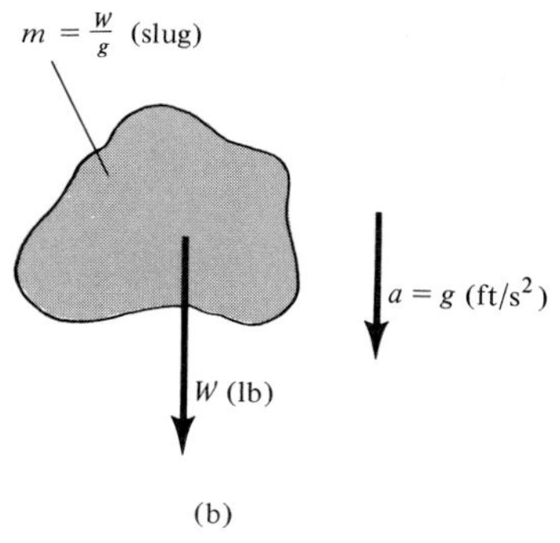

(b)

Fig. 13–1

*These systems of units are described in Appendix A.

13–2. The Equation of Motion

When more than one force acts on a particle, the resultant force $\mathbf{F}_R$ is determined by a vector summation of all the forces, i.e., $\mathbf{F}_R = \Sigma\mathbf{F}$. For this more general case, the equation of motion may be written as

$$\Sigma\mathbf{F} = m\mathbf{a} \tag{13-5}$$

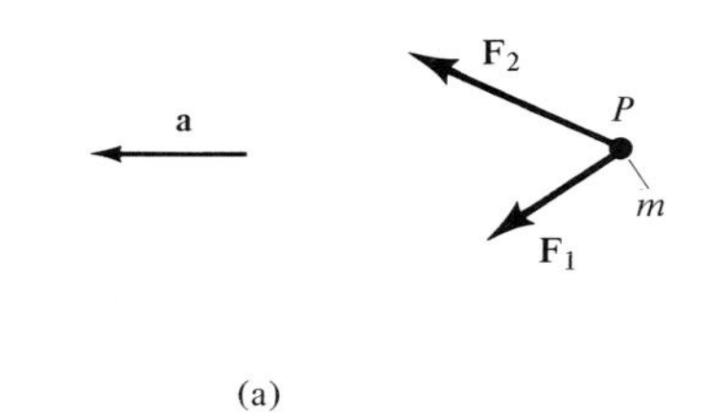

(a)

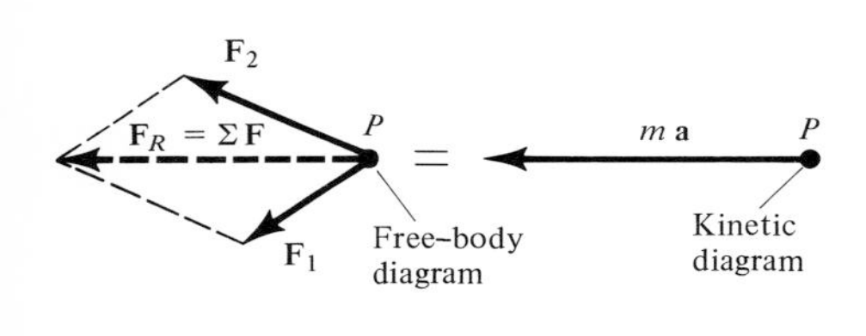

(b)

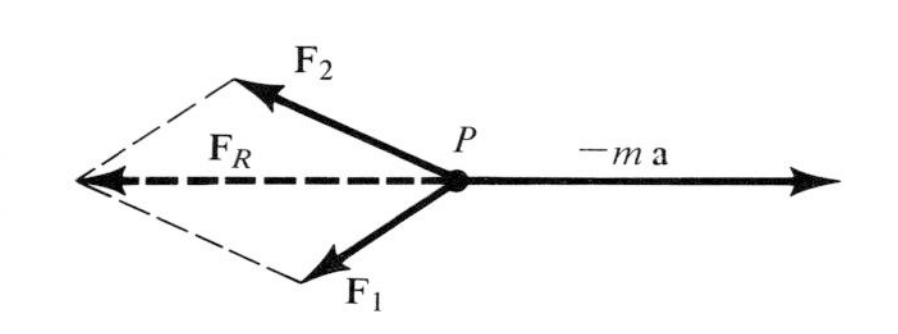

(c)

Fig. 13–2

To illustrate the use of this equation, consider the particle P, shown in Fig. 13–2*a,* which has a mass m and is subjected to the action of two forces, $\mathbf{F}_1$ and $\mathbf{F}_2$. The *free-body diagram,* constructed in Fig. 13–2*b*, graphically represents all the forces acting on the particle, $\Sigma\mathbf{F} = \mathbf{F}_1 + \mathbf{F}_2$, whereas the *kinetic diagram* graphically accounts for the vector $m\mathbf{a}$, Fig. 13–2*b*. As shown by these two diagrams, the particle accelerates in the direction of $\mathbf{F}_R = \mathbf{F}_1 + \mathbf{F}_2$, such that the magnitude of $a = F_R/m$. In particular, if $\mathbf{F}_R = \Sigma\mathbf{F} = \mathbf{0}$, the acceleration is also zero, in which case the particle will either remain at *rest* or move along a straight-line path with *constant velocity*. Such are the conditions of *static equilibrium,* Newton's first law of motion.

Dynamic Equilibrium. The equation of motion can be rewritten in the form $\Sigma\mathbf{F} - m\mathbf{a} = \mathbf{0}$ and the result illustrated on a diagram, which includes the "$m\mathbf{a}$ vector" acting in its reverse sense ($-m\mathbf{a}$), Fig. 13–2*c*. If $-m\mathbf{a}$ is treated in the same way as a "force vector," then the state of "equilibrium" created in Fig. 13–2*c* is referred to as *dynamic equilibrium*. This method for application of the equation of motion is often referred to as the *D'Alembert principle,* named after the French mathematician Jean le Rond d'Alembert (1717–1783).

The vector $-m\mathbf{a}$ is referred to as the *inertia-force vector*. The word "inertia" is incorporated in this terminology, because *mass m* is a quantitative measure of the inertia or resistance of a body to a *change* in its velocity (acceleration). It is important to realize, however, that the inertia-force vector is actually not the same as a force. The inertia of a body manifests itself as a force whenever an unbalanced force acts on the body and thereby causes an acceleration. For example, consider the passenger riding in a train that is accelerating, Fig. 13–3*a*. The forward motion of the train creates a horizontal force $\mathbf{F}$ which the seat exerts on her back, Fig. 13–3*b*. By the equation of motion, it is this unbalanced *force* which gives her a forward acceleration ($\mathbf{F} = m\mathbf{a}$). No force exists which pushes her back toward the seat ($-m\mathbf{a}$), although this is the sensation she receives, Fig. 13–3*c*.

Inertial Frame of Reference. Whenever the equation of motion is applied, it is required that measurements of the acceleration be made from a *Newtonian* or *inertial frame of reference*. Such a *coordinate system does not*

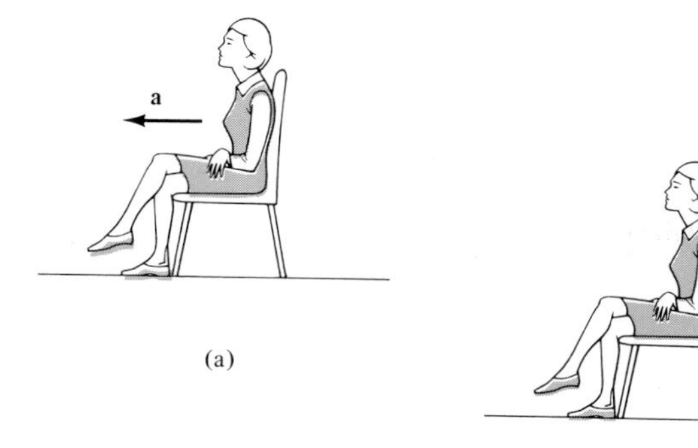

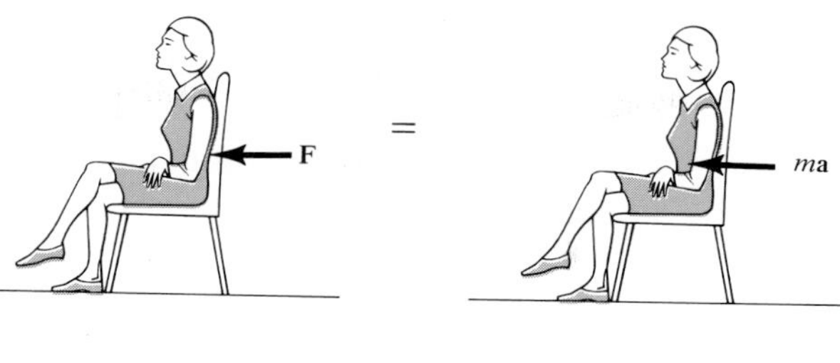

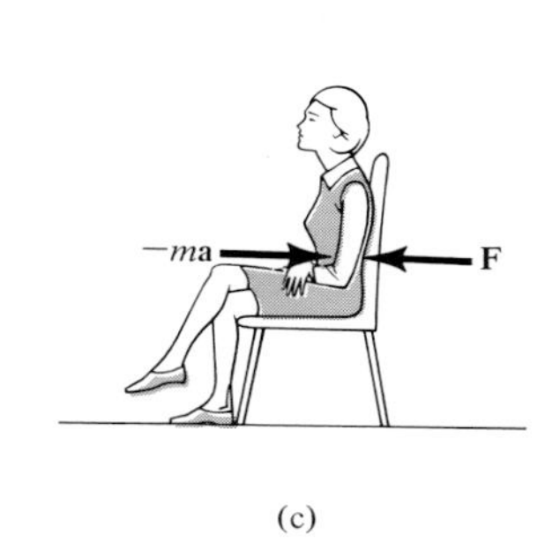

Fig. 13–3

rotate and is either fixed or translates in a given direction with a constant velocity (*zero acceleration*). This definition ensures that the particle *acceleration* measured by observers in two different inertial frames of reference will always be the *same*. When studying the motions of rockets and satellites it is justifiable to consider the inertial reference frame as fixed to the stars, whereas dynamics problems concerned with motions on or near the surface of the earth may be solved by using an inertial frame which is assumed fixed to the earth. Even though the earth *rotates* both about its own axis and about the sun, the acceleration created by these rotations can be neglected in most computations.

13–3. Equation of Motion for a System of Particles

The equation of motion will now be extended to include a system of n particles isolated within an enclosed region in space as shown in Fig. 13–4a. In particular, there is no restriction in the way the particles are connected, and as a result, the following analysis will apply equally well to the motion of a solid, liquid, or gas system. At the instant considered, the arbitrary ith particle, having a mass m_i, is subjected to a set of internal forces and a resultant external force. The *internal forces,* represented symbolically as $\Sigma \mathbf{f}_i$, are reactive forces which the other particles each exert on the ith particle. The *resultant external force* $\mathbf{F}_i$ represents the effect of gravitational, electrical, magnetic, or contact forces between adjacent bodies or particles not included within the system of n particles.

The free-body and kinetic diagrams for the ith particle are shown in Fig. 13–4b. Applying the equation of motion to the ith particle yields

$$\Sigma \mathbf{F} = m\mathbf{a}; \qquad \mathbf{F}_i + \Sigma \mathbf{f}_i = m_i \mathbf{a}_i$$

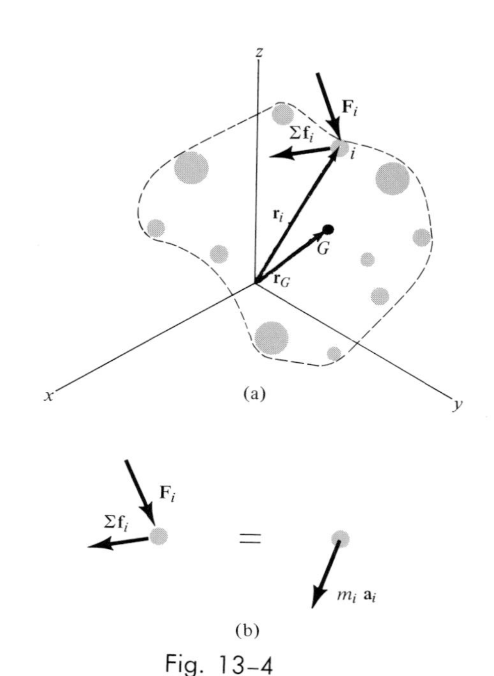

Fig. 13–4

When the equation of motion is applied to each of the other particles of the system, similar equations will result. However, when all these equations are added together *vectorially,* only the sum of the external forces, $\Sigma\mathbf{F} = \Sigma\mathbf{F}_i$, acting on the system of particles will remain. The vector sum of the internal forces acting within the system is zero, since, by Newton's third law of motion, all the internal forces occur in equal but opposite collinear pairs. Hence, the equation of motion, written for the *entire system of particles,* becomes

$$\Sigma\mathbf{F} = \Sigma m_i\mathbf{a}_i \tag{13-6}$$

If $\mathbf{r}_G$ is a position vector which locates the *center of mass* G of the particles, Fig. 13-4*a,* then, by definition of the center of mass,

$$m\mathbf{r}_G = \Sigma m_i\mathbf{r}_i$$

where $m = \Sigma m_i$ is the total mass of all n particles. Differentiating this equation twice with respect to time, assuming no mass is entering or leaving the system,* yields

$$m\mathbf{a}_G = \Sigma m_i\mathbf{a}_i$$

Substituting this result into Eq. 13-6, we obtain

$$\Sigma\mathbf{F} = m\mathbf{a}_G \tag{13-7}$$

This equation states that the sum of the external forces acting on the system of particles is equal to the mass $m = \Sigma m_i$ of a single "fictitious" particle times its acceleration. This fictitious particle is located at the center of mass G of all the particles.

Since, in reality, all particles must have finite size to possess mass, Eq. 13-7 justifies application of the equation of motion to a body that is represented as a single particle.

13-4. Equations of Motion for a Particle: Rectangular Coordinates

When a particle is moving relative to an inertial x, y, z frame of reference, the forces acting on it, as well as its acceleration, may be expressed in terms of their $\mathbf{i}, \mathbf{j}, \mathbf{k}$ components, Fig. 13-5. Applying the equation of motion, we have

$$\Sigma\mathbf{F} = m\mathbf{a}$$

$$\Sigma F_x\mathbf{i} + \Sigma F_y\mathbf{j} + \Sigma F_z\mathbf{k} = m(a_x\mathbf{i} + a_y\mathbf{j} + a_z\mathbf{k})$$

*A case in which m is a function of time (variable mass) is discussed in Sec. 15-9.

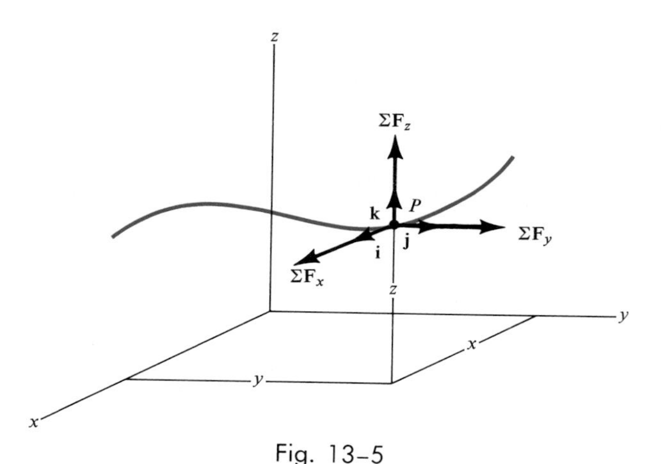

Fig. 13–5

Since the respective **i**, **j**, and **k** components must be equivalent, the solution of the above equation can be represented in terms of the following three scalar equations:

$$\Sigma F_x = ma_x \qquad \Sigma F_y = ma_y \qquad \Sigma F_z = ma_z \tag{13–8}$$

If all the forces acting on the particle lie in the x-y plane, the particle will only have motion in this plane. Hence, $\Sigma F_z = ma_z$ is not applicable and therefore only the first two of Eqs. 13–8 may be used to specify the motion.

PROCEDURE FOR ANALYSIS

The following three-step procedure should be applied when solving particle kinetic problems using the equations of motion:

Step 1: Establish the x, y, z inertial coordinate system and draw the particle's free-body and kinetic diagrams. As stated previously, the free-body diagram is a graphical representation of all the forces (Σ**F**) which act on the particle; whereas the kinetic diagram graphically accounts for m**a**, Fig. 13–2*b*.

Step 2: Apply the equations of motion. In most problems these equations can be applied in their scalar component form, provided the vector components of the forces and m**a** can be resolved directly from the free-body and kinetic diagrams. If the geometry of the problem appears to be complicated, which often occurs in three dimensions, Cartesian vector analysis should be used for the solution.

If the particle contacts a rough surface, it may be necessary to use the *frictional equation,* which relates the coefficient of kinetic friction μ_k to the frictional and normal forces $\mathbf{F}_f$ and $\mathbf{N}$ acting at the surfaces of contact,* i.e., $F_f = \mu_k N$.

If the particle is connected to an *elastic spring* having negligible mass, the magnitude of spring force $\mathbf{F}_s$ can be related to the stretch or compression x of the spring by the equation $F_s = kx$, where k is the spring's stiffness measured as a force per unit length.

Step 3: Use the equations of kinematics if a complete solution cannot be obtained strictly from the equation of motion. In particular, if the velocity or position of the particle is to be found, it will be necessary to apply the proper kinematic equations once the particle's acceleration $\mathbf{a}$ is determined from $\Sigma\mathbf{F} = m\mathbf{a}$.

If the *acceleration is a function of time,* use $a = dv/dt$ and $v = ds/dt$, which, when integrated, yield the particle's velocity and position, respectively.

If the *acceleration is a function of displacement,* integrate $a\,ds = v\,dv$ to obtain the velocity as a function of displacement.

If the *acceleration is constant,* use $v = v_1 + a_c t$, $s = s_1 + v_1 t + \frac{1}{2}a_c t^2$, or $v^2 = v_1^2 + 2a_c(s - s_1)$ to determine the position or velocity of the particle.

The following examples numerically illustrate application of this three-step procedure.

Example 13–1

The 50-kg crate shown in Fig. 13–6*a* rests on a horizontal plane for which the coefficient of kinetic friction is $\mu_k = 0.3$. If the crate is subjected to a 400-N towing force as shown, determine the velocity of the crate in 5 s starting from rest.

Solution

Step 1: From Eq. 13–3, the weight of the crate is

$$W = mg = 50\text{ kg}(9.81\text{ m/s}^2) = 490.5\text{ N}$$

The free-body and kinetic diagrams are shown in Fig. 13–6*b*. Note that the frictional force has a magnitude of $F = \mu_k N_C$ and acts to the left, since it opposes the motion of the crate.†

*A review of Sec. 8–1, on friction, of *Engineering Mechanics: Statics* is suggested *before* solving the problems.

†For problem solving, the normal force will be designated by the symbol *N with a subscript* in order to distinguish it from the symbol N used to specify its magnitude in newtons.

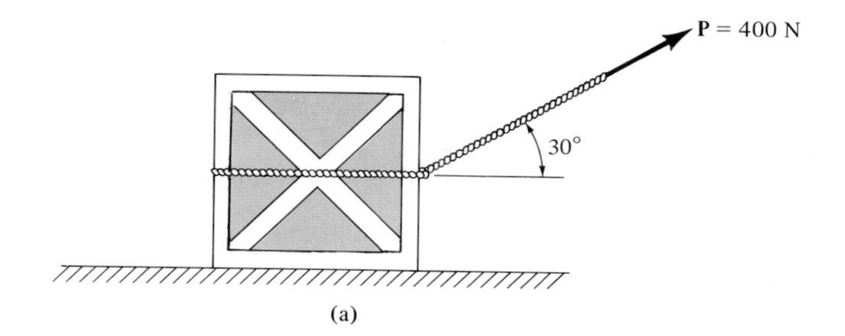

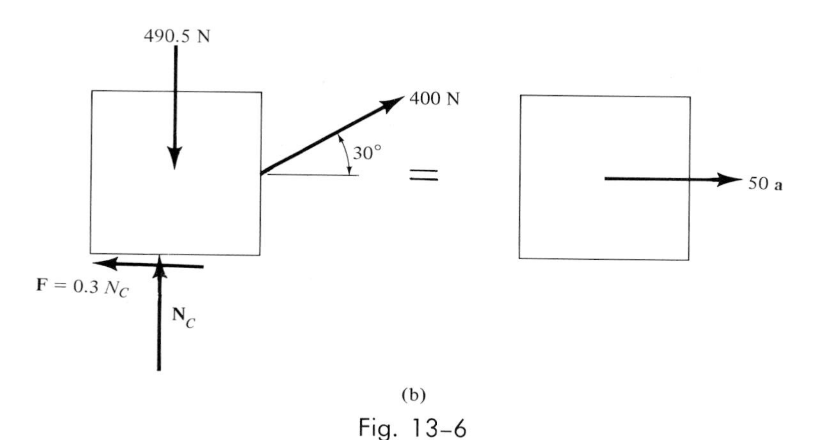

Fig. 13–6

Step 2: Applying the equations of motion, we have

$$\xrightarrow{+} \Sigma F_x = ma_x; \qquad 400 \cos 30° - 0.3\, N_C = 50a \qquad (1)$$

$$+\uparrow \Sigma F_y = ma_y; \qquad N_C - 490.5 + 400 \sin 30° = 0 \qquad (2)$$

Solving Eq. (2) for N_C, substituting the result into Eq. (1), and solving for a yields

$$N_C = 290.5 \text{ N}$$
$$a = 5.19 \text{ m/s}^2$$

Step 3: Since the acceleration is *constant,* and the initial velocity is zero, the velocity of the crate in 5 s is

$$(\xrightarrow{+}) \qquad v_2 = v_1 + a_c t$$
$$= 0 + 5.19(5)$$
$$= 26.0 \text{ m/s} \qquad \textit{Ans.}$$

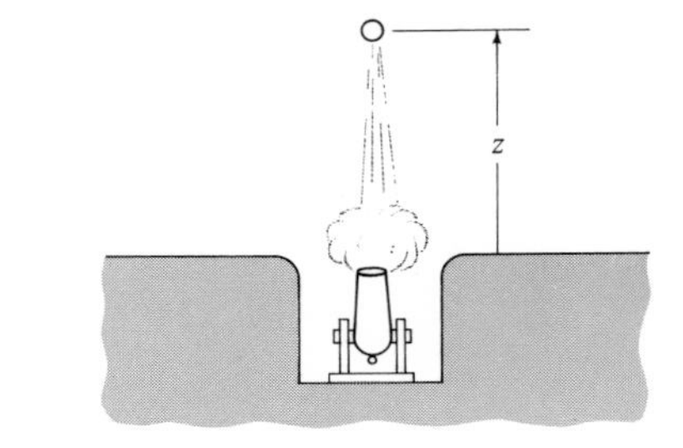

Fig. 13–7(a)

Example 13–2

A 10-kg projectile is fired vertically upward from the ground, with an initial velocity of 50 m/s, Fig. 13–7*a*. Determine the maximum height to which it will travel if (a) atmospheric resistance is neglected, and (b) the atmospheric resistance is measured as $F_D = (0.01v^2)$ N, where v is the speed at any instant, measured in m/s.

Fig. 13–7(b)

98.1 N = 10 a

Solution

Part (a). *Step 1:* The free-body and kinetic diagrams are shown in Fig. 13–7*b*. The weight is $W = mg = 10(9.81) = 98.1$ N.

Step 2: Applying the equation of motion,

$$+\uparrow \Sigma F_z = ma_z; \qquad -98.1 = 10a$$

$$a = -9.81 \text{ m/s}^2$$

The result indicates that the projectile, like every object having free-flight motion near the earth's surface, is subjected to a constant acceleration of 9.81 m/s². Because the answer is negative, the projectile is *decelerating* as it moves upward; that is, $m\mathbf{a}$ in Fig. 13–7*b* should act downward.

Step 3: Since the acceleration is *constant,* the maximum height h can be obtained using Eq. 12–10. Initially, $s_1 = 0$ and $v_1 = 50$ m/s, and at the maximum height, $v_2 = 0$; therefore,

$$(+\uparrow) \qquad v_2^2 = v_1^2 + 2a_c(s_2 - s_1)$$

$$0 = (50)^2 + 2(-9.81)(h - 0)$$

Thus,

$$h = 127.4 \text{ m} \qquad \textit{Ans.}$$

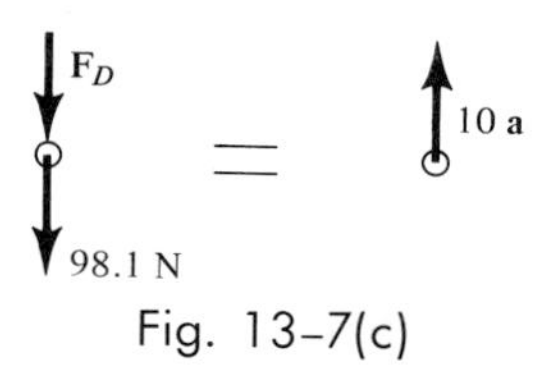

Fig. 13–7(c)

Part (b). *Step 1:* Since the force $F_D = (0.01v^2)$ N tends to retard the upward motion of the projectile, it acts downward as shown on the free-body diagram, Fig. 13–7*c*.

Step 2: Applying the equation of motion,

$$+\uparrow \Sigma F_z = ma_z; \qquad -0.01v^2 - 98.1 = 10a$$

$$a = -0.001v^2 - 9.81$$

Step 3: Here the acceleration is not constant; however, it can be related to the velocity and displacement by using $a\,dz = v\,dv$ or $a = v\,dv/dz$. Hence,

$$-0.001v^2 - 9.81 = \frac{v\,dv}{dz}$$

Separating the variables and integrating, realizing that initially $z_1 = 0$, $v_1 = 50$ m/s (positive upward), we have

$$dz = \frac{-v\,dv}{0.001v^2 + 9.81}$$

$$\int_0^z dz = -\int_{50}^{v} \frac{v\,dv}{0.001v^2 + 9.81}$$

$$z = -500 \ln (v^2 + 9810)\Big|_{50}^{v} = -500 \ln \left[\frac{v^2 + 9810}{(50)^2 + 9810}\right]$$

At the maximum height $z = h$ and $v = 0$; hence,

$$h = -500 \ln \left[\frac{9810}{(50)^2 + 9810}\right]$$

$$= 113.5 \text{ m} \qquad \textit{Ans.}$$

Comparison of the two answers (127.4 m and 113.5 m) reveals the retarding effect of air resistance $\mathbf{F}_D$ on projectile motion. When the speed is large yet less than the speed of sound, as illustrated here, the effect of air resistance is proportional to the *square* of the speed. The constant of proportionality, stated as 0.01, is dependent upon the density of air and the size and shape of the body.

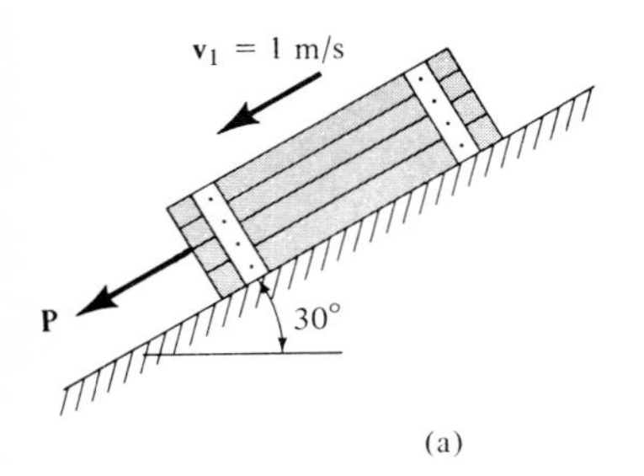

Example 13-3

The crate shown in Fig. 13-8*a* has a mass of 50 kg and is acted upon by a force having a variable magnitude of $P = 20t$, where P is in newtons and t is in seconds. Compute the crate's velocity 4 s after **P** has been applied. The crate's initial velocity is $v_1 = 1$ m/s down the plane, and the coefficient of kinetic friction between the crate and the plane is $\mu_k = 0.3$.

Solution

Step 1: The free-body and kinetic diagrams are shown in Fig. 13-8*b*. The frictional force is directed opposite to the crate's sliding motion and has a magnitude of $F = \mu_k N_C = 0.3 N_C$. The weight is $W = mg = 50(9.81) = 490.5$ N.

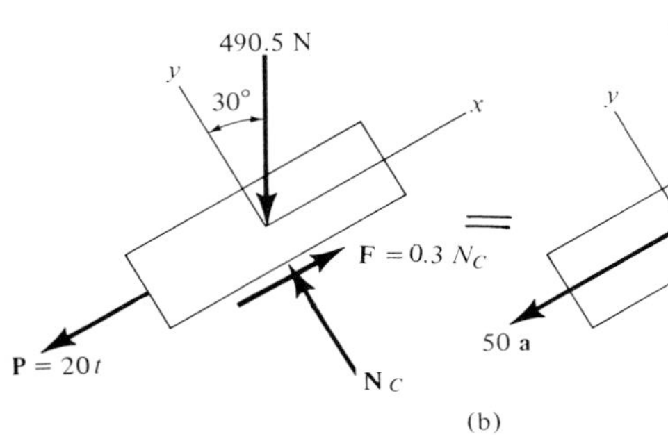

Fig. 13-8

Step 2: Applying the equations of motion, we have

$$+\nearrow \Sigma F^x = ma_x; \quad -20t + 0.3N_C - 490.5 \sin 30^\circ = -50a \qquad (1)$$

$$+\nwarrow \Sigma F_y = ma_y; \quad N_C - 490.5 \cos 30^\circ = 0 \qquad (2)$$

There are three unknowns, N_C, t, and a. Solving for N_C in Eq. (2) ($N_C = 424.7$ N), substituting into Eq. (1), and simplifying yields

$$a = 2.36 + 0.4t \qquad (3)$$

Step 3: Since the acceleration is a function of time and *not* constant, the velocity of the crate can be obtained by using $a = dv/dt$ with the initial condition that $v = 1$ m/s at $t = 0$. We have

$$dv = a\,dt$$

$$\int_1^v dv = \int_0^t (2.36 + 0.4t)\,dt$$

$$v = 2.36t + 0.2t^2 + 1$$

When $t = 4$ s,

$$v = 13.6 \text{ m/s} \qquad \textit{Ans.}$$

Example 13-4

A smooth 2-kg collar C, shown in Fig. 13-9*a*, is attached to a spring which has a stiffness of $k = 3$ N/m and an unstretched length of 0.75 m. If the collar is released from rest at A, determine the speed at which it is moving at the instant $s = 1$ m. Neglect friction between the collar and the shaft.

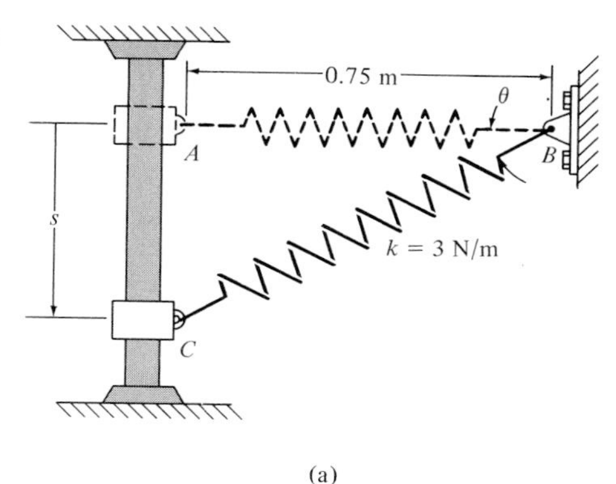

Solution

Step 1: The free-body and kinetic diagrams of the collar, when it is located at the arbitrary position s, are shown in Fig. 13–9b. Note that the collar has a weight of $W = 2(9.81) = 19.62$ N. Furthermore, the collar is assumed to be accelerating so that "2**a**" is shown acting downward.

Step 2: Writing the equations of motion,

$$\overset{+}{\rightarrow}\Sigma F_x = ma_x; \qquad -N_C + F_s \cos\theta = 0 \tag{1}$$

$$+\uparrow\Sigma F_y = ma_y; \qquad F_s \sin\theta - 19.62 = -2a \tag{2}$$

Equation (1) involves the unknown normal force N_C of the rod on the collar, which is of no consequence to the problem solution. From Eq. (2) it is seen that **a** is not constant; rather, it depends upon the magnitude and direction of $\mathbf{F}_s$.

Step 3: Since the speed of the collar is to be determined when $s = 1$ m (downward), the (downward) acceleration, velocity, and displacement can be related using $a\,ds = v\,dv$. Hence, Eq. (2) becomes

$$(F_s \sin\theta - 19.62)\,ds = -2(v\,dv) \tag{3}$$

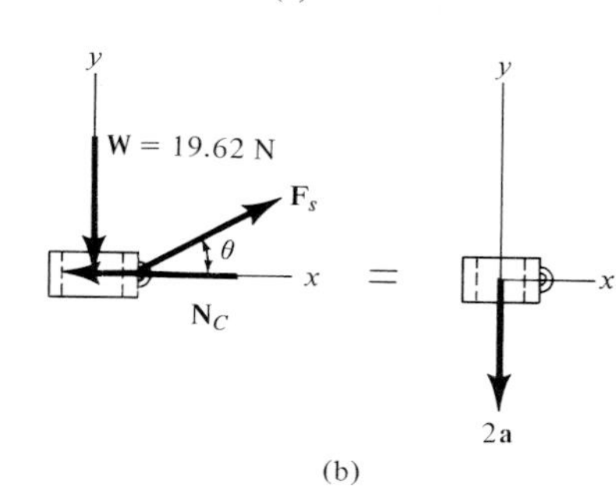

Fig. 13–9

Integration of this equation is possible provided the variables F_s and $\sin\theta$ can be expressed in terms of s. Recall that the magnitude of the spring force depends upon the stretch x of the spring, i.e., $F_s = kx$. Here the unstretched spring length is $AB = 0.75$ m, Fig. 13–9a; therefore, $x = CB - AB = \sqrt{s^2 + (0.75)^2} - 0.75$. Since $k = 3$ N/m, then

$$F_s = kx = 3(\sqrt{s^2 + (0.75)^2} - 0.75) \tag{4}$$

From Fig. 13–9a, the angle θ is related to s by trigonometry.

$$\sin\theta = \frac{s}{\sqrt{s^2 + (0.75)^2}} \tag{5}$$

Substituting Eqs. (4) and (5) into Eq. (3), simplifying, and integrating from $s_1 = 0$, $v_1 = 0$, yields

$$\int_0^s \left[3s - \frac{3(0.75)s}{\sqrt{(0.75)^2 + s^2}} - 19.62\right] ds = -2\int_0^v v\,dv$$

$$1.5s^2 - 2.25\sqrt{s^2 + (0.75)^2} - 19.62s\Big|_0^s = -v^2$$

$$1.5s^2 - 2.25\sqrt{s^2 + (0.75)^2} - 19.62s + 1.68 = -v^2$$

Setting $s = 1$ m and solving for v yields

$$v = 4.39 \text{ m/s} \qquad \textit{Ans.}$$

Example 13–5

The 100-kg block H shown in Fig. 13–10a is released from rest and travels down along the inclined plane. If the mass of the pulleys and the cord is neglected and the pulleys are frictionless, determine the speed of the 20-kg block E in 2 s.

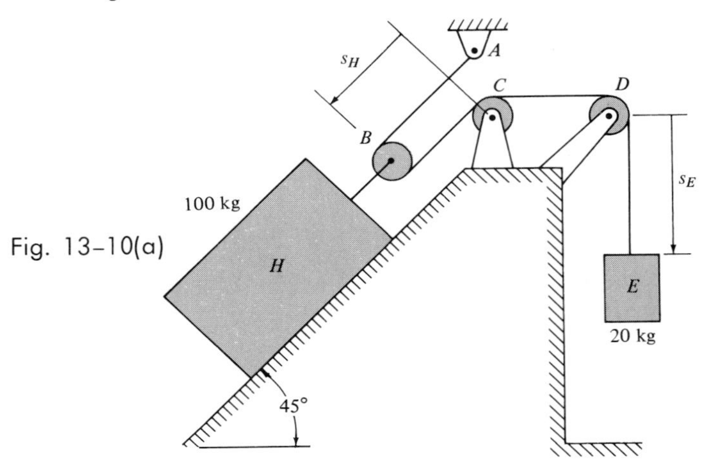

Fig. 13–10(a)

Solution

Motion of H and E will be analyzed separately.

Step 1: The necessary free-body and kinetic diagrams for each block are shown in Figs. 13–10b and 13–10c, respectively. Block H accelerates down the plane, causing block E to accelerate upward.

Step 2: Applying the equations of motion, we obtain for H,

$$+\swarrow \Sigma F_x = ma_x; \qquad -T_H + 981 \sin 45° = 100a_H \tag{1}$$

$$+\nwarrow \Sigma F_y = ma_y; \qquad N_H - 981 \cos 45° = 0 \tag{2}$$

and for E,

$$+\uparrow \Sigma F_y = ma_y; \qquad T_E - 196.2 = 20a_E \tag{3}$$

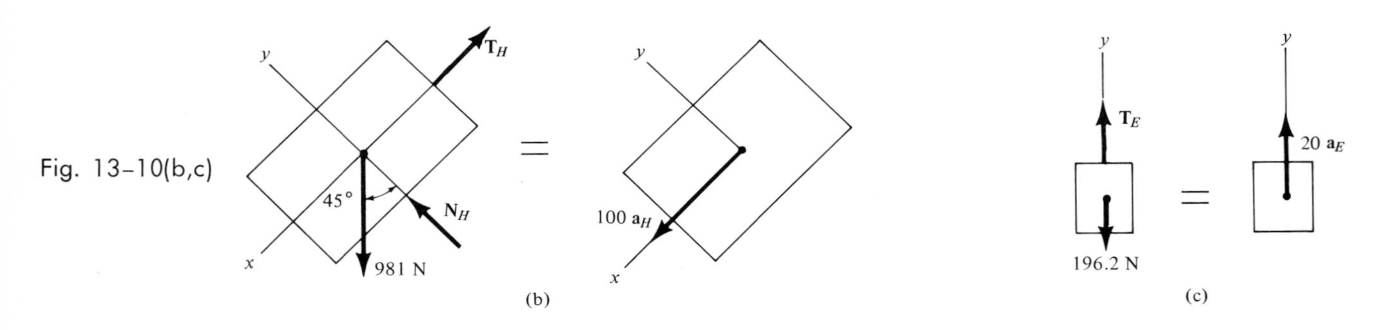

Fig. 13–10(b,c)

There are five unknowns (T_H, N_H, T_E, a_H, and a_E) in these three equations. Since the mass of the pulley is neglected and the pulley is frictionless, a relation between T_H and T_E may be obtained from the free-body diagram of the pulley at *B*, Fig. 13–10*d*, i.e.,

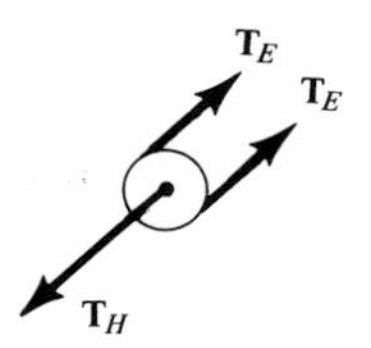

Fig. 13–10(d)

$$2T_E = T_H \qquad (4)$$

Step 3: A fifth equation is obtained by studying the kinematics of the pulley arrangement, thereby relating the accelerations a_E and a_H. Using the technique developed in Sec. 12–9, the coordinates s_H and s_E measure the positions of *E* and *H* from fixed points at *C* and *D*, Fig. 13–10*a*. It is seen that

$$2s_H + s_E = l$$

where l is *constant* and represents the *total length* of the segments of cable which *change* as the blocks displace. Differentiating this expression twice with respect to time yields

$$2a_H = -a_E$$

The sign in this equation indicates that if *H* accelerates down the incline ($+s_H$ direction), *E* will accelerate upward ($-s_E$ direction). Since it is necessary to obtain a *simultaneous solution* with Eqs. (1) through (4), it is important that the *directions* indicated by the above equation be consistent with those used for writing Eqs. (1) and (3). As shown on the kinetic *diagrams, Fig.* 13–10*b* and 13–10*c*, $100\mathbf{a}_H$ acts down the incline; whereas $20\mathbf{a}_E$ acts upward. Hence, to maintain a consistency with the directions of these vectors, we must write

$$2a_H = a_E \qquad (5)$$

Simultaneous solution of Eqs. (1) through (5) yields

$$T_H = 526.1 \text{ N}$$
$$T_E = 263.2 \text{ N}$$
$$N_H = 693.6 \text{ N}$$
$$a_H = 1.68 \text{ m/s}^2$$
$$a_E = 3.35 \text{ m/s}^2$$

Notice that $\mathbf{a}_H$ and $\mathbf{a}_E$ are constant, since the forces acting on *H* and *E* are constant. Because of this, the velocity of *E* may be obtained by using the kinematic equation 12–8. Since the block starts from rest, $(v_E)_1 = 0$, and therefore in 2 s, we have

$(+\uparrow)$
$$v_E = (v_E)_1 + a_E t$$
$$= 0 + (3.35)2$$
$$= 6.70 \text{ m/s} \qquad \textit{Ans.}$$

Problems

Except when stated otherwise, throughout this chapter assume that the coefficients of static and kinetic friction are equal; i.e., $\mu = \mu_s = \mu_k$.

13–1. Determine the gravitational attraction between two spheres which are just touching each other. Each sphere has a mass of 5 kg and a radius of 100 mm.

13–2. The planets Mars and Earth have diameters of 6775 and 12 755 km, respectively. The mass of Mars is 0.107 times that of the earth. If a body weighs 200 N on the surface of the earth, what would its weight be on Mars? Also, what is the mass of the body and the acceleration of gravity on Mars?

13–3. Determine the force which a boy of mass 34 kg exerts on the floor of an elevator when the elevator is (a) at rest, (b) descending with a constant velocity of 0.75 m/s, and (c) ascending with a constant acceleration of 2 m/s^2.

***13–4.** A block having a mass of 2 kg is placed on a spring scale located in an elevator that is moving downward. If the scale reading, which measures the force in the spring, is 16 N, determine the acceleration **a** of the elevator. Neglect the mass of the scale.

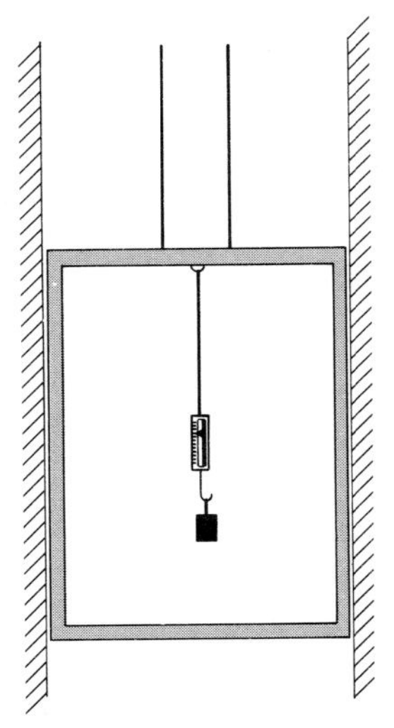

Prob. 13–4

13–5. The chain has a length of $l = 2$ m and a mass of $m = 3$ kg/m. If the coefficient of friction between the chain and the plane is $\mu = 0.2$, determine the velocity at which the end A will pass point B when the chain is released from rest.

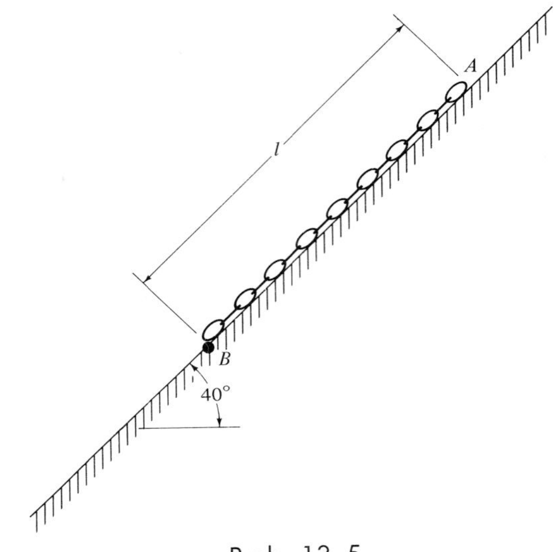

Prob. 13–5

13–5a. Solve Prob. 13–5 if $l = 5$ ft, $\mu = 0.3$, and the weight is $w = 0.5$ lb/ft.

13-6. A 0.75-kg brick is released from rest at A and slides down the inclined roof. If the coefficient of friction between the roof and the brick is $\mu = 0.3$, determine the speed at which the brick strikes the gutter G.

Prob. 13-6

13-7. The 200-kg ingot I, originally at rest, is being towed over a series of small rollers. Neglecting the mass of the cable, pulley P, and the rollers, compute the force in the cable when $t = 3$ s, if the motor M is drawing in the cable for a short time at a rate of $v = (0.2t^2)$ m/s, where t is in seconds ($0 \leqslant t \leqslant 4$ s). How far does the ingot move in 3 s?

Prob. 13-7

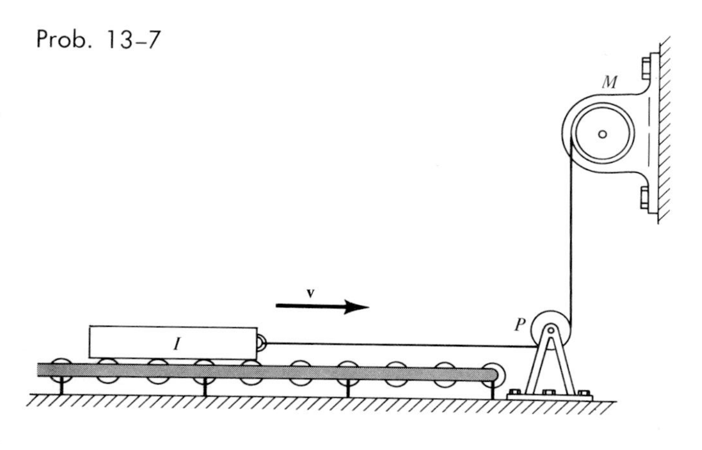

***13-8.** The conveyor belt delivers each 12-kg crate to the ramp at A such that the crate's velocity is $v_A = 2.5$ m/s, directed down *along* the ramp. If the kinetic coefficient of friction between each crate and the ramp is $\mu_k = 0.3$, determine the speed at which each crate slides off the ramp at B. Assume that no tipping occurs.

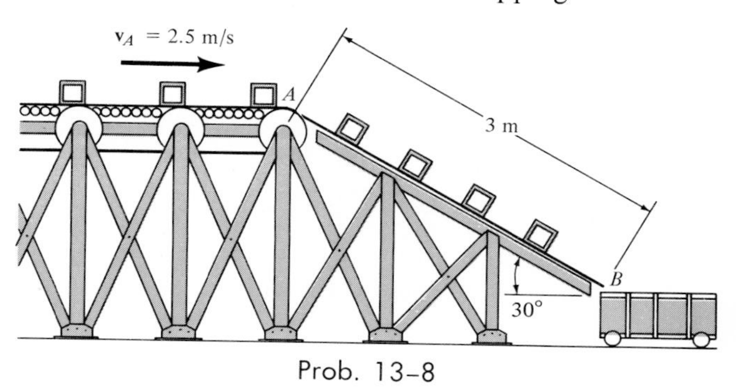

Prob. 13-8

13-9. A parachutist having a mass of 60 kg is falling at 8 m/s when he opens his parachute at a very high altitude. If the atmospheric drag resistance is $F_D = (30v^2)$ N, where v is in m/s, determine the velocity at which he lands on the ground. This velocity is referred to as the *terminal velocity,* which is found by letting the distance of fall $y \rightarrow \infty$.

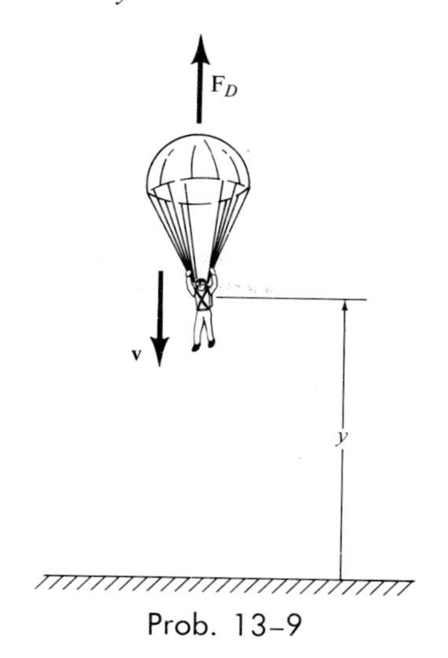

Prob. 13-9

13–10. The drag force of the water on a motor boat, having a mass of $m = 500$ kg, acts in the opposite direction of the boat's motion and has a magnitude of $F_D = Cv$, where v is the speed of the boat. Determine the value of the constant C if the boat is initially traveling forward at a speed of $v_1 = 40$ km/h, and when the engine is turned off it is observed that the speed drops to $v_2 = 20$ km/h in $t = 60$ s. How much horizontal force must be provided by the engine to propel the boat forward at a constant speed of $v_2 = 20$ km/h?

13–10a. Solve Prob. 13–10 if the boat weighs $W = 900$ lb, $v_1 = 30$ ft/s, $v_2 = 15$ ft/s, and $t = 18$ s.

13–11. A 50-kg log is given an initial forward velocity of $v_1 = 0.25$ m/s in still water. If the drag resistance of the water is proportional to the speed of the log at any instant and can be approximated by $F_D = (10v)$ N, where v is measured in m/s, determine the time needed for the log's speed to become 0.1 m/s. Given an initial velocity of $v_1 = 0.25$ m/s, through what distance does the log travel before it stops?

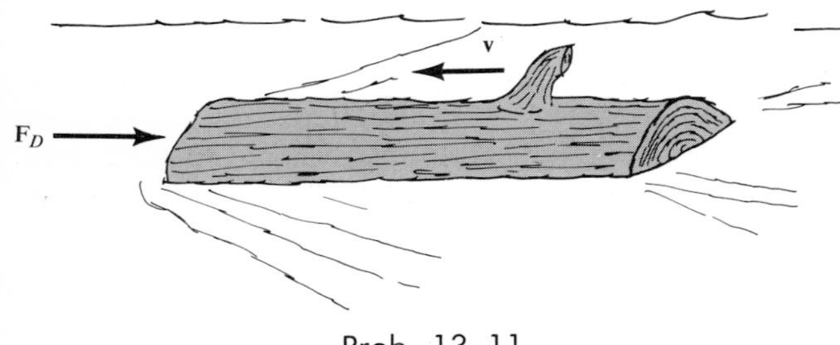

Prob. 13–11

***13–12.** The 5-kg shaft CA passes through a smooth journal bearing at B. Initially, the springs, which are coiled loosely around the shaft, are unstretched when no force is applied to the shaft. In this position $s = 250$ mm and the shaft is originally at rest. If a horizontal force of $F = 4$ kN is applied, determine the speed of the shaft at the instant $s = 50$ mm. The ends of the spring are attached to the bearing at B and the caps at C and A.

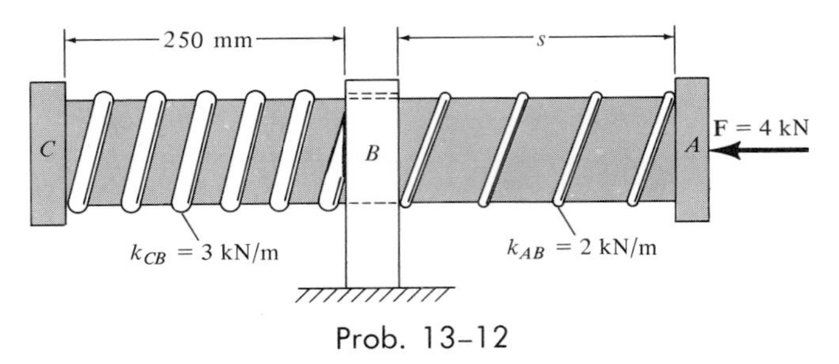

Prob. 13–12

13–13. The train shown consists of two 50-Mg cars B and C, and an 80-Mg locomotive A. If the train moves forward with an acceleration of 0.2 m/s² up the incline, determine the force exerted at the couplings D and E. Neglect the effect of rolling friction at the wheels of cars B and C, and calculate the resultant frictional tractive force $\mathbf{F}_t$ exerted on the tracks by the wheels of the locomotive A.

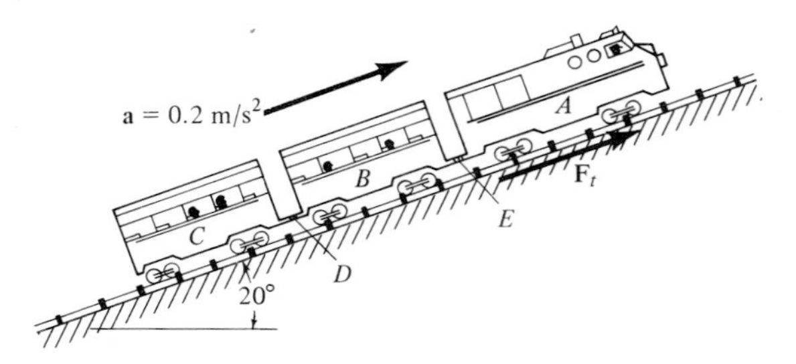

Prob. 13–13

13–14. Neglecting the effects of friction and the mass of the pulley and cord, determine the acceleration at which the block B will descend. What is the tension in the cord?

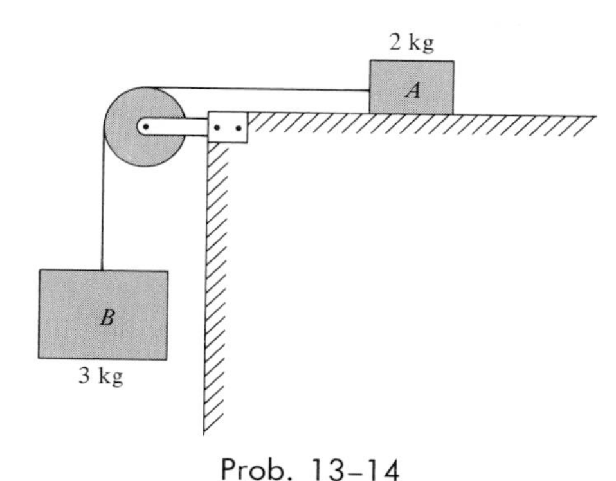

Prob. 13–14

13–15. Determine the required mass of block A so that when it is released from rest it moves block B $s_B = 0.75$ m up the smooth inclined plane in $t = 2$ s. Neglect the mass of the pulleys and cords. Block B has a mass of $m_B = 5$ kg.

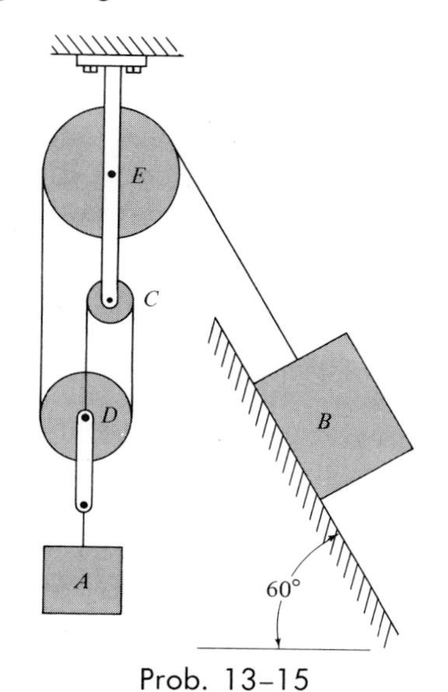

Prob. 13–15

13–15a. Determine the weight of block A in Prob. 13–15 if $s_B = 2$ ft, $t = 1.5$ s, and block B weighs $W_B = 10$ lb.

*** 13–16.** If the motor M pulls in its attached rope such that a particle P on the rope has an acceleration of $a_P = 6$ m/s², determine the towing force exerted by M on the rope in order to move the 50-kg crate C up the inclined plane. The kinetic coefficient of friction between the crate and the plane is $\mu_k = 0.3$. Neglect the mass of the pulleys and rope.

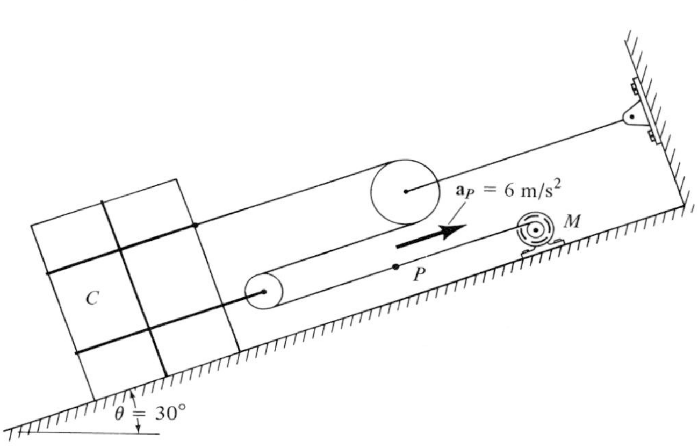

Prob. 13–16

13–17. Determine the tension developed in the two cords and the acceleration of each block. Neglect the mass of the pulleys and cords.

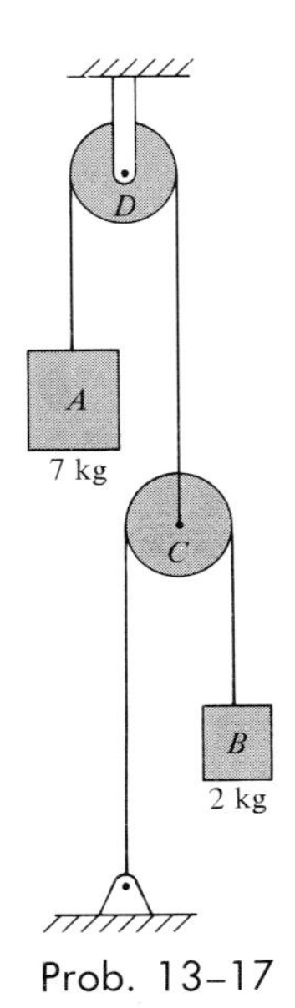

Prob. 13–17

13–18. Block B rests upon a smooth surface. If the coefficient of friction between A and B is $\mu = 0.5$, determine the acceleration of each block when (a) $F = 10$ N, and (b) $F = 20$ N.

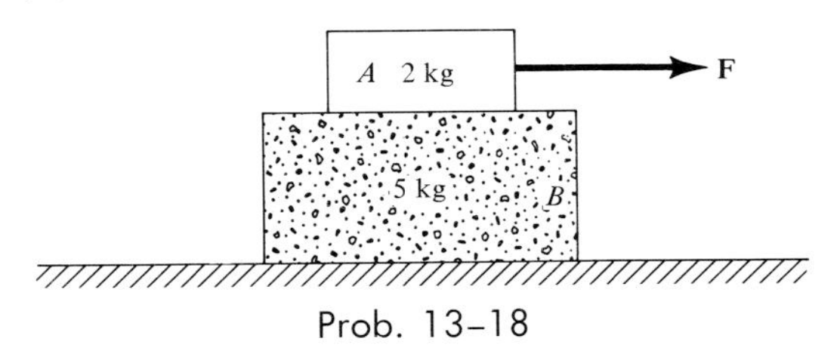

Prob. 13–18

13–19. The 20-kg block A rests on the 60-kg plate B in the position shown. Neglecting the mass of the rope and pulley, and using the coefficients of friction indicated, determine the time needed for block A to slide 0.5 m *on the plate* when the system is released from rest.

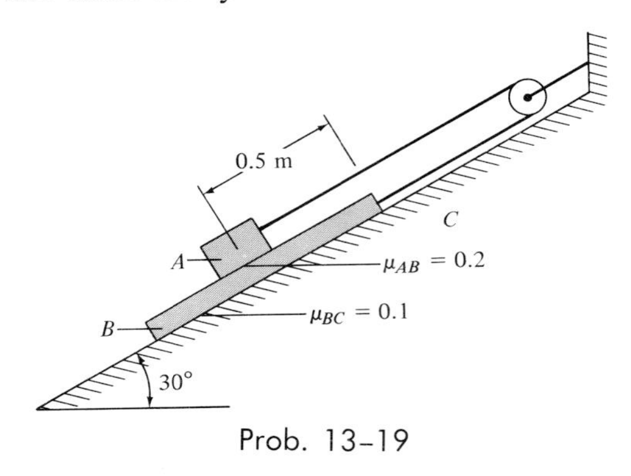

Prob. 13–19

*** 13–20.** Suppose it is possible to dig a smooth tunnel through the earth along a cord from a city at A to a city at B as shown. By the theory of gravitation, any vehicle C of mass m placed within the tunnel would be sub-

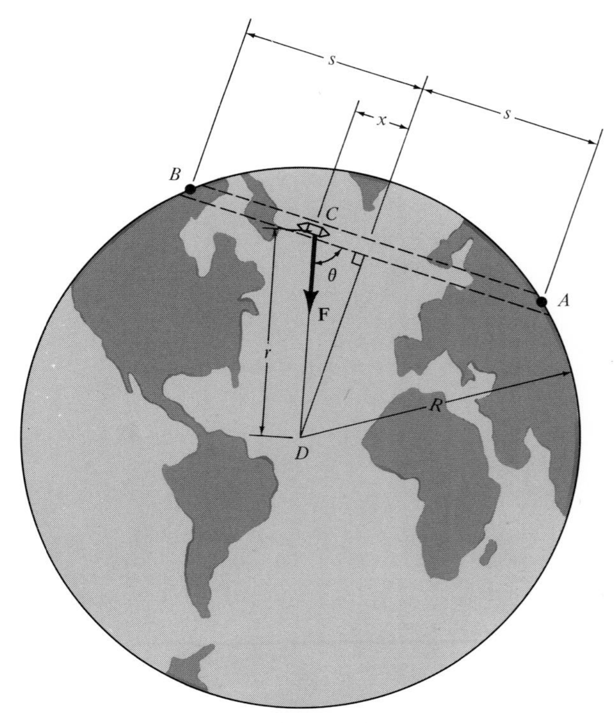

Prob. 13–20

jected to a gravitational force which is always directed toward the center of the earth D. This force $\mathbf{F}$ has a magnitude that is directly proportional to its distance r from the earth's center. Hence, if the vehicle has a weight of $W = mg$ when it is located on the earth's surface, then at an arbitrary location r, the magnitude of force $\mathbf{F}$ is $F = (mg/R)r$, where $R = 6328$ km, the radius of the earth. Provided the vehicle is released from rest when it is at B, $x = s = 2$ Mm, determine the time needed for it to reach A, and the maximum velocity it attains. Neglect the effect of the earth's rotation in the calculation. *Hint:* Write the equation of motion in the x direction, noting that $r \cos \theta = x$. Integrate, using the kinematic relation $v\,dv = a\,dx$, then integrate the result using $v = dx/dt$.

* **13–20a.** Solve Prob. 13–20 if $g = 32.2$ ft/s^2, $R = 3932$ mi, and $x = s = 1000$ mi, where 1 mi = 5280 ft.

13–5. Equations of Motion for a Particle: Cylindrical Coordinates

When all the forces acting on a particle are resolved into components along the unit-vector directions $\mathbf{u}_r$, $\mathbf{u}_\theta$, and $\mathbf{u}_z$, Fig. 13–11, the equation of motion may be expressed as

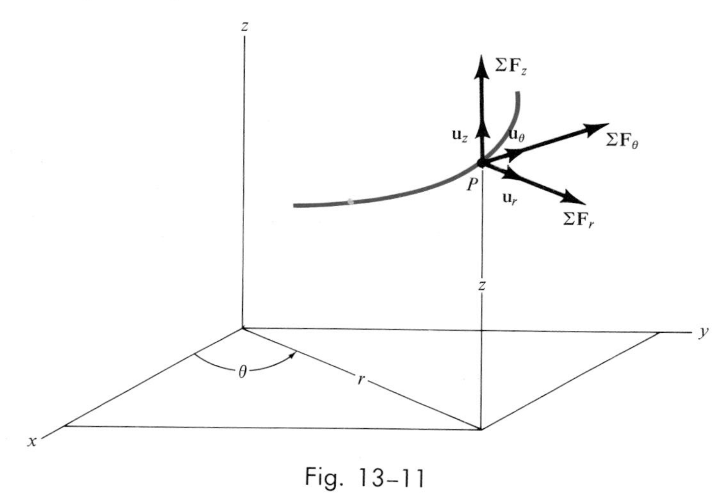

Fig. 13–11

$$\Sigma \mathbf{F} = m\mathbf{a}$$

$$\Sigma F_r \mathbf{u}_r + \Sigma F_\theta \mathbf{u}_\theta + \Sigma F_z \mathbf{u}_z = ma_r \mathbf{u}_r + ma_\theta \mathbf{u}_\theta + ma_z \mathbf{u}_z$$

Since the respective $\mathbf{u}_r$, $\mathbf{u}_\theta$, and $\mathbf{u}_z$ components must be equivalent, the solution of the above equation can be represented in terms of the following three scalar equations:

$$\begin{aligned} \Sigma F_r &= ma_r \\ \Sigma F_\theta &= ma_\theta \\ \Sigma F_z &= ma_z \end{aligned} \tag{13–9}$$

If all the forces acting on the particle lie in the r-θ plane, the particle will have motion in this plane. Hence, $\Sigma F_z = ma_z$ is not applicable, and therefore only the first two of Eqs. 13–9 may be used to specify the motion.

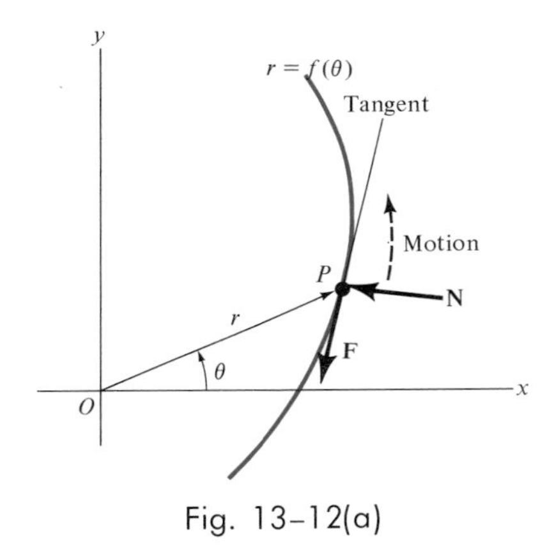

Fig. 13–12(a)

The most straightforward type of problem involving cylindrical coordinates requires the determination of the resultant force components $\Sigma \mathbf{F}_r$, $\Sigma \mathbf{F}_\theta$, and $\Sigma \mathbf{F}_z$, causing a particle to move with a *known* acceleration. If however the particle's accelerated motion is not completely specified at the given instant, then some information regarding the directions or magnitudes of the forces acting on the particle must be known or computed in order to solve Eqs. 13–9. For example, consider a particle P which is forced to move along a planar path defined in polar coordinates by $r = f(\theta)$, Fig. 13–12a. The normal force $\mathbf{N}$ acting on the particle is always perpendicular to the tangent of the path; whereas the frictional force $\mathbf{F}$ always acts along the tangent in the opposite direction of motion. The *directions* of these forces, can be determined by computing the angle ψ (psi), Fig. 13–12b, which is defined between the *extended* radial line $r = OP$ and the tangent to the curve. This angle, *always measured coun-*

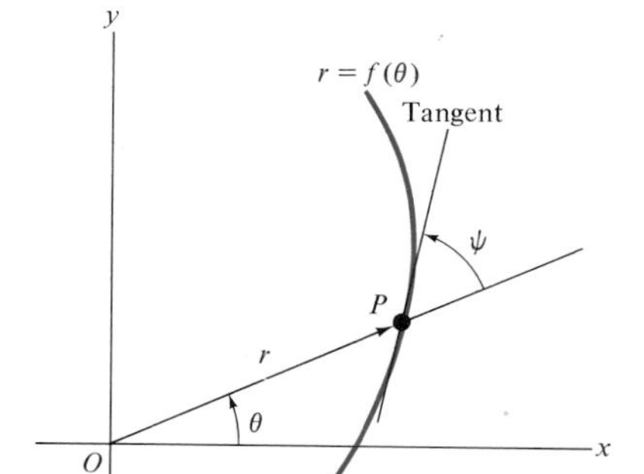

Fig. 13–12(b)

terclockwise, or in the *positive direction of θ,* is determined from the equation*

$$\tan \psi = r \Big/ \frac{dr}{d\theta} \tag{13–10}$$

Application is illustrated numerically in Example 13–8.

PROCEDURE FOR ANALYSIS

The basic three-step procedure outlined in Sec. 13–4 should be used when applying Eqs. 13–9. For problems involving cylindrical coordinates, these steps proceed as follows:

Step 1: Establish the r, θ, z inertial coordinate system and draw the particle's free-body and kinetic diagrams. When drawing the kinetic diagram always assume the vectors $m\mathbf{a}_r$, $m\mathbf{a}_\theta$, and $m\mathbf{a}_z$ act in the *positive direction* of r, θ, and z.

Step 2: Apply the equations of motion, Eqs. 13–9.

Step 3: Use the methods of Sec. 12–7 to determine r and the time derivatives $\dot{r}$, $\ddot{r}$, $\dot{\theta}$, $\ddot{\theta}$, and $\ddot{z}$, and evaluate the acceleration components, $a_r = \ddot{r} - r\dot{\theta}^2$, $a_\theta = r\ddot{\theta} + 2\dot{r}\dot{\theta}$, and $a_z = \ddot{z}$. If any of these three components are computed as negative quantities, it indicates that they act in their negative coordinate directions.

The following examples numerically illustrate application of these three steps. Further application of Eqs. 13–9 is given in Sec. 13–7, where problems involving rockets, satellites, and planetary bodies subjected to gravitational forces are considered.

*The derivation is given in any standard calculus text.

Example 13-6

From experimental measurements, the motion of a 2-kg particle is defined in terms of its cylindrical coordinates by the parametric equations $r = (t^2 + 2t)$ m, $\theta = (3t + 2)$ rad, and $z = (t^3 + 4)$ m, where t is in seconds. Determine the magnitude of the resultant force acting on the particle at the instant $t = 1$ s.

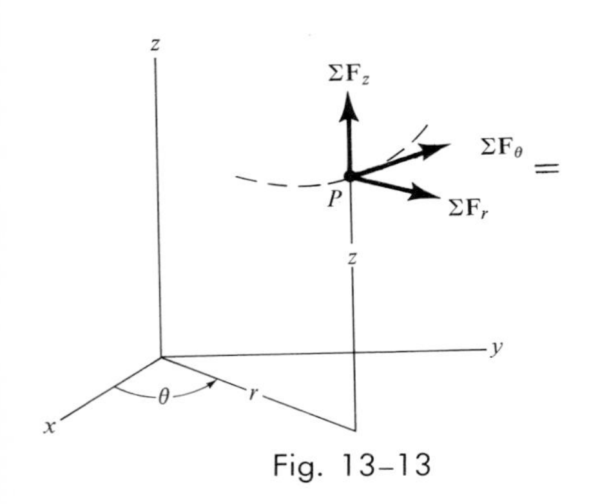

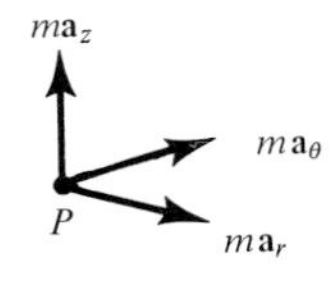

Fig. 13–13

Solution

Step 1: The free-body and kinetic diagrams of the particle, when it is located at an arbitrary point along the path, are shown in Fig. 13–13. The force components $\Sigma \mathbf{F}_r$, $\Sigma \mathbf{F}_\theta$, and $\Sigma \mathbf{F}_z$, which includes the particle's weight, are to be determined.

Step 2: Applying the equations of motion,

$$\Sigma F_r = ma_r; \qquad \Sigma F_r = 2(\ddot{r} - r\dot{\theta}^2) \qquad (1)$$

$$\Sigma F_\theta = ma_\theta; \qquad \Sigma F_\theta = 2(r\ddot{\theta} + 2\dot{r}\dot{\theta}) \qquad (2)$$

$$\Sigma F_z = ma_z; \qquad \Sigma F_z = 2\ddot{z} \qquad (3)$$

Step 3: Since the motion is specified, the radial coordinate r and the required time derivatives can be computed and evaluated at $t = 1$ s.

$$r = (t^2 + 2t)\Big|_{t=1\text{ s}} = 3 \text{ m}; \qquad \dot{\theta} = 3 \text{ rad/s}$$

$$\dot{r} = (2t + 2)\Big|_{t=1\text{ s}} = 4 \text{ m/s}; \qquad \ddot{\theta} = 0$$

$$\ddot{r} = 2\Big|_{t=1\text{ s}} = 2 \text{ m/s}^2; \qquad \ddot{z} = 6t\Big|_{t=1\text{ s}} = 6 \text{ m/s}^2$$

Substituting these results into Eqs. (1) through (3) yields

$$\Sigma F_r = 2[2 - 3(3)^2] = -50 \text{ N}$$

$$\Sigma F_\theta = 2[3(0) + 2(4)(3)] = 48 \text{ N}$$

$$\Sigma F_z = 2(6) = 12 \text{ N}$$

Hence, the magnitude of force acting on the particle is

$$F = \sqrt{(-50)^2 + (48)^2 + (12)^2}$$

$$= 70.3 \text{ N} \qquad \textit{Ans.}$$

Example 13–7

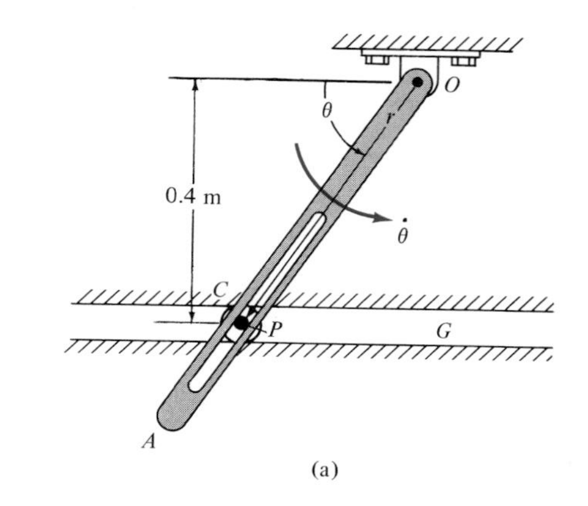

The 2-kg cylinder C shown in Fig. 13–14a has a peg P through its center which passes through the slot in arm OA. If the arm rotates in the *horizontal plane* at a constant rate of $\dot{\theta} = 0.5$ rad/s, determine the force that the arm exerts on the peg at the instant $\theta = 60°$. Neglect friction in the calculation and assume that the cylinder fits loosely in the straight horizontal groove G.

Solution

Step 1: The free-body and kinetic diagrams for the cylinder (top view) are shown in Fig. 13–14b. The force of the peg, $\mathbf{F}_P$, acts perpendicular to the slot in the arm. $\mathbf{N}_C$ represents the normal force of one of the walls of the groove on the cylinder. Both $2\mathbf{a}_r$ and $2\mathbf{a}_\theta$ are assumed to act in the direction of *positive* r and θ, respectively.

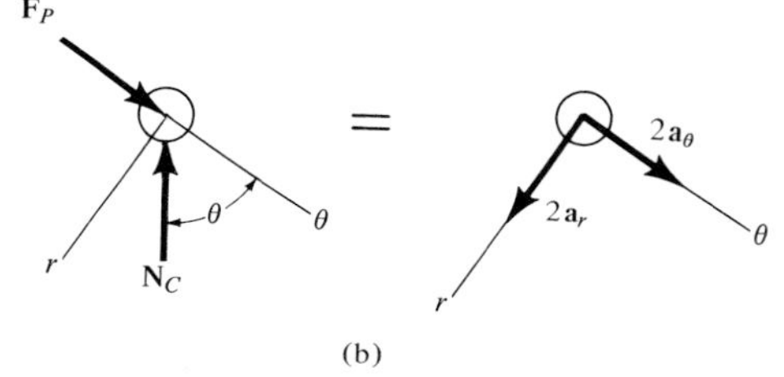

Fig. 13–14

Step 2: Applying the equations of motion,

$$+\swarrow \Sigma F_r = ma_r; \qquad -N_C \sin\theta = 2a_r \qquad (1)$$

$$+\searrow \Sigma F_\theta = ma_\theta; \qquad F_P - N_C \cos\theta = 2a_\theta \qquad (2)$$

Step 3: From Fig. 13–14a, r can be related to θ by the equation

$$r = \frac{0.4}{\sin\theta} = 0.4 \csc\theta$$

Since $d(\csc\theta) = -(\csc\theta \cot\theta)d\theta$ and $d(\cot\theta) = -(\csc^2\theta)\,d\theta$, then r and the necessary time derivatives become

$$\dot{\theta} = 0.5, \qquad \ddot{\theta} = 0$$

$$r = 0.4 \csc\theta$$

$$\dot{r} = -0.4 \csc\theta \cot\theta\dot{\theta}$$

$$= -0.2 \csc\theta \cot\theta$$

$$\ddot{r} = -0.2(-\csc\theta \cot\theta\dot{\theta})\cot\theta - 0.2\csc\theta(-\csc^2\theta\dot{\theta})$$

$$= 0.1 \csc\theta(\cot^2\theta + \csc^2\theta)$$

Evaluating these formulas at $\theta = 60°$, we get

$$\dot{\theta} = 0.5, \qquad \ddot{\theta} = 0$$

$$r = 0.462, \qquad \dot{r} = -0.133, \qquad \ddot{r} = 0.192$$

Thus,

$$a_r = \ddot{r} - r\dot{\theta}^2 = 0.192 - 0.462(0.5)^2 = 0.0765$$

$$a_\theta = r\ddot{\theta} + 2\dot{r}\dot{\theta} = 0 + 2(-0.133)(0.5) = -0.133$$

Note that since a_θ is calculated as a negative quantity, it indicates that $2\mathbf{a}_\theta$ acts in the opposite direction to that shown on the kinetic diagram in Fig. 13–14b. Substituting into Eqs. (1) and (2) with $\theta = 60°$ yields

$$-N_C \sin 60° = 2(0.0765)$$
$$F_P - N_C \cos 60° = 2(-0.133)$$

Solving,

$$N_C = -0.177 \text{ N}$$
$$F_P = -0.354 \text{ N} \qquad \textit{Ans.}$$

Since both of these forces are calculated as negative quantities, their directions are opposite to those shown on the free-body diagram in Fig. 13–14*b*.

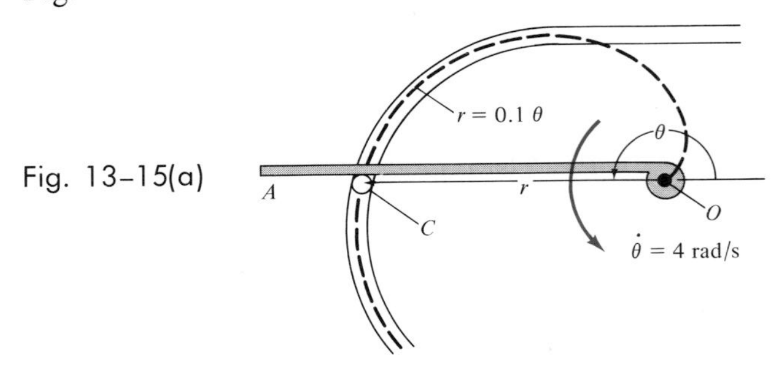

Fig. 13–15(a)

Example 13–8

A can C, having a mass of 0.5 kg, moves along a grooved horizontal path shown in Fig. 13–15*a*. The path is in the form of a spiral, which is defined by the equation $r = (0.1\theta)$ m, where θ is measured in radians. If the arm OA is rotating at a constant rate of $\dot{\theta} = 4$ rad/s in the horizontal plane, determine the force it exerts on the can at the instant $\theta = \pi$ rad. Neglect friction and the size of the can.

Solution

Step 1: The free-body and kinetic diagrams of the can are shown in Fig. 13–15*b*. As usual, the vectors $0.5\mathbf{a}_r$ and $0.5\mathbf{a}_\theta$ act in the *positive*

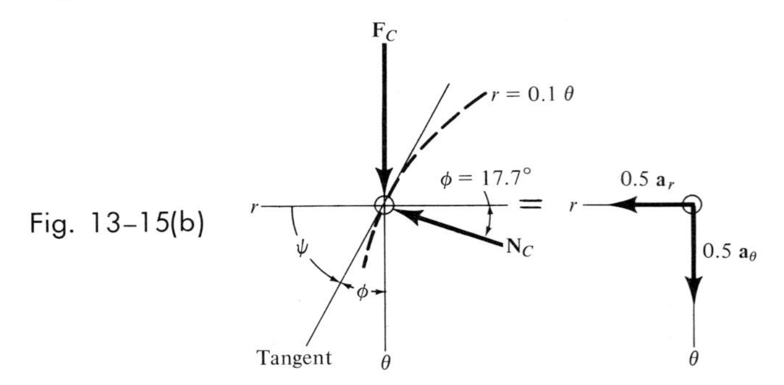

Fig. 13–15(b)

directions of r and θ, respectively. The driving force $\mathbf{F}_C$ acts perpendicular to the arm OA; whereas the force of the wall of the groove on the can, $\mathbf{N}_C$, acts perpendicular to the tangent to the curve at $\theta = \pi$ rad. Since the path is specified, the angle ψ which the extended radial line r makes with the tangent, Fig. 13–15b, can be determined from Eq. 13–10. We have $r = 0.1\theta$, so that $dr/d\theta = 0.1$, and therefore

$$\tan\psi = r\Big/\frac{dr}{d\theta} = \frac{0.1\theta}{0.1} = \theta$$

When $\theta = \pi$, $\psi = \tan^{-1}\pi = 72.3°$, so that $\phi = 90° - \psi = 17.7°$, as shown in Fig. 13–15$b$.

Step 2: Applying the equations of motion.

$$\overset{+}{\leftarrow}\Sigma F_r = ma_r; \qquad N_C\cos 17.7° = 0.5a_r \qquad (1)$$

$$+\downarrow\Sigma F_\theta = ma_\theta; \qquad F_C - N_C\sin 17.7° = 0.5a_\theta \qquad (2)$$

Step 3: Computing the time derivatives of r and θ, we have

$$\dot{\theta} = 4 \text{ rad/s}, \qquad \ddot{\theta} = 0$$

$$r = 0.1\theta, \qquad \dot{r} = 0.1\dot{\theta} = 0.1(4) = 0.4 \text{ m/s}, \qquad \ddot{r} = 0$$

At the instant $\theta = \pi$ rad,

$$a_r = \ddot{r} - r\dot{\theta}^2 = 0 - 0.1(\pi)(4)^2 = -5.03 \text{ m/s}^2$$

$$a_\theta = r\ddot{\theta} + 2\dot{r}\dot{\theta} = 0 + 2(0.4)(4) = 3.20 \text{ m/s}^2$$

Substituting these results into Eqs. (1) and (2) and solving yields

$$N_C = -2.64 \text{ N}$$

$$F_C = 0.797 \text{ N} \qquad \textit{Ans.}$$

Problems

13–21. Determine the magnitude of the unbalanced force acting on a 5-kg particle at the instant $t = 2$ s, if the particle is moving along a two-dimensional path defined by the equations $r = (2t + 10)$ m and $\theta = (1.5t^2 - 6t)$ rad, where t is measured in seconds.

13–22. The spool, which has a mass of 4 kg, slides along the rotating rod. At the instant shown, the angular rate of rotation of the rod is $\dot{\theta} = 6$ rad/s and the rotation is increasing at $\ddot{\theta} = 2$ rad/s^2. At this same instant, the spool has a velocity of 3 m/s and an acceleration of 1 m/s^2, both measured relative to the rod and directed away from the center O when $r = 0.5$ m. Determine the radial frictional force and the normal force both exerted by the rod on the spool at this instant.

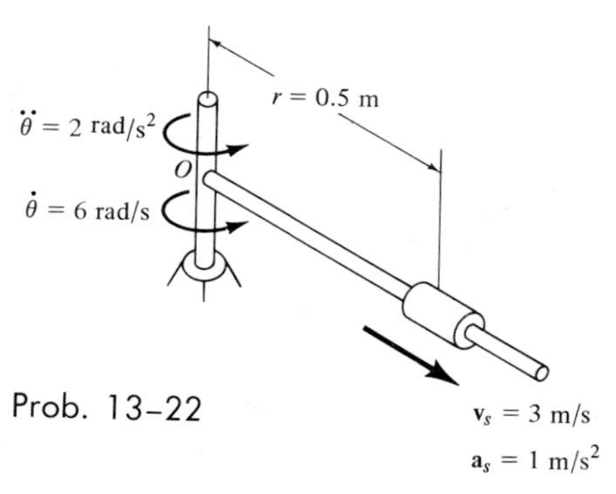

Prob. 13–22

13–23. The boy, of mass 40 kg, is sliding down the spiral chute at a constant speed such that his position, measured from the top of the chute, has components $r = 2$ m, $\theta = (0.7t)$ rad, and $z = (-0.5t)$ m, where t is measured in seconds. Determine the components of force $\mathbf{F}_r$, $\mathbf{F}_\theta$, and $\mathbf{F}_z$ which he exerts on the chute at the instant $t = 2$ s.

Prob. 13–23

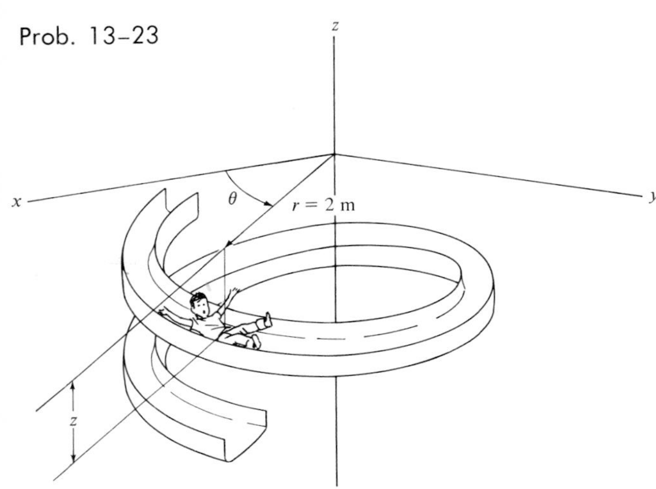

*** 13–24.** A smooth can C, having a mass of 3 kg, is lifted from a feed at A to a ramp at B by a forked rotating rod. If the rod maintains a constant angular motion of $\dot{\theta} = 0.5$ rad/s, determine the force which the rod exerts on the can at the instant $\theta = 30°$. Neglect the effects of friction in the calculation. The ramp from A to B is circular, having a radius of 600 mm.

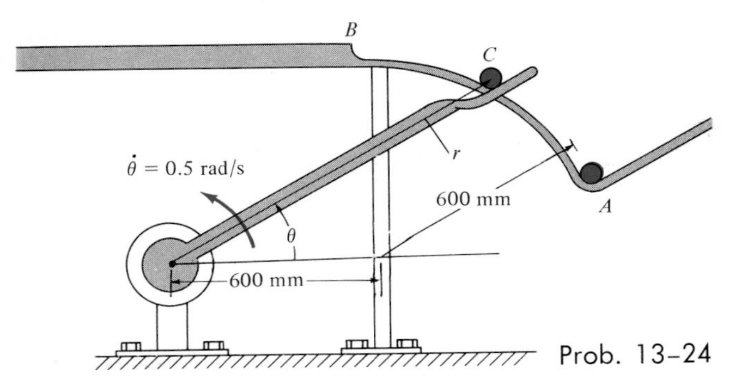

Prob. 13–24

13–25. A ride in an amusement park consists of a cart which is supported by small wheels. Initially, the cart is traveling in a circular path of radius $r_1 = 8$ m such that the angular rate of rotation is $\dot{\theta}_1 = 0.2$ rad/s. If the attached cable OC begins to shorten such that it is drawn inward at a constant speed of $\dot{r} = -0.5$ m/s, determine the tension it exerts on the cart at the instant $r = 4$ m. The cart and its passengers have a total mass of $m_t = 180$ kg. Neglect the effects of friction. *Hint:* The equation of motion in the θ direction can be written as $d/dt(r^2\dot{\theta}) = 0$, which, when integrated, yields $r^2\dot{\theta} =$ const.

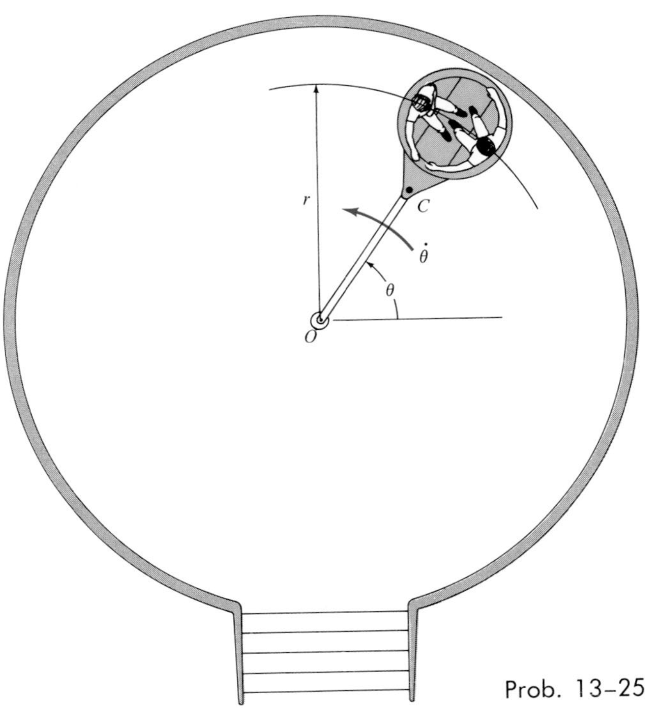

Prob. 13–25

13–25a. Solve Prob. 13–25 if $r_1 = 15$ ft, $\theta_1 = 0.3$ rad/s, $\dot{r} = -2$ ft/s, and $r = 6$ ft. The cart and passengers have a total weight of $W_t = 400$ lb.

13–26. A 0.2-kg spool slides down along a smooth rod. If the rod has a constant angular rate of rotation of $\dot{\theta} = 2$ rad/s in the vertical plane, show that the equations of motion for the spool are $\ddot{r} - 4r - 9.81 \sin\theta = 0$ and $0.8\dot{r} + N_s - 1.962 \cos\theta = 0$, where N_s is the magnitude of the normal force of the rod on the spool. Using the methods of differential equations, it can be shown that the solution of the first of these equations is $r = C_1e^{-2t} + C_2e^{2t} - (9.81/8)\sin 2t$. If r, $\dot{r}$, and θ are zero when $t = 0$, evaluate the constants C_1 and C_2 and determine r at the instant $\theta = \pi/4$ rad.

Prob. 13–26

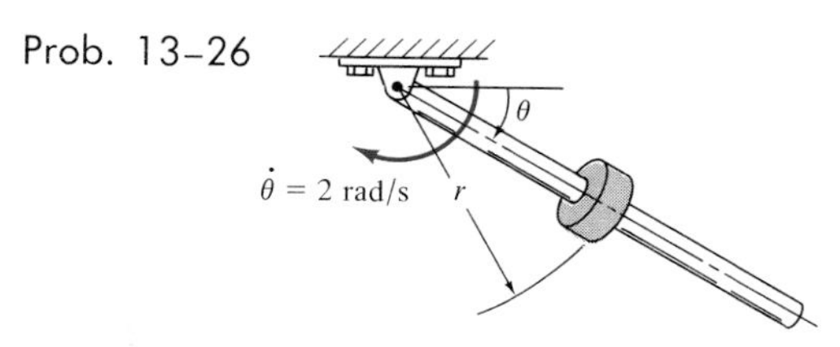

13–27. Using a forked rod, a smooth can C, having a mass of 0.5 kg, is forced to move along the *vertical* slotted path $r = (0.6\theta)$ m, where θ is measured in radians. If the can has a constant speed of $v_C = 2$ m/s, determine both the force of the forked rod on the can and the normal force of the slot on the can at the instant $\theta = \pi$ rad. Assume the can is in contact with only *one edge* of the rod and slot at any instant. *Hint:* To obtain the time derivatives necessary to compute the can's acceleration components, a_r and a_θ, take the first and second time derivatives of $r = 0.6\theta$. Then, for further information, use Eq. 12–45 to determine $\dot{r}$ and $\dot{\theta}$. Also, take the time derivative of Eq. 12–45, noting that $\dot{v}_C = 0$, to determine $\ddot{r}$ and $\ddot{\theta}$.

Prob. 13–27

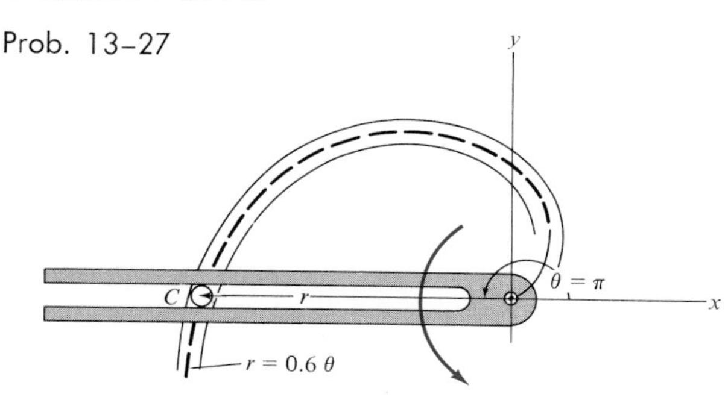

*** 13–28.** The tube rotates in the horizontal x-y plane at a constant angular rate of $\dot{\theta} = 4$ rad/s. If a 0.2-kg ball B starts at the origin O with an initial radial velocity of $\dot{r} = 1.5$ m/s and moves outward through the tube, determine the components of the ball's velocity, $\mathbf{v}_r$ and $\mathbf{v}_\theta$, at the instant it leaves the outer end at C, $r = 0.5$ m. *Hint:* Show that the equation of motion in the r direction is $\ddot{r} - 16r = 0$. The solution is of the form $r = Ae^{-4t} + Be^{4t}$. Evaluate the integration constants A and B, and determine the time t when $r = 0.5$ m. Proceed to obtain v_r and v_θ.

Prob. 13–28

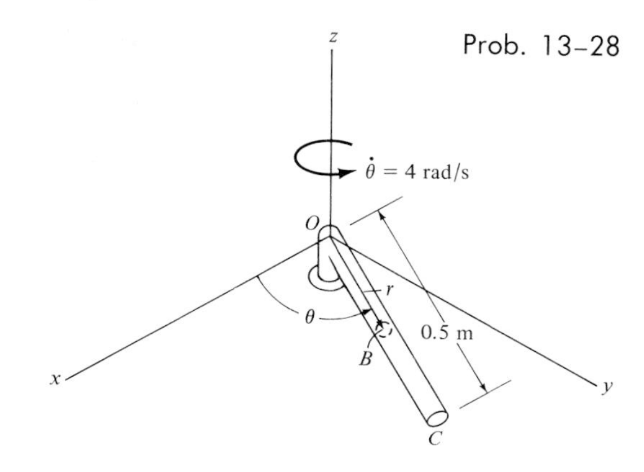

13–29. A chain is attached to a disk of mass m and the disk is given an initial velocity $\mathbf{v}_1$ perpendicular to the chain when $\theta = 0$ and $r = r_1$. At this instant, the chain is pulled downward through a hole in the center of a table with a constant speed v_2. Determine the equation that describes the path of the disk, i.e., $r = f(\theta)$. What is the magnitude of the force $\mathbf{F}$, measured as a function of time, which must be applied to the chain to maintain the motion? *Hint:* The equation of motion in the θ direction can be written as $d/dt(r^2\dot{\theta}) = 0$, which, when integrated, yields $r^2\dot{\theta} =$ const.

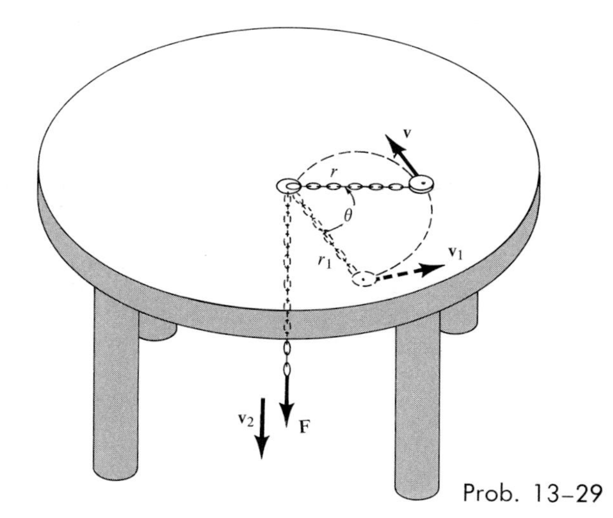

Prob. 13–29

13-30. A truck T has a mass of $m = 9$ Mg and is traveling along a portion of a road defined by the lemniscate $r^2 = (10^6)\cos 2\theta$, where r is measured in metres and θ is in radians. If the truck maintains a constant speed of $v_T = 10$ m/s, determine the magnitude of the resultant frictional force which must be exerted by all the wheels to maintain the motion when $\theta = 0$. *Hint:* To determine the time derivatives necessary to compute the truck's acceleration components a_r and a_θ, take the first and second time derivatives of $r^2 = 10^6 \cos 2\theta$. Then for further information, use Eq. 12–45 to determine $\dot{r}$ and $\dot{\theta}$. Also take the time derivative of Eq. 12–45, noting that $\dot{v}_T = 0$, to determine $\ddot{r}$ and $\ddot{\theta}$.

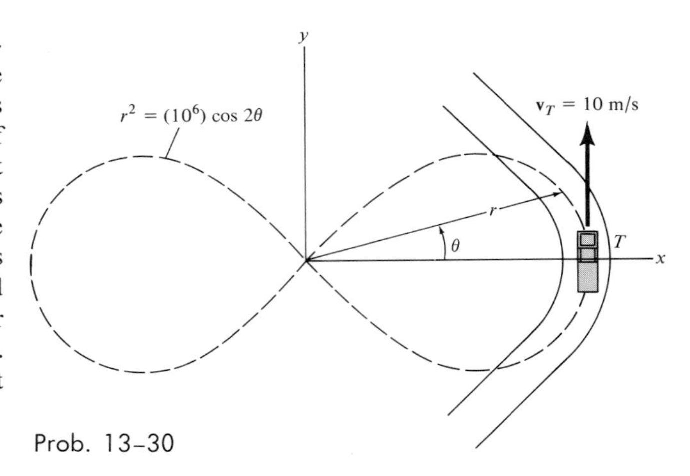

Prob. 13–30

13-30a. Solve Prob. 13–30 if the truck has a weight of $W = 8000$ lb, $v_T = 4$ ft/s, and the road is defined by $r^2 = 0.2(10^6)\cos 2\theta$, where r is in ft and θ is in radians.

13–6. Equations of Motion for a Particle: Normal and Tangential Coordinates

When a particle moves over a curved path which is known, the equation of motion for the particle may be expressed in terms of its normal and tangential components. Substituting Eq. 12–52 into Eq. 13–5 we have

$$\Sigma \mathbf{F} = m\mathbf{a}$$

$$\Sigma F_t \mathbf{u}_t + \Sigma F_n \mathbf{u}_n = m\mathbf{a}_t + m\mathbf{a}_n$$

Here ΣF_n and ΣF_t represent the sums of all the force components acting on the particle in the normal and tangential directions, respectively, Fig. 13–16. Since the respective $\mathbf{u}_t$ and $\mathbf{u}_n$ components must be equivalent, the solution of the above equation can be represented in terms of the following two scalar equations:

$$\Sigma F_t = ma_t \qquad (13\text{–}11)$$
$$\Sigma F_n = ma_n$$

The first of these equations indicates that the change in the magnitude of velocity ($a_t = dv/dt$) is caused by the sum of the tangential force components acting on the particle. Hence, if the resultant $\Sigma \mathbf{F}_t$ acts in

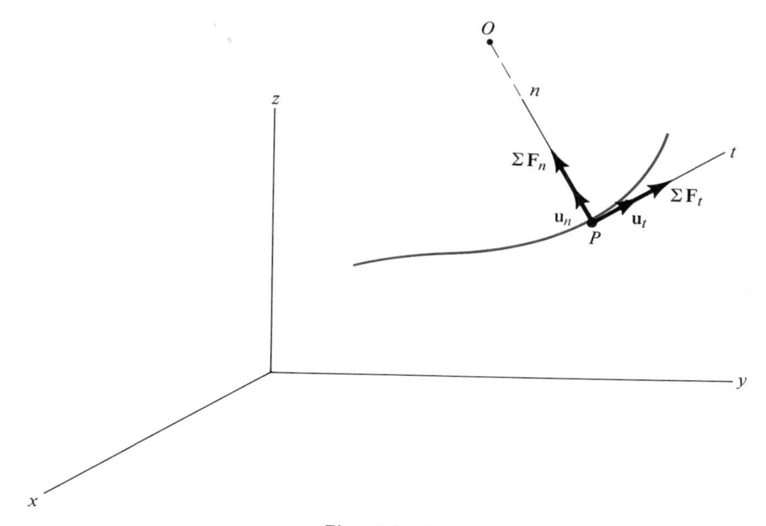

Fig. 13–16

the direction of motion, the speed is increasing; whereas if it acts in the opposite direction, the particle is slowing down.

The second of Eqs. 13–11 indicates that the change in the direction of velocity ($a_n = v^2/\rho$) is caused by the resultant of the normal force components acting on the particle. This resultant, $\Sigma\mathbf{F}_n$, will always act toward the center of curvature, O, of the path, Fig. 13–16, in the same direction as $\mathbf{a}_n$. In particular, when the particle is traveling on a constrained circular path with a constant speed, there is a normal force exerted on the particle by the constraint. This force is termed a *centripetal force*. The equal but opposite force exerted by the particle on the constraint is called a *centrifugal force*.

PROCEDURE FOR ANALYSIS

The three-step procedure given in Sec. 13–4 may be stated as follows when applied to problems involving n and t coordinates:

Step 1: Establish the inertial n-t coordinate system at the particle and draw the particle's free-body and kinetic diagrams.

Step 2: Apply the equations of motion, Eqs. 13–11.

Step 3: Formulate the normal and tangential components of acceleration using Eqs. 12–53 and 12–54; i.e., $a_t = dv/dt$ or $a_t = v\,dv/ds$ and $a_n = v^2/\rho$. If the path is defined as $y = f(x)$, the radius of curvature can be obtained from Eq. 12–56, namely, $\rho = |[1 + (dy/dx)^2]^{3/2}/(d^2y/dx^2)|$.

In most problems, *Steps 2* and *3* may be combined for the analysis.

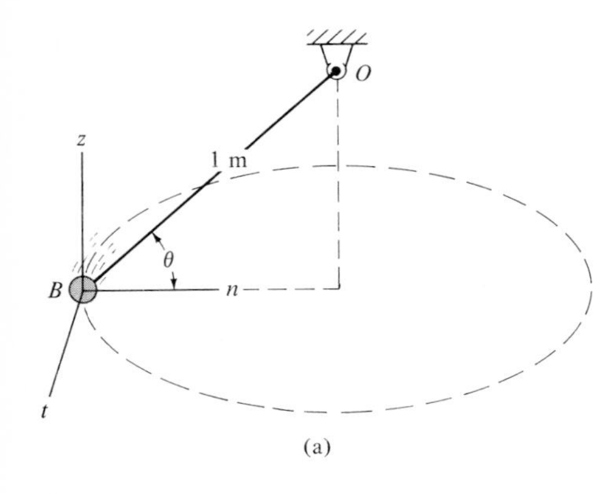

(a)

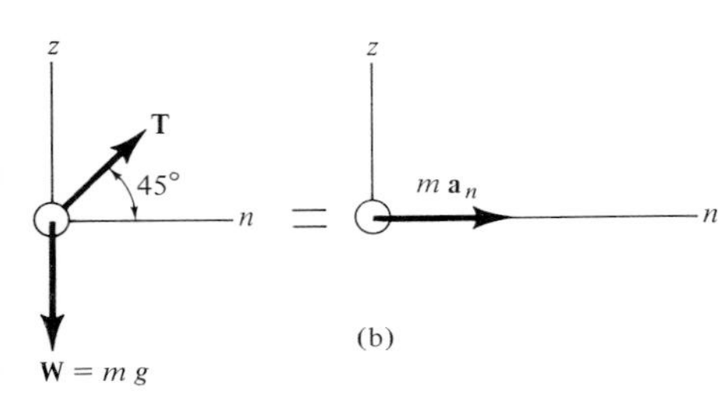

(b)

Fig. 13–17

Example 13–9

The ball B is fastened to the end of a 1-m-long string, Fig. 13–17a. If air resistance is neglected, then as the ball moves with a constant speed, it describes a horizontal circular path in which the string OB generates a locus of points defining the surface of a cone. Determine the speed of the ball along its circular path if $\theta = 45°$.

Solution

Step 1: The free-body and kinetic diagrams for the ball are shown in Fig. 13–17b. The weight is represented as $W = mg$, where $g = 9.81 \text{ m/s}^2$. Since the ball moves around the circle with constant speed, only a normal component of acceleration $\mathbf{a}_n$ (directed toward the center of curvature) is present. Why? This component has a magnitude of $a_n = v^2/\rho = v^2/(1 \cos 45°)$.

Step 2: Using the n-z axes shown, and applying the equations of motion, we have

$$+\uparrow \Sigma F_z = ma_z; \qquad T \sin 45° - m(9.81) = 0 \qquad (1)$$

$$\overset{+}{\rightarrow} \Sigma F_n = ma_n; \qquad T \cos 45° = m \frac{v^2}{1 \cos 45°} \qquad (2)$$

Application of the equation of motion in the tangential direction is of no consequence to the solution, since there are no forces in this direction. Solving Eq. (1) for T and substituting into Eq. (2), it is found that m cancels and the speed v of the ball is then

$$v = 2.63 \text{ m/s} \qquad \textit{Ans.}$$

Example 13–10

Determine the banking angle θ of the circular track so that the wheels of the sports car shown in Fig. 13–18a will not have to depend upon friction to prevent the car from sliding either up or down the curve. The car travels at a constant speed of 50 m/s. The radius of the track is 500 m.

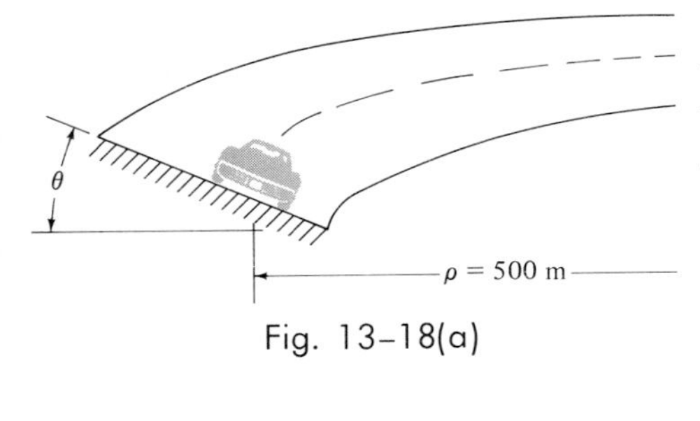

Fig. 13–18(a)

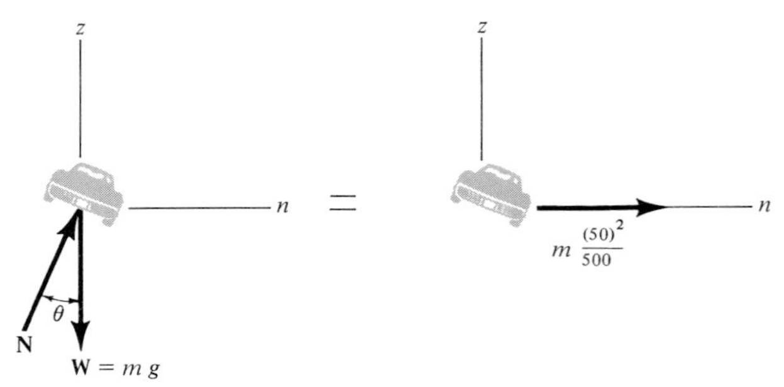

Fig. 13–18(b)

Solution

Step 1: Since the problem statement stipulates no friction, the free-body and kinetic diagrams for the car are shown in Fig. 13–18*b*. The car is assumed to have a mass m.

Step 2: Using the n-z axes shown, and applying the equations of motion, realizing that $a_n = v^2/\rho$, we have

$$+\uparrow \Sigma F_z = ma_z; \qquad N\cos\theta - m(9.81) = 0 \qquad (1)$$

$$\overset{+}{\rightarrow} \Sigma F_n = ma_n; \qquad N\sin\theta = m\frac{(50)^2}{500} \qquad (2)$$

Eliminating N and m from these equations by dividing Eq. (2) by Eq. (1), we obtain

$$\tan\theta = \frac{(50)^2}{9.81(500)}$$

$$\theta = \tan^{-1}(0.510) = 27.0° \qquad \textit{Ans.}$$

A force summation in the tangential direction of motion is of no consequence to the solution. If it were considered, note that $a_t = dv/dt = 0$, since the car moves with *constant speed.*

Example 13–11

Determine the constant speed of the satellite s in Fig. 13–19*a* so that it circles the earth with an orbit of radius $r = 15$ Mm. The mass of the earth is $5.976(10^{24})$ kg.

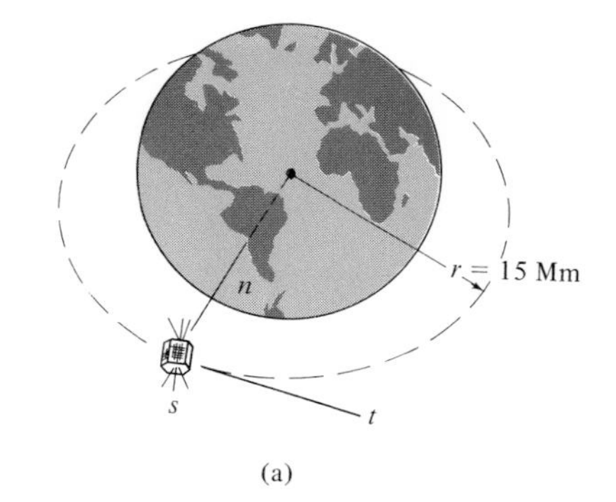

(a)

Solution

Step 1: The free-body and kinetic diagrams are shown in Fig. 13–19*b*. The only force acting on the satellite is the gravitational attraction of the earth. If m_s and m_e denote the mass of the satellite and earth, respectively, this force is determined by Eq. 13–2, $F = G(m_e m_s/r^2)$.

Step 2: Applying the second of Eqs. 13–11, we have

$$\Sigma F_n = ma_n; \qquad G\frac{m_e m_s}{r^2} = m_s\frac{v^2}{r}$$

Therefore,

$$v = \sqrt{\frac{Gm_e}{r}}$$

$$= \sqrt{\frac{6.673(10^{-11})[5.976(10^{24})]}{15(10^6)}}$$

$$= 5156 \text{ m/s} = 18.6 \text{ Mm/h} \qquad \textit{Ans.}$$

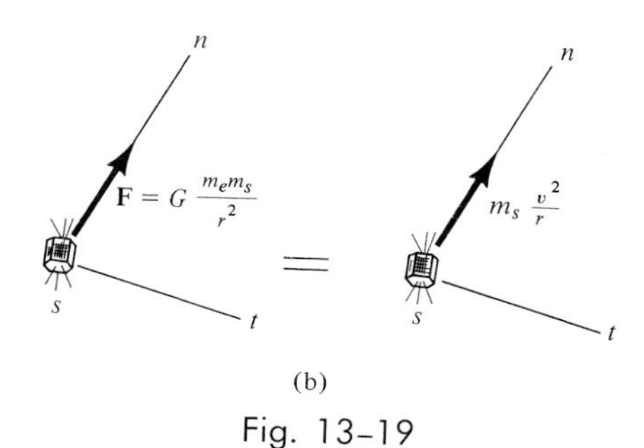

(b)

Fig. 13–19

Example 13–12

The 3-kg disk D is attached to the end of a cord as shown in Fig. 13–20*a*. The other end of the cord is attached to a ball-and-socket joint located at the center of a platform. If the platform is rotating at a very high rate of speed, and the disk is placed on it and released from rest, as shown, determine the time it takes for the disk to reach a speed great enough to break the cord. The maximum tension the cord can sustain is 100 N, and the coefficient of kinetic friction between the disk and the platform is $\mu_k = 0.1$.

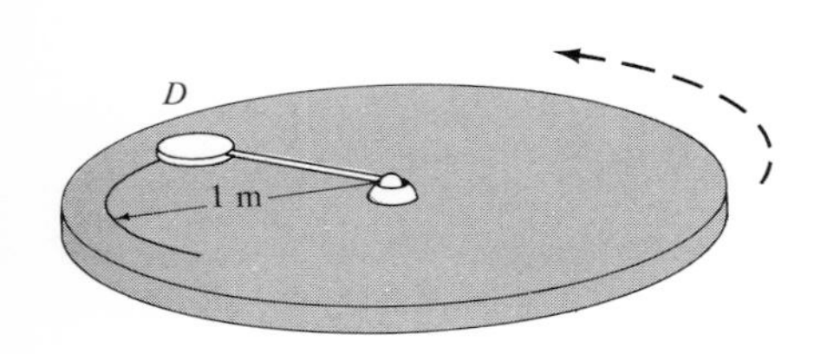

Fig. 13–20(a)

Solution

Step 1: The free-body and kinetic diagrams for the disk are shown in Fig. 13–20*b*. Note that the disk has *both* normal and tangential components of acceleration as a result of the unbalanced forces **T** and **F.** Since sliding occurs, the frictional force has a magnitude of $F = \mu_k N_D = 0.1 N_D$, and a direction that opposes the *relative motion* between the contacting surfaces, Fig. 13–20*b*. The weight of the disk is $W = 3(9.81) = 29.43$ N.

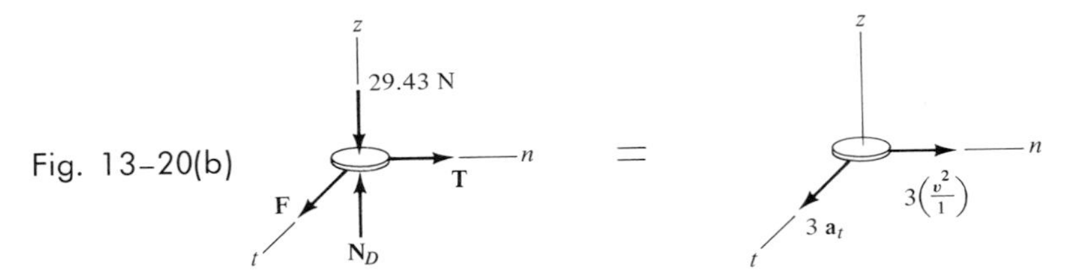

Fig. 13–20(b)

Step 2: Applying the equations of motion,

$\Sigma F_z = ma_z;$ $$N_D - 29.43 = 0 \qquad (1)$$

$\Sigma F_t = ma_t;$ $$0.1 N_D = 3a_t \qquad (2)$$

$\Sigma F_n = ma_n;$ $$T = 3\left(\frac{v^2}{1}\right) \qquad (3)$$

Since the maximum tension sustained by the cord is $T = 100$ N, Eq. (3) can be solved for the critical speed v_{cr} needed to break the cord. Solving all the equations, we get

$$N_D = 29.43 \text{ N}$$
$$a_t = 0.981 \text{ m/s}^2$$
$$v_{\text{cr}} = 5.77 \text{ m/s}$$

Step 3: Since a_t is *constant,* Eq. 12–8 can be applied to obtain the time needed to break the cord.

$$v = v_1 + a_t t$$
$$5.77 = 0 + (0.981)t$$
$$t = 5.89 \text{ s} \qquad \textit{Ans.}$$

Example 13-13

A block having a mass of 2 kg is given an initial velocity of 1 m/s when it is at the top surface of the smooth cylinder shown in Fig. 13–21*a*. If the block moves along a path of radius 0.5 m, determine the angle $\theta = \theta_{max}$ at which it begins to leave the cylinder's surface.

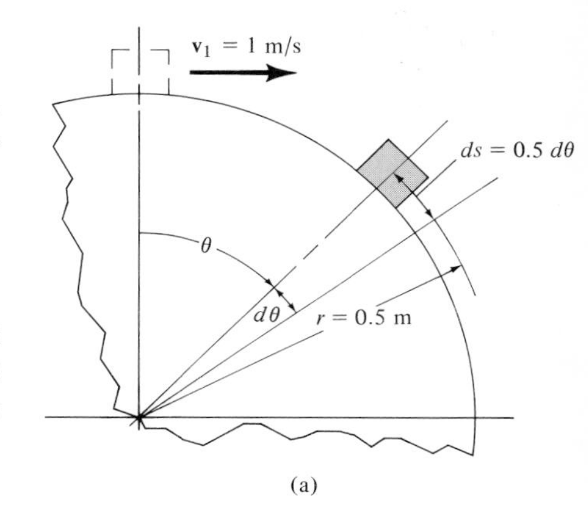

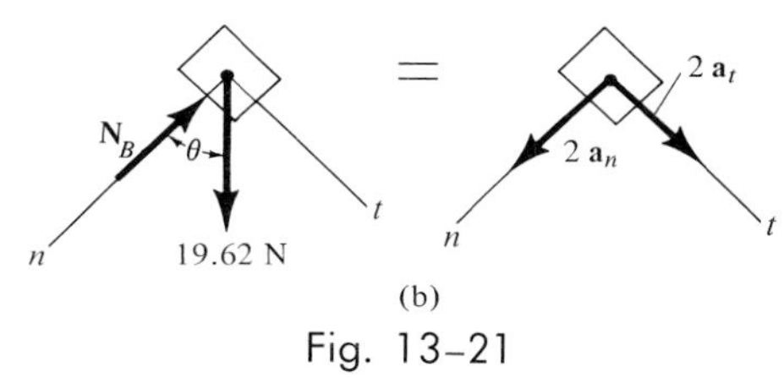

Fig. 13–21

Solution

Step 1: The free-body and kinetic diagrams for the block, when the block is located at the *general position* θ, are shown in Fig. 13–21*b*. The block must have a tangential acceleration $\mathbf{a}_t$, since its *speed* is always *increasing* as it slides downward. The weight is $W = 2(9.81) = 19.62$ N.

Step 2: Applying the equations of motion yields

$$+\swarrow \Sigma F_n = ma_n; \qquad -N_B + 19.62 \cos\theta = 2\frac{v^2}{0.5} \qquad (1)$$

$$+\searrow \Sigma F_t = ma_t; \qquad 19.62 \sin\theta = 2a_t \qquad (2)$$

These two equations contain four unknowns, N_B, v, a_t, and θ. At the instant $\theta = \theta_{max}$, however, the block leaves the surface of the cylinder so that $N_B = 0$.

Step 3: A third equation for the solution may be obtained by noting that the magnitude of tangential acceleration a_t may be related to the speed of the block v and the angle θ. From Eq. 12–7, $a_t\, ds = v\, dv$, but $ds = r\, d\theta = 0.5\, d\theta$, Fig. 13–21*a*. Thus,

$$a_t = \frac{v\, dv}{0.5\, d\theta} \qquad (3)$$

Substituting Eq. (3) into Eq. (2) and separating the variables, we have

$$v\, dv = 4.905 \sin\theta\, d\theta$$

Integrating both sides, realizing that when $\theta = 0°$, $v = 1$ m/s, yields

$$\int_1^v v\, dv = 4.905 \int_{0°}^{\theta} \sin\theta\, d\theta$$

$$\left.\frac{v^2}{2}\right|_1^v = -4.905 \cos\theta \Big|_{0°}^{\theta}$$

$$v^2 = 9.81(1 - \cos\theta) + 1$$

Substituting into Eq. (1) with $N_B = 0$ and solving for $\cos\theta_{max}$ gives

$$19.62 \cos\theta_{max} = \frac{2}{0.5}[9.81(1 - \cos\theta_{max}) + 1]$$

$$\cos\theta_{max} = \frac{43.24}{58.86}$$

$$\theta_{max} = 42.7° \qquad \textit{Ans.}$$

13-31. Compute the mass of the sun, knowing that the distance from the earth to the sun is $149.6(10^6)$ km. *Hint:* Represent the force of gravitation acting on the earth by using Eq. 13-2.

* **13-32.** Determine the maximum constant speed at which the car can coast freely over the crest (or top) of the hill without leaving the surface of the road. The radius of curvature at the crest measured to the center of mass of the car is $\rho = 80$ m, and the mass of the car is 1200 kg.

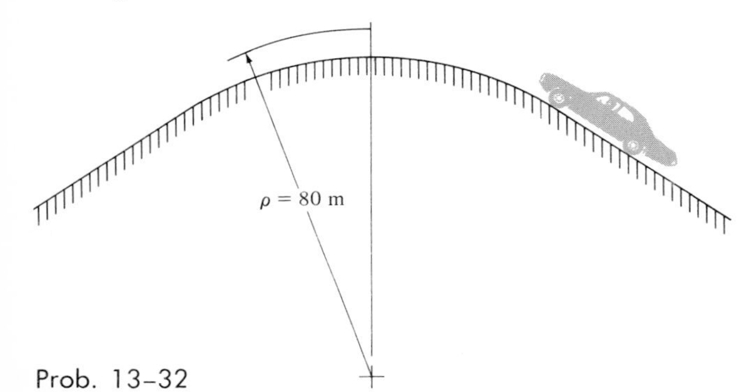

Prob. 13-32

13-33. The sports car, having a mass of 1700 kg, is traveling horizontally along a 20° banked track which is circular and has a radius of curvature of $\rho = 100$ m. If the coefficient of friction between the tires and the road is $\mu = 0.2$, determine the *maximum constant speed* at which the car can travel without sliding up the slope.

Prob. 13-33

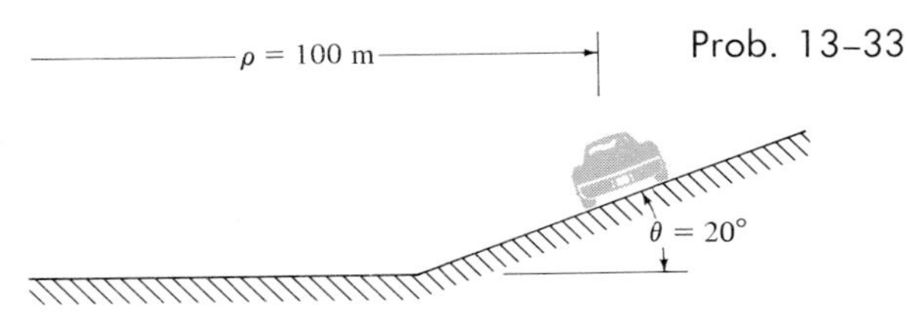

13-34. Using the data in Prob. 13-33, determine the *minimum speed* at which the car can travel around the track without sliding down the slope.

13-35. A pilot, having a mass of $m_p = 70$ kg, is traveling at a constant speed of $v_p = 150$ km/h. If he turns an inside vertical loop having a radius of $\rho = 90$ m, as shown, determine the vertical force he exerts on the seat of the plane at the instant he is upside down at A.

Prob. 13-35

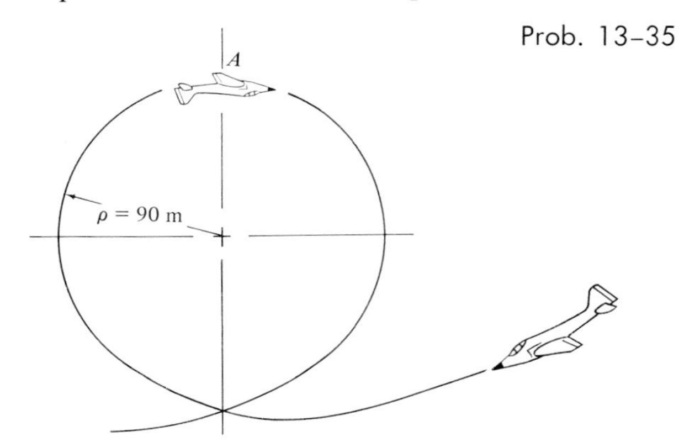

13-35a. Solve Prob. 13-35 if the pilot has a weight of $W_p = 170$ lb, and $v_p = 100$ ft/s, $\rho = 200$ ft.

* **13-36.** The rotational speed of the disk is controlled by a 20-g smooth contact arm AB which is spring-mounted on the disk. When the disk is *at rest,* the center of mass, G, of the arm is located 250 mm from the center O, and the preset compression in the spring is 18 mm. If the initial gap between B and the contact at C is 10 mm, determine the (controlling) speed of the arm's mass center, v_G, which will close the gap. The disk rotates in the horizontal plane. The spring has a stiffness of $k = 50$ N/m, and its ends are attached to the contact arm at D and to the disk at E.

Prob. 13-36

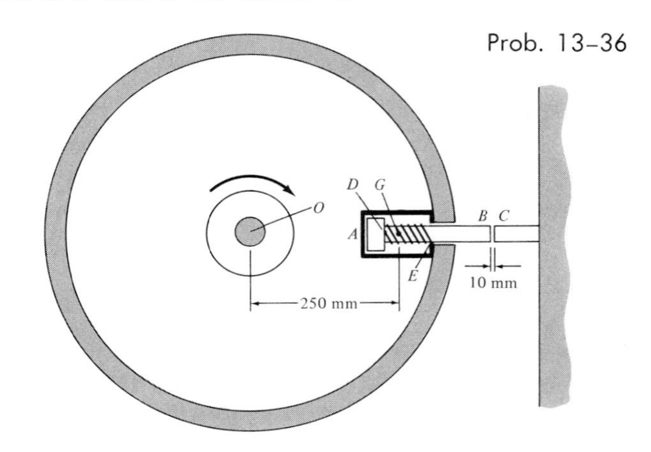

13–37. The block B, having a mass of 0.3 kg, is attached to the vertex A of the right circular cone using a light cord. The cone is rotating at a constant angular rate about the z axis such that the block attains a speed of 0.6 m/s. At this speed, determine the tension in the cord and the reaction which the cone exerts on the block. Neglect the size of the block in the computation.

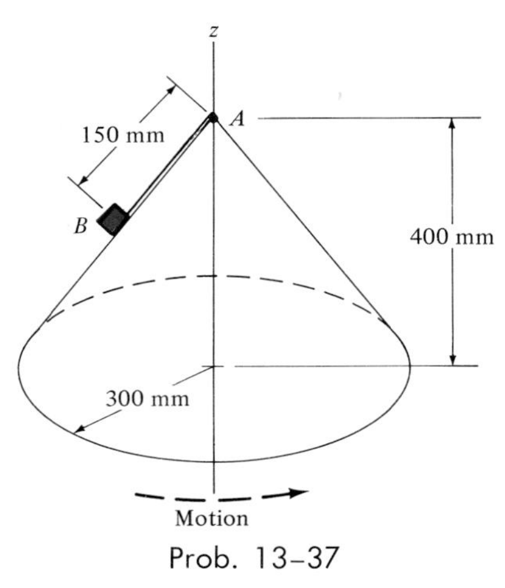

Prob. 13–37

13–38. Prove that if the block is released from rest at the top point B of a smooth path of *arbitrary shape,* the speed it attains when it reaches point A is equal to the speed it attains when it falls freely through a distance h.

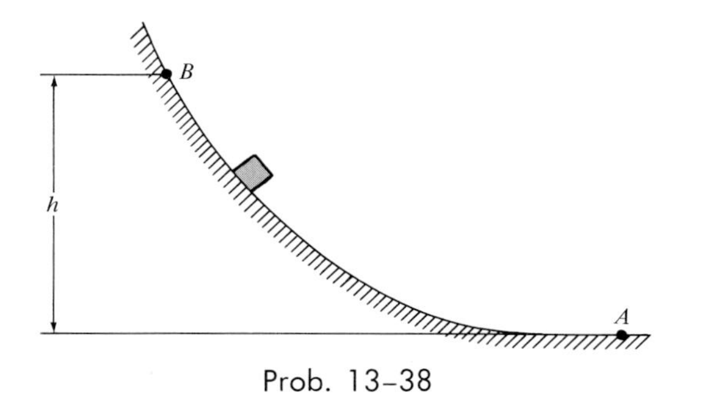

Prob. 13–38

13–39. A skier starts from rest at $A(10\text{ m}, 0)$ and descends the smooth slope, which may be approximated by a parabola. If she has a mass of 52 kg, determine the normal force she exerts on the ground at the instant she arrives at point B. *Hint:* Use the result of Prob. 13–38.

Prob. 13–39

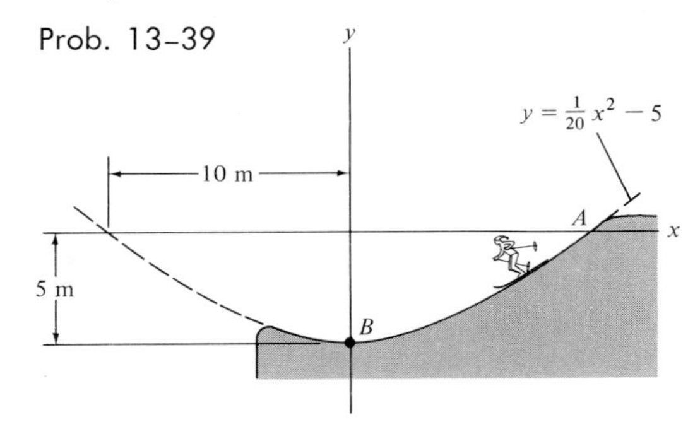

* **13–40.** The pendulum bob B has a mass of $m = 5$ kg and is released from rest in the position shown, $\theta = 0°$, by cutting the string AB. Determine the tension in string BC immediately after AB is cut and also at the instant the bob reaches point D, $\theta = 45°$, $r = 2$ m.

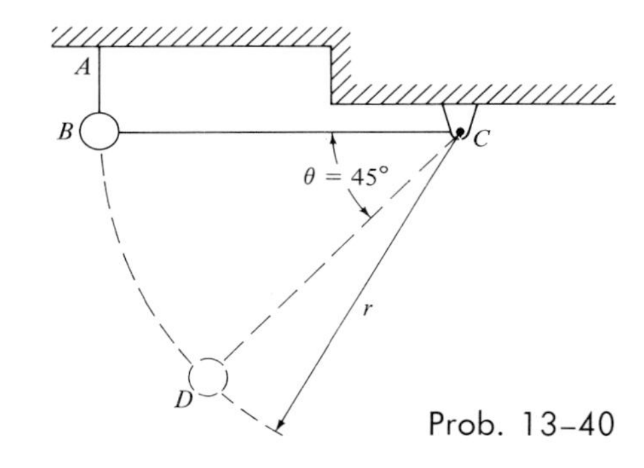

Prob. 13–40

* **13–40a.** Solve Prob. 13–40 if the bob has a weight of $W = 10$ lb and $r = 4$ ft.

13–41. A bucket filled with water has a total mass of 10 kg. If a cord is attached to the bucket and the bucket is whirled in a vertical circle having a radius of 1.5 m, determine the minimum speed the bucket must have when it reaches the top of the circle so that the cord does not slacken. Compute the tension in the cord when the bucket swings down and returns to the bottom of the circle after rounding the top.

13–42. A toboggan and passenger have a total mass of 90 kg and travel down along the (smooth) slope defined by the equation $y = 0.08x^2$. At the instant $x = 10$ m, the toboggan's speed is 5 m/s. At this point, determine the rate of increase in speed and the normal force which the toboggan exerts on the slope. Neglect the size of the toboggan and passenger for the calculation.

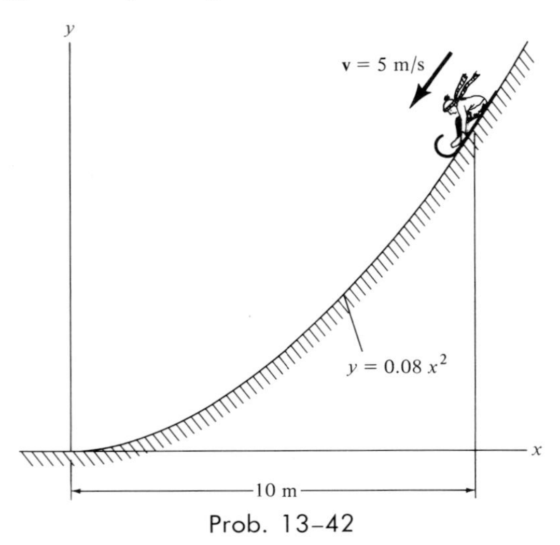

Prob. 13–42

13–43. The collar A, having a mass of 1.5 kg, is attached to a spring having a stiffness of $k = 200$ N/m. When rod BC rotates about the vertical axis, the collar slides outward along the smooth rod DE. If the spring is unstretched when $s = 0$, determine the constant speed of the collar in order that $s = 100$ mm.

Prob. 13–43

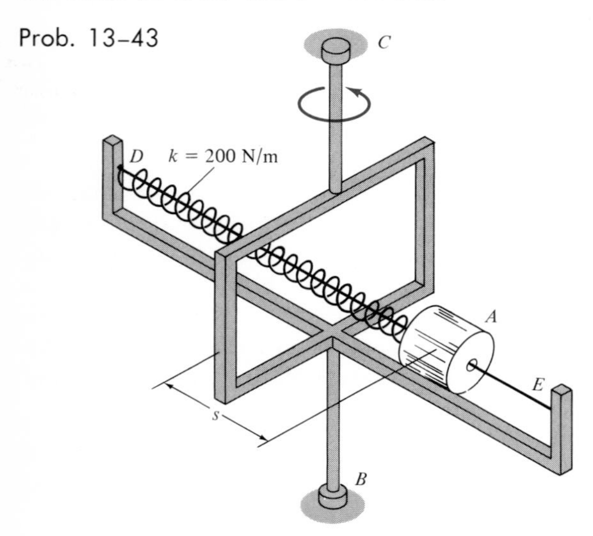

*** 13–44.** A 2-kg ball rolls within a vertical circular slot. If the ball is released from rest when $\theta = 10°$, determine the force which it exerts on the slot at the instants it arrives at points A and B. Neglect the rolling motion of the ball in the calculation.

Prob. 13–44

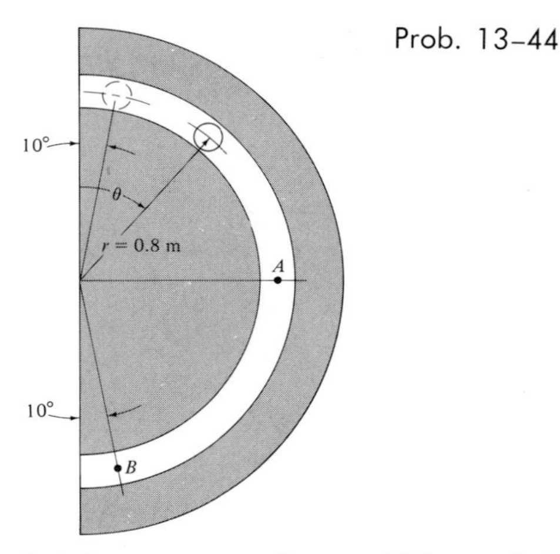

13–45. An acrobat has a mass of $m_a = 70$ kg and is sitting on a chair which is perched on top of a pole as shown. If by a mechanical drive, the pole rotates downward at a constant rate from $\theta = 0°$, such that the acrobat's center of mass G maintains a *constant speed* of $v_a = 3.5$ m/s, determine the angle θ at which he begins to "fly" out of the chair. Neglect friction and assume that the distance from the pivot O to G is $\rho = 3$ m.

Prob. 13–45

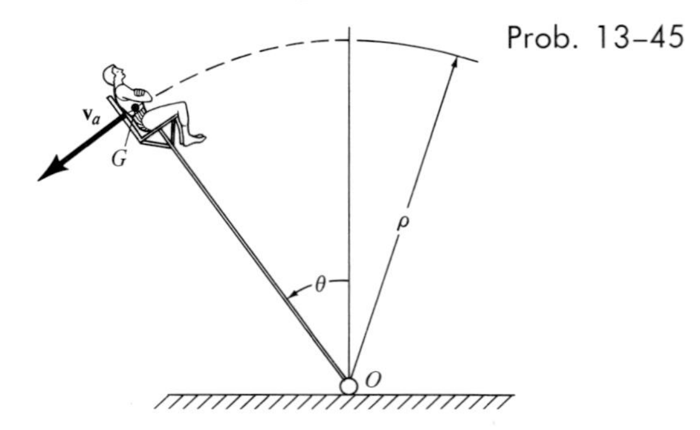

13–45a. Solve Prob. 13–45 if the acrobat weighs $W_a = 150$ lb, and $v_a = 9.5$ ft/s, $\rho = 12$ ft.

13–46. Solve Prob. 13–45 if the speed of the acrobat's center of mass is increased from $(v_a)_1 = 3.5$ m/s at $\theta = 0°$ by a constant rate of $\dot{v}_a = 0.1$ m/s².

13–47. A girl, having a mass of 50 kg, sits motionless relative to the surface of a horizontal platform, at a distance of $r = 5$ m from the platform's center. If the angular motion of the platform is *slowly* increased, so that the girl's tangential component of acceleration can be neglected, determine the maximum speed which the girl will have before she begins to slip off the platform. The coefficient of friction between the girl and the platform is $\mu = 0.3$.

Prob. 13–47

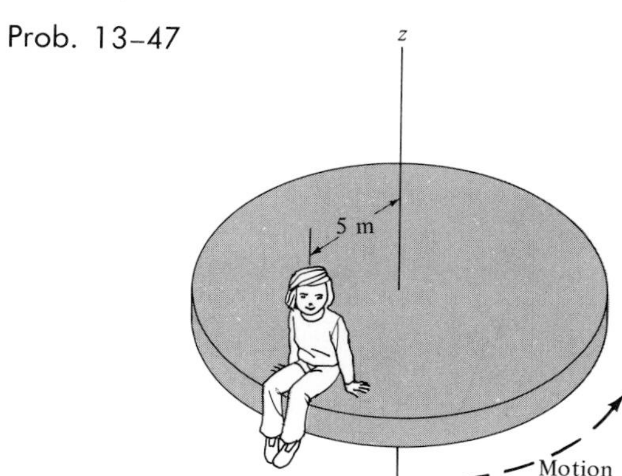

*** 13–48.** Solve Prob. 13–47 assuming that the platform starts rotating from rest, so that the girl's speed is increased uniformly at $\dot{v} = 1.25$ m/s².

13–49. A collar having a mass of 1.5 kg slides over the surface of a horizontal circular rod for which the coefficient of friction is $\mu = 0.3$. If the collar is given an initial speed of 2 m/s and then released at $\theta = 0°$, determine how far, *s*, it slides on the rod before coming to rest.

Prob. 13–49

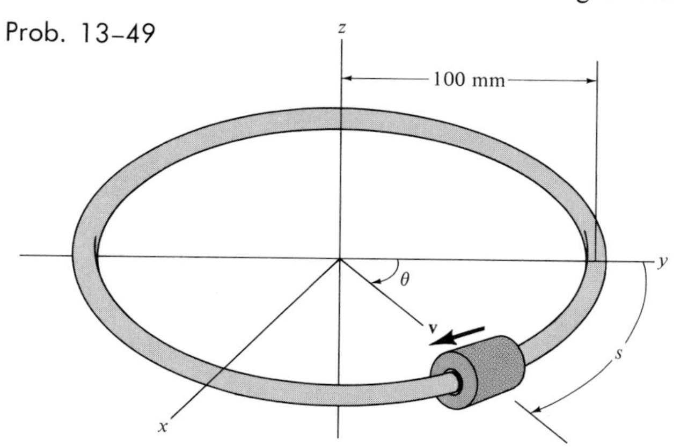

13–50. Packages having a mass of $m_p = 2.5$ kg ride on the surface of the conveyor belt. If the belt starts from rest and increases to a constant speed of $v_b = 0.75$ m/s in $t = 2$ s, determine the maximum angle of tilt, θ, so that none of the packages slip on the inclined surface *AB* of the belt. The coefficient of friction between the belt and each package is $\mu = 0.3$. At what angle ϕ do the packages first begin to slip off the surface of the belt after the belt is moving at a constant speed of $v_b = 0.75$ m/s? Set $r = 350$ mm.

Prob. 13–50

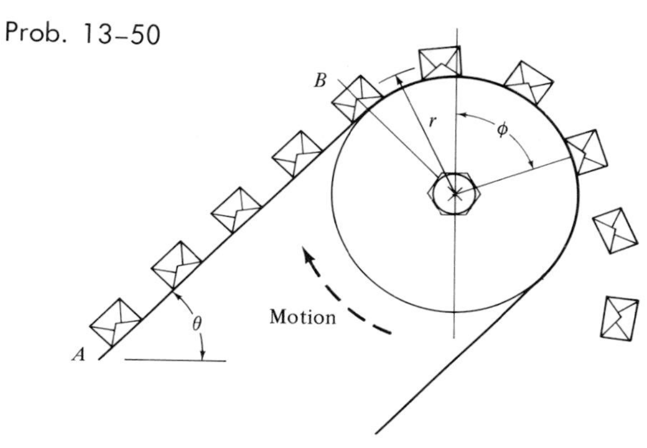

13–50a. Solve Prob. 13–50 if each package weighs $W_p = 2$ lb and $v_b = 1.5$ ft/s, $t = 3$ s, $\mu = 0.3$, $r = 0.75$ ft.

13–51. A thin rubber band *B*, having a mass of 0.9 kg/m, is placed over the smooth cylinder such that the tension in the band is 60 N. Determine the minimum speed which the cylinder can impart to the band before it begins to loosen on the cylinder.

Prob. 13–51

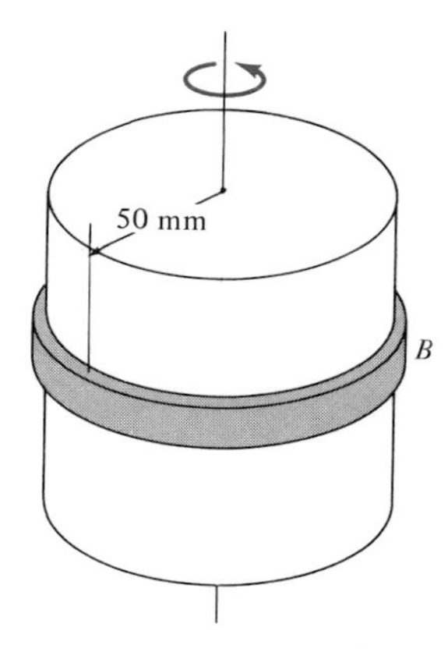

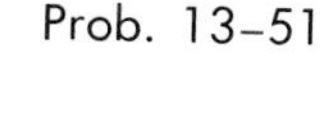

*13–7. Central-Force Motion and Space Mechanics

If a particle is moving only under the influence of a force having a line of action which is always directed toward a fixed point, the motion is called *central-force motion.* This type of motion is commonly caused by electrostatic and gravitational forces.

In Fig. 13–22*a*, the particle P having a mass m is acted upon only by the central force $\mathbf{F}$. The free-body and kinetic diagrams for the particle are shown in Fig. 13–22*b*. Using polar coordinates (r, θ), the equations of motion (Eq. 13–9) become

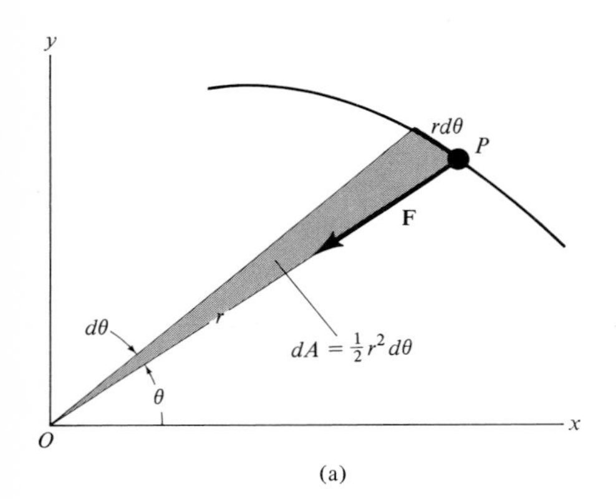

(a)

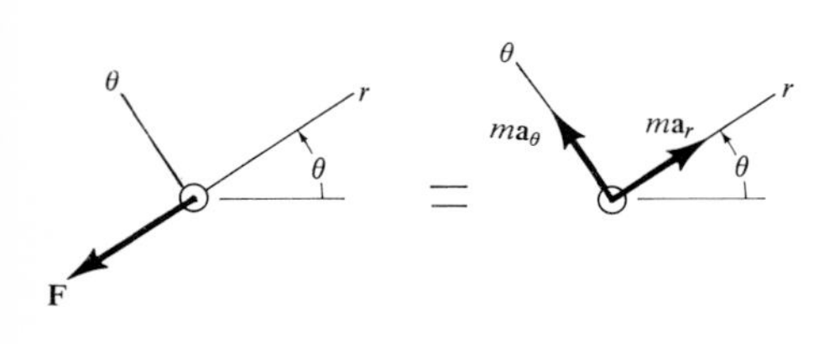

(b)

Fig. 13–22

$$-F = m\left[\frac{d^2r}{dt^2} - r\left(\frac{d\theta}{dt}\right)^2\right]$$
$$0 = m\left(r\frac{d^2\theta}{dt^2} + 2\frac{dr}{dt}\frac{d\theta}{dt}\right) \tag{13–12}$$

The second of these equations may be written in the form

$$\frac{1}{r}\left[\frac{d}{dt}\left(r^2\frac{d\theta}{dt}\right)\right] = 0$$

so that integrating yields

$$r^2\frac{d\theta}{dt} = C_1 \tag{13–13}$$

where C_1 is a constant of integration. From Fig. 13–22*a* it can be seen that the shaded area inscribed by the radius r, as r moves through an angle $d\theta$, is $dA = \frac{1}{2}r^2\,d\theta$. If the *areal velocity* is defined as

$$\frac{dA}{dt} = \frac{1}{2}r^2\frac{d\theta}{dt} \tag{13–14}$$

then, by comparison with Eq. 13–13, it is seen that the areal velocity for a particle subjected to central-force motion is *constant*. To obtain the *path of motion*, $r = f(\theta)$, the independent variable t must be eliminated from Eqs. 13–12. Using the chain rule of calculus and Eq. 13–13 the time derivatives of Eq. 13–12 may be replaced by

$$\frac{dr}{dt} = \frac{dr}{d\theta}\frac{d\theta}{dt} = \frac{C_1}{r^2}\frac{dr}{d\theta}$$

$$\frac{d^2r}{dt^2} = \frac{d}{dt}\left(\frac{C_1}{r^2}\frac{dr}{d\theta}\right) = \frac{d}{d\theta}\left(\frac{C_1}{r^2}\frac{dr}{d\theta}\right)\frac{d\theta}{dt} = \left[\frac{d}{d\theta}\left(\frac{C_1}{r^2}\frac{dr}{d\theta}\right)\right]\frac{C_1}{r^2}$$

Substituting a new dependent variable (xi) $\xi = 1/r$ into the second equation, it is seen that

$$\frac{d^2r}{dt^2} = -C_1^2\xi^2\frac{d^2\xi}{d\theta^2}$$

Also, the square of Eq. 13–13 becomes

$$\left(\frac{d\theta}{dt}\right)^2 = C_1^2\xi^4$$

Substituting these last two equations into the first of Eqs. 13–12, we have

$$-C_1^2\xi^2\frac{d^2\xi}{d\theta^2} - C_1^2\xi^3 = -\frac{F}{m}$$

or

$$\frac{d^2\xi}{d\theta^2} + \xi = \frac{F}{mC_1^2\xi^2} \qquad (13\text{–}15)$$

This differential equation defines the path over which the particle travels when it is subjected to the central force* **F**.

For application the force of gravitational attraction will be considered. Some common examples of central-force systems which depend upon gravitation include the motion of the moon and artificial satellites about the earth, and the motion of the planets about the sun. As a typical problem in space mechanics, consider the trajectory of a space satellite or space vehicle launched into orbit with an initial velocity $\mathbf{v}_o$, Fig. 13–23. It will be assumed that this velocity is initially *parallel* to the tangent at the surface of the earth, as shown in the figure.† Just after the satellite is released into free flight the only force acting upon it is the gravitational force of the earth. (Gravitational attractions involving other bodies such as the moon or sun will be neglected since for orbits close to the earth, their effect is small in comparison with the gravitational force of the earth.) According to Newton's law of gravitation, force **F** will always act between the mass centers of the earth and the satellite, Fig. 13–23. From Eq. 13–2, this force of attraction has a magnitude of

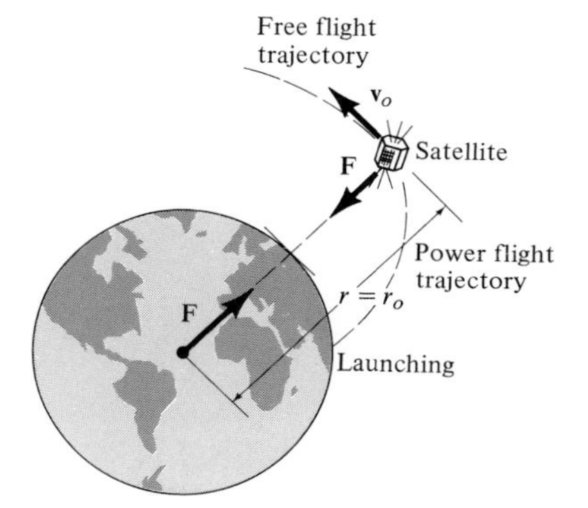

Fig. 13–23

$$F = G\frac{M_e m}{r^2}$$

where M_e and m represent the mass of the earth and the satellite, respectively, G is the gravitational constant, and r is the distance between the mass centers. Setting $\xi = 1/r$ in the above equation and substituting the result into Eq. 13–15, we obtain

$$\frac{d^2\xi}{d\theta^2} + \xi = \frac{GM_e}{C_1^2} \qquad (13\text{–}16)$$

*In the derivation, **F** is considered positive when it is directed toward point O. If **F** is oppositely directed, the right side of Eq. 13–15 should be negative.

†The case where $\mathbf{v}_o$ acts at some initial angle θ to the tangent is best described using the conservation of angular momentum. See Prob. 15–51.

This second-order ordinary differential equation has constant coefficients and is nonhomogeneous. The solution is represented as the sum of the complementary and particular solutions. The complementary solution is obtained when the term on the right is equal to zero. It is,

$$\xi_c = C_2 \cos(\theta - \phi)$$

where C_2 and ϕ are constants of integration. The particular solution is

$$\xi_p = \frac{GM_e}{C_1^2}$$

Thus, the complete solution to Eq. 13–16 is

$$\begin{aligned} \xi &= \xi_c + \xi_p \\ &= \frac{1}{r} = C_2 \cos(\theta - \phi) + \frac{GM_e}{C_1^2} \end{aligned} \tag{13–17}$$

The validity of this result may be checked by substitution into Eq. 13–16.

Equation 13–17 represents the *free-flight trajectory* of the satellite. It is the equation of a conic section expressed in terms of polar coordinates. As shown in Fig. 13–24, a *conic section* is defined as the locus of point P which moves in a plane in such a way that the ratio of its distance from a fixed point F to its distance from a fixed line is constant. The fixed point is called the *focus*, and the fixed line DD is called the *directrix*. The constant ratio is called the *eccentricity* of the conic and is denoted by e. Thus,

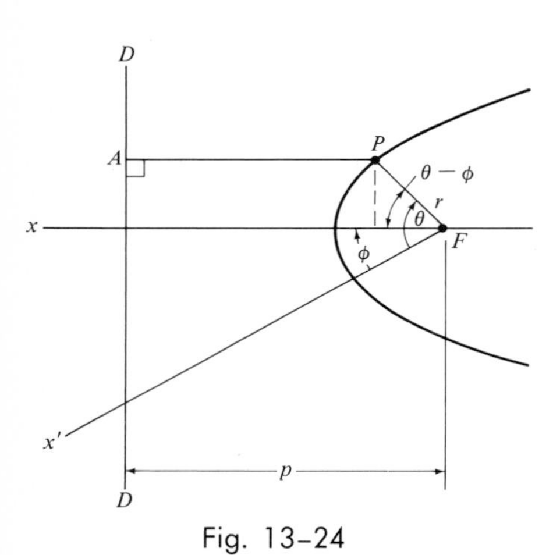

Fig. 13–24

$$e = \frac{FP}{PA}$$

which may be written in the form

$$FP = r = e(PA) = e[p - r\cos(\theta - \phi)]$$

or

$$\frac{1}{r} = \frac{1}{p}\cos(\theta - \phi) + \frac{1}{ep}$$

Comparing this equation with Eq. 13–17, it is seen that the eccentricity of the conic section for the trajectory is

$$e = \frac{C_2 C_1^2}{GM_e} \tag{13–18}$$

and the fixed distance from the focus to the directrix is

$$p = \frac{1}{C_2} \tag{13–19}$$

Provided the polar angle θ is measured from the x axis (an axis of symmetry since it is perpendicular to the directrix DD), the angle ϕ is zero,

Fig. 13-24, and therefore Eq. 13-17 reduces to

$$\frac{1}{r} = C_2 \cos\theta + \frac{GM_e}{C_1^2} \tag{13-20}$$

The constants C_1 and C_2 are determined from the data obtained for the position and velocity of the satellite at the end of the *power-flight trajectory*. For example, if the initial height or radius to the space vehicle is r_o (measured from the center of the earth) and its initial speed is v_o at the beginning of its free flight, Fig. 13-25, then the constant C_1 may be obtained from Eq. 13-13. When $\theta = \phi = 0°$, the velocity $\mathbf{v}_o$ has no radial component; therefore, from Eq. 12-43, $v_o = r_o(d\theta/dt)$, so that

$$C_1 = r_o^2 \frac{d\theta}{dt}$$

or

$$C_1 = r_o v_o \tag{13-21}$$

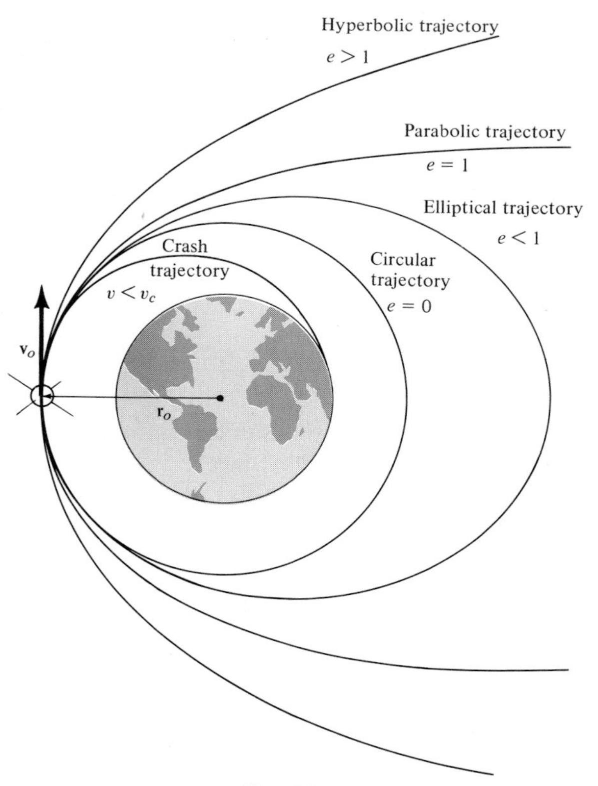

Fig. 13-25

To determine C_2, use Eq. 13–20 with $\theta = 0°$, $r = r_o$, and substitute Eq. 13–21 for C_1:

$$C_2 = \frac{1}{r_o}\left(1 - \frac{GM_e}{r_o v_o^2}\right) \tag{13–22}$$

The equation for the free-flight trajectory therefore becomes

$$\frac{1}{r} = \frac{1}{r_o}\left(1 - \frac{GM_e}{r_o v_o^2}\right)\cos\theta + \frac{GM_e}{r_o^2 v_o^2} \tag{13–23}$$

The type of path taken by the satellite is determined from the value of the eccentricity of the conic section as given by Eq. 13–18. If

$$\begin{aligned} e &= 0, && \text{free-flight trajectory is a circle} \\ e &= 1, && \text{free-flight trajectory is a parabola} \\ e &< 1, && \text{free-flight trajectory is an ellipse} \\ e &> 1, && \text{free-flight trajectory is a hyperbola} \end{aligned} \tag{13–24}$$

A graph of each of these trajectories is shown in Fig. 13–25. From the curves it is seen that when the satellite follows a parabolic path, it is "on the border" of not returning to its initial starting point. The initial velocity at launching, v_o, required for the satellite to follow a parabolic path is called the *escape velocity*. This velocity, v_e, can be determined by using the second of Eqs. 13–24 with Eqs. 13–18, 13–21, and 13–22. It is left as an exercise to show that

$$v_e = \sqrt{\frac{2GM_e}{r_o}} \tag{13–25}$$

The velocity v_c required to launch a satellite into a *circular orbit* can be found using the first of Eqs. 13–24. Since e is related to C_1 and C_2, Eq. 13–18, C_2 must be zero to satisfy this equation (from Eq. 13–21, C_1 cannot be zero); and, therefore, using Eq. 13–22, we have

$$v_c = \sqrt{\frac{GM_e}{r_o}} \tag{13–26}$$

Provided r_o represents a minimum height for launching, in which frictional resistance from the atmosphere is neglected, velocities of launching which are less than v_c will cause the satellite to reenter the earth's atmosphere and either burn up or crash, Fig. 13–25.

All the trajectories attained by planets and most satellites are elliptical, Fig. 13–26. For an orbit about the earth, the *minimum distance* from the

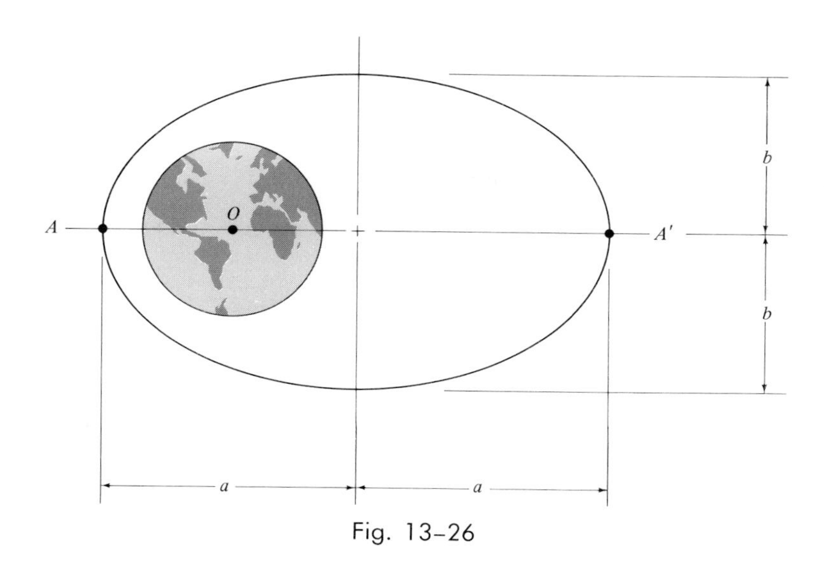

Fig. 13–26

orbit to the center of the earth O (which is located at one of the foci of the ellipse) is OA and can be found using Eq. 13–23 with $\theta = 0°$. Therefore,

$$OA = r_o \tag{13–27}$$

This distance is called the *perigee* of the orbit. The *apogee* or maximum distance OA' can be found using Eq. 13–23 with $\theta = 180°$. Thus,

$$OA' = \frac{r_o}{\dfrac{2GM_e}{r_o v_o^2} - 1} \tag{13–28}$$

With reference to Fig. 13–26, the semimajor axis a of the ellipse is

$$a = \frac{OA + OA'}{2} \tag{13–29}$$

Using analytical geometry it can be shown that the minor axis b is determined from the equation

$$b = \sqrt{(OA)(OA')} \tag{13–30}$$

Furthermore, by direct integration, the area of an ellipse is

$$A = \pi ab = \frac{\pi}{2}(OA + OA')\sqrt{(OA)(OA')} \tag{13–31}$$

The areal velocity has been defined by Eq. 13–14. Substituting this equation into Eq. 13–13, we have

$$\frac{dA}{dt} = \frac{C_1}{2}$$

Integrating yields

$$A = \frac{C_1}{2}T$$

where T is the *period* of time required to make one orbital revolution. From Eq. 13–31, the period is

$$T = \frac{\pi}{C_1}(OA + OA')\sqrt{(OA)(OA')} \quad (13\text{–}32)$$

In addition to predicting the orbital trajectory of earth satellites, the theory developed in this section is valid, as a surprisingly close approximation, in predicting the actual motion of the planets traveling around the sun. In this case the mass of the sun, M_s, should be substituted for M_e when using the appropriate formulas.

Example 13–14

A satellite is launched 600 km from the surface of the earth with an initial velocity of 30 Mm/h, acting parallel to the tangent at the surface of the earth, Fig. 13–27. Assuming that the radius of the earth is 6378 km and that its mass is $5.976(10^{24})$ kg, determine (a) the eccentricity of the orbital path, (b) the velocity of the satellite at apogee, and (c) the period of revolution.

Solution

Part (a). The eccentricity of the orbit is obtained using Eq. 13–18. The

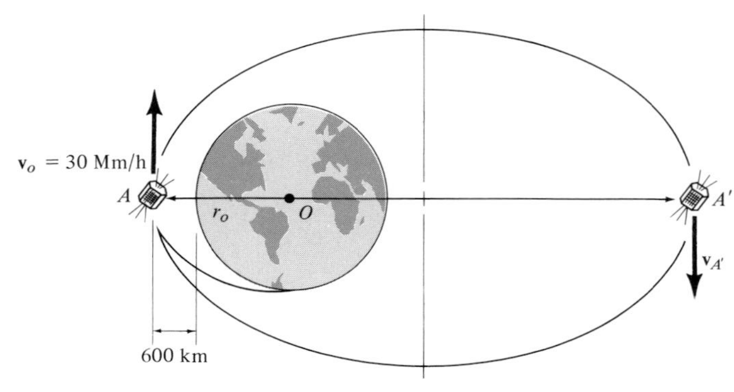

Fig. 13–27

constants C_1 and C_2 are first determined from Eqs. 13–21 and 13–22. Since

$$r_o = 6378 \text{ km} + 600 \text{ km} = 6.978(10^6) \text{ m}$$
$$v_o = 30 \text{ Mm/h} = 8333.3 \text{ m/s}$$

then

$$C_1 = r_o v_o = 6.978(10^6)(8333.3) = 5.815(10^{10}) \text{ m}^2/\text{s}$$

$$C_2 = \frac{1}{r_o}\left(1 - \frac{GM_e}{r_o v_o^2}\right)$$

$$= \frac{1}{6.978(10^6)}\left(1 - \frac{6.673(10^{-11})[5.976(10^{24})]}{6.978(10^6)(8333.3)^2}\right)$$

$$= 2.54(10^{-8}) \text{ m}^{-1}$$

Hence,

$$e = \frac{C_2 C_1^2}{GM_e} = \frac{2.54(10^{-8})[5.815(10^{10})]^2}{6.673(10^{-11})[5.976(10^{24})]}$$

$$= 0.215 < 1 \qquad \textit{Ans.}$$

From Eq. 13–24, observe that the orbit is an *ellipse.*

Part (b). If the satellite were launched at the apogee A', shown in Fig. 13–27, with a velocity $\mathbf{v}_{A'}$, the same orbit would be maintained provided

$$C_1 = r_o v_o = OA' v_{A'} = 5.815(10^{10}) \text{ m}^2/\text{s}$$

Using Eq. 13–28, we have

$$OA' = \frac{r_o}{\dfrac{2GM_e}{r_o v_o^2} - 1} = \frac{6.978(10^6)}{\dfrac{2[6.673(10^{-11})][5.976(10^{24})]}{6.978(10^6)(8333.3)^2} - 1}$$

$$= 10.804(10^6)$$

Thus,

$$v_{A'} = \frac{5.815(10^{10})}{10.804(10^6)}$$

$$= 5382.3 \text{ m/s} = 19.4 \text{ Mm/h} \qquad \textit{Ans.}$$

Part (c). The time for one revolution is determined from Eq. 13–32. Since $OA = r_o$,

$$T = \frac{\pi}{C_1}(OA + OA')\sqrt{(OA)(OA')}$$

$$= \frac{\pi}{5.815(10^{10})}[6.978(10^6) + 10.804(10^6)]\sqrt{6.978(10^6)[10.804(10^6)]}$$

$$= 8341.4 \text{ s} = 2.32 \text{ h} \qquad \textit{Ans.}$$

Problems

In the following problems, assume that the radius of the earth is 6378 km, the earth's mass is $5.976(10^{24})$ kg, the mass of the sun is $1.99(10^{30})$ kg, and the gravitational constant is $G = 6.673(10^{-11})\ \text{m}^3/(\text{kg} \cdot \text{s}^2)$.

***13-52.** Show that the velocity for a satellite launched into a circular orbit about the earth is given by Eq. 13-26. Determine the velocity of a satellite launched parallel to the surface of the earth, so that it travels in a circular orbit 700 km from the earth's surface.

13-53. If the orbit of an asteroid has an eccentricity of $e = 0.056$ about the sun, determine the perigee of the orbit. The orbit's apogee is $2.0(10^9)$ km.

13-54. The rocket is traveling in free flight along an elliptical trajectory $A'A$. The planet has no atmosphere, and its mass is 0.60 times that of the earth's. If the rocket has an apogee and perigee as shown in the figure, determine the speed of the rocket when it is at point A.

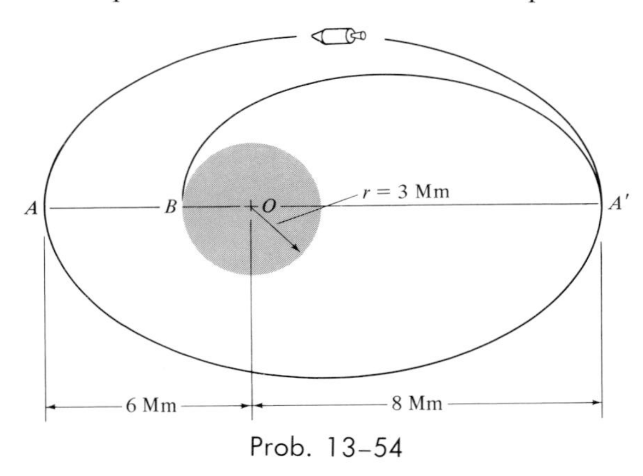

Prob. 13-54

13-55. The planet Jupiter travels around the sun in an elliptical orbit such that the eccentricity is $e = 0.048$. If the perigee between Jupiter and the sun is $r_o = 7.04(10^8)$ km, determine (a) Jupiter's speed at perigee, and (b) the apogee of the orbit.

13-55a. Solve Prob. 13-55 if $r_o = 4.40(10^8)$ mi, $G = 3.44(10^{-8})\ \text{lb} \cdot \text{ft}^2/\text{slug}^2$, $M_s = 1.97(10^{29})$ slug, and 1 mi = 5280 ft.

***13-56.** If the rocket in Prob. 13-54 is to land on the surface of the planet, determine the required free-flight speed it must have at A' so that the rocket strikes the planet at B. How long does it take for the rocket to land, in going from A' to B along the elliptical path?

13-57. A satellite is placed into orbit at a velocity of 6 km/s, parallel to the surface of the earth. Determine the proper altitude of the satellite above the earth's surface such that its orbit remains circular. What will happen to the satellite if its initial velocity is 3 km/s when placed into orbit at the calculated altitude?

13-58. The rocket shown is originally in a circular orbit 6 Mm above the surface of the earth. It is required that it travel in another circular orbit having an altitude of 12 Mm. To do this, the rocket is given a short pulse of power at A so that it travels in free flight along the dashed elliptical path from the first orbit to the second orbit. Determine the necessary speed it must have at A, just after the power pulse, and the time required to get to the outer orbit along the path AA'. What adjustment in speed must be made at A' to maintain the second circular orbit?

Prob. 13-58

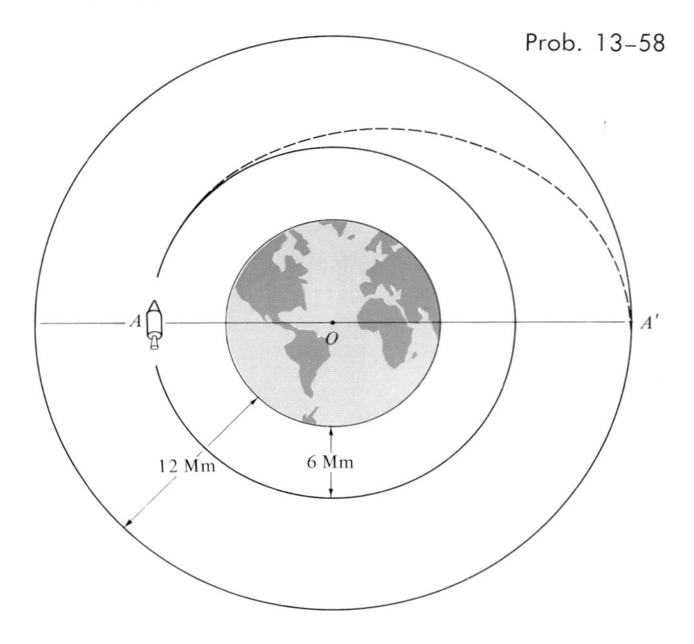

13-59. A rocket is in free-flight elliptical orbit around the planet Venus. Knowing that the perigee and apogee of the orbit are 8 Mm and 24 Mm, respectively, determine (a) the speed of the rocket at point A'; (b) the required speed which it must attain at A just after braking so that it undergoes an 8-Mm free-flight circular orbit around Venus; and (c) compute the periods of both the circular and elliptical orbits. The mass of Venus is 0.816 times the mass of the earth.

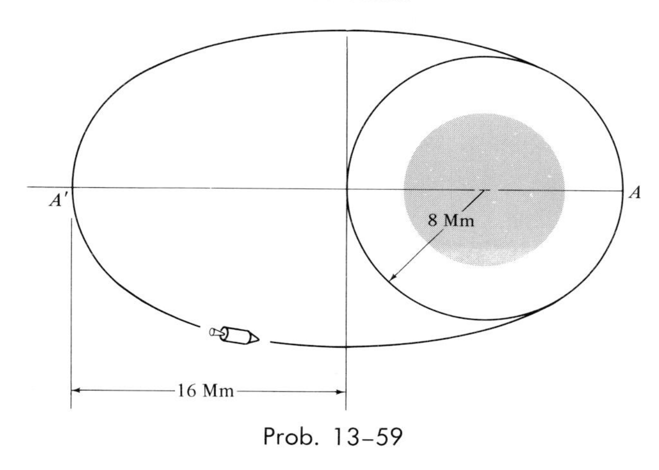

Prob. 13-59

* **13-60.** A satellite is launched with an initial velocity of $v_s = 4000$ km/h, parallel to the surface of the earth. Determine the required altitude (or range of altitudes) above the earth's surface for launching if the free-flight trajectory is to be (a) circular, (b) parabolic, (c) elliptical, and (d) hyperbolic.

* **13-60a.** Solve Prob. 13-60 if $v_s = 2500$ mi/h, $G = 3.44(10^{-8})$ (lb · ft²)/slug², $M_e = 4.09(10^{23})$ slug, the earth's radius is $r_e = 3960$ mi, and 1 mi = 5280 ft.

13-61. With what speed must the rocket in Prob. 13-59 be traveling so that it can leave its elliptical orbit at A and travel in free flight along a hyperbolic trajectory which has an eccentricity of $e = 1.5$?

13-62. The earth has an eccentricity ratio of $e = 0.0821$ in its orbit around the sun. Knowing that its farthest distance from the sun is $151.3(10^6)$ km, compute the speed at which it is traveling when it is at this distance. Determine the equation in polar coordinates which describes the orbit of the earth.

13-63. A rocket is docked next to a satellite located 12 Mm above the earth's surface. If the satellite is traveling in a circular orbit, determine the speed which must suddenly be given to the rocket, relative to the satellite, such that it travels in free flight away from the satellite along a parabolic trajectory as shown.

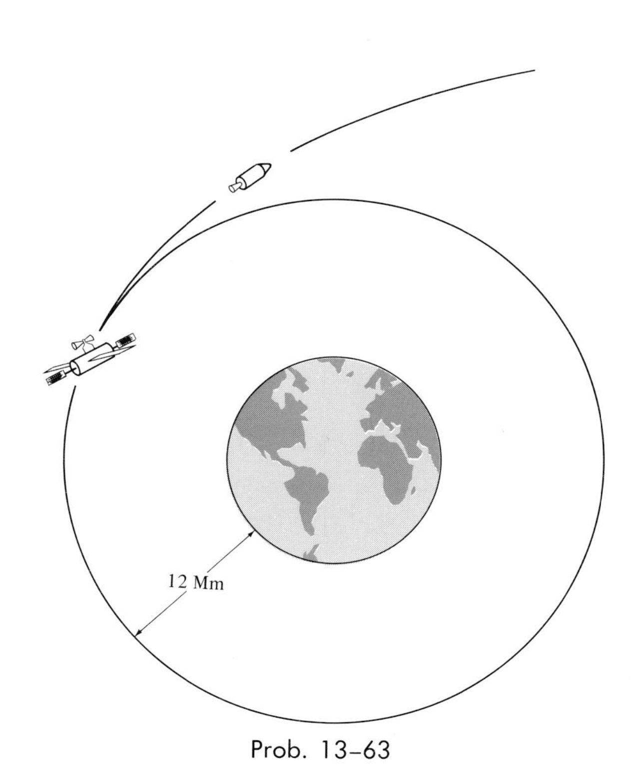

Prob. 13-63

14 Kinetics of a Particle: Work and Energy

14–1. The Work of a Force

In mechanics, a force **F** does an infinitesimal amount of work when it moves through a displacement $d\mathbf{s}$. In general, if θ is the angle formed between the tails of **F** and $d\mathbf{s}$, Fig. 14–1*a*, the *work dU* is a *scalar quantity,* defined by the dot product

$$dU = \mathbf{F} \cdot d\mathbf{s} = F\,ds\cos\theta \qquad (14\text{–}1)$$

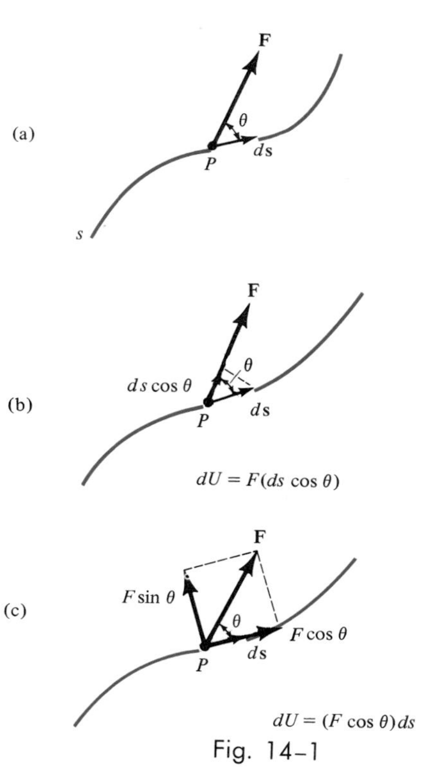

Fig. 14–1

Work, as expressed by this equation, may be interpreted in one of two ways: either as the product of F and the magnitude of the component of displacement in the direction of **F**, i.e., $ds\cos\theta$, Fig. 14–1*b;* or as the product of ds and the component of force magnitude in the direction of $d\mathbf{s}$, i.e., $F\cos\theta$, Fig. 14–1*c*. If $F\cos\theta$ and $d\mathbf{s}$ are in the *same direction,* the work is *positive;* if these vectors are in *opposite directions,* the work is *negative.* There are two cases for which a force does no work. As seen from Eq. 14–1, if **F** is *perpendicular* to $d\mathbf{s}$, $\cos 90° = 0$, so that the work is zero. The second case occurs when the force is applied at a *fixed point* in which case $d\mathbf{s} = \mathbf{0}$.

The basic unit for work, called a joule (J), combines the units of force and displacement. Specifically, 1 *joule* of work is done when a force of 1 newton moves 1 metre along its line of action (1 J = 1 N · m). The moment of a force has this same combination of units (N · m); however, the concepts of moment and work are in no way related. A moment is a vector quantity, whereas work is a scalar.

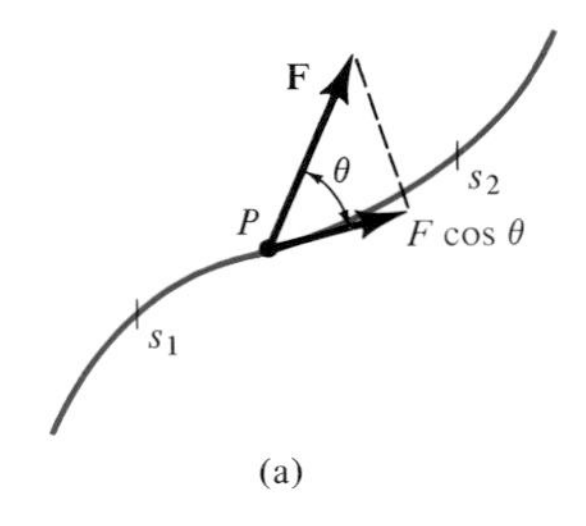

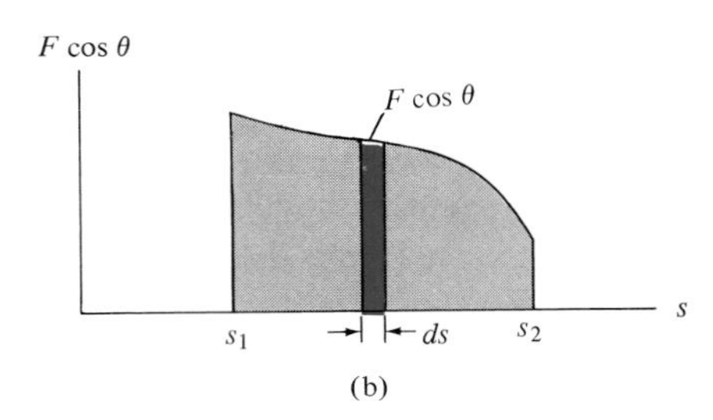

Fig. 14-2

Work of a Variable Force. If a force, acting on a particle P, undergoes a finite displacement along its path from s_1 to s_2, Fig. 14-2*a*, the work U_{1-2} done by $\mathbf{F}$ is determined by integrating Eq. 14–1. Provided $\mathbf{F}$ is a function of displacement, $\mathbf{F} = \mathbf{F}(s)$, we have

$$U_{1-2} = \int_{s_1}^{s_2} \mathbf{F} \cdot d\mathbf{s} = \int_{s_1}^{s_2} F \cos \theta \, ds \tag{14–2}$$

If the working component of the force, $F \cos \theta$, is plotted versus s, Fig. 14–2*b*, the integral represented in this equation can be interpreted as the *area under the curve* between the points s_1 and s_2.

Work of a Constant Force Moving Along a Straight Line. If the force $\mathbf{F}_c$ acting on particle P has a constant magnitude and direction and this force acts at an angle θ from its straight-lined path, Fig. 14–3*a*, then the component of $\mathbf{F}_c$ in the direction of motion is $F_c \cos \theta$. Hence, the work done by $\mathbf{F}_c$ when it is displaced from s_1 to s_2 is determined by Eq. 14–2, in which case

$$U_{1-2} = \int_{s_1}^{s_2} \mathbf{F}_c \cdot d\mathbf{s} = F_c \cos \theta \int_{s_1}^{s_2} ds$$

or

$$U_{1-2} = F_c \cos \theta (s_2 - s_1) \tag{14–3}$$

Here the work of $\mathbf{F}_c$ represents the *area under the rectangle* in Fig. 14–3*b*.

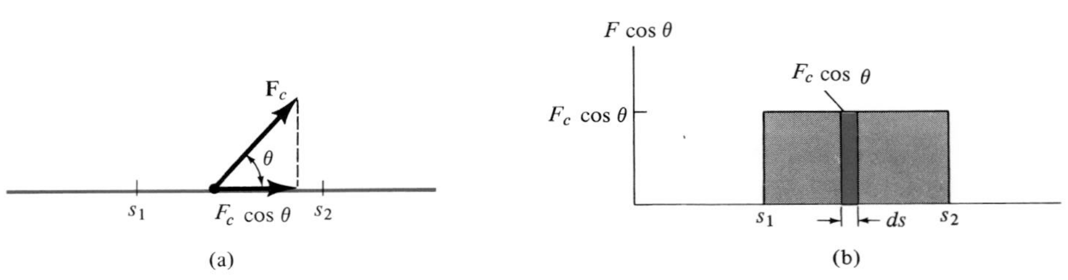

Fig. 14–3

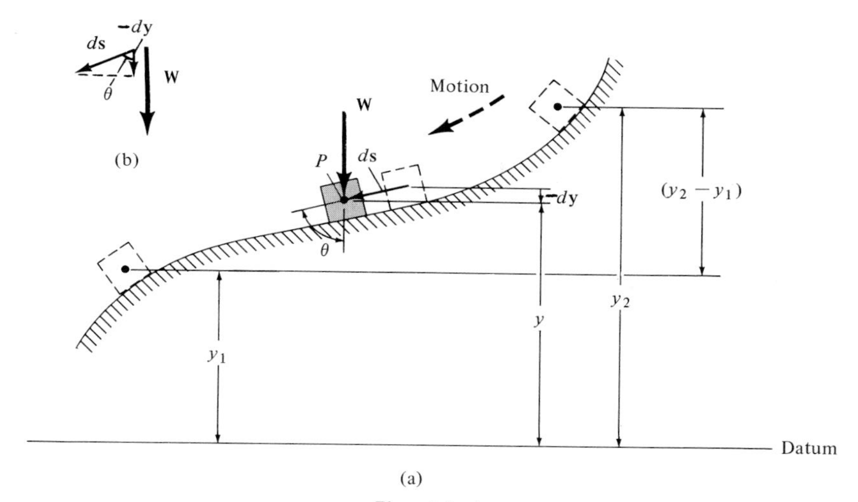

Fig. 14–4

Work of a Weight. Consider a particle (or block) P which moves down along the path shown in Fig. 14–4*a* from an initial elevation y_2 to a final elevation y_1, where both elevations are measured positive upward from a fixed horizontal reference plane or *datum*.* As the particle *descends* along the path by an amount $d\mathbf{s}$, the displacement component in the direction of $\mathbf{W}$ is $-d\mathbf{y}$, Fig. 14–4*a*. Hence, from Fig. 14–4*b*, we have $-dy = ds \cos \theta$, so that applying Eq. 14–2, realizing that $\mathbf{W}$ is constant, we obtain

$$U_{1-2} = \int_{y_2}^{y_1} -\mathbf{W} \cdot d\mathbf{y} = -W \int_{y_2}^{y_1} dy$$

or

$$U_{1-2} = W(y_2 - y_1) \qquad (14\text{–}4)$$

Thus, the work done is equal to the magnitude of $\mathbf{W}$ times the *difference* in elevation which defines the particle's vertical displacement.† In this case the total work is *positive,* since $\mathbf{W}$ and the vertical displacement $(y_2 - y_1)$ are both *downward,* that is, in the same direction. If the particle is displaced *upward* from y_1 to y_2, the work of the weight is *negative.* Why?

*Here the weight (force) is assumed to be *constant*. This assumption is suitable for small differences in elevation $(y_2 - y_1)$. If the elevation change is significant, however, a variation of weight with elevation must be taken into account.

†Note that the location of the datum plane is *arbitrary* since the results indicate that the work done depends *only upon the difference in elevation*.

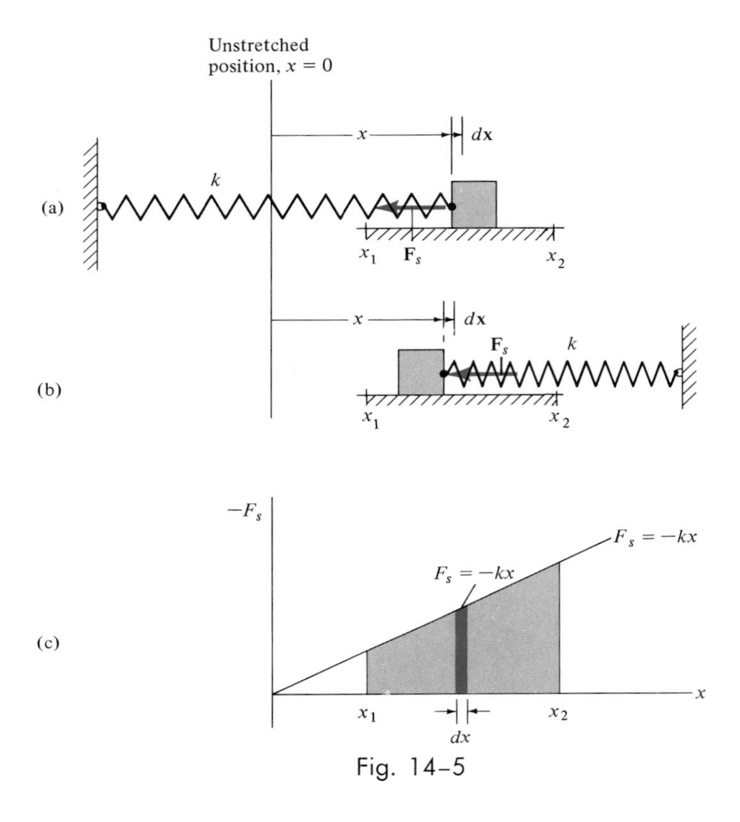

Fig. 14–5

Work of a Spring Force. The magnitude of force $\mathbf{F}_s$ developed by a linear elastic spring, when the spring is displaced a distance x from its unstretched position, is $F_s = kx$, where k is the spring stiffness. If a particle (or block) is attached to the spring and the spring is *elongated,* then *negative work* is done by $\mathbf{F}_s$ acting on the block.* This is because the differential displacement $d\mathbf{x}$ of the block is in the opposite direction of $\mathbf{F}_s$, Fig. 14-5*a*. Using Eq. 14–2 we can obtain the work U_{1-2} done by $\mathbf{F}_s$ when the block moves and therefore elongates the spring from an initial position x_1 to a final position x_2.

$$U_{1-2} = \int_{x_1}^{x_2} \mathbf{F}_s \cdot d\mathbf{x} = \int_{x_1}^{x_2} -kx\,dx$$

or

$$U_{1-2} = -(\tfrac{1}{2}kx_2^2 - \tfrac{1}{2}kx_1^2) \qquad (14\text{–}5)$$

*The mass of the spring is assumed to be small compared to the body to which it is connected.

In a similar way it can be shown that the same amount of work is done by $\mathbf{F}_s$ if the block moves such that it *compresses* the spring from an initial position x_1 to a final position x_2, Fig. 14–5*b*.

For either extension or compression, the work of the spring can be represented graphically as the shaded area under the line $F_s = -kx$ versus x, Fig. 14–5*c*.

As an example to illustrate the application of the above equations, consider the block of weight **W** shown in Fig. 14–6*a*, which is being pushed up along the rough path from A to B. The free-body diagram is shown in Fig. 14–6*b*. If the spring is unstretched when the block is at A, then the spring force $\mathbf{F}_s$ does negative work, $U_s = -\frac{1}{2}ks_{AB}^2$ (Eq. 14–5); the weight **W** does *negative work,* $U_W = -Wy_{AB}$ (Eq. 14–4); the applied constant force **P** does positive work, $U_P = Px_{AB}$ (Eq. 14–3); the constant frictional force $\mathbf{F}_f$ ($F_f = \mu_k N_B$) does negative work, $U_{F_f} = -F_f S_{AB}$ (Eq. 14–3); and the normal force $\mathbf{N}_B$ does no work, since it is always perpendicular to the displacement as the block moves from A to B.

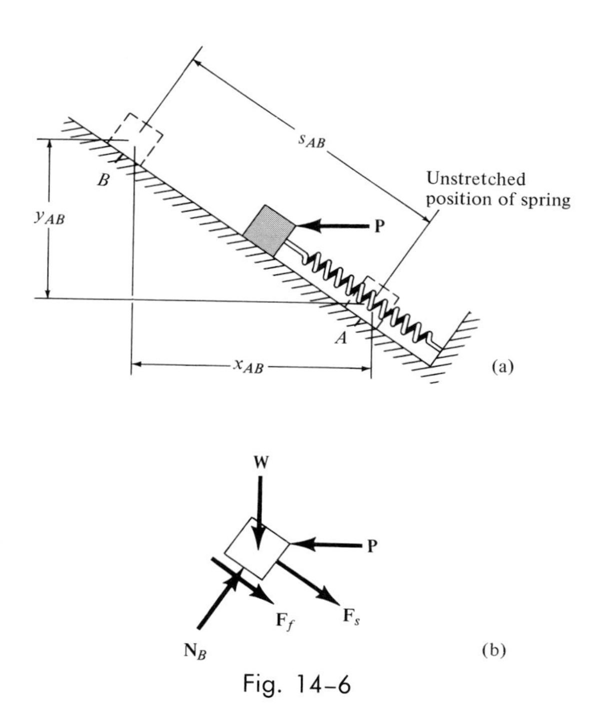

Fig. 14–6

14–2. Principle of Work and Energy

Consider a particle P which at an instant is located a distance s from one end of its path, as shown in Fig. 14–7. The vector $\mathbf{r}$ locates the position of P relative to the inertial frame of reference, x, y, z. At the instant considered, P has an instantaneous velocity $\mathbf{v}$ and it is subjected to a system of external forces, represented by the resultant $\mathbf{F}_R = \Sigma\mathbf{F}$. If $\mathbf{F}_R$, or all the external forces, are resolved into their normal and tangential components, then during an infinitesimal displacement $d\mathbf{s}$, the normal components $\Sigma F_n = \Sigma F \sin\theta$ *do no work,* since they do not displace in the normal direction; instead, only the tangential components of force, $\Sigma F_t = \Sigma F \cos\theta$, do work. As a result, consider writing the equation of motion for the particle in the tangential direction.

$$\Sigma F_t = ma_t; \qquad \Sigma F \cos\theta = ma_t$$

During the displacement $d\mathbf{s}$, $a_t = v\,dv/ds$ (Eq. 12–7). Hence,

$$\Sigma F \cos\theta\, ds = mv\, dv$$

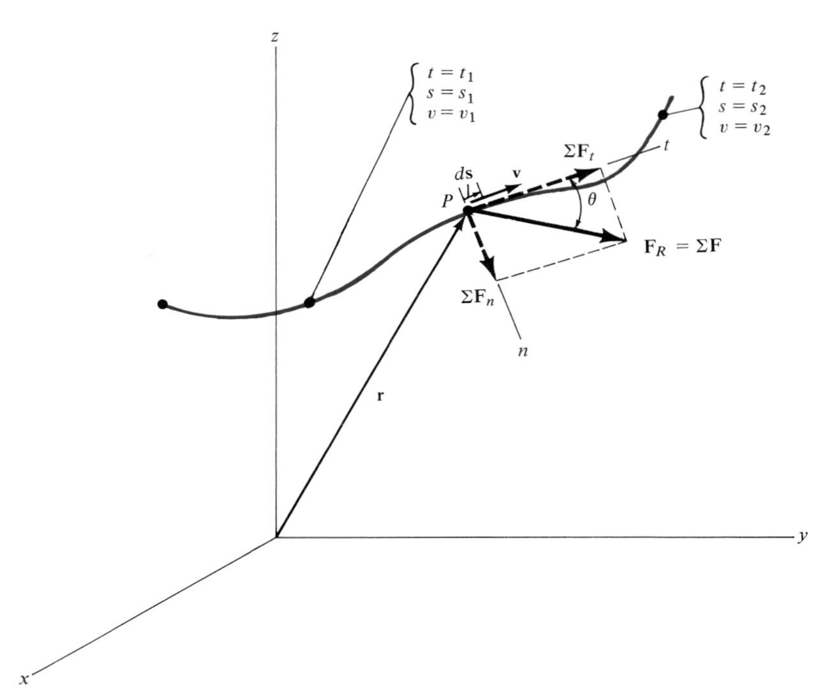

Fig. 14–7

Integrating both sides, assuming that initially the particle has a position $s = s_1$ and a speed $v = v_1$, and later $s = s_2$, $v = v_2$, Fig. 14–7, we have

$$\Sigma \int_{s_1}^{s_2} F \cos \theta \, ds = \int_{v_1}^{v_2} mv \, dv$$

or

$$\Sigma \int_{s_1}^{s_2} F \cos \theta \, ds = \tfrac{1}{2}mv_2^2 - \tfrac{1}{2}mv_1^2$$

Using Eq. 14–2, the final result may be written as

$$\Sigma U_{1-2} = \tfrac{1}{2}mv_2^2 - \tfrac{1}{2}mv_1^2 \tag{14–6}$$

This equation describes the *principle of work and energy* for the particle. The scalar term on the left represents the sum of the work done by *all the forces acting on the particle* as the particle moves from point 1 to point 2. The two terms on the right side, which are of the form $T = \frac{1}{2}mv^2$, define the particle's final and initial *kinetic energy*, respectively. These terms are positive scalar quantities since they do not depend on the direction of the particle's velocity. Because Eq. 14–6 is dimensionally homogeneous, the kinetic energy has the same units as work, e.g., joules (J).

As noted from the derivation, the principle of work and energy represents an integrated form of $\Sigma F_t = ma_t$, acquired by using the kinematic equation $a_t = v\,dv/ds$. Hence, if the particle's initial speed is known, and the work of all the forces acting on the particle can be computed, then Eq. 14–6 provides a *direct means* of obtaining the final speed v_2 of a particle after it undergoes a specified displacement. Notice that if v_2 is determined by means of the equation of motion, a two-step process is necessary; i.e., apply $\Sigma F_t = ma_t$ to obtain a_t, then integrate $a_t = v\,dv/ds$ to obtain v_2.

When applying Eq. 14–6 it is convenient to rewrite it in the form

$$T_1 + \Sigma U_{1-2} = T_2 \tag{14–7}$$

which states that the particle's initial kinetic energy, T_1, plus the work done by all the forces acting on the particle, ΣU_{1-2}, as the particle moves from its initial to its final position, is equal to the particle's final kinetic energy, T_2. It should be noted that Eq. 14–7 is a *scalar equation;* and therefore, only *one unknown* can be obtained by using this equation when it is applied to a single particle. The principle of work and energy cannot be used, for example, to determine forces directed *normal* to the path of motion since these forces do no work on the particle. For curved paths, however, the magnitude of the normal force is a function of velocity. Hence, it is generally easier to obtain the velocity using the principle of work and energy, and then to substitute this quantity into the equation of motion $\Sigma F_n = mv^2/\rho$ to obtain the normal force.

PROCEDURE FOR ANALYSIS

The principle of work and energy is used to solve kinetic problems that involve *velocity, force,* and *displacement,* since these terms are involved in the formulation. For applications, the following two-step procedure should be used:

Step 1: Draw a free-body diagram of the particle when it is located at an intermediate point along its path in order to account for all the forces that do work on the particle.

Step 2: Apply the principle of work and energy, $T_1 + \Sigma U_{1-2} = T_2$. The kinetic energy at the initial and final points is always positive, since it involves the speed squared ($T = \frac{1}{2}mv^2$). For the calculation v must be measured from an inertial reference frame. The work done by each force shown on the free-body diagram is computed by using the appropriate equations developed in Sec. 14–1. Since *algebraic addition* of the work terms is required, it is important that the proper sign of each term be specified. Specifically, work is positive when the force is in the same direction as its displacement, otherwise it is negative.

Numerical application of this two-step procedure is illustrated in the examples following Sec. 14–3.

14–3. Principle of Work and Energy for a System of Particles

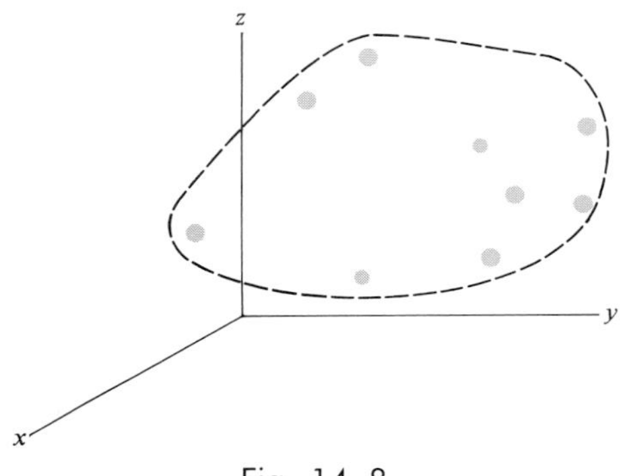

Fig. 14–8

The principle of work and energy may be applied to each particle of an enclosed system of particles, Fig. 14–8, and the results added algebraically. We can then write

$$\Sigma T_1 + \Sigma U_{1-2} = \Sigma T_2 \tag{14–8}$$

Here the system's initial kinetic energy, ΣT_1, plus the total work done by all the external and internal forces acting on the particles of the system, ΣU_{1-2}, is equal to the system's final kinetic energy, ΣT_2. In cases where all the particles of the system are connected by inextensible links or cables, or the particles are contained within a rigid body, the work created by the internal forces is zero. This is because these forces occur in equal but opposite pairs, and each pair of forces is displaced by an equal amount.

The procedure for analysis outlined in the previous section should be followed when applying Eq. 14–8. Since only one equation applies for the entire system, then if the particles are connected, further information can generally be obtained by using the kinematic principles outlined in Sec. 12–9 in order to *relate* the particles' speeds. See Example 14–4.

Example 14-1

The 1500-kg automobile shown in Fig. 14-9a is traveling up the 20° incline at a speed of 12 m/s. If the driver wishes to stop his car in a distance of 10 m, determine the frictional force at the pavement which must be supplied by the rear wheels.

Solution

Step 1: The free-body diagram of the auto is shown in Fig. 14-9b. Here it can be seen that the normal force $\mathbf{N}_A$ does no work since it is never displaced along its line of action. The friction force $\mathbf{F}_A$ is displaced 10 m, and the 1500(9.81)-N force is displaced 10 sin 20° m. Both forces do negative work since they act in the opposite direction to their displacement.

Step 2: Applying the principle of work and energy gives

$$T_1 + \Sigma U_{1-2} = T_2$$

$$\tfrac{1}{2}(1500)(12)^2 - 1500(9.81)(10 \sin 20°) - F_A(10) = 0$$

Solving for F_A yields

$$F_A = 5767 \text{ N} = 5.77 \text{ kN} \qquad \textit{Ans.}$$

In order to compare the solutions, try working the problem using the equation of motion.

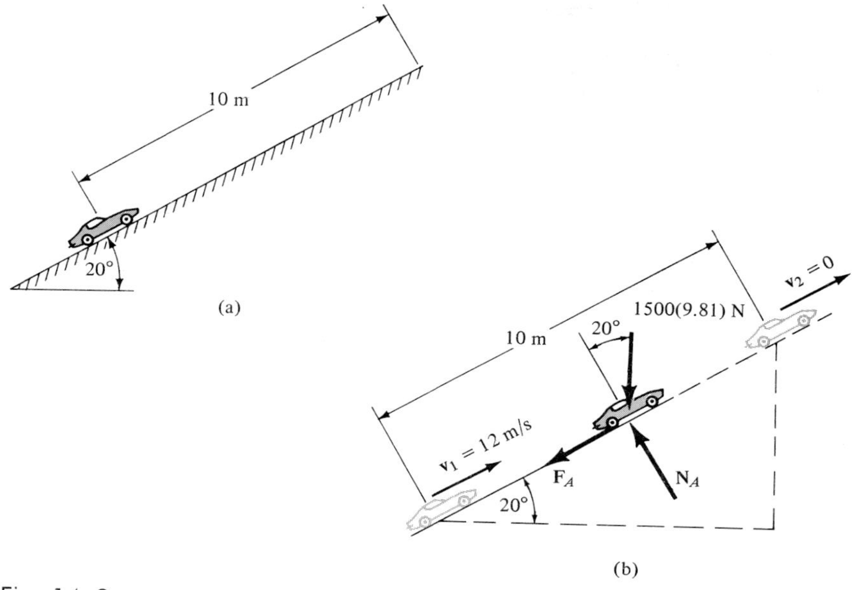

Fig. 14-9

Example 14–2

A 10-kg block rests on a horizontal surface, as shown in Fig. 14–10*a*. The spring, which is not attached to the block, has a stiffness of $k = 500$ N/m and is initially compressed 0.2 m from C to A. After the block is released from rest at A, determine (a) its velocity when it passes point D, and (b) the total distance d it moves before coming to rest. The coefficient of kinetic friction between the block and the plane is $\mu_k = 0.2$.

Solution

Part (a). *Step 1:* Two free-body diagrams for the block are shown in Fig. 14–10*b*. The block moves under the influence of the spring force $F_s = kx$, along the 0.2-m-long path AC, after which it continues to slide along the plane to point D. With reference to either free-body diagram, $\Sigma F_y = 0$; hence, $N_B = 98.1$ N. Only the spring and friction forces do work during the displacement—the spring force does positive work from A to C, whereas the frictional force does negative work from A to D. Why?

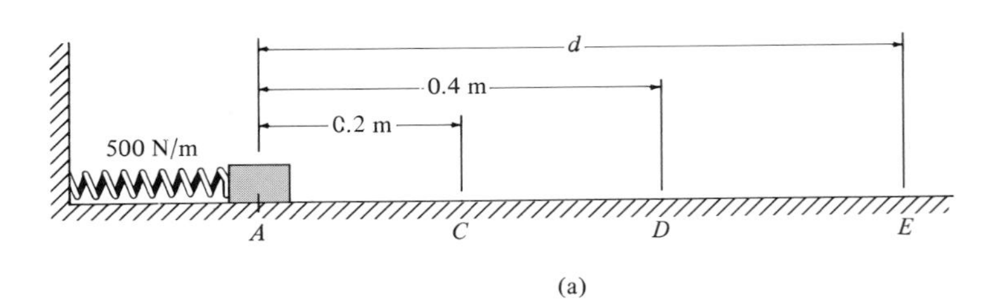

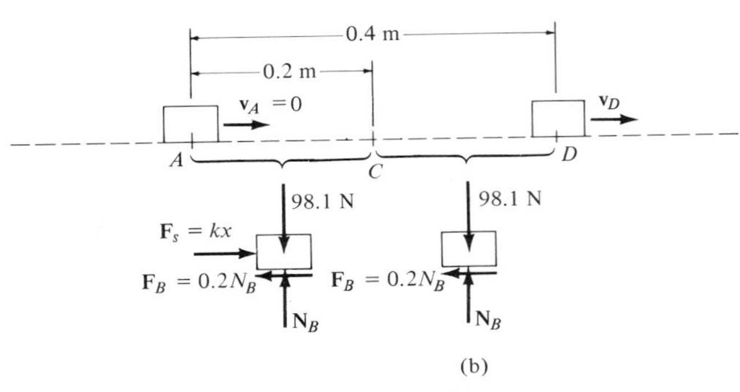

Fig. 14–10(a,b)

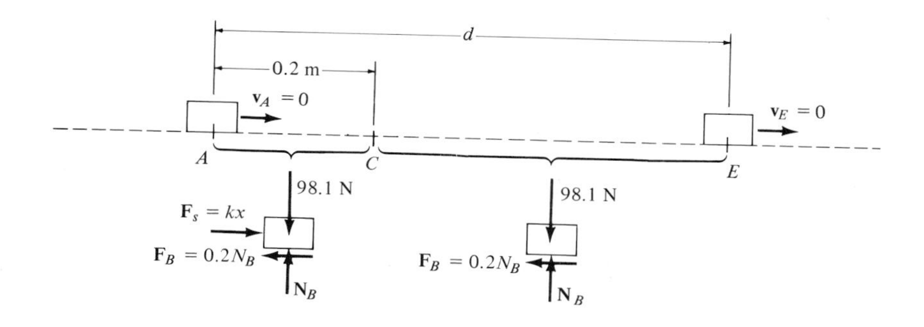

Fig. 14–10(c)

Step 2: Applying the principle of work and energy from A to D yields

$$T_A + \Sigma U_{A-D} = T_D$$

$$\tfrac{1}{2}m(v_A)^2 + \tfrac{1}{2}ks_{AC}^2 - (0.2N_B)s_{AD} = \tfrac{1}{2}m(v_D)^2$$

$$0 + \tfrac{1}{2}(500)(0.2)^2 - 0.2(98.1)(0.4) = \tfrac{1}{2}10(v_D)^2$$

Solving for v_D,

$$v_D = 0.656 \text{ m/s} \qquad \textit{Ans.}$$

Part (b). *Step 1:* When the block moves a distance d from its original position, the velocity $v_E = 0$. The free-body diagrams are shown in Fig. 14–10*c*.

Step 2: Applying the principle of work and energy between points A and E gives

$$T_A + \Sigma U_{A-E} = T_E$$

$$\tfrac{1}{2}m(v_A)^2 + \tfrac{1}{2}ks_{AC}^2 - (0.2N_B)s_{AE} = \tfrac{1}{2}m(v_E)^2$$

$$0 + \tfrac{1}{2}(500)(0.2)^2 - 0.2(98.1)d = 0$$

Thus,

$$d = 0.510 \text{ m} \qquad \textit{Ans.}$$

Example 14-3

A block having a mass of 2 kg is given an initial velocity of $v_1 = 1$ m/s when it is at the top surface of the smooth cylinder shown in Fig. 14–11*a*. If the block moves along a path of 0.5-m radius, determine the angle $\theta = \theta_{max}$ at which it begins to leave the cylinder's surface.

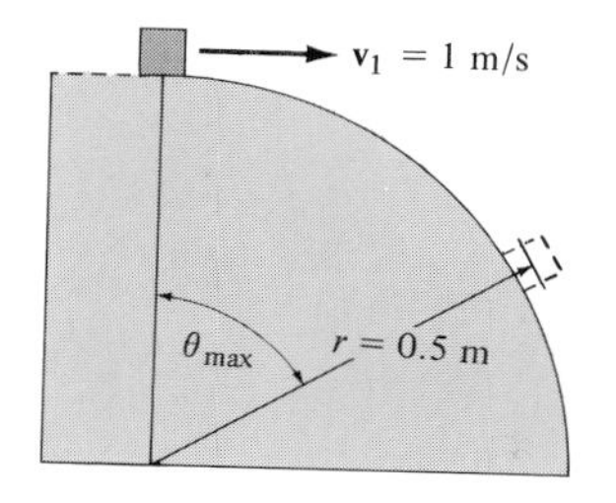

Fig. 14–11(a)

Solution

Step 1: The free-body diagram for the block is shown in Fig. 14–11*b*. Only the weight $W = 2(9.81) = 19.62$ N does work during the displacement. If the block is assumed to leave the surface when $\theta = \theta_{max}$ ($\mathbf{N}_B = \mathbf{0}$), then the weight moves through a vertical displacement of $0.5(1 - \cos\theta_{max})$ m, as shown in the figure.

Step 2: Applying the principle of work and energy yields

$$T_1 + \Sigma U_{1-2} = T_2$$

$$\tfrac{1}{2}(2)(1)^2 + 19.62(0.5)(1 - \cos\theta_{max}) = \tfrac{1}{2}(2)v_2^2$$

or

$$v_2^2 = 9.81(1 - \cos\theta_{max}) + 1 \qquad (1)$$

There are two unknowns in this equation, θ_{max} and v_2. A second equation relating these two variables may be obtained by applying the equation of motion in the *normal direction* to the forces acting on the free-body diagram. The kinetic diagram is shown in Fig. 14–11*c*. Thus,

$$+\swarrow \Sigma F_n = ma_n; \qquad -N_B + 19.62\cos\theta = 2\left(\frac{v^2}{0.5}\right)$$

When the block leaves the surface of the cylinder at $\theta = \theta_{max}$, $\mathbf{N}_B = \mathbf{0}$ and $v = v_2$; hence,

$$\cos\theta_{max} = \frac{v_2^2}{4.905} \qquad (2)$$

Eliminating the unknown v_2^2 between Eqs. (1) and (2) gives

$$4.905\cos\theta_{max} = 9.81(1 - \cos\theta_{max}) + 1$$

Solving for $\cos\theta_{max}$, we have

$$\cos\theta_{max} = 0.735$$

Thus,

$$\theta_{max} = 42.7° \qquad \textit{Ans.}$$

This problem has also been solved in Example 13–13. If the two methods of solution are compared it will be apparent that a work-energy approach yields a more direct solution.

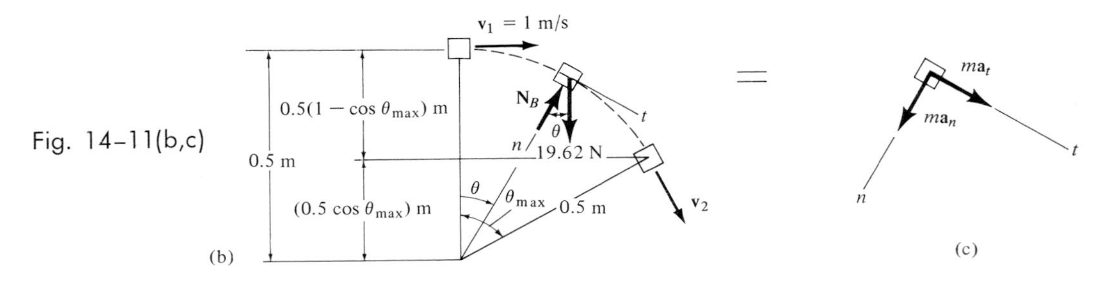

Fig. 14–11(b,c)

Example 14-4

The blocks A and B shown in Fig. 14–12a have a mass of 10 and 100 kg, respectively. If they are released from rest, determine the distance A travels at the instant the speed of B becomes 2 m/s.

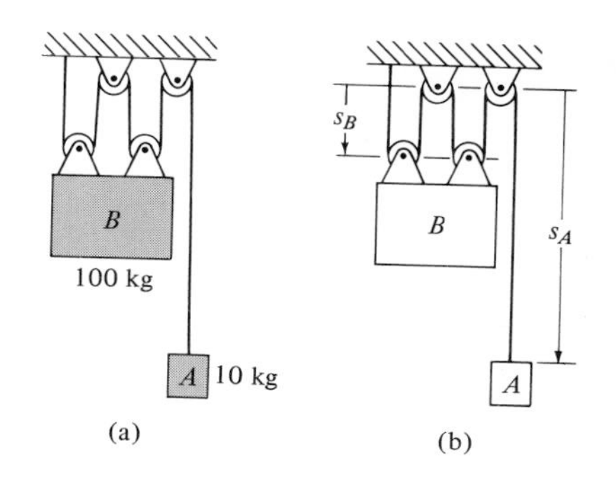

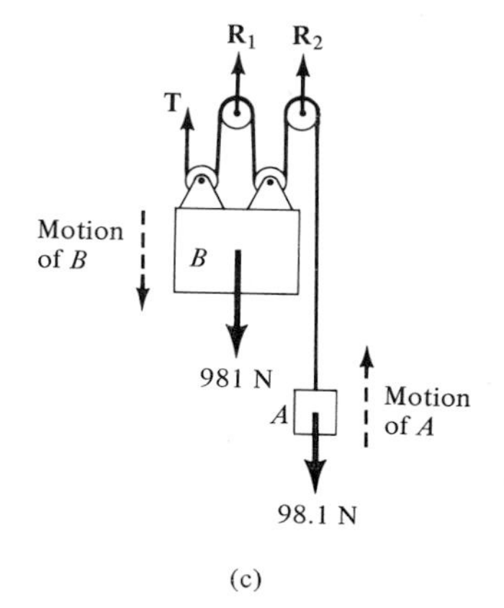

Fig. 14–12

Solution

This problem may be solved by considering blocks A and B separately and applying the principle of work and energy to each block. However, since the blocks are connected by an inextensible cable, blocks A and B will be considered together as a system.

If a free-body diagram of B were drawn and $\Sigma F_y = 0$ applied, it would be found that a cable tension of $\frac{1}{4}(100)(9.81) = 245.3$ N is required for *equilibrium*. Since A weighs $10(9.81) = 98.1$ N, which requires a cable tension of only 98.1 N for *equilibrium*, B will begin to accelerate downward while A accelerates *upward*. Using the methods of kinematics discussed in Sec. 12–9, it may be seen from Fig. 14–12b that at any given instant the total length l of all the vertical segments of cable may be expressed in terms of the coordinates s_A and s_B as

$$s_A + 4s_B = l$$

Hence, a displacement Δs_B (downward) of block B will cause a corresponding displacement $\Delta s_A = -4\Delta s_B$ (upward), i.e.,

$$|\Delta s_A| = |4\Delta s_B| \tag{1}$$

Taking the time derivative yields

$$|v_A| = |4v_B| \tag{2}$$

Step 1: Since the blocks start from rest, their initial velocities are zero. From Eq. (2), the final velocities are $(v_B)_2 = 2$ m/s (downward) and $(v_A)_2 = 8$ m/s (upward). The cable force $\mathbf{T}$ and reactions $\mathbf{R}_1$ and $\mathbf{R}_2$, shown on the free-body diagram, Fig. 14–12c, do *no work* since these forces represent the reactions at the supports and consequently do not move while the blocks are being displaced.

Step 2: Applying the principle of work and energy, using Eq. (1), yields

$$\Sigma T_1 + \Sigma U_{1-2} = \Sigma T_2$$

$$\tfrac{1}{2}m_A(v_A)_1^2 + \tfrac{1}{2}m_B(v_B)_1^2 - W_A\Delta s_A + W_B\Delta s_B = \tfrac{1}{2}m_A(v_A)_2^2 + \tfrac{1}{2}m_B(v_B)_2^2$$

$$0 + 0 - 98.1\,\Delta s_A + 981(\tfrac{1}{4}\Delta s_A) = \tfrac{1}{2}(10)(8)^2 + \tfrac{1}{2}(100)(2)^2$$

Solving for Δs_A gives

$$\Delta s_A = 3.53 \text{ m} \qquad \textit{Ans.}$$

Problems

Except when stated otherwise, throughout this chapter assume that the coefficients of static and kinetic friction are equal; i.e., $\mu = \mu_s = \mu_k$.

14–1. Solve Prob. 13–5 using the principle of work and energy.

14–2. Solve Prob. 13–6 using the principle of work and energy.

14–3. Solve Prob. 13–8 using the principle of work and energy.

* **14–4.** A car having a mass of 2 Mg strikes a smooth, rigid sign post with an initial speed of 40 km/h. To stop the car, the front end horizontally deforms 0.25 m. If the car is free to roll during the collision, determine the *average* horizontal collision force causing the deformation.

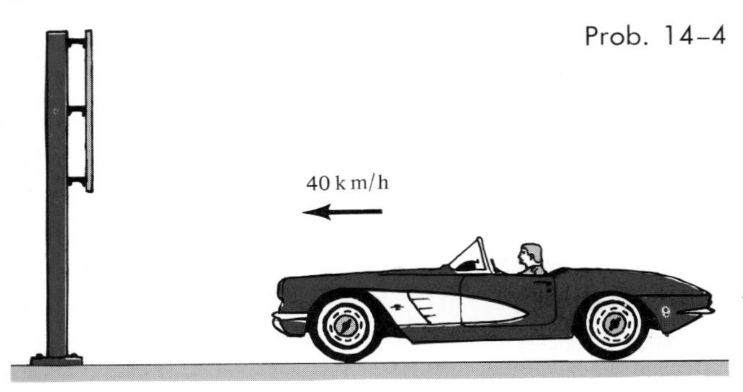

Prob. 14–4

14–5. The crate, having a mass of $m_c = 100$ kg, is subjected to the action of two forces, $F_1 = 800$ N and $F_2 = 1.5$ kN, as shown. If it is originally at rest, determine the distance it slides in order to attain a speed of $v_c = 6$ m/s. The coefficient of friction between the crate and the surface is $\mu = 0.2$.

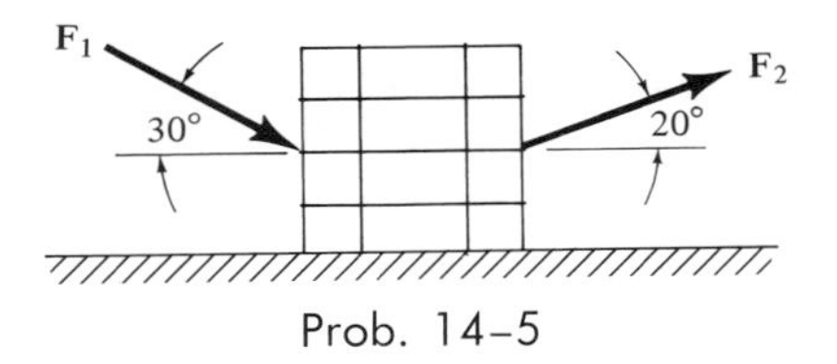

Prob. 14–5

14–5a. Solve Prob. 14–5 if the crate has a weight of $W_c = 50$ lb, and $F_1 = 40$ lb, $F_2 = 90$ lb, $v_c = 18$ ft/s, $\mu = 0.3$.

14–6. A freight car having a mass of 12 Mg is towed along a horizontal track. If the car starts from rest and attains a speed of 15 m/s after traveling a distance of 100 m, determine the constant horizontal towing force **T** applied to the car in this distance. Neglect friction and the mass of the wheels.

14–7. If a crate, having a mass of 50 kg, is released from rest at A, determine its speed when it has traveled 10 m down the plane. The coefficient of friction between the crate and plane is $\mu = 0.3$.

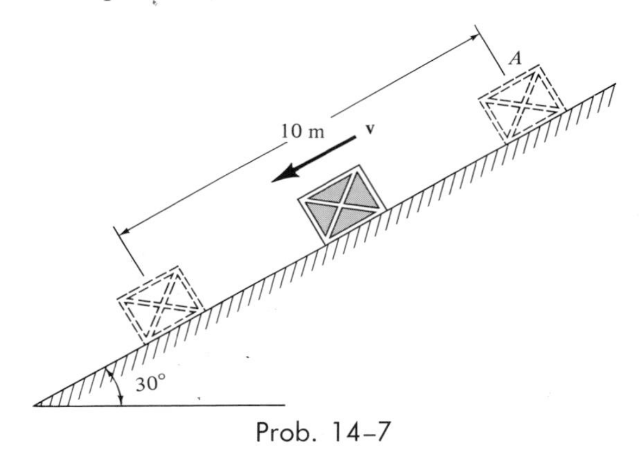

Prob. 14–7

* **14–8.** Coins are placed in a small container C. Using an elastic spring gun, a clerk "fires" the container along the smooth wire from A to a cashier located at B. If the spring in the gun has a stiffness of $k = 3$ kN/m and the spring is compressed 50 mm when the gun is fired, determine the greatest mass of coins which can be placed in the container and still allow it to reach the cashier. Neglect the effect of friction and assume the spring

becomes completely unstretched when the gun is fired. The empty container has a mass of 100 g.

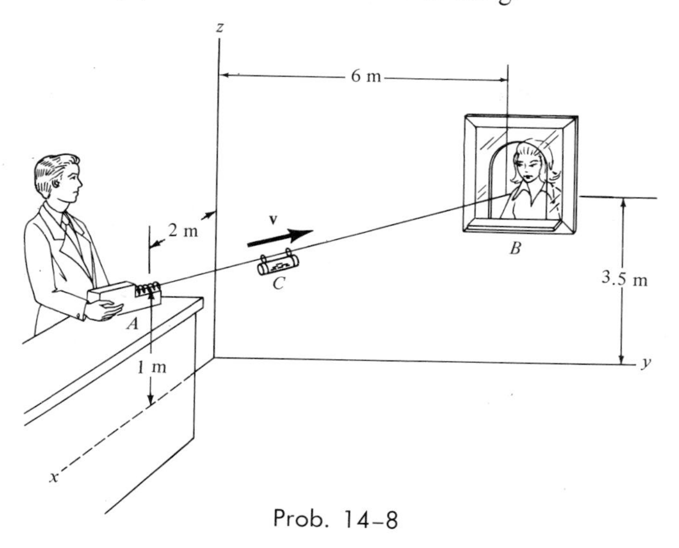

Prob. 14–8

14–9. If the cashier in Prob. 14–8 releases the *empty* 100-g container from rest at B, determine the speed at which it strikes the gun at A.

14–10. The block having a mass of $m = 1.5$ kg slides along a smooth plane and strikes a *nonlinear spring* with a speed of $v = 4$ m/s. The spring is termed "nonlinear," since it has a resistance of $F_s = kx^2$, where $k = 900$ N/m². Determine the speed of the block after it has compressed the spring $x = 0.2$ m.

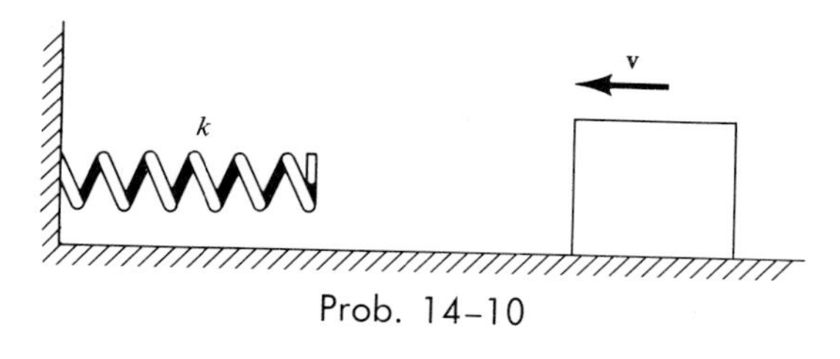

Prob. 14–10

14–10a. Solve Prob. 14–10 if the block weighs $W = 10$ lb, and $v = 6$ ft/s, $k = 8000$ lb/ft², $x = 0.1$ ft.

14–11. A car is equipped with a bumper B designed to absorb collisions. The bumper is mounted to the car by means of inexpensive pieces of flexible tubing T. Upon collision with a rigid barrier A, a constant horizontal force $\mathbf{F}$ is developed which causes a car deceleration of $3g = 29.43$ m/s² (the highest safe deceleration for a passenger without a seatbelt). If the car and passenger have a total mass of 1.5 Mg and the car is initially coasting with a speed of 2 m/s, compute the magnitude of $\mathbf{F}$ needed to stop the car and the deformation x of the bumper tubing.

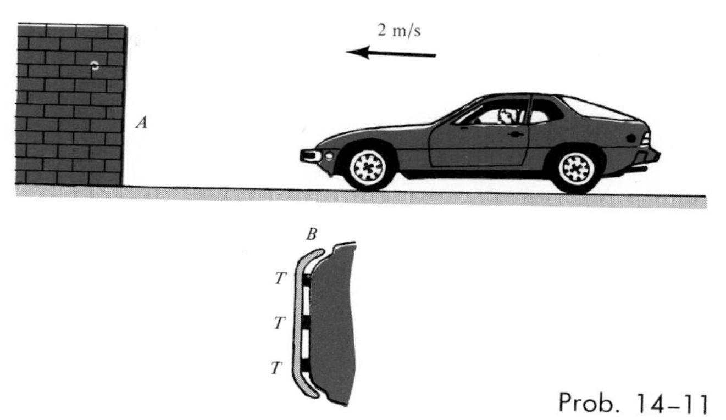

Prob. 14–11

***14–12.** When a 7-kg projectile is fired from a cannon barrel that has a length of 2 m, the explosive force exerted on the projectile, while it is in the barrel, varies in the manner shown. Determine the muzzle velocity of the projectile at the instant it leaves the barrel. Neglect the effects of friction inside the barrel and assume that the barrel is horizontal.

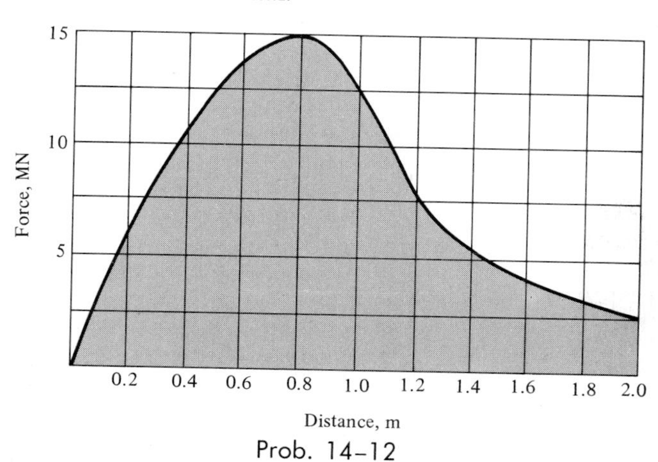

Prob. 14–12

14–13. The coefficient of friction between the 4-kg block and the surface is $\mu = 0.2$. The block is acted upon by a horizontal force of $P = 30$ N and has a speed of 5 m/s when it is at point A. Determine the maximum deformation of the outer spring B at the instant the block comes to rest. Spring B has a stiffness of $k_B = 2$ kN/m and the "nested" spring C has a stiffness of $k_C = 6$ kN/m.

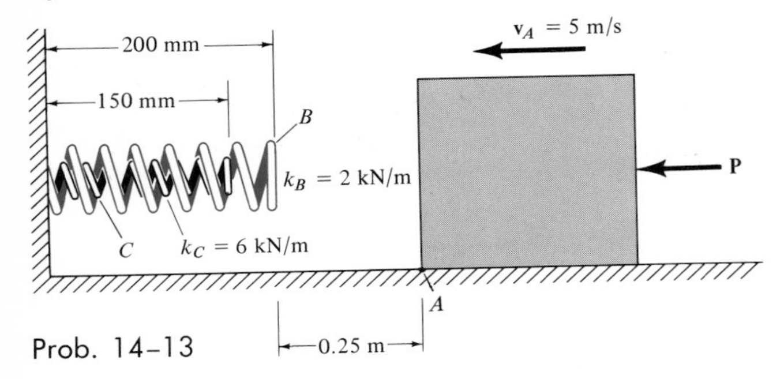

Prob. 14–13

14–14. If the 4-kg block in Prob. 14–13 is pressed against the two springs so that it is 100 mm from the wall and then released from rest, with **P** = **0**, determine how far from the wall the block slides before coming to rest. The coefficient of friction is $\mu = 0.2$.

14–15. The block has a mass of $m = 0.8$ kg and moves within the smooth vertical slot. If it starts from rest when the *attached* spring is in the unstretched position at A, determine the *constant* vertical force **F** which must be applied to the cord so that the block attains a speed of $v_B = 2.5$ m/s when it reaches B. Set $d = 0.3$ m, $s = 0.4$ m, $s_B = 0.15$ m, and $k = 100$ N/m.

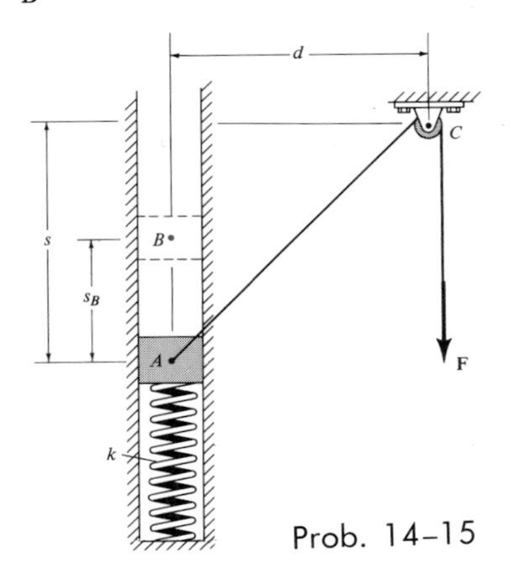

Prob. 14–15

14–15a. Solve Prob. 14–15 if the block weighs $W = 2.5$ lb, and $v_B = 2$ ft/s, $d = 1.5$ ft, $s = 2$ ft, $s_B = 1.25$ ft, $k = 20$ lb/ft.

* **14–16.** A car, assumed to be rigid and having a mass of 800 kg, strikes a barrel-barrier installation without the driver applying the brakes. From experiments, the magnitude of the force of resistance $\mathbf{F}_r$, created by deforming the barrels successively, is shown as a function of vehicle penetration. If the car strikes the barrier traveling at $v_c = 60$ km/h, determine approximately the distance s to which the car penetrates the barrier.

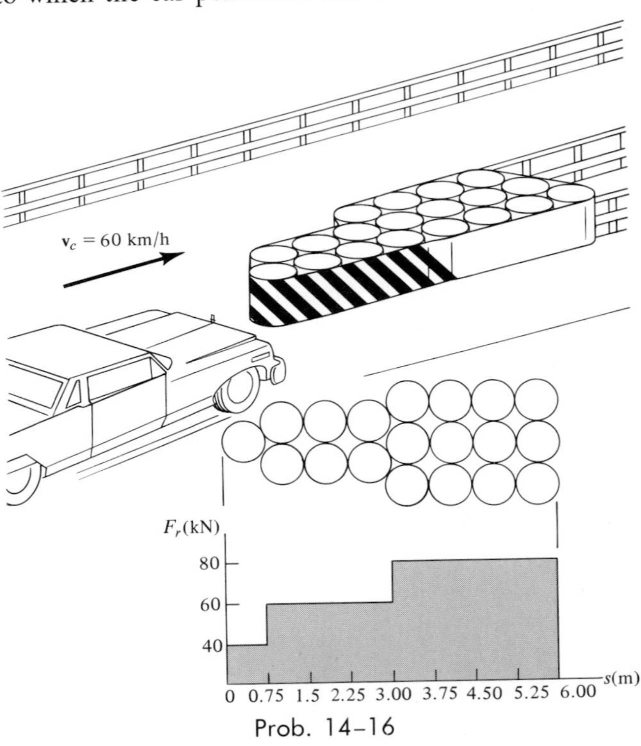

Prob. 14–16

14–17. The catapulting mechanism is used to propel the 10-kg slider A to the right along the smooth track. The propelling action is obtained by drawing the pulley attached to rod BC rapidly to the left by means of a piston P. If the piston applies a constant force of $F = 20$ kN on BC such that it moves the rod 0.2 m, determine the speed attained by the slider if it was originally at rest. Neglect the mass of the pulleys, cable, the piston, and rod BC.

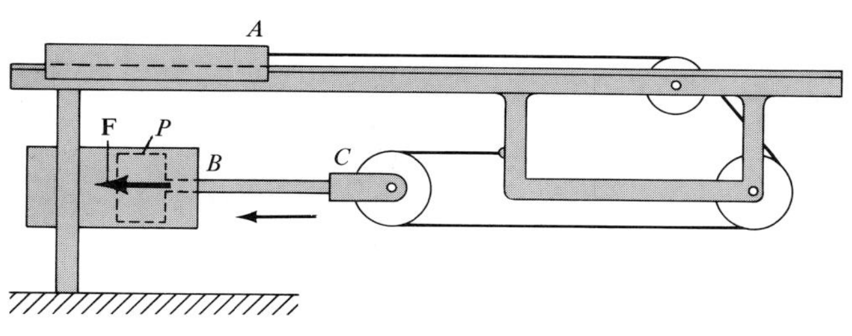

Prob. 14–17

14–18. The block has a mass of 1.5 kg and is traveling with a speed of $v_A = 3$ m/s when it reaches point A. If the spring stiffness is $k = 900$ N/m, determine the maximum compression in the spring at the instant the block stops. Neglect the mass of the spring.

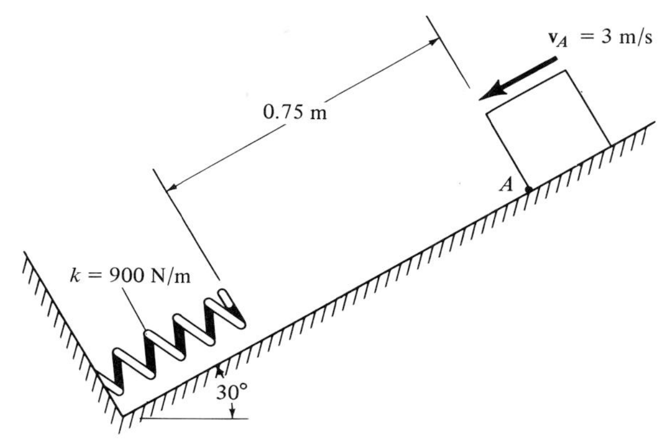

Prob. 14–18

14–19. Rework Prob. 14–18 assuming that the coefficient of friction between the block and plane is $\mu = 0.25$.

* **14–20.** The "flying car" is a ride at an amusement park, which consists of a car having wheels which roll along a track mounted on a rotating drum. Motion of the car is created by applying the car's brake, thereby gripping the car to the track and allowing it to move with a speed of $v_t = 2$ m/s. If the rider applies the brake when going from B to A and then releases it at the top of the drum, A, so that the car coasts freely down along the track to B ($\theta = \pi$ rad), determine the speed of the car at B and the normal reaction which the track exerts on the car at B. Neglect friction during the motion from A to B. The rider and car have a total mass of $m = 200$ kg and the center of mass of the car and rider moves along a circular path of radius $r = 8$ m.

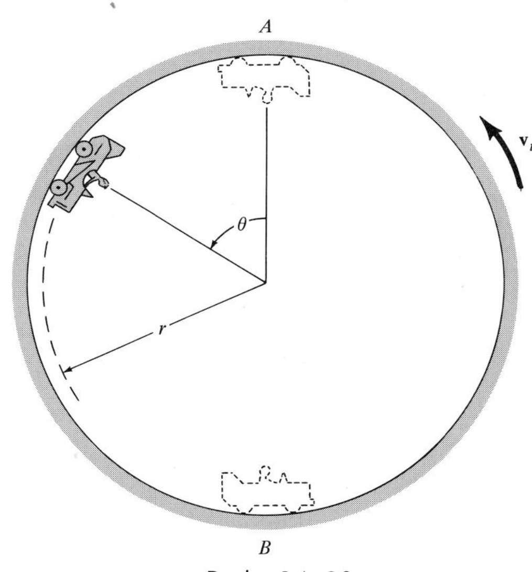

Prob. 14–20

* **14–20a.** Solve Prob. 14–20 if $v_t = 3$ ft/s, the total weight of the car and rider is $W = 200$ lb, and $r = 8$ ft.

14–21. A rocket of mass m is fired vertically from the surface of the earth, i.e., at $r = r_1$. Assuming that no mass is lost as it travels upward, determine the work it must do against gravity to reach a distance r_2. The force of gravity is $F = GM_e m/r^2$ (Eq. 13–2), where M_e is the mass of the earth and r the distance between the rocket and the center of the earth.

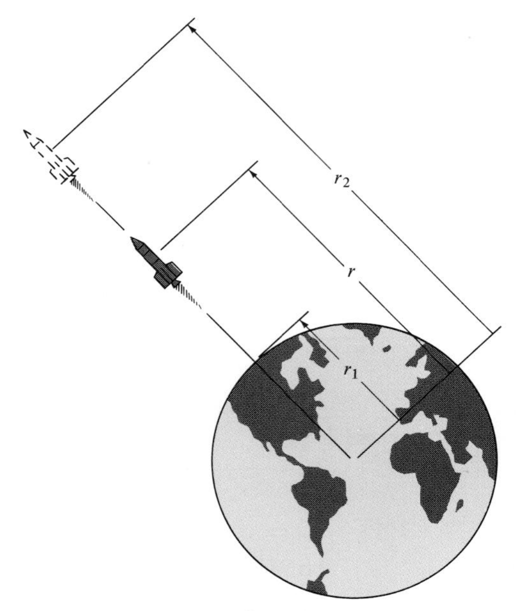

Prob. 14–21

14–4. Power and Efficiency

Power. *Power* is defined as the amount of work performed per unit of time. Hence, the *average power* generated by a machine or engine that performs an amount of work ΔU within the time interval Δt is

$$P_{\text{avg}} = \frac{\Delta U}{\Delta t} \tag{14–9}$$

If the time $\Delta t \to dt$, and consequently $\Delta U \to dU$, then the *instantaneous power* is defined as

$$P = \frac{dU}{dt} \tag{14–10}$$

Provided the work dU is expressed by $dU = \mathbf{F} \cdot d\mathbf{s}$, then it is also possible to write

$$P = \frac{dU}{dt} = \frac{\mathbf{F} \cdot d\mathbf{s}}{dt}$$

or

$$P = \mathbf{F} \cdot \mathbf{v} \qquad (14\text{–}11)$$

Hence, power is a *scalar,* where in the formulation, **v** represents the instantaneous velocity of the object which is acted upon by the unbalanced force **F**. The basic unit used to measure power is the watt (W), which is equivalent to 1 joule of work completed per second, i.e., $1 \text{ W} = 1 \text{ J/s} = 1 \text{ N} \cdot \text{m/s}$.

The term "power" provides a useful basis for determining the type of motor or machine which is required to do a certain amount of work in a given time. For example, two pumps may each be able to empty a reservoir if given enough time; however, the pump having the larger power will complete the job sooner.

Efficiency. The *mechanical efficiency* of a machine is defined as the ratio of the output of useful power created by the machine to the input of power supplied to the machine. Hence,

$$e = \frac{\text{power output}}{\text{power input}} \qquad (14\text{–}12)$$

If work is being done by the machine at a *constant rate,* then the efficiency may be expressed in terms of the ratio of output energy to input energy, i.e.,

$$e = \frac{\text{energy output}}{\text{energy input}} \qquad (14\text{–}13)$$

Since machines consist of a series of moving parts, frictional forces will always be developed within the machine, and as a result, extra work or power is needed to overcome these forces. Consequently, *the efficiency of a machine is always less than 1.* In all machines the transformation of mechanical energy into thermal energy owing to frictional forces is unavoidable; however, in some situations this is desirable. For example, the kinetic energy of a moving car is dissipated into thermal energy utilizing the frictional forces developed by the brakes.

PROCEDURE FOR ANALYSIS

When computing the power supplied to a body, one must first determine the unbalanced external force **F** acting on the body which causes the motion. This force is usually developed by a machine or engine placed either within or external to the body. If the body is accelerating, it may be necessary to draw the proper free-body and kinetic diagrams and apply the equation of motion ($\Sigma \mathbf{F} = m\mathbf{a}$) to determine **F**. Once **F** has been computed, the power is determined by multiplying the force magnitude by the instantaneous velocity at which **F** is moving along its line of action, i.e., $P = \mathbf{F} \cdot \mathbf{v} = Fv \cos \theta$.

In some problems the power may also be computed by calculating the work done per unit of time ($P_{avg} = \Delta U/\Delta t$, or $P = dU/dt$). Depending upon the problem, this work is done either by the external or internal force of a machine or engine, by the weight of the body, or by an elastic spring force acting on the body.

Example 14-5

The motor M of the hoist shown in Fig. 14–13a operates with an efficiency of 0.85. Determine the power of the motor which is required to lift the crate C having a mass of 50 kg, if the cable is being drawn in with an acceleration of 2 m/s² and at the instant shown the cable speed is 5 m/s. Neglect the mass of the pulley and cable.

Solution

In order to compute the power supplied by the motor to the cable, it is first necessary to determine the tension force **T** in the cable. Since the crate is subjected to an acceleration, this requires application of the equation of motion.

The free-body and kinetic diagrams for the crate are shown in Fig. 14–13b. Hence,

$$+\uparrow \Sigma F_y = ma_y; \qquad 2T - 50(9.81) = 50a_C \qquad (1)$$

The acceleration of the crate can be obtained by using kinematics to relate the motion of the crate to the known motion of a point P located on the cable, Fig. 14–13a. Hence, by the methods of Sec. 12–9, the coordinates s_C and s_P in Fig. 14–13a can be related to a constant portion of cable length l which is changing in the vertical and horizontal directions. We have

$$2s_C + s_P = l$$

Taking the second time derivative of this equation yields

$$2a_C = -a_P$$

From the problem data, $a_P = 2\ \text{m/s}^2$; thus, $a_C = (-2\ \text{m/s}^2)/2 = -1\ \text{m/s}^2$. Substituting this result into Eq. (1), *neglecting* the negative sign since it indicates that the acceleration is upward, in accordance with the direction of $m\mathbf{a}_C$ in Fig. 14–13b, we have

$$2T - 50(9.81) = 50(1)$$
$$T = 270.2\ \text{N}$$

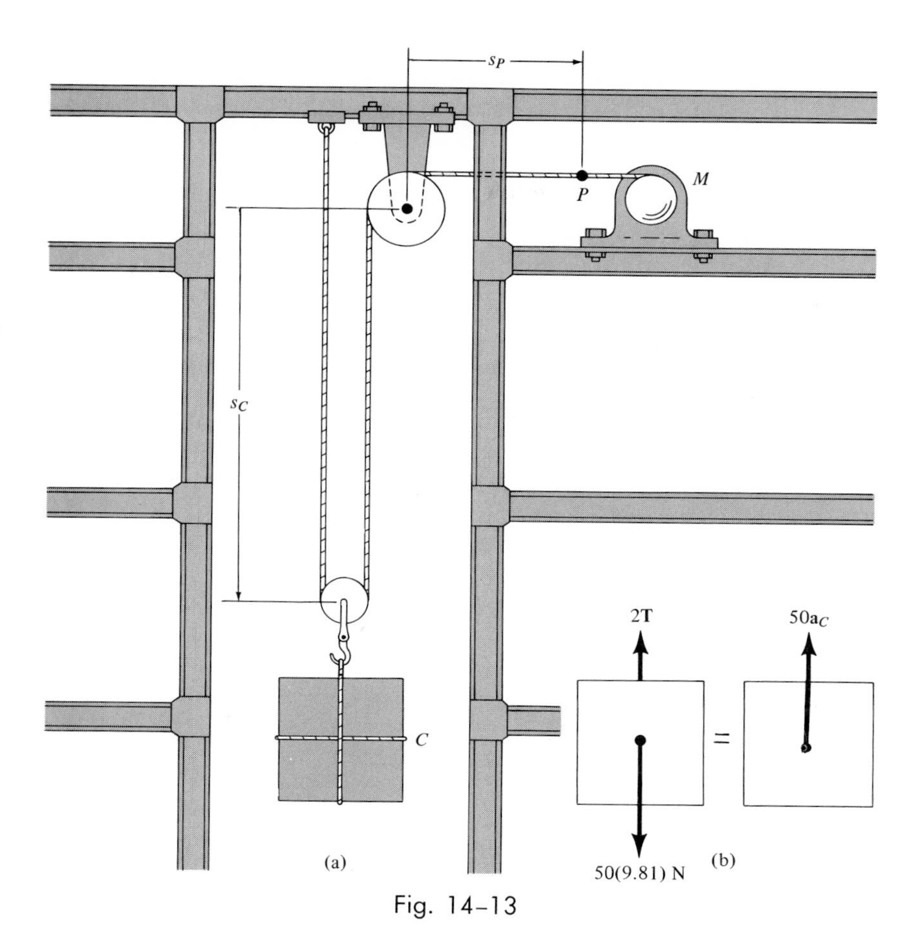

Fig. 14–13

The power of the motor required to draw the cable in at a rate of 5 m/s is, therefore,

$$P = \mathbf{T} \cdot \mathbf{v} = (270.2 \text{ N})(5 \text{ m/s}) = 1351 \text{ W} = 1.35 \text{ kW}$$

This *power output* requires that the motor provide a *power input* of

$$\text{Power input} = \frac{1}{e}(\text{power output})$$

$$= \left(\frac{1}{0.85}\right)1.35 = 1.59 \text{ kW} \qquad \textit{Ans.}$$

Since the velocity of the crate is constantly changing, notice that this power requirement is *instantaneous*.

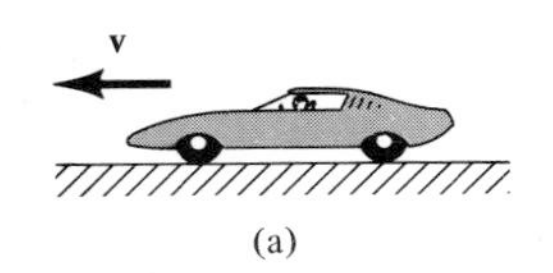

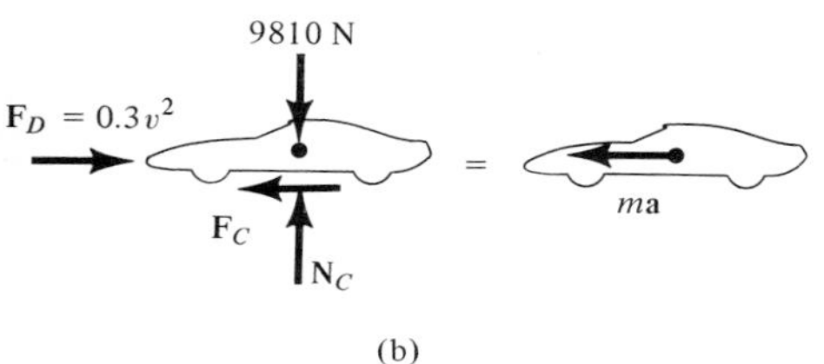

Fig. 14–14

Example 14-6

The car shown in Fig. 14–14*a* has a mass of 1000 kg and an engine running efficiency of $e = 0.63$. As it moves forward, the wind creates a drag resistance on the car of $F_D = 0.3v^2$ N, where v is the instantaneous velocity in m/s. Assuming that the engine supplies power to *all the wheels,* determine the maximum power that can be supplied by the engine. The coefficient of friction between the wheels and the pavement is $\mu = 0.25$.

Solution

Maximum power is attained when the car reaches its maximum velocity. To determine this velocity it is necessary to apply the equation of motion. (Notice that if the principle of work and energy were applied, it would involve the *unknown displacement* of the car, which is of no consequence to the solution of the problem.)

The free-body and kinetic diagrams for the car are shown in Fig. 14–14*b*. The normal force $\mathbf{N}_C$ and frictional force $\mathbf{F}_C$, shown on the free-body diagram, represent the *resultant forces* of all four wheels. In particular, the unbalanced frictional force drives or pushes the car *forward.* This effect is, of course, created by the rotating motion of the wheels on the pavement.

Applying the equations of motion, we have

$$\overset{+}{\rightarrow}\Sigma F_x = ma_x; \qquad -F_C + 0.3v^2 = -1000\frac{dv}{dt} \tag{1}$$

$$+\uparrow\Sigma F_y = ma_y; \qquad N_C - 9810 = 0 \qquad N_C = 9810 \text{ N}$$

At *maximum power* the drag resistance from the wind will *balance* the maximum frictional force which can be developed at the wheels, thereby preventing the car from accelerating further. When this happens the car is in *equilibrium* since it moves with *constant velocity,* i.e., $dv/dt = 0$. The frictional force reaches its maximum value when $F_C = 0.25N_C = 0.25(9810) = 2452.5$ N. Hence, Eq. (1) becomes

$$-2452.5 + 0.3\,v^2 = 0$$

Solving for v, we get

$$v = 90.4 \text{ m/s}$$

The power output of the car is created by the driving (frictional) force $\mathbf{F}_C$. Thus,

$$P = \mathbf{F}_C \cdot \mathbf{v} = (2452.5)90.4 = 221.7 \text{ kW}$$

The power supplied by the engine (power input) is therefore

$$\text{power input} = \frac{1}{e}(\text{power output}) = \frac{1}{0.63}(221.7) = 351.9 \text{ kW} \qquad \textit{Ans.}$$

14–22. An electrically powered train car draws 300 kW of power. If the car has a mass of 18 Mg and travels along a horizontal track, determine the speed it attains in 20 s starting from rest. The mechanical efficiency is $e = 0.8$.

14–23. An electric train car, having a mass of 30 Mg, travels up a 10° incline with a constant speed of 60 km/h. Determine the power required to overcome the force of gravity.

* **14–24.** A spring having a stiffness of 6 kN/m is compressed 600 mm. The stored energy in the spring is used to drive a machine which requires 50 W of power. Determine how long the spring can supply energy at the required rate.

14–25. An automobile having a mass of $m = 2$ Mg travels up a 7° slope at a constant speed of $v = 100$ km/h. If friction and wind resistance are to be neglected, determine the power developed by the engine if the automobile has a mechanical efficiency of $e = 0.65$.

14–25a. Solve Prob. 14–25 if the automobile has a weight of $W = 4300$ lb, $v = 80$ ft/s, and $e = 0.52$. Express the answer in units of *horsepower* (hp), where 1 hp = 550 ft · lb/s.

14–26. A motor hoists a crate that has a mass of 60 kg to a height of $h = 5$ m in 2 s. If the indicated power of the motor is 3.2 kW, determine the motor's efficiency. The crate is hoisted at a constant speed.

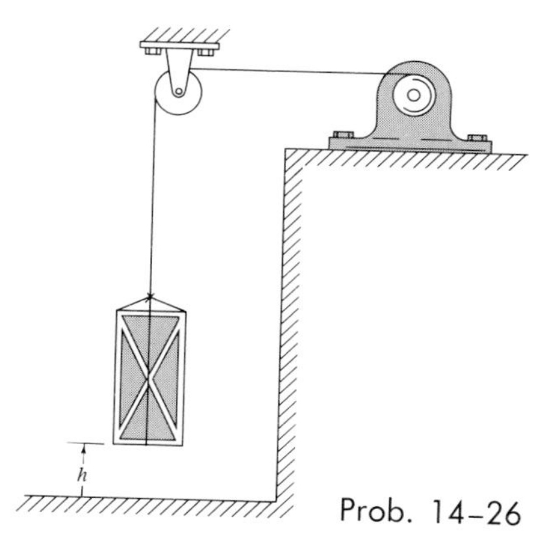

Prob. 14–26

14–27. A truck has a mass of 12 Mg and an engine which transmits a power of 260 kW to *all* the wheels. Assuming that the wheels do not slip on the ground, determine the angle θ of the largest incline the truck can climb at a constant speed of $v = 8$ m/s.

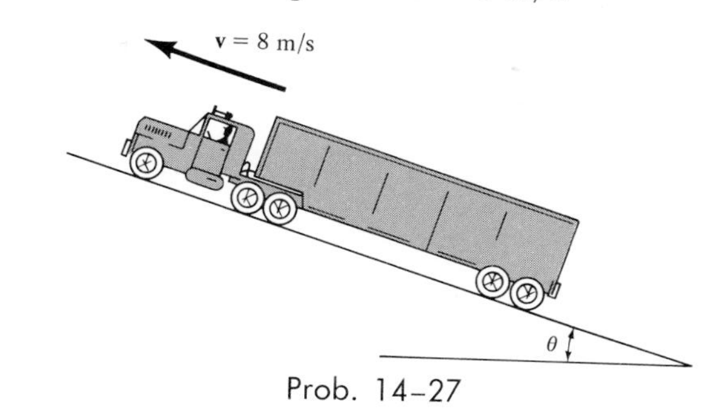

Prob. 14–27

* **14–28.** The escalator steps move with a constant speed of 0.6 m/s. If the steps are 125 mm high and 250 mm in length, determine the horsepower of a motor needed to lift an average mass of 150 kg per step. There are 32 steps.

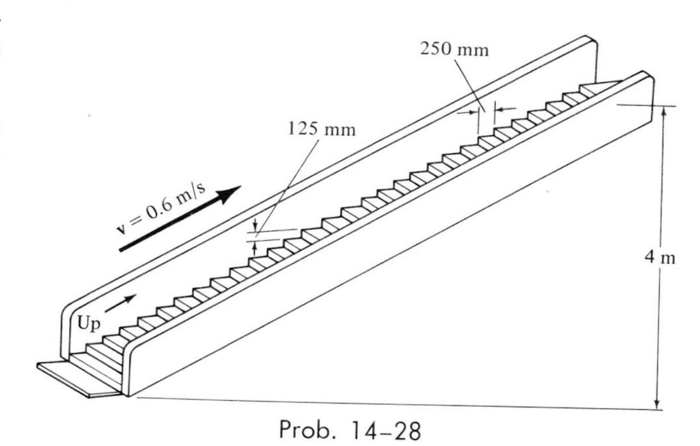

Prob. 14–28

14–29. If the escalator in Prob. 14–28 is *not moving*, determine the constant speed at which a man having a mass of 80 kg must walk up the steps to generate 100 W of power—the same amount that is needed to power a standard light bulb.

14-30. The crate, having a mass of $m_c = 50$ kg, is hoisted up the 30° incline by the pulley system and motor M. If the crate starts from rest and, by constant acceleration, attains a speed of $v_c = 4$ m/s after traveling $s_c = 8$ m along the plane, determine the power that must be supplied to the motor at the instant the crate has moved $s_c = 8$ m. Neglect friction along the plane. The motor has an efficiency of $e = 0.74$.

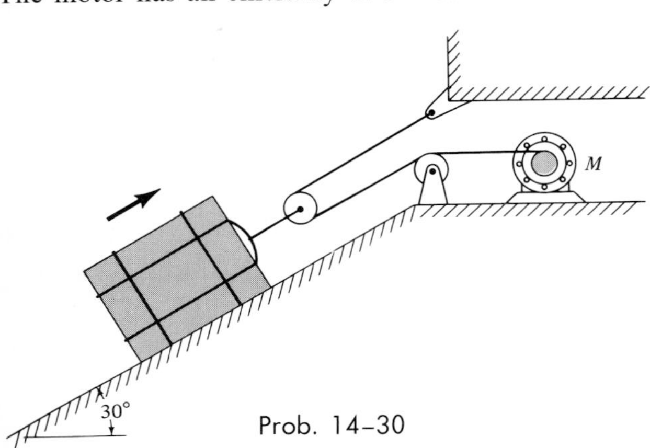

Prob. 14–30

14-30a. Solve Prob. 14–30 if the crate has a weight of $W_c = 75$ lb, and $v_c = 5$ ft/s, $s_c = 10$ ft, $e = 0.70$. Express the answer in units of *horsepower* (hp), where 1 hp = 550 ft · lb/s.

14-31. Solve Prob. 14–30 ($e = 0.74$) if the coefficient of friction between the plane and the crate is $\mu = 0.30$.

* **14-32.** The elevator E and its freight have a total mass of 300 kg. Hoisting is provided by the motor M and the 40-kg block C. If the motor has an efficiency of $e = 0.6$, determine the power that must be supplied to the motor when the elevator is hoisted upward at a constant speed of $v_E = 3$ m/s.

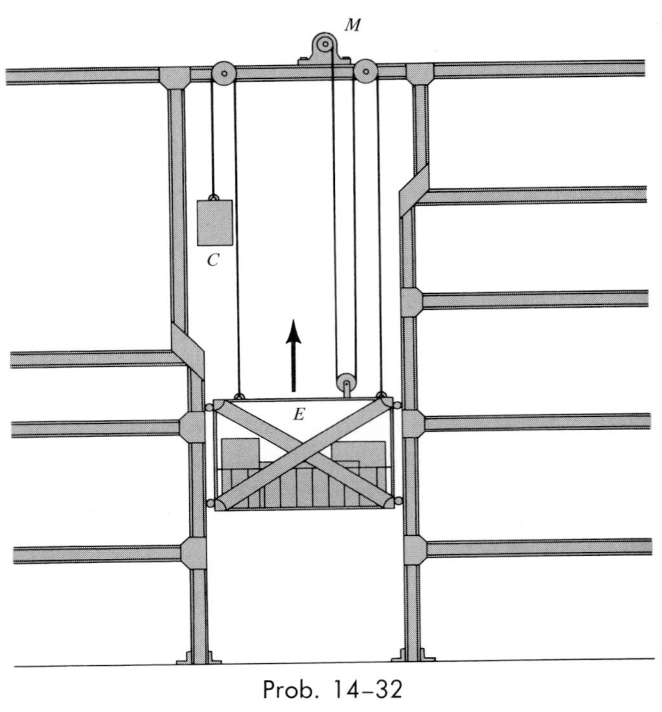

Prob. 14–32

14–5. Conservative Forces and Potential Energy

Conservative Force. A particularly simple type of force acting on a particle is one that depends *only* on the particle's position and is independent of the particle's velocity and acceleration. Furthermore, if the work done by this force in moving the particle from one point to another is *independent of the path* followed by the particle, this force is called a *conservative force*. The weight of a particle and the force of an elastic spring are two examples of conservative forces often encountered in mechanics.

Weight. The work done by the weight **W** of a particle when the particle is displaced downward along an *arbitrary path* is computed from Eq. 14–4,

$$U_{1-2} = W(y_2 - y_1)$$

This equation is independent of the path; rather it depends only on the particle's *vertical displacement*.

Elastic Spring. The work done by a spring force $\mathbf{F}_s$ acting on a particle is defined from Eq. 14–5,

$$U_{1-2} = -(\tfrac{1}{2}kx_2^2 - \tfrac{1}{2}kx_1^2)$$

Here x_1 is the initial spring position, from which the spring is either extended or compressed to a *further position* x_2, $|x_2| > |x_1|$. Notice that the work depends only upon the initial and final lengths x_1 and x_2 of the spring. Whether the spring is extended or compressed, the work of the spring force acting on a particle is negative, since this force is always opposite in direction to the particle's displacement.

Friction. In contrast to a conservative force, consider the force of friction exerted on a moving object by a fixed surface. The work done by the frictional force *depends upon the path*—the longer the path, the greater the work. Consequently, *frictional forces are nonconservative*. The work is dissipated from the body in the form of heat.

Potential Energy. When a conservative force acts on a particle, it gives the particle the capacity to do work. This capacity, measured as *potential energy, V,* depends only upon the *location* of the particle when acted upon by the force.

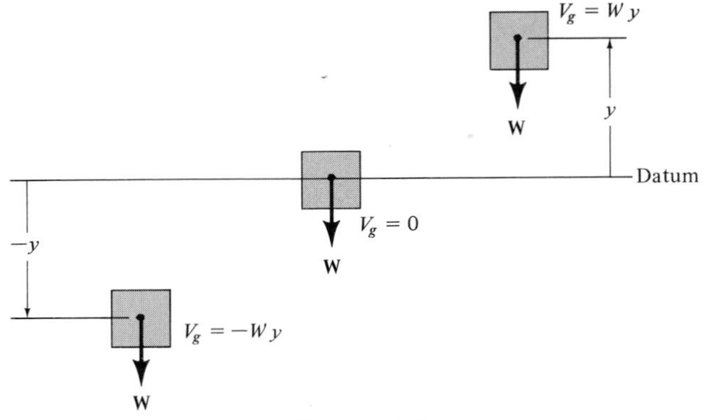

Fig. 14–15

Gravitational Potential Energy. If a particle (or block) is located a distance *y above* a datum, as shown in Fig. 14–15, the particle's weight **W** has positive *gravitational potential energy,* V_g, since **W** has the capacity of doing positive work when the particle is moved back down to the datum. This energy can be expressed mathematically as

$$V_g = Wy \tag{14–14}$$

If the particle is located a distance *y below* the datum, V_g is negative,

$$V_g = -Wy \tag{14–15}$$

since the weight does negative work when the particle is moved back up to the datum.

Elastic Potential Energy. When an elastic spring is elongated or compressed a distance x from its unstretched position, the elastic potential energy V_e which the spring imparts to an attached particle (or block) can be expressed as

$$V_e = \tfrac{1}{2}kx^2 \tag{14–16}$$

Here V_e is positive since, in the deformed position, the spring has the *capacity* for doing positive work when the particle is returned to its original position x, Fig. 14–16.

Potential-Energy Function. In the general case, if a particle is subjected to both gravitational and elastic forces, the particle's potential energy can be expressed as a *potential-energy function,* which is the algebraic sum

$$V = V_g + V_e \tag{14–17}$$

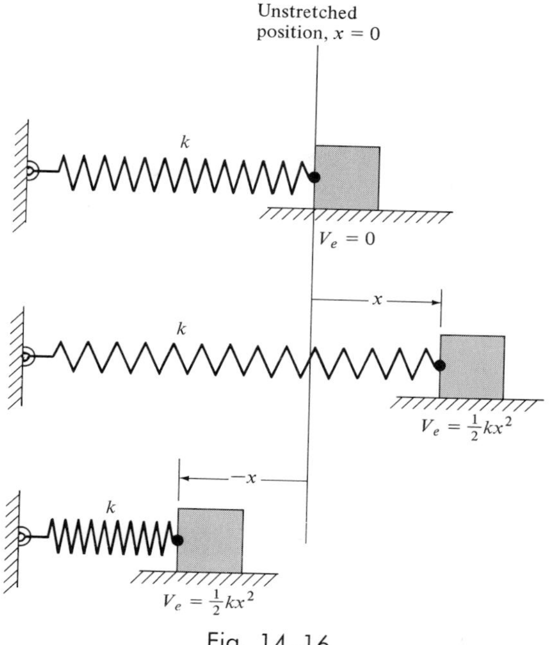

Fig. 14–16

Measurement of V depends upon the location of the particle with respect to a selected datum in accordance with Eqs. 14–14, 14–15, and 14–16.

In general, if a particle is located at an arbitrary point (x, y, z) in space, its potential energy can be defined using a potential-energy function $V = V(x, y, z)$. The work done by a conservative force in moving the particle from point (x_1, y_1, z_1) to point (x_2, y_2, z_2) is then measured by the *difference* of this function, i.e.,

$$U_{1-2} = V_1(x_1, y_1, z_1) - V_2(x_2, y_2, z_2) \qquad (14\text{–}18)$$

For example, the potential-energy function for a block of weight **W** suspended from a spring can be expressed in terms of its position y, measured from a datum located at the unstretched length of the spring, Fig. 14–17. We have

$$\begin{aligned} V &= V_g + V_e \\ &= -Wy + \tfrac{1}{2}ky^2 \end{aligned}$$

If the block moves from y_1 to a further downward position y_2, then applying Eq. 14–18, the work is

$$\begin{aligned} U_{1-2} = V_1 - V_2 &= -W(y_1 - y_2) + (\tfrac{1}{2}ky_1^2 - \tfrac{1}{2}ky_2^2) \\ &= W(y_2 - y_1) - (\tfrac{1}{2}ky_2^2 - \tfrac{1}{2}ky_1^2) \end{aligned}$$

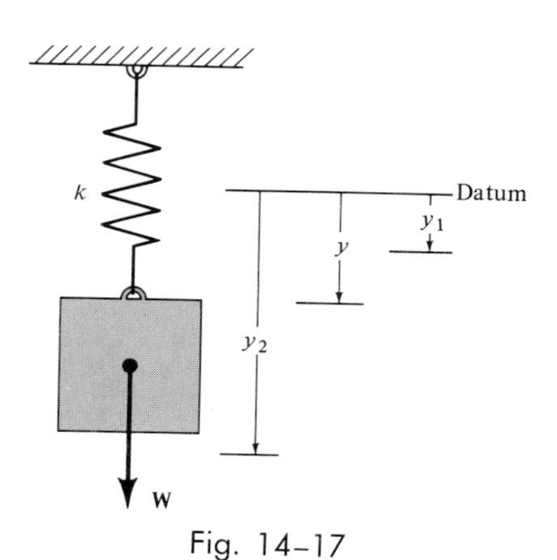

Fig. 14–17

When the displacement path is infinitesimal, i.e., from point (x, y, z) to $(x + dx, y + dy, z + dz)$, Eq. 14–18 becomes

$$\begin{aligned} dU &= V(x, y, z) - V(x + dx, y + dy, z + dz) \\ &= -dV(x, y, z) \end{aligned} \tag{14–19}$$

Provided both the force $\mathbf{F}$, causing the work, and displacement $d\mathbf{s}$ are defined using rectangular coordinates, then the work can also be expressed as

$$\begin{aligned} dU &= \mathbf{F} \cdot d\mathbf{s} = (F_x\mathbf{i} + F_y\mathbf{j} + F_z\mathbf{k}) \cdot (dx\mathbf{i} + dy\mathbf{j} + dz\mathbf{k}) \\ &= F_x\,dx + F_y\,dy + F_z\,dz \end{aligned}$$

Substituting this result into Eq. 14–19 and expressing the total derivative of $dV(x, y, z)$ in terms of its partial derivatives yields

$$F_x\,dx + F_y\,dy + F_z\,dz = -\left(\frac{\partial V}{\partial x}dx + \frac{\partial V}{\partial y}dy + \frac{\partial V}{\partial z}dz\right)$$

Since changes in x, y, and z are all independent of one another, this equation is satisfied provided

$$F_x = -\frac{\partial V}{\partial x}, \qquad F_y = -\frac{\partial V}{\partial y}, \qquad F_z = -\frac{\partial V}{\partial z} \tag{14–20}$$

Thus,

$$\begin{aligned} \mathbf{F} &= -\frac{\partial V}{\partial x}\mathbf{i} - \frac{\partial V}{\partial y}\mathbf{j} - \frac{\partial V}{\partial z}\mathbf{k} \\ &= -\left(\frac{\partial}{\partial x}\mathbf{i} + \frac{\partial}{\partial y}\mathbf{j} + \frac{\partial}{\partial z}\mathbf{k}\right)V \end{aligned}$$

or

$$\mathbf{F} = -\boldsymbol{\nabla} V \tag{14–21}$$

where $\boldsymbol{\nabla}$ (del) represents the vector operator $\boldsymbol{\nabla} = (\partial/\partial x)\mathbf{i} + (\partial/\partial y)\mathbf{j} + (\partial/\partial z)\mathbf{k}$.

Equation 14–21 relates a force $\mathbf{F}$ to its potential-energy function, V, and thereby provides a mathematical criterion for proving that $\mathbf{F}$ is conservative. For example, the gravitational potential-energy function for a weight $\mathbf{W}$, located a distance y above a datum, is $V_g = Wy$, Eq. 14–14. To prove that $\mathbf{W}$ is conservative, it is necessary to show that it satisfies Eq. 14–21 (or Eq. 14–20), in which case

$$F_y = -\frac{\partial V}{\partial y}; \qquad F = -\frac{\partial}{\partial y}(Wy) = -W$$

The negative sign indicates that $\mathbf{W}$ acts downward, opposite to the positive y elevation.

14–6. Conservation-of-Energy Theorem

When a particle is acted upon by a *system* of conservative forces, using Eq. 14–18 ($\Sigma U_{1-2} = V_1 - V_2$), the principle of work and energy ($T_1 + \Sigma U_{1-2} = T_2$) may be written in the form,

$$T_1 + V_1 = T_2 + V_2 \tag{14–22}$$

This equation is often referred to as the *conservation-of-energy theorem.* It states that the sum of the particle's initial kinetic and potential energy is equal to the sum of the particle's final kinetic and potential energy. In other words, at any instant the sum of the particle's kinetic and potential energy remains *constant*. For example, consider the energy involved in dropping a ball of weight **W** from a height h above the ground (datum), Fig. 14–18. The potential energy of the ball is maximum before it is dropped, at which time its kinetic energy is zero. The total energy of the ball in its initial position is thus

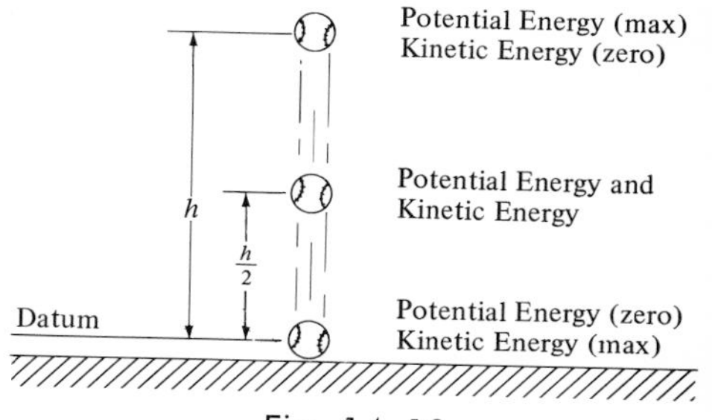

Fig. 14–18

$$E = T_1 + V_1 = 0 + Wh = Wh$$

When the ball has fallen a distance $h/2$, its speed can be determined by using $v^2 = v_1^2 + 2a_c(y - y_1)$, which yields $v = \sqrt{2g(h/2)} = \sqrt{gh}$. The energy of the ball at the midheight position is therefore

$$E = V_2 + T_2 = W\frac{h}{2} + \frac{1}{2}\frac{W}{g}(\sqrt{gh})^2 = Wh$$

Just before the ball strikes the ground, its potential energy is zero, and its speed is $v = \sqrt{2gh}$. Here, again, the total energy of the ball is

$$E = V_3 + T_3 = 0 + \frac{1}{2}\frac{W}{g}(\sqrt{2gh})^2 = Wh$$

When the ball comes in contact with the ground, it deforms somewhat and, provided the ground is hard enough, the ball will rebound off the surface, reaching a new height h', which will be less than the height h from which it was first released. The difference in height accounts for an energy loss, $E_l = W(h - h')$, occurring at the moment of collision. Portions of this loss produce noise, deformation of the ball and ground, vibrations, and heat.

System of Particles. An equation similar to Eq. 14–22 can be written for a system of particles, which is based on Eq. 14–8 ($\Sigma T_1 + \Sigma U_{1-2} = \Sigma T_2$). We have

$$\Sigma T_1 + \Sigma V_1 = \Sigma T_2 + \Sigma V_2 \tag{14–23}$$

Here, the sum of the system's initial kinetic and potential energies is equal to the sum of the system's final kinetic and potential energies.

PROCEDURE FOR ANALYSIS

The conservation-of-energy theorem is used to solve problems involving velocity, displacement, and conservative force systems. For applications, the following two-step procedure should be used.

Step 1: Draw two diagrams showing the particle located at its initial and final points along the path. If the particle is subjected to a vertical displacement, determine where to establish the fixed horizontal datum from which to measure the particle's gravitational potential energy, V_g. Although this position can be selected arbitrarily, it is best to locate the datum either at the initial or final point of the path, since at the datum $V_g = 0$. Data pertaining to the elevation of the particle from the datum, and the extension or compression of any connecting springs, can be determined from the geometry associated with the two diagrams.

Step 2: Apply the conservation-of-energy theorem, $T_1 + V_1 = T_2 + V_2$. When computing the kinetic energy, $T = \frac{1}{2}mv^2$, the particle's speed v must be measured from an inertial reference frame. If the problem involves a system of *connected particles,* the speed of each of the particles may be related by using the kinematic principles outlined in Sec. 12–9. The potential energy $V = V_g + V_e$ is formulated on the basis of applying $V_g = \pm Wy$ (Eqs. 14–14 and 14–15) and $V_e = \frac{1}{2}kx^2$ (Eq. 14–16).

It is important to remember that only problems involving conservative force systems may be solved by using the conservation-of-energy theorem. As stated previously, friction or other drag-resistant forces, which depend upon velocity or acceleration, are nonconservative. The work done by such forces is transformed into thermal energy used to heat up the surfaces of contact, and consequently this energy dissipates into the surroundings and may not be recovered. Therefore, problems involving frictional forces should either be solved by using the principle of work and energy if it applies, or the equation of motion.

The following example problems numerically illustrate application of the two-step procedure given above.

Example 14–7

A small block having a weight **W** starts from rest at point A and slides down the smooth curved path shown in Fig. 14–19*a*. Determine the velocity of the block when it reaches points B and C.

Solution

Step 1: Diagrams showing the block at points A, B, and C are given in Fig. 14–19*b*. For convenience, the potential-energy datum has been established through the center of mass of the block when it is located at B.

Step 2: Applying Eq. 14–22 between points A and B yields

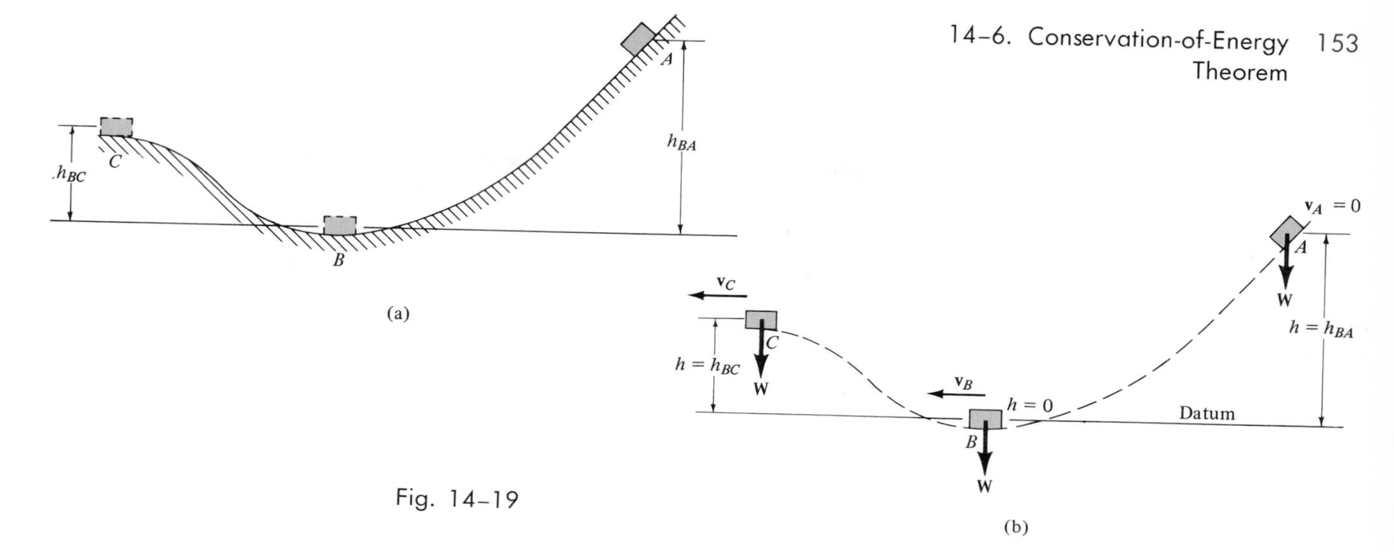

Fig. 14-19

$$V_A + T_A = V_B + T_B$$

$$Wh_{BA} + 0 = 0 + \frac{1}{2}\frac{W}{g}(v_B)^2$$

$$v_B = \sqrt{2gh_{BA}} \qquad \textit{Ans.}$$

Applying Eq. 14-22 between A and C gives

$$V_A + T_A = V_C + T_C$$

$$Wh_{BA} + 0 = Wh_{BC} + \frac{1}{2}\frac{W}{g}(v_C)^2$$

Hence,

$$v_C = \sqrt{2g(h_{BA} - h_{BC})} \qquad \textit{Ans.}$$

This same result may also be obtained by applying Eq. 14-22 between B and C, i.e.,

$$V_B + T_B = V_C + T_C$$

$$0 + \frac{1}{2}\frac{W}{g}(\sqrt{2gh_{BA}})^2 = Wh_{BC} + \frac{1}{2}\frac{W}{g}(v_C)^2$$

Solving for v_C gives

$$v_C = \sqrt{2g(h_{BA} - h_{BC})} \qquad \textit{Ans.}$$

The results of this problem indicate that the block attains a speed that is *independent* of the path. Furthermore, this speed is the *same* as that computed by Eq. 12-10 [$v^2 = v_1^2 + 2a_c(y - y_1)$] when the block *falls freely* from the vertical height of the path.

Example 14-8

A smooth 2-kg collar C, shown in Fig. 14-20*a*, fits loosely on the vertical shaft. If the spring is unextended when the collar is in the dashed position A, determine the speed at which the collar is moving when $s = 1$ m, if (a) it is released from rest at A, (b) it is released at A with an *upward* velocity of $v_A = 2$ m/s.

Solution

Part a. *Step 1:* The two diagrams indicating the collar in its initial and final positions are shown in Fig. 14-20*b*. For convenience, the datum is established through AB. When the collar is at C, the gravitational potential energy is $-(mg)s$, since the collar is *below* the datum; and the elastic potential energy is $\frac{1}{2}kx_{CB}^2$. Here $x_{CB} = 0.5$ m, which represents the *stretch* in the spring as computed on the figure.

Step 2: Applying Eq. 14-22, we have

$$T_A + V_A = T_C + V_C$$

$$0 + 0 = \tfrac{1}{2}mv_C^2 - mgs + \tfrac{1}{2}kx_{CB}^2$$

$$0 + 0 = \tfrac{1}{2}(2)v_C^2 - 2(9.81)(1) + \tfrac{1}{2}(3)(0.5)^2$$

$$0 = v_C^2 - 19.62 + 0.375$$

$$v_C = 4.39 \text{ m/s} \qquad \textit{Ans.}$$

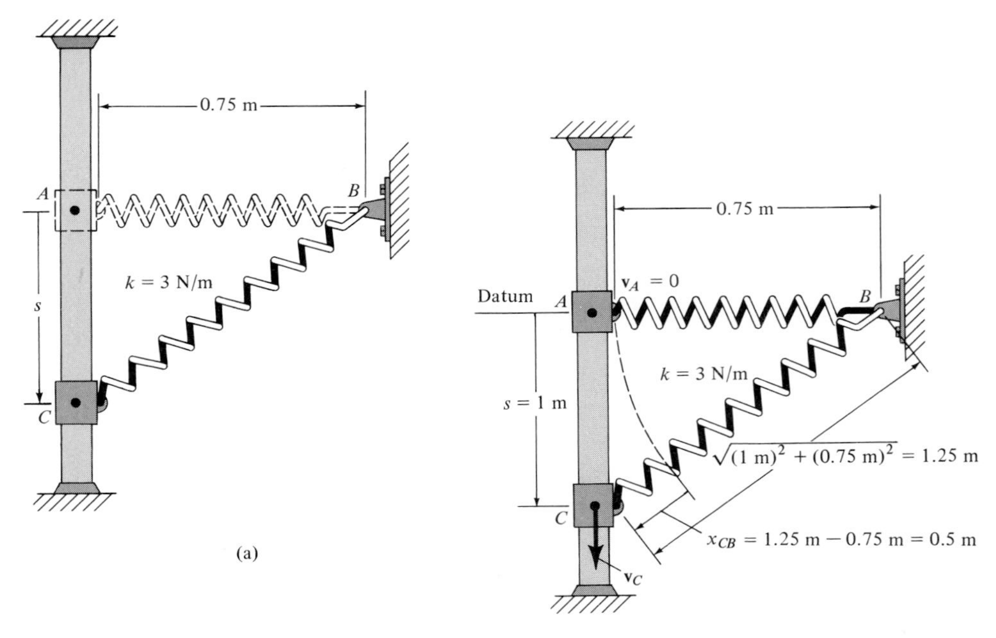

Fig. 14-20

This problem was solved using the equation of motion in Example 13–4. Comparing the solutions, it is seen that the conservation of energy is clearly advantageous since the calculations depend *only* on data calculated at the initial and final points of the path.

Part b. Applying Eq. 14–22 for the case when $v_A = 2$ m/s, using the data in Fig. 14–20*b*, we have

$$T_A + V_A = T_C + V_C$$
$$\tfrac{1}{2}mv_A^2 + 0 = \tfrac{1}{2}mv_C^2 - mgs + \tfrac{1}{2}kx_{CB}^2$$
$$\tfrac{1}{2}(2)(2)^2 + 0 = \tfrac{1}{2}(2)v_C^2 - 2(9.81)(1) + \tfrac{1}{2}(3)(0.5)^2$$
$$4 = v_C^2 - 19.62 + 0.375$$
$$v_C = 4.82 \text{ m/s} \qquad \textit{Ans.}$$

Note that the kinetic energy of the collar depends only on the *magnitude* of velocity and therefore it is immaterial if the collar is moving up or down at 2 m/s when released at *A*.

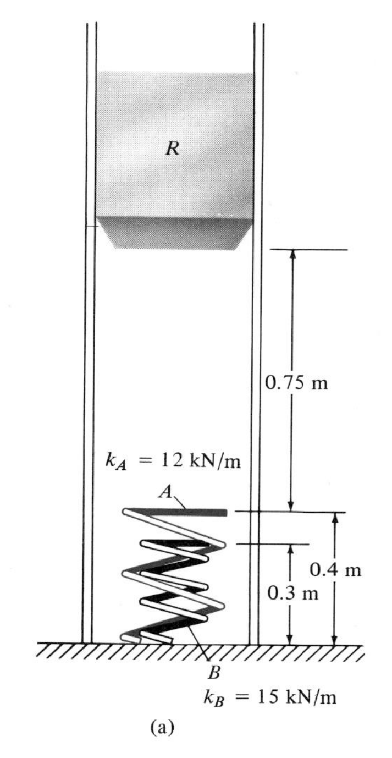

(a)

Example 14–9

The ram *R* shown in Fig. 14–21*a* has a mass of 100 kg and is released from rest 0.75 m from the top of a spring *A* that has a stiffness of $k_A = 12$ kN/m. If a second spring *B*, having a stiffness of $k_B = 15$ kN/m, is "nested" in *A*, determine the maximum deflection of *A* needed to stop the downward motion of the ram. The unstretched length of each spring is indicated in the figure.

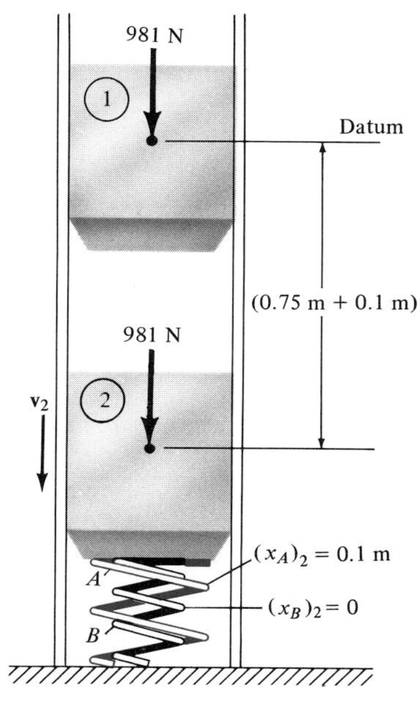

(b)

Fig. 14–21(a,b)

Solution

Step 1: In order to obtain the answer, it is first necessary to determine if the ram compresses *both* springs. To do this the datum is located through the center of gravity of the ram, at its initial position, Fig. 14–21*b*. After striking spring *A*, and compressing it 0.1 m, the ram is assumed to be moving downward with a speed $v_2 > 0$.

Step 2: Applying Eq. 14–22, we have

$$T_1 + V_1 = T_2 + V_2$$
$$0 + 0 = \tfrac{1}{2}m(v_2)^2 - Wh_{1-2} + \tfrac{1}{2}k_A(x_A)_2^2$$
$$0 + 0 = \tfrac{1}{2}(100)(v_2^2) - 981(0.75 + 0.1) + \tfrac{1}{2}(12\,000)(0.1)^2$$
$$\tfrac{1}{2}(100)v_2^2 = +773.85$$
$$v_2 = 3.93 \text{ m/s}$$

Since the final kinetic energy T_2 is *positive* (+773.85 J), the answer indicates that both springs will be compressed. Why?

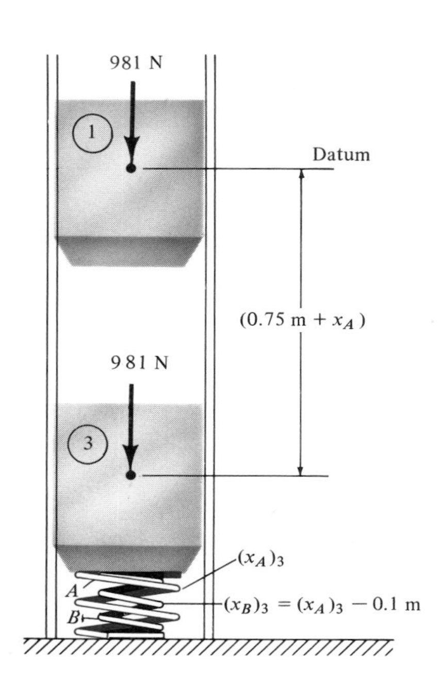

Fig. 14–21(c)

Step 1: When the kinetic energy is reduced to zero ($v_3 = 0$), it will be assumed that A is compressed a distance $(x_A)_3$ so that B compresses $(x_B)_3 = (x_A)_3 - 0.1$ m, Fig. 14–21*c*.
Step 2: Applying Eq. 14–22, we have

$$T_1 + V_1 = T_3 + V_3$$

$$0 + 0 = 0 - Wh_{1-3} + \tfrac{1}{2}k_A(x_A)_3^2 + \tfrac{1}{2}k_B[(x_A)_3 - 0.1]^2$$

$$0 + 0 = 0 - 981[0.75 + (x_A)_3] + \tfrac{1}{2}(12\,000)(x_A)_3^2 + \tfrac{1}{2}(15\,000)[(x_A)_3 - 0.1]^2$$

Rearranging the terms,

$$13\,500(x_A)_3^2 - 2481(x_A)_3 - 660.75 = 0$$

Using the quadratic formula, and solving for the positive root, we have

$$(x_A)_3 = 0.331 \text{ m} \qquad \textit{Ans.}$$

The second root, $(x_A)_3 = -0.148$ m, does not represent the physical situation, since it indicates that the spring A is *extended* instead of compressed. (Positive x has been measured downward.)

Example 14–10

A chain has a mass density of 10 kg/m and hangs across the surface of a smooth peg, as shown in Fig. 14–22*a*. If it is released from rest in the position shown, determine its speed when end A passes point B. Neglect the radius of the peg.

Solution

Step 1: The two diagrams showing the chain at its initial and final positions are indicated in Fig. 14–22*b*. The datum has been located at the peg. Since each link or particle of the chain moves with the same speed as its connecting links, the total energy for the *system of links* can be found by determining the mechanical energy for a given length of chain and adding (algebraically) the results. In particular, the potential energy of a chain length can be computed on the basis of knowing the position of the

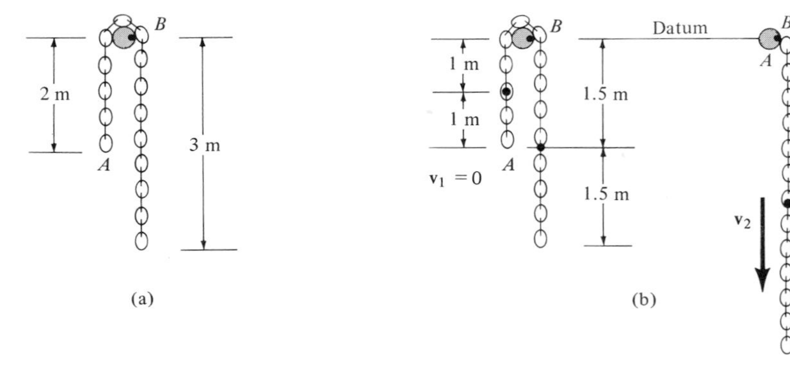

Fig. 14–22

mass center of the chain length. The mass centers for the two lengths of chain on the right and left sides of the peg are located with respect to the datum as shown in the figure. In all cases the potential energy of a given length of chain is negative, since the mass center of each length lies below the datum.

Step 2: Applying Eq. 14–22, noting that the chain has a weight of $10(9.81) = 98.1$ N/m, we have

$$V_1 + T_1 = V_2 + T_2$$

$$-98.1\ \text{N/m}(2\ \text{m})(1\ \text{m}) - 98.1\ \text{N/m}(3\ \text{m})(1.5\ \text{m}) + 0 = -98.1\ \text{N/m}(5\ \text{m})(2.5\ \text{m}) + \tfrac{1}{2}(10\ \text{kg/m})(5\ \text{m})v_2^2$$

so that

$$25v_2^2 = 588.6$$

Solving,

$$v_2 = 4.85\ \text{m/s} \qquad \textit{Ans.}$$

Problems

14-33. Solve Prob. 14–8 using the conservation-of-energy theorem.

14-34. Using the conservation-of-energy theorem, determine the speed of the car in Prob. 14–20 when it reaches point B.

14-35. The block has a mass of $m_b = 0.2$ kg and slides along the smooth chute AB. It is released from rest at A, which has coordinates of $A(0.75\ \text{m}, 0, 2.25\text{m})$. Determine the speed at which it slides off at B, which has coordinates of $B(0, 1.75\ \text{m}, 0)$.

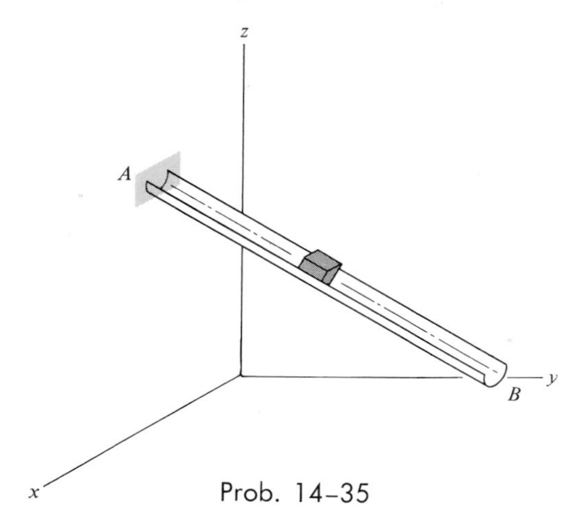

Prob. 14–35

14-35a. Solve Prob. 14–35 if the block weighs $W_b = 1.5$ lb, and $A(5\ \text{ft}, 0, 10\ \text{ft})$, $B(10\ \text{ft}, 0, 0)$.

* **14-36.** The firing mechanism of a pinball machine consists of a plunger P having a mass of 0.2 kg and a spring of stiffness $k = 300$ N/m. When $s = 0$, the spring is compressed 50 mm. If the arm is pulled back such that $s = 100$ mm and released, determine the speed of the 0.3-kg pinball B *just before* the plunger strikes the stop, i.e., $s = 0$. Assume all surfaces of contact to be smooth. The ball moves in the horizontal plane. Neglect friction and the rolling motion of the ball.

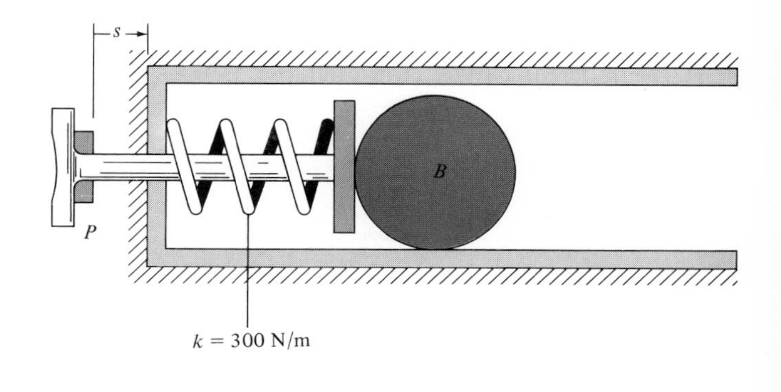

Prob. 14–36

14–37. The bob of the pendulum has a mass of 0.2 kg and is released from rest when it is in the horizontal position shown. Determine its speed and the tension in the cord at the instant the bob passes through its lowest position.

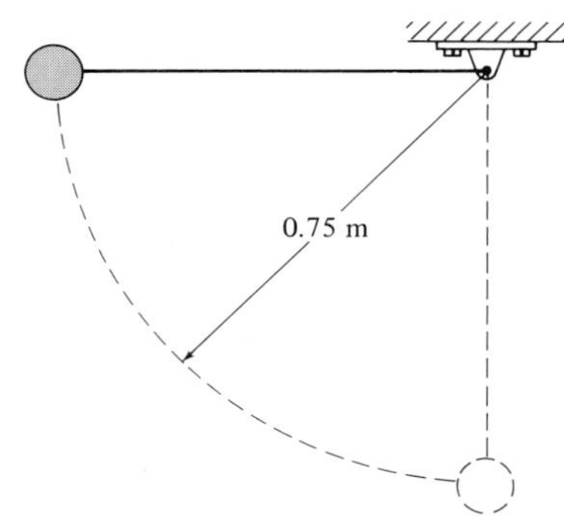

Prob. 14–37

14–38. Four inelastic cables C are attached to a plate P and hold the spring 80 mm in compression when *no force* acts on the plate. If a block B, having a mass of 0.5 kg, is placed on the plate and the plate is pushed down 40 mm and released from rest, determine how high the block rises from the point where it was released. Neglect the mass of the plate.

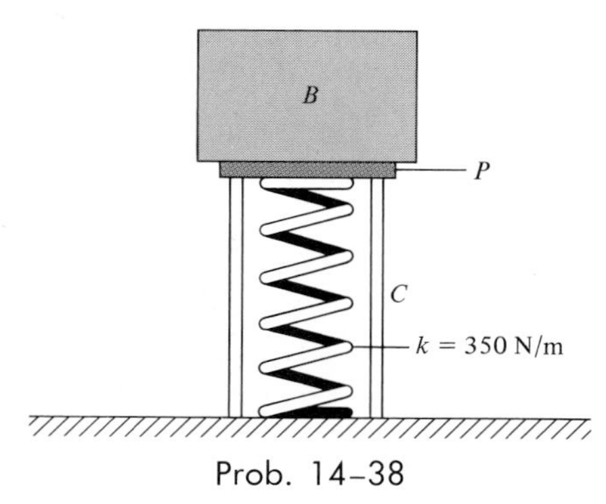

Prob. 14–38

14–39. The collar of negligible size has a mass of 0.25 kg and is attached to a spring having an unstretched length of 100 mm. If it is released from rest at A and travels along the smooth guide, determine the speed at which it strikes B.

Prob. 14–39

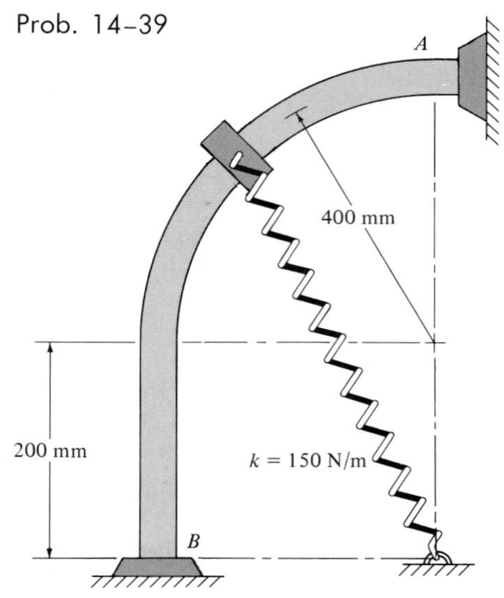

***14–40.** The block A having a mass of $m_A = 0.5$ kg slides in the smooth horizontal slot, where $d = 0.4$ m. If the block is drawn back so that $s = 0.3$ m, and released from rest, determine its speed at the instant $s = 0$. Each of the two springs has a stiffness of $k = 400$ N/m and an unstretched length of $l = 0.15$ m.

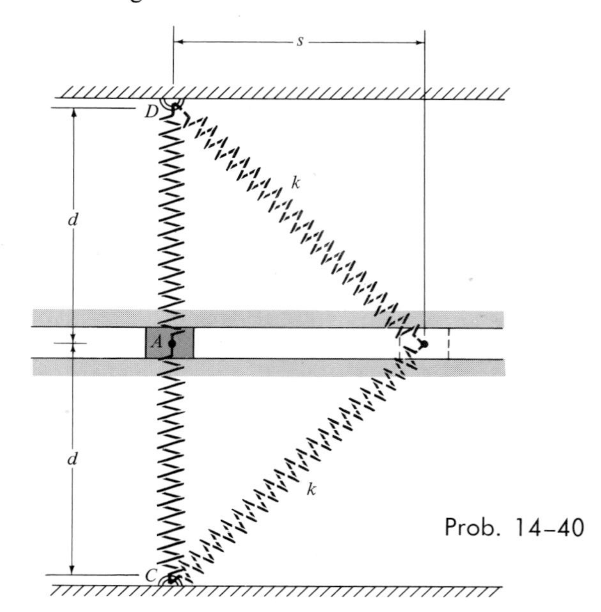

Prob. 14–40

***14–40a.** Solve Prob. 14–40 if the block has a weight of $W_A = 1.25$ lb, and $d = 2$ ft, $s = 1.5$ ft, $k = 150$ lb/ft, $l = 0.75$ ft.

14–41. The toy car has a mass of 200 g. Determine the minimum height h from which it can be released from rest so that it travels around the loop without leaving the track. Neglect friction and the size of the car in the calculation.

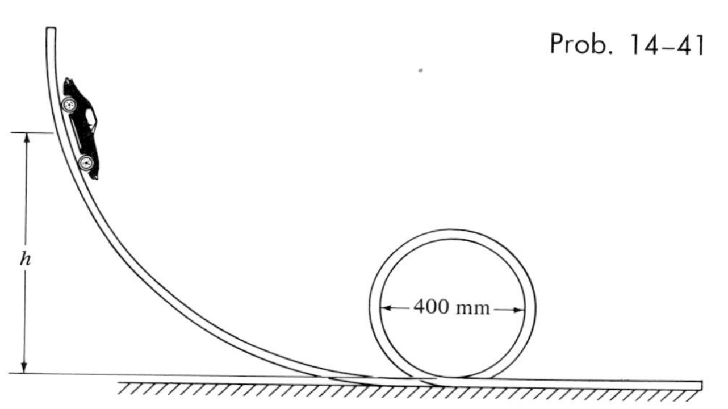

Prob. 14–41

14–42. A block having a mass of 15 kg is attached to four springs. If each spring has a stiffness of $k = 2$ kN/m and an unstretched length of 150 mm, determine the *maximum* downward vertical displacement s_{max} of the block if it is released from rest when $s = 0$.

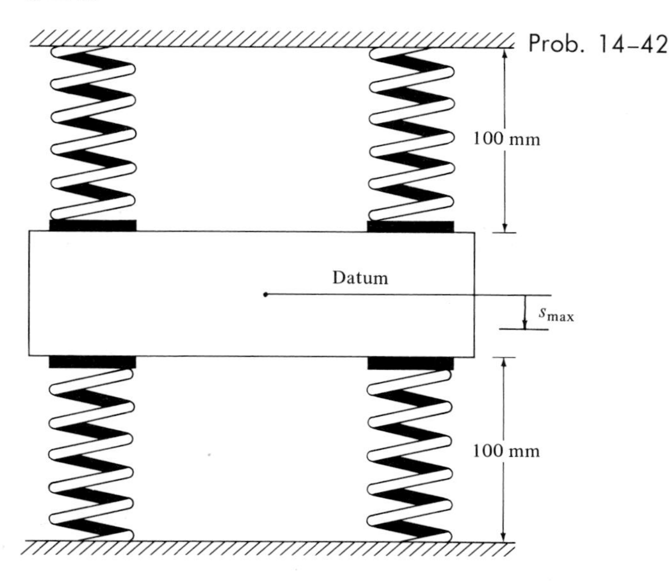

Prob. 14–42

14–43. The bob of a pendulum has a mass of 0.75 kg. It is fired from position A by a spring which has a stiffness of $k = 6$ kN/m and is compressed 125 mm. Determine the speed of the bob and the tension in the cord when it is at positions B and C. Point B is located on the path where the radius of curvature is still 0.6 m.

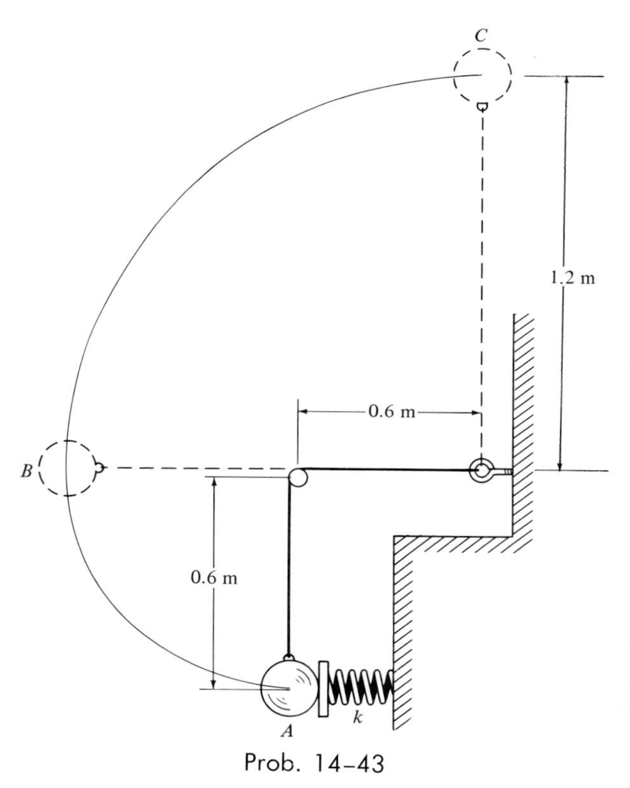

Prob. 14–43

***14–44.** If the spring in Prob. 14–43 is compressed 50 mm and released, determine the stiffness k so that (a) the speed of the bob at B is zero, and (b) the tension in the cord at C is zero.

14–45. The car C and its contents have a mass of $m_C = 400$ kg, whereas block B has a mass of $m_B = 75$ kg. If the car is released from rest, determine its speed when it travels $s_C = 10$ m down the 20° incline. *Suggestion:* Establish separate datums at the initial elevations of B and C.

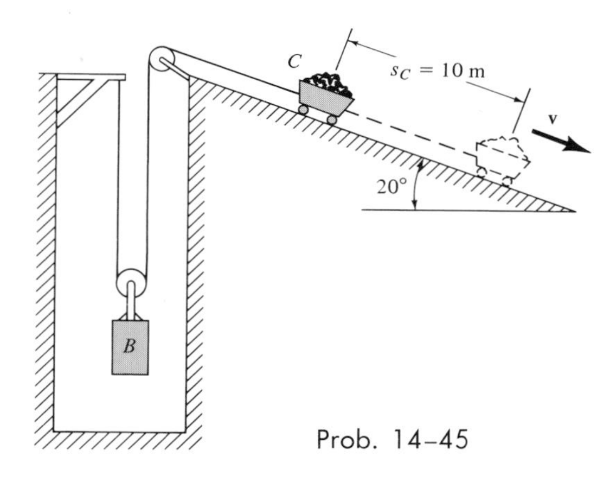

Prob. 14–45

14–45a. Solve Prob. 14–45 if the car and its contents weigh $W_C = 600$ lb, block B weighs $W_B = 175$ lb, and $s_C = 30$ ft.

14–46. A chain is placed within a smooth tube as shown. If it has a mass of 1.5 kg/m and is released from rest, determine the speed of end B when it comes through the other end of the tube at A.

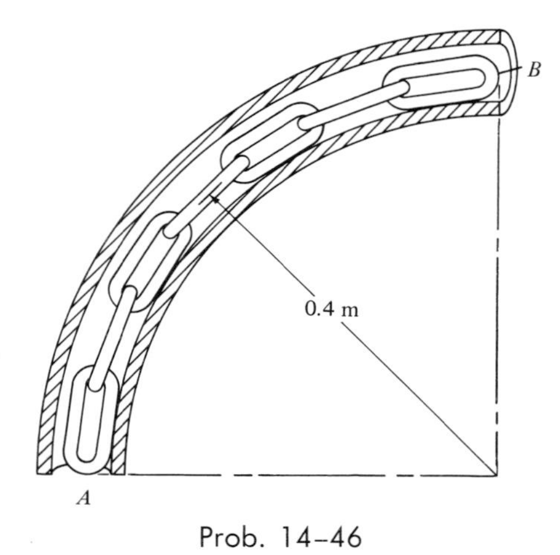

Prob. 14–46

14–47. If the mass of the earth is M_e, show that the gravitational potential energy of a body of mass m, located a distance r from the center of the earth, is $V = -GM_e m/r$. Recall that the gravitational force acting between the earth and the body is $F = G(M_e m/r^2)$ (Eq. 13–2). For the calculation, locate the datum at $r \to \infty$. Also, prove that $\mathbf{F}$ is a conservative force.

* **14–48.** A 60-kg satellite is traveling in free flight along an elliptical orbit such that at A, where $r_A = 20$ Mm, it has a speed of $v_A = 40$ Mm/h. What is the speed of the satellite when it reaches point B, where $r_B = 80$ Mm? *Hint:* See Prob. 14–47, where $M_e = 5.976(10^{24})$ kg and $G = 6.673(10^{-11})\ \text{m}^3/(\text{kg} \cdot \text{s}^2)$.

Prob. 14–48

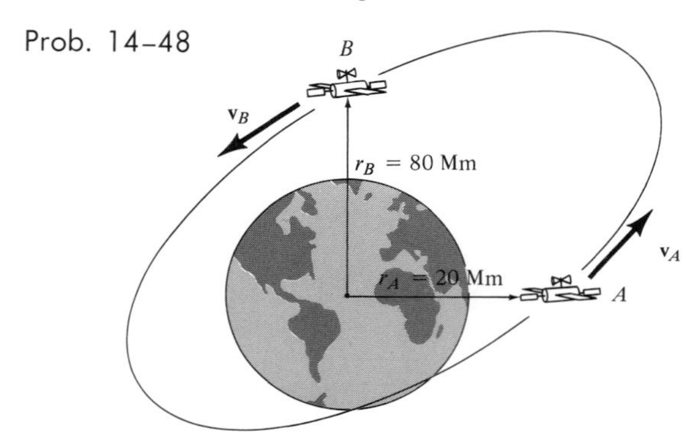

14–49. A toboggan and its two riders have a total mass of 125 kg. Determine the greatest initial speed v_A the toboggan can have at A so that it arrives at C in the shortest time without leaping off the path. The path consists of a 30° sector of a circular arc which has an inflection point at B. Neglect friction. *Hint:* The greatest speed is reached when the toboggan tends to leave the path at B, where the radius of curvature is still 10 m.

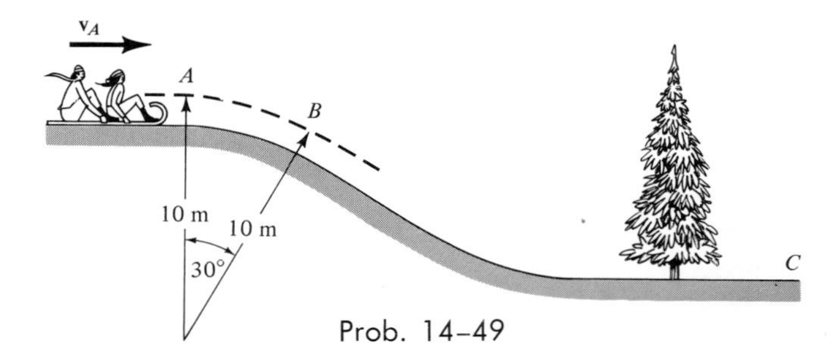

Prob. 14–49

14-50. A tank car is stopped by two spring bumpers A and B, having a stiffness of $k_A = 20$ kN/m and $k_B = 40$ kN/m, respectively. Bumper A is attached to the car, whereas bumper B is attached to the wall. If the car has a mass of $m_c = 50$ Mg and is freely coasting at $v_c = 0.15$ m/s, compute the maximum deflection of each spring at the instant the bumpers stop the car.

Prob. 14-50

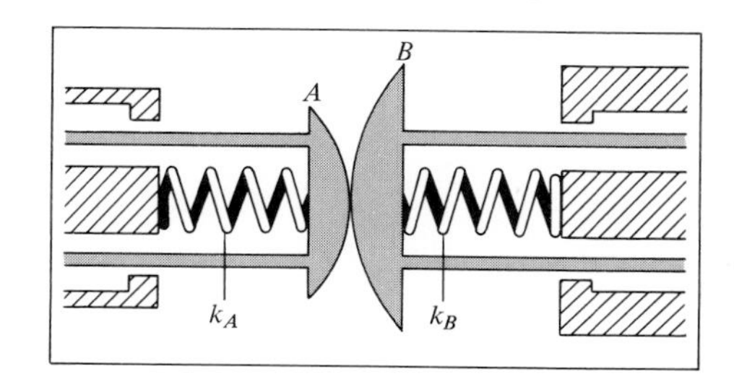

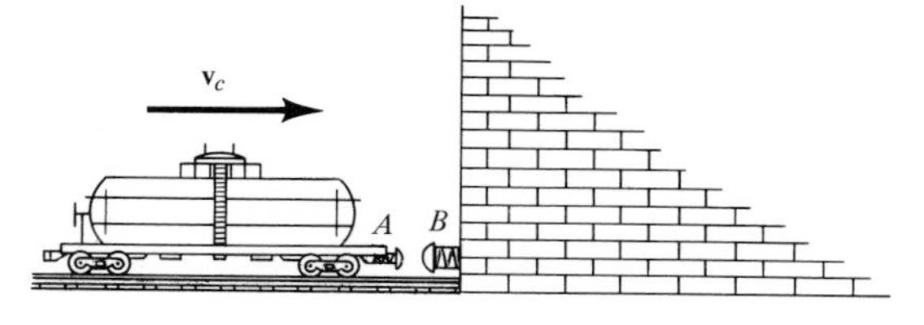

14-50a. Solve Prob. 14-50 if $k_A = 1.5(10^4)$ lb/ft, $k_B = 3(10^4)$ lb/ft, the car has a weight of $W_c = 2.5(10^4)$ lb, and $v_c = 3$ ft/s.

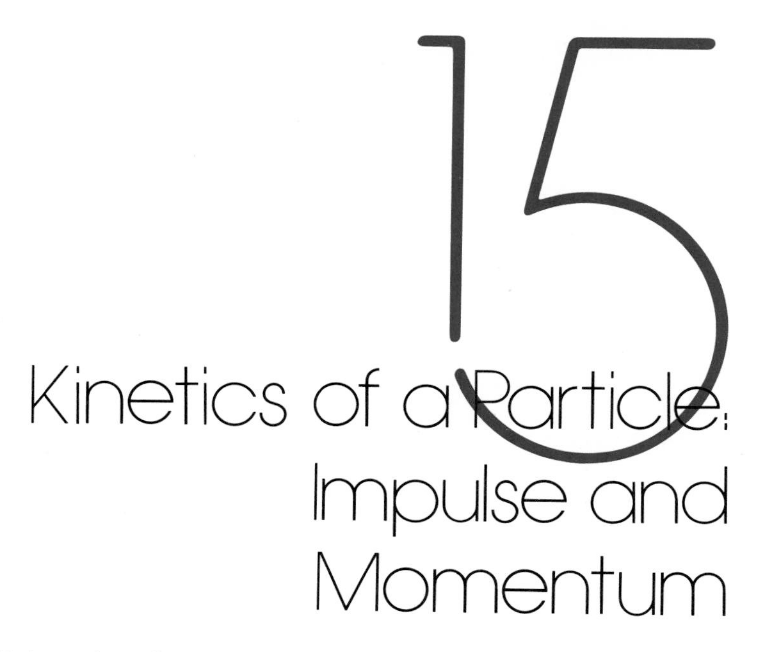

15 Kinetics of a Particle: Impulse and Momentum

15–1. Principle of Linear Impulse and Momentum for a Particle

Consider a particle of mass m which is subjected to several forces $\Sigma\mathbf{F}$. The equation of motion for the particle can be written as

$$\Sigma\mathbf{F} = m\mathbf{a} = m\frac{d\mathbf{v}}{dt} \qquad (15\text{–}1)$$

where $\mathbf{a}$ and $\mathbf{v}$ indicate the particle's instantaneous acceleration and velocity, respectively. Rearranging the terms and integrating between the limits $\mathbf{v} = \mathbf{v}_1$ at $t = t_1$ and $\mathbf{v} = \mathbf{v}_2$ at $t = t_2$, we have

$$\Sigma\int_{t_1}^{t_2} \mathbf{F}\,dt = m\int_{\mathbf{v}_1}^{\mathbf{v}_2} d\mathbf{v}$$

or

$$\Sigma\int_{t_1}^{t_2} \mathbf{F}\,dt = m\mathbf{v}_2 - m\mathbf{v}_1 \qquad (15\text{–}2)$$

This equation, which is referred to as the *principle of linear impulse and momentum,* provides a *direct means* of obtaining the particle's final velocity $\mathbf{v}_2$ after a specified time period when the particle's initial velocity is known and the forces acting on the particle are either constant or can be expressed as functions of time. Notice from the derivation that if $\mathbf{v}_2$ is determined using the equation of motion, a two-step process is necessary; i.e., apply $\Sigma\mathbf{F} = m\mathbf{a}$ to obtain $\mathbf{a}$, then integrate $\mathbf{a} = d\mathbf{v}/dt$ to obtain $\mathbf{v}_2$.

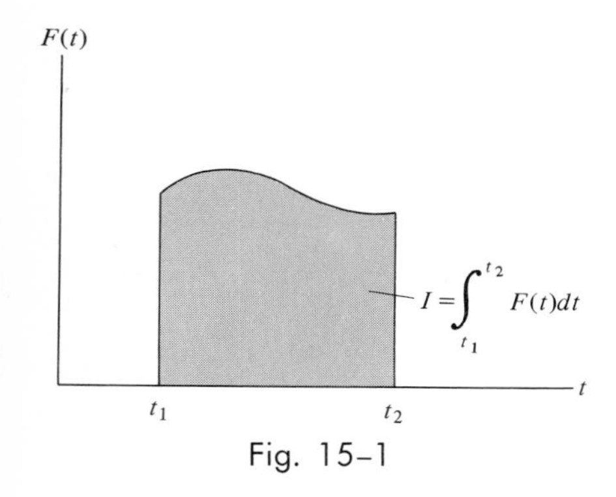

Fig. 15–1

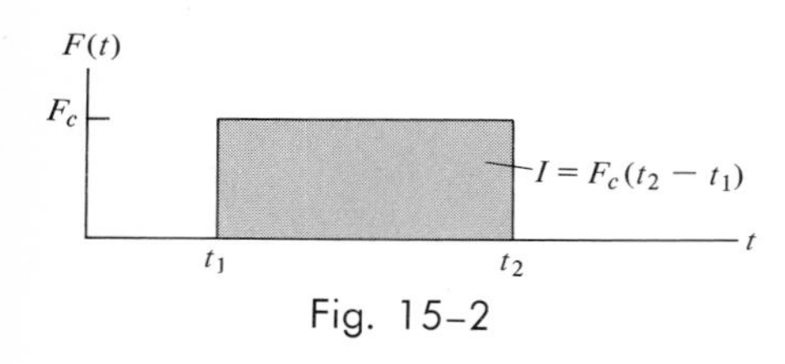

Fig. 15–2

Linear Impulse. The integral $\mathbf{I} = \int_{t_1}^{t_2} \mathbf{F}\,dt$ in Eq. 15–2 is defined as the *linear impulse*. This term is a vector quantity which measures the effect of a force during the time the force acts. The impulse vector acts in the same direction as the force, and its magnitude has units of force–time, e.g., N · s. Provided the force is expressed as a function of time, the impulse given to the particle may be determined by direct evaluation of the integral. In particular, if **F** acts in a constant direction during the time period t_1 to t_2, the magnitude of the impulse $\mathbf{I} = \int_{t_1}^{t_2} \mathbf{F}\,dt$ can be represented experimentally by the shaded area under the curve of force versus time, Fig. 15–1. If the force is constant in both magnitude and direction, $\mathbf{F} = \mathbf{F}_c$, the resulting impulse becomes $\mathbf{I} = \int_{t_1}^{t_2} \mathbf{F}_c\,dt = \mathbf{F}_c(t_2 - t_1)$, which represents the shaded rectangular area shown in Fig. 15–2.

Linear Momentum. Each of the two vectors of the form $\mathbf{L} = m\mathbf{v}$ in Eq. 15–2 is defined as the *linear momentum* of the particle. Since m is a scalar, the linear momentum vector acts in the same direction as **v**, and its magnitude mv has units of either mass–velocity, e.g., kg · m/s, or force–time, e.g., N · s, the same units used to measure the linear impulse.

Principle of Linear Impulse and Momentum. For problem solving, Eq. 15–2 will be rewritten in the form

$$m\mathbf{v}_1 + \Sigma \int_{t_1}^{t_2} \mathbf{F}\,dt = m\mathbf{v}_2 \tag{15–3}$$

which states that the initial momentum of the particle at t_1, $m\mathbf{v}_1$, plus the vector sum of all the impulses applied to the particle during the time interval t_1 to t_2, $\Sigma \int_{t_1}^{t_2} \mathbf{F}\,dt$, is equivalent to the final momentum of the particle at t_2, $m\mathbf{v}_2$. These three terms are illustrated graphically on the *momentum- and impulse-vector diagrams* shown in Fig. 15–3.

Provided the velocities $\mathbf{v}_1$ and $\mathbf{v}_2$ and the forces $\Sigma\mathbf{F}$ can be expressed by their Cartesian vector components, Eq. 15–3 may be written as

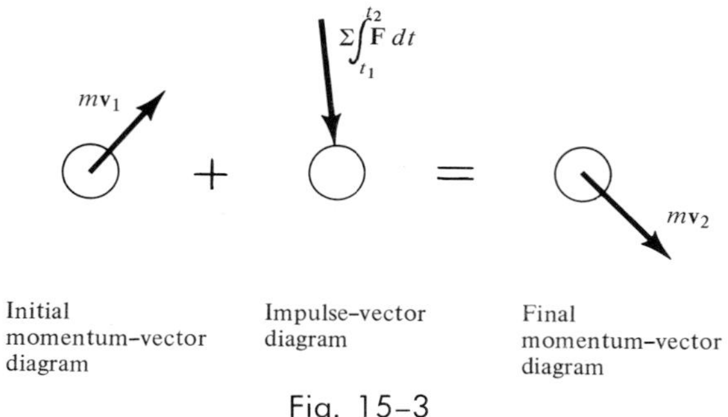

Fig. 15–3

$$m(v_x)_1\mathbf{i} + m(v_y)_1\mathbf{j} + m(v_z)_1\mathbf{k} + \left(\Sigma\int_{t_1}^{t_2} F_x\,dt\right)\mathbf{i} + \left(\Sigma\int_{t_1}^{t_2} F_y\,dt\right)\mathbf{j} + \left(\Sigma\int_{t_1}^{t_2} F_z\,dt\right)\mathbf{k} = m(v_x)_2\mathbf{i} + m(v_y)_2\mathbf{j} + m(v_z)_2\mathbf{k}$$

The solution is determined by equating the respective **i**, **j**, and **k** components, which yields the following three scalar equations:

$$\begin{aligned} m(v_x)_1 + \Sigma\int_{t_1}^{t_2} F_x\,dt &= m(v_x)_2 \\ m(v_y)_1 + \Sigma\int_{t_1}^{t_2} F_y\,dt &= m(v_y)_2 \\ m(v_z)_1 + \Sigma\int_{t_1}^{t_2} F_z\,dt &= m(v_z)_2 \end{aligned} \qquad (15\text{–}4)$$

These equations represent the principle of linear impulse and momentum for the particle in the x, y, and z directions, respectively.

PROCEDURE FOR ANALYSIS

The principle of linear impulse and momentum is used to solve problems involving force, time, and velocity, since these terms are involved in the formulation. For applications, the following two-step procedure should be used.

Step 1: Draw the momentum- and impulse-vector diagrams for the particle. Each of these diagrams graphically accounts for the vectors in Eq. 15–3, that is, $m\mathbf{v}_1 + \Sigma\int_{t_1}^{t_2}\mathbf{F}\,dt = m\mathbf{v}_2$. The two *momentum-vector diagrams* are simply outlined shapes of the particle which indicate the direction and magnitude of the particle's initial and final momentum, $m\mathbf{v}_1$ and $m\mathbf{v}_2$, respectively, Fig. 15–3. Similar to the free-body diagram, the *impulse-vector diagram* is an outlined shape of the particle, showing all the impulses that act on the particle when it is located at some intermediate point along its path. In general, whenever the magnitude or direction of a force $\mathbf{F}$ varies, the impulse of the force is determined by integration and represented on the impulse-vector diagram as $\mathbf{I} = \int_{t_1}^{t_2}\mathbf{F}\,dt$. If the force $\mathbf{F} = \mathbf{F}_c$ is *constant* for the time interval $(t_2 - t_1)$, the impulse applied to the particle is $\mathbf{I} = \mathbf{F}_c(t_2 - t_1)$, acting in the same direction as $\mathbf{F}_c$.

Step 2: Apply the principle of linear impulse and momentum, $m\mathbf{v}_1 + \Sigma\int_{t_1}^{t_2}\mathbf{F}\,dt = m\mathbf{v}_2$. If motion occurs in the x-y plane, the two scalar component equations can be formulated by resolving the vector components *directly* from each of the momentum- and impulse-vector diagrams.

The following examples numerically illustrate application of this two-step procedure.

Example 15–1

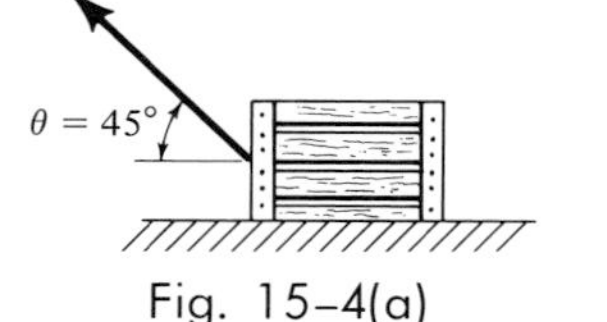

Fig. 15–4(a)

The 100-kg crate shown in Fig. 15–4a is originally at rest on the smooth horizontal surface. If a force $F = 200$ N, acting at an angle of $\theta = 45°$, is applied to the crate for 10 s, determine the final velocity of the crate and the normal force which the surface exerts on the crate during the time interval.

Solution

Step 1: The momentum- and impulse-vector diagrams for the crate are shown in Fig. 15–4b. It has been assumed that during the application of **F**, the crate remains on the surface and after 10 s, the crate moves to the left with a velocity $\mathbf{v}_2$. Since all the forces acting on the crate are *constant,* the respective impulses are simply the product of the force magnitude and 10 s, i.e., $\mathbf{I} = \mathbf{F}_c(t_2 - t_1)$.

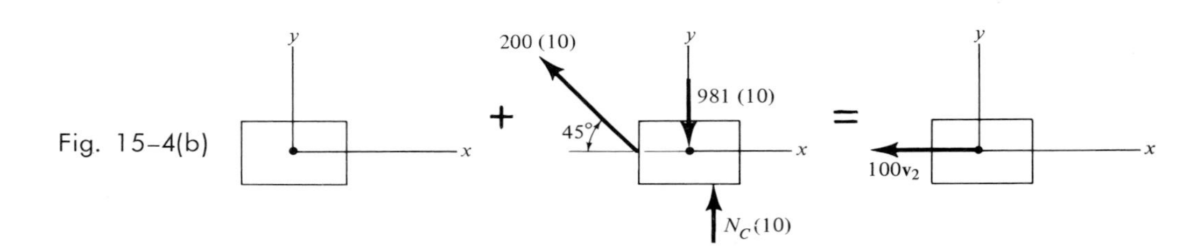

Fig. 15–4(b)

Step 2: Applying the principle of linear impulse and momentum by resolving the vectors along the x,y axes shown in the figure, we have

$(\overset{+}{\leftarrow})$

$$m(v_x)_1 + \Sigma \int_{t_1}^{t_2} F_x\,dt = m(v_x)_2$$

$$0 + 200(10)\cos 45° = 100v_2$$

$$v_2 = 14.1 \text{ m/s} \qquad \textit{Ans.}$$

$(+\uparrow)$

$$m(v_y)_1 + \Sigma \int_{t_1}^{t_2} F_y\,dt = m(v_y)_2$$

$$0 + N_C(10) - 981(10) + 200(10)\sin 45° = 0$$

$$N_C = 839.6 \text{ N} \qquad \textit{Ans.}$$

Example 15–2

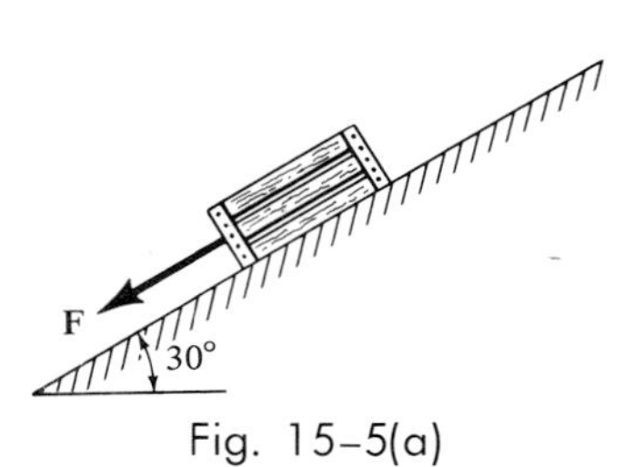

Fig. 15–5(a)

The crate shown in Fig. 15–5a has a mass of 50 kg and is acted upon by a force having a variable magnitude of $F = (20t)$ N, where t is in seconds. Compute the crate's velocity 4 s after **F** has been applied. The crate has an initial velocity of $v_1 = 1$ m/s down the plane and the coefficient of kinetic friction between the crate and the plane is $\mu_k = 0.3$.

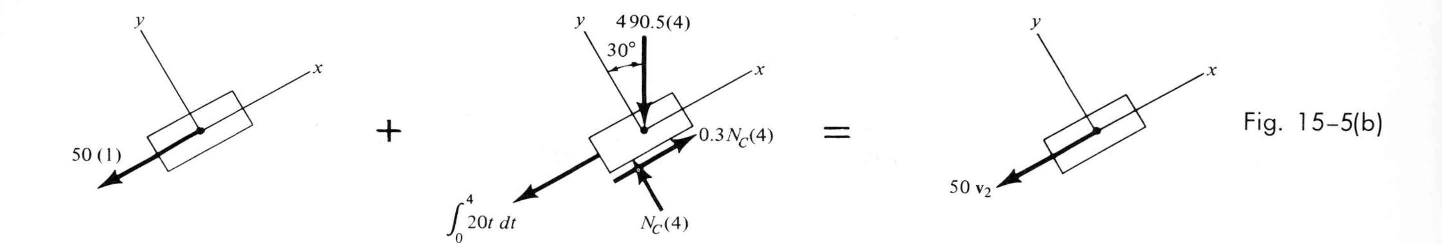

Fig. 15-5(b)

Solution

Step 1: The momentum- and impulse-vector diagrams of the crate are shown in Fig. 15-4*b*. Since the magnitude of force $F = 20t$ *varies* with time, the impulse of **F** must be determined by integrating over the 4-s time interval ($\mathbf{I} = \int_{t_1}^{t_2} \mathbf{F}\,dt$). The weight ($9.81(50) = 490.5$ N), normal force, and frictional force (which acts opposite to the direction of motion) are all *constant,* so that the impulse created by each of these forces is simply the magnitude of the force times 4 s ($\mathbf{I} = \mathbf{F}_c(t_2 - t_1)$).

Step 2: Applying the principle of impulse and momentum, with respect to the x and y axes, yields

$$(+\nearrow) \qquad m(v_x)_1 + \Sigma \int_{t_1}^{t_2} F_x\,dt = m(v_x)_2$$

$$-50(1) - \int_0^4 20t\,dt + 0.3N_C(4) - (490.5)(4)\sin 30^\circ = -50v_2$$

$$-50 - 160 + 1.2N_C - 981 = -50v_2$$

or

$$v_2 = 23.8 - 0.024N_C \tag{1}$$

Also,

$$(+\nwarrow) \qquad m(v_y)_1 + \Sigma \int_{t_1}^{t_2} F_y\,dt = m(v_y)_2$$

$$0 + N_C(4) - 490.5(4)\cos 30^\circ = 0 \tag{2}$$

Solving Eq. (2) yields

$$N_C = 424.8 \text{ N}$$

Substituting into Eq. (1) and solving for v_2, we have

$$v_2 = 13.6 \text{ m/s} \qquad \textit{Ans.}$$

This problem has also been solved using the equation of motion in Example 13–3. The two methods of solution should be compared. Since *force, velocity,* and *time* were involved in the problem, application of the principle of impulse and momentum eliminates the need for using kinematics ($a = dv/dt$) and thereby yields an easier method for solution.

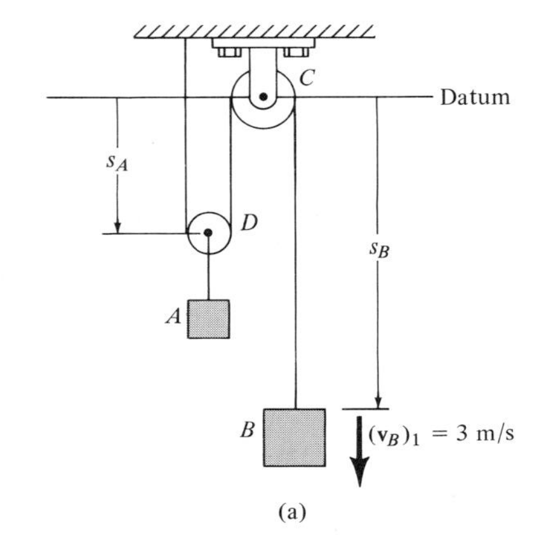

(a)

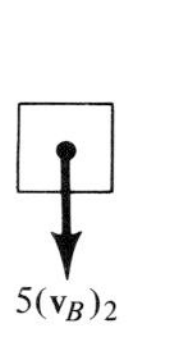

(b)

(c)

Fig. 15–6

Example 15–3

Blocks A and B shown in Fig. 15–6a have a mass of 3 kg and 5 kg, respectively. If B is initially moving downward with a velocity of 3 m/s, determine its velocity in 6 s. Neglect the mass of the pulleys and cord.

Solution

Since the blocks are subjected to dependent motion, the velocity of A may be related to that of B by using the kinematic analysis discussed in Sec. 12–9. A horizontal datum is established through the fixed point at C, Fig. 15–6a, and the changing positions of the blocks, s_A and s_B, are related to the constant total length l of the vertical segments of the cord by the equation

$$2s_A + s_B = l$$

Taking the time derivative yields

$$2v_A = -v_B$$

As indicated by the negative sign, when B moves downward at 3 m/s, as shown, A moves upward at 1.5 m/s.

Step 1: Using the above equation, the momentum- and impulse-vector diagrams for each block are shown in Fig. 15–6b. Since the impulses of the blocks' weights are constant, the impulses of the cord tensions, $\mathbf{T}_A$ and $\mathbf{T}_B$, are also constant. Furthermore, since the mass of pulley D is neglected, the cord tension $T_A = 2T_B$, Fig. 15–6c.

Step 2: Applying the principle of linear impulse and momentum in the vertical direction, we have

Block A:

$$(+\uparrow) \qquad m(v_A)_1 + \Sigma \int_{t_1}^{t_2} F_y\,dt = m(v_A)_2$$

$$3(1.5) + (2T_B)(6) - 3(9.81)(6) = 3\left(\frac{(v_B)_2}{2}\right) \qquad (1)$$

Block B:

$$(+\downarrow) \qquad m(v_B)_1 + \Sigma \int_{t_1}^{t_2} F_y\,dt = m(v_B)_2$$

$$5(3) + 5(9.81)(6) - T_B(6) = 5(v_B)_2 \qquad (2)$$

Solving Eqs. (1) and (2) yields

$$(v_B)_2 = 38.8 \text{ m/s}$$

$$T_B = 19.2 \text{ N} \qquad \textit{Ans.}$$

Problems

Except when stated otherwise, throughout this chapter assume that the coefficients of static and kinetic friction are equal, i.e., $\mu = \mu_s = \mu_k$.

15–1. A cannonball having a mass of 10 kg is fired upward, in the vertical direction, with a muzzle velocity of 200 m/s. Determine how long it takes before its velocity is reduced to zero, which occurs when it reaches a maximum height. Use the principle of impulse and momentum.

15–2. Using the principle of impulse and momentum, determine the tension in the cable and the velocity of the blocks in Prob. 13–14 in $t = 2$ s if they are released from rest.

15–3. A train consists of a 50-Mg engine and three cars, each having a mass of 30 Mg. If it takes 80 s for the train to increase its speed uniformly to 40 km/h, starting from rest, determine the force **T** developed at the coupling between the engine E and the first car A. The wheels of the engine provide the resultant frictional tractive force **F** which gives the train forward motion, whereas the car wheels roll freely. Determine **F** acting on the engine wheels.

Prob. 15–3

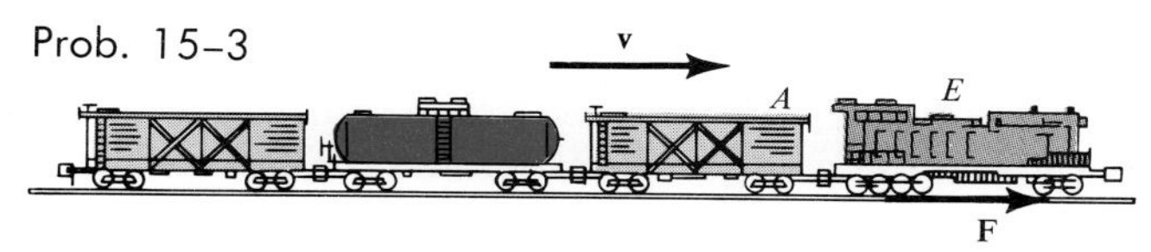

*** 15–4.** Assuming that the impulse acting on a 2-g bullet, as it passes horizontally through the barrel of a rifle, varies with time in the manner shown, determine the maximum net force $\mathbf{F}_O$ applied to the bullet when it is fired. The muzzle velocity is 500 m/s when $t = 0.75$ ms. Neglect friction between the bullet and the rifle barrel.

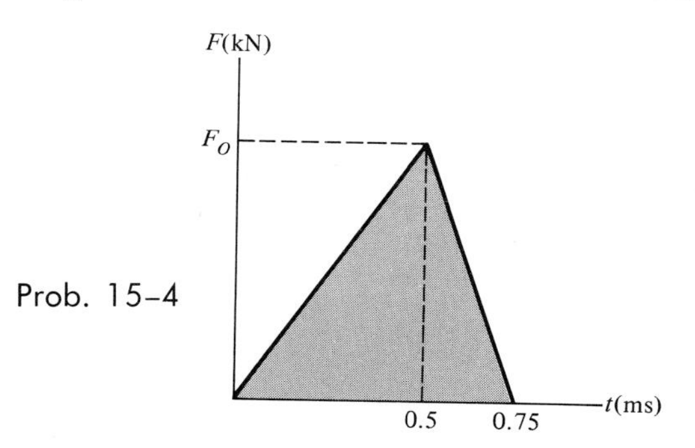

Prob. 15–4

15–5. A projectile having a mass of $m = 15$ kg is fired horizontally from a cannon located $h = 20$ m from the ground. If an average force of $F = 600$ kN is exerted on the projectile for $\Delta t = 0.02$ s while it is being fired through the cannon barrel, determine the range R, measured from the end of the cannon, A, to where the projectile strikes the ground at B.

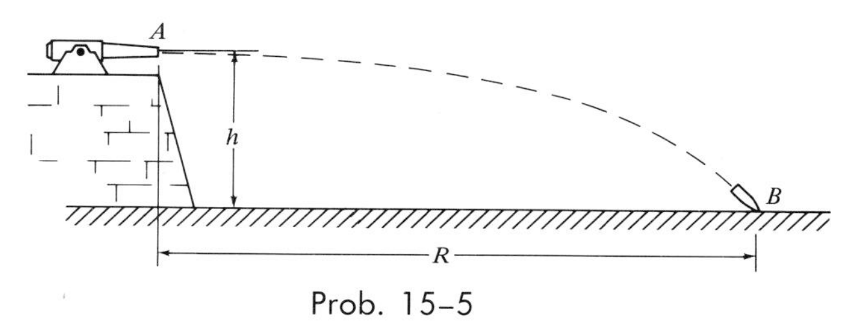

Prob. 15–5

15–5a. Solve Prob. 15–5 if the projectile has a weight of $W = 30$ lb, and $h = 60$ ft, $F = 13(10^4)$ lb, $\Delta t = 0.015$ s.

15–6. A 15-kg block is initially moving along a smooth horizontal surface with a speed of $v_1 = 3$ m/s to the left. If it is acted upon by a force **F**, which varies in the manner shown, determine the velocity of the block in 15 s. The argument for the cosine is in radians.

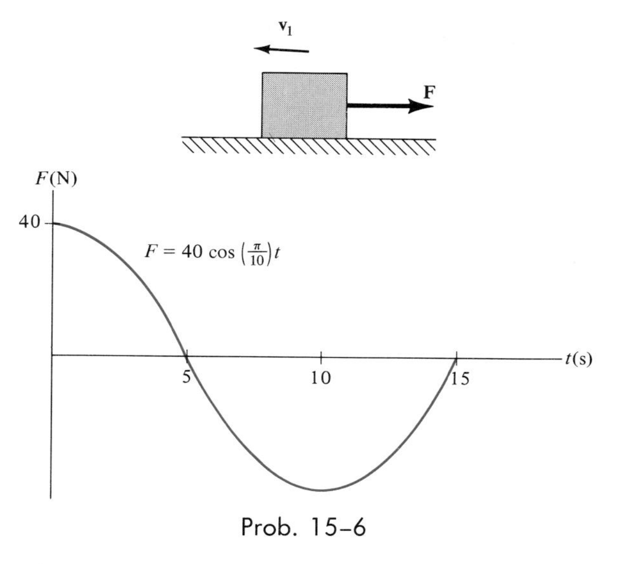

Prob. 15–6

15-7. The fuel-element assembly of a nuclear reactor has a mass of 3 Mg. Suspended in a vertical position and initially at rest, it is given an upward speed of 200 mm/s in 0.3 s using a crane hook H. Determine the average tension in cables AC and AB during this time interval.

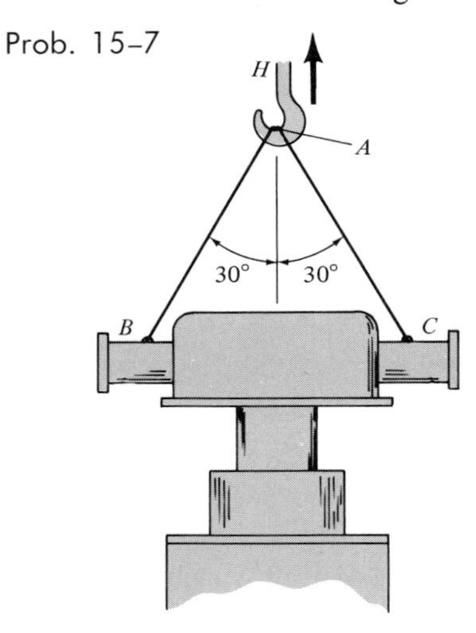

Prob. 15-7

*** 15-8.** In cases of emergency, the gas actuator can be used to move a 100-kg block B by exploding a charge C near a pressurized cylinder of negligible mass. As a result of the explosion, the cylinder fractures and the released gas forces the front part of the cylinder, A, to move B forward, giving it a speed of 200 mm/s in 0.4 s. If the coefficient of friction between B and the floor is $\mu = 0.5$, determine the impulse that the actuator must impart to B.

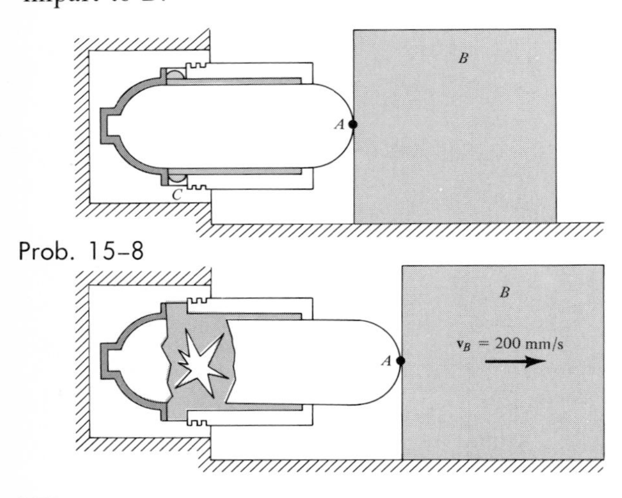

Prob. 15-8

15-9. A jet plane having a mass of 8 Mg is to be launched forward from a stationary position on an aircraft carrier using a catapult that exerts a horizontal force on the plane which varies as shown on the graph. If the carrier is traveling forward with a speed of 40 km/h, and the plane is to achieve an airspeed of 200 km/h after 5 s, determine the peak force $\mathbf{F}_O$ which must be exerted on the plane. While the catapult is in operation, the jet on the plane exerts a constant horizontal thrust of 60 kN.

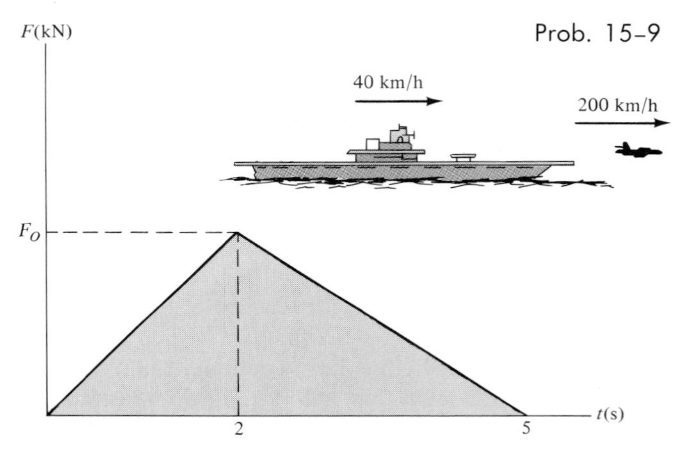

Prob. 15-9

15-10. Determine the speed of each block when $t = 2$ s after the blocks are released from rest. What is the tension in the cord? Neglect the mass of the cord and pulleys in the calculation. The blocks have a mass of $m_A = 4$ kg and $m_B = 8$ kg.

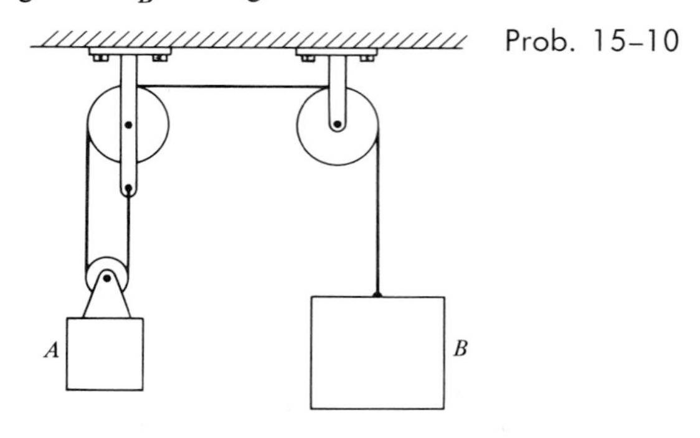

Prob. 15-10

15-10a. Solve Prob. 15-10 if $t = 1.5$ s, and the blocks weigh $W_A = 10$ lb and $W_B = 30$ lb.

15-11. A tank car has a mass of 20 Mg and is freely rolling to the right with a speed of 0.75 m/s. If it strikes the barrier, determine the horizontal impulse needed to stop the car if the spring in the bumper B has a stiffness of (a) $k \rightarrow \infty$ (bumper is rigid) and (b) $k = 15$ kN/m.

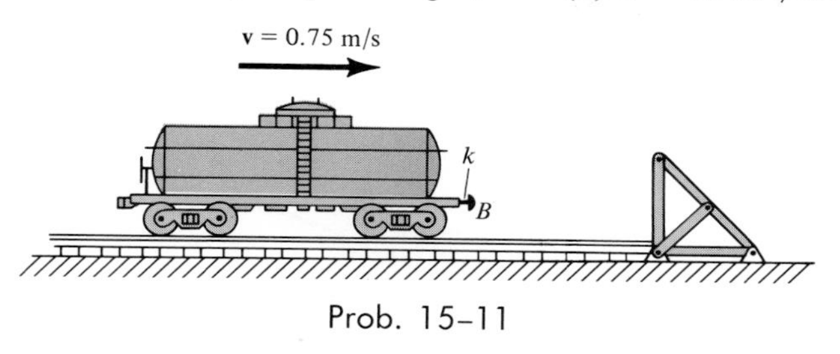

Prob. 15-11

* **15-12.** The 40-kg slider block is moving to the right with a speed of 1.5 m/s when it is acted upon by the forces $\mathbf{F}_1$ and $\mathbf{F}_2$. These loadings vary in the manner shown on the graph. Determine the speed of the block at $t = 6$ s. Neglect friction and the mass of the pulleys and cords.

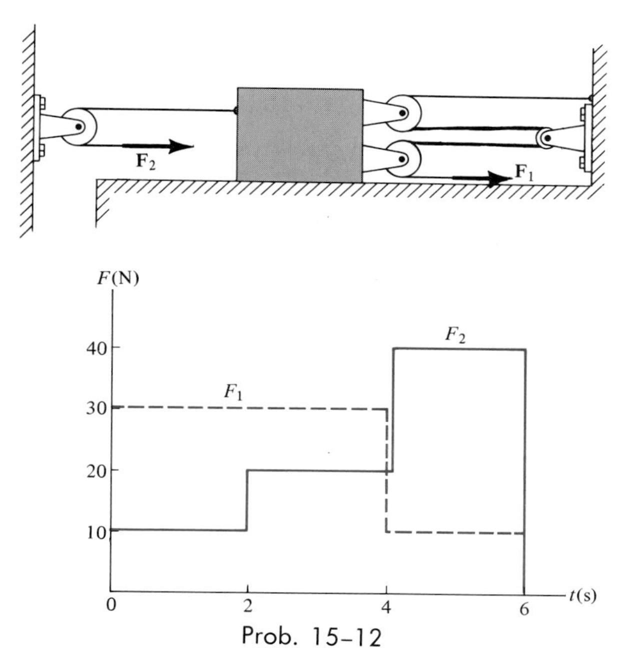

Prob. 15-12

15-13. The hoisting block and motor M have a total mass of 50 kg. When the motor is turned on, it draws in its attached cable at the rate of 0.75 m/s in 3 s, starting from rest. Neglecting the mass of the pulleys and cable, determine the force developed in the cable during this time.

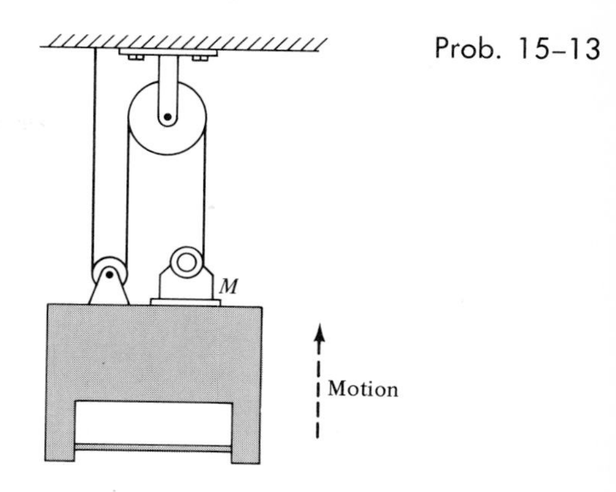

Prob. 15-13

15-14. A 20-kg crate is originally at rest on a horizontal surface for which the coefficient of friction is $\mu = 0.6$. If a horizontal force $\mathbf{F}$ is applied such that it varies with time as shown, determine the speed of the crate in 10 s. *Hint:* First determine the time needed to overcome friction and start the crate moving.

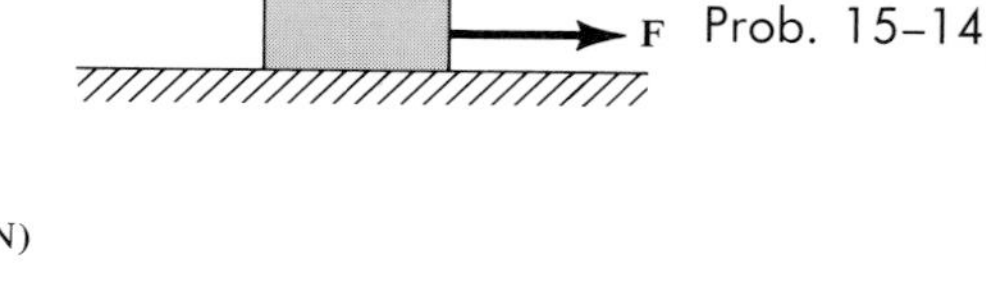

Prob. 15-14

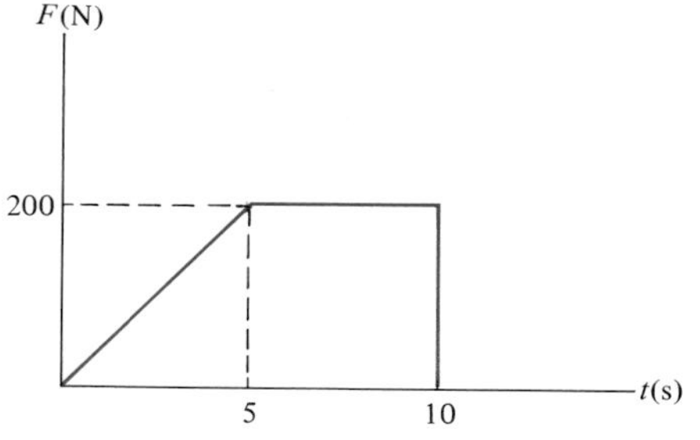

15-15. As a sphere, having a mass of $m = 1.2$ kg, falls vertically from rest through a liquid, the drag force exerted on it is $F_D = (0.6v)$ N, where v is the speed measured in m/s. Using the differential form of the impulse and momentum principle ($\Sigma F\,dt = m\,dv$), determine the time required for the sphere to attain one fourth of its terminal velocity (the terminal velocity occurs when $t \rightarrow \infty$).

15-15a. Solve Prob. 15-15 if the sphere has a weight of $W = 4$ lb and $F_D = (0.052v)$ lb, where v is measured in ft/s.

*** 15-16.** The motor M pulls on the cable with a force $\mathbf{F}$ that has a magnitude which varies as shown on the graph. If the 20-kg crate is originally resting on the floor such that the cable tension is zero when the motor is turned on, determine the speed of the crate when $t = 6$ s. *Hint:* First determine the time needed to begin lifting the crate.

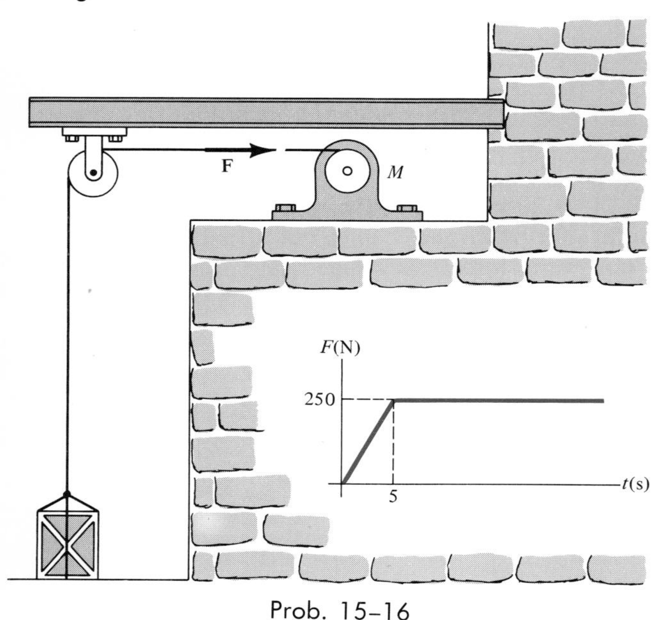

Prob. 15-16

15-17. A crate having a mass of 50 kg rests against a stop block s which prevents the crate from moving down the plane. If the coefficient of friction between the plane and the crate is $\mu = 0.3$, determine the time needed for the force $\mathbf{F}$ to give the crate a speed of 2 m/s up the plane. $\mathbf{F}$ always acts parallel to the plane and has a magnitude of $F = (300t)$ N, where t is measured in seconds. *Hint:* First determine the time needed to overcome friction and start the crate moving.

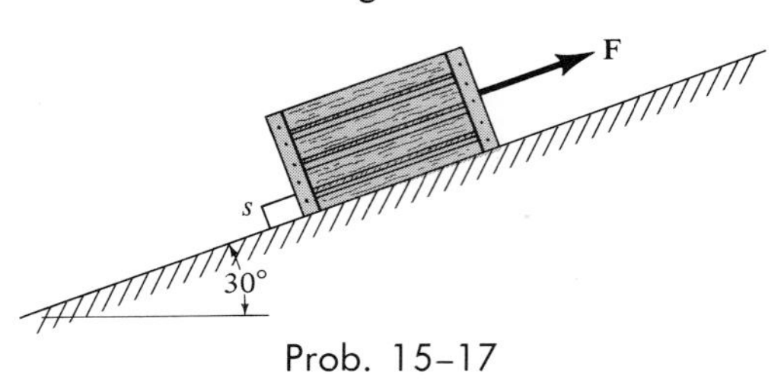

Prob. 15-17

15-2. Principle of Linear Impulse and Momentum for a System of Particles

The principle of linear impulse and momentum for a system of n particles, Fig. 15-7, may be obtained from Eq. 13-6 ($\Sigma \mathbf{F} = \Sigma m_i \mathbf{a}_i$), which may be rewritten as

$$\Sigma \mathbf{F} = \Sigma m_i \frac{d\mathbf{v}_i}{dt} \qquad (15\text{-}5)$$

The term on the left side represents the sum of all the *external forces* acting on the system of particles. The internal forces between particles do not appear with this summation, since they occur in equal but opposite collinear pairs and therefore cancel out. On the right side $\mathbf{v}_i$ defines the instantaneous velocity of the ith particle. Multiplying both sides of Eq. 15–5 by dt and integrating between the limits $t = t_1$, $\mathbf{v}_i = (\mathbf{v}_i)_1$, and $t = t_2$, $\mathbf{v}_i = (\mathbf{v}_i)_2$, yields

$$\Sigma \int_{t_1}^{t_2} \mathbf{F}\, dt = \Sigma m_i \int_{(\mathbf{v}_i)_1}^{(\mathbf{v}_i)_2} d\mathbf{v}_i$$

$$\Sigma \int_{t_1}^{t_2} \mathbf{F}\, dt = \Sigma m_i (\mathbf{v}_i)_2 - \Sigma m_i (\mathbf{v}_i)_1$$

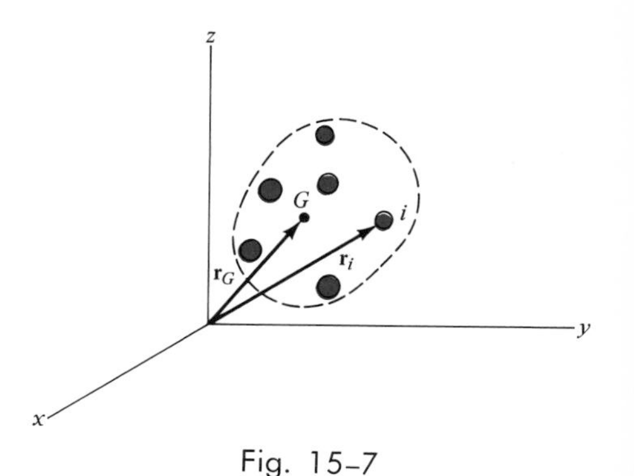

Fig. 15–7

or

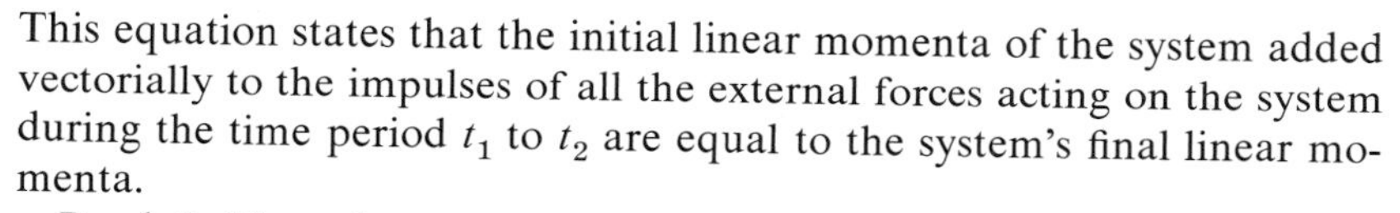

$$\Sigma m_i (\mathbf{v}_i)_1 + \Sigma \int_{t_1}^{t_2} \mathbf{F}\, dt = \Sigma m_i (\mathbf{v}_i)_2 \qquad (15\text{–}6)$$

This equation states that the initial linear momenta of the system added vectorially to the impulses of all the external forces acting on the system during the time period t_1 to t_2 are equal to the system's final linear momenta.

By definition, the mass center G of the system is determined from

$$m\mathbf{r}_G = \Sigma m_i \mathbf{r}_i$$

where $m = \Sigma m_i$ is the total mass of all the particles, and $\mathbf{r}_G$ and $\mathbf{r}_i$ are defined in Fig. 15–7. Taking the time derivative, we have

$$m\mathbf{v}_G = \Sigma m_i \mathbf{v}_i \qquad (15\text{–}7)$$

which states that the total linear momentum of the system of particles is equivalent to the linear momentum of a "fictitious" aggregate particle of mass $m = \Sigma m_i$ which is moving with the velocity of the mass center G of the system. Substituting Eq. 15–7 into Eq. 15–6, yields

$$m(\mathbf{v}_G)_1 + \Sigma \int_{t_1}^{t_2} \mathbf{F}\, dt = m(\mathbf{v}_G)_2$$

This equation states that the initial linear momentum of the aggregate particle plus (vectorially) the external impulses acting on the system of particles during the time interval t_1 to t_2 is equal to the aggregate particle's final linear momentum. Since in reality all particles must have finite size to possess mass, the above equation justifies application of the principle of linear impulse and momentum to a rigid body represented as a single particle.

15–3. Conservation of Linear Momentum for a System of Particles

When the sum of the external impulses acting on a system of particles is zero, Eq. 15–6 reduces to a simplified form,

$$\Sigma m_i(\mathbf{v}_i)_1 = \Sigma m_i(\mathbf{v}_i)_2 \tag{15–8}$$

This equation is referred to as the *conservation of linear momentum.* It states that the linear momenta for a system of particles remain constant throughout the time period t_1 to t_2. Using Eq. 15–7 ($m\mathbf{v}_G = \Sigma m_i\mathbf{v}_i$), we can also write

$$(\mathbf{v}_G)_1 = (\mathbf{v}_G)_2 \tag{15–9}$$

which indicates that the velocity $\mathbf{v}_G$ of the mass center for the system of particles does not change.

To illustrate a situation for which the conservation of linear momentum applies, consider the missile shown in Fig. 15–8*a* which has an intended

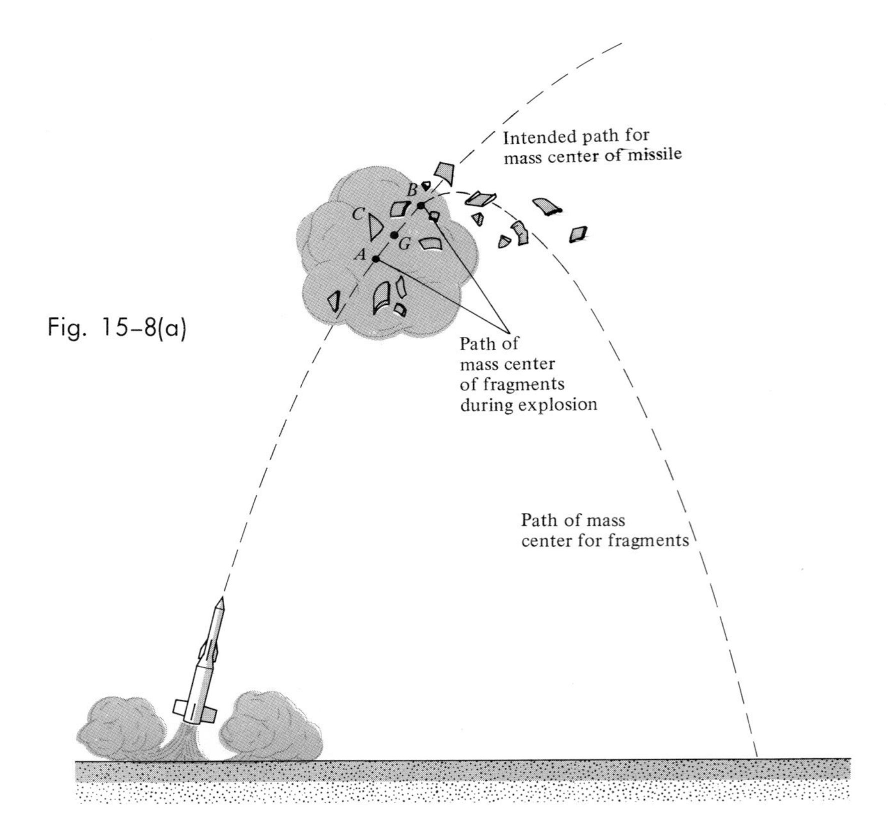

Fig. 15–8(a)

trajectory shown by the dashed path. Suppose that when the missile reaches point A, it suddenly blows up. Just before the explosion the velocity of the fragments (or all the particles Σm_i) is the same, since the missile is intact. Just after the explosion, however, the velocity of each fragment changes. Even so, the *total momentum of the system is conserved during the time Δt of the explosion.* There are two reasons for this. First, the fragments are blown apart only by impulses which occur from equal but opposite *internal forces* acting on the system; and, second, the time for explosion Δt is very *short,* so that the *external impulse* created by gravity is negligible in comparison to the large internal explosive impulses given to the system. For these reasons the mass center G of the fragments will continue to follow the intended path AB of the missile for the time Δt, Fig. 15–8*a*. As the time *after the explosion* becomes longer, however, the gravitational impulse becomes influential and the path of the mass center of the fragments changes, as shown in the figure.

If a study is made of the motion of just *one* of the missile fragments instead of the entire system of fragments, the explosive impulse on the fragment is considered to be "external" and hence it must be included in the analysis involving impulse and momentum. For example, consider fragment C of mass m in Fig. 15–8*a*. The momentum- and impulse-vector diagrams are shown in Fig. 15–8*b*. Initially, the fragment has a momentum $m\mathbf{v}_1$ *just before* the explosion, where $\mathbf{v}_1$ is the velocity of the missile. During the time Δt of the explosion, both the explosive impulse $\int \mathbf{F}\,dt$ and the weight impulse $\mathbf{W}(\Delta t)$ act on the fragment. However, as stated above, since the weight is much smaller than the explosive force and Δt is very small, $\mathbf{W}(\Delta t) \approx \mathbf{0}$ and it can be neglected in the analysis. Assuming the final momentum of the fragment to be $m\mathbf{v}_2$, then applying the principle of impulse and momentum, we have

$$m\mathbf{v}_1 + \int \mathbf{F}\,dt = m\mathbf{v}_2$$

If the final velocity of the fragment is known, the explosive impulse is

$$\mathbf{I} = \int \mathbf{F}\,dt = m(\mathbf{v}_2 - \mathbf{v}_1)$$

Generally, the variation of the explosive force $\mathbf{F}$ with time t is *not known.* However, it is suspected that during the explosion time Δt the magnitude of $\mathbf{F}$ will increase sharply to some value F_{max}, as shown in Fig. 15–8*c*, then decrease sharply to zero. Consequently, the impulse is represented as the *colored area* under the graph. Provided Δt and $\mathbf{I}$ are *known,* then for purposes of calculation, the impulsive loading may be *approximated* by an *equivalent rectangular area,* shown shaded in Fig. 15–8*c*. In this case, the *average impulsive force,* $\mathbf{F}_{avg}$, can be determined since $\mathbf{I} = \mathbf{F}_{avg}\,\Delta t$. Using the above equation,

$$\mathbf{F}_{avg} = \frac{m}{\Delta t}(\mathbf{v}_2 - \mathbf{v}_1)$$

$\int \mathbf{F}dt = \mathbf{F}_{avg}\,\Delta t$

$\Delta m\mathbf{v}_1$ + = $m\mathbf{v}_2$

$\mathbf{W}\Delta t \approx 0$

(b)

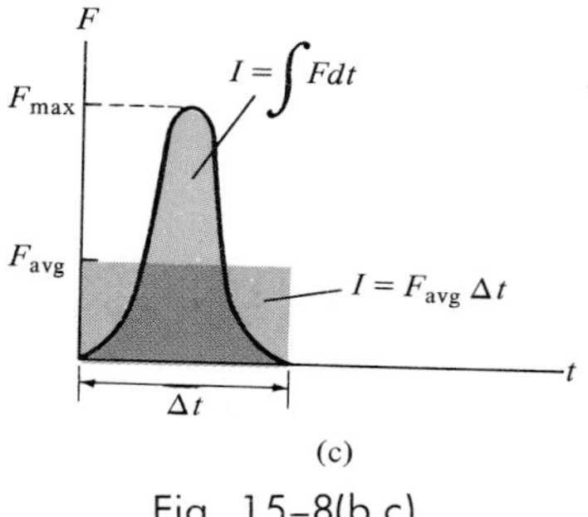

(c)

Fig. 15–8(b,c)

Impulsive and Nonimpulsive Forces. A force that is relatively very large and acts for a very short time, such that it produces a significant change in momentum, is called an *impulsive force*. In the above example, the average force of the explosion, $\mathbf{F}_{avg}$, acting on the missile fragment, may be classified as an impulsive force. By comparison, the weight $\mathbf{W}$ of the fragment is a *nonimpulsive force,* since it creates no significant change in the momentum of the fragment during the short time Δt. Consequently, to simplify the analysis of a problem, since the impulse of a nonimpulsive force is much smaller than that of an impulsive force, the effects of nonimpulsive forces are *neglected* when applying the principles of impulse and momentum. In general, *impulsive forces occur from the striking of one body against another, whereas nonimpulsive forces include the weight of a body, the force imparted by a slightly deformed spring having a relatively small stiffness, or any force that is small compared to other larger (impulsive) forces.*

PROCEDURE FOR ANALYSIS

Generally the principle of linear impulse and momentum or the conservation of linear momentum is applied to a system of particles in order to determine the final velocities of the particles *just after* the time period considered. By applying these equations to the entire system, the internal impulses acting within the system, which may be unknown, are eliminated from the analysis, since they occur in equal but opposite collinear pairs. For application, the following two-step procedure should be used.

Step 1: Draw the momentum- and impulse-vector diagrams for the system of particles. The conservation of linear momentum applies to the system in a given direction when *no external impulsive forces* act on the system in that direction. To determine if this condition applies, one should investigate the impulse-vector diagram in order to clearly distinguish the external impulsive and nonimpulsive forces from the system's internal impulses.

Step 2: Apply the principle of linear impulse and momentum or the conservation of linear momentum in the appropriate directions. Most often the scalar component equations can be formulated by resolving the vector components *directly* from each of the momentum- and impulse-vector diagrams. If the particles are subjected to dependent motion, kinematics, as discussed in Sec. 12–9, can be used to relate the velocities.

The following examples numerically illustrate application of this two-step procedure.

Example 15–4

The 600-kg cannon shown in Fig. 15–9*a* fires a 5-kg projectile with a muzzle velocity of 525 m/s. If firing takes place in 0.008 s, determine (a) the velocity of the cannon just after firing, and (b) the average impulsive force acting on the projectile. The cannon support is firmly fixed to the ground and the horizontal recoil of the cannon is absorbed by two springs.

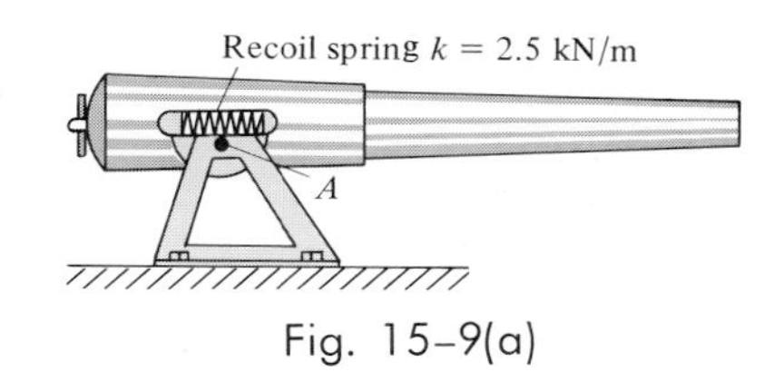

Fig. 15–9(a)

Solution

Part (a). *Step 1:* The momentum- and impulse-vector diagrams for the projectile and cannon, when considered as a single system, are shown in Fig. 15–9*b*. The impulsive forces, $\int \mathbf{F}\,dt$, between the cannon and projectile are *internal* to the system and are therefore not included in the momentum-impulse analysis of the system. Furthermore, during the time

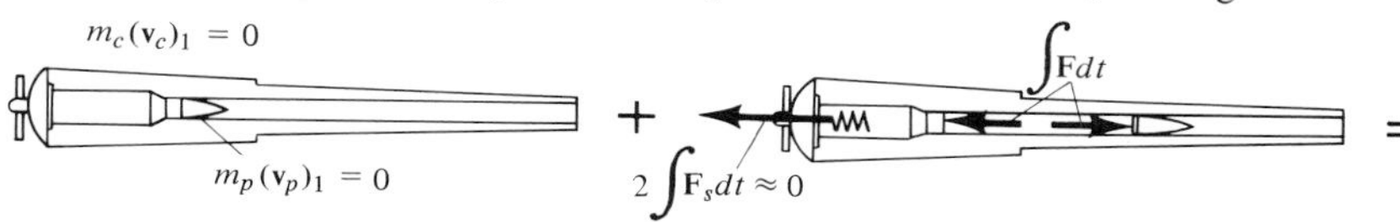

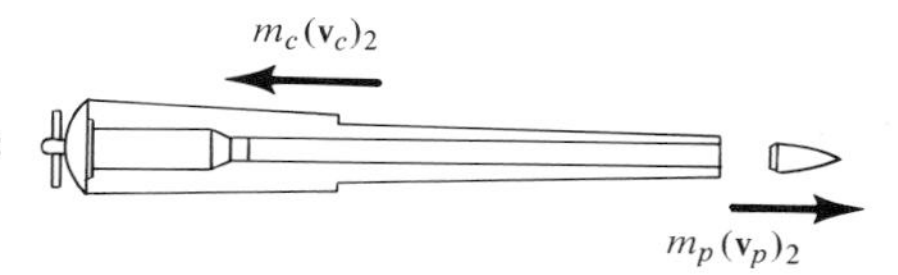

Fig. 15–9(b)

$\Delta t = 0.008$ s, the two recoil springs, which are attached to the support, each exert a nonimpulsive force $\mathbf{F}_s$ on the cannon. This is because Δt is very short, so that during this time the cannon only moves through a very small distance* x. Consequently, $F_s = kx \approx 0$. Hence it may be concluded that momentum for the system is conserved in the *horizontal direction*.
Step 2: Applying the conservation of linear momentum to the system, using the data on the momentum-vector diagrams, we have

$$\Sigma m\mathbf{v}_1 = \Sigma m\mathbf{v}_2$$

$$(\overset{+}{\rightarrow}) \qquad m_c(v_c)_1 + m_p(v_p)_1 = m_c(v_c)_2 + m_p(v_p)_2$$

$$0 + 0 = -600(v_c)_2 + 5(525)$$

$$(v_c)_2 = 4.37 \text{ m/s} \qquad \textit{Ans.}$$

Part (b). The average impulsive force exerted by the cannon on the projectile is determined by applying the principle of linear impulse and momentum to the projectile. The momentum- and impulse-vector diagrams are shown in Fig. 15–9*c*. Since $\int F\,dt = F_{\text{avg}}\,\Delta t = F_{\text{avg}}(0.008)$, we have

$$(\overset{+}{\rightarrow}) \qquad m(v_p)_1 + \Sigma \int F\,dt = m(v_p)_2$$

$$0 + F_{\text{avg}}(0.008) = 5(525)$$

$$F_{\text{avg}} = 328.1 \text{ kN} \qquad \textit{Ans.}$$

$\int \mathbf{F}dt$ 5(525)

Fig. 15–9(c)

*If the cannon is firmly fixed to its support (no springs), the reactive force of the support on the cannon must be considered as an external impulse to the system, since the support allows no movement of the cannon.

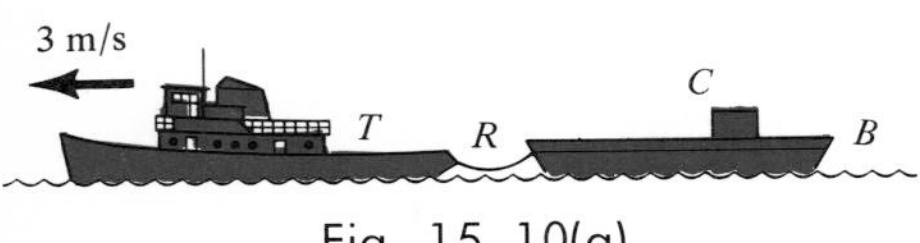

Fig. 15–10(a)

Example 15–5

The 350-Mg tugboat T shown in Fig. 15–10a is used to pull the 50-Mg barge B with a rope R. A 10-Mg crate C rests on top of the barge, where the coefficient of friction between the barge and the crate is $\mu = 0.3$. If the initial velocity of the tugboat is $(v_T)_1 = 3$ m/s while the rope is slack, determine the velocity of the tugboat *directly after* towing occurs. Assume that when towing occurs, the rope does not stretch. Neglect the frictional effects of the water.

Solution

Step 1: Since the rope does not stretch, the impulse created between the tugboat and the barge is *instantaneous*. As a result, the barge is pulled (or jerked) from under the crate so fast that the crate has effectively *zero velocity* and hence slips on the surface of the barge. The impulse- and momentum-vector diagrams for the entire system (tugboat, crate, and barge) are shown in Fig. 15–10b. The impulses created by the rope R and the frictional force between the crate and the barge are *internal* to the system, and therefore momentum of the system is conserved during the instant of towing.

Step 2: Noting that $(v_B)_2 = (v_T)_2$, we have

$$(\overset{+}{\leftarrow}) \quad m_T(v_T)_1 + m_B(v_B)_1 + m_C(v_C)_1 = m_T(v_T)_2 + m_B(v_B)_2 + m_C(v_C)_2$$

$$350(10^3)(3) + 0 + 0 = 350(10^3)(v_T)_2 + 50(10^3)(v_T)_2 + 0$$

Solving,

$$(v_T)_2 = 2.63 \text{ m/s} \qquad \textit{Ans.}$$

This value represents the tugboat's velocity *just after* the towing impulse. At this instant, the crate, while slipping on the barge, still has zero velocity, as explained previously. In time, however, the frictional force developed between the crate and the barge creates an impulse on the crate which eventually gives it a velocity of 2.63 m/s.

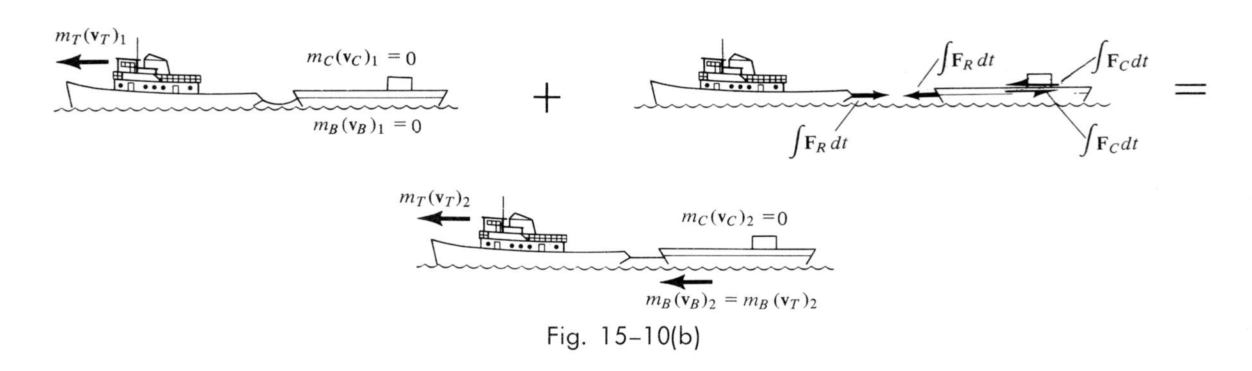

Fig. 15–10(b)

Example 15-6

A rigid pile P shown in Fig. 15–11a has a mass of 800 kg, and is driven into the ground using a hammer H that has a mass of 300 kg. The hammer falls from rest from a height of $y_0 = 0.5$ m and strikes the top of the pile. Determine the initial impulse which the hammer imparts on the pile if (a) the bottom of the pile is resting on rigid bedrock at B, and (b) the pile is surrounded entirely by loose sand so that after striking, the hammer does *not* rebound off the pile.

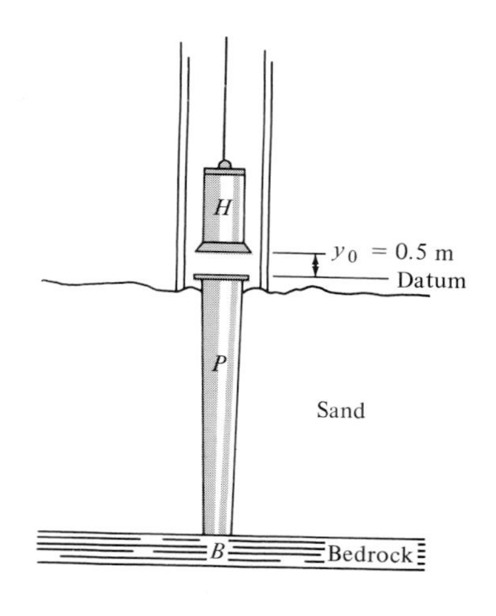

(a)

Solution

The velocity at which the hammer strikes the pile can be determined using the conservation-of-energy theorem applied to the hammer. With the datum at the top of the pile, Fig. 15–11a, we have

$$T_0 + V_0 = T_1 + V_1$$
$$\tfrac{1}{2}m_H(v_H)_0^2 + W_H y_0 = \tfrac{1}{2}m_H(v_H)_1^2 + W_H y_1$$
$$0 + 300(9.81)(0.5) = \tfrac{1}{2}(300)(v_H)_1^2 + 0$$
$$(v_H)_1 = 3.13 \text{ m/s}$$

Part (a). When the hammer strikes the pile with an impulse $\int \mathbf{R}\,dt$, the bedrock imparts an equal but opposite impulsive reaction $\int \mathbf{R}\,dt$, as shown in Fig. 15–11b. Here the weight of the pile, $\mathbf{W}_P$, is a nonimpulsive force, since $R \gg W_P$. Since no downward movement of the pile occurs, the hammer is brought momentarily to rest shortly after striking the pile. Under these conditions, the momentum- and impulse-vector diagrams of the hammer are shown in Fig. 15–11c.

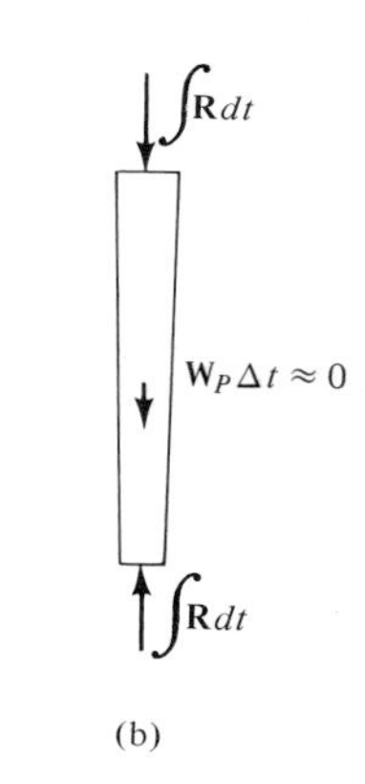

(b)

Applying the principle of linear impulse and momentum to the hammer, realizing that $W_H \ll R$, we have

$$(+\uparrow) \qquad m_H(v_H)_1 + \Sigma \int_{t_1}^{t_2} F_y\,dt = m_H(v_H)_2$$
$$-300(3.13) + \int R\,dt = 0$$
$$\int R\,dt = 939 \text{ N} \cdot \text{s} \qquad \textit{Ans.}$$

(c) 300(3.13) + $\mathbf{W}_H \Delta t \approx 0$, $\int \mathbf{R}\,dt$ =

Part (b). When the soil is loose, both the pile and hammer move downward with a *common velocity* $(\mathbf{v})_2$ just after contact. Hence, in this case the momentum- and impulse-vector diagrams *for the hammer* are shown in Fig. 15–11d. Consequently,

$$(+\uparrow) \qquad m_H(v_H)_1 + \Sigma \int_{t_1}^{t_2} F_y\,dt = m_H(v_H)_2$$
$$-300(3.13) + \int R\,dt = -300\,v_2 \qquad (1)$$

(d) 300(3.13) + $\mathbf{W}_H \Delta t \approx 0$, $\int \mathbf{R}\,dt$ = $300\mathbf{v}_2$

Fig. 15–11(a–d)

A second equation, used to determine v_2, is obtained by considering the hammer and pile as a single system.

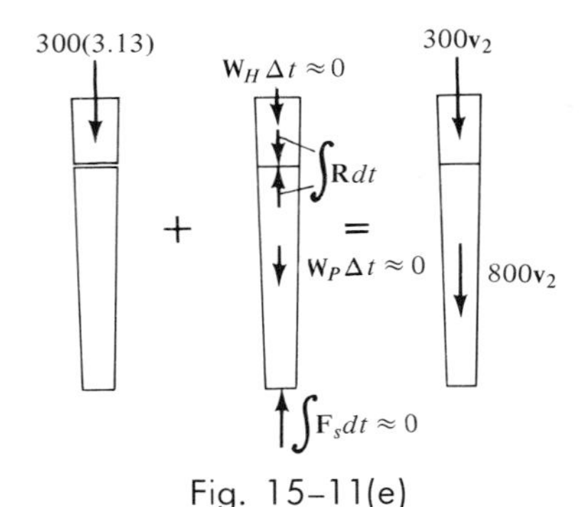

Fig. 15–11(e)

Step 1: From the physical aspects of the problem, the impulse-vector diagram indicates that during the short time occurring just before to just after the collision, the weights of the hammer and pile, and the resistance force $\mathbf{F}_s$ of the soil are all *nonimpulsive*. Furthermore, the impulse $\int \mathbf{R}\, dt$ is internal to the system and therefore cancels. Consequently, momentum is conserved.

Step 2: Applying the conservation of linear momentum, we have

$$(+\downarrow) \qquad m_H(v_H)_1 + m_P(v_P)_1 = m_H v_2 + m_P v_2$$

$$300(3.13) + 0 = 300v_2 + 800v_2$$

$$v_2 = 0.854 \text{ m/s}$$

Substituting into Eq. (1) and solving for the impulse yields

$$\int R\, dt = 682.9 \text{ N} \cdot \text{s} \qquad \textit{Ans.}$$

Since in reality all materials deform, the actual impulse given to the pile has a magnitude somewhere between the upper and lower bounds determined in parts (a) and (b) of this problem.

Example 15–7

A boy having a mass of 40 kg stands on the back of a 15-kg toboggan which is originally at rest, Fig. 15–12*a*. If he walks to the front *B* and stops, determine the distance the toboggan moves. Assume that the toboggan is resting on ice, so that friction on the bottom of the toboggan may be neglected.

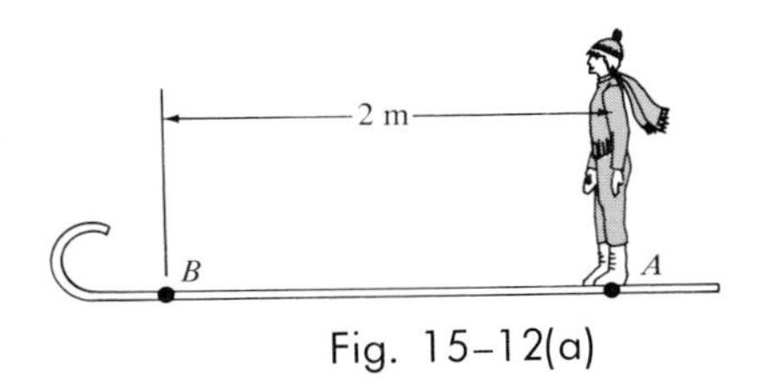

Fig. 15–12(a)

Solution I

The unknown frictional force of the boy's shoes on the bottom of the toboggan can be *excluded* from the analysis if the toboggan and boy on it are considered as a single system. In this way the frictional force becomes internal and the conservation of momentum applies. Since both the initial and final momentum of the system are zero (because the initial and final velocities are zero), the momentum must also be zero when the boy is at some intermediate point between *A* and *B*. From the momentum diagram, Fig. 15–12*b*, we can therefore write

$$(\overset{+}{\rightarrow}) \qquad -m_b v_b + m_t v_t = 0 \qquad (1)$$

Here the two unknowns v_b and v_t represent the velocities of the boy and the toboggan measured from a *fixed inertial reference* on the ground. The *positions* of the toboggan and boy are determined by integration. Assuming the initial position of the system to be zero, we have

$$-m_b s_b + m_t s_t = 0$$

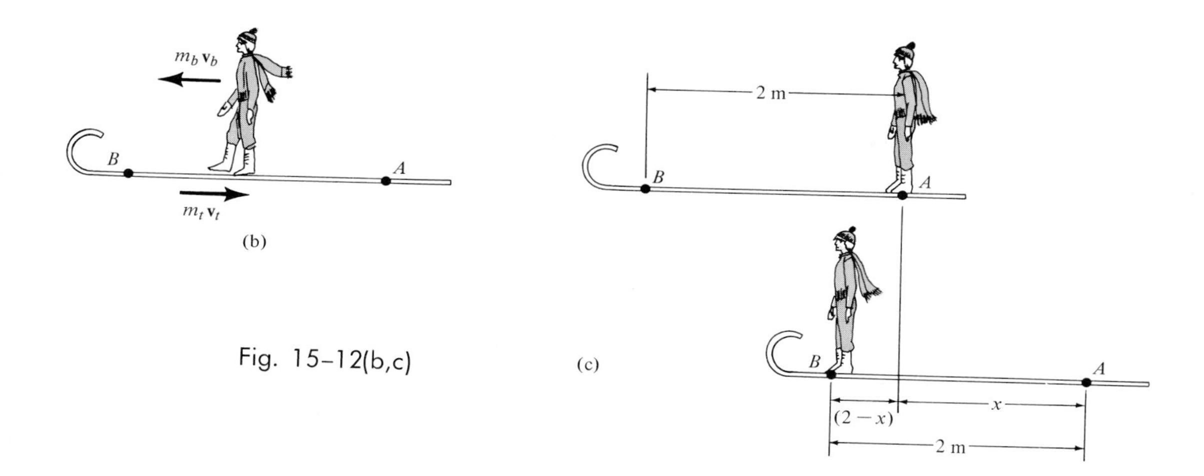

(b)

Fig. 15–12(b,c)

(c)

If in the final position point A on the toboggan is located $s_t = x$ to the right (in the same direction as $\mathbf{v}_t$), Fig. 15–12c, then when the boy is at B, this point has the position $s_b = (2 - x)$ m to the left (in the same direction as $\mathbf{v}_b$). Hence,

$$-m_b(2 - x) + m_t x = 0$$

$$x = \frac{2m_b}{m_b + m_t} = \frac{2(40)}{40 + 15} = 1.45 \text{ m} \qquad \textit{Ans.}$$

Solution II

The problem may also be solved by considering the relative motion of the boy with respect to the toboggan, $\mathbf{v}_{b/t}$. This velocity is related to the velocities of the boy and toboggan by the equation $\mathbf{v}_b = \mathbf{v}_t + \mathbf{v}_{b/t}$, Eq. 12–58. Since positive motion is assumed to be to the right in Eq. (1), $\mathbf{v}_b$ and $\mathbf{v}_{b/t}$ are negative, since the boy's motion is to the left. Hence, in scalar form, $-v_b = v_t - v_{b/t}$ and Eq. (1) then becomes

$$m_b(v_t - v_{b/t}) + m_t v_t = 0$$

Integrating gives

$$m_b(s_t - s_{b/t}) + m_t s_t = 0$$

Assuming that the toboggan moves a distance x to the right, Fig. 15–12c, realizing $s_{b/t} = 2$ m, we have

$$m_b(x - 2) + m_t x = 0$$

$$x = \frac{2m_b}{m_b + m_t} = \frac{2(40)}{40 + 15} = 1.45 \text{ m} \qquad \textit{Ans.}$$

15–18. A rifle has a mass of 2.3 kg. If it is loosely gripped and a 1.4-g bullet is fired from it with a muzzle velocity of 1300 m/s, determine the recoil velocity of the rifle just after firing.

15–19. A railroad car having a mass of 15 Mg is coasting at 1.5 m/s on a horizontal track. At the same time another car having a mass of 12 Mg is coasting at 0.75 m/s in the opposite direction. If the cars meet and couple together, determine the speed of both cars just after the coupling. Compute the difference between the total kinetic energy before and after coupling has occurred, and explain qualitatively what happened to this energy.

***15–20.** The barge has a mass of $m_B = 20$ Mg and supports the automobile, which has a mass of $m_A = 1.7$ Mg. If the barge is not tied to the pier P, and the automobile is driven forward at a constant speed of $v' = 5$ m/s measured relative to the barge, determine the speed at which the barge moves away from the pier. Neglect the resistance of the water.

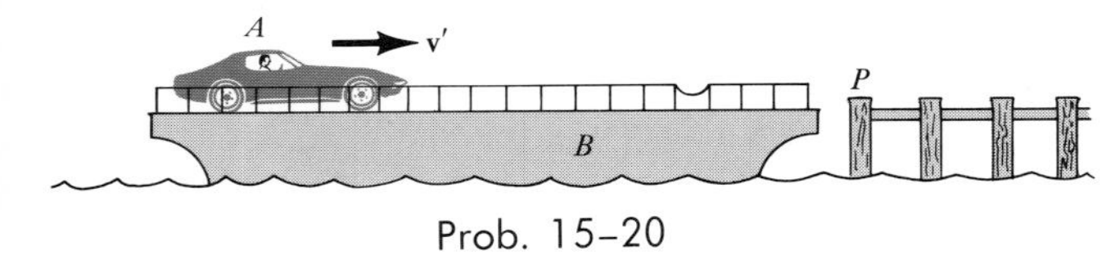

Prob. 15–20

***15–20a.** Solve Prob. 15–20 if the barge weighs $W_B = 4(10^4)$ lb, the automobile weighs $W_A = 3500$ lb, and $v' = 7$ ft/s.

15–21. A girl having a mass of 50 kg slides down the smooth slide onto the surface of a wagon having a mass of 10 kg. Determine the speed of the wagon at the instant the girl stops sliding on the wagon. If someone ties the wagon to the slide at B, determine the horizontal impulse the girl will exert at C in order to stop her motion. Neglect friction and assume that the girl starts from rest at the top of the slide, A.

Prob. 15–21

15–22. When the 4-kg wooden block is at rest, $\theta = 0°$, a 2-g bullet strikes and becomes embedded in it. If it is observed that the block swings upward to a maximum angle of $\theta = 6°$, estimate the initial speed of the bullet.

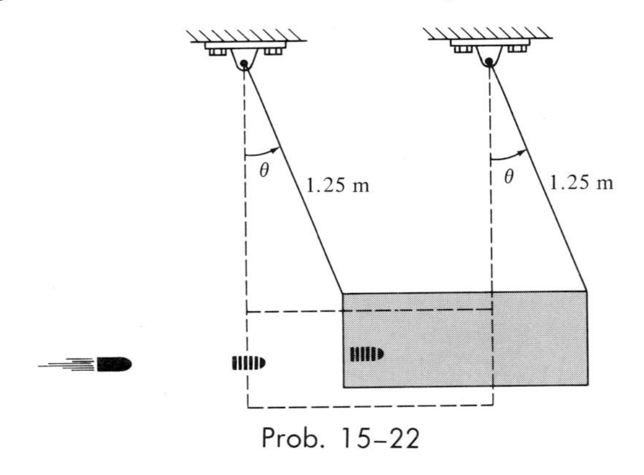

Prob. 15–22

15–23. Two men, each having a mass of 70 kg, are standing in the back of a 100-kg boat, which is originally at rest. Each man runs to the front of the boat with a constant speed of 3 m/s, measured relative to the boat, and dives off horizontally. Determine the final velocity of the boat if (a) both men dive off together, and (b) the men dive off one at a time.

***15-24.** A man wearing ice skates, throws a 20-kg block with an initial velocity of 4 m/s measured relative to him in the direction shown. If he is originally at rest, and completes the throw in 1.5 s while keeping his legs rigid, determine the horizontal velocity of the man just after releasing the block. What is the vertical reaction of both his skates on the ice during the throw? The man has a mass of 70 kg. Neglect friction and the motion of his arms.

Prob. 15-24

15-25. A boy, having a mass of $m_b = 50$ kg, jumps off a wagon with a velocity of $v_{b/w} = 3$ m/s, measured relative to the wagon. If the angle of jump, measured from the wagon, is 30°, determine the horizontal velocity $(\mathbf{v}_w)_2$ of the wagon just after the jump. Originally both the wagon and the boy are at rest. Also, compute the total average impulsive force that all four wheels of the wagon exert on the ground if the boy jumps off in $\Delta t = 0.8$ s. The wagon has a mass of 20 kg.

Prob. 15-25

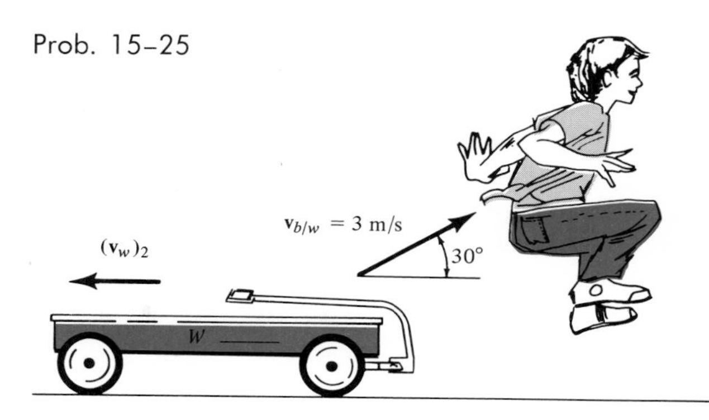

15-25a. Solve Prob. 15-25 if the boy weighs $W_b =$ 90 lb, the wagon weighs $W_w = 30$ lb, $v_{b/w} = 7$ ft/s, and $\Delta t = 0.8$ s.

15-26. The barge B has a mass of 15 Mg and supports an automobile having a mass of 2 Mg. If the barge is not tied to the pier P and someone drives the automobile 60 m to the other side for unloading, determine how far the barge moves away from the pier just after the automobile stops. Neglect the resistance of the water.

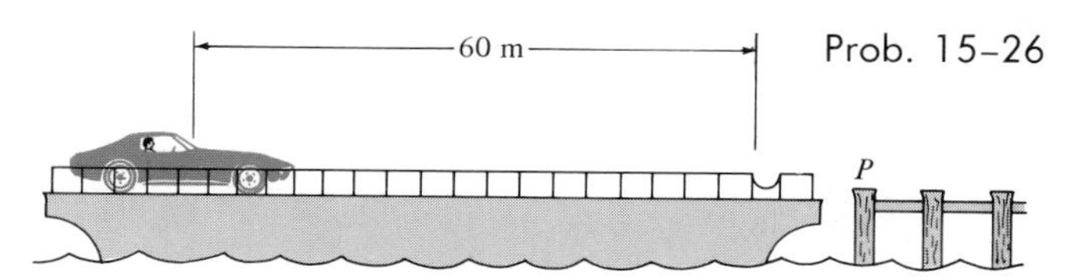

Prob. 15-26

15-27. A tugboat T having a mass of 19 Mg is tied to a barge B having a mass of 75 Mg. If the rope is "elastic" such that it has a stiffness of $k = 500$ kN/m, determine the maximum stretch in the rope as the barge is beginning to be towed. Originally both the tugboat and barge are moving in the same direction with speeds of $(v_T)_1 = 15$ km/h and $(v_B)_1 = 10$ km/h, respectively. Neglect the resistance of the water.

Prob. 15-27

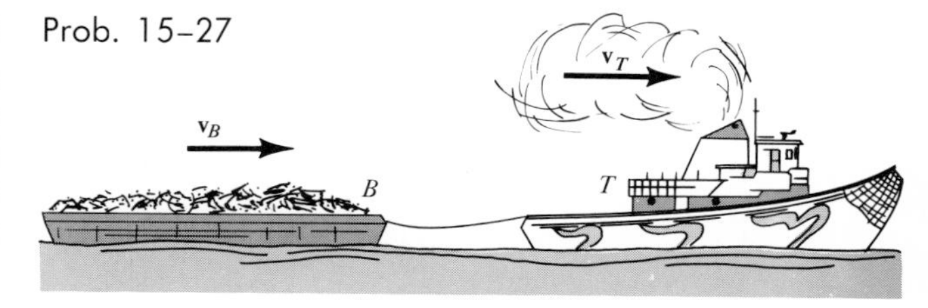

***15–28.** A box A, having a mass of 20 kg, is released from rest at the position shown and slides freely down the smooth inclined ramp. When it reaches the bottom of the ramp, it slides horizontally onto the surface of a 10-kg cart for which the coefficient of friction between the cart and the box is $\mu = 0.6$. If $h = 0.2$ m, determine the final velocity of the cart once the block comes to rest on it. Also, determine the position s of the box on the cart after it comes to rest on the cart.

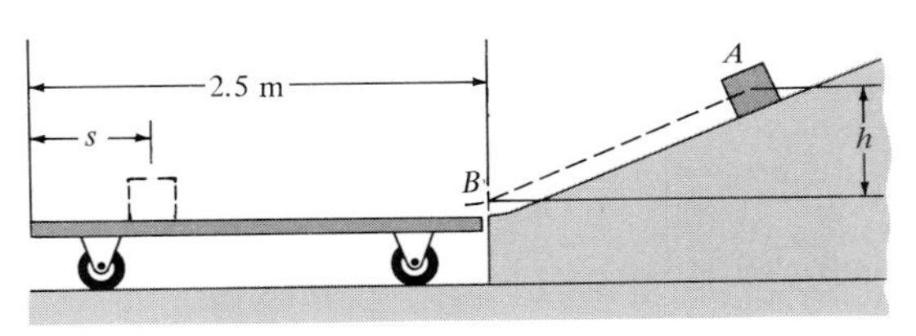

Prob. 15–28

15–29. The free-rolling ramp has a mass of 40 kg. A 10-kg crate is released from rest at A and slides down 3.5 m to point B. If the surface of the ramp is smooth, determine the ramp's speed when the crate reaches B.

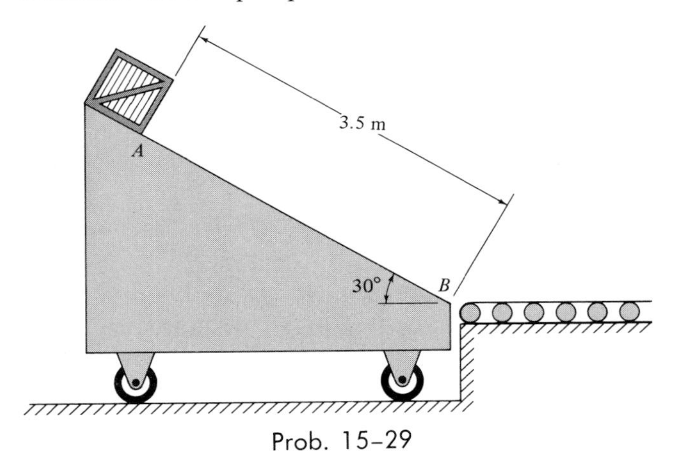

Prob. 15–29

15–30. A toboggan having a mass of $m_t = 10$ kg starts from rest at A and carries a girl and boy having a mass of $m_g = 40$ kg and $m_b = 45$ kg, respectively. When the toboggan reaches the bottom of the slope at B, $h = 3$ m, the boy is pushed off from the back with a horizontal velocity of $v_{b/t} = 2$ m/s, measured relative to the toboggan. Determine the velocity of the toboggan afterwards. Neglect friction in the calculation.

Prob. 15–30

15–30a. Solve Prob. 15–30 if the weights of the toboggan, girl, and boy are $W_t = 20$ lb, $W_g = 80$ lb, and $W_b = 110$ lb, respectively; $h = 10$ ft, $v_{b/t} = 4$ ft/s.

15–4. Impact

Impact occurs when two bodies collide with each other during a very *short* interval of time, causing relatively large (impulsive) forces to be exerted between the bodies. The striking of a hammer and nail, or a golf club and ball, are common examples of impact loadings.

In general, there are two types of impact. *Central impact* occurs when the direction of motion of the mass centers of the two colliding particles is along the *line of impact,* Fig. 15–13*a*. When the motion of one or both of the particles is at an angle with the line of impact, Fig. 15–13*b,* the impact is said to be *oblique impact*.

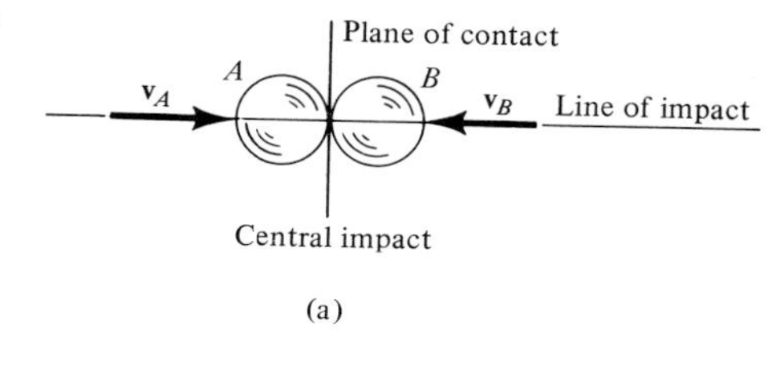

(a)

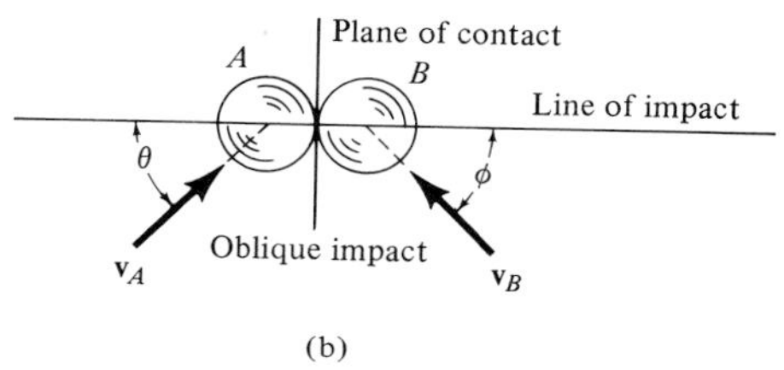

(b)

Fig. 15–13

Central Impact. To illustrate the method for analyzing the mechanics of impact, consider the case involving the central impact of particles *A* and *B,* shown in Fig. 15–14*a*. These particles have mass m_A and m_B and initial velocities $(\mathbf{v}_A)_1$ and $(\mathbf{v}_B)_1$. Provided $(v_A)_1 > (v_B)_1$, collision will eventually occur. When this happens, the particles undergo a *period of deformation* such that at the instant of *maximum deformation,* they both have the *same* velocity $\mathbf{v}$, Fig. 15–14*b*. A *period of restitution* then occurs, in which case the particles will either return to their original shape or remain permanently deformed. After restitution, the particles will have final velocities $(\mathbf{v}_A)_2$ and $(\mathbf{v}_B)_2$, where $(v_B)_2 > (v_A)_2$, Fig. 15–14*c*. The mechanics of the impact just described depends primarily upon the type of materials involved, and to a lesser extent, upon the particles' size as well as their relative velocities.

The final velocity of each of the particles after collision can be obtained by applying the conservation of momentum to the *system of particles* and the principle of impulse and momentum to *each* of the particles. Momentum for the *system* is conserved just before and just after collision, since during collision the (internal) impulses of deformation and restitution between the particles are equal but opposite and therefore *cancel*. Hence, referring to Figs. 15–14*a* and *c,* we have

$$(\overset{+}{\rightarrow}) \qquad m_A(v_A)_1 + m_B(v_B)_1 = m_A(v_A)_2 + m_B(v_B)_2 \qquad (15\text{–}10)$$

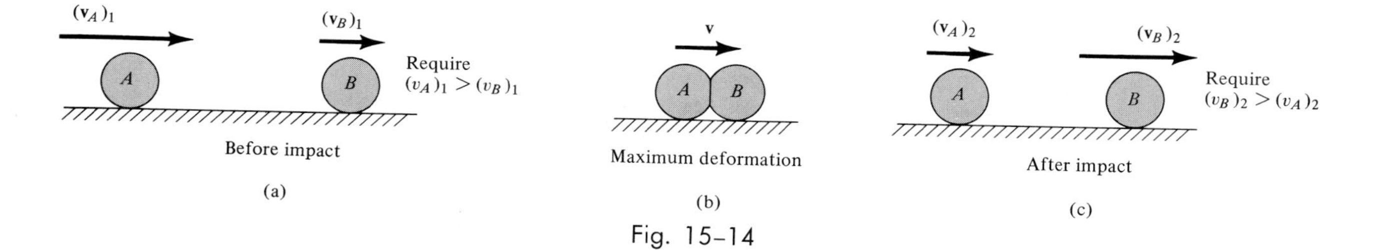

Fig. 15–14

The momentum- and impulse-vector diagrams for each particle during the deformation and restitution phases of the collision are shown in Fig. 15–15. The *deformation impulse* $\int \mathbf{P}\,dt$ is applied during the time of deformation, whereas a different *restitution impulse* $\int \mathbf{R}\,dt$ is applied during the time of restitution. In reality, the physical properties of any two bodies are such that the force of deformation, **P**, is *always greater* than the force of restitution, **R**. Hence, $\int P\,dt > \int R\,dt$. Furthermore, each of these impulses acts with equal magnitude but opposite direction on each of the particles.

Referring to the momentum- and impulse-vector diagrams for the deformation phase, we have

$$(\overset{+}{\rightarrow}) \qquad m_A(v_A)_1 - \int P\,dt = m_A(v) \qquad \text{(particle } A\text{)} \qquad (15\text{–}11)$$

$$(\overset{+}{\rightarrow}) \qquad m_B(v_B)_1 + \int P\,dt = m_B(v) \qquad \text{(particle } B\text{)} \qquad (15\text{–}12)$$

Likewise, for the restitution phase,

$$(\overset{+}{\rightarrow}) \qquad m_A(v) - \int R\,dt = m_A(v_A)_2 \qquad \text{(particle } A\text{)} \qquad (15\text{–}13)$$

$$(\overset{+}{\rightarrow}) \qquad m_B(v) + \int R\,dt = m_B(v_B)_2 \qquad \text{(particle } B\text{)} \qquad (15\text{–}14)$$

An equation involving the unknowns $(v_A)_2$ and $(v_B)_2$ can be obtained from the above four equations in the following manner: combine Eqs. 15–11 and 15–13, which yields

$$\frac{\int R\,dt}{\int P\,dt} = \frac{v - (v_A)_2}{(v_A)_1 - v} \qquad (15\text{–}15)$$

Also, from Eqs. 15–12 and 15–14 we have

$$\frac{\int R\,dt}{\int P\,dt} = \frac{(v_B)_2 - v}{v - (v_B)_1} \qquad (15\text{–}16)$$

The *coefficient of restitution, e,* is defined as the ratio of the impulse of restitution to the impulse of deformation, i.e.,

$$e = \frac{\int R\,dt}{\int P\,dt} \qquad (15\text{–}17)$$

Substituting "e" into Eqs. 15–15 and 15–16, and solving for v in each of these equations, we get

$$v = \frac{(v_A)_1 e + (v_A)_2}{1 + e} \quad \text{and} \quad v = \frac{(v_B)_1 e + (v_B)_2}{1 + e}$$

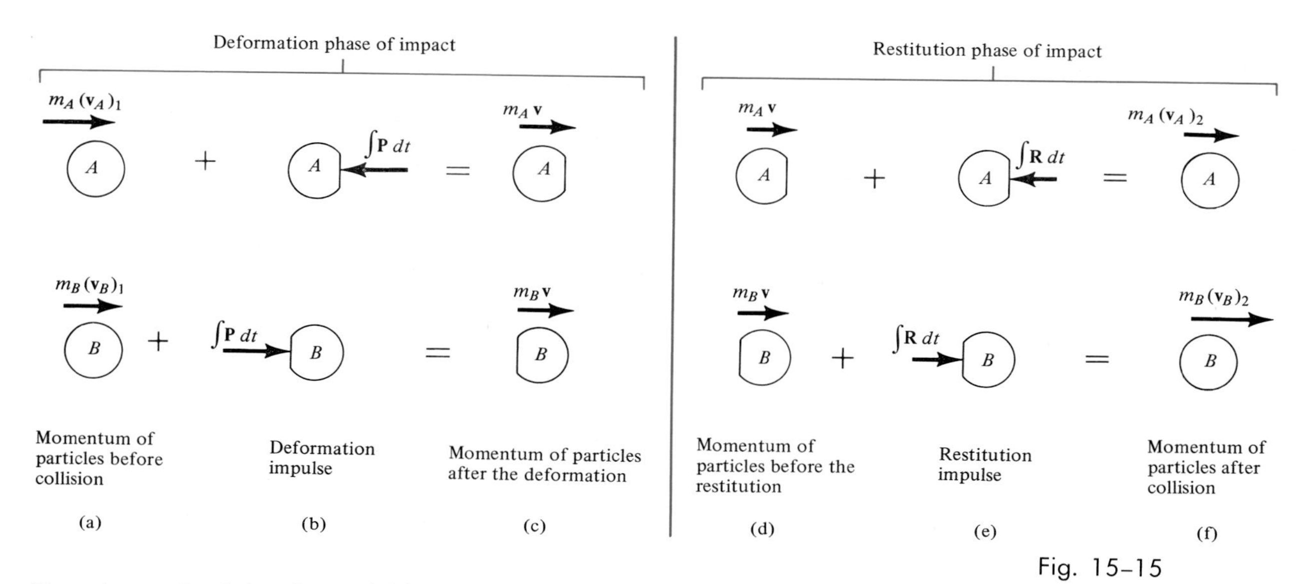

Fig. 15–15

Equating and solving for e yields

$$e = \frac{(v_B)_2 - (v_A)_2}{(v_A)_1 - (v_B)_1} \tag{15–18}$$

Provided a value for e is specified, Eqs. 15–10 and 15–18 may be solved simultaneously to obtain $(v_A)_2$ and $(v_B)_2$.

Coefficient of Restitution. With reference to Figs. 15–14*c* and 15–14*a*, it is seen that Eq. 15–18 indicates that the coefficient of restitution, e, is equal to the ratio of the relative velocity of the particle's separation *just after impact* $(v_{rel})_2 = (v_B)_2 - (v_A)_2$, to the relative velocity of the particle's approach *just before impact*, $(v_{rel})_1 = (v_A)_1 - (v_B)_1$. Hence, Eq. 15–18 may be written in a more generalized form as

$$e = \frac{(v_{rel})_2}{(v_{rel})_1} \quad \text{(along line of impact)} \tag{15–19}$$

When this equation is applied, the relative velocities are obtained from the directions of velocity of the particles *along the line of central impact* just before or just after collision. For example, if *both* particles are moving *toward* one another before collision, Fig. 15–16*a*, their relative velocity is $(v_A)_1 + (v_B)_1$. After collision, if it is *assumed* that both particles move to the left, the relative velocity of *separation* is $(v_A)_2 - (v_B)_2$, since $(v_A)_2 > (v_B)_2$, Fig. 15–16*a*. Hence, for this case

$$e = \frac{(v_{rel})_2}{(v_{rel})_1} = \frac{(v_A)_2 - (v_B)_2}{(v_A)_1 + (v_B)_1}$$

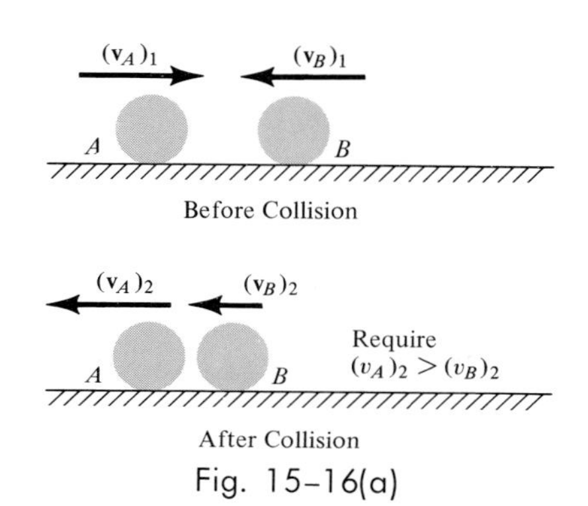

Fig. 15–16(a)

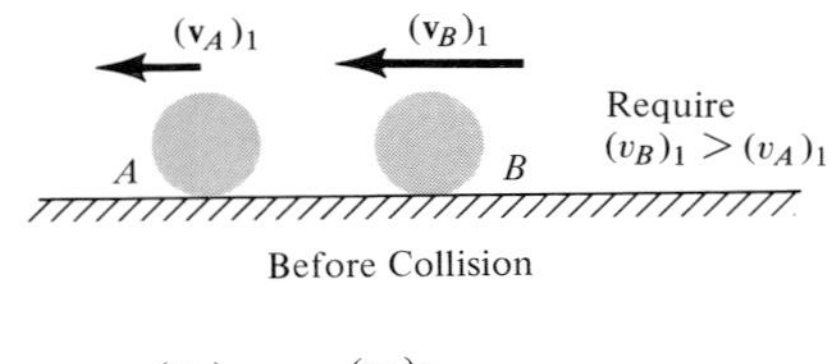

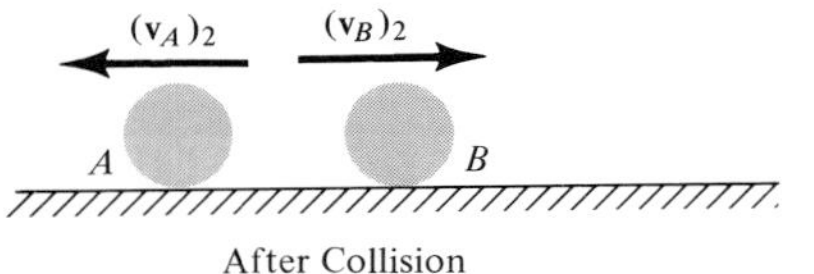

Fig. 15–16(b)

In a similar manner, particle B collides with particle A in Fig. 15–16b, only if $(v_B)_1 > (v_A)_1$, so that the relative velocity is $(v_B)_1 - (v_A)_1$. If it is *assumed* that after collision the particles move in opposite directions, Fig. 15–16b, the relative velocity is $(v_A)_2 + (v_B)_2$. Hence,

$$e = \frac{(v_{\text{rel}})_2}{(v_{\text{rel}})_1} = \frac{(v_A)_2 + (v_B)_2}{(v_B)_1 - (v_A)_1}$$

Equation 15–19 provides a relatively simple means for an experimental determination of e. By measuring the relative velocities, it has been found that e varies appreciably with impact velocity as well as with the size and shape of the colliding bodies. Differences in measurement occur because some of the initial kinetic energy of the bodies is transformed into heat energy as well as creating sound and elastic shock waves when the collision occurs. For these reasons the coefficient of restitution is reliable only when used under conditions which closely approximate those which were known to exist when measurements were made.

Elastic Impact $(e = 1)$**.** If the collision between the two particles is *perfectly elastic,* the deformation impulse ($\int \mathbf{P}\,dt$) is equal and opposite to the restitution impulse ($\int \mathbf{R}\,dt$). Although in reality this can never be achieved, from Eq. 15–17, $e = 1$ for an elastic collision. Under these conditions no energy is lost in the collision. (See Prob. 15–39.)

Plastic Impact $(e = 0)$**.** The impact is said to be *inelastic or plastic* when $e = 0$. In this case there is no restitution impulse given to the particles ($\int \mathbf{R}\,dt = \mathbf{0}$), so that after collision both particles couple *together* and move with a common velocity. In this case the energy lost during collision is a maximum.

PROCEDURE FOR ANALYSIS

In most cases the *final velocities* of two colliding particles are to be determined *just after* the particles are subjected to direct central impact. Provided the coefficient of restitution e, the mass of each particle, and each particle's initial velocity *just before* impact are known, the two equations available for solution are:

1. The conservation of momentum applies to the system of particles, $\Sigma m v_1 = \Sigma m v_2$.
2. The coefficient of restitution, $e = (v_{\text{rel}})_2/(v_{\text{rel}})_1$, relates the relative velocities of the particles from just before to just after impact.

When applying these two equations, the sense of direction of an unknown velocity can be assumed. If the solution yields a negative magnitude, the velocity acts in the opposite sense of direction.

Oblique Impact. When oblique impact occurs, the particles move away from each other with velocities having unknown directions as well as unknown magnitudes. Provided the initial velocities are known, four unknowns are present in the problem. As shown in Fig. 15–17*a*, these unknowns may be represented as $(v_A)_2$, $(v_B)_2$, θ_2, and ϕ_2.

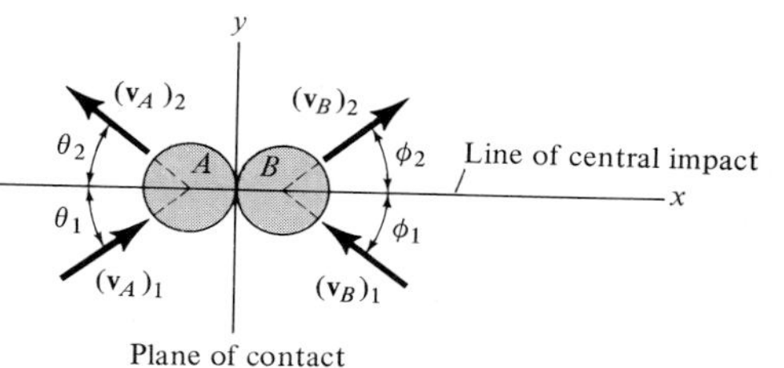

Fig. 15–17(a)

PROCEDURE FOR ANALYSIS

Provided the y axis is established along the plane of contact and the x axis along the line of central impact, the impulsive forces of deformation and restitution act *only in the x direction*, Fig. 15–17*b*. Resolving the velocities or momentum vectors into components along the x and y axes, Fig. 15–17*b*, it is possible to write four independent scalar equations in order to determine $(v_{Ax})_2$, $(v_{Ay})_2$, $(v_{Bx})_2$, and $(v_{By})_2$.

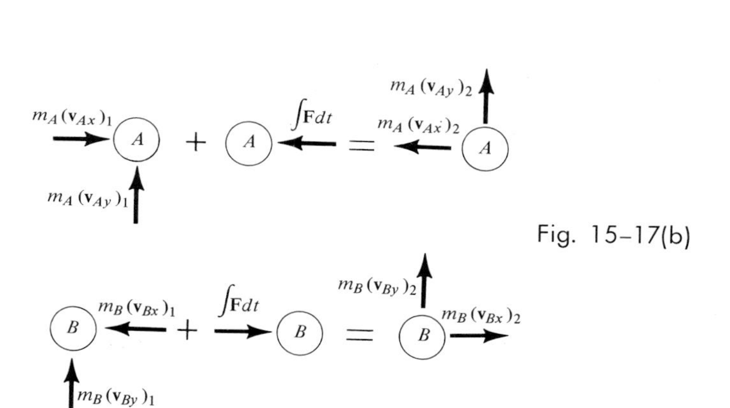

Fig. 15–17(b)

1. Momentum of the system is conserved *along the line of impact, x* axis, so that $\Sigma m(v_x)_1 = \Sigma m(v_x)_2$.
2. The coefficient of restitution, $e = (v_{\text{rel}})_2/(v_{\text{rel}})_1$, relates the relative-velocity *components* of the particles *along the line of impact, x* axis, from just before to just after impact.
3. Momentum of particle A is conserved perpendicular to the line of central impact, y axis, since no impulse acts on the particle in this direction.
4. Momentum of particle B is conserved perpendicular to the line of central impact, y axis, since no impulse acts on the particle in this direction.

Application of these four equations is illustrated numerically in Example 15–9.

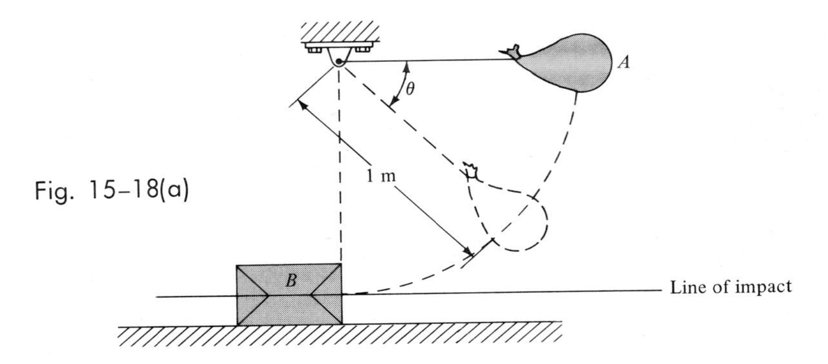

Fig. 15–18(a)

Example 15–8

The bag *A,* having a mass of 2 kg, is released from rest at the position $\theta = 0°$, as shown in Fig. 15–18*a*. It strikes a 20-kg box *B* when $\theta = 90°$. If the coefficient of restitution between the bag and box is $e = 0.3$, determine the velocities of the bag and box just after impact.

Solution

This problem involves central impact. Why? Before analyzing the mechanics of the impact, however, it is first necessary to obtain the velocity of the bag *just before* it strikes the box. This can be done by applying the principle of work and energy to the bag, from $\theta = 0°$ to $\theta = 90°$. As shown in Fig. 15–18*b*, work is done only by the bag's weight, which moves through a vertical displacement of 1 m. (The tension **T** does

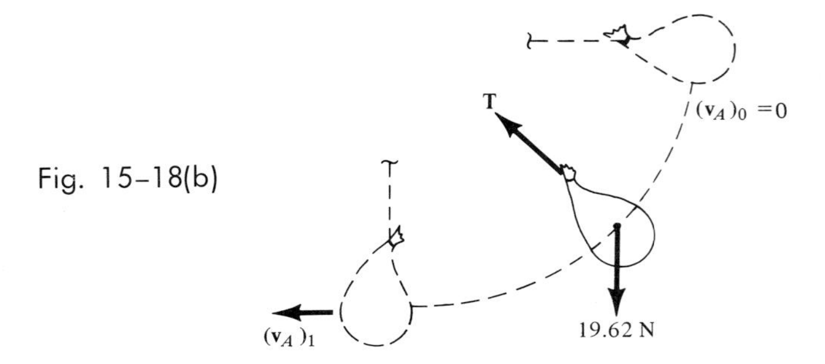

Fig. 15–18(b)

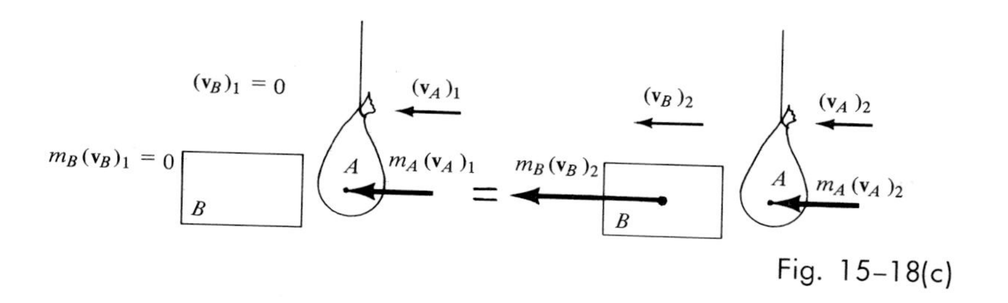

Fig. 15–18(c)

no work, since this force is always perpendicular to the displacement.) Hence,

$$T_0 + \Sigma U_{0-1} = T_1$$

$$0 + 19.62(1) = \frac{1}{2}(2)(v_A)_1^2$$

Thus,

$$(v_A)_1 = 4.43 \text{ m/s}$$

The momentum diagrams of A and B, just before and just after impact, are shown in Fig. 15–18c. Here it is assumed that after collision both bodies continue to travel to the left. Applying the conservation of momentum to the system, we have

$$(\overset{+}{\leftarrow}) \qquad m_B(v_B)_1 + m_A(v_A)_1 = m_B(v_B)_2 + m_A(v_A)_2$$

$$0 + 2(4.43) = 20(v_B)_2 + 2(v_A)_2$$

or

$$(v_A)_2 = 4.43 - 10(v_B)_2 \tag{1}$$

Using the definition of the coefficient of restitution and realizing that for separation to occur after collision $(v_B)_2 > (v_A)_2$, Fig. 15–18c, we have

$$e = \frac{(v_{\text{rel}})_2}{(v_{\text{rel}})_1} = \frac{(v_B)_2 - (v_A)_2}{(v_A)_1 - (v_B)_1}$$

$$0.3 = \frac{(v_B)_2 - (v_A)_2}{4.43 - 0}$$

or

$$(v_A)_2 = (v_B)_2 - 1.33 \tag{2}$$

Solving Eqs. (1) and (2) simultaneously yields

$$(v_A)_2 = -0.806 \text{ m/s} \quad \text{and} \quad (v_B)_2 = 0.524 \text{ m/s} \qquad \textit{Ans.}$$

The negative sign for $(v_A)_2$ indicates that the bag moves to the *right* after impact instead of to the left as shown in Fig. 15–18c.

Example 15–9

Two spheres A and B, having a mass of 1 and 2 kg, respectively, collide with initial velocities as shown in Fig. 15–19a. If the coefficient of restitution for the spheres is $e = 0.75$, determine the final velocity of each sphere after collision, and the loss in kinetic energy due to the collision.

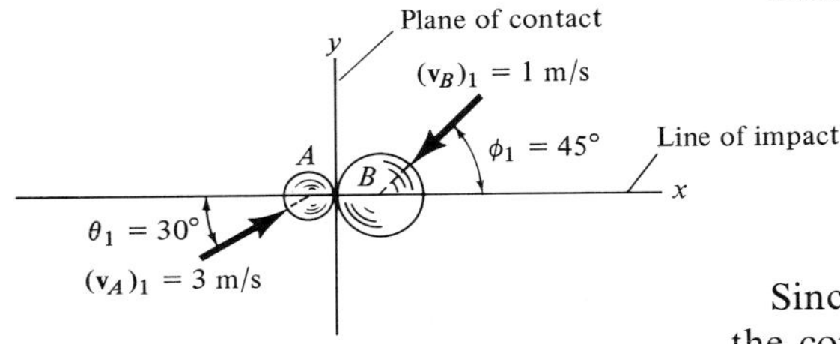

Fig. 15–19(a)

Solution

The problem involves *oblique impact*. Why? In order to seek a solution, the x and y axes have been established along the line of central impact and the plane of contact, respectively, Fig. 15–19a.

Resolving each of the initial velocities into x and y components, we have

$$(v_{Ax})_1 = 3 \cos 30° = 2.60 \text{ m/s}$$
$$(v_{Ay})_1 = 3 \sin 30° = 1.50 \text{ m/s}$$
$$(v_{Bx})_1 = -1 \cos 45° = -0.707 \text{ m/s}$$
$$(v_{By})_1 = -1 \sin 45° = -0.707 \text{ m/s}$$

Since the impact occurs only in the x direction (line of central impact), the conservation of momentum for *both* spheres can be applied in this direction. Why? Thus, in reference to the momentum diagrams in Fig. 15–19b,

$$(\overset{+}{\rightarrow}) \qquad m_A(v_{Ax})_1 + m_B(v_{Bx})_1 = m_A(v_{Ax})_2 + m_B(v_{Bx})_2$$
$$1(2.60) + 2(-0.707) = 1(v_{Ax})_2 + 2(v_{Bx})_2$$
$$(v_{Ax})_2 + 2(v_{Bx})_2 = 1.18 \qquad (1)$$

Since both spheres are *assumed* to have components of velocity in the $+x$ direction after collision, Fig. 15–19b, it is necessary that $(v_{Bx})_2 > (v_{Ax})_2$ for separation to occur along this axis. Hence, from the direction of the x components of velocity, we can write

$$e = \frac{(v_{\text{rel}})_{2x}}{(v_{\text{rel}})_{1x}} = \frac{(v_{Bx})_2 - (v_{Ax})_2}{(v_{Ax})_1 + (v_{Bx})_1}$$
$$0.75 = \frac{(v_{Bx})_2 - (v_{Ax})_2}{2.60 + 0.707}$$

Fig. 15–19(b)

$$(v_{Bx})_2 - (v_{Ax})_2 = 2.48 \qquad (2)$$

Solving Eqs. (1) and (2) for $(v_{Ax})_2$ and $(v_{Bx})_2$ yields

$$(v_{Ax})_2 = -1.26 \text{ m/s} \quad (\leftarrow)$$
$$(v_{Bx})_2 = 1.22 \text{ m/s} \quad (\rightarrow)$$

The momentum of *each sphere* is *conserved* in the *y* direction (plane of contact), since *no impact* occurs in this direction. Hence, in reference to Fig. 15–19*b*,

$(+\uparrow)$
$$m_A(v_{Ay})_1 = m_A(v_{Ay})_2$$
$$(v_{Ay})_2 = 1.50 \text{ m/s} \quad (\uparrow)$$

and

$(+\uparrow)$
$$m_B(v_{By})_1 = m_B(v_{By})_2$$
$$(v_{By})_2 = -0.707 \text{ m/s} \quad (\downarrow)$$

Summing the vector components, we have

$$(v_A)_2 = \sqrt{(v_{Ax})_2^2 + (v_{Ay})_2^2} = \sqrt{(-1.26)^2 + (1.50)^2} = 1.96 \text{ m/s} \qquad \textit{Ans.}$$

$$\theta_2 = \tan^{-1}\frac{(v_{Ay})_2}{(v_{Ax})_2} = \tan^{-1}\frac{1.50}{1.26} = 50.0^\circ \quad \theta_2 \measuredangle (\mathbf{v}_A)_2 \qquad \textit{Ans.}$$

$$(v_B)_2 = \sqrt{(v_{Bx})_2^2 + (v_{By})_2^2} = \sqrt{(1.22)^2 + (-0.707)^2} = 1.41 \text{ m/s} \qquad \textit{Ans.}$$

$$\phi_2 = \tan^{-1}\frac{(v_{By})_2}{(v_{Bx})_2} = \tan^{-1}\frac{0.707}{1.22} = 30.1^\circ \quad \phi_2 \measuredangle (\mathbf{v}_B)_2 \qquad \textit{Ans.}$$

These results are shown in Fig. 15–19*c*.

Since the speed of each sphere both before and after the collision is known, the loss in kinetic energy becomes

$$\begin{aligned}\Delta T &= T_1 - T_2 \\ &= \left[\frac{1}{2}m_A(v_A)_1^2 + \frac{1}{2}m_B(v_B)_1^2\right] - \left[\frac{1}{2}m_A(v_A)_2^2 + \frac{1}{2}m_B(v_B)_2^2\right] \\ &= \left[\frac{1}{2}(1)(3)^2 + \frac{1}{2}(2)(1)^2\right] - \left[\frac{1}{2}(1)(1.96)^2 + \frac{1}{2}(2)(1.41)^2\right] \\ &= 1.59 \text{ J}\end{aligned} \qquad \textit{Ans.}$$

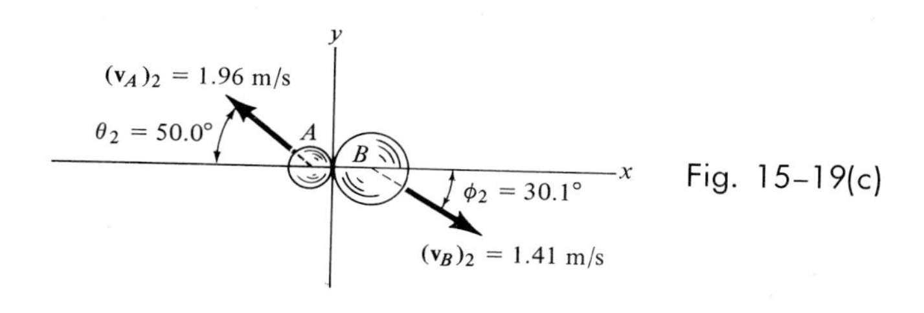

Fig. 15–19(c)

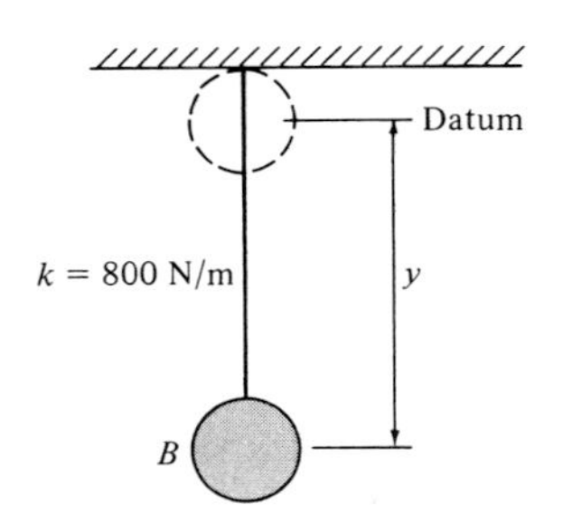

Fig. 15–20(a)

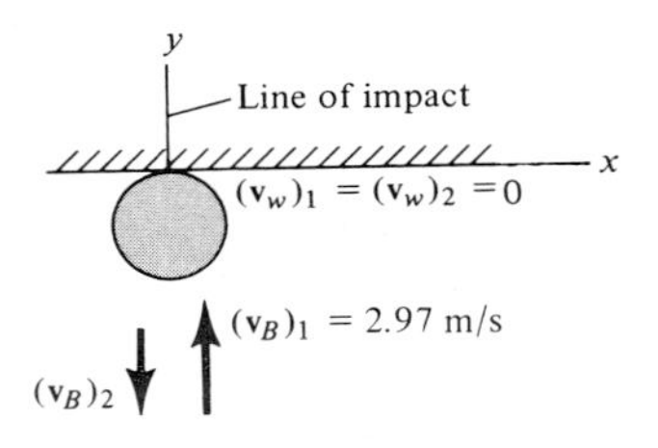

Fig. 15–20(b)

Example 15–10

The ball B shown in Fig. 15–20a has a mass of 1.5 kg and is suspended from a ceiling by a 1-m-long elastic cord. If the cord is *stretched* downward 250 mm and the ball is released from rest, determine how far the cord stretches after the ball rebounds from the ceiling. The stiffness of the cord is $k = 800$ N/m and the coefficient of restitution between the ball and ceiling is $e = 0.8$. The ball makes a central impact with the ceiling.

Solution

It is necessary to first obtain the initial velocity of the ball *just before* it strikes the ceiling. This may be done by applying the conservation of energy theorem to the ball, in which case the gravitational potential energy of the ball and the elastic potential energy of the cord are transformed into kinetic energy of the ball. With the datum located as shown in Fig. 15–20a, realizing that initially $y = y_0 = (1 + 0.25)$ m $= 1.25$ m, we have

$$T_0 + V_0 = T_1 + V_1$$

$$\tfrac{1}{2}m(v_B)_0^2 - W_B y_0 + \tfrac{1}{2}kx^2 = \tfrac{1}{2}m(v_B)_1^2 + 0$$

$$0 - 1.5(9.81)(1.25) + \tfrac{1}{2}(800)(0.25)^2 = \tfrac{1}{2}(1.5)(v_B)_1^2$$

Solving for $(v_B)_1$ yields

$$(v_B)_1 = 2.97 \text{ m/s}$$

The interaction of the ball with the ceiling will now be considered using the principles of impact. Note that the y axis, Fig. 15–20b, represents the line of central impact for the ball. Since an indeterminate portion of the mass of the ceiling is involved in the impact, the conservation of momentum for the ball–ceiling system will not be written. The "velocity" of this portion of ceiling remains at rest *both* before and after impact, so that Eq. 15–19 will be applied, i.e.,

$$e = \frac{(v_{\text{rel}})_2}{(v_{\text{rel}})_1} = \frac{0 - (v_B)_2}{(v_B)_1 - 0} \qquad 0.8 = \frac{-(v_B)_2}{2.97}$$

Solving gives

$$(v_B)_2 = -2.37 \text{ m/s}$$

The maximum stretch x in the cord may be determined by again applying the conservation-of-energy theorem to the ball just after collision. In this case the ball's kinetic energy is transformed into both elastic potential energy of the cord and gravitational potential energy of the ball. Assuming $y = y_3 = (1 + x)$ m, Fig. 15–20a, then

$$T_2 + V_2 = T_3 + V_3$$

$$\tfrac{1}{2}m(v_B)_2^2 + 0 = \tfrac{1}{2}m(v_B)_3^2 - W_B y_3 + \tfrac{1}{2}kx_3^2$$

$$\tfrac{1}{2}(1.5)(2.37)^2 = 0 - 9.81(1.5)(1 + x_3) + \tfrac{1}{2}(800)x_3^2$$

or

$$400x_3^2 - 14.72x_3 - 18.94 = 0$$

Solving this quadratic equation for the positive root yields

$$x_3 = 0.237 \text{ m} = 237 \text{ mm} \qquad \textit{Ans.}$$

Problems

15-31. Ball A has a mass of 200 g and an initial velocity of $(v_A)_1 = 2$ m/s. As it rolls on a horizontal plane, it makes a direct collision with ball B, which has a mass of 300 g and is originally at rest. If both balls have the same size and the collision is perfectly elastic ($e = 1$), determine the velocity of each ball after the collision. Show that the kinetic energy of the balls before and after collision is the same.

*** 15-32.** Blocks A and B have a mass of 10 and 15 kg, respectively. After striking block B, A slides 55 mm to the right, and B slides 300 mm to the right. If the coefficient of friction between the blocks and the surface is $\mu = 0.35$, determine the coefficient of restitution between the blocks. Block B is originally at rest.

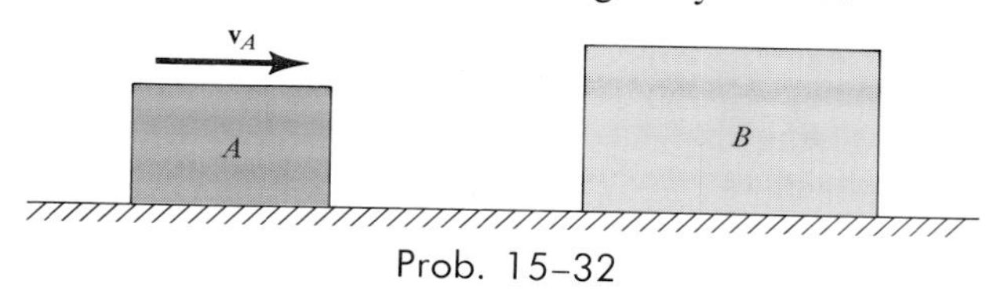

Prob. 15-32

15-33. An ivory ball having a mass of 200 g is released from rest at a height of 400 mm above a very large fixed metal surface. If the ball rebounds to a height of 325 mm above the surface, determine the coefficient of restitution between the ball and the surface.

15-34. The ball B has a mass of 0.75 kg and is moving forward with a velocity of $(v_B)_1 = 4$ m/s when it strikes the 1.5-kg block A, which is originally at rest. If the coefficient of restitution between the ball and the block is $e = 0.7$, compute the distance block A slides before coming to rest. The coefficient of friction between the block and the surface is $\mu_A = 0.4$.

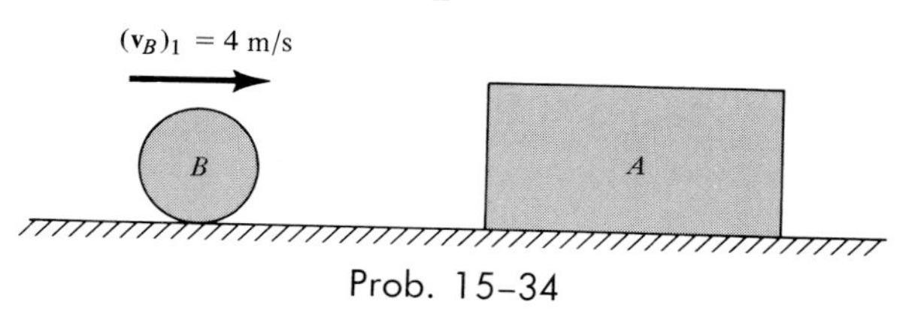

Prob. 15-34

15-35. The drop hammer H has a mass of $m_H = 10$ Mg and falls from rest $h = 0.75$ m onto a forged anvil plate P that has a mass of $m_P = 6$ Mg. The plate is mounted on a set of springs which have a combined stiffness of $k_T = 3$ MN/m. Determine the maximum compression in the springs caused by the impact if the coefficient of restitution between the hammer and the plate is $e = 0.5$. Neglect friction along the vertical guide posts A and B.

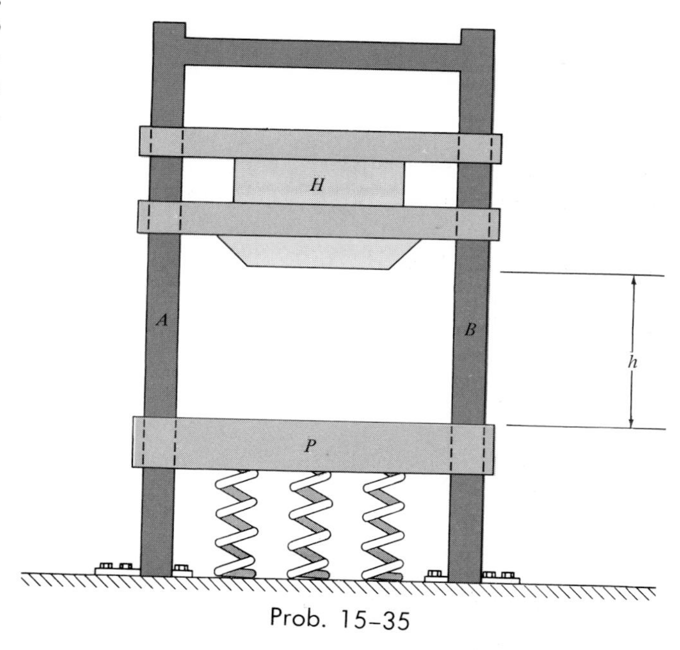

Prob. 15-35

15-35a. Solve Prob. 15-35 if the hammer and plate have weights of $W_H = 900$ lb and $W_P = 400$ lb, respectively; $h = 3$ ft, $k_T = 500$ lb/ft, and $e = 0.6$.

***15–36.** A stunt driver in car A travels in free flight off the edge of a ramp at C. At the point of maximum height he strikes car B. If the direct collision is perfectly plastic, determine the required ramp speed v_C at the end of the ramp, C, and the approximate distance s where both cars strike the ground. Each car has a mass of 4 Mg. Neglect the size of the cars in the calculation.

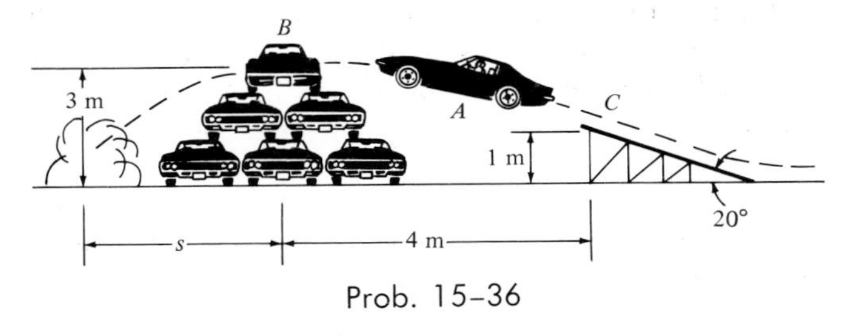

Prob. 15–36

15–37. Plates A and B have a mass of $m_A = 3$ kg and $m_B = 2$ kg, respectively, and are restricted to move along the frictionless guides. If the coefficient of restitution between the plates is $e = 0.7$, determine the maximum deflection of the spring if A has a velocity of 4 m/s just before striking B. Plate B is originally at rest.

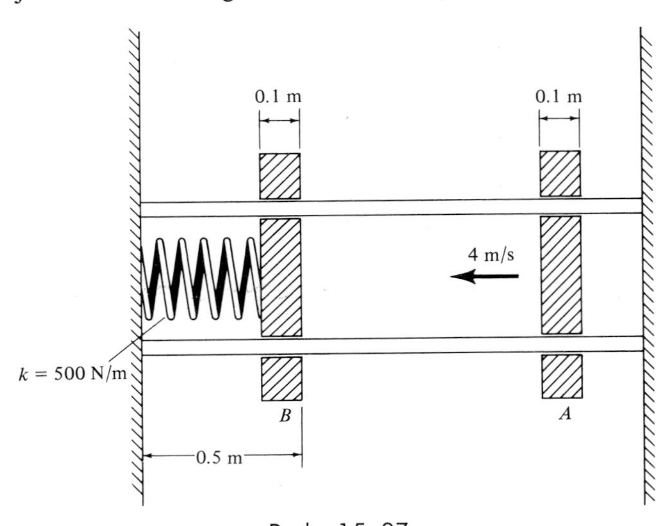

Prob. 15–37

15–38. Solve Prob. 15–37 assuming that the coefficient of restitution between the plates is (a) $e = 1$, and (b) $e = 0$.

15–39. If two spheres A and B are subjected to direct central impact and the collision is perfectly elastic ($e = 1$), prove that the kinetic energy before collision equals the kinetic energy after collision.

***15–40.** Two billiard balls A and B have an equal mass of $m = 200$ g. If A strikes B with a velocity of $(v_A)_1 = 1.5$ m/s as shown, determine their final velocities just after collision. Ball B is originally at rest and the coefficient of restitution is $e = 0.85$.

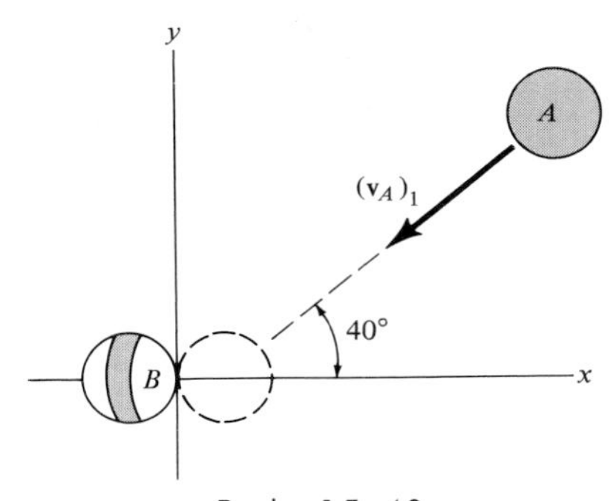

Prob. 15–40

***15–40a.** Solve Prob. 15–40 if each ball weighs $W = 0.4$ lb, and $(v_A)_1 = 8$ ft/s, $e = 0.85$.

15–41. Two cars A and B, each having a mass of 1.6 Mg, collide on the icy pavement of an intersection. The direction of motion of each car after collision is measured from snow tracks as shown. If the driver in car A states that he was going at 50 km/h just before collision and that after collision he applied the brakes, so that his car skidded 4 m before stopping, determine the approximate speed of car B just before the collision. Assume that the coefficient of friction between the car wheels and the pavement is $\mu = 0.15$. *Note:* The line of impact has not been defined; furthermore, this information is not needed for the solution.

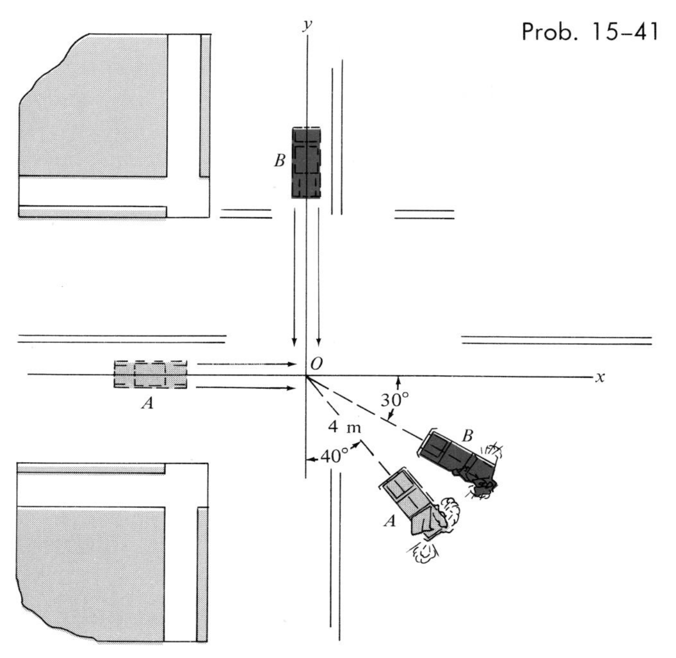

Prob. 15–41

15–42. The billiard ball has a mass of 200 g and is moving with a speed of 2.5 m/s when it strikes the side of the pool table at A. If the coefficient of restitution between the ball and the side of the table is $e = 0.6$, determine the speed of the ball just after striking the table twice, i.e., at A, then at B. Neglect the effects of friction while the ball is rolling.

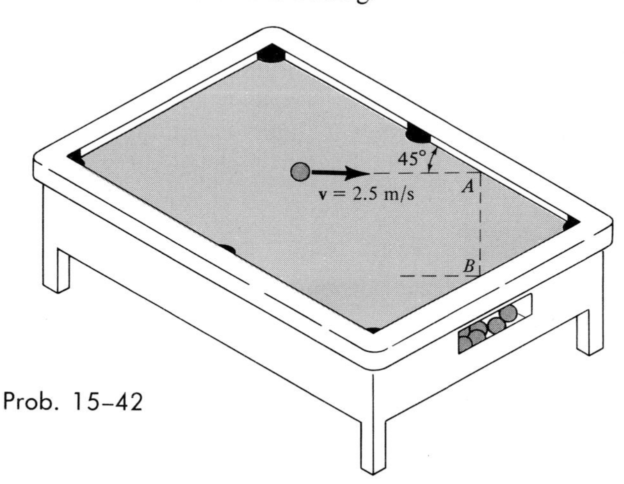

Prob. 15–42

15–43. The two hockey pucks A and B each have a mass of 250 g. If they collide at O and are deflected along the dashed paths, determine their speeds just after impact. Assume that the icy surface over which they slide is smooth. *Hint:* Since the y axis is *not* along the line of impact, apply the conservation of momentum along the x and along the y axis.

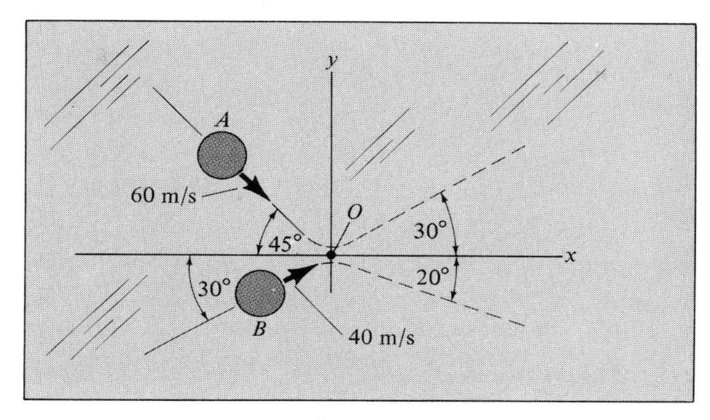

Prob. 15–43

*** 15–44.** The two disks A and B have a mass of 3 and 5 kg, respectively. If they collide with the initial velocities shown, determine their velocities just after impact. The coefficient of restitution is $e = 0.65$.

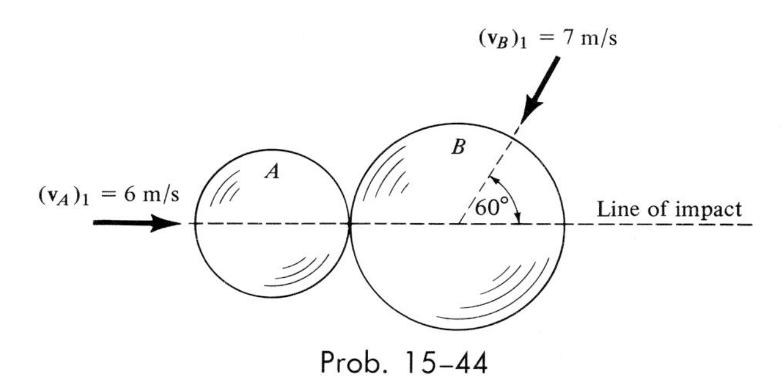

Prob. 15–44

15–5. Angular Momentum of a Particle

The *angular momentum* $\mathbf{H}_O$ of a particle is defined as the "moment" of the particle's linear momentum $\mathbf{L} = m\mathbf{v}$ about an axis passing through point O. Since this concept is analogous to finding the moment of a force about an axis, $\mathbf{H}_O$ is sometimes referred to as the *moment of momentum.*

Scalar Formulation. If a particle P is moving along a *planar curve,* Fig. 15–21, the angular momentum can generally be computed about an axis passing through the origin O of an x, y, z reference frame by using a scalar formulation. The *magnitude* of $\mathbf{H}_O$ is then

$$H_O = (d)(mv) \tag{15-20}$$

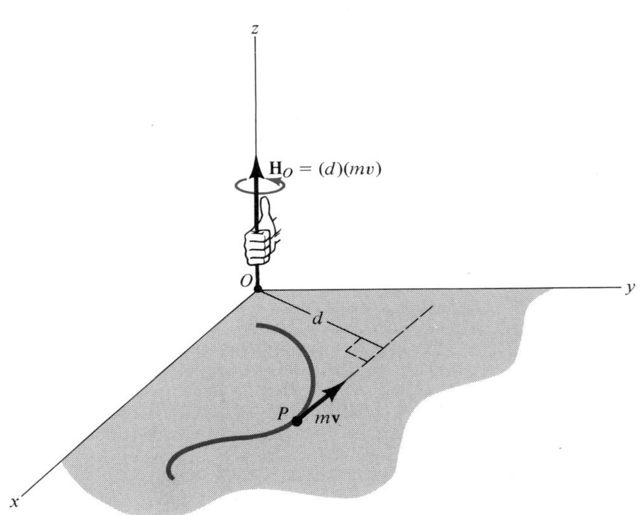

Fig. 15–21

Here d is the moment arm or perpendicular distance from O to the line of action of $m\mathbf{v}$. Common units for this magnitude are $\text{kg} \cdot \text{m}^2/\text{s}$ (or $\text{N} \cdot \text{m} \cdot \text{s}$). The *direction* of $\mathbf{H}_O$ is defined by the right-hand rule. As shown in Fig. 15–21, the curl of the fingers of the right hand indicates the sense of rotation of $m\mathbf{v}$ about O, so that the thumb or $\mathbf{H}_O$ is directed perpendicular to the x-y plane along the $+z$ axis.

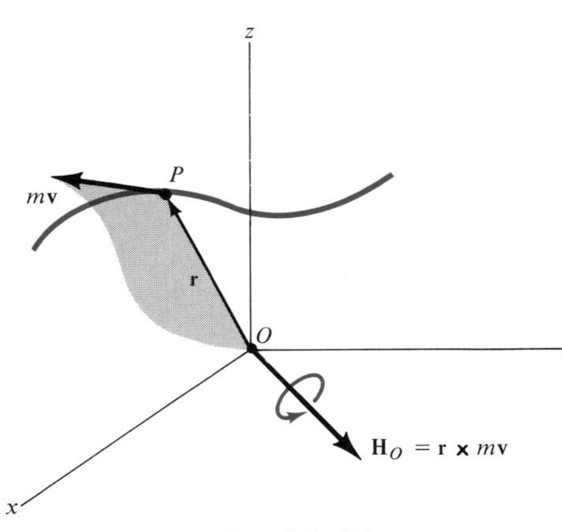

Fig. 15–22

Vector Formulation. If the particle is moving along a space curve, Fig. 15–22, the vector cross product can conveniently be used to define the *angular momentum.* In this case

$$\mathbf{H}_O = \mathbf{r} \times m\mathbf{v} \tag{15-21}$$

where $\mathbf{r}$ denotes a position vector drawn from O to P. As shown in the figure, $\mathbf{H}_O$ is *perpendicular* to the shaded plane containing $\mathbf{r}$ and $m\mathbf{v}$.

In order to conveniently use the cross product, $\mathbf{r}$ and $m\mathbf{v}$ should be expressed in terms of their Cartesian components, so that the angular momentum is determined from the evaluation of the determinant:

$$\mathbf{H}_O = \begin{vmatrix} \mathbf{i} & \mathbf{j} & \mathbf{k} \\ r_x & r_y & r_z \\ mv_x & mv_y & mv_z \end{vmatrix} \tag{15–22}$$

Relation Between Moment of a Force and Angular Momentum. The moments about point O of all the forces acting on the particle may be related to the particle's angular momentum by using the equation of motion. Since the mass of the particle is constant, we may write

$$\Sigma\mathbf{F} = m\dot{\mathbf{v}}$$

Performing a cross-product multiplication of each side of this equation by the position vector $\mathbf{r}$ to obtain the moments $\Sigma\mathbf{M}_O$ of the forces about point O of an inertial frame of reference, we have

$$\Sigma\mathbf{M}_O = \mathbf{r} \times \Sigma\mathbf{F} = \mathbf{r} \times m\dot{\mathbf{v}} \tag{15–23}$$

By definition of the derivative of the cross product of two vectors, Appendix B, we can write

$$\frac{d}{dt}(\mathbf{r} \times m\mathbf{v}) = \dot{\mathbf{r}} \times m\mathbf{v} + \mathbf{r} \times m\dot{\mathbf{v}}$$

The first term on the right side, $\dot{\mathbf{r}} \times m\mathbf{v} = m(\dot{\mathbf{r}} \times \dot{\mathbf{r}}) = \mathbf{0}$, since the cross product of a vector with itself is zero. Hence, substituting into Eq. 15–23 yields

$$\Sigma\mathbf{M}_O = \frac{d}{dt}(\mathbf{r} \times m\mathbf{v})$$

Using Eq. 15–21, the final result is

$$\Sigma\mathbf{M}_O = \dot{\mathbf{H}}_O \tag{15–24}$$

Therefore, *the moments about point O of all the forces acting on the particle are equal to the time rate of change of the angular momentum of the particle about point O.* This result is similar to Eq. 15–1, which can be written as

$$\Sigma\mathbf{F} = \dot{\mathbf{L}} \tag{15–25}$$

Here $\mathbf{L} = m\mathbf{v}$, so that *the resultant force acting on the particle is equal to the time rate of change of the linear momentum of the particle.*

15–6. Angular Momentum of a System of Particles

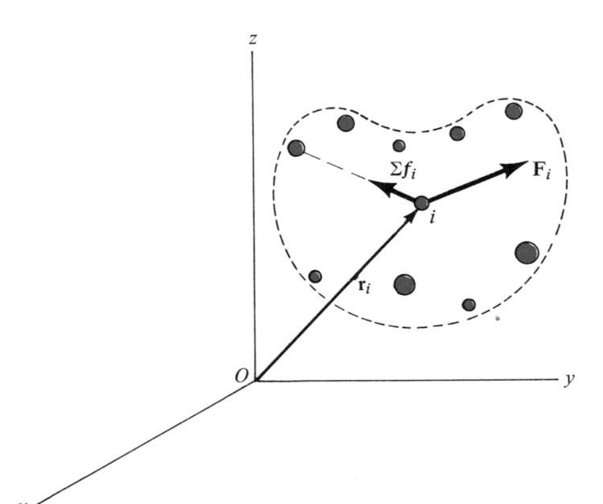

Fig. 15–23

An equation having the same form as Eq. 15–24 may be derived for the system of n particles shown in Fig. 15–23. The forces acting on the arbitrary ith particle of the system consist of a resultant *external force* $\mathbf{F}_i$ and the internal forces $\Sigma\mathbf{f}_i$ which act between the ith particle and all other particles within the system. Expressing the moments of these forces about point O, using the form of Eq. 15–24, we have

$$(\mathbf{r}_i \times \mathbf{F}_i) + (\mathbf{r}_i \times \Sigma\mathbf{f}_i) = (\dot{\mathbf{H}}_i)_O$$

Here $\mathbf{r}_i$ represents the position vector drawn from the origin O of an inertial frame of reference to the ith particle and $(\mathbf{H}_i)_O$ is the angular momentum of the ith particle about O. Similar equations can be written for each of the other particles of the system. However, when all results are summed vectorially, the moments created by the internal forces will cancel out. This is because corresponding pairs of internal forces are equal in magnitude, opposite in direction, and collinear. Hence, the final result may be written as

$$\Sigma\mathbf{M}_O = \dot{\mathbf{H}}_O \tag{15–26}$$

which states that *the sum of the moments about point O of all the external forces acting on a system of n particles is equal to the time rate of change of the total angular momentum of the system of particles about point O.*

When moments of the external forces acting on the system of particles are summed about the system's *mass center G,* one again obtains the same simple form of Eq. 15–26, relating the moment summation $\Sigma\mathbf{M}_G$ to the angular momentum $\mathbf{H}_G$. To show this, consider the system of n particles in Fig. 15–24, where x, y, z represents an inertial frame of reference and the x', y', z' axes, with origin at G, *translate* with respect to this frame. In general, G is *accelerating,* so by definition the translating frame is *not* an inertial reference. The angular momentum of the ith particle with respect to this frame is, however,

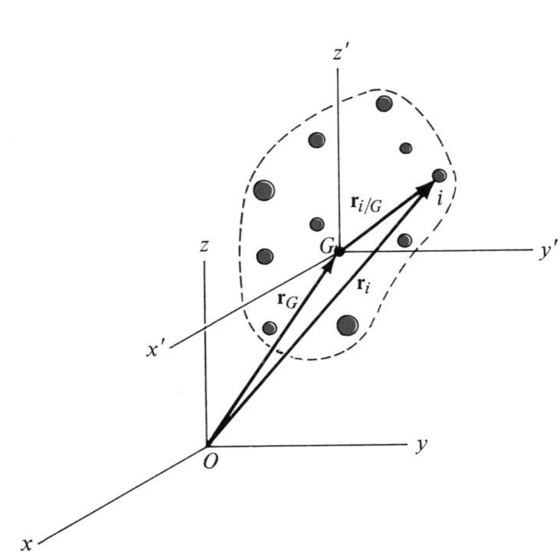

Fig. 15–24

$$(\mathbf{H}_i)_G = \mathbf{r}_{i/G} \times m_i\mathbf{v}_{i/G}$$

where $\mathbf{r}_{i/G}$ and $\mathbf{v}_{i/G}$ represent the relative position and relative velocity of the ith particle with respect to G. Taking the time derivative gives

$$(\dot{\mathbf{H}}_i)_G = \dot{\mathbf{r}}_{i/G} \times m_i\mathbf{v}_{i/G} + \mathbf{r}_{i/G} \times m_i\dot{\mathbf{v}}_{i/G}$$

By definition, $\mathbf{v}_{i/G} = \dot{\mathbf{r}}_{i/G}$. Thus, the first term on the right side is zero since the cross product of parallel vectors is zero. Also, $\mathbf{a}_{i/G} = \dot{\mathbf{v}}_{i/G}$, so that

$$(\dot{\mathbf{H}}_i)_G = \mathbf{r}_{i/G} \times m_i\mathbf{a}_{i/G}$$

Similar expressions can be written for the other particles of the system. When the results are summed for all n particles, we get

$$\dot{\mathbf{H}}_G = \Sigma(\mathbf{r}_{i/G} \times m_i\mathbf{a}_{i/G})$$

The relative acceleration for the ith particle is defined by the equation $\mathbf{a}_{i/G} = \mathbf{a}_i - \mathbf{a}_G$, where $\mathbf{a}_i$ and $\mathbf{a}_G$ represent, respectively, the accelerations of the ith particle and point G measured with respect to the *inertial frame of reference*. Substituting and expanding, using the distributive property of the vector cross product, yields

$$\dot{\mathbf{H}}_G = \Sigma(\mathbf{r}_{i/G} \times m_i\mathbf{a}_i) - (\Sigma m_i\mathbf{r}_{i/G}) \times \mathbf{a}_G$$

By definition of the mass center, the sum $(\Sigma m_i\mathbf{r}_{i/G}) = (\Sigma m_i)\bar{\mathbf{r}}$ is equal to zero, since the position vector $\bar{\mathbf{r}}$ relative to G is zero. Hence, the last term in the above equation is zero. Using the equation of motion, the product $m_i\mathbf{a}_i$ may be replaced by the resultant *external force* $\mathbf{F}_i$ acting on the ith particle. Denoting $\Sigma\mathbf{M}_G = \Sigma\mathbf{r}_{i/G} \times \mathbf{F}_i$, the final result may be written as

$$\Sigma\mathbf{M}_G = \dot{\mathbf{H}}_G \qquad (15\text{–}27)$$

It will be shown in Chapter 21 that this equation represents one of the most important formulations in rigid-body mechanics, since it forms the basis for obtaining the rotational equations of motion for the body.

15–7. Angular Impulse and Momentum Principles for a Particle

Principle of Angular Impulse and Momentum for a Particle. If Eq. 15–24 is rewritten in the form $\Sigma\mathbf{M}_O\,dt = d\mathbf{H}_O$ and integrated, we have, assuming that at time $t = t_1$, $\mathbf{H}_O = (\mathbf{H}_O)_1$ and at time $t = t_2$, $\mathbf{H}_O = (\mathbf{H}_O)_2$,

$$\Sigma\int_{t_1}^{t_2} \mathbf{M}_O\,dt = (\mathbf{H}_O)_2 - (\mathbf{H}_O)_1$$

or

$$(\mathbf{H}_O)_1 + \Sigma\int_{t_1}^{t_2} \mathbf{M}_O\,dt = (\mathbf{H}_O)_2 \qquad (15\text{–}28)$$

This equation is referred to as the *principle of angular impulse and momentum*. The initial and final angular momenta $(\mathbf{H}_O)_1$ and $(\mathbf{H}_O)_2$ in Eq. 15–28 are defined by the moment of the linear momentum of the particle ($\mathbf{H}_O = \mathbf{r} \times m\mathbf{v}$) at the instants t_1 and t_2, respectively. The second term on the left side, $\Sigma\int_{t_1}^{t_2}\mathbf{M}_O\,dt$, is called the *angular impulse*. It is computed on the basis of integrating, with respect to time, the moments of all the forces acting on the particle over the time interval t_1 to t_2. Since the moment of a force about point O is defined as $\mathbf{M}_O = \mathbf{r} \times \mathbf{F}$, the angular impulse may be expressed in vector form as

$$\text{angular impulse} = \int_{t_1}^{t_2} \mathbf{M}_O\,dt = \int_{t_1}^{t_2} (\mathbf{r} \times \mathbf{F})\,dt \qquad (15\text{–}29)$$

Here $\mathbf{r}$ is a position vector which extends from point O to any point on the line of action of $\mathbf{F}$.

In a similar manner, using Eq. 15–26, the principle of angular impulse and momentum for a system of n particles may be written as

$$\Sigma(\mathbf{H}_O)_1 + \Sigma \int_{t_1}^{t_2} \mathbf{M}_O \, dt = \Sigma(\mathbf{H}_O)_2 \qquad (15\text{–}30)$$

Here the first and third terms represent the angular momenta of the system of particles ($\Sigma \mathbf{H}_O = \Sigma(\mathbf{r}_i \times m\mathbf{v}_i)$) at the instants t_1 and t_2. The second term in Eq. 15–30 is the vector sum of the angular impulses given to all the particles during the time period t_1 to t_2. These impulses are created only by the moments of external forces acting on the system where, for the ith particle, $\mathbf{M}_O = \mathbf{r}_i \times \mathbf{F}_i$.

Using impulse and momentum principles, it is therefore possible to write two vector equations which define the particle motion; namely, Eqs. 15–3 and 15–28, restated as

$$m\mathbf{v}_1 + \Sigma \int_{t_1}^{t_2} \mathbf{F} \, dt = m\mathbf{v}_2$$
$$(\mathbf{H}_O)_1 + \Sigma \int_{t_1}^{t_2} \mathbf{M}_O \, dt = (\mathbf{H}_O)_2 \qquad (15\text{–}31)$$

In general, these two vector equations may be expressed in x, y, z component form, yielding a total of six independent scalar equations. In particular, if the particle is confined to move in the x-y plane, three independent scalar equations may be written to express the motion,

$$m(v_x)_1 + \Sigma \int_{t_1}^{t_2} F_x \, dt = m(v_x)_2$$
$$m(v_y)_1 + \Sigma \int_{t_1}^{t_2} F_y \, dt = m(v_y)_2 \qquad (15\text{–}32)$$
$$(H_O)_1 + \Sigma \int_{t_1}^{t_2} M_O \, dt = (H_O)_2$$

The first two of these equations represent the principle of linear impulse and momentum in the x and y directions, and the third equation represents the principle of angular impulse and momentum about the z axis.

Conservation of Angular Momentum. When the angular impulses acting on a particle are all zero during the time t_1 to t_2, Eq. 15–28 reduces to the following simplified form,

$$(\mathbf{H}_O)_1 = (\mathbf{H}_O)_2 \qquad (15\text{–}33)$$

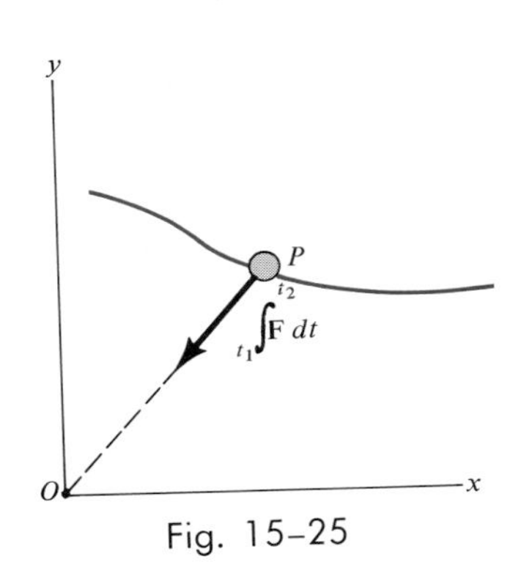

Fig. 15–25

This equation is known as the *conservation of angular momentum*. It states that from t_1 to t_2 the particle's angular momentum remains constant. Obviously, if no external impulse is applied to the particle, both linear and angular momentum will be conserved. In some cases, however, the particle's angular momentum will be conserved and linear momentum may not. An example of this occurs when the particle is subjected only to a *central force* (Sec. 13–7). As shown in Fig. 15–25, the impulsive central force **F** is always directed toward point O as the particle moves along the path. Hence, the angular impulse (moment) created by this force about the z axis passing through point O is always zero, and therefore angular momentum of the particle is conserved about this axis. As a second example of angular-momentum conservation, consider the ball of mass m shown in Fig. 15–26*a,* which is moving along a spiral path as its attached cord wraps around the thin pole. The momentum- and impulse-vector diagrams for the ball, as it moves from A to C, are shown in Fig. 15–26*b.* Because the pole is thin, **T** is assumed always to be directed to a point on the z axis, and **W** is always parallel to the z axis. Hence, these forces do not create an angular impulse about the z axis, and therefore the angular momentum of the ball is only conserved about this axis, i.e., $(H_z)_A = (H_z)_C$.

On the basis of Eq. 15–30, we can also write the conservation of angular momentum for a system of n particles, namely,

$$\Sigma(\mathbf{H}_O)_1 = \Sigma(\mathbf{H}_O)_2 \tag{15–34}$$

In this case the summation must include the angular momenta of all the particles in the system.

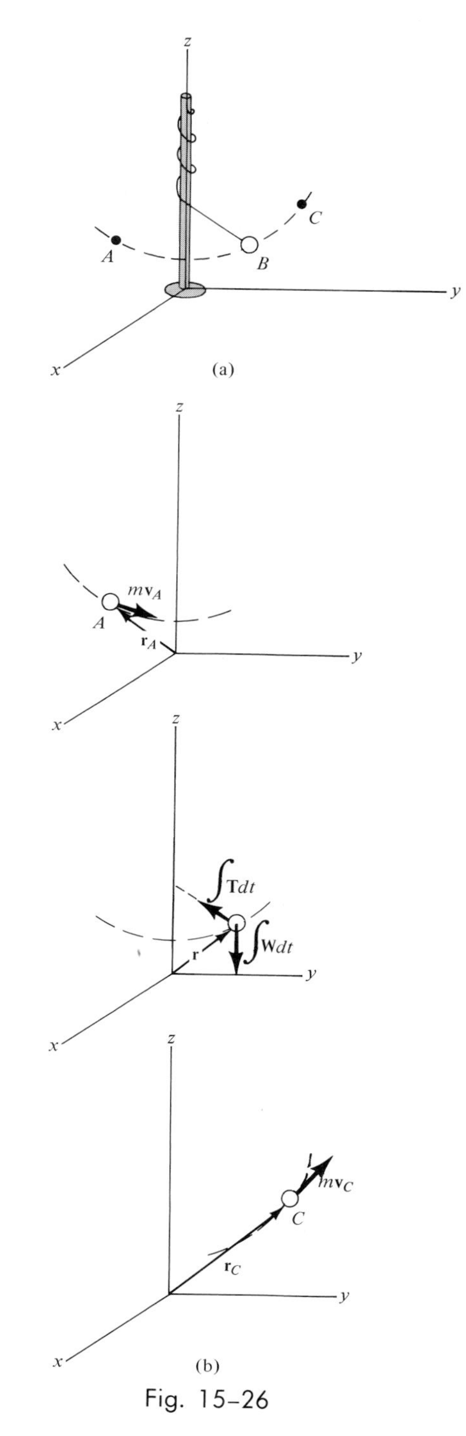

Fig. 15–26

PROCEDURE FOR ANALYSIS

When applying the principles of angular impulse and momentum, or the conservation of angular momentum, it is suggested that the following two-step procedure be used.

Step 1: Draw the momentum- and impulse-vector diagrams for the particle. In particular, the impulse-vector diagram provides a convenient means for determining any axis about which angular momentum may be conserved. For this to occur, the moments of the linear impulses about the axis must be zero throughout the time period t_1 to t_2.

Step 2: Apply the principle of angular impulse and momentum, $(\mathbf{H}_O)_1 + \Sigma \int_{t_1}^{t_2} \mathbf{M}_O\, dt = (\mathbf{H}_O)_2$ or, if appropriate, the conservation of angular momentum, $(\mathbf{H}_O)_1 = (\mathbf{H}_O)_2$. Each of the terms in these equations can be formulated directly from the data shown on the momentum- and impulse-vector diagrams.

If other equations are needed for the problem solution, when appropriate, use the principle of linear impulse and momentum, the principle of work and energy, the equations of motion, or kinematics.

The following examples numerically illustrate application of the above procedure.

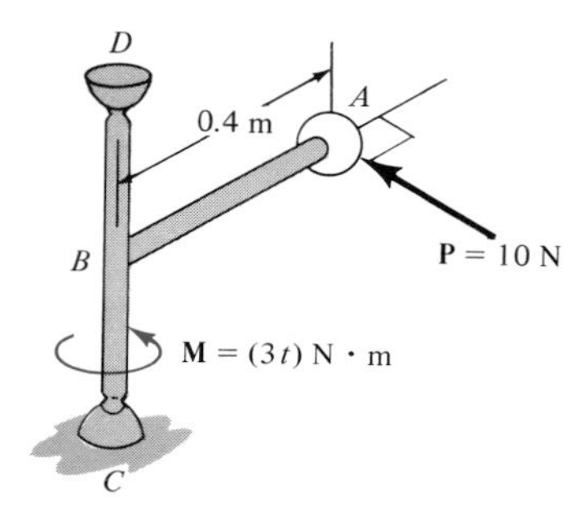

Fig. 15–27(a)

Example 15–11

The frame shown in Fig. 15–27*a* is originally at rest and supports the 5-kg sphere A. A moment of $M = (3t)$ N · m, where t is measured in seconds, is applied to shaft CD and a force **P** of 10 N is applied perpendicular to arm AB. If the mass of the frame is neglected, determine the speed of the sphere in 4 s.

Solution

Step 1: The momentum- and impulse-vector diagrams of the frame are shown in Fig. 15–27*b*. The impulses $\int \mathbf{F}_D\,dt$ and $\int \mathbf{F}_C\,dt$ are unknown in both magnitude and direction and represent the effect of the supports on the frame. These impulses can be eliminated from the analysis by applying the principle of angular impulse and momentum about the z axis. Why? If this is done, the angular impulse created by the weight of sphere A is also eliminated from this equation, since it acts parallel to the z axis and therefore creates zero moment about this axis.

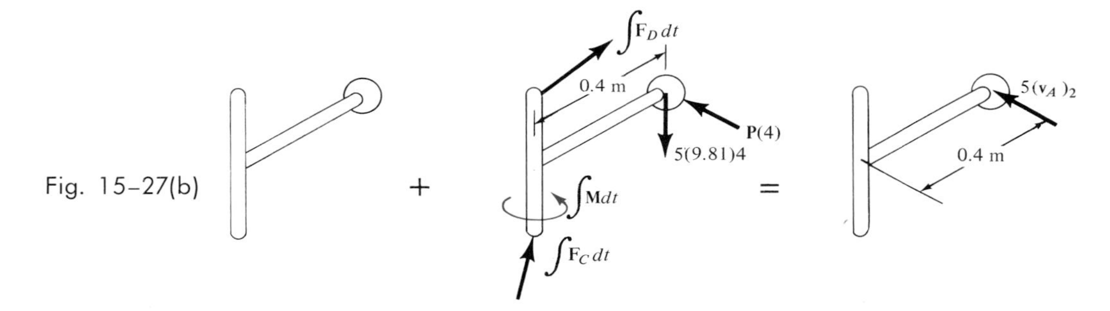

Fig. 15–27(b)

Step 2: Applying Eq. 15–30 about the z axis, we have

$$(H_z)_1 + \Sigma \int M_z\,dt = (H_z)_2$$

$$(H_{Az})_1 + \int M\,dt + \int Pr_{AB}\,dt = (H_{Az})_2$$

$$0 + \int_0^4 3t\,dt + (10)(0.4)(4) = 5(v_A)_2(0.4)$$

$$24 + 16 = 2(v_A)_2$$

$$(v_A)_2 = 20.0 \text{ m/s} \qquad \textit{Ans.}$$

Example 15–12

The 2-kg block shown in Fig. 15–28*a* rests on a smooth *horizontal surface* and is attached to an elastic cord that has a stiffness of $k_c = 20$ N/m and is initially unstretched. If it is given a velocity of $(v_B)_1 = 1.5$ m/s in the direction shown, determine the rate at which the cord is being stretched and the speed of the block at the instant the cord is stretched 0.2 m.

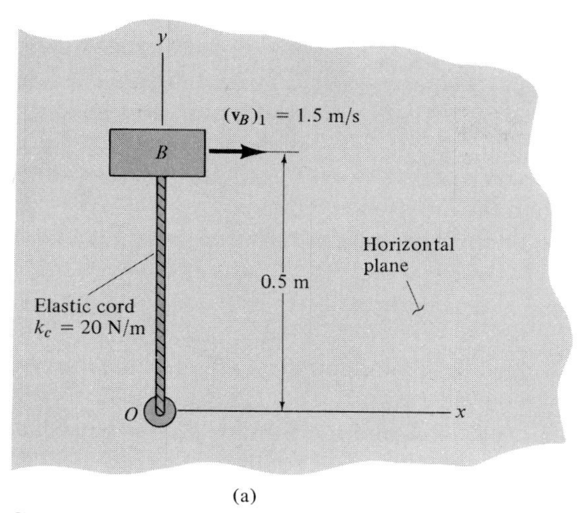

(a)

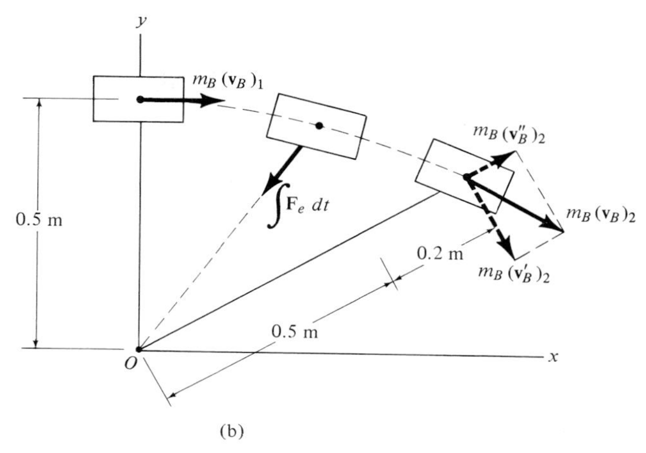

(b)

Fig. 15–28

Solution

Step 1: After the block has been launched, it slides along the dashed path shown in Fig. 15–28*b*. By inspection of the impulse-vector diagram, angular momentum about point O (or the z axis) is *conserved,* since the moment of the linear impulse $\int \mathbf{F}_e\, dt$ about O is always zero. ($\mathbf{F}_e$ is a central force.) Also, when the distance is 0.7 m, only the component of momentum $m_B(\mathbf{v}'_B)_2$ is effective in producing angular momentum of the block about O.

Step 2: The component $(\mathbf{v}'_B)_2$ can be obtained by applying the conservation of angular momentum about O (the z axis), i.e.,

$$(\mathbf{H}_O)_1 = (\mathbf{H}_O)_2$$

$$r_1 m_B (v_B)_1 = r_2 m_B (v'_B)_2$$

$$\circlearrowleft + \qquad 0.5(2)(1.5) = 0.7(2)(v'_B)_2$$

$$(v'_B)_2 = 1.07 \text{ m/s}$$

The speed of the block, $(v_B)_2$, may be obtained by applying the conservation-of-energy theorem before the block was launched and at the instant the cord is stretched 0.2 m.

$$T_1 + V_1 = T_2 + V_2$$

$$\tfrac{1}{2}(2)(1.5)^2 + 0 = \tfrac{1}{2}(2)(v_B)_2^2 + \tfrac{1}{2}(20)(0.2)^2$$

Thus,

$$(v_B)_2 = 1.36 \text{ m/s} \qquad \textit{Ans.}$$

Having determined $(v_B)_2$ and its component $(v'_B)_2$, the rate of stretch of the cord $(v''_B)_2$ is determined from the Pythagorean theorem,

$$(v''_B)_2 = \sqrt{(v_B)_2^2 - (v'_B)_2^2}$$

$$= \sqrt{(1.36)^2 - (1.07)^2}$$

$$= 0.838 \text{ m/s} \qquad \textit{Ans.}$$

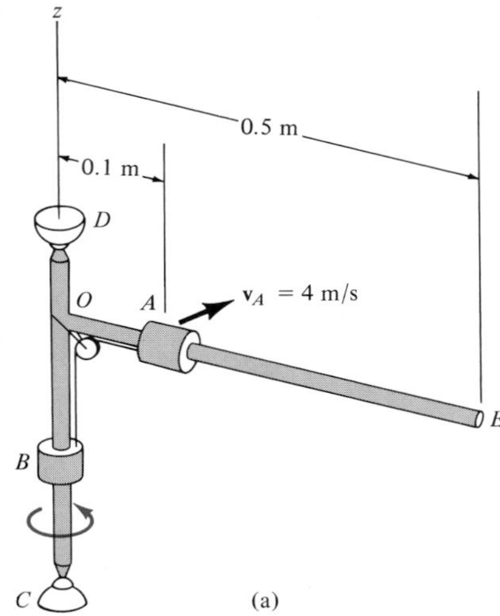

(a)

Example 15–13

Block A shown in Fig. 15–29*a* has a mass of 2 kg and is free to slide along the smooth horizontal rod OE. This block is connected to the 1-kg block B, which slides freely along rod CD. Holding B so that it cannot travel up rod CD, a moment is applied to CD until the velocity of A reaches 4 m/s, as shown. When the moment is removed and B is released, determine the upward speed of B when A slides outward until it is 0.3 m from the axis of rotation, CD. Neglect the mass of the rods.

Solution

Step 1: The impulse- and momentum-vector diagrams for the frame, cords, and two blocks, considered as a system, are shown in Figs. 15–29*b* and 15–29*c*. From the impulse-vector diagram, Fig. 15–29*b*, it is seen that the angular impulses of the block weights and the support reactions at C and D are zero about the z axis, since they do not create a "moment" about this axis. Hence, angular momentum of the system is conserved about the z axis.

Step 2: By applying the conservation of angular momentum in scalar form about the z axis, it is possible to determine the component of velocity $(v_A')_2$.

$$(H_z)_{A1} = (H_z)_{A2}$$
$$(r_A)_1 m_A (v_A)_1 = (r_A)_2 m_A (v_A')_2$$
$$0.1(2)(4) = 0.3(2)(v_A')_2$$

(b)

Fig. 15–29

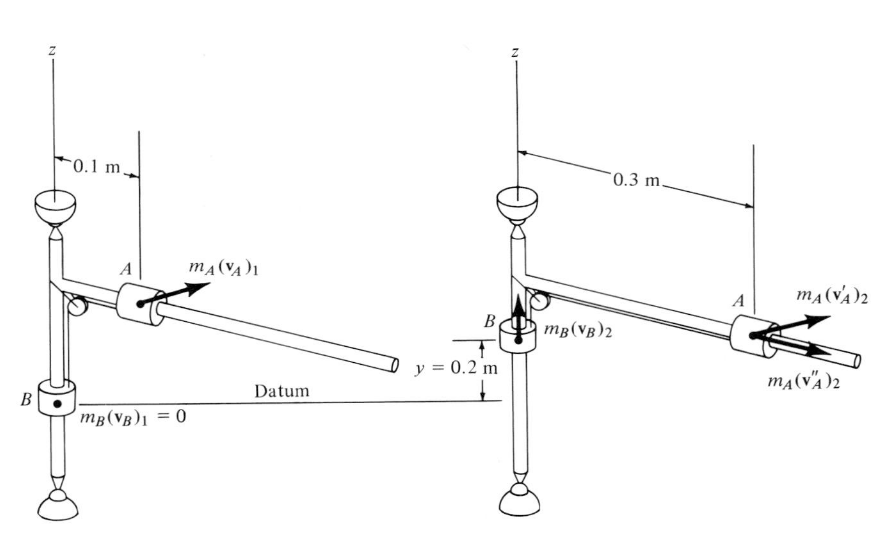

(c)

Thus,

$$(v_A')_2 = 1.33 \text{ m/s}$$

The velocity of B may be determined by applying the conservation-of-energy principle to the system. Locating the potential-energy datum at the lowest position of block B, Fig. 15–29c, and noting that the potential energy of block A does not change, we have

$$\{T_1\} + \{V_1\} = \{T_2\} + \{V_2\}$$

$$\{\tfrac{1}{2}m_A(v_A)_1^2 + \tfrac{1}{2}m_B(v_B)_1^2\} + \{0\} = \{\tfrac{1}{2}m_A(v_A')_2^2 + \tfrac{1}{2}m_A(v_A'')_2^2 + \tfrac{1}{2}m_B(v_B)_2^2\} + \{W_B y\}$$

Substituting the computed value of $(v_A')_2$, realizing that $(v_A'')_2 = (v_B)_2$ because of the cord, and noting that B moves upward, $y = (0.3 - 0.1)$ m $= 0.2$ m, we have

$$\{\tfrac{1}{2}(2)(4)^2 + 0\} + \{0\} = \{\tfrac{1}{2}(2)(1.33)^2 + \tfrac{1}{2}(2)(v_B)_2^2 + \tfrac{1}{2}(1)(v_B)_2^2\} + \{(1)(9.81)0.2\}$$

Solving for $(v_B)_2$ yields

$$(v_B)_2 = 2.86 \text{ m/s} \qquad \textit{Ans.}$$

Problems

15-45. The projectile having a mass of $m = 2$ kg is fired from a cannon with a muzzle velocity of $v_O = 500$ m/s. Determine the projectile's angular momentum about point O at the instant it is at the maximum height of its trajectory.

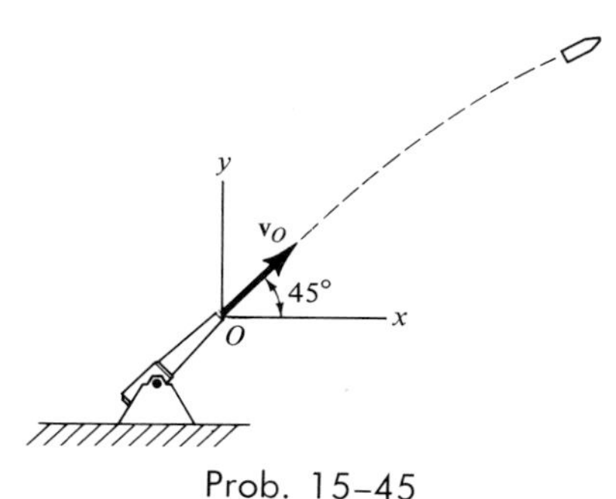

Prob. 15-45

15-45a. Solve Prob. 15–45 if the projectile has a weight of $W = 15$ lb and $v_O = 800$ ft/s.

15-46. Determine the total angular momentum $\mathbf{H}_O$ for the system of three particles about point O. All the particles are moving in the x-y plane.

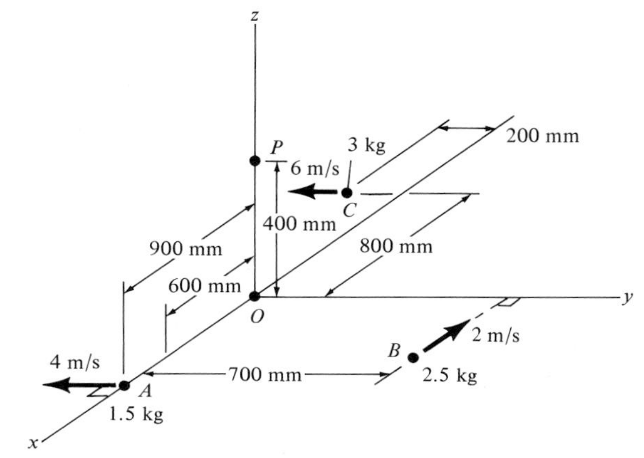

Prob. 15-46

15–47. Determine the total angular momentum $\mathbf{H}_P$ for the system of three particles in Prob. 15–46 about the point P.

*** 15–48.** The four spheres each have a mass of 2 kg and are rigidly attached to a crossbar frame of negligible mass. If a couple $M = (5t + 2)$ N · m, where t is in seconds, is applied as shown, determine the speed of each of the spheres in 4 s, starting from rest.

Prob. 15–48

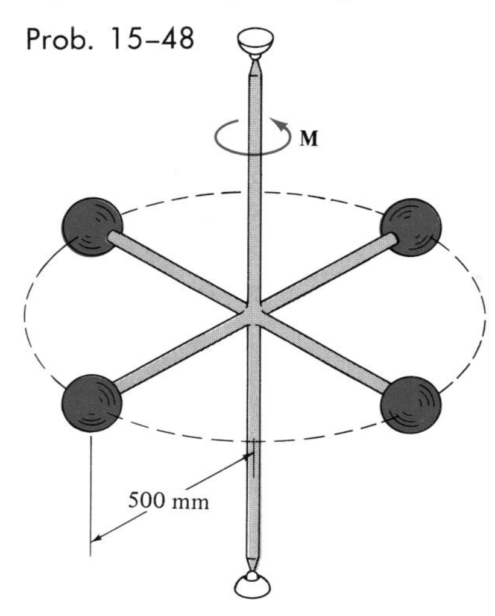

15–49. The two blocks A and B each have a mass of 400 g. The blocks are fixed to the horizontal rods and their initial velocity is 2 m/s in the direction shown. If a couple moment of $M = 0.6$ N · m is applied about CD of the frame, determine the speed of the blocks in 3 s. The mass of the supporting frame is negligible and it is free to rotate about CD.

Prob. 15–49

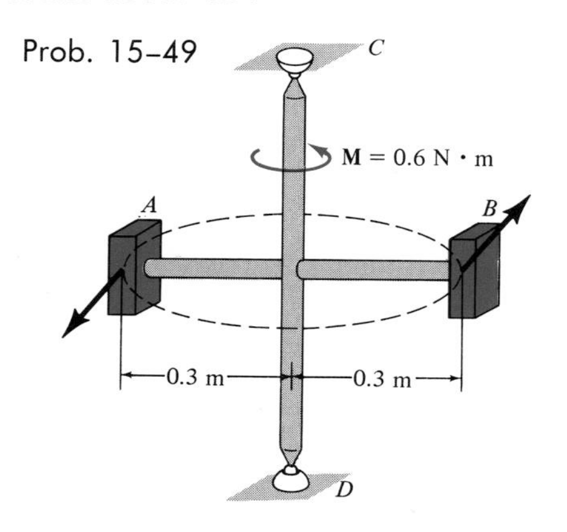

15–50. A basket and its contents have a mass of $m_b = 8$ kg. If the end of a rope is attached to the basket and the rope passes over a pulley for which $r = 0.1$ m, determine the speed at which the basket rises when $t = 3$ s, if initially a monkey, having a mass of $m_m = 10$ kg, begins to climb upward along the other end of the rope with a constant speed of $v_{m/r} = 2$ m/s, measured relative to the rope. Neglect the mass of the pulley and rope.

Prob. 15–50

15–50a. Solve Prob. 15–50 if the basket and its contents weigh $W_b = 10$ lb, the monkey weighs $W_m = 18$ lb, $t = 2$ s, $r = 0.4$ ft, and $v_{m/r} = 1$ ft/s.

15–51. An earth satellite of mass 800 kg is launched into a free-flight trajectory about the earth with an initial velocity of $v_A = 12$ km/s when the distance from the center of the earth is $r_A = 15$ Mm. If the launch angle at this position is $\phi_A = 60°$, determine the velocity v_B of the satellite and its closest distance r_B from the center of the earth. The earth has a mass of $M_e = 5.976(10^{24})$ kg.

Hint: Under these conditions, the satellite is subjected only to the earth's gravitational force, $F = GM_e m_s/r^2$ (Eq. 13–2). For part of the solution, use the conservation of energy; see Prob. 14–47.

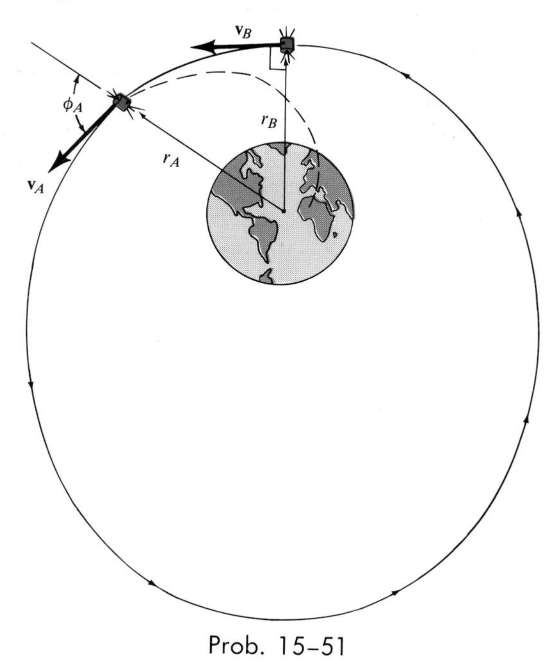

Prob. 15–51

*** 15–52.** A girl having a mass of 40 kg is sitting on a swing such that when she is at A, her velocity is equal to zero and her center of mass is 2.5 m from the center of rotation D. Upon swinging downward, she *suddenly* changes the position of her center of mass at B by moving her legs upward, so that her center of mass is 2.3 m from the center of rotation. If she maintains this position from B to C, determine the angle θ to which she rises before stopping. Neglect the mass of the swing. *Hint:* In practice, the transition in leg movement is uniform from A to C. By assuming *rigidity* from A to B, then from B to C, potential and kinetic energy are conserved along these segments of the path. Angular momentum is conserved at the instant *just before* and *just after* passing point B. Why?

Prob. 15–52

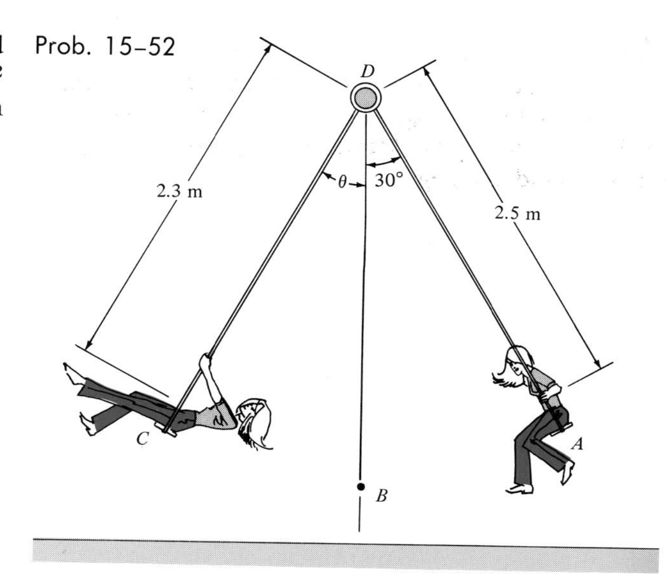

15–53. The elastic cord, having an unstretched length of 300 mm and a stiffness of $k = 400$ N/m, is attached to a fixed point at A and a block at B, which has a mass of 1.5 kg. If the block is given an initial velocity of $v_B = 2$ m/s along the x axis, from the position shown, determine the speed of the block at the instant the cord becomes slack at C. How close, d, does the block come within approaching A? The block slides on the smooth horizontal plane.

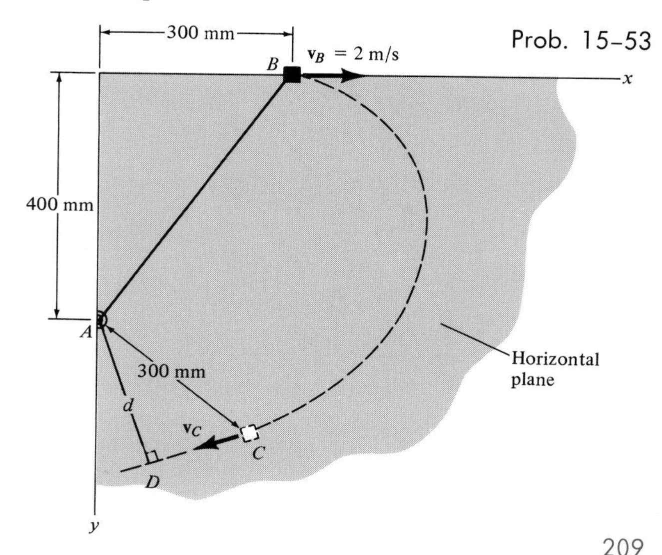

Prob. 15–53

15–54. A cord of length l is attached to a ball of mass m and a peg having a radius $a \ll l$. If the ball is given an initial velocity v_1, perpendicular to the cord, determine the ball's velocity just after it has wrapped the cord four times around the peg.

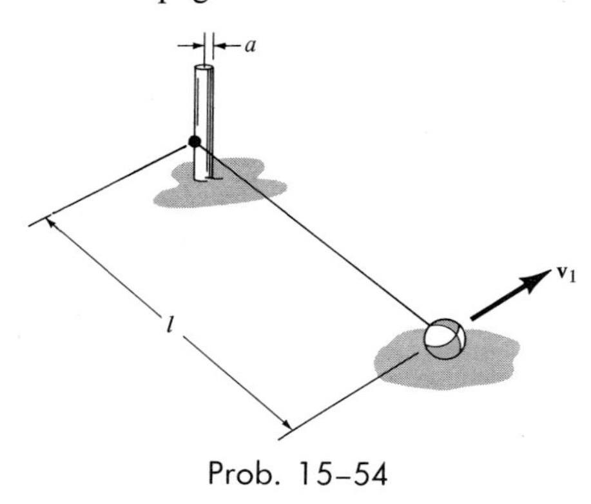

Prob. 15–54

15–55. A ball B has a mass of $m_B = 0.5$ kg and is attached to a cord which passes through a hole A in a smooth table. When the ball is $r_1 = 400$ mm from the hole, it is rotating around the hole in a circle such that its speed is $(v_B)_1 = 1.5$ m/s. If, by applying the force $\mathbf{F}$, the cord is then pulled downward through the hole, with a constant speed of $v_c = 2$ m/s, determine the speed of the ball at the instant it is $r_2 = 200$ mm from the hole. How much work is done by the force $\mathbf{F}$ in shortening the cord?

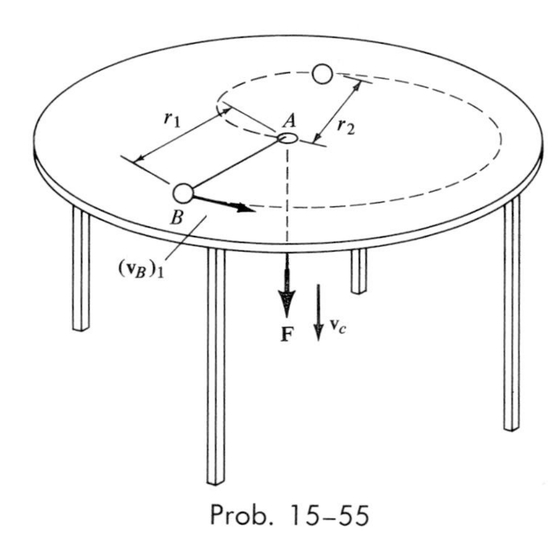

Prob. 15–55

15–55a. Solve Prob. 15–55 if the ball has a weight of $W_b = 0.8$ lb, and $r_1 = 1.75$ ft, $(v_B)_1 = 4$ ft/s, $v_c = 6$ ft/s, $r_2 = 0.6$ ft.

***15–56.** A toboggan and rider, having a total mass of 150 kg, enter horizontally tangent to a 90° circular curve with a velocity of $v_A = 80$ km/h. If the track is flat and banked at an angle of 60°, determine the velocity $\mathbf{v}_B$ and the angle θ_B of "descent," measured from the horizontal in the vertical $(y - z)$ plane, at which the toboggan exits at B. Neglect friction in the calculation. The radius r_B equals 57 m.

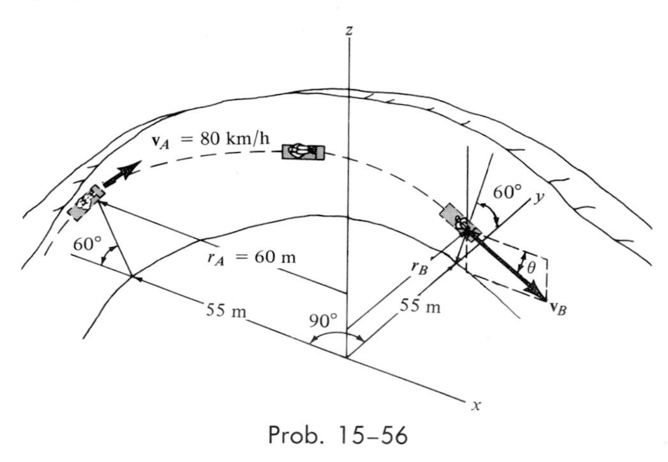

Prob. 15–56

15–57. A small particle having a mass m is placed inside the semicircular tube. The particle is displaced to the position shown and released. Applying the principle of angular momentum about point O $(\Sigma M_O = \dot{H}_O)$, show that the motion of the particle is governed by the differential equation $\ddot{\theta} + (g/R)\sin\theta = 0$.

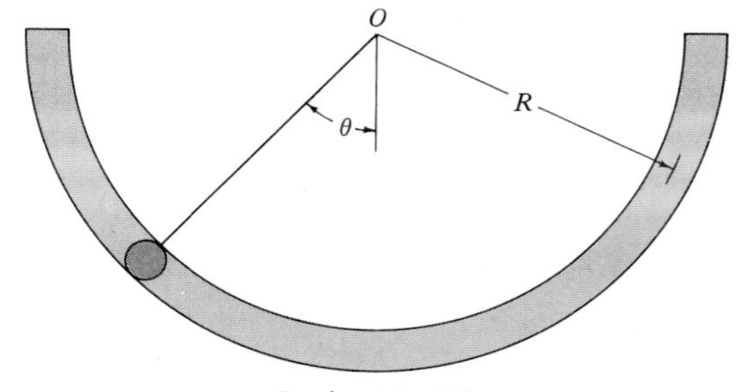

Prob. 15–57

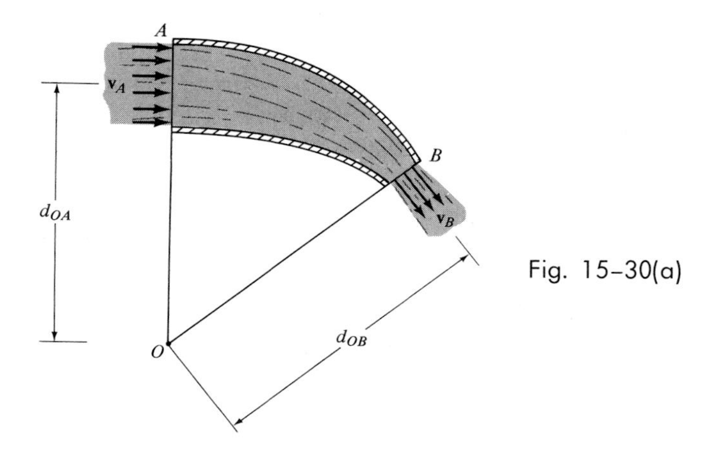

Fig. 15–30(a)

*15–8. Steady Fluid Streams

Knowledge of the forces developed by steadily moving fluid streams is of importance in the design and analysis of turbines, pumps, blades, and fans. To illustrate how the principle of impulse and momentum may be used to determine these forces, consider the diversion of a steady stream of fluid (liquid or gas) by a fixed pipe, Fig. 15–30*a*. The fluid enters the pipe with a velocity $\mathbf{v}_A$ and exits with a velocity $\mathbf{v}_B$. The momentum- and impulse-vector diagrams for the fluid stream are shown in Fig. 15–30*b*. The force $\Sigma\mathbf{F}$, shown acting on the impulse-vector diagram, represents the resultant force of all the external forces acting on the fluid stream. It is this loading which gives the fluid stream an impulse whereby the original momentum of the fluid is changed in both its magnitude and direction. Since the flow is steady, this force $\Sigma\mathbf{F}$ will be *constant* during the time interval dt. During this time, the fluid stream is in motion, and as a result a small amount of fluid, having a mass dm, enters the pipe with a velocity $\mathbf{v}_A$ at time t. If this element of mass and the mass of fluid in the pipe is considered as a "closed system," then at time $t + dt$ a corresponding element of mass dm must leave the pipe with a velocity $\mathbf{v}_B$. The fluid stream *within* the pipe section has a mass m and an *average velocity* $\mathbf{v}$

Fig. 15–30(b)

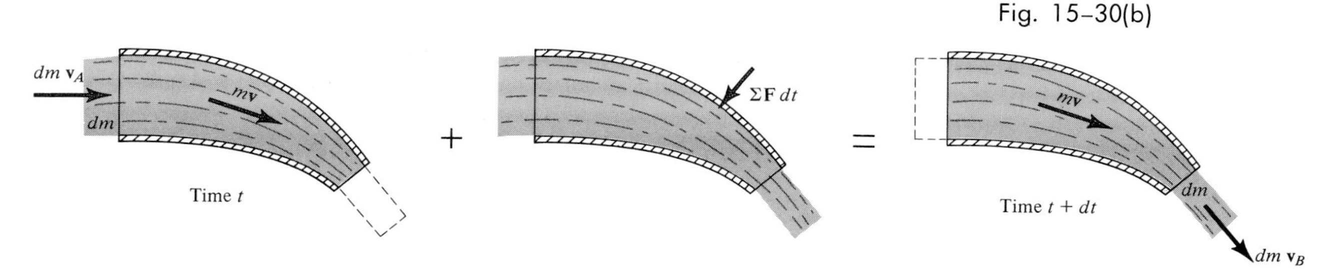

which is constant during the time interval dt. Applying the principle of linear impulse and momentum to the fluid stream, we have

$$dm\,\mathbf{v}_A + m\mathbf{v} + \Sigma\mathbf{F}\,dt = dm\,\mathbf{v}_B + m\mathbf{v}$$

Solving for the resultant force yields

$$\Sigma\mathbf{F} = \frac{dm}{dt}(\mathbf{v}_B - \mathbf{v}_A) \tag{15-35}$$

Provided the motion of the fluid can be represented in the x-y plane, it is usually convenient to express this vector equation in the form of two scalar component equations, i.e.,

$$\begin{aligned} \Sigma F_x &= \frac{dm}{dt}(v_{Bx} - v_{Ax}) \\ \Sigma F_y &= \frac{dm}{dt}(v_{By} - v_{Ay}) \end{aligned} \tag{15-36}$$

The term dm/dt is called the *mass flow* and indicates the constant amount of fluid which flows either into or out of the pipe per unit of time. If the cross-sectional areas and densities of the fluid at the entrance A and exit B are ρ_A, A_A, and ρ_B, A_B, respectively, Fig. 15–30*c*, then continuity of mass requires that $dm = \rho\,dV = \rho_A(ds_A\,A_A) = \rho_B(ds_B\,A_B)$. Hence, during the time dt, since $\mathbf{v}_A = d\mathbf{s}_A/dt$ and $\mathbf{v}_B = d\mathbf{s}_B/dt$, we have

$$\begin{aligned} \frac{dm}{dt} &= \rho_A v_A A_A = \rho_B v_B A_B \\ &= \rho_A Q_A = \rho_B Q_B \end{aligned} \tag{15-37}$$

Here $Q = vA$ is the volumetric *flow rate* which measures the volume of fluid flowing per unit of time, e.g., m^3/s.

In some cases it is necessary to obtain the support reactions on the fluid-carrying device. If Eq. 15–35 does not provide enough information to do this, the principle of angular impulse and momentum must be used. The formulation of this principle applied to fluid streams, can be obtained from Eq. 15–26, $\Sigma\mathbf{M}_O = \dot{\mathbf{H}}_O$, which states that the moment of all the external forces acting on the system about point O is equal to the time rate of change of angular momentum about O. In the case of the pipe shown in Fig. 15–30*a*, the flow is steady in the x-y plane; hence, we have

$$(\circlearrowleft +) \qquad \Sigma M_O = \frac{dm}{dt}(d_{OB}v_B - d_{OA}v_A) \tag{15-38}$$

where the moment arms d_{OB} and d_{OA} are directed from O to the *center* of the openings at A and B.

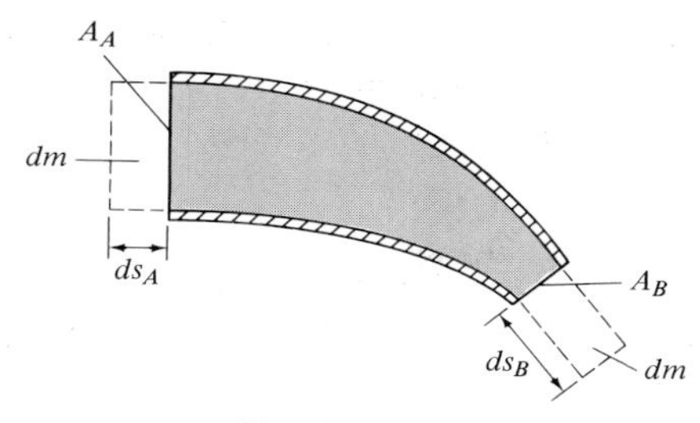

Fig. 15–30(c)

PROCEDURE FOR ANALYSIS

The following three-step procedure should be used when applying the above equations to the solution of problems involving steady flow.

Step 1: In problems where the device is *moving*, a *kinematic diagram* may be helpful for determining the absolute entrance and exit velocities of the fluid flowing onto the device, since a *relative-motion analysis* of velocity will be involved. The *kinematic diagram* in this case is simply a graphical representation of the velocities showing the vector addition of the relative-motion components. Once the absolute velocity of the fluid flowing onto the device is determined, the mass flow is calculated using Eq. 15–37.

Step 2: Draw a free-body diagram of the device which is directing the fluid in order to establish the forces $\Sigma \mathbf{F}$ acting on it. These external forces will include the support reactions, the weight of the device and the fluid contained within it, and the static (gauge) pressure forces of the fluid at the entrance and exit sections of the device.*

Step 3: Apply the equations of steady flow, Eqs. 15–36 and 15–38, using the appropriate components of velocity and force shown on the diagrams in *Steps 1* and *2*.

The following examples numerically illustrate application of this procedure.

Example 15–14

Determine the reaction components which the pipe joint at A exerts on the elbow in Fig. 15–31*a*, if water flowing through the pipe is subjected to

*Pressure is measured using the *pascal* (Pa) as the basic unit, where $1 \text{ Pa} = 1 \text{ N/m}^2$.

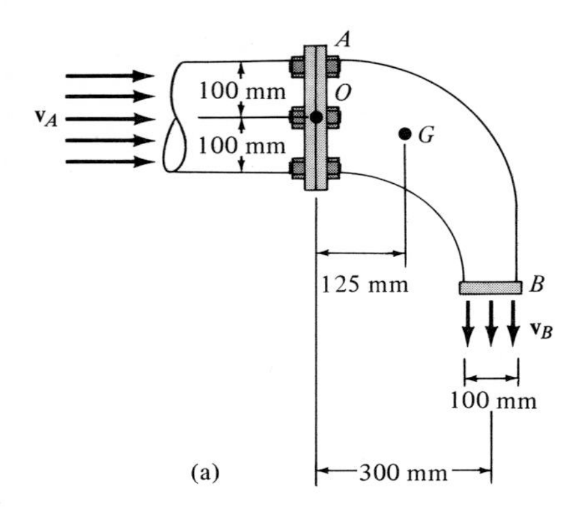

a static pressure of 100 kPa at A. The discharge at B is $Q_B = 0.2\ \text{m}^3/\text{s}$. Water is assumed to have a constant density of $\rho_w = 1000\ \text{kg/m}^3$. The water-filled elbow has a mass of 20 kg and center of gravity at G.

Solution

Step 1: The velocity of flow at A and B and the mass flow rate can be obtained from Eq. 15–37. Since the density of water is constant, $Q_B = Q_A = Q$. Hence,

$$\frac{dm}{dt} = \rho_w Q = (1000\ \text{kg/m}^3)(0.2\ \text{m}^3/\text{s}) = 200\ \text{kg/s}$$

$$v_B = \frac{Q}{A_B} = \frac{0.2\ \text{m}^3/\text{s}}{\pi(0.05\ \text{m})^2} = 25.46\ \text{m/s}$$

$$v_A = \frac{Q}{A_A} = \frac{0.2\ \text{m}^3/\text{s}}{\pi(0.1\ \text{m})^2} = 6.37\ \text{m/s}$$

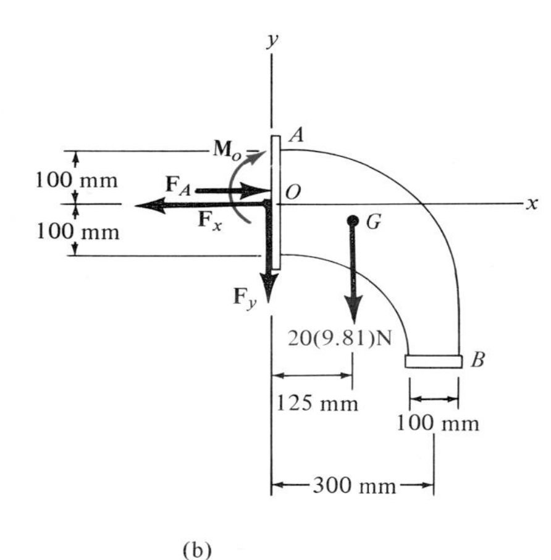

Fig. 15–31

Step 2: As shown on the free-body diagram, Fig. 15–31*b*, the fixed connection at A exerts a resultant couple $\mathbf{M}_O$ and force components $\mathbf{F}_x$ and $\mathbf{F}_y$ on the elbow. Due to the static pressure of water in the pipe, the pressure force acting on the fluid at A is $F_A = p_A A_A$. Since 1 kPa = 1000 N/m²,

$$F_A = p_A A_A; \quad F_A = (100(10^3)\ \text{N/m}^2)[\pi(0.1\ \text{m})^2] = 3141.6\ \text{N}$$

There is no static pressure acting at B, since the water is discharged at atmospheric pressure, i.e., the pressure measured by a gauge at B is equal to zero, $p_B = 0$.

Step 3: Applying Eqs. 15–36 to obtain the force components,

$$(\overset{+}{\rightarrow}) \qquad \Sigma F_x = \frac{dm}{dt}(v_{Bx} - v_{Ax})$$

$$-F_x + 3141.6\ \text{N} = 200\ \text{kg/m}^3(0 - 6.37\ \text{m/s})$$

$$F_x = 4415.6\ \text{N} = 4.42\ \text{kN} \qquad \textit{Ans.}$$

$$(+\uparrow) \qquad \Sigma F_y = \frac{dm}{dt}(v_{By} - v_{Ay})$$

$$-F_y - 20(9.81)\ \text{N} = 200\ \text{kg/m}^3(-25.46\ \text{m/s} - 0)$$

$$F_y = 4895.8\ \text{N} = 4.90\ \text{kN} \qquad \textit{Ans.}$$

The couple $\mathbf{M}_O$ is obtained by applying Eq. 15–38. If moments are summed about point O, Fig. 15–31*b*, $\mathbf{F}_x$, $\mathbf{F}_y$, and the static pressure $\mathbf{F}_A$ are eliminated, as well as the moment of momentum of the water entering at A, Fig. 15–31*a*. Hence,

$$(\curvearrowleft +) \qquad \Sigma M_O = \frac{dm}{dt}(d_{OB}v_B - d_{OA}v_A)$$

$$M_O + 20(9.81)\ \text{N}(0.125\ \text{m}) = 200\ \text{kg/m}^3[(0.3\ \text{m})(25.46\ \text{m/s}) - 0]$$

$$M_O = 1503.1\ \text{N}\cdot\text{m} = 1.50\ \text{kN}\cdot\text{m} \qquad \textit{Ans.}$$

Example 15–15

A 100-mm-diameter water jet having a velocity of 25 m/s impinges upon a single moving blade, Fig. 15–32*a*. If the blade is moving at 5 m/s away from the jet, determine the horizontal and vertical components of force which the blade is exerting on the water, $\rho_w = 1000 \text{ kg/m}^3$. What power does the fluid generate on the blade?

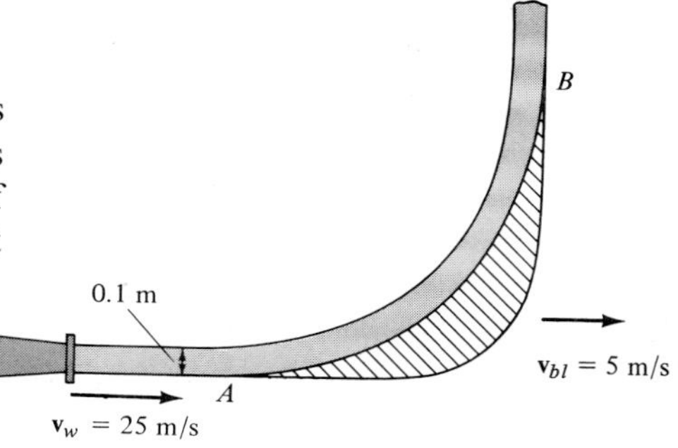

(a)

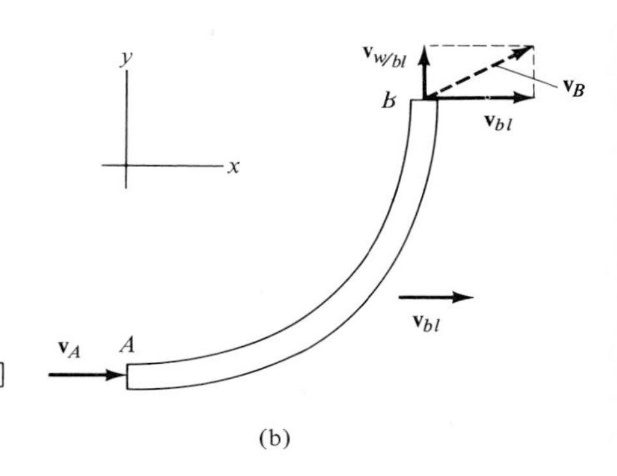

(b)

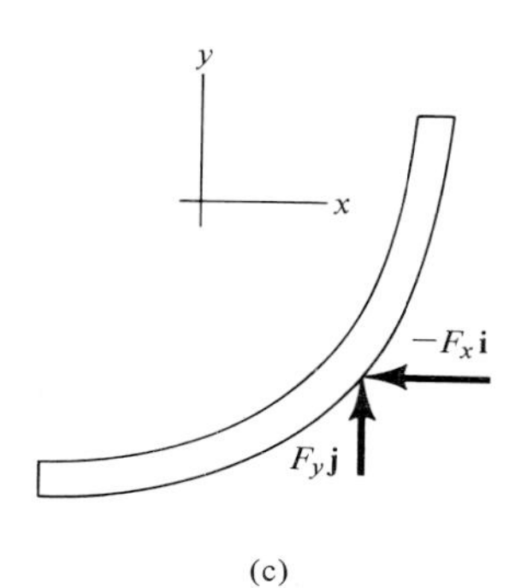

(c)

Fig. 15–32

Solution

Step 1: The kinematic diagram of velocity is shown in Fig. 15–32*b*. The rate at which water enters the blade is

$$\mathbf{v}_A = \mathbf{v}_w - \mathbf{v}_{bl}$$
$$= 25\mathbf{i} - 5\mathbf{i} = \{20\mathbf{i}\} \text{ m/s}$$

This same *relative-flow velocity* is directed vertically upward on the blade at *B*. Hence, $\mathbf{v}_{w/bl} = (v_A)\mathbf{j} = \{20\mathbf{j}\}$ m/s. Since the blade is moving with a velocity of $\mathbf{v}_{bl} = \{5\mathbf{i}\}$ m/s, the velocity of flow at *B* is the vector sum, shown in Fig. 15–32*b*. Thus,

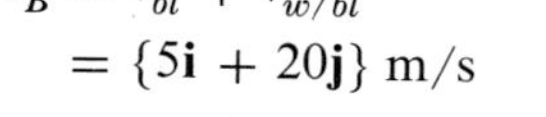

$$\mathbf{v}_B = \mathbf{v}_{bl} + \mathbf{v}_{w/bl}$$
$$= \{5\mathbf{i} + 20\mathbf{j}\} \text{ m/s}$$

The mass flow of water onto the blade is

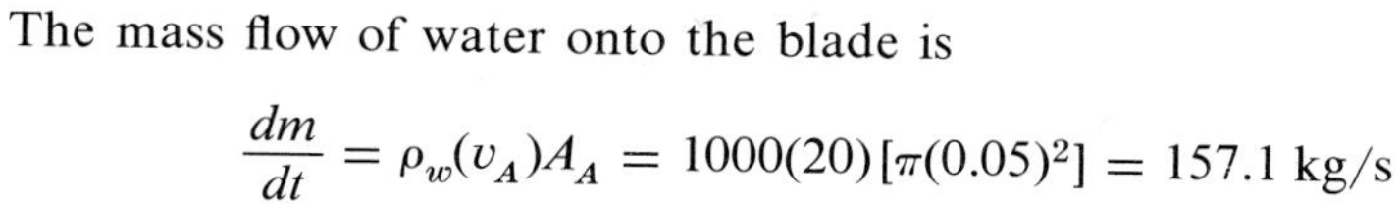

$$\frac{dm}{dt} = \rho_w(v_A)A_A = 1000(20)[\pi(0.05)^2] = 157.1 \text{ kg/s}$$

Step 2: The free-body diagram of a section of fluid acting on the blade is shown in Fig. 15–32*c*. The weight of the fluid will be neglected in the calculation, since this force is small compared to the reactive components $\mathbf{F}_x$ and $\mathbf{F}_y$.

Step 3: Applying Eq. 15–35 yields

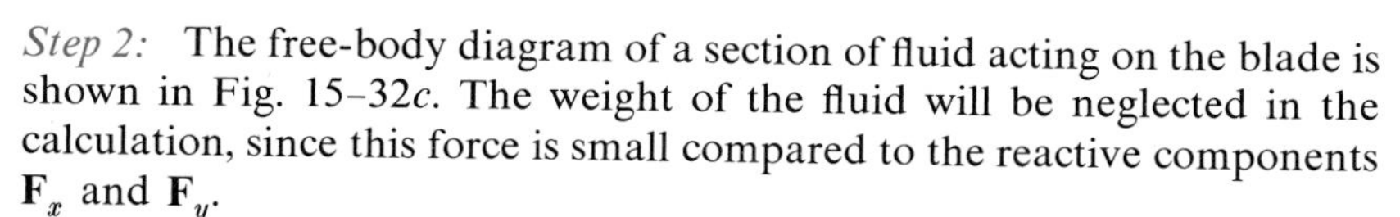

$$\Sigma\mathbf{F} = \frac{dm}{dt}(\mathbf{v}_B - \mathbf{v}_A)$$
$$-F_x\mathbf{i} + F_y\mathbf{j} = 157.1(5\mathbf{i} + 20\mathbf{j} - 20\mathbf{i})$$

Equating the respective **i** and **j** components gives

$$F_x = 157.1(15) = 2.36 \text{ kN} \qquad \textit{Ans.}$$
$$F_y = 157.1(20) = 3.14 \text{ kN} \qquad \textit{Ans.}$$

The water exerts equal but opposite forces on the blade.

Since the fluid force which causes the blade to move forward horizontally with a velocity of 5 m/s is $F_x = 2.36$ kN, then from Eq. 14–11, the power is

$$P = \mathbf{F} \cdot \mathbf{v}; \qquad P = 2.36 \text{ kN}(5 \text{ m/s}) = 11.8 \text{ kW} \qquad \textit{Ans.}$$

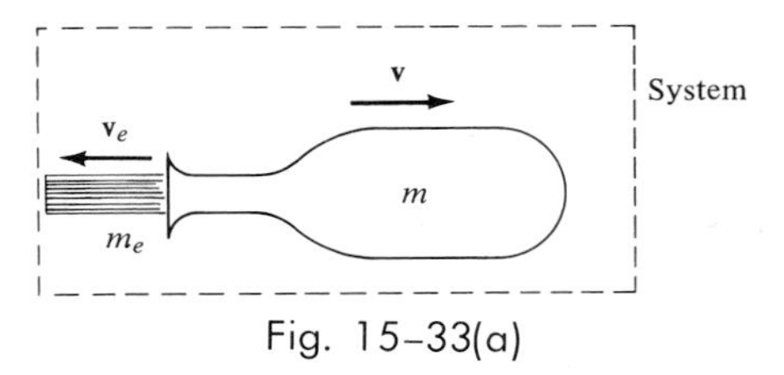

Fig. 15–33(a)

★15–9. Propulsion with Variable Mass

In the previous section the case in which a *constant* amount of mass dm entered and left a "*closed system*" was considered. There are, however, two other important cases involving mass flow, which are represented by a system which is either gaining or losing mass. In this section these cases will be discussed separately.

A System That Loses Mass. Consider a device which at an instant of time has a mass m and a forward velocity $\mathbf{v}$, Fig. 15–33*a*. At this same instant the device is expelling an amount of mass m_e such that the mass flow velocity $\mathbf{v}_e$ is *constant* when the measurement is made a small distance away from the device. For the analysis, consider the "*closed system*" at an instant to include *both the mass m and the expelled mass* m_e, as shown by the dashed line in the figure. The momentum- and impulse-vector diagrams for the system are shown in Fig. 15–33*b*. During the time interval dt, the velocity of the device is increased from $\mathbf{v}$ to $\mathbf{v} + d\mathbf{v}$. This increase in forward velocity, however, does not change the velocity $\mathbf{v}_e$ of the expelled mass, since this mass moves at a constant speed once it has been ejected. To increase the velocity of the device during the time dt, an amount of mass dm_e has been ejected and thereby gained in the exhaust. The impulses are created by $\Sigma \mathbf{F}_s$, which represents the resultant of all the external forces which *act on the system* in the direction of motion. This force resultant *does not include* the force which causes the device to move

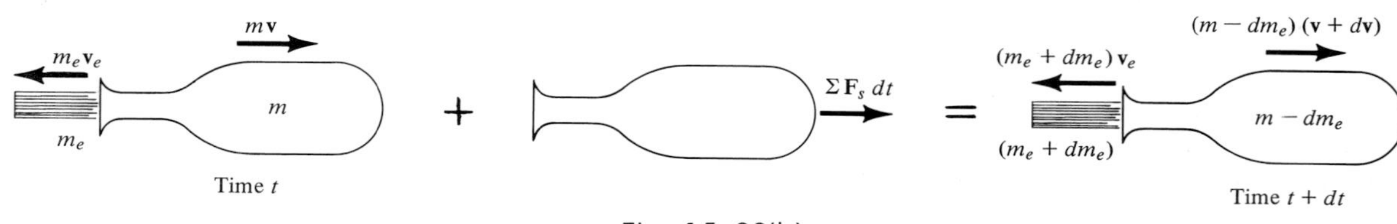

Fig. 15–33(b)

forward, since this force (called a *thrust*) is *internal to the system;* that is, the thrust acts with equal magnitude but opposite direction on the mass m of the device and the expelled exhaust mass m_e.* Applying the principle of impulse and momentum to the system, in reference to Fig. 15–33*b,* we have

$$(\overset{+}{\rightarrow}) \quad mv - m_e v_e + \Sigma F_s dt = (m - dm_e)(v + dv) - (m_e + dm_e)v_e$$

or

$$\Sigma F_s dt = -v\, dm_e + m\, dv - dm_e\, dv - v_e\, dm_e$$

Without loss of accuracy, the third term on the right side of this equation may be neglected since it is a "second-order" differential. Dividing by dt gives

$$\Sigma F_s = m\frac{dv}{dt} - (v + v_e)\frac{dm_e}{dt}$$

Noting that the relative velocity of the device as seen by an observer moving with the particles of the ejected mass is $v_{D/e} = (v + v_e)$, the final result can be written as

$$\Sigma F_s = m\frac{dv}{dt} - v_{D/e}\frac{dm_e}{dt} \qquad (15\text{–}39)$$

Here the term dm_e/dt represents the rate at which mass is being ejected.

To illustrate an application of Eq. 15–39, consider the rocket shown in Fig. 15–34, which has a weight $\mathbf{W}$ and is moving upward against an atmospheric drag force $\mathbf{F}_D$. The system to be considered consists of the mass of the rocket and the mass of ejected gas m_e. Applying Eq. 15–39 to this system gives

$$(+\uparrow) \qquad -F_D - W = \frac{W}{g}\frac{dv}{dt} - v_{D/e}\frac{dm_e}{dt}$$

The last term of this equation represents the *thrusting force* $\mathbf{T}$ which the engine exhaust exerts on the rocket, Fig. 15–34. Recognizing that $dv/dt = a$, we may therefore write

$$(+\uparrow) \qquad T - F_D - W = \frac{W}{g}a$$

If a free-body diagram of the rocket is drawn, it becomes obvious that this latter equation represents an application of $\Sigma\mathbf{F} = m\mathbf{a}$ for the rocket.

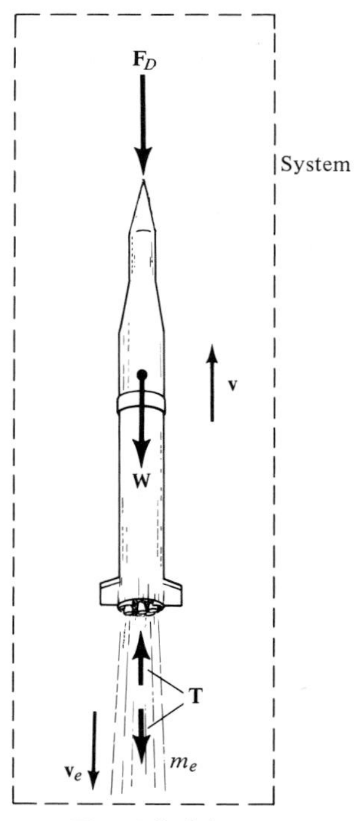

Fig. 15–34

*$\Sigma\mathbf{F}_s$ represents the external resultant force *acting on the system,* which is different from $\Sigma\mathbf{F}$, the resultant force acting only on the device.

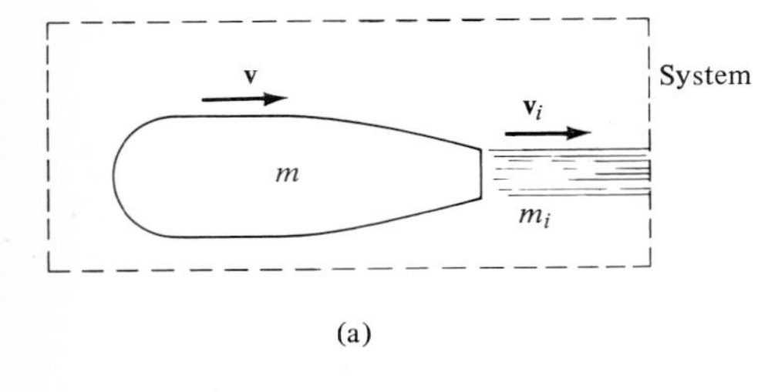

(a)

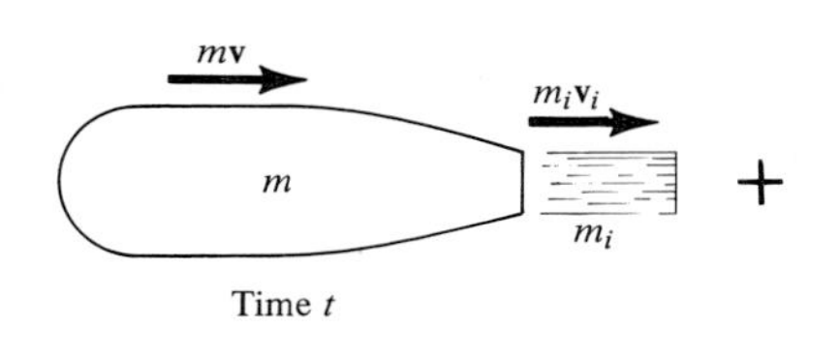

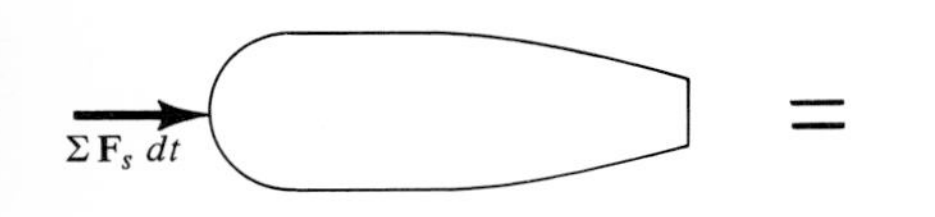

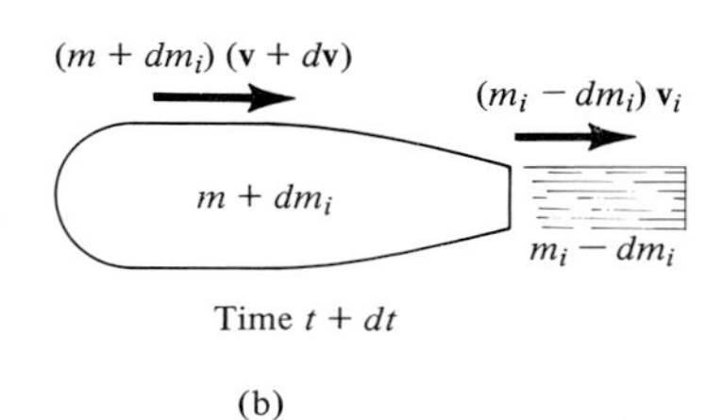

(b)

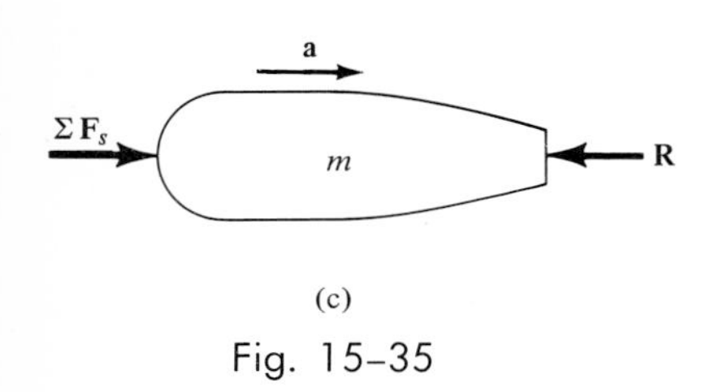

(c)

Fig. 15–35

A System That Gains Mass. A device such as a scoop or a shovel may gain mass as it moves forward. Consider, for example, the device shown in Fig. 15–35*a*, which at an instant of time has a mass m and is moving forward with a velocity $\mathbf{v}$. At the same instant, the device is collecting a particle stream of mass m_i. The flow velocity $\mathbf{v}_i$ of this injected mass is constant and independent of the velocity $\mathbf{v}$. It is required that $v > v_i$. The system to be considered at this instant includes both the mass of the device and the mass of the injected fluid, as shown by the dashed line in the figure. The momentum- and impulse-vector diagrams for this system are shown in Fig. 15–35*b*. With an increase in mass dm_i gained by the device, there is an assumed increase in velocity $d\mathbf{v}$ during the time interval dt. This increase is caused by the impulse created by the resultant of all the external forces *acting on the system* in the direction of motion, $\Sigma \mathbf{F}_s$. The force summation does not include the retarding force of the injected mass acting on the device. Why? Applying the principle of impulse and momentum to the system, we have

$$(\overset{+}{\rightarrow}) \quad mv + m_i v_i + \Sigma F_s \, dt = (m + dm_i)(v + dv) + (m_i - dm_i)v_i$$

Using the same procedure as in the previous case, we may write this equation as

$$\Sigma F_s = m\frac{dv}{dt} + (v - v_i)\frac{dm_i}{dt}$$

Since the relative velocity of the device as seen by an observer moving with the particles of the injected mass is $v_{D/i} = (v - v_i)$, the final result can be written as

$$\Sigma F_s = m\frac{dv}{dt} + v_{D/i}\frac{dm_i}{dt} \qquad (15\text{–}40)$$

where dm_i/dt is the rate of mass injected into the device. The last term in this equation therefore represents the magnitude of force $\mathbf{R}$ which the injected mass *exerts on the device*. Since $dv/dt = a$, Eq. 15–40 becomes

$$\Sigma F_s - R = ma$$

which represents the application of $\Sigma \mathbf{F} = m\mathbf{a}$, Fig. 15–35*c*.

As in the case of steady flow, problems which are solved using Eqs. 15–39 and 15–40 should be accompanied by the necessary free-body diagram. With this diagram one can then determine ΣF_s *for the system* and isolate the force exerted on the device by the particle stream.

Example 15–16

The initial combined mass of a rocket and its fuel is $m_0 = 2.5(10^6)$ kg. If a total of $1.5(10^6)$ kg of fuel is consumed at a constant rate of $7.5(10^3)$ kg/s, and expelled at a constant speed of 2500 m/s, determine the velocity of the rocket at the instant the fuel runs out. Neglect the change in gravitational force with altitude and the drag resistance of the air. The rocket is fired vertically from rest.

Solution

Since the rocket is losing mass as it moves upward, Eq. 15–39 can be used for the solution. The only *external force* acting on the *system* consisting of the rocket and a portion of the expelled mass is the weight **W**, Fig. 15–36. Hence,

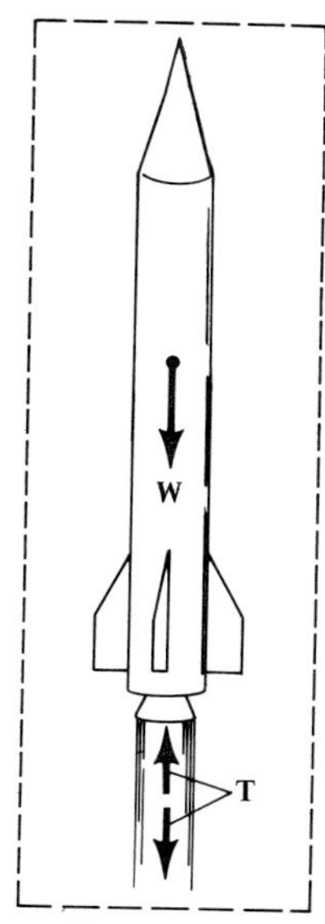

Fig. 15–36

$$(+\uparrow) \qquad \Sigma F_s = m\frac{dv}{dt} - v_{D/e}\frac{dm_e}{dt}$$

$$-W = m\frac{dv}{dt} - v_{D/e}\frac{dm_e}{dt} \qquad (1)$$

From the problem data,

$$\frac{dm_e}{dt} = 7.5(10^3)\text{ kg/s} \quad \text{and} \quad v_{D/e} = 2500\text{ m/s}$$

At any given time t during the flight, the mass of the rocket can be expressed as

$$m = m_O - (dm_e/dt)t = 2.5(10^6) - 7.5(10^3)t$$

Since $W = mg$, Eq. (1) becomes

$$-[2.5(10^6) - 7.5(10^3)t](9.81)$$
$$= [2.5(10^6) - 7.5(10^3)t]\frac{dv}{dt} - 2500[7.5(10^3)]$$

Rearranging the terms and integrating, realizing that $v_1 = 0$ at $t_1 = 0$, we have

$$\int_0^v dv = \int_0^t \left[\frac{18.75(10^6)}{2.5(10^6) - 7.5(10^3)t} - 9.81\right] dt$$

$$v = -2500 \ln [2.5(10^6) - 7.5(10^3)t] - 9.81\, t \Big|_0^t$$

$$= 2500 \ln \frac{2.5(10^6)}{2.5(10^6) - 7.5(10^3)t} - 9.81\, t \qquad (2)$$

This equation gives the speed of the rocket at any instant of time. The time t_2 needed to consume all the fuel is

$$m_f = (dm_e/dt)t_2$$
$$1.5(10^6) = [7.5(10^3)]t_2$$

Hence,

$$t_2 = 200\text{ s}$$

Substituting into Eq. (2) and solving yields

$$v = 328.7\text{ m/s} \qquad \textit{Ans.}$$

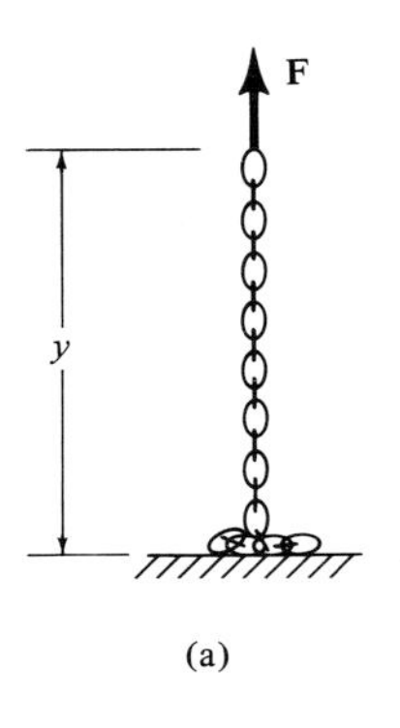

(a)

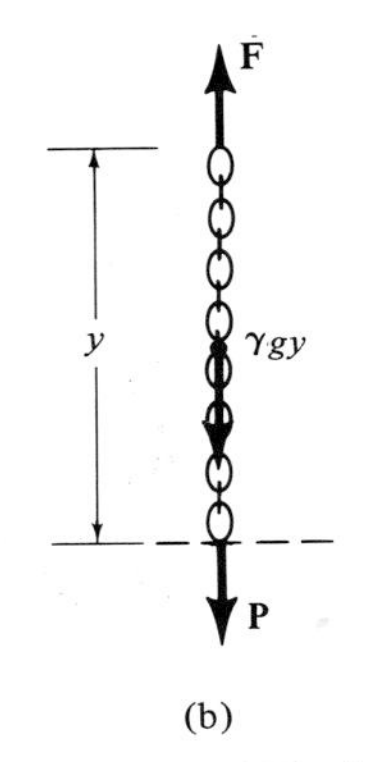

(b)

Fig. 15–37(a,b)

Example 15–17

A chain of length l, Fig. 15–37*a*, has a mass per unit length of γ. Determine the magnitude of force **F** required to (a) raise the chain with a constant speed v_c, starting from rest when $y = 0$, and (b) lower the chain with a constant speed v_c, starting from rest when $y = l$.

Solution

Part (a). As the chain is raised, all the suspended links are given a sudden impulse downward by each added link which is lifted off the ground. Thus, the *suspended portion* of the chain may be considered as a device which is *gaining mass*. A free-body diagram of a portion of the chain which is located at an arbitrary height y above the ground is shown in Fig. 15–37*b*. The system to be considered is the length of chain y which is suspended by **F** at any instant, including the next link which is about to be added but is still at rest. The forces acting on this system *exclude* the internal force **P** which the added link exerts on the suspended portion of the chain. Hence, $\Sigma F_s = F - \gamma g y$.

It now becomes necessary to find the rate at which mass is being added to the system. The velocity $\mathbf{v}_c$ of the chain at a given instant is equivalent to $\mathbf{v}_{D/i}$. Why? Since v_c is constant, $dv_c/dt = 0$ and $dy/dt = v_c$. Integrating, using the initial condition that $y = 0$ at $t = 0$, gives

$$y = v_c t$$

Thus, the mass of the system at any instant is

$$m = \gamma y = \gamma v_c t$$

and therefore the *rate* at which mass is *added* to the suspended chain is

$$\frac{dm_i}{dt} = \frac{dm}{dt} = \gamma v_c$$

Applying Eq. 15–40 to the system, using this data we have

$$(+\uparrow) \qquad \Sigma F_s = m\frac{dv_c}{dt} + v_{D/i}\frac{dm_i}{dt}$$

$$F - \gamma g y = 0 + v_c(\gamma v_c)$$

Hence,

$$F = \gamma(gy + v_c^2) \qquad \textit{Ans.}$$

Part (b). When the chain is being lowered, the links which are expelled (given zero velocity) *do not* impart an impulse to the *remaining* suspended links. Why? Thus, the system in part (a) cannot be considered. Instead, the equation of motion will be used to obtain the solution. At time t the portion of chain still off the floor is y. The free-body and kinetic diagrams

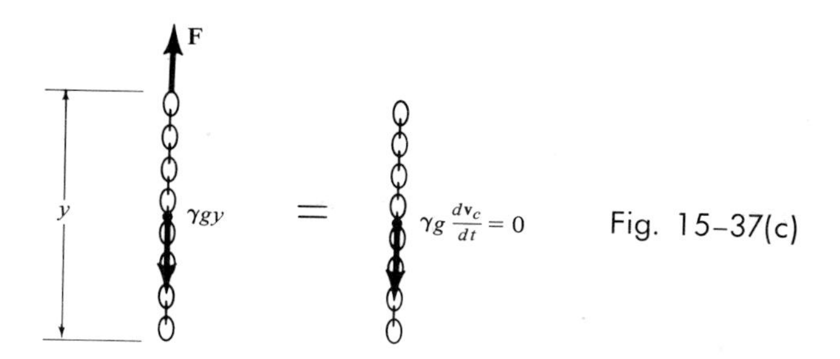

Fig. 15-37(c)

for a suspended portion of the chain are shown in Fig. 15-37*c*. Thus,

$$+\uparrow \Sigma F = ma \qquad F - \gamma g y = 0$$

$$F = \gamma g y \qquad \textit{Ans.}$$

Note: In determining the *reaction on the ground* in part (a), the equation of motion should be used for the solution. As the chain is raised, each link taken from the resting portion of the chain *does not* exert an impulse on the ground. However, as the chain is lowered, part (b), the links added to the ground *do* cause an impulsive reaction. For this case the pile of links represents a "device which is gaining mass" and therefore Eq. 15-40 should be used.

Problems

15-58. A jet of water has a cross-sectional area of 200 mm^2. If it strikes the fixed blade with a speed of 15 m/s, determine the magnitude of force which the water exerts on the blade; $\rho_w = 1$ Mg/m^3.

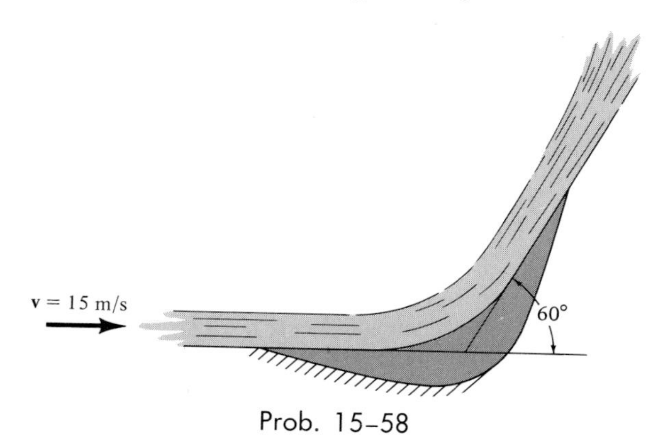

Prob. 15-58

15-59. The fan draws air essentially from rest at A through a vent with a speed of $v_B = 12$ m/s at B. If the cross-sectional area in the vent is 0.09 m^2, determine the horizontal force exerted on the fan blade. The density of the air is $\rho_a = 1.22$ kg/m^3.

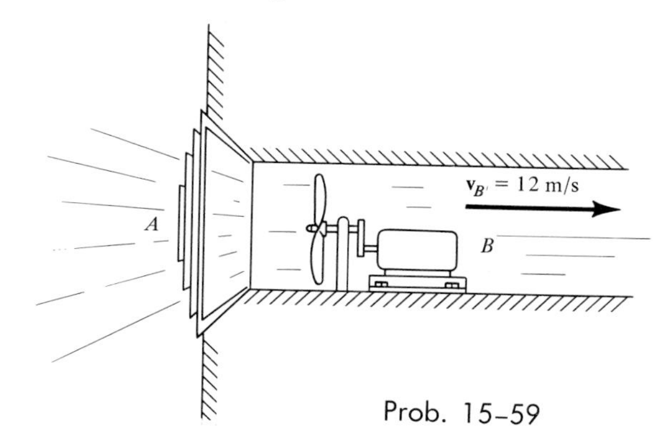

Prob. 15-59

*** 15–60.** The nozzle discharges water at the rate of $v_1 = 15$ m/s against a shield. If the cross-sectional area of the water stream is $A = 300$ mm², determine the force **F** required to hold the shield motionless. $\rho_w = 1$ Mg/m³.

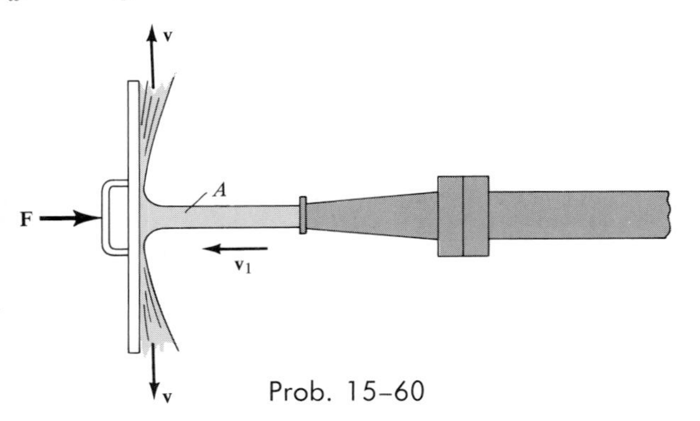

Prob. 15–60

*** 15–60a.** Solve Prob. 15–60 if $v_1 = 50$ ft/s, $A = 2$ in.², and $\rho_w = 62.4$ lb/ft³.

15–61. What force is required to move the shield in Prob. 15–60 forward against the nozzle with a speed of 3 m/s?

15–62. The nozzle has a diameter of 40 mm. If it discharges water uniformly with a downward velocity of 20 m/s against the fixed blade, determine the vertical force exerted by the water on the blade. $\rho_w = 1$ Mg/m³.

Prob. 15–62

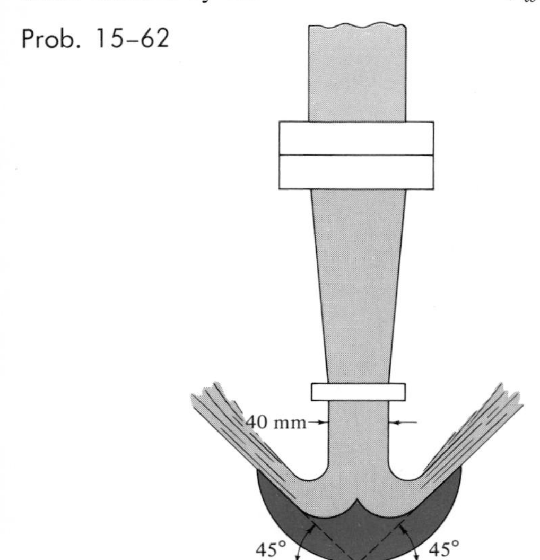

15–63. Solve Prob. 15–62 assuming that the blade has a downward velocity of 5 m/s.

*** 15–64.** A power lawn mower hovers very close over the ground. This is done by drawing in air at a speed of 8 m/s through an intake unit A, which has a cross-sectional area of $A_A = 0.25$ m², and discharging it at the ground, B, where the cross-sectional area $A_B = 0.4$ m². If air at A is subjected only to atmospheric pressure, determine the air pressure which the lawn mower exerts on the ground when the weight of the mower is freely supported and no load is placed on the handle. The mower has a mass of 15 kg with center of mass at G. Assume that air has a constant density of $\rho_a = 1.22$ kg/m³.

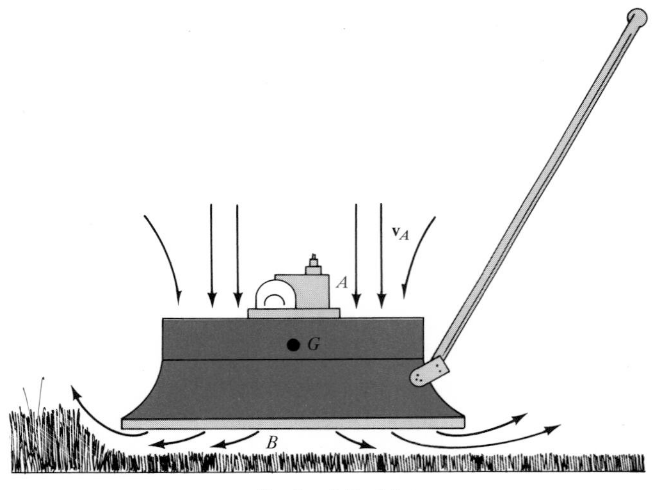

Prob. 15–64

15–65. A snow blower having a scoop S with a cross-sectional area of $A_S = 0.12$ m² is pushed into a snow drift with a speed of $v_S = 0.5$ m/s. The machine discharges the snow through a tube T that has a cross-sectional area of $A_T = 0.03$ m² and is directed 60° from the horizontal. If the density of snow is $\rho_s = 104$ kg/m³, determine the horizontal force **P** required to push the blower forward, and the resultant frictional force **F** of the tires on the ground, necessary to prevent the blower from moving sideways. The tires roll freely.

Prob. 15–65

15–65a. Solve Prob. 15–65 if $A_S = 1.5\ \text{ft}^2$, $v_S = 3\ \text{ft/s}$, $A_T = 0.25\ \text{ft}^2$, and $\rho_s = 6.5\ \text{lb/ft}^3$.

15–66. The 150-kg boat is powered by a fan F which develops a slipstream having a diameter of 0.75 m. If the fan ejects air with a speed of 15 m/s, measured relative to the boat, determine the acceleration of the boat if it is initially at rest. Assume that air has a constant density of $\rho_a = 1.22\ \text{kg/m}^3$ and that the entering air is essentially at rest. Neglect the drag resistance of the water to the forward motion of the boat.

Prob. 15–66

15–67. Water is flowing from the 150-mm-diameter fire hydrant with a velocity of $v_B = 10\ \text{m/s}$. Determine the horizontal and vertical components of force and the moment developed at the base joint A if the static (gauge) pressure at A is 50 kPa and the density of water is 1 Mg/m³.

Prob. 15–67

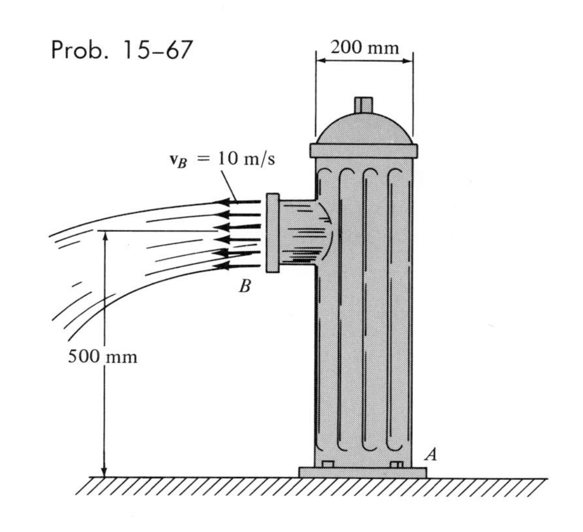

***15–68.** When operating, the air-jet fan discharges air with a speed of $v_B = 20\ \text{m/s}$ into a slipstream having a diameter of 0.5 m. If air has a density of $1.22\ \text{kg/m}^3$, determine the horizontal and vertical components of reaction at C and the vertical reaction at each of the two wheels, D, when the fan is in operation. The fan and its frame have a mass of 20 kg and a center of mass at G. Due to symmetry, both of the wheels support an equal load. Note that air entering the fan at A is essentially at rest.

Prob. 15–68

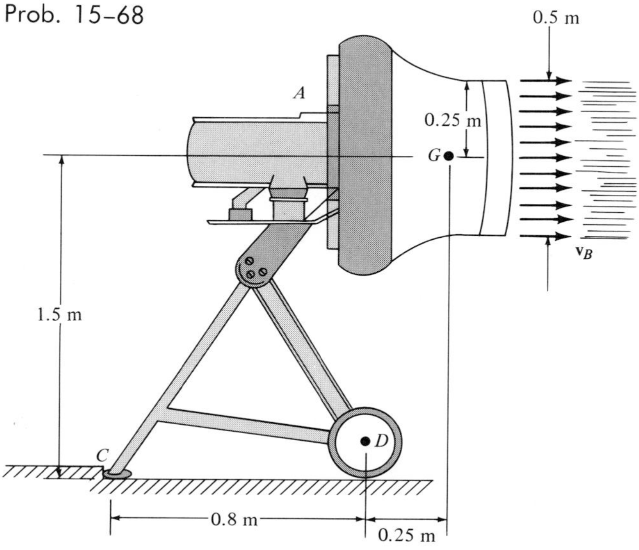

15–69. The truck has a mass of 40 Mg when empty. When it is unloading 5 m^3 of sand at a constant rate of 0.85 m^3/s, the sand flows out the back at a speed of 6 m/s measured relative to the truck in the direction shown. If the truck is free to roll, determine its initial acceleration just as the load begins to empty. Neglect the mass of the wheels and any frictional resistance to motion. The density of sand is $\rho_s = 1520$ kg/m^3.

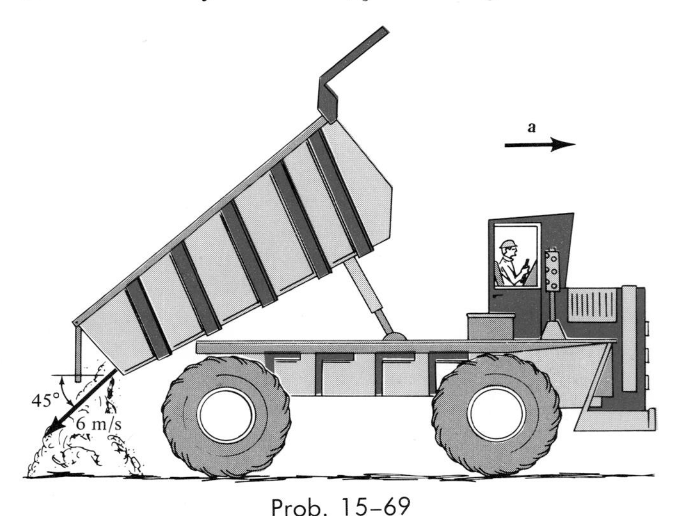

Prob. 15–69

15–70. A rocket burns $m_f = 600$ kg of fuel at a constant rate for $t = 120$ s. If the velocity of the exhaust gas is $v_{e/r} = 2.5$ km/s, measured with respect to the rocket, determine the thrust of the engine.

15–70a. Solve Prob. 15–70 if the weight of fuel is $W_f = 1000$ lb, $t = 40$ s, and $v_{e/r} = 7500$ ft/s.

15–71. The earthmover initially carries 12 m^3 of sand having a density of 1520 kg/m^3. The sand is unloaded horizontally through a 2-m^2 dumping port P at a rate of 900 kg/s measured relative to the port. If the trailer maintains a constant resultant tractive force of 3 kN at its wheels to provide forward motion, determine the acceleration of the earthmover when half the sand is dumped. The earthmover, when empty, has a mass of 31 Mg. Neglect any resistance to forward motion and assume that the rear wheels are free to roll.

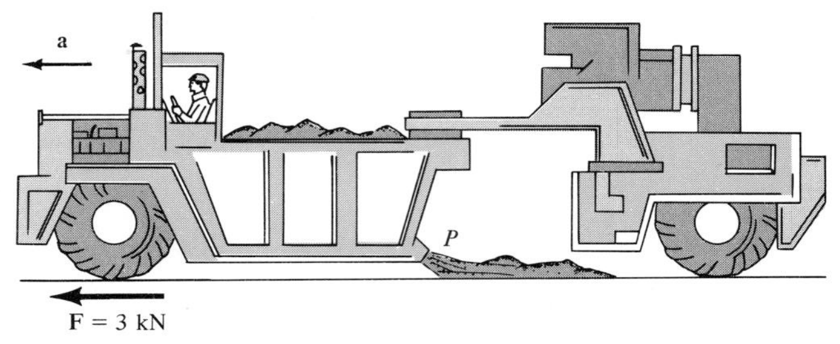

Prob. 15–71

***15–72.** The second stage S_2 of the two-stage rocket has a mass of 1.2 Mg (empty) and is launched from the first stage S_1 such that its velocity is 800 km/h. The fuel in the second stage has a mass of 0.6 Mg and it is consumed at the rate of 5 kg/s. If it is ejected from the rocket at the rate of 2.5 km/s, measured relative to S_2, determine the acceleration of S_2 at the instant the engine is fired, and just after all the fuel is consumed. Neglect the effects of gravitation and air resistance.

Prob. 15–72

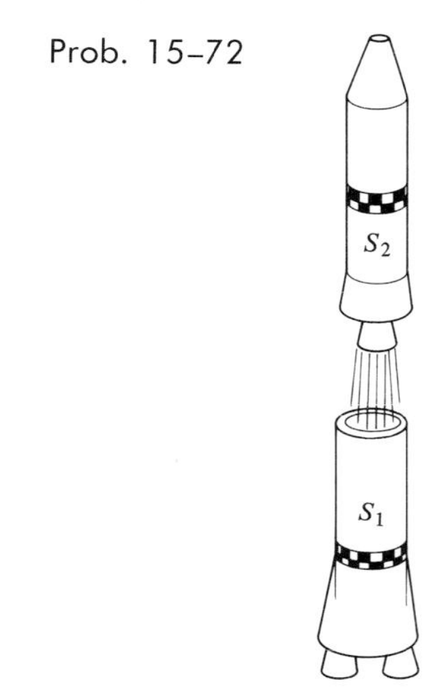

15–73. The 15-Mg jet airplane has a constant speed of 950 km/h when it is flying along the horizontal straight-line path shown. Air enters the intake scoops S at the rate of 54 m^3/s. If the engine burns fuel at the rate of 0.4 kg/s and the gas is exhausted relative to the plane with a speed of 500 m/s, determine the resultant drag force exerted on the plane by air resistance. Assume that air has a constant density of 1.22 kg/m³. *Hint:* Since mass both enters and exits the plane, Eqs. 15–39 and 15–40 must be combined to yield

$$\Sigma F_s = m\frac{dv}{dt} - v_{D/e}\frac{dm_e}{dt} + v_{D/i}\frac{dm_i}{dt}$$

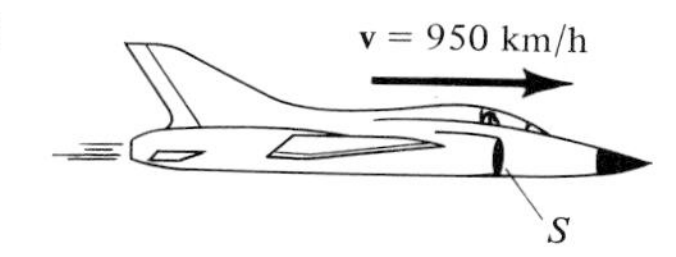

Prob. 15–73

15–74. A rocket has a mass of 900 kg (empty) and carries 200 kg of fuel. If the fuel is burned at the rate of 5 kg/s and ejected from the rocket with a relative velocity of 2 km/s, determine the maximum speed the rocket attains starting from rest. Neglect the effects of gravitation and air resistance.

15–75. The missile has a mass of $m_m = 1.5$ Mg (empty) and carries a total of $m_f = 800$ kg of fuel used for the *two* rocket boosters *B*. The fuel for *each* booster (400 kg) is burned at a constant rate of $\dot{m}_f = 20$ kg/s and exhausted with a relative velocity of $v_{e/m} = 1.2$ km/s. A constant thrust of $T = 40$ kN is provided by the turbojet engine, which has negligible fuel loss. If the missile is traveling with an initial speed of $v_1 = 500$ km/h when the boosters are turned on, determine the missile's speed $t = 10$ s later. Assume that the missile maintains a horizontal flight and neglect the effect of air resistance.

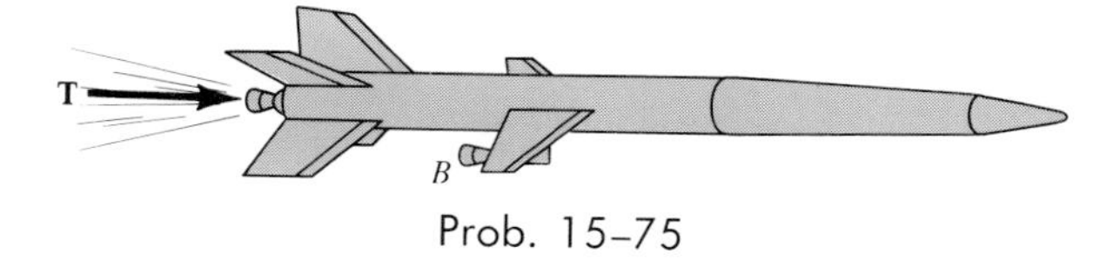

Prob. 15–75

15–75a. Solve Prob. 15–75 if the empty weight of the missile is $W_m = 4000$ lb, the weight of fuel is $W_f = 600$ lb, the burning rate is $\dot{W}_f = 20$ lb/s, $v_{e/m} = 3000$ ft/s, $T = 2500$ lb, $v_1 = 300$ ft/s, and $t = 5$ s.

***15–76.** The rocket car has a mass of 2 Mg (empty) and carries 120 kg of fuel. If the fuel is consumed at a constant rate of 6 kg/s and ejected from the car with a relative velocity of 300 m/s, determine the maximum speed attained by the car starting from rest. The drag resistance due to the atmosphere is $F_D = (40v^2)$ N, where v is the speed measured in m/s.

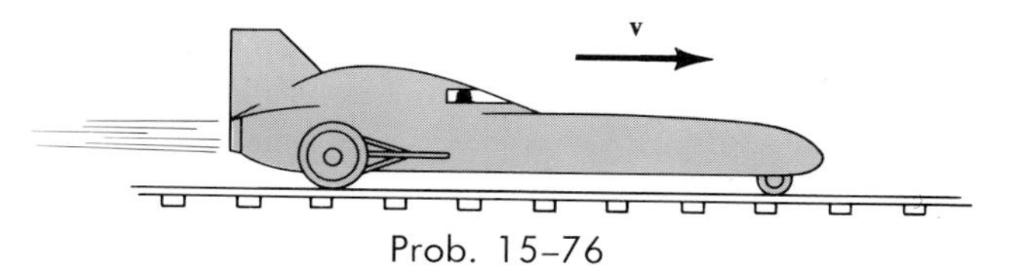

Prob. 15–76

15–77. The rocket has an initial mass m_0, including the fuel. For practical reasons desired for the passengers, it is required that it maintain a constant upward acceleration a_0. If the fuel is expelled from the rocket at a relative speed $v_{e/r}$, determine the rate at which the fuel should be consumed to maintain the motion. Neglect air resistance, and assume that the gravitational acceleration is constant.

Prob. 15–77

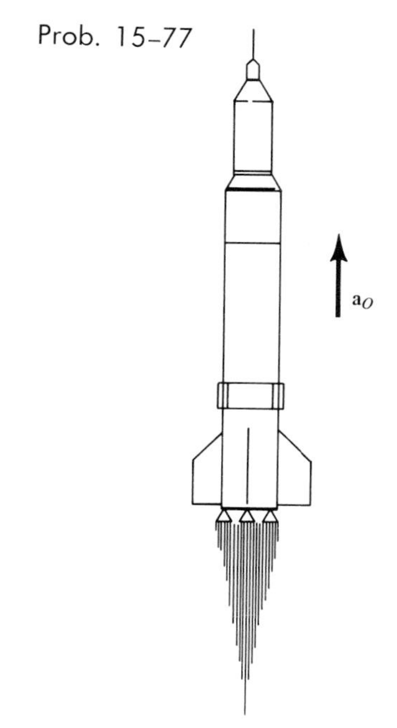

15–78. If the chain is lowered at a constant velocity of $v = 400$ mm/s by the force **F**, determine the normal reaction on the floor as a function of time. The chain has a mass of $m = 2$ kg/m and a total length of $l = 3$ m.

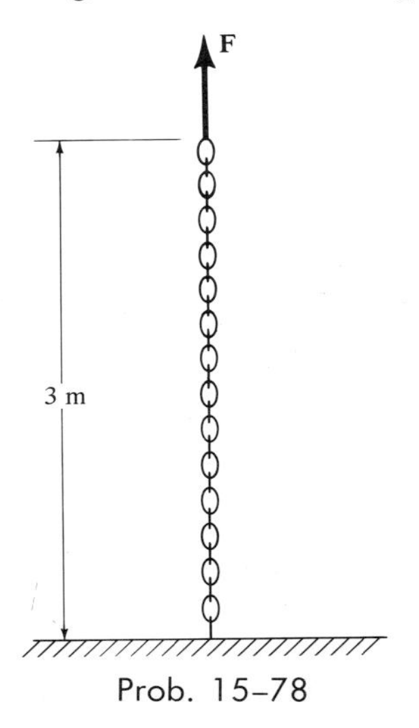

Prob. 15–78

15–79. The chain has a total length of $l = 3$ m and a mass of $m = 2$ kg/m. Determine the magnitude of force **F** as a function of time which must be applied to the end of the chain to raise it with a constant velocity of $v = 0.5$ m/s. Initially, the entire chain is at rest on the ground.

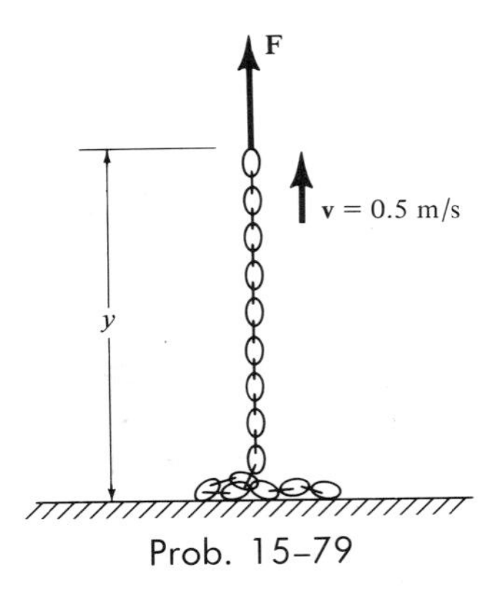

Prob. 15–79

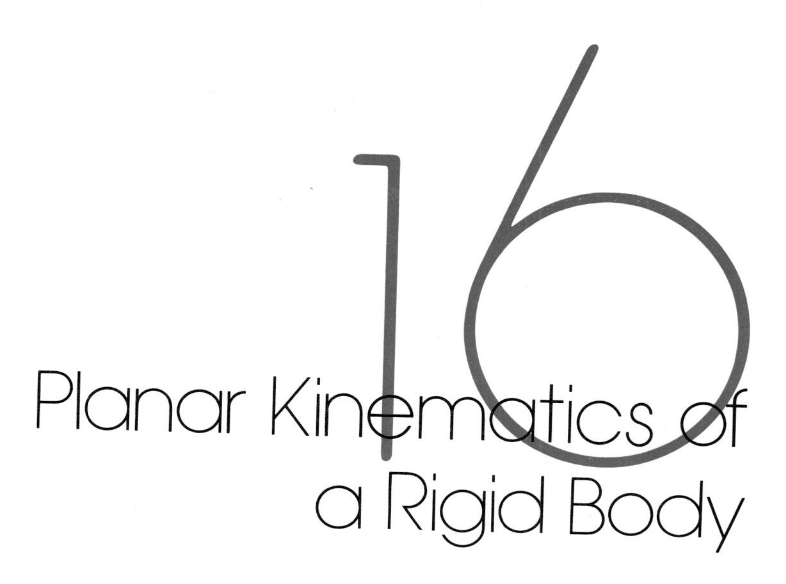

16 Planar Kinematics of a Rigid Body

16–1. Rigid-Body Motion

In this chapter the *kinematics,* or the geometry of motion, for a rigid body will be discussed. This study is important since in many cases the design of gears, cams, and mechanisms used for machine operations depends upon the geometry of their motions. Furthermore, once the kinematics of a rigid body is thoroughly understood, it will be possible to apply the equations of motion, which relate the forces on the body to the body's motion.

Rigid Body. A *rigid body* is considered as a combination of a large number of particles in which all the particles remain at a fixed distance from one another both before and after applying a load. This assumption is an important idealization in mechanics; however, one should be clearly aware of its limitations, because in reality, all bodies deform. In many cases the deformations occurring in engineering structures, machines, mechanisms, and so on, are relatively small, and therefore the rigid-body assumption is suitable for the analysis of the body's motion. Hence, with the exception of springs, the principles of mechanics as discussed in this and subsequent chapters are based on the assumption that all materials are rigid.

Planar Motion of a Rigid Body. When each of the particles of a rigid body moves along a path which is equidistant from a fixed plane, the body is said to undergo *planar motion.** There are three types of planar motion; in order of increasing complexity, they are:

* When a body is subjected to *spatial motion,* all moving particles of the body follow paths which cannot be represented in the same plane. This case is considered in Chapter 20.

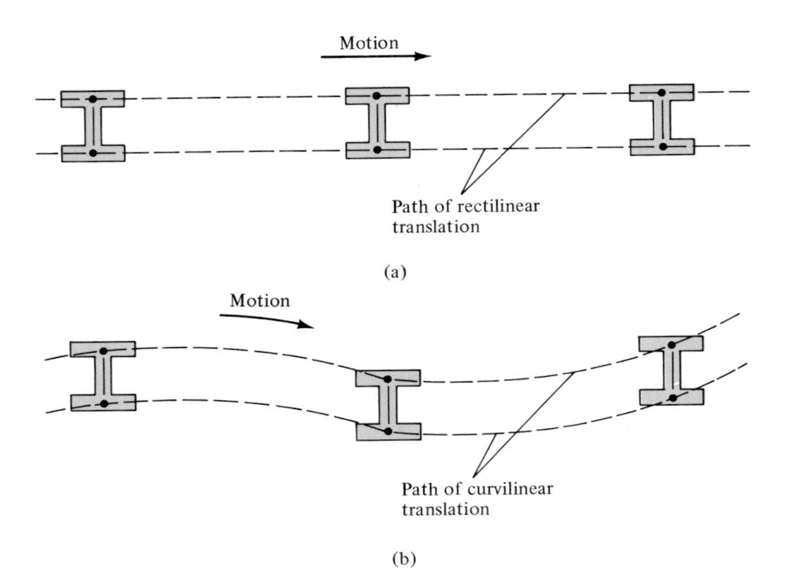

Fig. 16–1

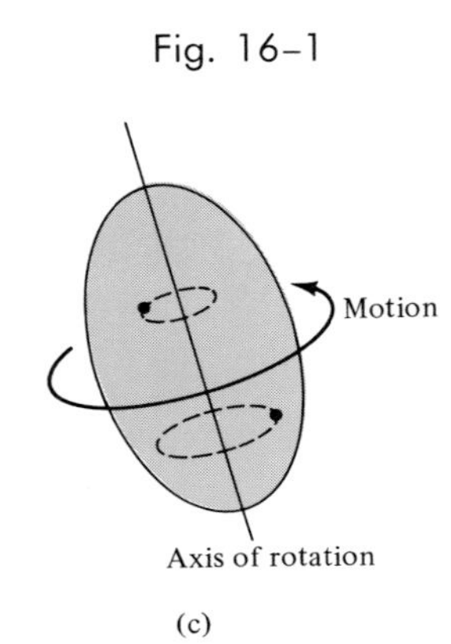

(c)

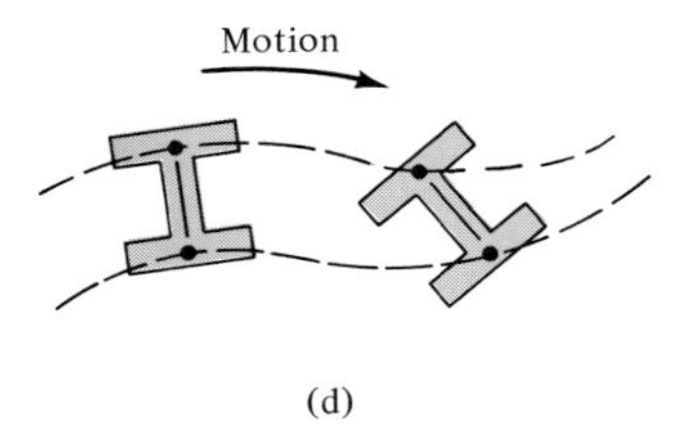

(d)

1. *Translation:* This type of motion occurs if any line segment on the body remains parallel to its original direction during the motion. When the paths of motion for all the particles of the body are along parallel straight lines, Fig. 16–1*a*, the motion is called *rectilinear translation*. However, if the paths of motion are along curved lines which are all parallel, Fig. 16–1*b*, the motion is called *curvilinear translation*.
2. *Rotation about a fixed axis:* When a rigid body rotates about a fixed axis, all the particles of the body, except those which lie on the axis of rotation, move along circular paths, Fig. 16–1*c*.
3. *General plane motion:* When a body is subjected to general plane motion, it undergoes a combination of translation *and* rotation, Fig. 16–1*d*. The translation occurs within a reference plane, and the rotation occurs about an axis perpendicular to the reference plane.

The above planar motions are exemplified by the motion of each part of the crank mechanism shown in Fig. 16–2. The piston A has *rectilinear translation*, the connecting rod BC has *curvilinear translation*, the wheels D and E rotate about a fixed axis, and the connecting rod AB has *general plane motion*.

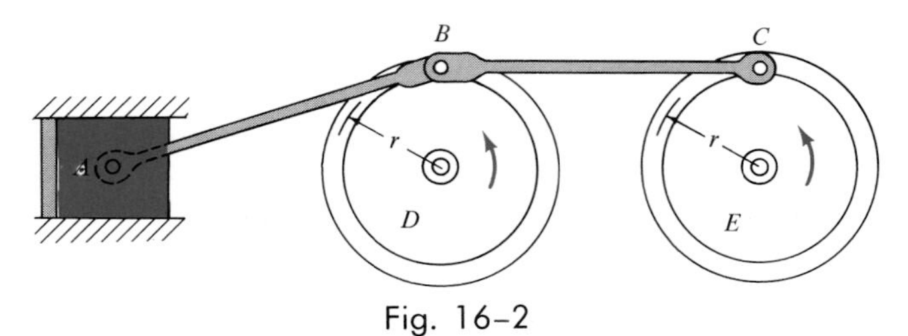

Fig. 16–2

In this chapter each of the three types of planar motion will be discussed separately. In all cases it will be shown that rigid-body planar motion is completely specified provided the motions of any two points on the body are known.

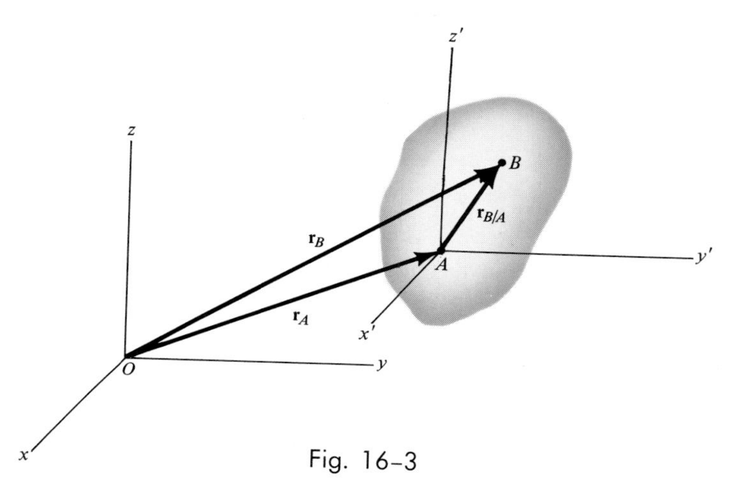

Fig. 16–3

16–2. Translation of a Rigid Body

Consider a rigid body which is subjected to either rectilinear or curvilinear translation, Fig. 16–3.

Position. The locations of points A and B in the body are defined from the fixed x, y, z reference frame by using *position vectors* $\mathbf{r}_A$ and $\mathbf{r}_B$. The translating x', y', z' coordinate system is *fixed in the body* and has its origin located at A, hereafter referred to as the *base point*. The position of B with respect to A is denoted by the *relative-position vector* $\mathbf{r}_{B/A}$ ("$\mathbf{r}$ of B with respect to A"). By vector addition, the three vectors shown in the figure are related by the equation

$$\mathbf{r}_B = \mathbf{r}_A + \mathbf{r}_{B/A} \tag{16–1}$$

Velocity. A relationship between the instantaneous velocities of A and B is obtained by taking the time derivative of Eq. 16–1, which yields

$$\mathbf{v}_B = \mathbf{v}_A + \frac{d\mathbf{r}_{B/A}}{dt}$$

Here $\mathbf{v}_A$ and $\mathbf{v}_B$ denote *absolute velocities* since these vectors are measured from the x, y, z axes. The term $d\mathbf{r}_{B/A}/dt = \mathbf{0}$, since the *magnitude* of $\mathbf{r}_{B/A}$ is constant, by definition of a rigid body; and because the body is translating, the *direction* of $\mathbf{r}_{B/A}$ is constant. Therefore,

$$\mathbf{v}_B = \mathbf{v}_A \tag{16–2}$$

Acceleration. Taking the time derivative of Eq. 16–2 yields a similar relationship between the instantaneous accelerations of A and B,

$$\mathbf{a}_B = \mathbf{a}_A \tag{16-3}$$

The above two equations indicate that *all points* in a rigid body subjected to either curvilinear or rectilinear *translation* move with the *same velocity and acceleration*. As a result, the kinematics of particle motion, discussed in Chapter 12, may be applied to specify the kinematics of translating rigid bodies.

16–3. Rotation of a Rigid Body About a Fixed Axis

Angular Motion of the Body. The characteristics of motion of any point P located in a body which is rotating about a fixed axis depend upon the angular motion of the body about the axis. To study the angular motion, consider the body shown in Fig. 16–4*a*, where the fixed x, y, z coordinate system has an origin O located on the axis of rotation.

Angular Position. At the instant shown, the radial line r, directed perpendicular from the axis at A to point P, is contained in a plane which is *fixed to the body*, and is located at an *angular position* θ, measured from the fixed x-z plane.

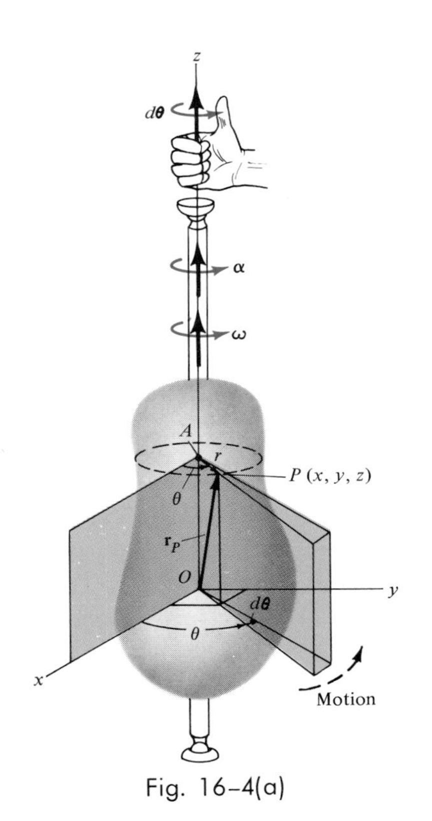

Fig. 16–4(a)

Angular Displacement. The change in the angular position θ, often measured as a differential $d\boldsymbol{\theta}$, is called the *angular displacement*, Fig. 16–4*a*. This vector has a *magnitude* $d\theta$, which is commonly measured in radians or revolutions, where 1 rev $= 2\pi$ rad. The *direction* of $d\boldsymbol{\theta}$ is determined by the right-hand rule; that is, the fingers of the right hand are curled with the sense of rotation, from the x to the y axis, so that the thumb, or $d\boldsymbol{\theta}$, points along the positive z axis.

Angular Velocity. The time rate of change in the angular position of r is called the *angular velocity* $\boldsymbol{\omega}$ (omega). The *magnitude* of this vector is

$$\omega = \frac{d\theta}{dt} \tag{16-4}$$

which is often measured in rad/s. The *direction* of $\boldsymbol{\omega}$ is the same as that of the angular displacement $d\boldsymbol{\theta}$, Fig. 16–4*a*.

Angular Acceleration. The *angular acceleration* $\boldsymbol{\alpha}$ (alpha) measures the time rate of change of the angular velocity. Hence, the *magnitude* of this vector may be written as

$$\alpha = \frac{d\omega}{dt} \tag{16-5}$$

Or taking the second time derivative of Eq. 16–4, it is possible to express α as

$$\alpha = \frac{d^2\theta}{dt^2} \tag{16–6}$$

The line of action of $\boldsymbol{\alpha}$ is the same as that for $\boldsymbol{\omega}$, Fig. 16–4*a*; however, its sense of *direction* depends upon whether $\boldsymbol{\omega}$ is increasing or decreasing with time. In particular, if $\boldsymbol{\omega}$ is decreasing, $\boldsymbol{\alpha}$ is called an *angular deceleration* and therefore it has a sense of direction which is opposite to $\boldsymbol{\omega}$.

By eliminating dt from Eqs. 16–4 and 16–5, it is possible to obtain a differential relation between the angular acceleration, angular velocity, and angular displacement,

$$dt = \frac{d\theta}{\omega} = \frac{d\omega}{\alpha}$$

or

$$\alpha\, d\theta = \omega\, d\omega \tag{16–7}$$

The similarity between the differential relations for angular motion and those developed for rectilinear motion of a particle ($v = ds/dt$, $a = dv/dt$, and $a\, ds = v\, dv$) should be apparent.

Constant Angular Acceleration. If the angular acceleration of the body is constant, $\boldsymbol{\alpha} = \boldsymbol{\alpha}_c$, Eqs. 16–4, 16–5, and 16–7, when integrated, yield a set of formulas which relate the body's angular velocity, its angular position, and time. These equations are similar to Eqs. 12–8, 12–9, and 12–10 used for rectilinear motion. The results are

$$\omega = \omega_1 + \alpha_c t \tag{16–8}$$

$$\theta = \theta_1 + \omega_1 t + \tfrac{1}{2}\alpha_c t^2 \tag{16–9}$$

$$\omega^2 = \omega_1^2 + 2\alpha_c(\theta - \theta_1) \tag{16–10}$$

Here θ_1 and ω_1 are the initial values of angular position and angular velocity, respectively.

Motion of Point *P*. As the rigid body in Fig. 16–4*a* rotates, the arbitrary point P travels along the dashed *circular path* of radius r and with center at point A.

Position. Point P in Fig. 16–4*a* may be located with reference to the origin O of the fixed x, y, z coordinate system by using the position vector $\mathbf{r}_P$. If at the instant shown P has coordinates x, y, z, then expressing $\mathbf{r}_P$ as a Cartesian vector yields

$$\mathbf{r}_P = x\mathbf{i} + y\mathbf{j} + z\mathbf{k} \tag{16–11}$$

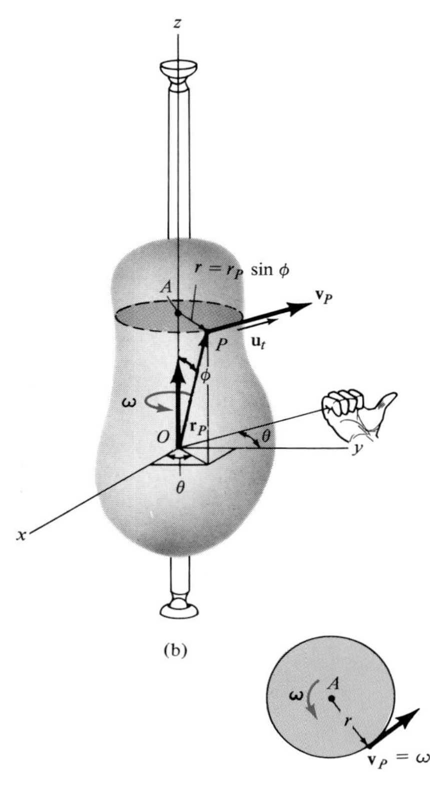

(b)

(c)

Velocity. The *magnitude* of the velocity of P can be determined by using Eqs. 12–44, which describe the circular motion of P in terms of its polar coordinate components $\mathbf{v}_r$ and $\mathbf{v}_\theta$. Since $r(=AP)$ is constant, Fig. 16–4*b*, the radial component $v_r = \dot{r} = 0$, so that $v_P = v_\theta = r\dot{\theta}$. Because $\omega = \dot{\theta}$ (Eq. 16–4), the result is

$$v_P = \omega r \tag{16–12}$$

As shown in Fig. 16–4*b*, the *direction* of $\mathbf{v}_P$ is tangent to the circular path, which can be defined either by $\mathbf{u}_\theta$ or the tangential unit vector $\mathbf{u}_t$.

Since $r = r_P \sin\phi$, Fig. 16–4*a*, then $v_P = \omega r_P \sin\phi$, and therefore $\mathbf{v}_P$ may be expressed in vector form as

$$\mathbf{v}_P = (\omega r_P \sin\phi)\mathbf{u}_t$$

Realizing that this equation defines the same result as the cross product of $\boldsymbol{\omega}$ and $\mathbf{r}_P$, we have*

$$\mathbf{v}_P = \boldsymbol{\omega} \times \mathbf{r}_P \tag{16–13}$$

The order of the vectors in this formulation is important, since the cross product is not commutative, i.e., $\boldsymbol{\omega} \times \mathbf{r}_P \neq \mathbf{r}_P \times \boldsymbol{\omega}$. In this regard, notice in Fig. 16–4*b* how the correct direction of $\mathbf{v}_P$ is established by the right-hand rule. The fingers of the right hand are curled from $\boldsymbol{\omega}$ to $\mathbf{r}_P$ ($\boldsymbol{\omega}$ "cross" $\mathbf{r}_P$). The thumb indicates the correct direction of $\mathbf{v}_P$, that is, tangent to the path in the direction of motion.

The motion of point P can be viewed in two dimensions if a thin slab of the body is considered which incorporates the dashed circular path of P as shown in Fig. 16–4*c*. For the projection, point O is chosen to be coincident with point A in Fig. 16–4*b* and $\mathbf{r}_P$ has a magnitude of r, the radius of the slab.

Acceleration. The acceleration of P may be determined from its polar-coordinate components $\mathbf{a}_r$ and $\mathbf{a}_\theta$, Eqs. 12–47. Since $r(=AP)$ is constant, Fig. 16–4*d*, $\dot{r} = 0$, $\dot{\theta} = \omega$ (Eq. 16–4), and $\ddot{\theta} = \alpha$ (Eq. 16–6), then $(a_P)_r = \ddot{r} - r(\dot{\theta})^2 = -r\omega^2$ and $(a_P)_\theta = r\ddot{\theta} + 2\dot{r}\dot{\theta} = \alpha r$. The negative sign for the radial component indicates that $(\mathbf{a}_P)_r$ acts *toward* point A, Fig. 16–4*d*, i.e., in the *direction* defined by the *normal unit vector* $\mathbf{u}_n$; whereas $(\mathbf{a}_P)_\theta$ acts in the direction defined by the *tangential unit vector* $\mathbf{u}_t$ (or $\mathbf{u}_\theta$). Hence, it is also possible to express the acceleration in terms of its normal and tangential components,

$$(a_P)_t = \alpha r \tag{16–14}$$

$$(a_P)_n = \omega^2 r \tag{16–15}$$

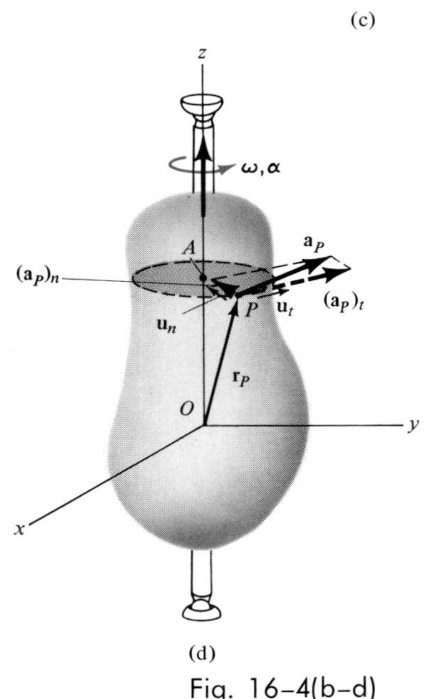

(d)

Fig. 16–4(b–d)

*See Appendix B.

The *tangential component of acceleration,* $(\mathbf{a}_P)_t$, Fig. 16–4*d,* represents the change in the magnitude of $\mathbf{v}_P$. If the speed of P is increasing, then $(\mathbf{a}_P)_t$ acts in the same *direction* as $\mathbf{v}_P$; however, if the speed is decreasing, $(\mathbf{a}_P)_t$ acts in the opposite direction of $\mathbf{v}_P$.

The *normal component of acceleration,* $(\mathbf{a}_P)_n$, represents the change in the direction of $\mathbf{v}_P$. The *direction* of $(\mathbf{a}_P)_n$ is always toward the center of the circular path, Fig. 16–4*d.*

Like the velocity, the acceleration of point P may be expressed in terms of the vector cross product. Taking the time derivative of Eq. 16–13 yields

$$\mathbf{a}_P = \frac{d\mathbf{v}_P}{dt} = \frac{d\boldsymbol{\omega}}{dt} \times \mathbf{r}_P + \boldsymbol{\omega} \times \frac{d\mathbf{r}_P}{dt}$$

Recalling that $\boldsymbol{\alpha} = d\boldsymbol{\omega}/dt$, and using Eq. 16–13 ($d\mathbf{r}_P/dt = \mathbf{v}_P = \boldsymbol{\omega} \times \mathbf{r}_P$), we have

$$\mathbf{a}_P = (\mathbf{a}_P)_t + (\mathbf{a}_P)_n = \boldsymbol{\alpha} \times \mathbf{r}_P + \boldsymbol{\omega} \times (\boldsymbol{\omega} \times \mathbf{r}_P) \qquad (16\text{–}16)$$

Using the right-hand rule as applied to the cross products, verify that the direction of each of the two component accelerations in Eq. 16–16 is in accordance with that shown in Fig. 16–4*d.* The magnitudes of these components are defined by Eqs. 16–14 and 16–15.

By choosing point O to be coincident with point A so that $r_P = r$, Fig. 16–4*b*, the accelerated motion of P can be viewed in two dimensions, as shown on the projected slab in Fig. 16–4*e*. The equations listed in this figure define the accelerated motion.

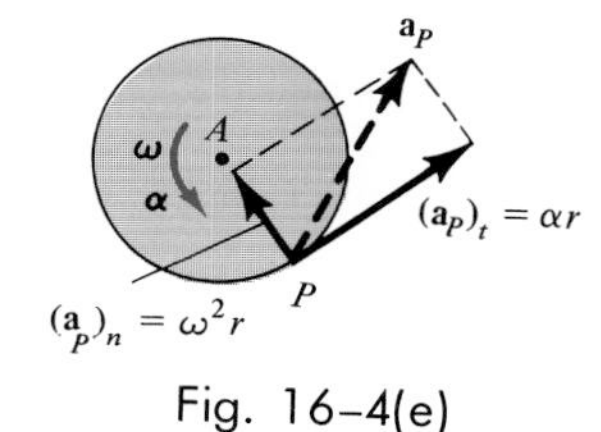

Fig. 16–4(e)

PROCEDURE FOR ANALYSIS

In order to determine the velocity and acceleration of a point located in a rigid body that is rotating about a fixed axis, it is necessary to know the body's angular velocity $\boldsymbol{\omega}$ and angular acceleration $\boldsymbol{\alpha}$.

Angular Motion. If $\boldsymbol{\alpha}$ or $\boldsymbol{\omega}$ are unknown, and $\boldsymbol{\alpha}$ is *not constant,* then the appropriate differential equation $\omega = d\theta/dt$, $\alpha = d\omega/dt$, or $\alpha\, d\theta = \omega\, d\omega$ must be used to obtain relationships between the angular motions. If the body's angular acceleration is *constant,* $\boldsymbol{\alpha}_c$, the equations $\theta = \theta_1 + \omega_1 t + \frac{1}{2}\alpha_c t^2$, $\omega = \omega_1 + \alpha_c t$, and $\omega^2 = \omega_1^2 + 2\alpha_c(\theta - \theta_1)$ may be used for this purpose.

Motion of P. When the motion of a point P in the body is to be determined, it is suggested that a *kinematic diagram* accompany the problem solution. This diagram is simply a graphical representation showing the motion of the point in the body (Fig. 16–4*b* or Fig. 16–4*d*). If the spatial geometry of the problem is simple, a slab of the body may be used for this purpose (Fig. 16–4*c* or Fig. 16–4*e*).

In most cases the velocity and the two components of acceleration can be determined from the scalar equations $v_P = \omega r$, $(a_P)_t = \alpha r$, and $(a_P)_n = \omega^2 r$. However, if the geometry of the problem appears complex, the vector equations $\mathbf{v}_P = \boldsymbol{\omega} \times \mathbf{r}_P$, $(\mathbf{a}_P)_t = \boldsymbol{\alpha} \times \mathbf{r}$, and $(\mathbf{a}_P)_n = \boldsymbol{\omega} \times (\boldsymbol{\omega} \times \mathbf{r})$ should be used. For application, each of the vectors in these equations can be expressed in terms of their **i, j, k** components and, if necessary, the cross products computed by using a determinant expansion (see Eq. B–12 of Appendix B).

The following examples numerically illustrate application of these principles.

Example 16–1

A cord is wrapped around a wheel, which is initially at rest, as shown in Fig. 16–5*a*. If a force **F** is applied to the cord and gives it an acceleration of $a = (4t)$ m/s^2, where t is in seconds, determine (a) the angular velocity of the wheel at $t = 1$ s, (b) the magnitudes of velocity and acceleration of point A at $t = 1$ s, and (c) the number of revolutions the wheel makes during the first second.

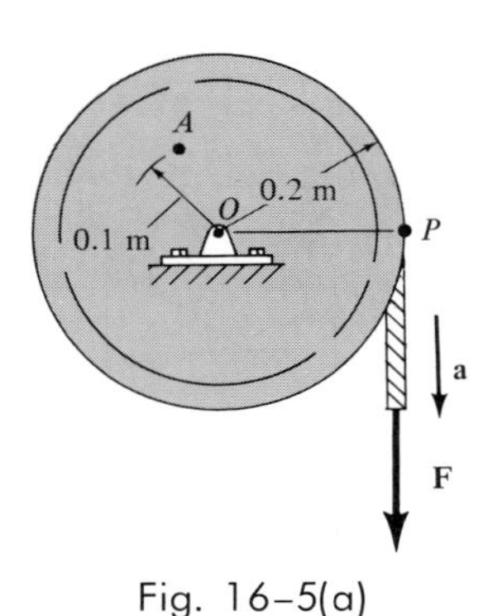

Fig. 16–5(a)

Solution

Part (a). The wheel is subjected to rotation about a fixed axis passing through point O, Fig. 16–5*a*. Thus, point P on the wheel has motion about a circular path, and therefore the acceleration of this point has both tangential and normal components. In particular, $(a_P)_t = (4t)$ m/s^2, since the cord is connected to the wheel and *tangent* to it at P. Hence, from Eq. 16–14 the angular acceleration of the wheel is

$$(a_P)_t = \alpha r$$

$$4t = \alpha(0.2)$$

$$\alpha = 20t \text{ rad/s}^2 \qquad (1)$$

The wheel's angular velocity $\boldsymbol{\omega}$ can be determined by using $\alpha = d\omega/dt$, since this equation relates α, t, and ω. Integrating, with the initial condition that $\omega_1 = 0$ at $t_1 = 0$, yields

$$\alpha = \frac{d\omega}{dt} = 20t$$

$$\int_0^\omega d\omega = \int_0^t 20t\, dt$$

$$\omega = 10t^2 \text{ rad/s} \qquad (2)$$

Thus, for $t = 1$ s,

$$\omega = 10(1)^2 = 10 \text{ rad/s} \quad \text{Ans.}$$

Why is it not possible to use Eq. 16–8 ($\omega = \omega_1 + \alpha_c t$) to obtain this result?

Part (b). The velocity of A is shown on the kinematic diagram in Fig. 16–5b. It is, of course, *tangent to the path of motion.* When $t = 1$ s, v_A can be obtained by using Eq. 16–12 with $\omega = 10$ rad/s. Hence,

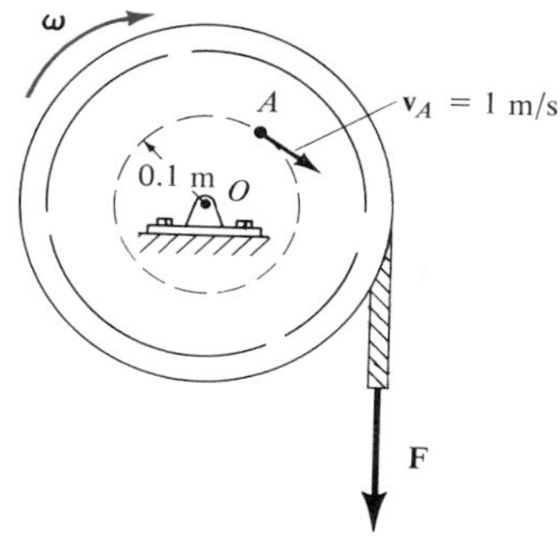

Fig. 16–5(b)

$$v_A = \omega r$$
$$= (10 \text{ rad/s})(0.1 \text{ m}) = 1 \text{ m/s} \qquad \textit{Ans.}$$

The acceleration of point A, at $t = 1$ s, has both normal and tangential components, as shown on the kinematic diagram in Fig. 16–5c. These two components may be obtained by using Eq. 16–14, Eq. (1), and Eq. 16–15; i.e.,

$$(a_A)_t = \alpha r = (20t)r = [20(1)](0.1) = 2 \text{ m/s}^2$$

and

$$(a_A)_n = \omega^2 r = (10)^2(0.1) = 10 \text{ m/s}^2$$

Note that $\mathbf{a}_A$ is *not* tangent to the path, Fig. 16–5c. The *magnitude* of acceleration is thus

$$a_A = \sqrt{(a_A)_t^2 + (a_A)_n^2}$$
$$= \sqrt{(2)^2 + (10)^2} = 10.2 \text{ m/s}^2 \qquad \textit{Ans.}$$

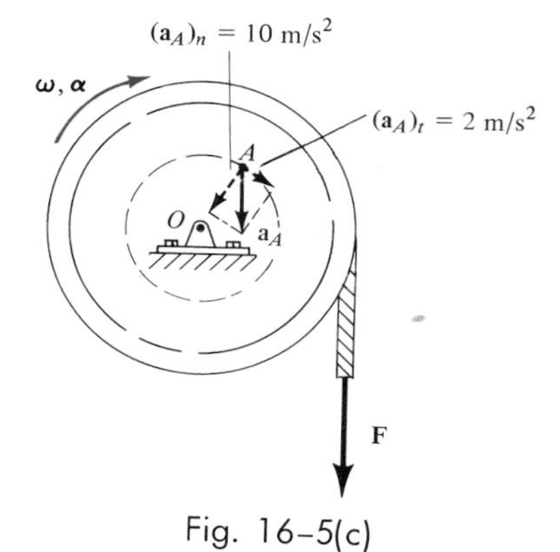

Fig. 16–5(c)

Part (c). The angular velocity is related to time by Eq. (2). Hence, the number of revolutions the wheel makes in 1 s can be computed using $\omega = d\theta/dt$, since this equation relates θ, ω, and t. Using the initial condition, $\theta_1 = 0$ at $t_1 = 0$, we have

$$\frac{d\theta}{dt} = \omega = 10t^2$$

$$\int_0^\theta d\theta = \int_0^t 10t^2\, dt$$

or

$$\theta = \tfrac{10}{3}t^3 \text{ rad}$$

Hence, for $t = 1$ s,

$$\theta = \tfrac{10}{3}(1)^3 = \tfrac{10}{3} \text{ rad}$$

Since there are 2π radians to one revolution,

$$\theta = \tfrac{10}{3} \text{ rad}\left(\frac{1 \text{ rev}}{2\pi \text{ rad}}\right) = 0.531 \text{ rev} \qquad \textit{Ans.}$$

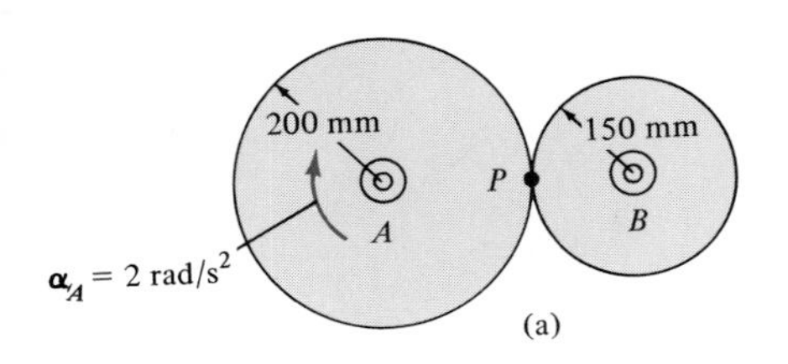

(a)

Example 16–2

Disk A, shown in Fig. 16–6a, starts from rest and rotates with a constant angular acceleration of $\alpha_A = 2\ \text{rad/s}^2$. How much time is needed for it to turn 10 revolutions? If disk A is in contact with disk B and no slipping occurs between the disks, determine the angular velocity and angular acceleration of B just after A turns 10 revolutions.

Solution

The rotational motion of A can be determined using the equations of *constant* angular acceleration since $\alpha_A = 2\ \text{rad/s}^2$. Since there are 2π radians to one revolution,

$$\theta_A = 10\ \text{rev}\left(\frac{2\pi\ \text{rad}}{1\ \text{rev}}\right) = 62.83\ \text{rad}$$

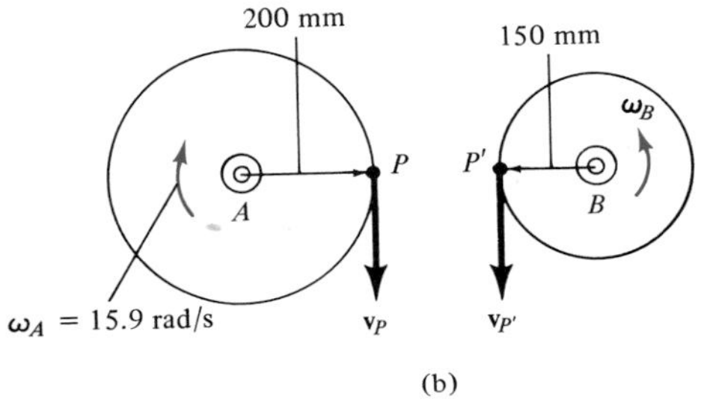

(b)

Since it is known that disk A starts from rest, its angular velocity when $\theta_A = 62.83$ rad can be determined from Eq. 16–10, i.e.,

$$(\circlearrowleft +) \qquad \begin{aligned} \omega^2 &= \omega_1^2 + 2\alpha_c(\theta - \theta_1) \\ \omega_A^2 &= 0 + 2(2)(62.83 - 0) \\ \omega_A &= 15.9\ \text{rad/s} \ \circlearrowleft \end{aligned}$$

The time can be found by solving the quadratic equation 16–9 $(\theta = \theta_1 + \omega_1 t + \frac{1}{2}\alpha_c t^2)$. An easier solution, however, is obtained using the computed result for ω_A and Eq. 16–8, in which case

$$(\circlearrowleft +) \qquad \begin{aligned} \omega &= \omega_1 + \alpha_c t \\ 15.9 &= 0 + 2t \\ t &= 7.95\ \text{s} \end{aligned} \qquad \textit{Ans.}$$

As shown in Fig. 16–6b, the speed of the contacting point P on the rim of A is

$$v_P = \omega_A r_A = (15.9)(0.2) = 3.18\ \text{m/s} \ \downarrow$$

The velocity is always tangent to the path of motion, so that the speed of point P' on B is the *same* as the speed of P on A. The angular velocity of B is therefore

$$\omega_B = \frac{v_{P'}}{r_B} = \frac{3.18}{0.15} = 21.2\ \text{rad/s} \ \circlearrowright \qquad \textit{Ans.}$$

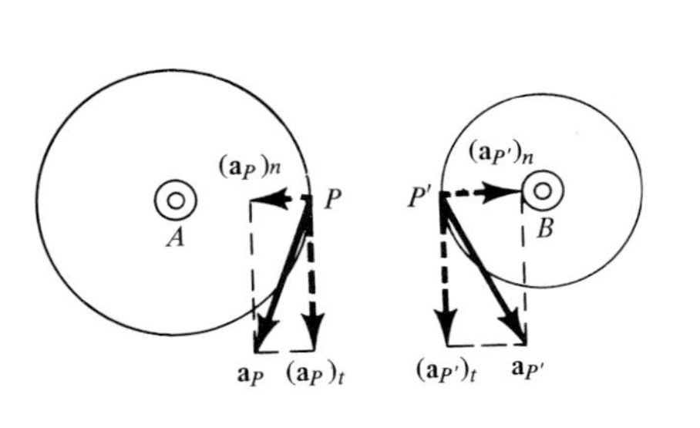

(c)

Fig. 16–6

As shown in Fig. 16–6c, the *tangential components* of acceleration of both disks are equal, since the disks are in contact with one another. Hence,

$$(a_P)_t = (a_{P'})_t \quad \text{or} \quad \alpha_A r_A = \alpha_B r_B$$

$$\alpha_B = \alpha_A\left(\frac{r_A}{r_B}\right) = 2\left(\frac{0.2}{0.15}\right) = 2.67\ \text{rad/s}^2 \ \circlearrowright \qquad \textit{Ans.}$$

Notice that the normal components of acceleration $(a_P)_n$ and $(a_{P'})_n$ act in *opposite directions,* since the paths of motion for both points are *different*. Furthermore, $(a_P)_n \neq (a_{P'})_n$, since the *magnitudes* of these components depend upon the radius and angular velocity of each disk, i.e., $(a_P)_n = \omega_A^2 r_A$ and $(a_{P'})_n = \omega_B^2 r_B$.

Problems

16–1. A wheel has an initial clockwise angular velocity of 10 rad/s and a constant angular acceleration of 3 rad/s². Determine the number of revolutions it must undergo to acquire a clockwise angular velocity of 15 rad/s. What time is required?

16–2. The spin drier of a washing machine has a constant angular acceleration of 3 rev/s², starting from rest. Determine how many revolutions it makes (a) in 11 s, and (b) from the fifth to the sixth second.

16–3. During a gust of wind, the blades of the windmill are given an acceleration of $\alpha = (0.2\,\theta)$ rad/s², where θ is measured in radians. If initially the blades have an angular velocity of 6 rad/s, determine the speed of point P located at the tip of one of the blades just after the blade has turned two revolutions.

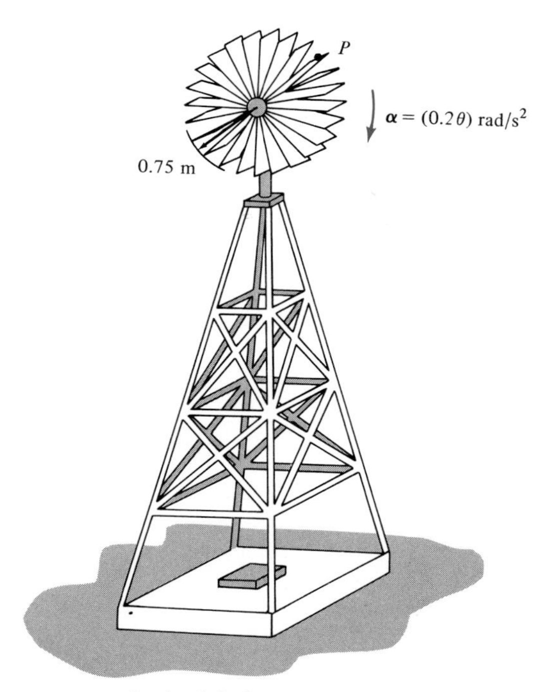

Prob. 16–3

***16–4.** The disk is partially submerged in a heavy liquid and is driven by a motor in the counterclockwise direction. It is found that the angular motion can be described by the relation $\theta = (0.5t + 0.2t^2)$ rad, where t is measured in seconds. Determine the number of revolutions turned, the angular velocity, and angular acceleration of the disk when $t = 20$ s.

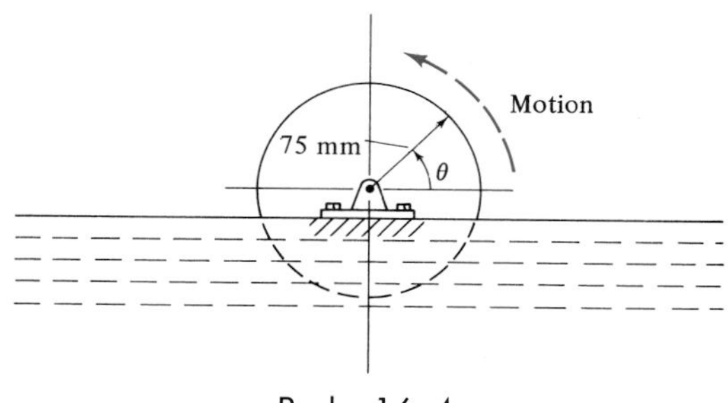

Prob. 16–4

16–5. Gear A is in mesh with gear B as shown. If A starts from rest and has a constant angular acceleration of $\alpha_A = 2 \text{ rad/s}^2$, determine the time needed for B to attain an angular velocity of $\omega_B = 50 \text{ rad/s}$. $r_A = 25 \text{ mm}$, $r_B = 100 \text{ mm}$.

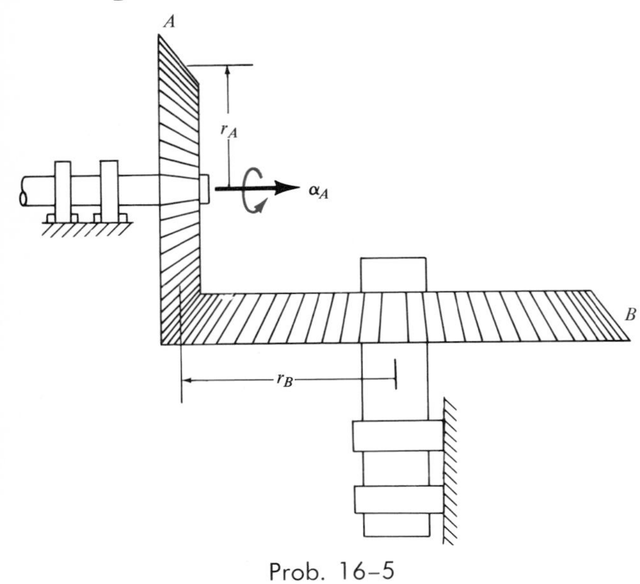

Prob. 16–5

16–5a. Solve Prob. 16–5 if $\alpha_A = 4 \text{ rad/s}^2$, $\omega_B = 60 \text{ rad/s}$, $r_A = 0.2 \text{ ft}$, and $r_B = 0.5 \text{ ft}$.

16–6. A stamp S, located on the revolving drum, is used to label canisters. If the canisters are centered 200 mm apart on the conveyor, determine the radius r_A of the driving wheel A and the radius r_B of the conveyor belt drum so that for each revolution of the stamp it marks the top of a canister. How many canisters are marked per minute if the drum at B is rotating at $\omega_B = 0.2 \text{ rad/s}$?

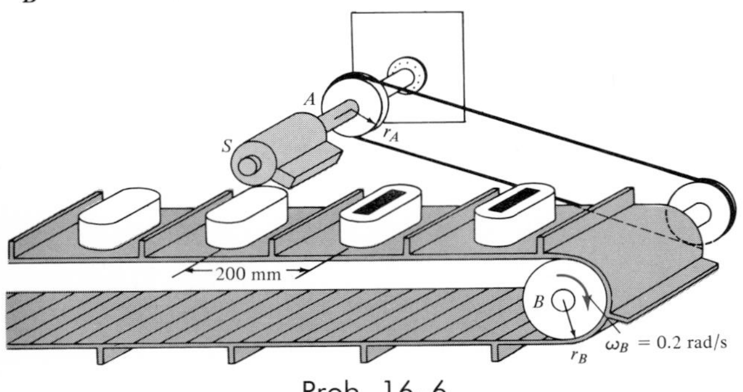

Prob. 16–6

16–7. The rod is bent into the shape of a sine curve and is forced to rotate about the y axis by connecting the spindle S to a motor. If the rod starts from rest in the position shown and a motor drives it for a short time with an angular acceleration of $\alpha = (1.5\, e^t) \text{ rad/s}^2$, where t is measured in seconds, determine the magnitude of the angular velocity and the angular displacement of the rod when $t = 3$ s. Locate the point on the rod which has the greatest velocity and acceleration, and compute the magnitudes of the velocity and acceleration of this point when $t = 3$ s. The curve defining the rod is $z = 0.25 \sin(\pi y)$, where the argument for the sine is given in radians when y is in metres.

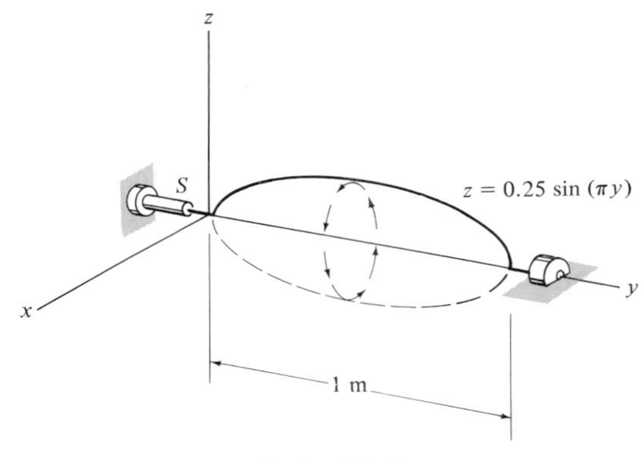

Prob. 16–7

***16–8.** The power of a bus engine is transmitted using the belt-and-pulley arrangement shown. If the engine turns pulley A at 60 rad/s, compute the angular velocities of the generator pulley B and the air-conditioning pulley C. The hub at D is rigidly *connected* to pulley B and turns with it.

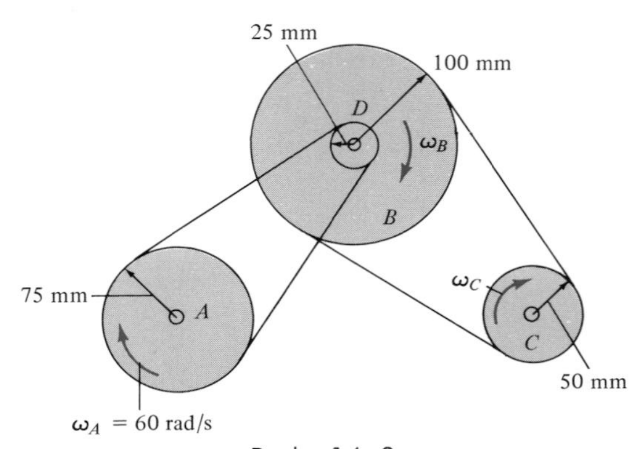

Prob. 16–8

16–9. The motor M is used via the gear arrangement to lift block E. If the motor shaft is turning gear A at a constant rate of $\omega_A = 4$ rad/s, determine the velocity of E.

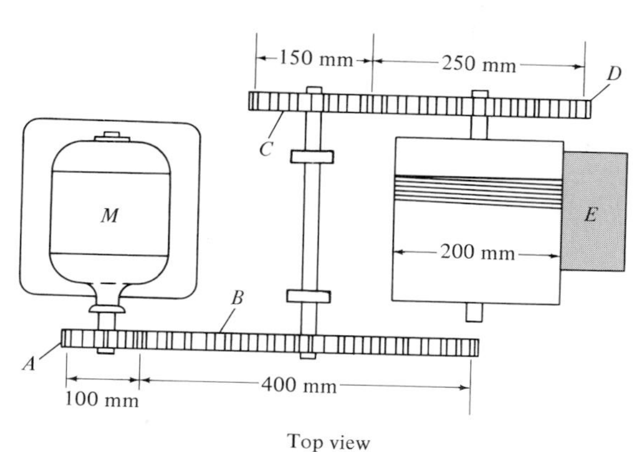

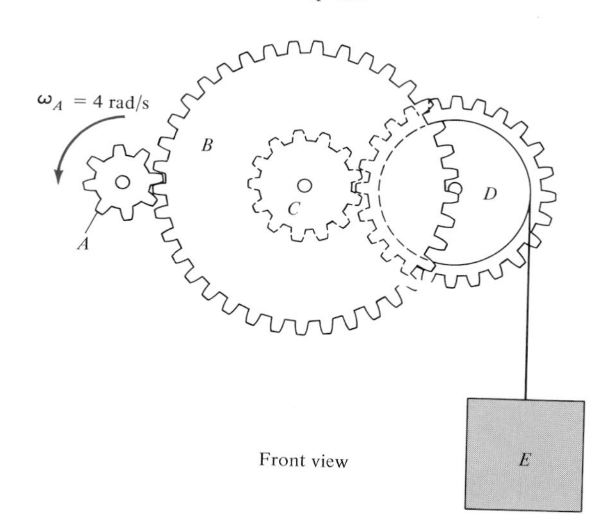

Prob. 16–9

16–10. If the hoisting gear A has an initial angular velocity of $\omega_A = 8$ rad/s and a deceleration of $\alpha_A = -1.5$ rad/s^2, determine the velocity and acceleration of block C when $t = 2$ s. $r_A = 100$ mm, $r_B = 200$ mm, and $r_C = 50$ mm.

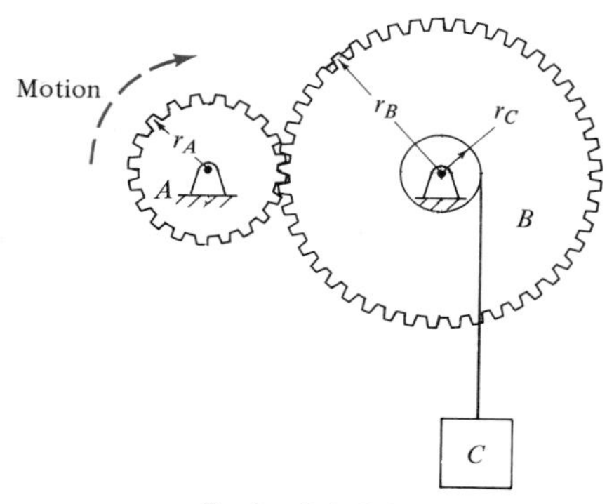

Prob. 16–10

16–10a. Solve Prob. 16–10 if $\omega_A = 10$ rad/s, $\alpha_A = -2$ rad/s^2, $t = 2$ s, $r_A = 0.25$ ft, $r_B = 0.5$ ft, and $r_C = 0.15$ ft.

16–11. Solve Prob. 16–10 assuming that the angular deceleration of gear A is defined by the relation $\alpha_A = (-0.5e^{-0.4t})$ rad/s^2, where t is measured in seconds.

* **16–12.** The concrete cylinder C is subjected to rotational motion caused by the driving wheel at A. Roller B maintains contact with the inner surface of the cylinder and is free to rotate. Likewise, the roller at D is free to rotate. If the angular acceleration of A is $\alpha_A = (0.3t^2)$ rad/s^2, when the cylinder is originally at rest, determine the angular velocity of B when $t = 5$ s.

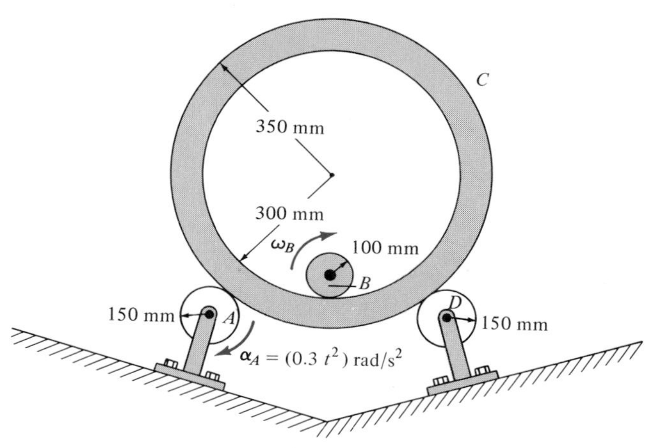

Prob. 16–12

16–13. The board rests on the surface of two drums. At the instant shown, it has a constant acceleration of $a = 0.5\ \text{m/s}^2$ to the right while, at the same instant, points on the periphery of the drums have an acceleration with a magnitude of $3\ \text{m/s}^2$. If the board does not slip on the drums, determine its speed due to the motion.

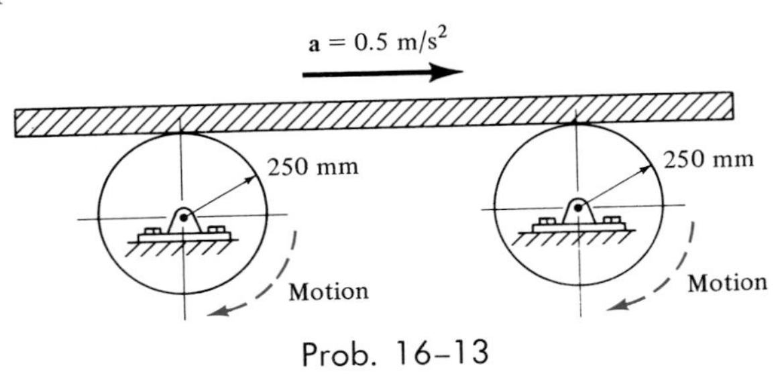

Prob. 16–13

16–14. The drive wheel A of the tape recorder has a constant angular velocity $\boldsymbol{\omega}_A$. If, at a particular instant, the radius of tape wound on each wheel is as shown, determine the angular acceleration of wheel B. The tape has a thickness t. *Hint:* First show that during the time dt the volume of tape exchanged between both wheels results in $r_B\, dr_B = -r_A\, dr_A$. By relating ω_B to ω_A and using $\alpha_A = \dot{\omega}_A$, show that $\alpha_B = (\omega_A/r_B^3)(r_A^2 + r_B^2)\dot{r}_A$. $\dot{r}_A = t\omega_A/2\pi$ is then obtained by computing the volume of tape coming off A per unit of time.

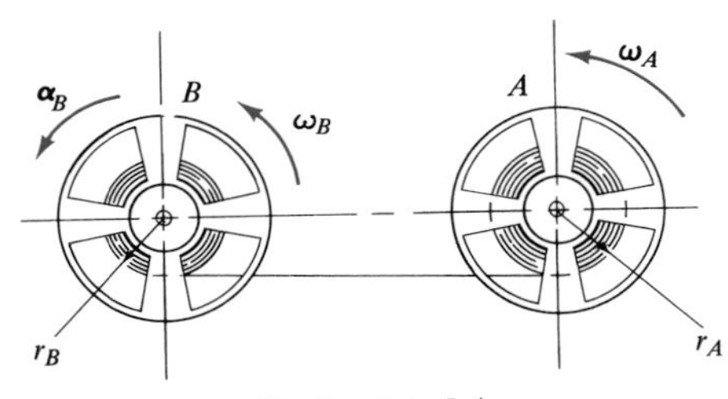

Prob. 16–14

16–15. The rod assembly is supported by ball-and-socket joints at A and B. If it is rotating about the y axis with an angular velocity of $\omega = 5\ \text{rad/s}$ and it has an angular acceleration of $\alpha = 8\ \text{rad/s}^2$, determine the magnitudes of the velocity and acceleration of point C at the instant shown. Set $l = 1\ \text{m}$, $d = 0.4\ \text{m}$, $s = 0.3\ \text{m}$. *Suggestion:* Solve the problem using Cartesian vectors and Eqs. 16–13 and 16–16.

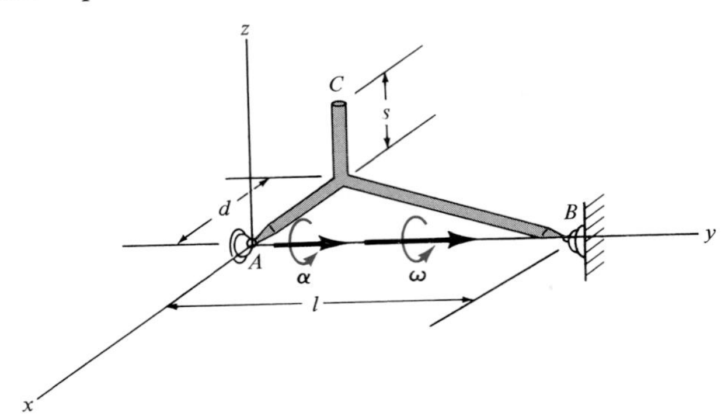

Prob. 16–15

16–15a. Solve Prob. 16–15 if $\omega = 6\ \text{rad/s}$, $\alpha = 10\ \text{rad/s}^2$, $l = 2\ \text{ft}$, $d = 0.8\ \text{ft}$, and $s = 0.6\ \text{ft}$.

16–4. Absolute-General-Plane-Motion Analysis of a Rigid Body

During an instant of time dt, a body subjected to *general plane motion* undergoes a *simultaneous* translation and rotation. If the body is represented by a thin slab, the slab translates in the plane and rotates about an axis perpendicular to the plane. In some problems, the position of a *point* located on a body may be easily related to the body's angular position. Then, by direct application of the time-differential equations $v = ds/dt$, $a = dv/dt$, $\omega = d\theta/dt$, and $\alpha = d\omega/dt$, the motion of the point and the angular motion of the body can be specified.

PROCEDURE FOR ANALYSIS

The following two-step procedure can be used to relate the absolute motion of a point P on a body to the body's angular motion.

Step 1: Relate the position s of the point P on the body to the body's angular position θ. It is necessary that the coordinate s be measured from a *fixed origin* and that it have the *same direction* as the specified path of motion of point P. The relationship $s = f(\theta)$, is obtained in terms of the dimensions of the body by using geometry and/or trigonometry.

Step 2: The first derivative of $s = f(\theta)$ with respect to time yields a relationship between v and ω, whereas the second time derivative yields a relationship between a and α.

This procedure is illustrated in the following two example problems.

Example 16–3

At a given instant, the cylinder of radius r, shown in Fig. 16–7a, has an angular velocity $\boldsymbol{\omega}$ and angular acceleration $\boldsymbol{\alpha}$. Determine the velocity and acceleration of its center G if it rolls without slipping.

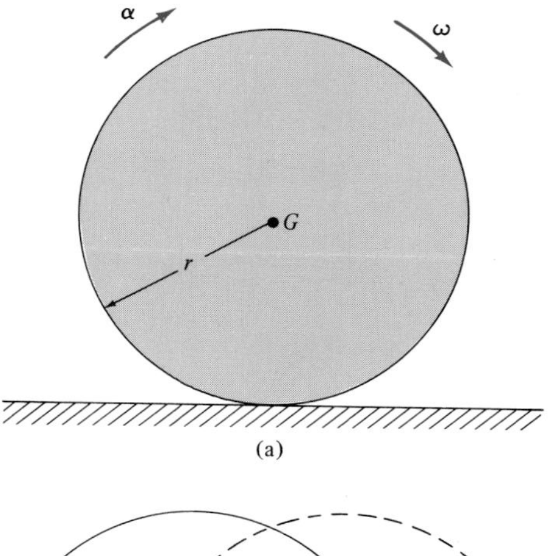

(a)

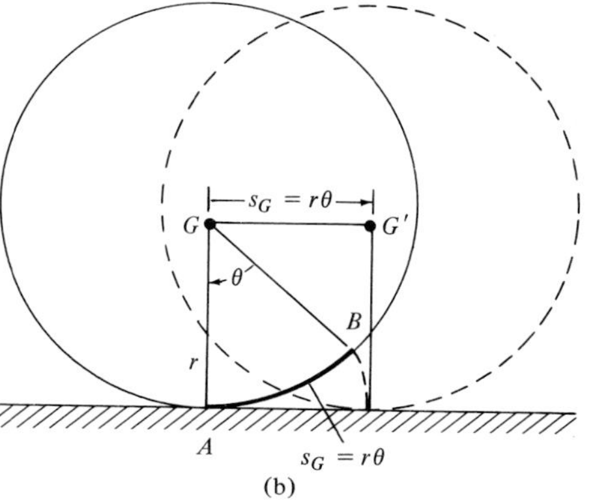

(b)

Fig. 16–7

Solution

Step 1: Coordinates s_G and θ are chosen for the analysis since the rectilinear motion of G is defined by $v_G = ds_G/dt$ and the angular motion of the cylinder is defined by $\omega = d\theta/dt$. The origin of the s_G axis is located at a point coincident with G at some instant, Fig. 16–7b, then as the cylinder rotates through an angle θ, succeeding points on the arc length $AB = r\theta$ contact the ground, such that G travels a distance $s_G = r\theta$.

Step 2: Taking successive time derivatives of this equation, realizing that r is constant, $\omega = d\theta/dt$, and $\alpha = d\omega/dt$, gives the necessary relationships

$$s_G = r\theta$$

$$v_G = r\omega \qquad \textit{Ans.}$$

$$a_G = r\alpha \qquad \textit{Ans.}$$

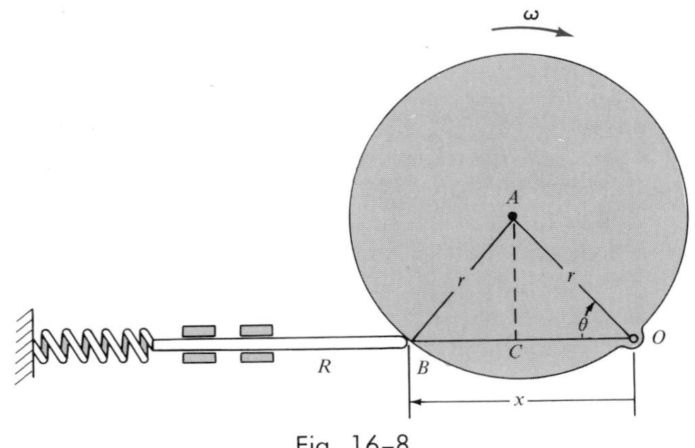

Fig. 16–8

Example 16–4

The end of rod R shown in Fig. 16–8 maintains contact with the cam by means of a spring. If for a short time the cam rotates about an axis through point O, with a constant angular velocity $\boldsymbol{\omega}$, compute the velocity of R for any angle θ of the cam.

Solution

Step 1: Coordinates x and θ are chosen for the analysis since the angular motion of the cam is defined by $\omega = d\theta/dt$, and the rectilinear motion of the rod (or horizontal component of the motion of point B) is $v = dx/dt$. These coordinates are measured from the fixed point O and may be related to each other using trigonometry. Since $OC = CB = r\cos\theta$, Fig. 16–8, then

$$x = 2r\cos\theta$$

Step 2: Taking the time derivative to obtain the necessary relation between the linear and angular motions yields

$$\frac{dx}{dt} = -2r\sin\theta\frac{d\theta}{dt}$$

Since $\omega = d\theta/dt$, then

$$v = -2r\,\omega\sin\theta \qquad \textit{Ans.}$$

***16–16.** The mechanism is used to convert the *constant* circular motion ω of rod AB into translating motion of rod CD. Compute the velocity and acceleration of CD for any angle θ of AB.

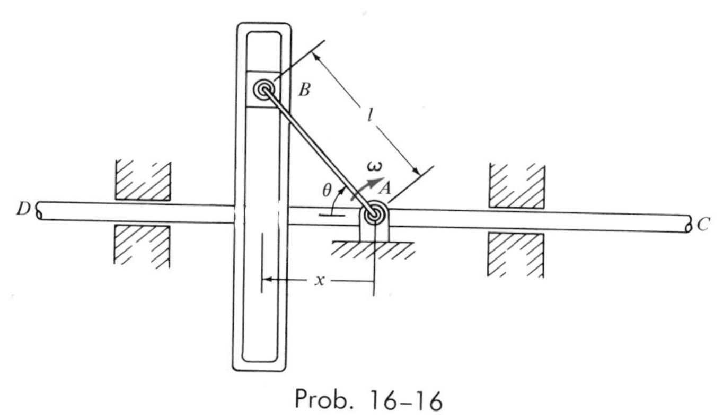

Prob. 16–16

16–17. Solve Prob. 16–16 assuming that rod AB has an angular acceleration $\boldsymbol{\alpha}$ and an angular velocity $\boldsymbol{\omega}$ at any instant.

16–18. Rod CD presses against AB and thereby gives it an angular velocity. If ω is to be constant, determine the required speed v of CD for any angle θ of rod AB.

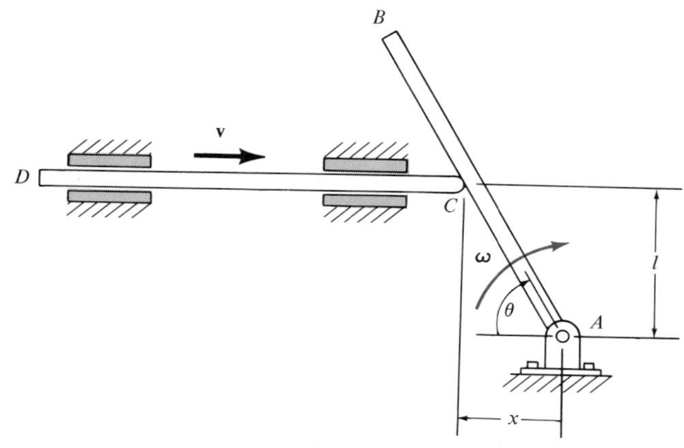

Prob. 16–18

16–19. The scaffold S is raised hydraulically by moving the rollers at A toward the pin at B. If A is approaching B with a speed v_A, determine the speed at which the platform is rising as a function of θ. Each link is pin-connected at its midpoint and end points and has a length l.

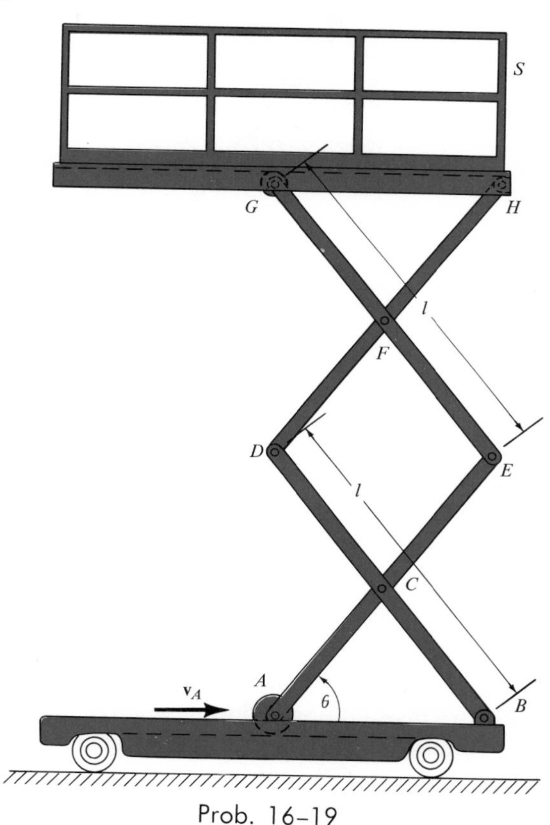

Prob. 16–19

***16–20.** The link OA is pinned at O and rotates because of the sliding action of rod R along the horizontal groove. If R starts from rest when $\theta = 0°$ and has a constant acceleration of $a_R = 60\ \text{mm/s}^2$ to the right, determine the angular velocity and angular acceleration of OA when $t = 2$ s. Set $d = 400$ mm.

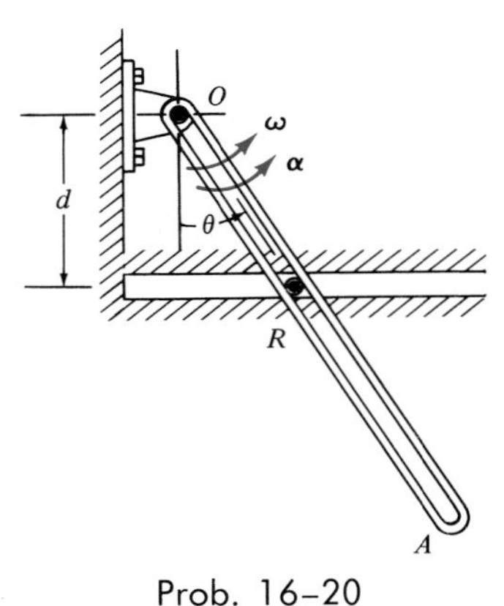

Prob. 16–20

***16–20a.** Solve Prob. 16–20 if $a_R = 0.25\ \text{ft/s}^2$, $t = 3$ s, and $d = 1.5$ ft.

16–21. At the instant shown, $\theta = 60°$ and rod AB is subjected to a deceleration of $5\ \text{m/s}^2$ when the velocity is $10\ \text{m/s}$. Determine the angular velocity and angular acceleration of link CD at this instant.

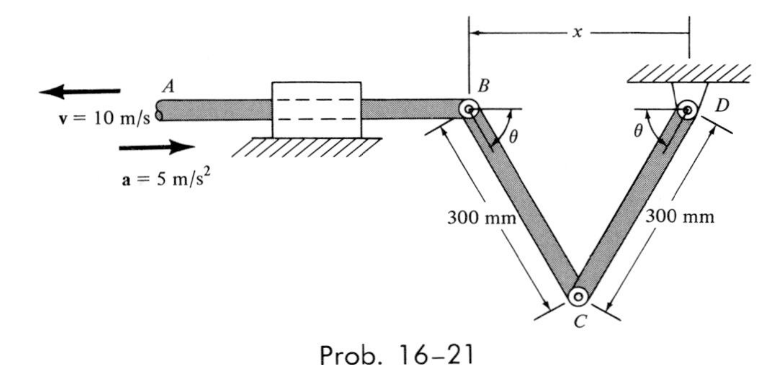

Prob. 16–21

16–22. The safe is transported on a platform which rests on rollers, each having a radius r. If the rollers do not slip, determine their angular velocity if the platform moves forward with a velocity **v**.

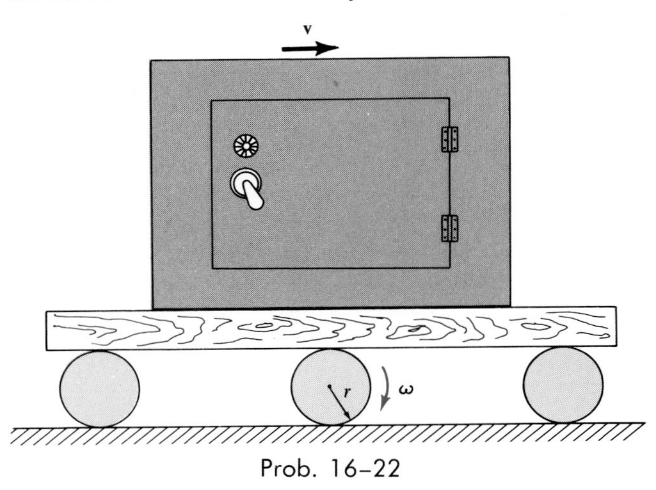

Prob. 16–22

16–23. Wheel A rolls without slipping over the surface of the *fixed* cylinder B. Determine the angular velocity of A if its center C has a speed of $v_C = 2\ \text{m/s}$. How many revolutions will A make about its center after link DC completes one revolution?

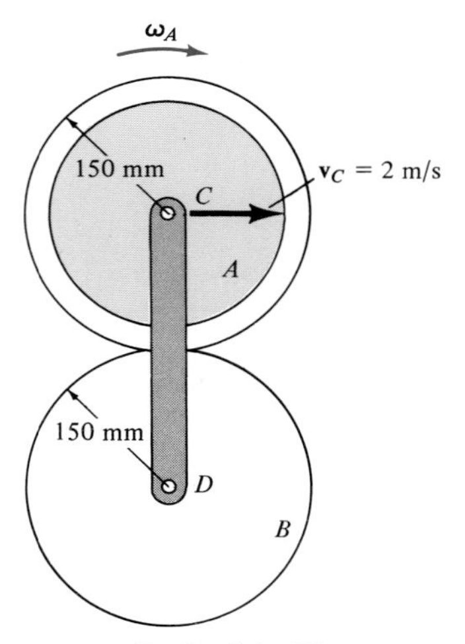

Prob. 16–23

*** 16–24.** Compute the velocity of the rod R for any angle θ of the cam C if the cam rotates with a constant angular velocity $\boldsymbol{\omega}$. The pin connection at O does not cause an interference with the motion of A on C.

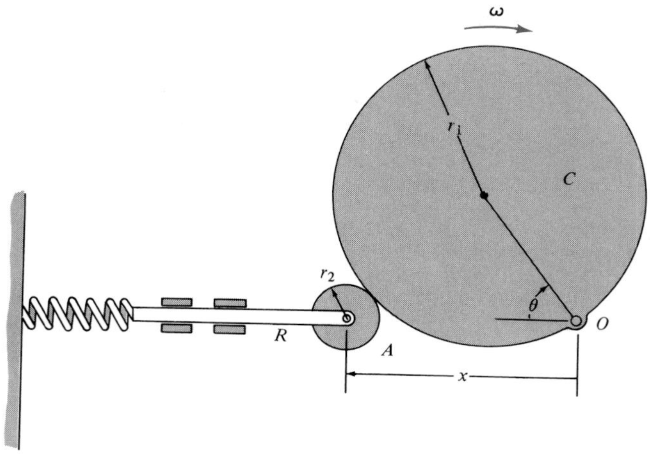

Prob. 16–24

16–25. At the instant $\theta = 30°$, the angular velocity of the crankshaft AB is $\omega = 250$ rad/s. Determine the velocity of the piston C at this instant. Set $l = 150$ mm and $d = 300$ mm.

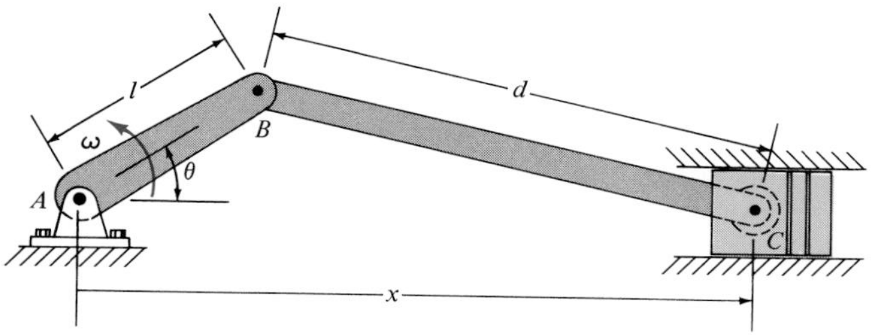

Prob. 16–25

16–25a. Solve Prob. 16–25 if $\theta = 30°$, $\omega = 200$ rad/s, $l = 0.5$ ft, and $d = 1$ ft.

16–26. Compute the velocity of the rod R for any angle θ of the cam C if the cam rotates with a constant angular velocity $\boldsymbol{\omega}$. The pin connection at O does not cause an interference with the motion of A on C.

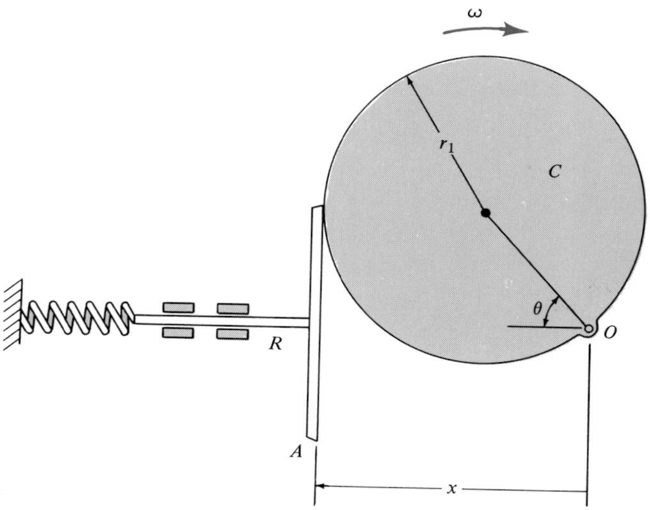

Prob. 16–26

16–27. The Geneva wheel A provides intermittent rotary motion $\boldsymbol{\omega}_A$ for continuous motion $\omega_D = 2$ rad/s of disk D. By choosing $d = 100\sqrt{2}$ mm, the wheel has zero angular velocity at the instant pin B enters or leaves one of the four slots. Determine the magnitude of the angular velocity $\boldsymbol{\omega}_A$ of the Geneva wheel at any angle θ for which pin B is in contact with the slot.

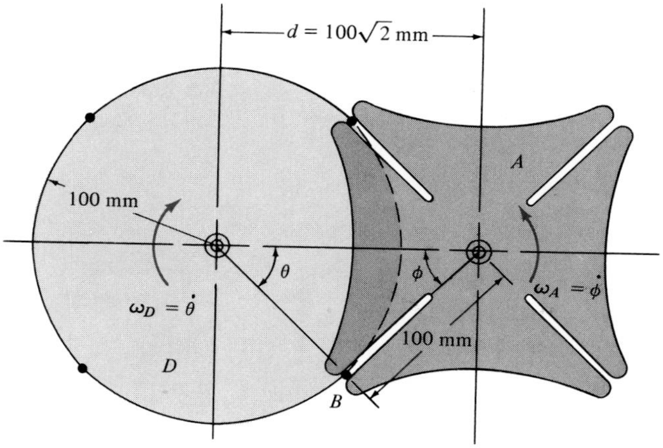

Prob. 16–27

16–5. Relative-General-Plane-Motion Analysis of a Rigid Body Using Translating Axes. Velocity

Since general plane motion consists of a combination of translation and rotation of the body during an instant of time, it is often convenient to view each of these "component" motions *separately* using a *relative-motion analysis*. Two sets of coordinate axes will be used to do this. The x, y, z coordinate system is fixed and therefore it will be used to measure *absolute* positions, velocities, and accelerations of points in the rigid body, Fig. 16–9*a*. The origin of the x', y', z' coordinate system will be fixed to a selected "base point" A in the body. The axes of this coordinate system are *not* fixed to the body; rather they will only be allowed to *translate* with respect to the fixed frame.

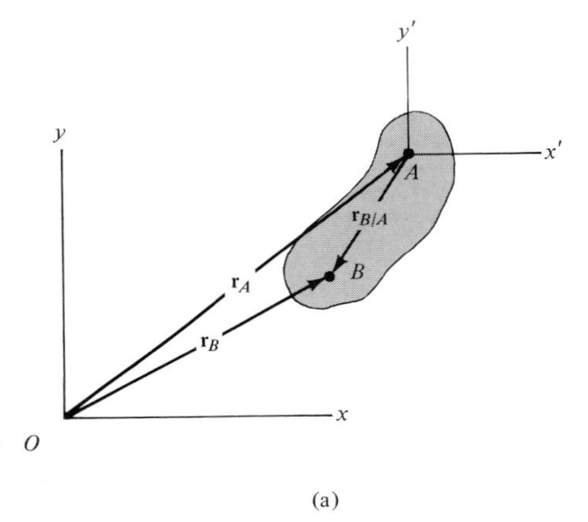

(a)

Position. The *position vector* $\mathbf{r}_A$ in Fig. 16–9*a* specifies the location of the "base point" A; whereas the *relative-position vector* $\mathbf{r}_{B/A}$ locates an arbitrary point B in the body with respect to A. By vector addition, the *absolute position* of B can be determined from the equation

$$\mathbf{r}_B = \mathbf{r}_A + \mathbf{r}_{B/A} \tag{16–17}$$

Velocity. To determine the relationship between the instantaneous velocities of points A and B, it is necessary to take the time derivative of Eq. 16–17, i.e.,

$$\frac{d\mathbf{r}_B}{dt} = \frac{d\mathbf{r}_A}{dt} + \frac{d\mathbf{r}_{B/A}}{dt}$$

The terms $d\mathbf{r}_B/dt = \mathbf{v}_B$ and $d\mathbf{r}_A/dt = \mathbf{v}_A$ are measured from the fixed x, y, z axes and represent the *absolute velocity* of points A and B, Fig. 16–9*b*. The time rate of change in the relative position $\mathbf{r}_{B/A}$ is denoted as the *relative velocity*, $\mathbf{v}_{B/A}$, which represents the velocity of B measured by an observer stationed at A and fixed to the translating x', y', z' axes. Since the body is *rigid*, this observer sees point B move along the dashed *circular arc* that has a radius of curvature of $\mathbf{r}_{B/A}$, Fig. 16–9*d*. In other words, the body moves as if it were *pinned* at A. Consequently, $\mathbf{v}_{B/A}$ has a *magnitude* of

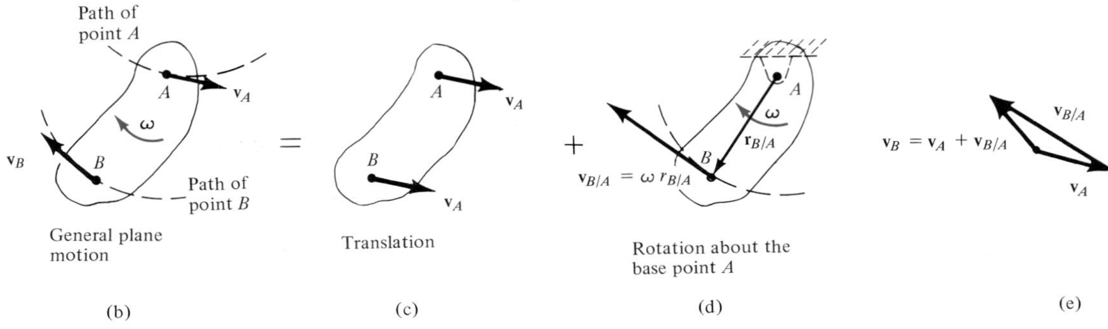

Fig. 16–9

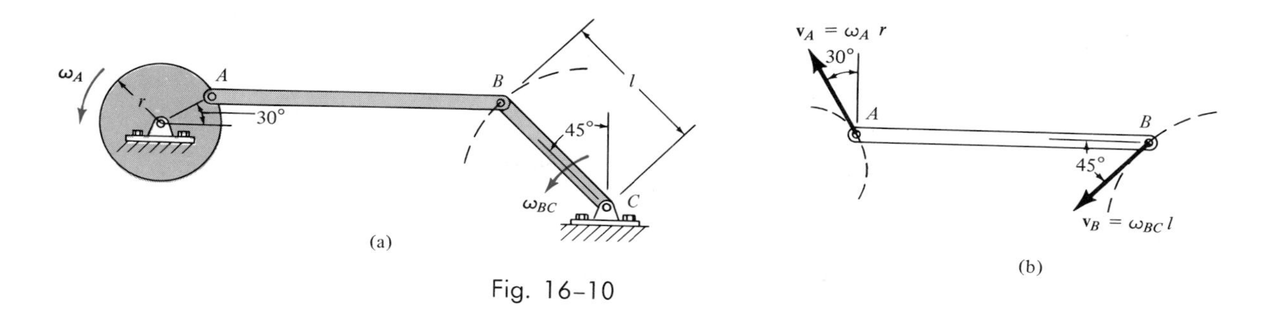

Fig. 16–10

$v_{B/A} = \omega r_{B/A}$ (Eq. 16–12) and a *direction* which is tangent to the curved path at B. Hence, the above equation can be written in the form

$$\mathbf{v}_B = \mathbf{v}_A + \mathbf{v}_{B/A} \tag{16–18}$$

where

$\mathbf{v}_B$ = absolute velocity of point B
$\mathbf{v}_A$ = absolute velocity of the base point A
$\mathbf{v}_{B/A}$ = relative velocity of "B with respect to A" as measured by a *translating* observer; the *magnitude* is $v_{B/A} = \omega r_{B/A}$ and the *direction* is perpendicular to the line of action of $\mathbf{r}_{B/A}$
$\boldsymbol{\omega}$ = absolute angular velocity of the body
$\mathbf{r}_{B/A}$ = relative-position vector drawn from A to B

Each of the three terms in Eq. 16–18 is represented graphically on the *kinematic diagrams* in Figs. 16–9*b*, 16–9*c*, and 16–9*d*. Here it is seen that at a given instant the velocity of B, Fig. 16–9*b*, is determined by considering the entire body to translate with a velocity $\mathbf{v}_A$, Fig. 16–9*c*, and rotate about the base point A with an instantaneous angular velocity $\boldsymbol{\omega}$, Fig. 16–9*d*. Vector addition of these two effects, applied to B, yields $\mathbf{v}_B$, as shown in Fig. 16–9*e*.

The velocity equation 16–18 may be used in a practical manner to study the motion of a rigid body which is either pin-connected to or in contact with other moving bodies. To obtain the necessary data when applying this equation, points A and B should generally be selected at joints which are pin-connected, or at points in contact with adjacent bodies which have a *known motion*. For example, both points A and B on link AB, Fig. 16–10*a*, have circular paths of motion. Hence, at the instant shown, the magnitudes of their velocities are determined by the angular motion of the wheel and link BC so that $v_A = \omega_A r$ and $v_B = \omega_{BC} l$. The *directions* of $\mathbf{v}_A$ and $\mathbf{v}_B$ are always tangent to their paths of motion, Fig. 16–10*b*. In the case of the wheel shown in Fig. 16–11, point A can be selected at the ground. Provided the wheel rolls *without slipping*, this point has zero velocity, since it is in contact with the ground at this instant. Furthermore, the center of the wheel, B, moves along a horizontal path, so that $\mathbf{v}_B$ is horizontal.

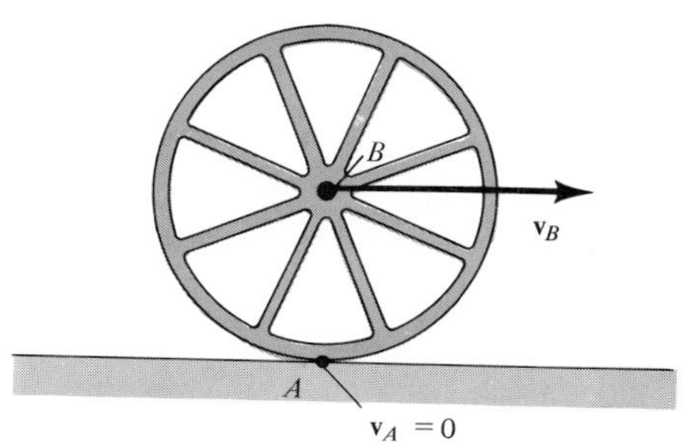

Fig. 16–11

PROCEDURE FOR ANALYSIS

The velocity equation $\mathbf{v}_B = \mathbf{v}_A + \mathbf{v}_{B/A}$ can be applied to any two points located in the *same* rigid body. For application, the following three-step procedure should be used.

Step 1: Draw a kinematic diagram of the body such as shown in Fig. 16–9*b*, and indicate on it the absolute velocities $\mathbf{v}_A$ and $\mathbf{v}_B$ of points A and B, and the angular velocity $\boldsymbol{\omega}$. Usually the base point A is selected as a point having a known velocity.

Step 2: Draw a kinematic diagram of the body such as shown in Fig. 16–9*d* and indicate on it the relative velocity $\mathbf{v}_{B/A}$. Since the body is considered to be pinned at the base point A, the *magnitude* of $\mathbf{v}_{B/A}$ is $v_{B/A} = \omega r_{B/A}$, where $r_{B/A}$ is calculated as the distance from A to B. The *direction* is established from the diagram, such that $\mathbf{v}_{B/A}$ acts perpendicular to $\mathbf{r}_{B/A}$, in accordance with the rotational motion $\boldsymbol{\omega}$ of the body.*

Step 3: Write the velocity equation in symbolic form: $\mathbf{v}_B = \mathbf{v}_A + \mathbf{v}_{B/A}$, and underneath each of the terms represent the respective vectors by their magnitudes and directions. To do this, use the data shown on the kinematic diagrams in *Steps 1* and *2*. Since motion occurs in the plane, this "vector" equation can be expressed in terms of *two scalar equations* which represent the addition of vector components in two directions. Once established, these two equations provide a solution for at most two unknown scalar quantities.

The following example problems numerically illustrate this scalar method of application.

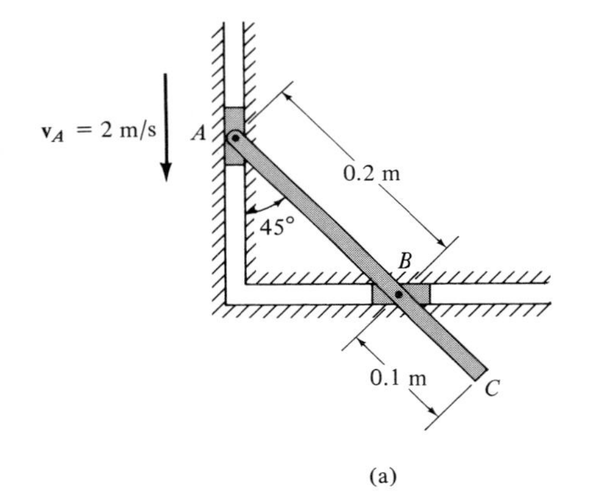

(a)

Example 16–5

The link shown in Fig. 16–12*a* is guided by two blocks at A and B, which move in the fixed slots. At the instant shown, the velocity of A is 2 m/s downward. Determine the velocity of (a) point B, and (b) point C located on the link at this same instant.

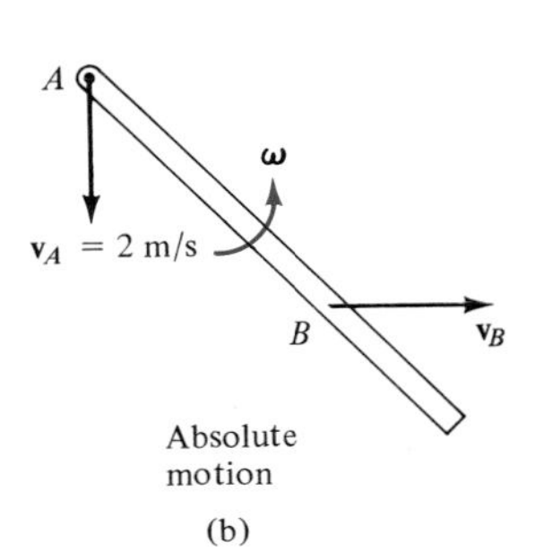

(b)

Fig. 16–12(a,b)

Solution

Part (a). *Step 1:* Since points A and B are restricted to move along the fixed slots, and $\mathbf{v}_A$ is directed downward, the velocity $\mathbf{v}_B$ must be directed horizontally to the right, Fig. 16–12*b*. This motion causes the link to rotate counterclockwise; that is, by the right-hand rule the angular velocity $\boldsymbol{\omega}$ is directed outward, perpendicular to the page (the plane of motion). Knowing the magnitude and direction of $\mathbf{v}_A$ and the lines of action of $\mathbf{v}_B$ and $\boldsymbol{\omega}$, it is possible to apply the velocity equation, $\mathbf{v}_B = \mathbf{v}_A + \mathbf{v}_{B/A}$, to points A and B in order to solve for the two unknown magnitudes v_B and ω.

*Perhaps the notation $\mathbf{v}_B = \mathbf{v}_A + \mathbf{v}_{B/A(\text{pin})}$ helps in recalling that A is pinned.

Step 2: As shown in Fig. 16–12*c*, the relative velocity $\mathbf{v}_{B/A}$ acts up to the right, perpendicular to $\mathbf{r}_{B/A}$ in accordance with the angular velocity $\boldsymbol{\omega}$. The magnitude of $\mathbf{v}_{B/A}$ is

$$v_{B/A} = \omega r_{B/A} = \omega(0.2 \text{ m})$$

Step 3: Applying Eq. 16–18, yields

$$\mathbf{v}_B = \mathbf{v}_A + \mathbf{v}_{B/A}$$
$$\underset{\rightarrow}{v_B} = \underset{\downarrow}{2 \text{ m/s}} + \underset{\angle^{45^\circ}}{\omega(0.2 \text{ m})}$$

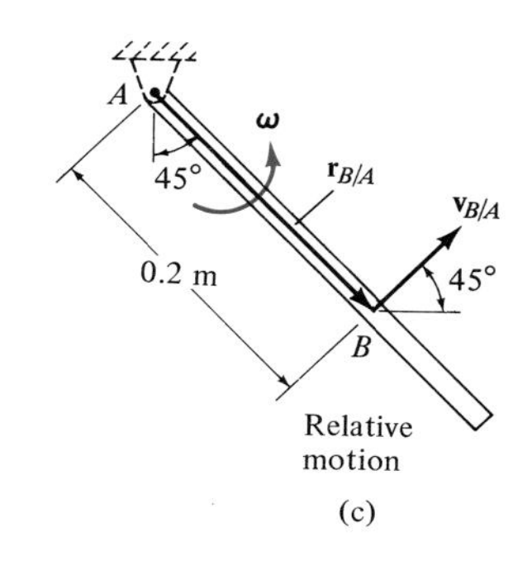

Equating the horizontal and vertical components, with the assumed positive directions to the right and upward, we have

$(\overset{+}{\rightarrow})$ $\quad v_B = 0 + 0.2\,\omega\cos 45^\circ$

$(+\uparrow)$ $\quad 0 = -2 + 0.2\,\omega\sin 45^\circ$

Thus,

$$\omega = 14.14 \text{ rad/s}$$
$$v_B = 2 \text{ m/s} \qquad \textit{Ans.}$$

Since both results are *positive,* the *directions* of $\mathbf{v}_B$ and $\boldsymbol{\omega}$ are indeed *correct,* as shown in Fig. 16–12*b*.

Part (b). *Step 1:* Knowing the angular velocity of the link, it is possible to obtain the magnitude and direction of the velocity of C (the two unknowns) by applying Eq. 16–18 to points A and C, i.e., $\mathbf{v}_C = \mathbf{v}_A + \mathbf{v}_{C/A}$, Fig. 16–12*d*.

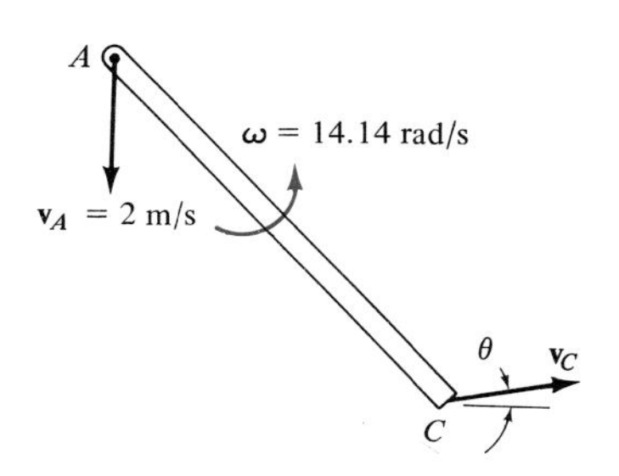

Step 2: The relative velocity $\mathbf{v}_{C/A}$ acts as shown in Fig. 16–12*e* and has a magnitude of

$$v_{C/A} = \omega r_{C/A} = 14.14 \text{ rad/s}(0.3 \text{ m}) = 4.24 \text{ m/s}$$

Step 3: Applying Eq. 16–18, we have

$$\mathbf{v}_C = \mathbf{v}_A + \mathbf{v}_{C/A}$$
$$\underset{\angle^{\theta}}{v_C} = \underset{\downarrow}{2 \text{ m/s}} + \underset{\angle^{45^\circ}}{4.24 \text{ m/s}}$$

Equating the horizontal and vertical components yields

$(\overset{+}{\rightarrow})$ $\quad v_C\cos\theta = 0 + 4.24\cos 45^\circ$

$(+\uparrow)$ $\quad v_C\sin\theta = -2 + 4.24\sin 45^\circ$

Dividing one equation by the other eliminates v_C and allows a solution for θ. v_C is obtained by resubstitution of θ into one of the equations. The results are

$$v_C = 3.16 \text{ m/s}$$
$$\theta = 18.4^\circ \quad \angle_{\theta}^{\mathbf{v}_C} \qquad \textit{Ans.}$$

Using the results for v_B and ω obtained in part (a), try to apply Eq. 16–18 to points B and C to obtain the same results for $\mathbf{v}_C$.

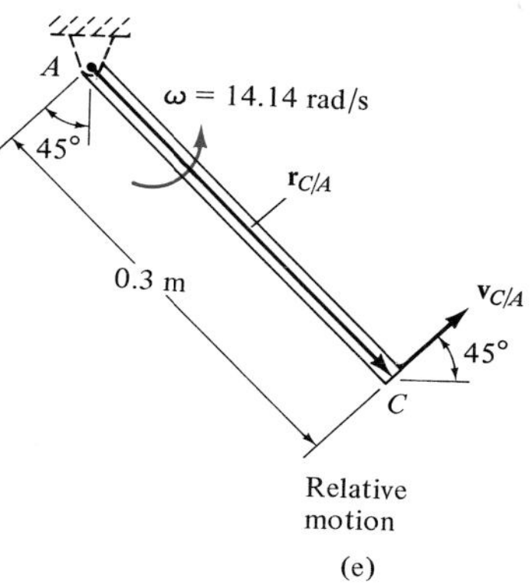

Fig. 16–12(c–e)

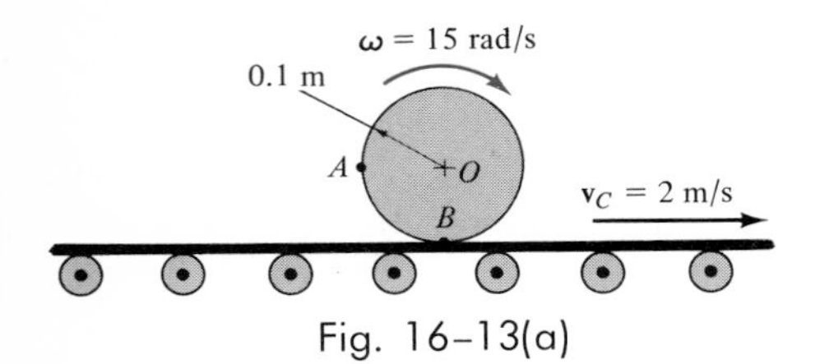

Fig. 16–13(a)

Example 16-6

The cylinder shown in Fig. 16–13*a* rolls freely on the surface of a conveyor belt which is moving at 2 m/s. Assuming that no slipping occurs between the cylinder and the belt, determine the velocity of point *A*. The cylinder has a clockwise angular velocity of $\omega = 15$ rad/s at the instant shown.

Solution

Step 1: Since no slipping occurs, point *B* on the cylinder has the same velocity as the conveyor, Fig. 16–13*b*. Knowing the angular velocity of the cylinder, it is possible to apply Eq. 16–18 to *B*, the base point, and *A* to determine the magnitude and direction of $\mathbf{v}_A$, i.e., $\mathbf{v}_A = \mathbf{v}_B + \mathbf{v}_{A/B}$.

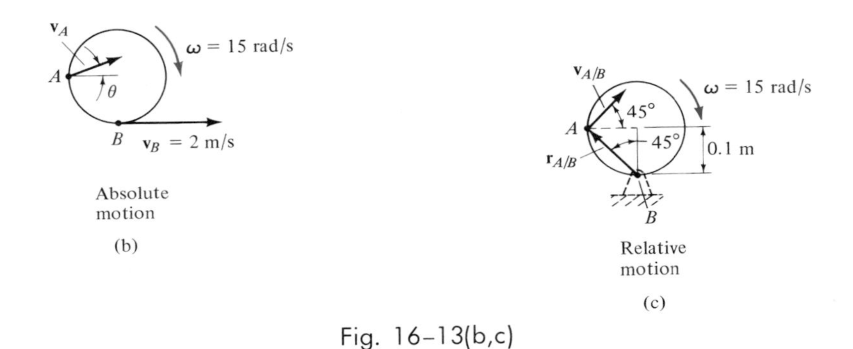

Fig. 16–13(b,c)

Step 2: The relative velocity $\mathbf{v}_{A/B}$ has a direction as shown in Fig. 16–13*c* and a magnitude of

$$v_{A/B} = \omega r_{A/B} = (15 \text{ rad/s})\left(\frac{0.1 \text{ m}}{\cos 45^\circ}\right) = 2.12 \text{ m/s}$$

Step 3: Applying Eq. 16–18 yields

$$\mathbf{v}_A = \mathbf{v}_B + \mathbf{v}_{A/B}$$

$$\underset{\angle^{\theta}}{v_A} = \underset{\rightarrow}{2 \text{ m/s}} + \underset{\angle^{45^\circ}}{2.12 \text{ m/s}}$$

Equating the horizontal and vertical components gives

$(\xrightarrow{+})$ $\qquad v_A \cos\theta = 2 + 2.12 \cos 45^\circ$

$(+\uparrow)$ $\qquad v_A \sin\theta = 0 + 2.12 \sin 45^\circ$

Solving,

$$v_A = 3.81 \text{ m/s}$$

$$\theta = 23.2^\circ \quad \angle^{\theta}\mathbf{v}_A \qquad \textit{Ans.}$$

Example 16-7

The bar AB of the linkage shown in Fig. 16–14a has a clockwise angular velocity of 30 rad/s when $\theta = 60°$. Compute the angular velocity of members BC and DC at this instant.

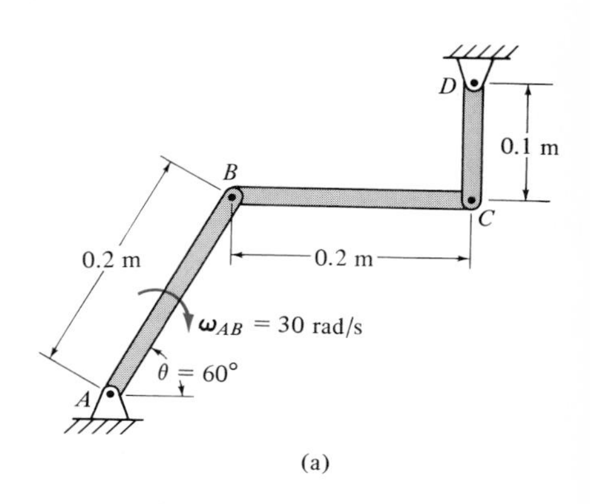

(a)

Solution

From the figure, AB and DC are subjected to rotation about a fixed axis, and BC is subjected to general plane motion. The magnitude and direction of $\mathbf{v}_B$ may be determined from the kinematic diagram shown in Fig. 16–14b. Because of the rotation, $\mathbf{v}_B$ *always* acts perpendicular to $\mathbf{r}_{B/A}$. The magnitude of $\mathbf{v}_B$ is

$$v_B = \omega_{AB} r_{B/A} = 30 \text{ rad/s}(0.2 \text{ m}) = 6 \text{ m/s}$$

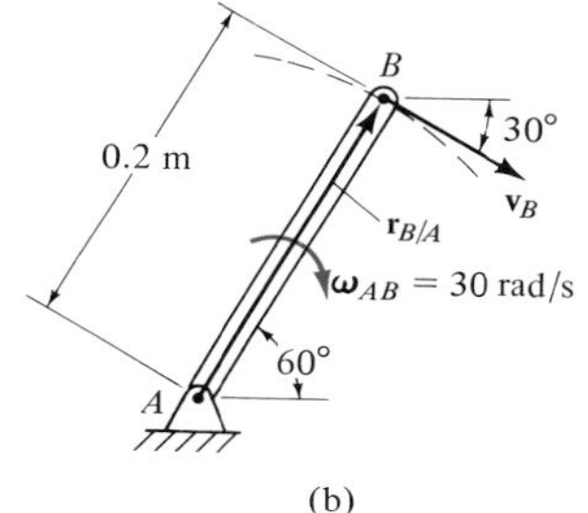

(b)

From the kinematic diagram of bar DC, Fig. 16–4c, $\mathbf{v}_C$ acts perpendicular to $\mathbf{r}_{C/D}$ and has a magnitude of

$$\begin{aligned} v_C &= \omega_{DC} r_{C/D} \\ &= \omega_{DC}(0.1 \text{ m}) \end{aligned} \tag{1}$$

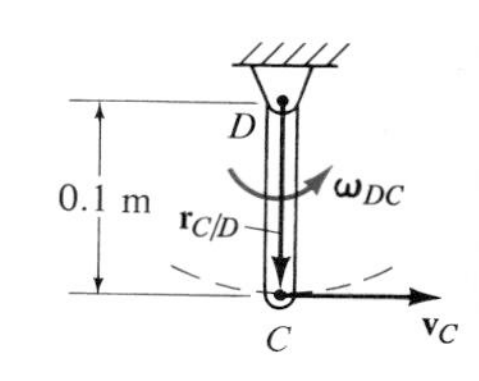

(c)

Step 1: Knowing $\mathbf{v}_B$ and the direction of $\mathbf{v}_C$, Fig. 16–14d, we can determine the two unknown magnitudes v_C and ω_{BC} by applying Eq. 16–18 between B, the base point, and C lying on rod BC, i.e., $\mathbf{v}_C = \mathbf{v}_B + \mathbf{v}_{C/B}$.
Step 2: As shown in Fig. 16–14e, the relative velocity $\mathbf{v}_{C/B}$ acts upward and has a magnitude of

$$\begin{aligned} v_{C/B} &= \omega_{BC}(r_{C/B}) \\ &= \omega_{BC}(0.2 \text{ m}) \end{aligned}$$

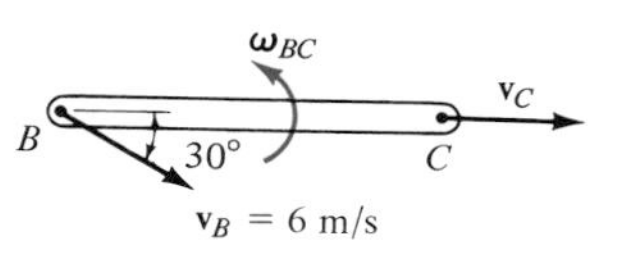

Absolute motion

(d)

Step 3: Applying Eq. 16–18, we have

$$\mathbf{v}_C = \mathbf{v}_B + \mathbf{v}_{C/B}$$

$$\underset{\rightarrow}{v_C} = \underset{\searrow 30°}{6 \text{ m/s}} + \underset{\uparrow}{\omega_{BC}(0.2 \text{ m})}$$

Equating the horizontal and vertical components yields

$(\overset{+}{\rightarrow})$ $\qquad v_C = 6 \cos 30° + 0 = 5.20 \text{ m/s}$

$(+\uparrow)$ $\qquad 0 = -6 \sin 30° + 0.2\omega_{BC}$

$$\omega_{BC} = 15 \text{ rad/s} \circlearrowleft \qquad \textit{Ans.}$$

Relative motion

(e)

Fig. 16–14

Using Eq. (1),

$$\omega_{DC}(0.1) = 5.20 \text{ m/s}$$

$$\omega_{DC} = 52 \text{ rad/s} \circlearrowleft \qquad \textit{Ans.}$$

16–6. Method for Determining Velocity Using Cartesian Vectors

In the previous section the velocity equation $\mathbf{v}_B = \mathbf{v}_A + \mathbf{v}_{B/A}$ was applied using a scalar method for solution. It is also possible to apply this equation using a Cartesian vector analysis.

Since the relative velocity $\mathbf{v}_{B/A}$ represents the effect of *circular motion*, observed from translating axes passing through the base point A, this term can be expressed by the cross product $\mathbf{v}_{B/A} = \boldsymbol{\omega} \times \mathbf{r}_{B/A}$ (Eq. 16–13). Hence, for application,

$$\mathbf{v}_B = \mathbf{v}_A + (\boldsymbol{\omega} \times \mathbf{r}_{B/A}) \tag{16–19}$$

where

$\mathbf{v}_B$ = absolute velocity of B
$\mathbf{v}_A$ = absolute velocity of the base point A
$\boldsymbol{\omega}$ = absolute angular velocity of the body
$\mathbf{r}_{B/A}$ = relative-position vector drawn from A to B

Applying Eq. 16–19 in Cartesian vector form is very methodical and therefore has an advantage for solving problems involving mechanisms consisting of *several connected members*. When using this method, however, the physical aspects of the kinematics are often lost. Furthermore, because of the additional vector algebra which must be included in the solution, generally more computations than the scalar method must be made to obtain the complete solution.

PROCEDURE FOR ANALYSIS

The following two-step procedure should be used when applying Eq. 16–19 to two points A and B located in the *same* rigid body.

Step 1: Draw a kinematic diagram of the body, which is used to indicate the absolute velocities $\mathbf{v}_A$ and $\mathbf{v}_B$ of points A and B, the angular velocity $\boldsymbol{\omega}$, and the relative-position vector $\mathbf{r}_{B/A}$.
Step 2: Establish the fixed x, y, z axes and express the vectors shown on the kinematic diagram in terms of their **i, j, k** components. Substitute these vectors into Eq. 16–19 and evaluate the cross product. Then equate the respective **i** and **j** components to obtain two scalar equations which may be solved for at most two unknowns.

The following two examples numerically illustrate this procedure. Refer also to *Steps 1* and *2* of Examples 16–15 and 16–16, where this technique is used in part of the solution.

Example 16–8

The link shown in Fig. 16–15*a* is guided by two blocks at A and B which move in the fixed slots. At the instant shown, the velocity of the block at A is 2 m/s downward. Determine the velocity of point B at this instant.

Solution

Step 1: A kinematic diagram showing the velocities of points A and B, and the angular velocity of the link, is given in Fig. 16–15*b*. The two unknowns are represented by the magnitudes of $\boldsymbol{\omega}$ and $\mathbf{v}_B$.

Step 2: Expressing each of the vectors in Fig. 16–15*b* in terms of their **i**, **j**, and **k** components, we have

$$\mathbf{v}_A = \{-2\mathbf{j}\}\text{ m/s}, \qquad \mathbf{v}_B = v_B\mathbf{i}, \qquad \boldsymbol{\omega} = \omega\mathbf{k}$$

$$\mathbf{r}_{B/A} = \{0.2 \sin 45^\circ\mathbf{i} - 0.2 \cos 45^\circ\mathbf{j}\}\text{ m}$$

Applying Eq. 16–19 to A, the base point, and B, we have

$$\mathbf{v}_B = \mathbf{v}_A + (\boldsymbol{\omega} \times \mathbf{r}_{B/A})$$

$$v_B\mathbf{i} = -2\mathbf{j} + [\omega\mathbf{k} \times (0.2 \sin 45^\circ\mathbf{i} - 0.2 \cos 45^\circ\mathbf{j})]$$

or

$$v_B\mathbf{i} = -2\mathbf{j} + 0.2\omega \sin 45^\circ\mathbf{j} + 0.2\omega \cos 45^\circ\mathbf{i}$$

Equating the **i** and **j** components gives

$$v_B = 0.2\omega \cos 45^\circ, \qquad 0 = -2 + 0.2\omega \sin 45^\circ$$

Thus,

$$\omega = 14.14\text{ rad/s}$$

$$v_B = 2\text{ m/s} \qquad \textit{Ans.}$$

Now that $\boldsymbol{\omega}$ has been determined, apply Eq. 16–19 to points A and C to determine $\mathbf{v}_C$ as in Example 16–5.

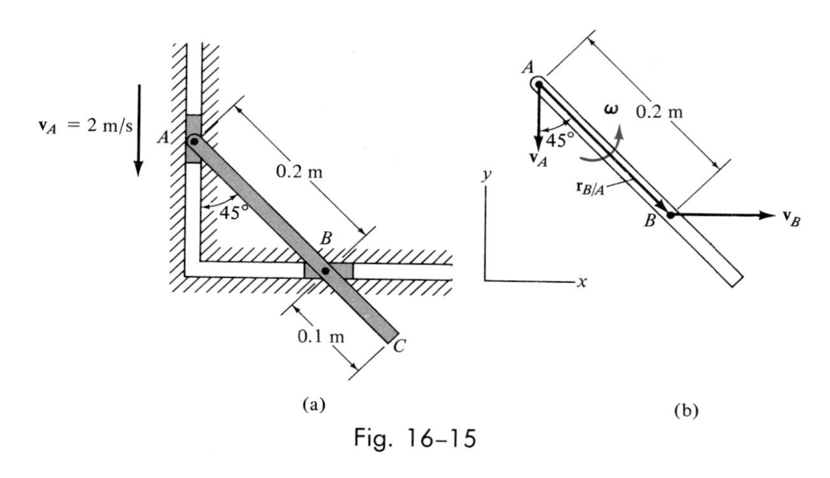

Fig. 16–15

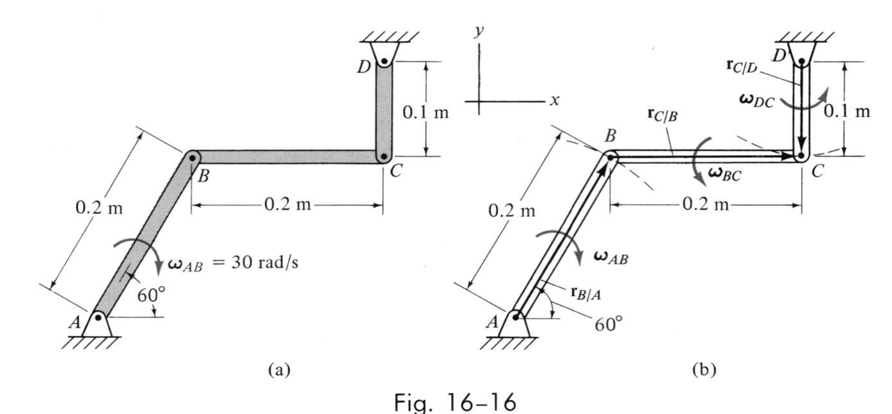

Fig. 16–16

Example 16–9

The bar AB of the linkage shown in Fig. 16–16a has a clockwise angular velocity of 30 rad/s when $\theta = 60°$. Compute the angular velocity of members BC and DC at this instant.

Solution

Step 1: The velocities of points B and C are defined by the rotation of links AB and DC about their fixed axes. The relative-position vectors and the angular velocity of each link are shown on the kinematic diagram in Fig. 16–16b.

Step 2: Expressing each vector in Fig. 16–16b in Cartesian vector form, we have

$$\boldsymbol{\omega}_{AB} = \{-30\mathbf{k}\}\ \text{rad/s}, \qquad \boldsymbol{\omega}_{BC} = \omega_{BC}\mathbf{k}, \qquad \boldsymbol{\omega}_{DC} = \omega_{DC}\mathbf{k}$$
$$\mathbf{r}_{B/A} = \{0.2 \cos 60°\mathbf{i} + 0.2 \sin 60°\mathbf{j}\}\ \text{m}, \qquad \mathbf{r}_{C/B} = \{0.2\mathbf{i}\}\ \text{m},$$
$$\mathbf{r}_{C/D} = \{-0.1\mathbf{j}\}\ \text{m}$$

For link AB:

$$\begin{aligned}\mathbf{v}_B &= \boldsymbol{\omega}_{AB} \times \mathbf{r}_{B/A}\\ &= (-30\mathbf{k}) \times (0.2 \cos 60°\mathbf{i} + 0.2 \sin 60°\mathbf{j})\\ &= \{5.20\mathbf{i} - 3.0\mathbf{j}\}\ \text{m/s}\end{aligned}$$

For link BC:

$$\begin{aligned}\mathbf{v}_C &= \mathbf{v}_B + (\boldsymbol{\omega}_{BC} \times \mathbf{r}_{C/B})\\ &= 5.20\mathbf{i} - 3.0\mathbf{j} + [(\omega_{BC}\mathbf{k}) \times (0.2\mathbf{i})]\\ &= 5.20\mathbf{i} + (0.2\omega_{BC} - 3.0)\mathbf{j}\end{aligned}$$

For link DC:

$$\mathbf{v}_C = \boldsymbol{\omega}_{DC} \times \mathbf{r}_{C/D}$$
$$5.20\mathbf{i} + (0.2\omega_{BC} - 3.0)\mathbf{j} = (\omega_{DC}\mathbf{k}) \times (-0.1\mathbf{j})$$
$$5.20\mathbf{i} + (0.2\omega_{BC} - 3.0)\mathbf{j} = 0.1\omega_{DC}\mathbf{i}$$

Equating the respective **i** and **j** components and solving, we have

$$\omega_{DC} = 52 \text{ rad/s} \quad \circlearrowleft \qquad \textit{Ans.}$$
$$\omega_{BC} = 15 \text{ rad/s} \quad \circlearrowleft \qquad \textit{Ans.}$$

Problems

***16-28.** The rigid body of arbitrary shape moves in the plane. If h and θ are known, and the speed of A and B is $v_A = v_B = v$, compute the angular velocity $\boldsymbol{\omega}$ of the body and the direction ϕ of $\mathbf{v}_B$.

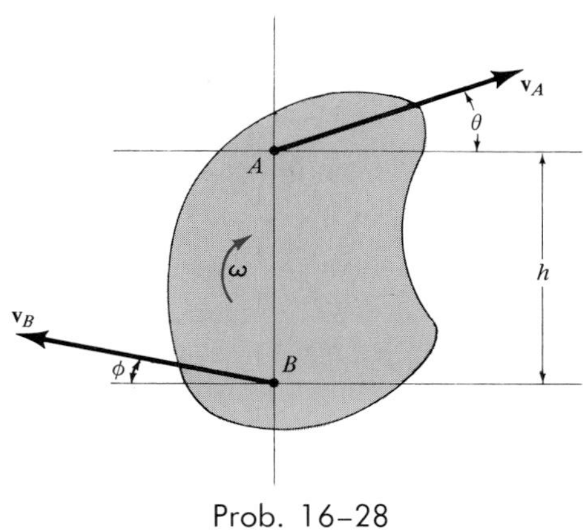
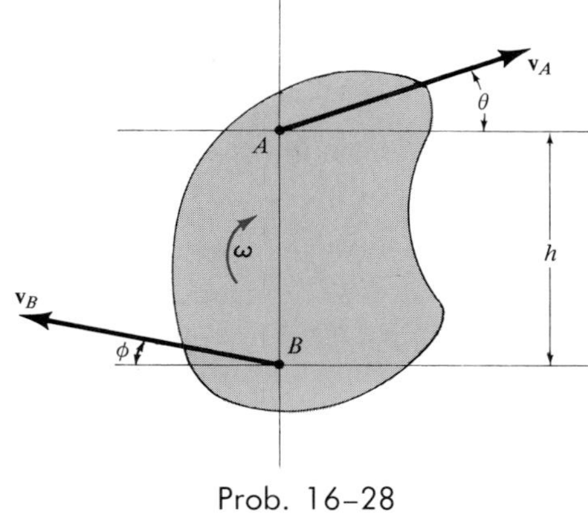

Prob. 16-28

16-29. When the slider block C is in the position shown, link AB has a clockwise angular velocity of $\omega_{AB} = 5$ rad/s. Determine the velocity of C at this instant.

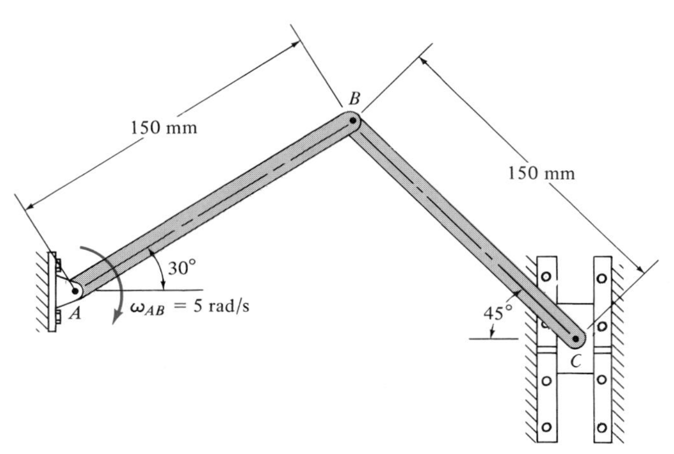

Prob. 16-29

16-30. The center O of the wheel is moving to the right with a speed of $v_O = 3$ m/s. If no slipping occurs at the ground A, determine the velocities of points B and C. Set $r = 250$ mm. *Hint:* To determine the wheel's angular velocity $\boldsymbol{\omega}$, use $v_O = \omega r$ as indicated by the result of Example 16-3.

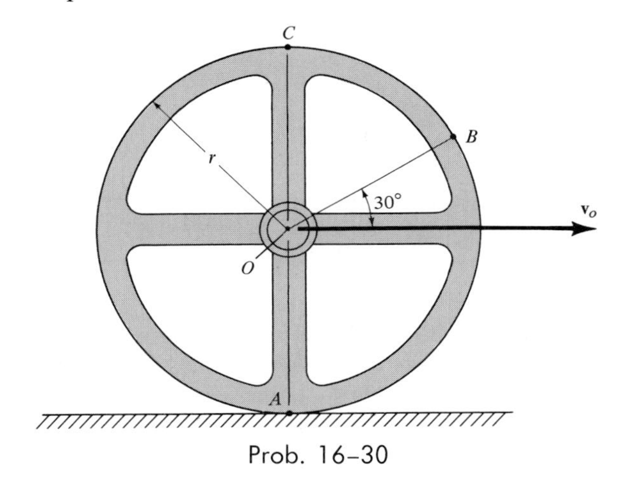

Prob. 16-30

16-30a. Solve Prob. 16-30 if $v_O = 4$ ft/s and $r = 0.75$ ft.

16–31. If rod CD has a downward velocity of 90 mm/s at the instant shown, determine the velocity of the gear rack A at this instant. The rod is pinned at C to gear B.

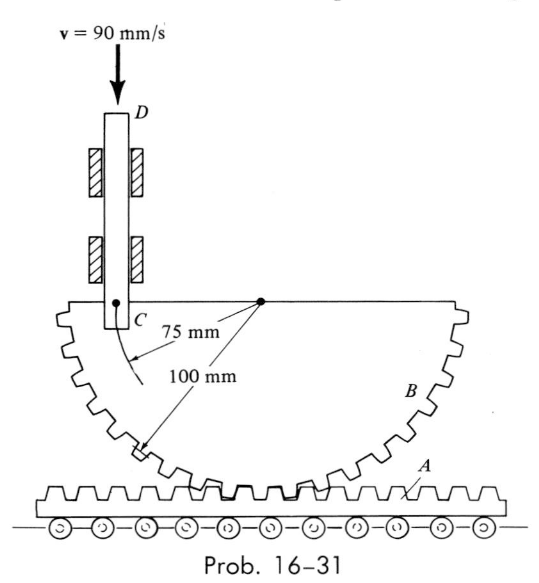

Prob. 16–31

***16–32.** When the crank on the Chinese windlass is turning, the rope on shaft A unwinds while that on shaft B winds up. Determine the speed at which the bucket lowers if the crank is turning with an angular velocity of $\omega = 2$ rad/s. What is the angular velocity of the pulley at C?

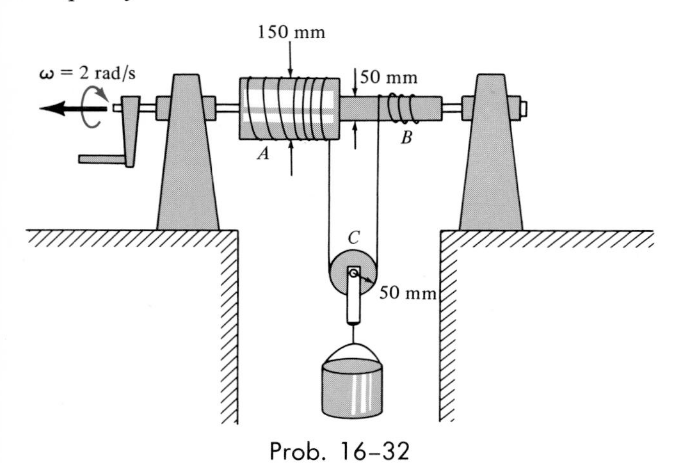

Prob. 16–32

16–33. The spool gear rests on the fixed horizontal rack and a cord is wrapped around the inner core so that it remains horizontally tangent at A. If the cord is pulled with a constant speed of 150 mm/s, determine the velocity of the center of the gear, C. *Hint:* Note that point D, considered to be a hypothetical extension of the gear, has zero velocity.

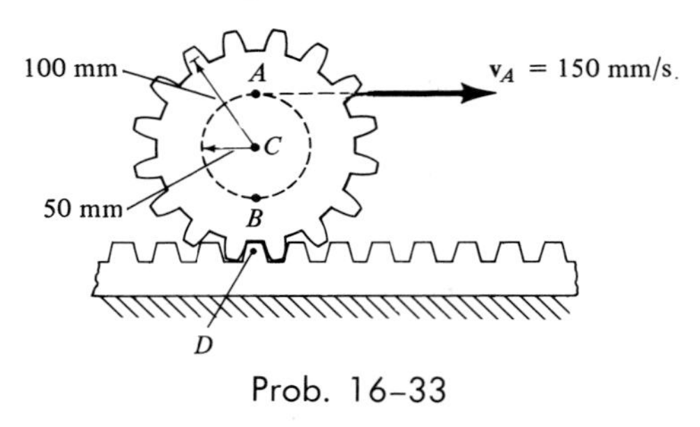

Prob. 16–33

16–34. Solve Prob. 16–33 assuming that the cord is wrapped around the core in the opposite sense, so that the end of the cord remains horizontally tangent to the core at B, and is pulled to the right at $v_B = 150$ mm/s.

16–35. If, at a given instant, point B has a downward velocity of $v_B = 3$ m/s, determine the velocity of point A at this instant. Notice that for this motion to occur, the wheel must slip on the surface of the horizontal plane. Set $r_o = 0.4$ m, $r_i = 0.15$ m.

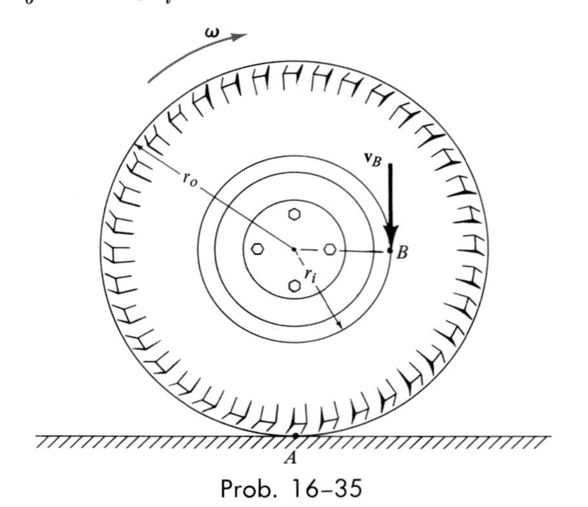

Prob. 16–35

16–35a. Solve Prob. 16–35 if $v_B = 6$ ft/s, $r_o = 2$ ft, and $r_i = 1.25$ ft.

*** 16–36.** If the piston P is moving upward with a speed of 50 m/s at the instant shown, determine the angular velocity of the crankshaft AB at this instant.

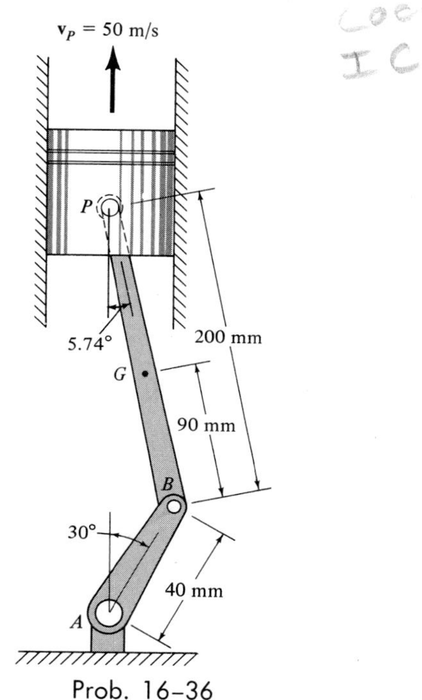

Prob. 16–36

16–37. Determine the velocity of the center of gravity G of the connecting rod in Prob. 16–36 at the instant shown. The piston is moving upward with a speed of 50 m/s.

16–38. If the end of the cable is pulled downward at a speed of $v_C = 120$ mm/s, determine the angular velocities of pulleys A and B and the upward speed of the block at D. Assume that the cable does not slip on the pulleys.

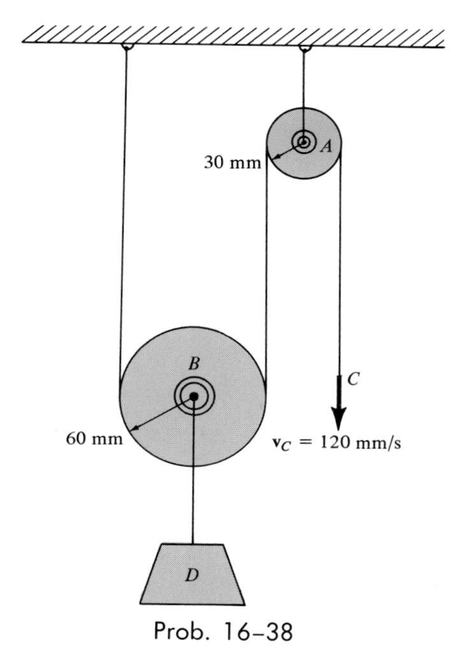

Prob. 16–38

16–39. The inner hub of the roller bearing is rotating with an angular velocity of $\omega_i = 6$ rad/s, while the outer hub is rotating in the opposite direction at $\omega_o = 4$ rad/s. Determine the angular velocity of each of the rollers if they roll on the hubs without slipping.

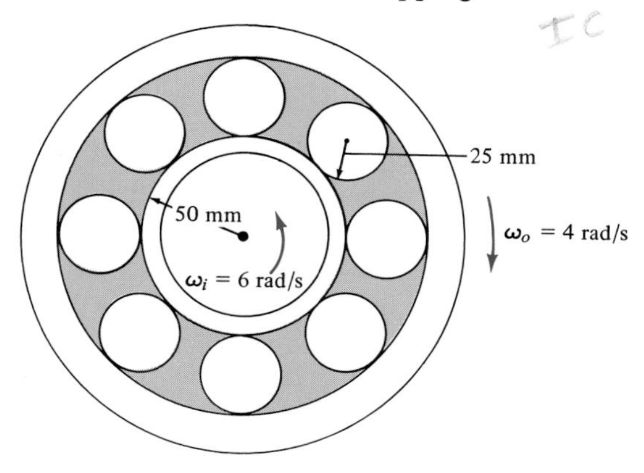

Prob. 16–39

IC

*** 16-40.** Link AB rotates clockwise about the fixed point A with an angular velocity of $\omega_{AB} = 3$ rad/s. If the frame F is stationary, determine the angular velocity of gear C. Set $r_C = 40$ mm and $r_D = 80$ mm.

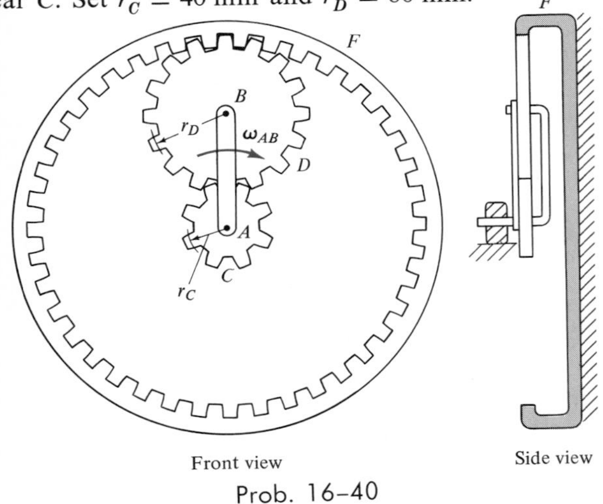

Prob. 16-40

*** 16-40a.** Solve Prob. 16-40 if $\omega_{AB} = 4$ rad/s, $r_C = 0.2$ ft, and $r_D = 0.4$ ft.

16-41. The device is used to indicate the safe load acting at the end of the boom B when it is in any angular position. It consists of a fixed dial plate D and an indicator arm ACE which is pinned to the plate at C and to a short link EF. If the boom is pin-connected to the trunk frame at G and is rotating downward at $\omega_B = 3$ rad/s, determine the velocity of the dial pointer A at the instant shown, i.e., when EF and AC are in the vertical position.

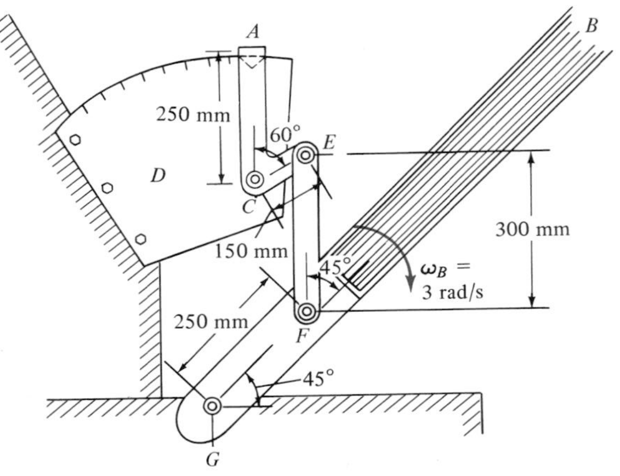

Prob. 16-41

16-42. From a side view, both sides (BC) of a window are pinned to links (AB). If the ends A, B, and C are confined to move in the grooved slots, and C is given a velocity of 0.75 m/s, determine the angular velocities of BC and AB at the instant $\theta = 30°$.

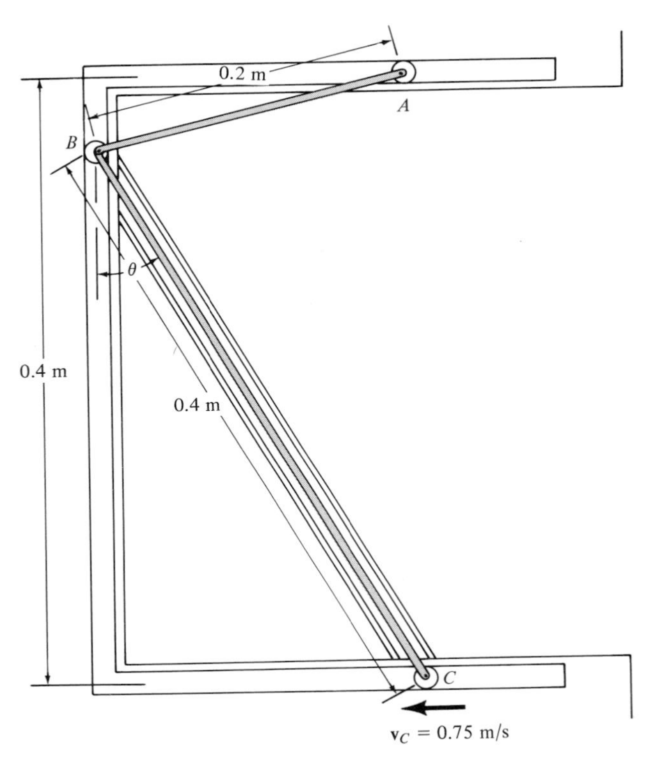

Prob. 16-42

16–43. The rotation of link AB creates an oscillating movement of gear F. If AB has an angular velocity of $\omega_{AB} = 6$ rad/s, determine the angular velocity of gear F at the instant shown. Gear E is rigidly attached to arm CD and pinned at D to a fixed point.

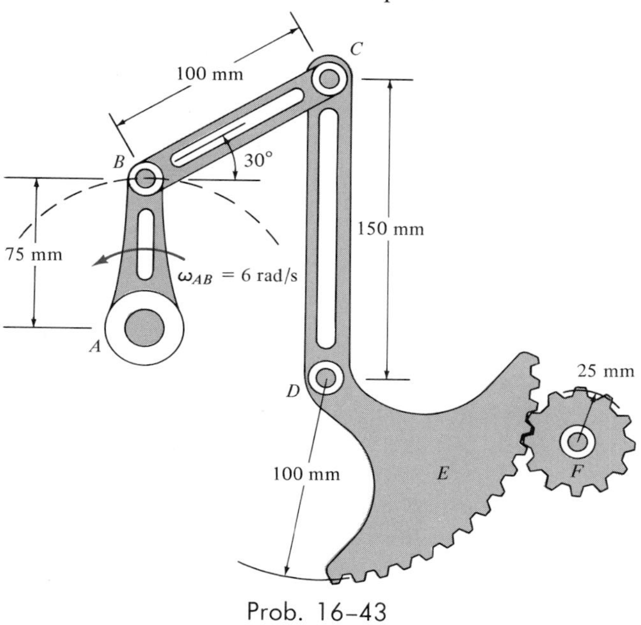

Prob. 16–43

***16–44.** The mechanism shown is used to give the printing table T oscillating motion. If the crank AB is rotating at a constant rate of 3 rad/s, determine the velocity of the table at the instant shown. The gears D and E rotate on the stationary gear rack F and drive the movable top rack attached to the table.

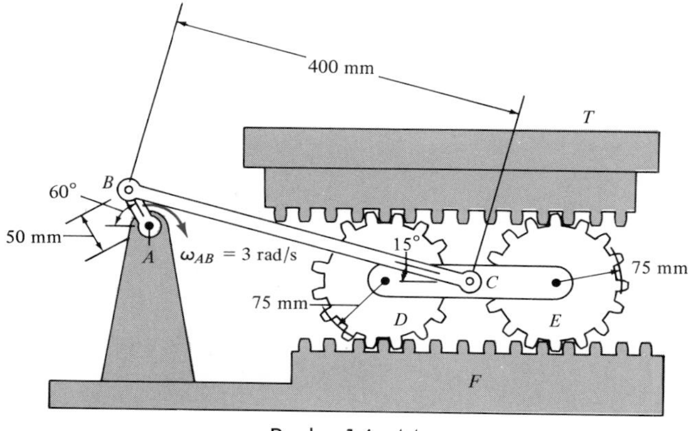

Prob. 16–44

16–45. The slider mechanism is used to increase the stroke of travel of one slider with respect to that of another. As shown, when the slider A is moving forward, the attached pinion F rolls on the *fixed* rack D, forcing slider C to move forward. This in turn causes the attached pinion at G to roll on the *fixed* rack E, thereby moving slider B. If A has a velocity of $v_A = 6$ m/s at the instant shown, determine the velocity of B. $r = 50$ mm.

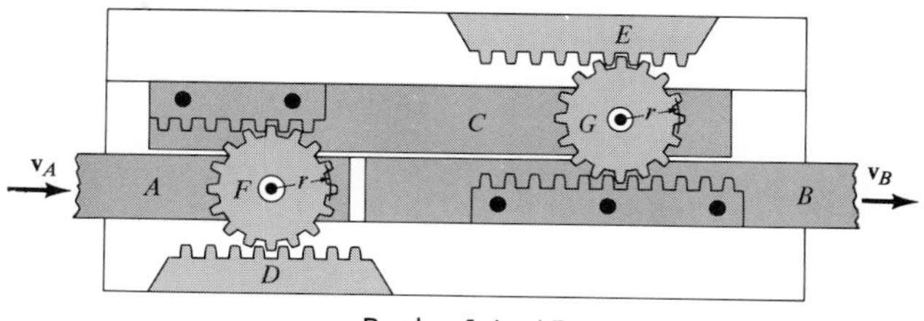

Prob. 16–45

16–45a. Solve Prob. 16–45 if $v_A = 2$ ft/s and $r = 0.2$ ft.

16–46. The mechanism is used on a machine for the manufacturing of a wire product. Because of the rotational motion of link AB and the sliding of block F, the segmental gear lever DE undergoes general plane motion. If AB is rotating at $\omega_{AB} = 5$ rad/s, determine the velocity of point E at the instant shown.

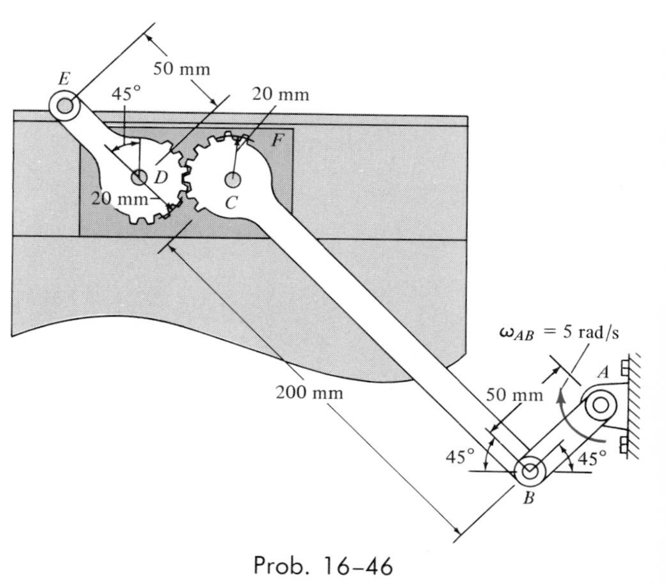

Prob. 16–46

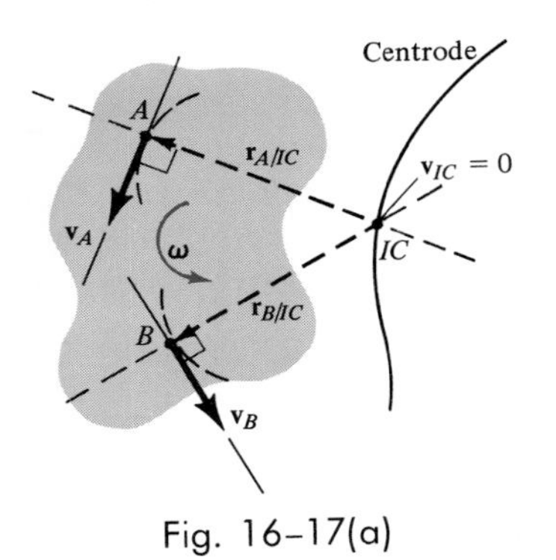

Fig. 16–17(a)

16–7. Instantaneous Center of Zero Velocity

The velocity of any point B, located in a rigid body, can be obtained in a simple way if one chooses the base point IC to be a point that has *zero velocity* at the instant considered. In this case, $\mathbf{v}_{IC} = \mathbf{0}$, and therefore the velocity equation (Eq. 16–18) becomes $\mathbf{v}_B = \mathbf{v}_{IC} + \mathbf{v}_{B/IC} = \mathbf{v}_{B/IC}$. For a body having general plane motion, point IC, so chosen, is called the *instantaneous center of zero velocity* (IC), and it lies on the *instantaneous axis of zero velocity* (IA). Hence, the IA is always perpendicular to the plane used to represent the motion, and the intersection of the IA with this plane defines the IC. Provided the location of the IC is *known,* Fig. 16–17a, the relative-position vector $\mathbf{r}_{B/IC}$, which is drawn from the IC to point B, can be established. Since $\mathbf{v}_B = \mathbf{v}_{B/IC}$, point B appears to move momentarily in a *circular path* about the IC such that the *magnitude* of $\mathbf{v}_B$ is

$$v_B = \omega r_{B/IC} \tag{16–20}$$

where ω is the magnitude of the angular velocity of the body. The *direction* of $\mathbf{v}_B$ is perpendicular to the line of action of $\mathbf{r}_{B/IC}$, as shown in Fig. 16–17a.

Location of the *IC*. If the location of the IC is unknown, it may be determined provided one knows the lines of action for the velocities $\mathbf{v}_A$ and $\mathbf{v}_B$ of any two points A and B in the body, Fig. 16–17a. From each of these lines construct at points A and B *perpendicular* line segments which then define the lines of action of $\mathbf{r}_{A/IC}$ and $\mathbf{r}_{B/IC}$, respectively. Extending these perpendiculars to their *point of intersection* as shown, locates the IC at the instant considered. As a special case, if the lines of action of the velocities $\mathbf{v}_A$ and $\mathbf{v}_B$ are *parallel* to one another, Fig. 16–17b or 16–17c, then provided one knows the magnitudes of $\mathbf{v}_A$ and $\mathbf{v}_B$, the IC can be located using proportional right triangles.

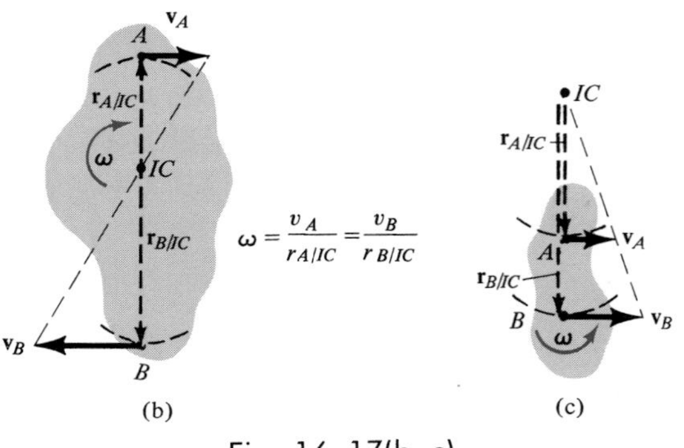

Fig. 16–17(b–c)

Centrode. When a body is subjected to general plane motion, the point selected to represent the instantaneous center of zero velocity for the body can only be used for an instant of time. Since the body changes its position from one instant to the next, then for each position of the body, a unique instantaneous center must be determined. The locus of points which defines the *IC* for the body during various instants of time is called a *centrode,* Fig. 16–17*a*. Hence, each point on the centrode acts as the *IC* for the body only for an instant of time.

PROCEDURE FOR ANALYSIS

In order to determine the velocity of a point in a body which is subjected to general plane motion, it is first necessary to locate the *IC* of the body by using the method described above. Drawing a kinematic diagram, such as shown in Fig. 16–18, the body can be imagined as "extended and pinned" at the *IC* such that the body rotates about this pin with its instantaneous angular velocity $\boldsymbol{\omega}$. The *magnitude* of velocity for the arbitrary points *A*, *B*, and *C* on the body can then be determined by using the equation $v = \omega r$ (Eq. 16–12), where r is the radial line drawn from the *IC* to the point. The line of action of each velocity vector is *perpendicular* to the radial line, and the velocity has a *direction* which tends to move the point in a manner consistent with the angular rotation of the radial line, Fig. 16–18.

The following examples illustrate the method of determining the *IC* and computing the velocities of points in a rigid body subjected to general plane motion. See *Step 1* of Examples 16–12, 16–13, and 16–14, which further illustrates this method of analysis.

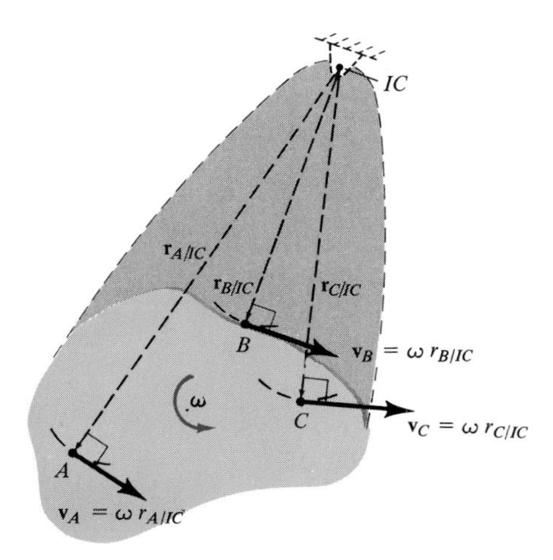

Fig. 16–18

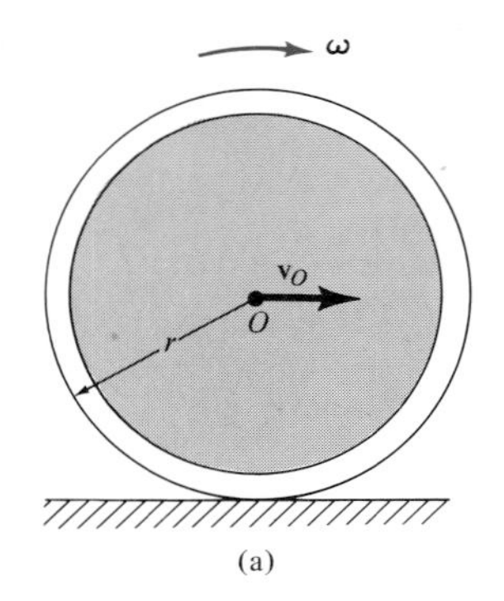

(a)

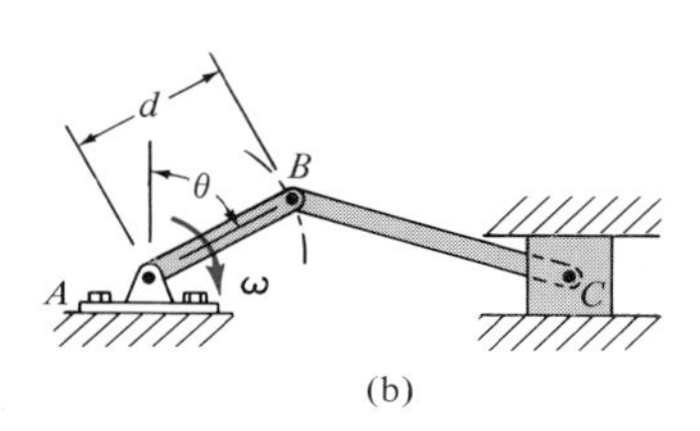

(b)

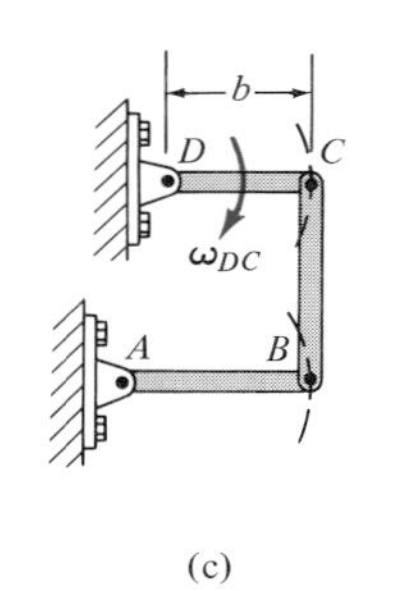

(c)

Example 16–10

Determine the location of the instantaneous center of zero velocity for (a) the wheel shown in Fig. 16–19*a* which is rolling without slipping along the ground, (b) the crankshaft BC shown in Fig. 16–19*b*, and (c) the link CB shown in Fig. 16–19*c*.

Solution

Part (a). The wheel rolls without slipping, and therefore the point of *contact* of the wheel with the ground has *zero velocity*. Hence, this point represents the IC for the wheel, Fig. 16–19*d*. If it is imagined that the wheel is momentarily pinned at this point, the velocity of points A, B, O, and so on, can be found using Eq. 16–20. As shown, the distances $r_{A/IC}$, $r_{B/IC}$, and $r_{O/IC}$ are determined from the geometry of the wheel, and the angular velocity $\boldsymbol{\omega}$ of the wheel must be known.

Part (b). As shown in Fig. 16–19*b*, link AB rotates about the fixed pin A. Thus, point B has a speed of $v_B = \omega d$, caused by the clockwise rotation of AB. Also, the velocity is perpendicular to AB, so that it acts at an angle θ from the horizontal as shown in Fig. 16–19*e*. The motion of point B causes the piston to move forward *horizontally* with a velocity $\mathbf{v}_C$. Consequently, point C on the rod moves horizontally with this same velocity. When lines are drawn perpendicular to $\mathbf{v}_B$ and $\mathbf{v}_C$, Fig. 16–19*e*, they intersect at the IC. The magnitudes of $\mathbf{r}_{B/IC}$ and $\mathbf{r}_{C/IC}$ are determined strictly from the geometry of construction.

Part (c). Points B and C follow circular paths of motion since rods AB and DC are each subjected to rotation about a fixed axis, Fig. 16–19*c*. In particular, $v_C = \omega_{DC} b$. Since the velocity is always tangent to the path, at the instant considered, the velocities of point C on rod DC and point B on rod AB are both directed vertically downward, along the axis of link CB, Fig. 16–19*f*. Furthermore, since CB is *rigid,* no relative displacement occurs between points B and C, so that $\mathbf{v}_B = \mathbf{v}_C$. Radial lines drawn perpendicular to these two velocities form parallel lines which intersect at "infinity," i.e., $r_{C/IC} \rightarrow \infty$ and $r_{B/IC} \rightarrow \infty$. Since $v_C = r_{C/IC}\omega_{CB}$, then $\omega_{CB} = v_C / r_{C/IC} = (\omega_{DC} b)/\infty = 0$. As a result, rod CB momentarily *translates* with a speed of $v_C = \omega_{DC} b$. An instant later, however, CB will move to a new position, causing the instantaneous center to *change* to some finite location.

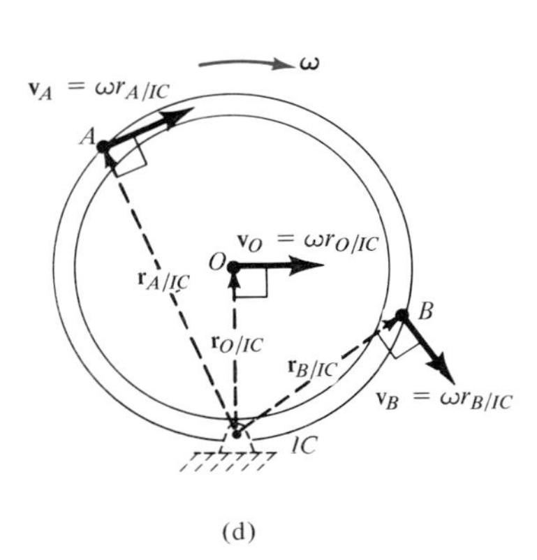

(d)

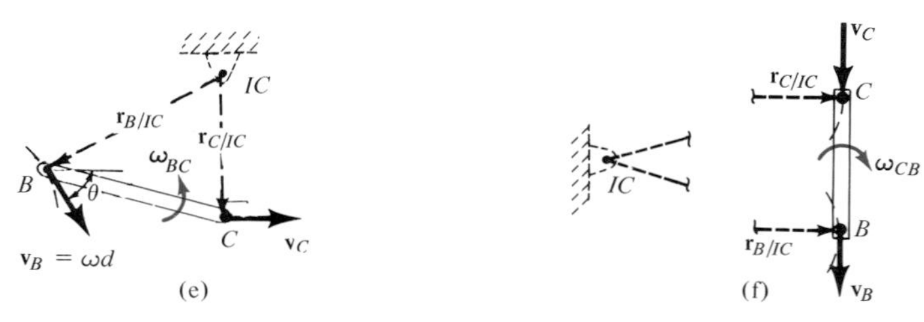

(e) (f)

Fig. 16–19

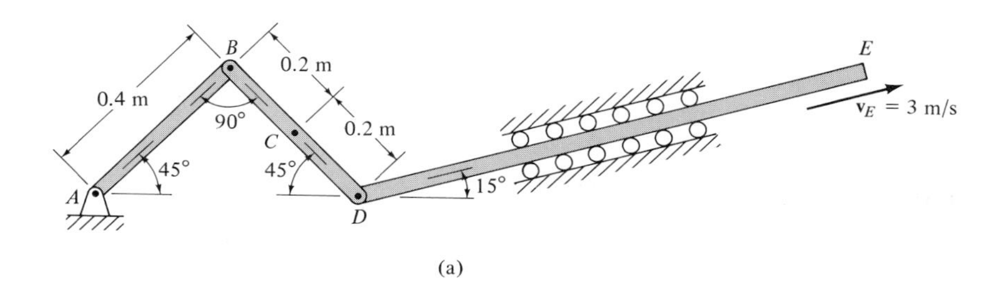

(a)

Example 16–11

Rod DE, shown in Fig. 16–20a, moves with a speed of 3 m/s. Determine the angular velocity of link AB and the velocity of point C on link BD at the instant shown.

Solution

Due to the constraint, DE is subjected to rectilinear translation. Hence, the speed of point D (and any other point on DE) is 3 m/s directed upward along the axis of the rod. This movement causes arm AB to rotate about point A in a clockwise direction, Fig. 16–20c, so that the velocity of point B is directed perpendicular to AB, Fig. 16–20b. The instantaneous center of zero velocity for BD is located at the intersection of the line segments drawn perpendicular to the velocities $\mathbf{v}_B$ and $\mathbf{v}_D$, Fig. 16–20b. From the geometry,

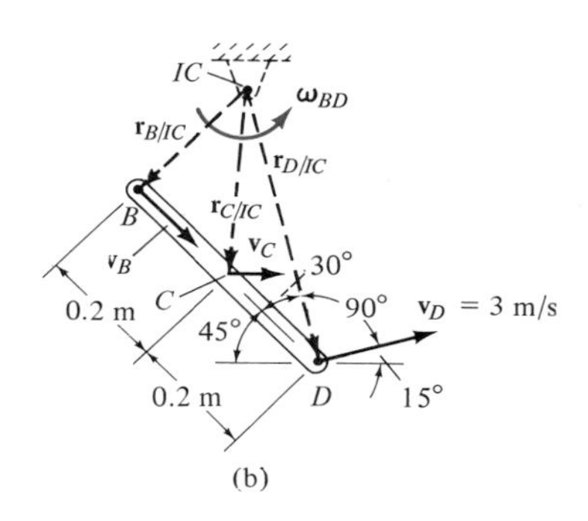

(b)

$$r_{B/IC} = 0.4 \tan 30° \text{ m} = 0.231 \text{ m}$$

$$r_{D/IC} = \frac{0.4 \text{ m}}{\cos 30°} = 0.462 \text{ m}$$

and

$$r_{C/IC} = \sqrt{(r_{B/IC})^2 + (0.2)^2}$$
$$= \sqrt{(0.231)^2 + (0.2)^2} = 0.306 \text{ m}$$

Since link BD can be considered momentarily pinned at the IC, the angular velocity of the link is

$$\omega_{BD} = \frac{v_D}{r_{D/IC}} = \frac{3 \text{ m/s}}{0.462 \text{ m}} = 6.49 \text{ rad/s} \circlearrowleft$$

The velocities of B and C are therefore

$$v_B = \omega_{BD}(r_{B/IC}) = 6.49 \text{ rad/s } (0.231 \text{ m}) = 1.50 \text{ m/s}$$

and

$$v_C = \omega_{BD}(r_{C/IC}) = 6.49 \text{ rad/s } (0.306 \text{ m}) = 1.99 \text{ m/s} \qquad \textit{Ans.}$$

Since link AB is subjected to rotation about a fixed axis passing through A, Fig. 16–20c, and the velocity of point B is known, the angular velocity of the link is

$$\omega_{AB} = \frac{v_B}{r_{B/A}} = \frac{1.50 \text{ m/s}}{0.4 \text{ m}} = 3.75 \text{ rad/s} \circlearrowright \qquad \textit{Ans.}$$

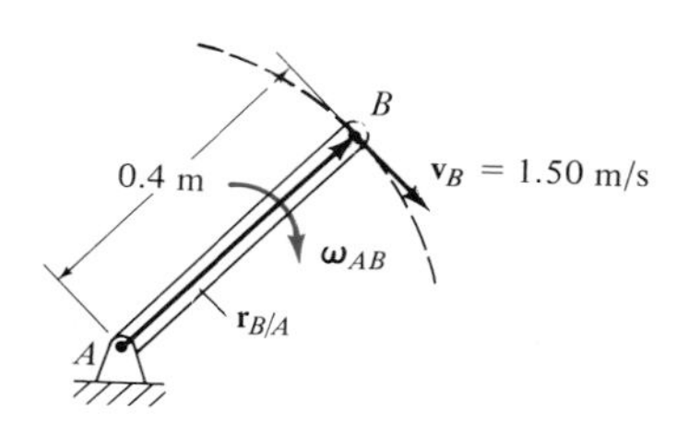

(c)

Fig. 16–20

16–47. Solve Prob. 16–31 using the method of instantaneous center of zero velocity.

*** 16–48.** Solve Prob. 16–33 using the method of instantaneous center of zero velocity.

16–49. Solve Prob. 16–38 using the method of instantaneous center of zero velocity.

16–50. Solve Prob. 16–35 using the method of instantaneous center of zero velocity.

16–50a. Solve Prob. 16–35*a* using the method of instantaneous center of zero velocity.

16–51. Solve Prob. 16–39 using the method of instantaneous center of zero velocity.

*** 16–52.** Solve Prob. 16–40 using the method of instantaneous center of zero velocity.

16–53. Solve Prob. 16–42 using the method of instantaneous center of zero velocity.

16–54. The instantaneous center of zero velocity for the body is located at point *IC* (0.5 m, 2 m). If the body has an angular velocity of 4 rad/s, as shown, determine the relative velocity of *B* with respect to *A*.

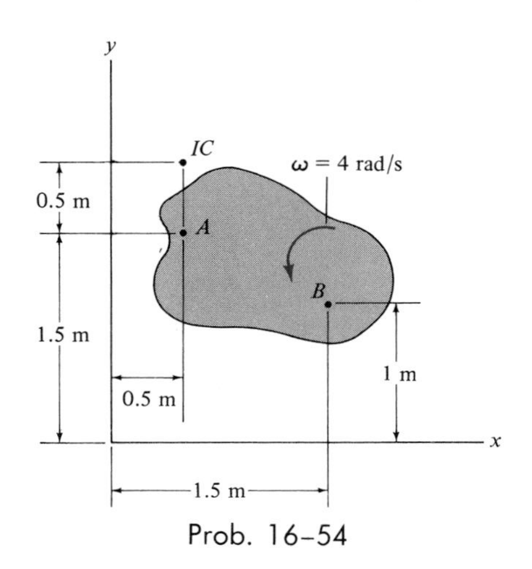

Prob. 16–54

16–55. The center of the wheel has a velocity of $v_C = 0.5$ m/s up the inclined plane and at the same instant it has a clockwise angular velocity of $\omega = 4$ rad/s which causes the wheel to slip at its contact point *A*. Determine the velocity of point *A* if $r = 0.4$ m.

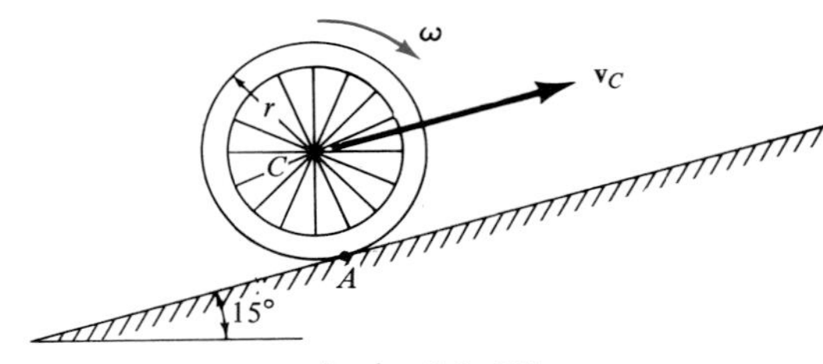

Prob. 16–55

16–55a. Solve Prob. 16–55 if $v_C = 2.5$ ft/s, $\omega = 5$ rad/s, and $r = 1.25$ ft.

*** 16–56.** Show that if the rim of the wheel and its hub maintain contact with the three tracks as the wheel rolls, it is necessary that slipping occurs at the hub *A* if no slipping occurs at *B*. Under these conditions, what is the speed at *A* if the wheel has an angular velocity **ω**?

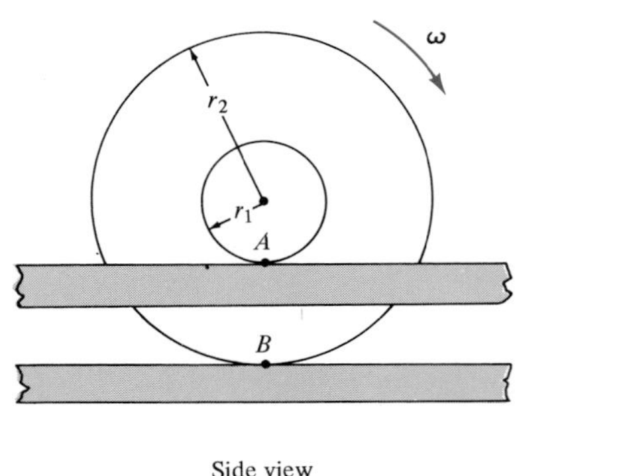

Prob. 16–56

16–57. The wheel rolls on its hub without slipping on the horizontal surface S. If the velocity of the center of the wheel is $v_C = 250$ mm/s to the right, determine the velocities of points A and B at the instant shown.

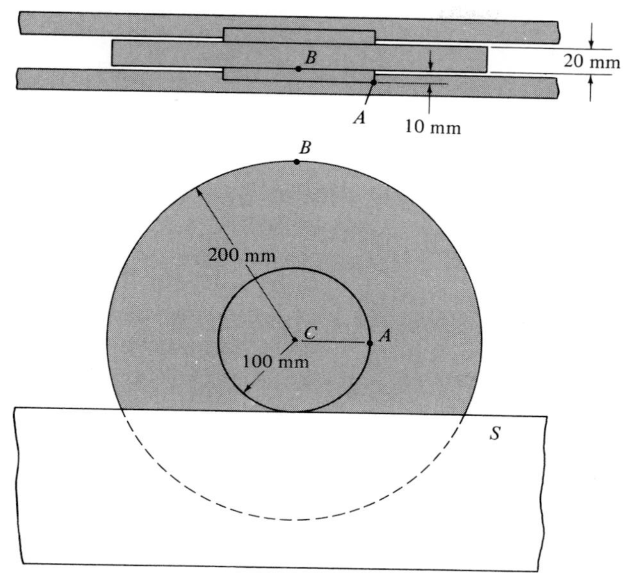

Prob. 16–57

16–58. Knowing that the angular velocity of link CD is $\omega_{CD} = 6$ rad/s, determine the velocity of point E on link BC and the angular velocity of link AB at the instant shown.

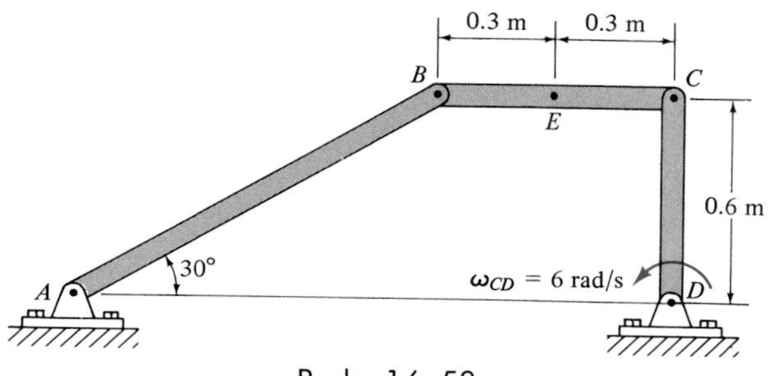

Prob. 16–58

16–59. If link BC rotates clockwise with an angular velocity of 12 rad/s, while the outer gear rack rotates counterclockwise with an angular velocity of 3 rad/s, determine the angular velocity of gear A.

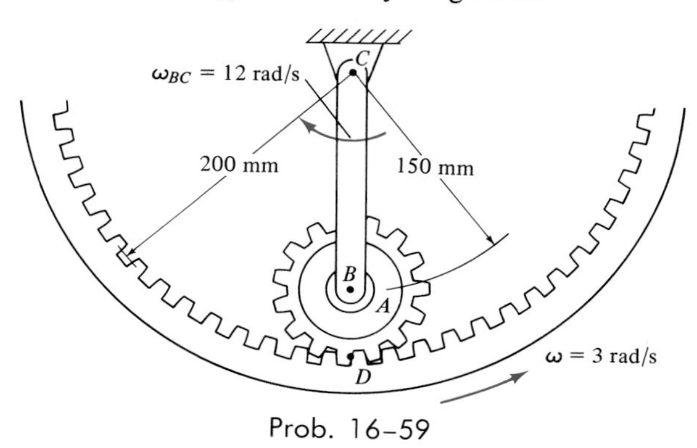

Prob. 16–59

*** 16–60.** If gear C rotates clockwise with an angular velocity of $\omega_C = 1.5$ rad/s, while the connecting link AB rotates counterclockwise at $\omega_{AB} = 6$ rad/s, determine the angular velocity of gear D. Set $r_C = 40$ mm and $r_D = 30$ mm.

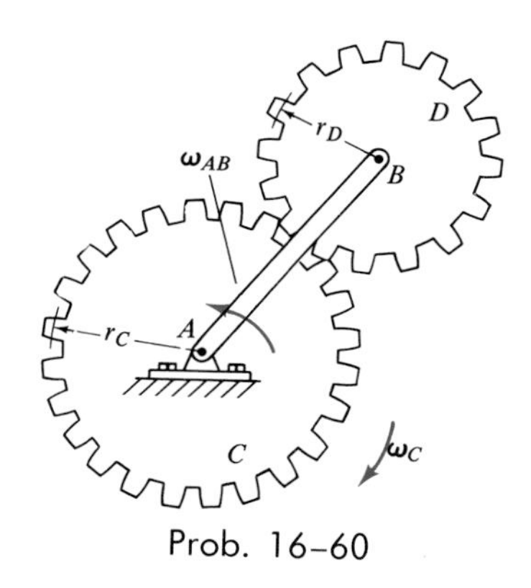

Prob. 16–60

*** 16–60a.** Solve Prob. 16–60 if $\omega_C = 4$ rad/s, $\omega_{AB} = 2.5$ rad/s, $r_C = 0.5$ ft, and $r_D = 0.25$ ft.

16-61. The mechanism shown is used in a riveting machine. It consists of a driving piston A, three links, and a riveter which is attached to the slider block D. Determine the velocity of D at the instant shown, when the piston at A is traveling at $v_A = 20$ m/s.

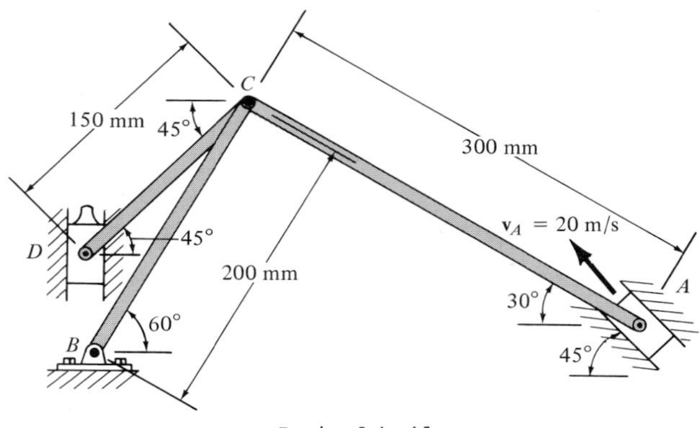

Prob. 16-61

16-8. Relative-General-Plane-Motion Analysis of a Rigid Body Using Translating Axes. Acceleration

An equation that relates the accelerations of two points in a rigid body subjected to general plane motion may be determined by differentiating Eq. 16-18 ($\mathbf{v}_B = \mathbf{v}_A + \mathbf{v}_{B/A}$) with respect to time. Thus,

$$\frac{d\mathbf{v}_B}{dt} = \frac{d\mathbf{v}_A}{dt} + \frac{d\mathbf{v}_{B/A}}{dt}$$

The terms $d\mathbf{v}_B/dt = \mathbf{a}_B$ and $d\mathbf{v}_A/dt = \mathbf{a}_A$ are measured from a set of *fixed x, y, z axes* and represent the *absolute accelerations* of points B and A, Fig. 16-21*a*. The time rate of change of the relative velocity $\mathbf{v}_{B/A}$ is denoted as the *relative acceleration,* $\mathbf{a}_{B/A}$. This vector represents the acceleration of B measured by an observer stationed at A and fixed to a set of *translating axes*. Since the body is rigid, to this observer point B appears to move along the dashed *circular arc* that has a radius of curvature of $r_{B/A}$, Fig. 16-21*c*. In other words, the body moves as if it were *pinned* at A. Consequently, $\mathbf{a}_{B/A}$ can be expressed in terms of its tangential and normal components of motion, i.e., $\mathbf{a}_{B/A} = (\mathbf{a}_{B/A})_t + (\mathbf{a}_{B/A})_n$, where $(a_{B/A})_t = \alpha r_{B/A}$ (Eq. 16-14) and $(a_{B/A})_n = \omega^2 r_{B/A}$ (Eq. 16-15). Hence, the above equation can be written in the form

$$\mathbf{a}_B = \mathbf{a}_A + (\mathbf{a}_{B/A})_t + (\mathbf{a}_{B/A})_n \qquad (16\text{-}21)$$

where

$\mathbf{a}_B$ = absolute acceleration of point B
$\mathbf{a}_A$ = absolute acceleration of point A
$(\mathbf{a}_{B/A})_t$ = relative tangential acceleration component of "B with respect to

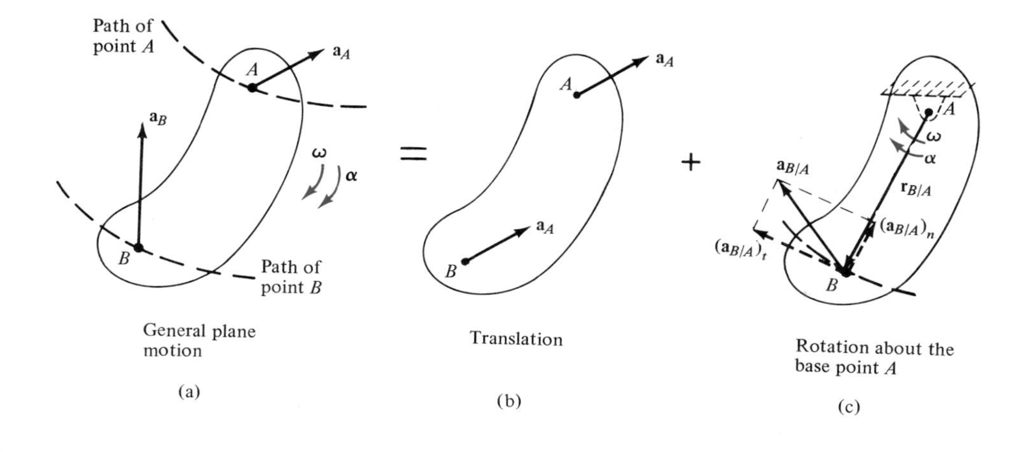

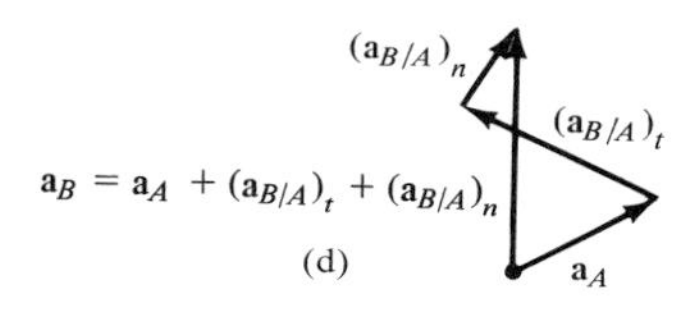

Fig. 16–21

A;" the *magnitude* is $(a_{B/A})_t = \alpha r_{B/A}$ and the *direction* is perpendicular to the line of action of $\mathbf{r}_{B/A}$

$(\mathbf{a}_{B/A})_n$ = relative normal acceleration component of "B with respect to A;" the *magnitude* is $(a_{B/A})_n = \omega^2 r_{B/A}$ and the *direction* is always from B toward A

$\boldsymbol{\alpha}$ = absolute angular acceleration of the body

$\boldsymbol{\omega}$ = absolute angular velocity of the body

$\mathbf{r}_{B/A}$ = relative-position vector drawn from B to A

Each of the four terms in Eq. 16–21 is represented graphically on the *kinematic diagrams* shown in Fig. 16–21. Here it is seen that at a given instant, the acceleration of B, Fig. 16–21*a*, is determined by considering the body to translate with an acceleration $\mathbf{a}_A$, Fig. 16–21*b*, and rotate about the base point A with an instantaneous angular velocity $\boldsymbol{\omega}$ and angular acceleration $\boldsymbol{\alpha}$, Fig. 16–21*c*. Vector addition of these two effects, applied to B, yields $\mathbf{a}_B$, as shown in Fig. 16–21*d*. It should be noted from Fig. 16–21*a* that points A and B are both shown to move along *curved paths* and as a result, the accelerations of these points have *both tangential and normal components*. Recall that the acceleration of a point is *tangent to the path* only when the path is *rectilinear*.

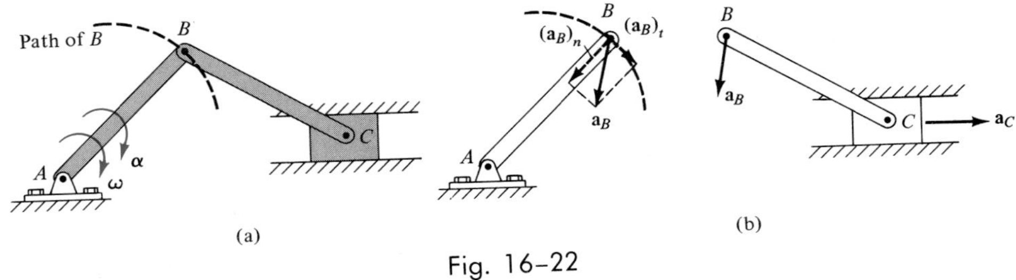

Fig. 16–22

If Eq. 16–21 is applied in a practical manner to study the accelerated motion of a rigid body which is pin-connected to two other bodies, it should be realized that points which are *coincident at the pin* move with the *same acceleration,* since the path of motion over which they travel is the *same*. For example, point B lying on either rod AB or BC of the crank mechanism shown in Fig. 16–22a has the same acceleration, since the rods are pin-connected at B. Here the motion of B is along a *curved path,* so that $\mathbf{a}_B$ is calculated on the basis of its tangential and normal components $(a_B)_t = \alpha r_{B/A}$ and $(a_B)_n = \omega^2 r_{B/A}$, which are defined by the angular motion of AB. At the other end of rod BC, however, point C moves along a *rectilinear path,* which is defined by the piston. Hence, in this case, the acceleration $\mathbf{a}_C$ is directed along the path, Fig. 16–22b.

If two bodies contact one another *without slipping,* and the *points in contact* move along *different paths,* the *tangential components* of acceleration of the points will be the *same;* however, the *normal components* will *not* be the same. For example, consider the two meshed gears in Fig. 16–23a. Point A is located on gear B and A' is located on gear C. Due to the rotational motion, $(\mathbf{a}_A)_t = (\mathbf{a}_{A'})_t$; however, since both points follow different curved paths, $(\mathbf{a}_A)_n \neq (\mathbf{a}_{A'})_n$ and therefore $\mathbf{a}_A \neq \mathbf{a}_{A'}$, Fig. 16–23$b$.

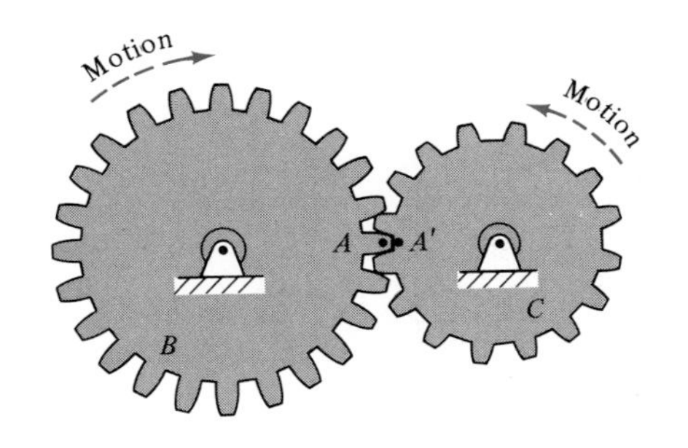

(a)

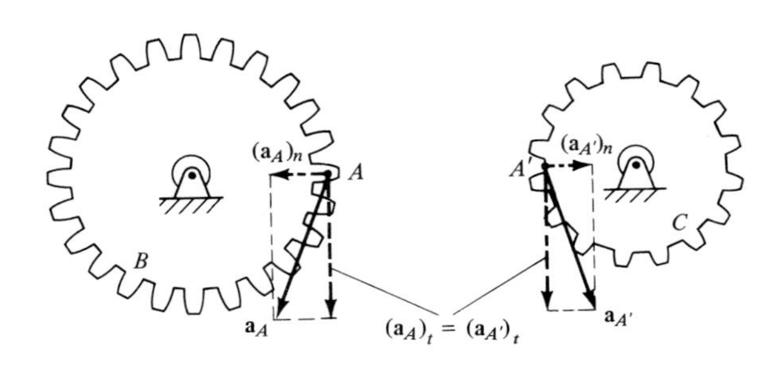

(b)

Fig. 16–23

PROCEDURE FOR ANALYSIS

The acceleration equation $\mathbf{a}_B = \mathbf{a}_A + (\mathbf{a}_{B/A})_t + (\mathbf{a}_{B/A})_n$ applies between any two points located on the *same* rigid body. For application, the following four-step procedure should be used.

Step 1: If the angular velocity $\boldsymbol{\omega}$ of the body is unknown, determine it by using a velocity analysis as discussed in Sec. 16–5 or 16–7. Determine also the velocities $\mathbf{v}_A$ and $\mathbf{v}_B$ of points A and B *if these points move along curved paths.*

Step 2: Draw a kinematic diagram of the body such as shown in Fig. 16–21*a*; and indicate on it the absolute accelerations of points A and B, $\mathbf{a}_A$ and $\mathbf{a}_B$, the angular velocity $\boldsymbol{\omega}$, and the angular acceleration $\boldsymbol{\alpha}$. Usually the base point A is selected as a point having a known acceleration. In particular, if points A and B move along *curved paths,* their accelerations should be expressed in terms of their tangential and normal components of motion, i.e., $\mathbf{a}_A = (\mathbf{a}_A)_t + (\mathbf{a}_A)_n$ and $\mathbf{a}_B = (\mathbf{a}_B)_t + (\mathbf{a}_B)_n$. Here $(a_A)_n = (v_A)^2/\rho_A$ and $(a_B)_n = (v_B)^2/\rho_B$, where ρ_A and ρ_B define the radii of curvature of the paths of points A and B, respectively (Eq. 12–54).

Step 3: Draw a kinematic diagram of the body such as shown in Fig. 16–21*c* and indicate on it the two components of the relative acceleration, $(\mathbf{a}_{B/A})_t$ and $(\mathbf{a}_{B/A})_n$. Since the body is considered to be pinned at the base point A, these components have a *magnitude* of $(a_{B/A})_t = \alpha r_{B/A}$ and $(a_{B/A})_n = \omega^2 r_{B/A}$; and their *directions* are established from the diagram such that $(\mathbf{a}_{B/A})_t$ acts perpendicular to $\mathbf{r}_{B/A}$, in accordance with the rotational motion $\boldsymbol{\alpha}$ of the body, and $(\mathbf{a}_{B/A})_n$ is directed from B toward A.*

Step 4: Write Eq. 16–22 in symbolic form: $\mathbf{a}_B = \mathbf{a}_A + (\mathbf{a}_{B/A})_t + (\mathbf{a}_{B/A})_n$, and underneath each of the terms, represent the respective vectors by their magnitudes and directions. To do this use the data shown on the kinematic diagrams in *Steps 2* and *3.* Since motion occurs in the plane, this "vector" equation can be expressed in terms of two scalar equations which represent the addition of vector components in two directions. Once established, these two equations provide a solution for at most two unknown scalar quantities.

The following example problems numerically illustrate this scalar method of application.

*Perhaps the notation $\mathbf{a}_B = \mathbf{a}_A + (\mathbf{a}_{B/A(\text{pin})})_t + (\mathbf{a}_{B/A(\text{pin})})_n$ helps in recalling that A is pinned.

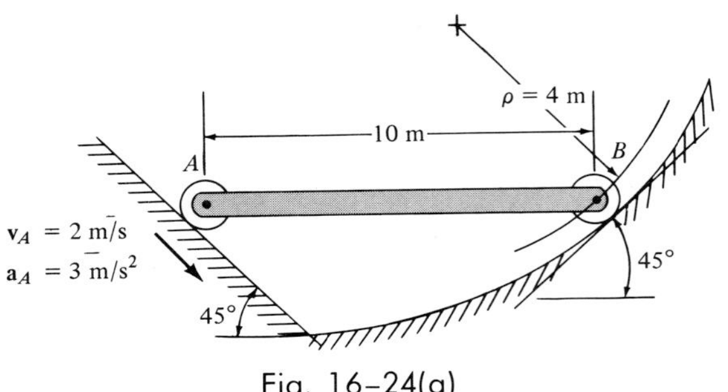

Fig. 16–24(a)

Example 16–12

The rod AB shown in Fig. 16–24a is confined to move along the inclined plane at A and the circular arc at B. If point A has an acceleration of 3 m/s^2 and a velocity of 2 m/s, both directed down the plane at the instant the rod becomes horizontal, determine the angular acceleration of the rod at this instant.

Solution

Step 1: Since the velocity of A is known and the velocity of B is tangent to the curved path of motion at B, the instantaneous center of zero velocity for the rod is located as shown in Fig. 16–24b. From the geometry,

$$r_{A/IC} = r_{B/IC} = 10 \cos 45° = 07.07 \text{ m}$$

Thus,

$$\omega = \frac{v_A}{r_{A/IC}} = \frac{2 \text{ m/s}}{07.07 \text{ m}} = 0.283 \text{ rad/s}$$

Also, since point B follows a *curved path,*

$$v_B = \omega(r_{B/IC}) = 2.83 \text{ rad/s}(0.707 \text{ m}) = 2 \text{ m/s}$$

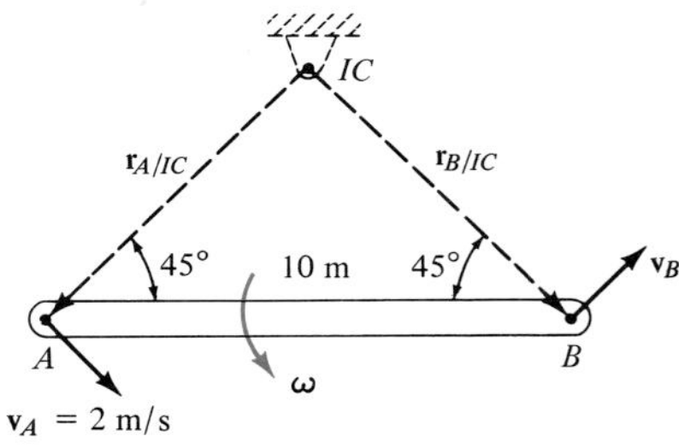

Fig. 16–24(b)

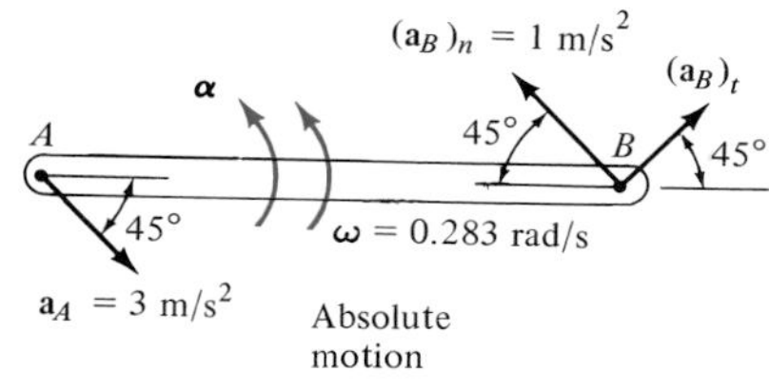

(c)

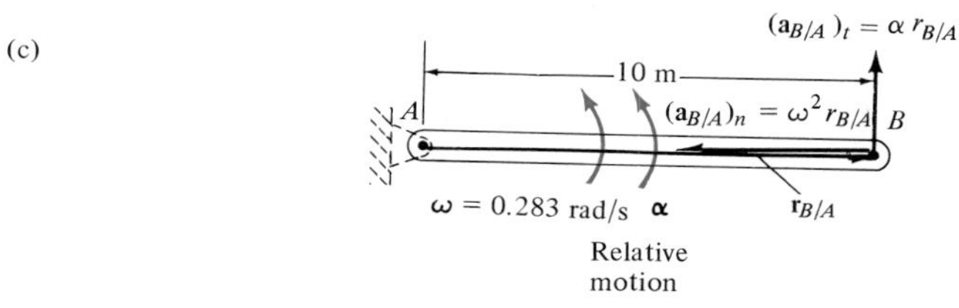

(d)

Fig. 16–24(c,d)

Step 2: The accelerated motion of points A and B is shown on the kinematic diagram in Fig. 16–24*c*. Since B moves along a *curved path,* the acceleration of B is represented in terms of its tangential and normal components. The magnitude of the normal component is determined from Eq. 12–54, i.e.,

$$(a_B)_n = \frac{v_B^2}{\rho_B} = \frac{(2\ \text{m/s})^2}{4\ \text{m}} = 1\ \text{m/s}^2$$

There are two unknowns in Fig. 16–24*c*, i.e., $(a_B)_t$ and α. Point A will be chosen as the base point, so that $\mathbf{a}_B = \mathbf{a}_A + \mathbf{a}_{B/A}$.
Step 3: The relative accelerated motion of B with respect to A is shown on the kinematic diagram in Fig. 16–24*d*.
Step 4: Applying Eq. 16–21 to points A and B, we have

$$(\mathbf{a}_B)_t + (\mathbf{a}_B)_n = \mathbf{a}_A + (\mathbf{a}_{B/A})_t + (\mathbf{a}_{B/A})_n$$

$$\underset{\measuredangle 45^\circ}{(a_B)_t} + \underset{45^\circ\nwarrow}{1\ \text{m/s}^2} = \underset{\searrow 45^\circ}{3\ \text{m/s}^2} + \underset{\uparrow}{\alpha(10\ \text{m})} + \underset{\leftarrow}{(0.283\ \text{rad/s})^2(10\ \text{m})}$$

Equating the horizontal and vertical components yields

$$(\overset{+}{\rightarrow}) \qquad (a_B)_t \cos 45^\circ - 1 \cos 45^\circ = 3 \cos 45^\circ + 0 - (0.283)^2(10)$$

Hence,

$$(a_B)_t = -2.87\ \text{m/s}^2$$

$$(+\uparrow) \qquad (a_B)_t \sin 45^\circ + 1 \sin 45^\circ = -3 \sin 45^\circ + \alpha(10) + 0$$

Substituting for $(a_B)_t$ and solving for α, we have $\alpha = -2.35\ \text{rad/s}^2$ so that

$$\alpha = 0.486\ \text{rad/s}^2 \ \circlearrowright \qquad \textit{Ans.}$$

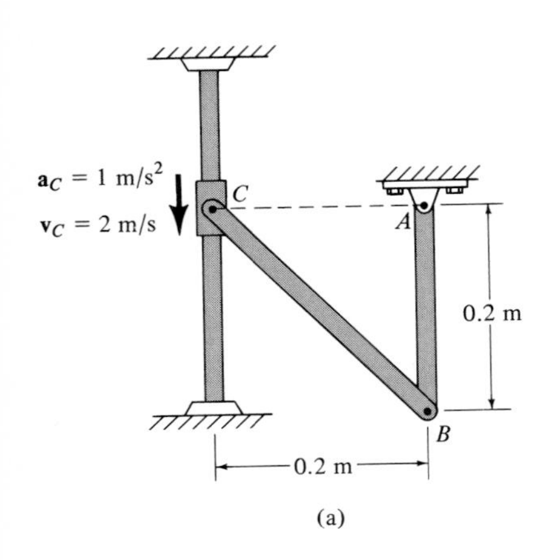

(a)

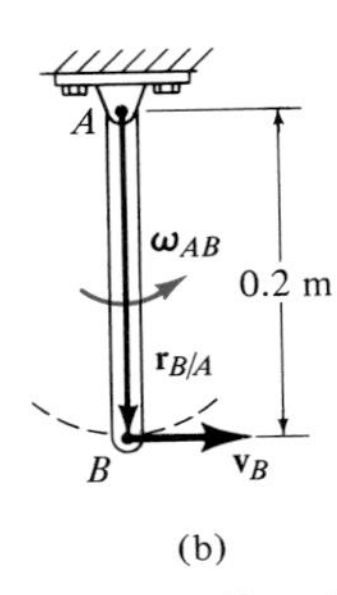

(b)

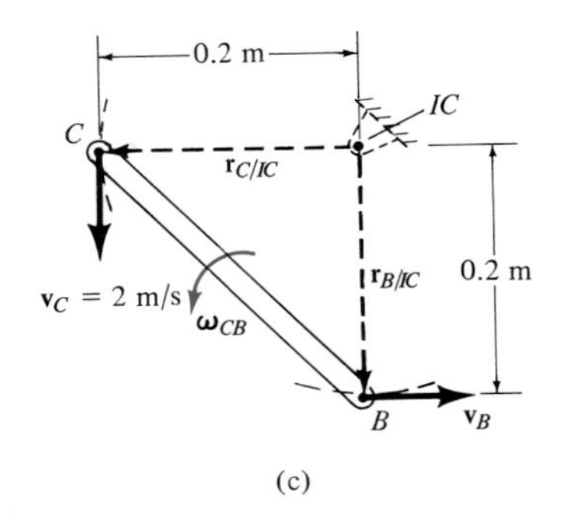

(c)

Fig. 16–25(a–c)

Example 16–13

The collar C in Fig. 16–25*a* is moving downward with an acceleration of 1 m/s², and at the instant shown it has a speed of 2 m/s. Determine the angular acceleration of links CB and AB at this instant.

Solution

Step 1: The velocity of point C is directed downward, and due to the rotational motion of link AB about the fixed point A, the velocity of B is directed horizontally (to the right), Fig. 16–25*b*. Knowing this, the instantaneous center of zero velocity for link CB is located as shown in Fig. 16–25*c*. From the figure, $r_{C/IC} = r_{B/IC} = 0.2$ m and ω_{CB} is assumed to act counterclockwise. Hence,

$$v_C = \omega_{CB}(r_{C/IC})$$
$$2 \text{ m/s} = \omega_{CB}(0.2 \text{ m})$$
$$\omega_{CB} = 10 \text{ rad/s}$$

Computing also the velocity of point B since this point moves along a curved path,

$$v_B = \omega_{CB}(r_{B/IC}) = 10 \text{ rad/s}(0.2 \text{ m}) = 2 \text{ m/s}$$

Furthermore, from Fig. 16–25*b*,

$$\omega_{AB} = \frac{v_B}{r_{B/A}} = \frac{2 \text{ m/s}}{0.2 \text{ m}} = 10 \text{ rad/s}$$

Step 2: As shown on the kinematic diagram in Fig. 16–25*d*, the acceleration of point C is directed downward since it moves along a straight-line path. This point will be chosen as the base point since $\mathbf{a}_C$ is known. Point B moves along a *curved path* and therefore the acceleration $\mathbf{a}_B$ is represented by its normal and tangential components. In particular, the normal component of acceleration, Fig. 16–25*f*, can be computed using either Eq. 12–54 or 16–15, i.e.,

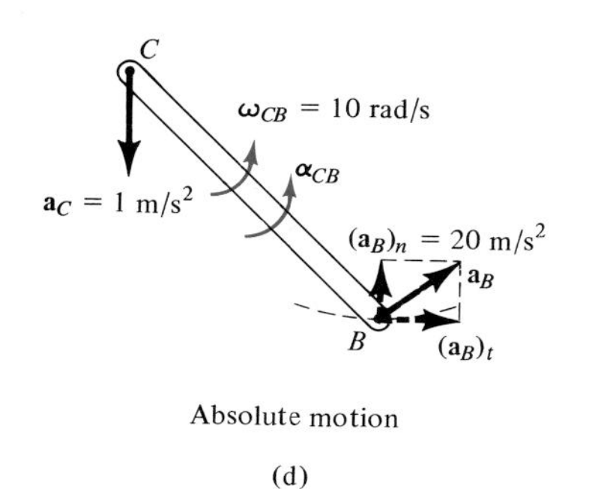

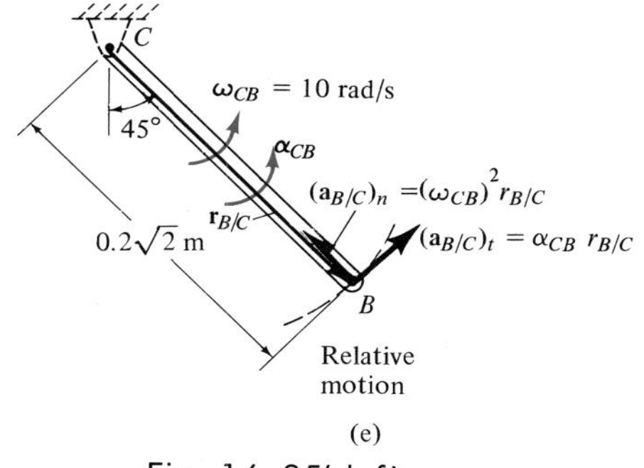

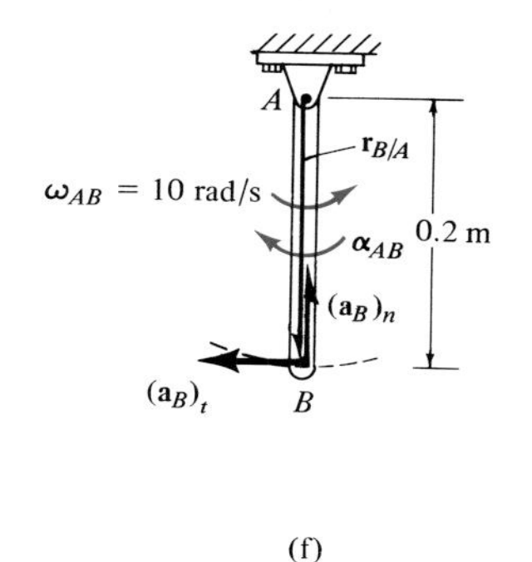

Fig. 16–25(d–f)

$$(a_B)_n = \frac{(v_B)^2}{r_{B/A}} = (\omega_{AB})^2(r_{B/A}) = (10 \text{ rad/s})^2(0.2 \text{ m}) = 20 \text{ m/s}^2$$

There are two unknowns shown on the kinematic diagram in Fig. 16–25*d*, α_{CB} and $(a_B)_t$.

Step 3: The relative accelerated motion of B with respect to C is shown on the kinematic diagram in Fig. 16–25*e*.

Step 4: Applying Eq. 16–21 to points B and C, we have

$$(\mathbf{a}_B)_t + (\mathbf{a}_B)_n = \mathbf{a}_C + (\mathbf{a}_{B/C})_t + (\mathbf{a}_{B/C})_n$$

$$\underset{\rightarrow}{(a_B)_t} + \underset{\uparrow}{20 \text{ m/s}^2} = \underset{\downarrow}{1 \text{ m/s}^2} + \underset{\measuredangle 45^\circ}{\alpha_{CB}(0.2\sqrt{2} \text{ m})} + \underset{45^\circ \searrow}{(10 \text{ rad/s})^2(0.2\sqrt{2} \text{ m})}$$

Equating components in the horizontal and vertical directions, gives

$$(\overset{+}{\rightarrow}) \quad (a_B)_t + 0 = 0 + \alpha_{CB}(0.2\sqrt{2}) \cos 45^\circ - (10)^2(0.2\sqrt{2}) \cos 45^\circ$$

or

$$(a_B)_t = 0.2\alpha_{CB} - 20$$

$$(+\uparrow) \quad 0 + 20 = -1 + \alpha_{CB}(0.2\sqrt{2}) \sin 45^\circ + (10)^2(0.2\sqrt{2}) \sin 45^\circ$$

or

$$1 = 0.2\alpha_{CB}$$

Solving,

$$\alpha_{CB} = 5 \text{ rad/s}^2$$

$$(a_B)_t = -19 \text{ m/s}^2$$

The negative sign indicates that $(\mathbf{a}_B)_t$ acts in the opposite direction to that shown in Fig. 16–25*d*.

A kinematic diagram of link AB is shown in Fig. 16–25*f*, where $(\mathbf{a}_B)_t$ acts to the left and consequently $\boldsymbol{\alpha}_{AB}$ is clockwise. Hence,

$$\alpha_{AB} = \frac{(a_B)_t}{r_{B/A}} = \frac{19}{0.2} = 95 \text{ rad/s}^2 \qquad \textit{Ans.}$$

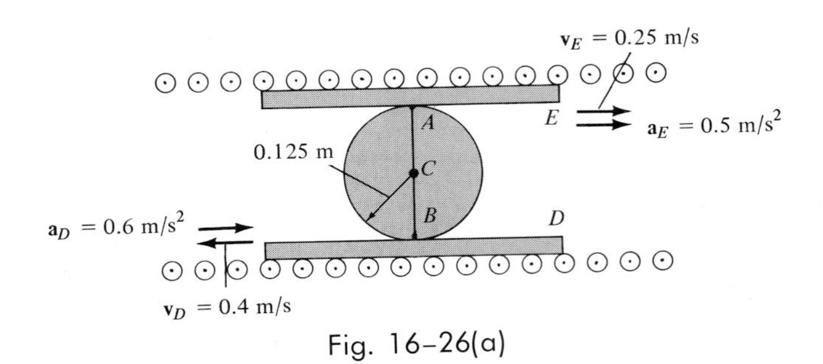

Fig. 16–26(a)

Example 16–14

The cylinder shown in Fig. 16–26*a* rolls without slipping between the two moving plates *E* and *D*. Determine the angular acceleration of the cylinder and the magnitude of acceleration of point *B*, at the instant shown.

Solution

Step 1: Since no slipping occurs, the contact points *B* and *A* on the cylinder have the same velocity as the plates *E* and *D*. Furthermore, the velocities $\mathbf{v}_A$ and $\mathbf{v}_B$ are *parallel,* so that by the proportionality of right triangles, the *IC* is located at a point on line *AB*, Fig. 16–26*b*. Assuming this point to be a distance *x* from *B*, we have

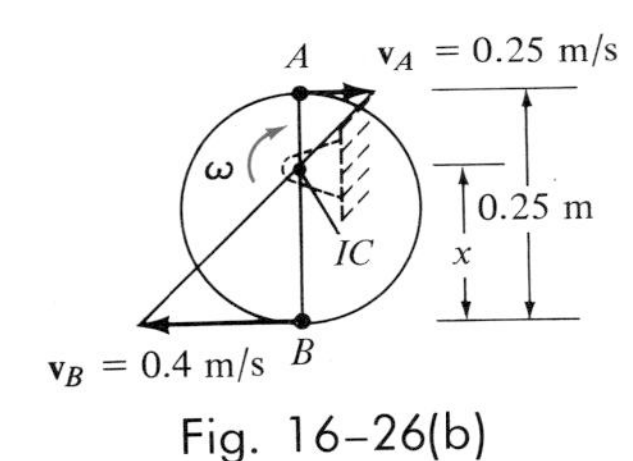

Fig. 16–26(b)

$$v_B = \omega x; \qquad 0.4 = \omega x$$

$$v_A = \omega(0.25 - x); \qquad 0.25 = \omega(0.25 - x)$$

Dividing one of these equations into the other eliminates ω and yields

$$0.4(0.25 - x) = 0.25x$$

$$x = \frac{0.1}{0.65} = 0.154 \text{ m}$$

Hence, the angular velocity is

$$\omega = \frac{v_B}{x} = \frac{0.4}{0.154} = 2.60 \text{ rad/s}$$

Step 2: Since both points *A* and *B* move along *curved paths* they have normal and tangential components of motion as shown on the kinematic diagram in Fig. 16–26*c*. The tangential components are specified by the deceleration of plate *D* and the acceleration of plate *E*; whereas the normal components have unknown magnitudes since the radii of curvature of the paths at *A* and *B* are *unknown.** Point *A* will be chosen as the base point so that $\mathbf{a}_B = \mathbf{a}_A + \mathbf{a}_{B/A}$.

*These paths do *not* have a radius of curvature defined by the radius of the cylinder, since the cylinder is *not* rotating about point *C*. Furthermore, the radii of curvature are *not* defined from the *IC*, since the location of the *IC* depends only on the velocity of points *A* and *B* and *not* the geometry of their paths.

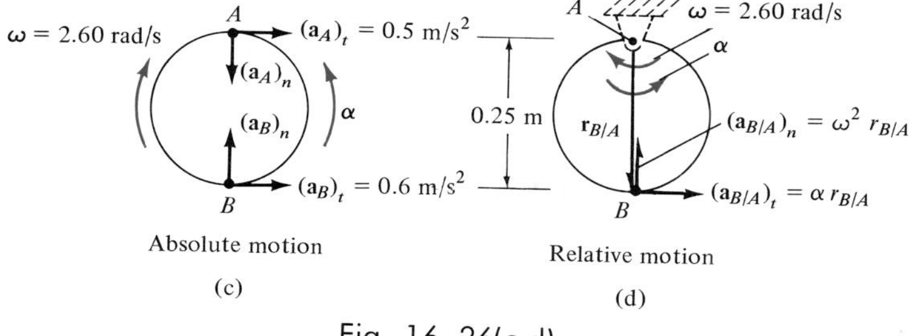

Fig. 16–26(c,d)

Step 3: The relative accelerated motion of B with respect to A is shown on the kinematic diagram in Fig. 16–25d.
Step 4: Applying Eq. 16–21 to points A and B, we have

$$\mathbf{(a_B)}_t + \mathbf{(a_B)}_n = \mathbf{(a_A)}_t + \mathbf{(a_A)}_n + \mathbf{(a_{B/A})}_t + \mathbf{(a_{B/A})}_n$$

$$\underset{\rightarrow}{0.6 \text{ m/s}^2} + \underset{\uparrow}{(a_B)_n} = \underset{\rightarrow}{0.5 \text{ m/s}^2} + \underset{\downarrow}{(a_A)_n} + \underset{\rightarrow}{\alpha(0.25 \text{ m})} + \underset{\uparrow}{(2.60 \text{ rad/s})^2(0.25 \text{ m})}$$

A solution for α can be obtained from the horizontal components, i.e.,

$$(\overset{+}{\rightarrow}) \qquad 0.6 + 0 = 0.5 + 0 + \alpha(0.25) + 0$$

$$\alpha = 0.40 \text{ rad/s}^2 \circlearrowright \qquad \textit{Ans.}$$

In order to obtain $(a_B)_n$ it is necessary to apply the acceleration equation between B and some other point. The center point C will be chosen since it moves with rectilinear translation; that is, the path is known. The two necessary kinematic diagrams, where C is the base point, are shown in Figs. 16–26e and 16–26f. There are two unknowns, a_C and $(a_B)_n$. Hence,

$$\mathbf{(a_B)}_t + \mathbf{(a_B)}_n = \mathbf{a}_C + \mathbf{(a_{B/C})}_t + \mathbf{(a_{B/C})}_n$$

$$\underset{\rightarrow}{0.6 \text{ m/s}^2} + \underset{\uparrow}{(a_B)_n} = \underset{\rightarrow}{a_C} + \underset{\rightarrow}{(0.40 \text{ rad/s}^2)(0.125 \text{ m})} + \underset{\uparrow}{(2.60)^2(0.125 \text{ m})}$$

Equating the vertical components yields

$$(+\uparrow) \qquad 0 + (a_B)_n = 0 + 0 + (2.60)^2(0.125)$$

$$(a_B)_n = 0.845 \text{ m/s}^2$$

Hence,

$$a_B = \sqrt{(a_B)_n^2 + (a_B)_t^2} = \sqrt{(0.845)^2 + (0.6)^2} = 1.04 \text{ m/s}^2 \qquad \textit{Ans.}$$

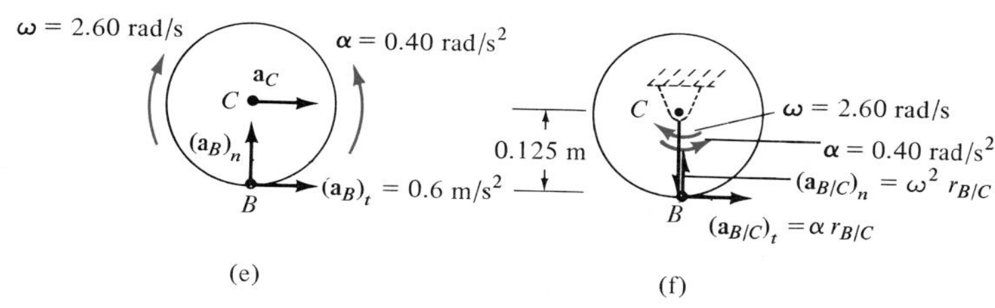

Fig. 16–26(e,f)

16–9. Method for Determining Acceleration Using Cartesian Vectors

In the previous section, the acceleration equation $\mathbf{a}_B = \mathbf{a}_A + (\mathbf{a}_{B/A})_t + (\mathbf{a}_{B/A})_n$ was applied using a scalar method for solution. It is also possible to apply this equation using a Cartesian vector analysis.

Since the relative-acceleration components represent the effect of *circular motion* observed from translating axes passing through the base point A, these terms can be expressed by the cross products $(\mathbf{a}_{B/A})_t = \boldsymbol{\alpha} \times \mathbf{r}_{B/A}$ and $(\mathbf{a}_{B/A})_n = \boldsymbol{\omega} \times (\boldsymbol{\omega} \times \mathbf{r}_{B/A})$ (Eq. 16–16). Since $\boldsymbol{\omega}$ is directed perpendicular to the plane of motion, $(\mathbf{a}_{B/A})_n$ *always* acts from B toward A, Fig. 16–21*c*. Noting that this *direction* is the same as $-\mathbf{r}_{B/A}$, and that the *magnitude* is $(a_{B/A})_n = \omega^2 r_{B/A}$, we can express $(\mathbf{a}_{B/A})_n$ in a much simpler vector form as $(\mathbf{a}_{B/A})_n = -\omega^2 \mathbf{r}_{B/A}$. Hence, for Cartesian vector application, Eq. 16–21 becomes

$$\mathbf{a}_B = \mathbf{a}_A + \boldsymbol{\alpha} \times \mathbf{r}_{B/A} - \omega^2 \mathbf{r}_{B/A} \tag{16–22}$$

where

$\mathbf{a}_B$ = absolute acceleration of point B

$\mathbf{a}_A$ = absolute acceleration of the base point A

$\boldsymbol{\omega}$ = absolute angular velocity of the body

$\boldsymbol{\alpha}$ = absolute angular acceleration of the body

$\mathbf{r}_{B/A}$ = relative-position vector drawn from A to B

Applying Eq. 16–22 in Cartesian vector form is very methodical and thus it has an advantage for solving problems involving mechanisms having *several connected members*. When applying this method, however, the physical aspects of the kinematics are often lost. Furthermore, because of the additional vector algebra, which must be included in the solution, in general more computations than the scalar method must be made to obtain a complete solution.

PROCEDURE FOR ANALYSIS

The following two-step procedure should be used when applying Eq. 16–22 to two points A and B located on the *same* rigid body.

Step 1: Draw a kinematic diagram of the body such as shown in Fig. 16–21*a* and indicate on it the absolute accelerations of points A and B, $\mathbf{a}_A$ and $\mathbf{a}_B$, if they are specified, the angular acceleration $\boldsymbol{\alpha}$, the angular velocity $\boldsymbol{\omega}$, and the relative-position vector $\mathbf{r}_{B/A}$. If $\boldsymbol{\omega}$ is to be determined, also indicate the absolute velocities $\mathbf{v}_A$ and $\mathbf{v}_B$ of points A and B on the diagram.

Step 2: Establish the fixed x, y, z axes and express each of the vectors on the kinematic diagram in terms of their **i**, **j**, **k** components. If $\boldsymbol{\omega}$ is to be determined use Eq. 16–19, $\mathbf{v}_B = \mathbf{v}_A + (\boldsymbol{\omega} \times \mathbf{r}_{B/A})$. Afterward, apply the acceleration equation, Eq. 16–22. In each case evaluate the cross product and equate the respective **i** and **j** components. This yields two scalar equations which may be solved for at most two unknowns.

The following two examples numerically illustrate this procedure.

Example 16–15

The collar C in Fig. 16–27a is moving downward with an acceleration of 1 m/s², and at the instant shown, it has a speed of 2 m/s. Determine the angular acceleration of links CB and AB at this instant.

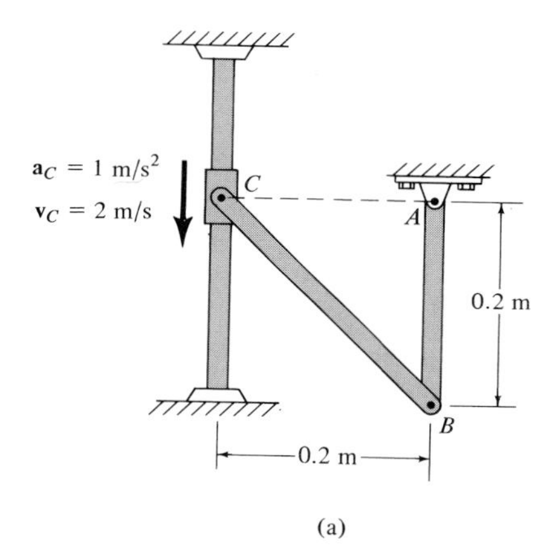

(a)

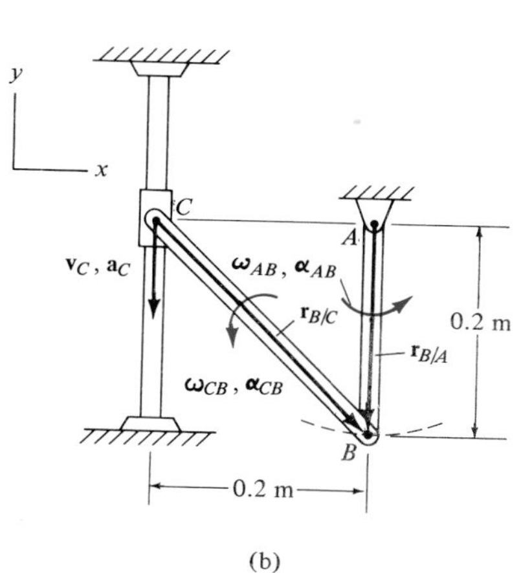

(b)

Fig. 16–27

Solution

Step 1: The kinematic diagrams of both links AB and CB are shown in Fig. 16–27b.

Step 2: Expressing each of the vectors in Cartesian vector form, we have

$$\mathbf{r}_{B/A} = \{-0.2\mathbf{j}\} \text{ m} \qquad \mathbf{r}_{B/C} = \{0.2\mathbf{i} - 0.2\mathbf{j}\} \text{ m}$$
$$\boldsymbol{\omega}_{AB} = \omega_{AB}\mathbf{k} \qquad \boldsymbol{\alpha}_{AB} = \alpha_{AB}\mathbf{k}$$
$$\boldsymbol{\omega}_{CB} = \omega_{CB}\mathbf{k} \qquad \boldsymbol{\alpha}_{CB} = \alpha_{CB}\mathbf{k}$$
$$\mathbf{v}_C = \{-2\mathbf{j}\} \text{ m/s} \qquad \mathbf{a}_C = \{-1\mathbf{j}\} \text{ m/s}^2$$

Velocity Analysis. Since AB is subjected to rotational motion about a *fixed axis* passing through point A, using Eq. 16–13 we have

$$\mathbf{v}_B = \boldsymbol{\omega}_{AB} \times \mathbf{r}_{B/A}$$
$$\mathbf{v}_B = (\omega_{AB}\mathbf{k}) \times (-0.2\mathbf{j}) = 0.2\omega_{AB}\mathbf{i}$$

Link CB undergoes general plane motion; thus, using Eq. 16–19 with C as the base point, we have

$$\mathbf{v}_B = \mathbf{v}_C + \boldsymbol{\omega}_{CB} \times \mathbf{r}_{B/C}$$
$$0.2\omega_{AB}\mathbf{i} = -2\mathbf{j} + (\omega_{CB}\mathbf{k}) \times (0.2\mathbf{i} - 0.2\mathbf{j})$$
$$0.2\omega_{AB}\mathbf{i} = -2\mathbf{j} + 0.2\omega_{CB}\mathbf{j} + 0.2\omega_{CB}\mathbf{i}$$

Equating the respective **i** and **j** components gives

$$0.2\omega_{AB} = 0.2\omega_{CB}$$
$$0 = -2 + 0.2\omega_{CB}$$

Thus,

$$\omega_{AB} = \omega_{CB} = 10 \text{ rad/s}$$

Acceleration Analysis. Applying Eq. 16–16 to points A and B on link AB, we have

$$\mathbf{a}_B = \boldsymbol{\alpha}_{AB} \times \mathbf{r}_{B/A} - \omega_{AB}^2 \mathbf{r}_{B/A}$$
$$\mathbf{a}_B = (\alpha_{AB}\mathbf{k}) \times (-0.2\mathbf{j}) - (10)^2(-0.2\mathbf{j})$$
$$\mathbf{a}_B = 0.2\alpha_{AB}\mathbf{i} + 20\mathbf{j}$$

Applying Eq. 16–22 to points B and C on rod BC, using the above value for $\mathbf{a}_B$, yields

$$\mathbf{a}_B = \mathbf{a}_C + \boldsymbol{\alpha}_{CB} \times \mathbf{r}_{B/C} - \omega_{CB}^2 \mathbf{r}_{B/C}$$
$$0.2\alpha_{AB}\mathbf{i} + 20\mathbf{j} = -1\mathbf{j} + (\alpha_{CB}\mathbf{k}) \times (0.2\mathbf{i} - 0.2\mathbf{j}) - (10)^2(0.2\mathbf{i} - 0.2\mathbf{j})$$
$$0.2\alpha_{AB}\mathbf{i} + 20\mathbf{j} = -1\mathbf{j} + 0.2\alpha_{CB}\mathbf{j} + 0.2\alpha_{CB}\mathbf{i} - 20\mathbf{i} + 20\mathbf{j}$$

Equating the respective $\mathbf{i}$ and $\mathbf{j}$ components,

$$0.2\alpha_{AB} = 0.2\alpha_{CB} - 20 \qquad (1)$$
$$20 = -1 + 0.2\alpha_{CB} + 20 \qquad (2)$$

Hence,

$$\alpha_{CB} = 5 \text{ rad/s}^2$$
$$\alpha_{AB} = -95 \text{ rad/s}^2$$

Thus,

$$\boldsymbol{\alpha}_{CB} = \{5\mathbf{k}\} \text{ rad/s}^2 \qquad \textit{Ans.}$$
$$\boldsymbol{\alpha}_{AB} = \{-95\mathbf{k}\} \text{ rad/s}^2 \qquad \textit{Ans.}$$

Example 16–16

The crankshaft AB of an engine turns with a clockwise angular acceleration of 20 rad/s², Fig. 16–28*a*. Determine the acceleration of the piston at the instant when AB is in the position shown and has a clockwise angular velocity of 10 rad/s.

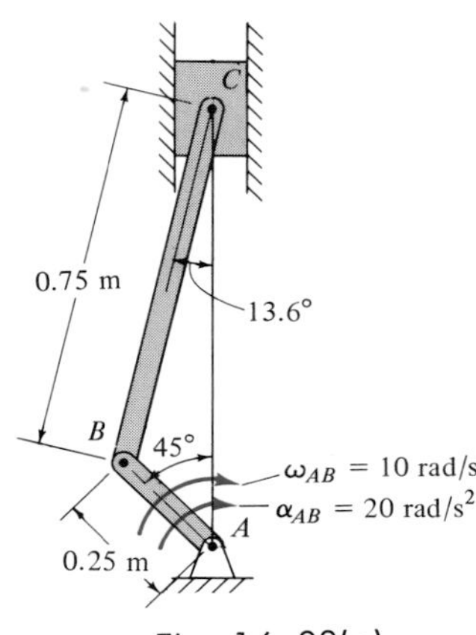

Fig. 16–28(a)

Solution

Step 1: The kinematic diagrams of AB and BC are shown in Fig. 16–28*b*.
Step 2: Expressing each of the vectors in Cartesian vector form yields

$$\mathbf{r}_{B/A} = \{-0.25 \sin 45°\mathbf{i} + 0.25 \cos 45°\mathbf{j}\} \text{ m} = \{-0.177\mathbf{i} + 0.177\mathbf{j}\} \text{ m}$$
$$\mathbf{r}_{C/B} = \{0.75 \sin 13.6°\mathbf{i} + 0.75 \cos 13.6°\mathbf{j}\} \text{ m} = \{0.176\mathbf{i} + 0.729\mathbf{j}\} \text{ m}$$
$$\boldsymbol{\omega}_{AB} = \{-10\mathbf{k}\} \text{ rad/s} \qquad \boldsymbol{\alpha}_{AB} = \{-20\mathbf{k}\} \text{ rad/s}^2$$
$$\boldsymbol{\omega}_{BC} = \omega_{BC}\mathbf{k} \qquad \boldsymbol{\alpha}_{BC} = \alpha_{BC}\mathbf{k}$$
$$\mathbf{v}_C = v_C\mathbf{j} \qquad \mathbf{a}_C = a_C\mathbf{j}$$

Velocity Analysis. Since AB is subjected to rotational motion about a *fixed axis,* using Eq. 16–13, $\mathbf{v}_B$ can be determined.

$$\mathbf{v}_B = \boldsymbol{\omega}_{AB} \times \mathbf{r}_{B/A}$$
$$= (-10\mathbf{k}) \times (-0.177\mathbf{i} + 0.177\mathbf{j})$$
$$= 1.77\mathbf{j} + 1.77\mathbf{i}$$

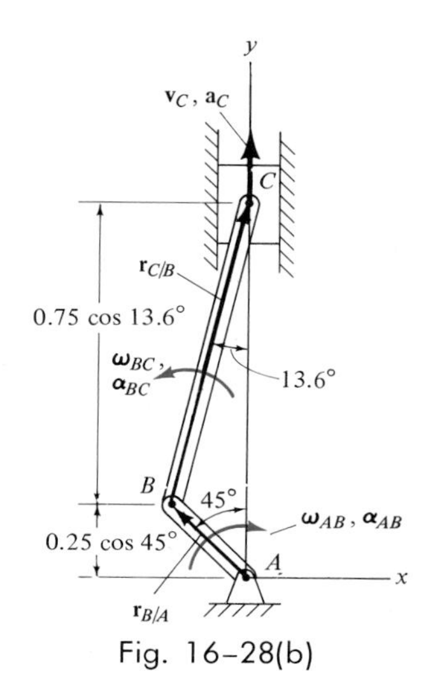

Fig. 16–28(b)

The angular velocity of rod BC is determined by applying Eq. 16–19 to points B and C on the rod and using the above result.

$$\mathbf{v}_C = \mathbf{v}_B + \boldsymbol{\omega}_{BC} \times \mathbf{r}_{C/B}$$
$$v_C\mathbf{j} = (1.77\mathbf{j} + 1.77\mathbf{i}) + (\omega_{BC}\mathbf{k}) \times (0.176\mathbf{i} + 0.729\mathbf{j})$$
$$v_C\mathbf{j} = 1.77\mathbf{j} + 1.77\mathbf{i} + 0.176\omega_{BC}\mathbf{j} - 0.729\omega_{BC}\mathbf{i}$$

Equating the respective **i** and **j** components,

$$0 = 1.77 - 0.729\omega_{BC}$$
$$v_C = 1.77 + 0.176\omega_{BC}$$

Hence,

$$\omega_{BC} = 2.43 \text{ rad/s}, \qquad v_C = 2.20 \text{ m/s}$$

Acceleration Analysis. Point B moves in a *circular* path because of the rotational motion of AB about A. Using Eq. 16–16, the acceleration of this point is

$$\mathbf{a}_B = \boldsymbol{\alpha}_{AB} \times \mathbf{r}_{B/A} - \omega_{AB}^2\mathbf{r}_{B/A}$$
$$= (-20\mathbf{k}) \times (-0.177\mathbf{i} + 0.177\mathbf{j}) - (10)^2(-0.177\mathbf{i} + 0.177\mathbf{j})$$
$$= \{21.24\mathbf{i} - 14.16\mathbf{j}\} \text{ m/s}^2$$

Applying Eq. 16–22 to points B and C on rod BC, yields

$$\mathbf{a}_C = \mathbf{a}_B + \boldsymbol{\alpha}_{BC} \times \mathbf{r}_{C/B} - \omega_{BC}^2\mathbf{r}_{C/B}$$
$$a_C\mathbf{j} = 21.24\mathbf{i} - 14.16\mathbf{j} + (\alpha_{BC}\mathbf{k}) \times (0.176\mathbf{i} + 0.729\mathbf{j}) - (2.43)^2(0.176\mathbf{i} + 0.729\mathbf{j})$$
$$a_C\mathbf{j} = 21.24\mathbf{i} - 14.16\mathbf{j} + 0.176\alpha_{BC}\mathbf{j} - 0.729\alpha_{BC}\mathbf{i} - 1.04\mathbf{i} - 4.30\mathbf{j}$$

Equating the respective **i** and **j** components,

$$0 = 20.20 - 0.729\alpha_{BC}$$
$$a_C = 0.176\alpha_{BC} - 18.46$$

Solving,

$$\alpha_{BC} = 27.7 \text{ rad/s}^2$$
$$a_C = -13.6 \text{ m/s}^2 \qquad \textit{Ans.}$$

The negative sign indicates that the piston is decelerating; i.e., $\mathbf{a}_C = \{-13.6\mathbf{j}\}$ m/s². This causes the speed v_C of the piston to decrease until the crankshaft AB and the connecting rod BC become vertical, at which time the piston is momentarily at rest.

16–62. At a given instant, the slider block A has the velocity and deceleration shown. Determine the acceleration of block B and the angular acceleration of the link at this same instant.

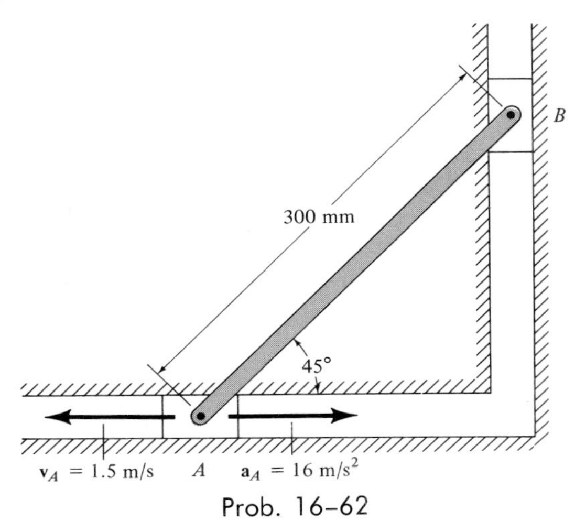

Prob. 16–62

16–63. The center of the wheel moves to the right with a velocity of 2 m/s and has an acceleration of 10 m/s² at the instant shown. Assuming that the wheel does not slip at A, determine the accelerations of points A and B at this instant. *Hint:* To determine α, use $a_C = \alpha r$, as indicated by the result of Example 16–3.

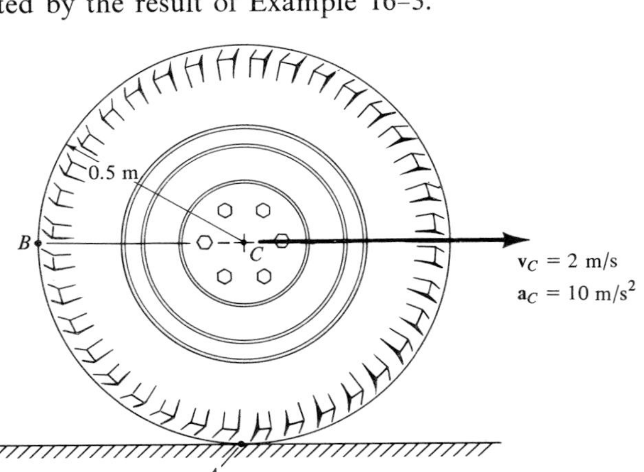

Prob. 16–63

***16–64.** Determine the accelerations of points B and D located on the planetary gear A in Prob. 16–59. Use the data given in Prob. 16–59.

16–65. Determine the angular accelerations of links AB and BC at the instant $\theta = 90°$ if the collar C has an instantaneous velocity of $v_C = 3$ m/s and deceleration $a_C = 2$ m/s² as shown. Set $b = 200$ mm.

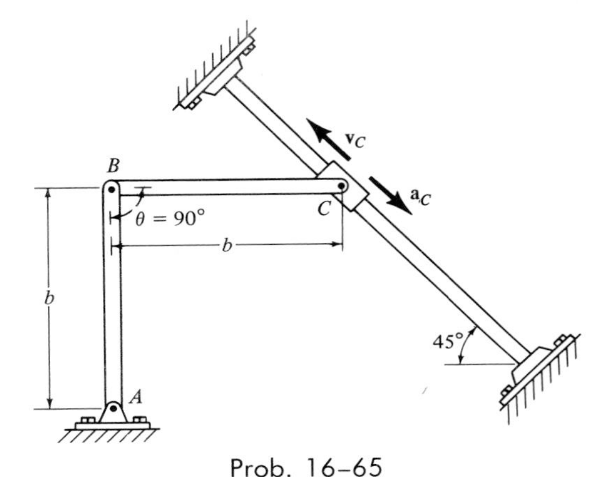

Prob. 16–65

16–65a. Solve Prob. 16–65 if $\theta = 90°$, $v_C = 4$ ft/s, $a_C = 3$ ft/s² (deceleration), and $b = 0.5$ ft.

16–66. The center of the pulley is being lifted vertically with an acceleration of 4 m/s^2 at the instant it has a velocity of 2 m/s. If the cable does not slip on the pulley's surface, determine the accelerations of the cylinder B and point C on the pulley. *Hint:* Note that the pulley "rolls" upward along the cord at D, without slipping. Hence, to determine α, use $a_A = \alpha r$, as indicated by the result in Example 16–3.

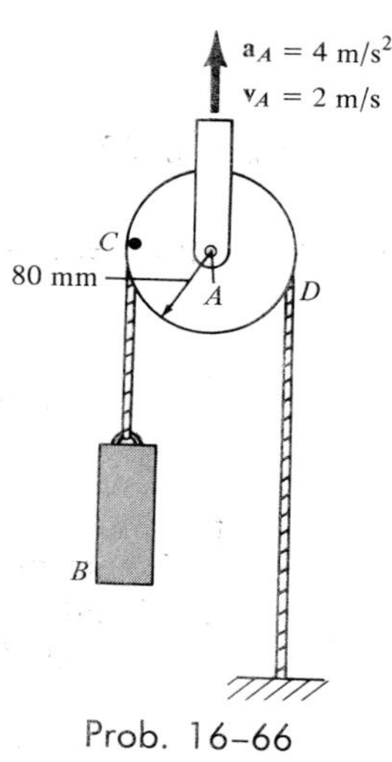

Prob. 16–66

16–67. Determine the acceleration of point A and the angular acceleration of the gear if at the given instant the gear racks have the velocities and accelerations shown.

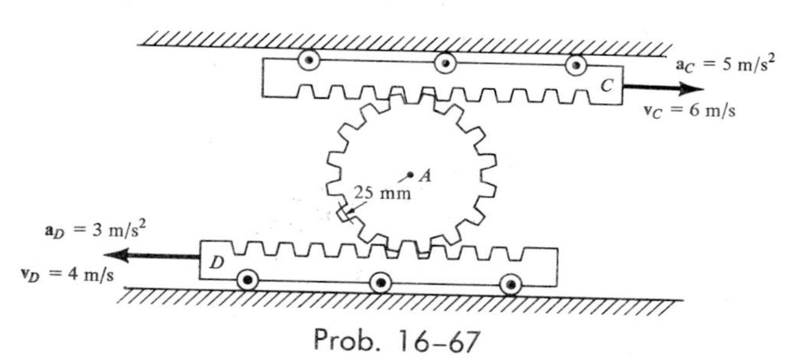

Prob. 16–67

***16–68.** At the instant shown, link AB has a clockwise angular velocity of 0.5 rad/s and a counterclockwise angular deceleration of 2 rad/s^2. Determine the acceleration of point C and the angular acceleration of link CD at this instant.

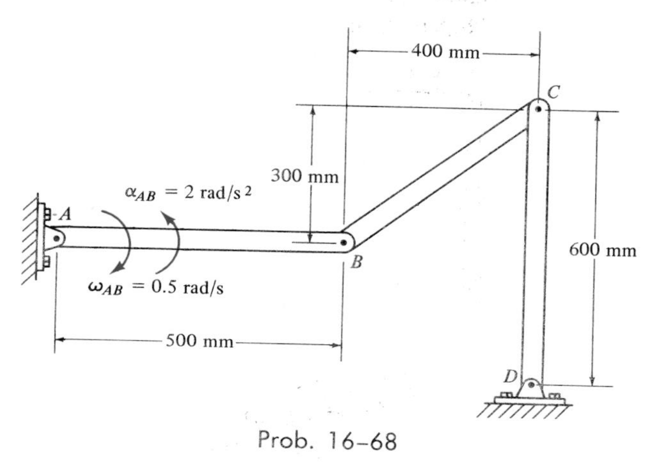

Prob. 16–68

16-69. A single pulley having both an inner and outer rim is pin-connected to block B at A. As cord CF unwinds from the inner rim of the pulley with the motion shown, cord DE unwinds from the outer rim. Determine the angular acceleration of the pulley and the acceleration of block B at the instant shown.

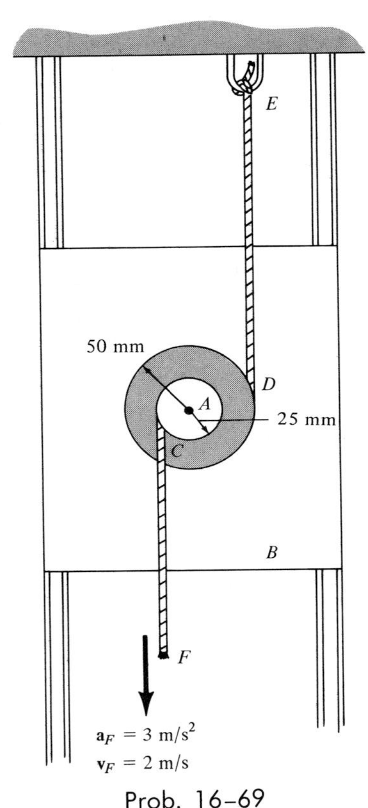

Prob. 16-69

16-70. Gear C is rotating with a constant angular velocity of $\omega_C = 3$ rad/s. Determine the acceleration of the piston, A, and the angular acceleration of rod AB at the instant $\theta = 90°$. Set $r_C = 25$ mm, $r_D = 50$ mm, and $l = 130$ mm.

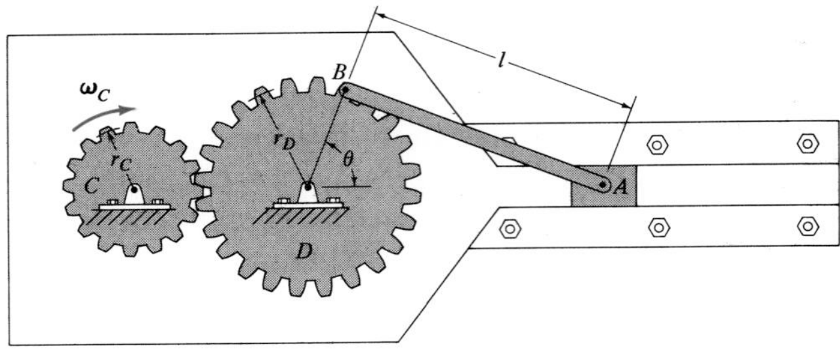

Prob. 16-70

16-70a. Solve Prob. 16-70 if $\omega_C = 5$ rad/s, $r_C = 0.2$ ft, $r_D = 0.3$ ft, and $l = 1.5$ ft.

16-71. Determine the acceleration of point A and the angular acceleration of rod AB at the instant $\theta = 0°$ in Prob. 16-70.

***16-72.** The mechanism produces intermittent motion of link AB. If the sprocket S is turning with an angular acceleration of $\alpha_S = 1.5$ rad/s² and has an angular velocity of $\omega_S = 6$ rad/s at the instant shown, determine the angular velocity and angular acceleration of link AB at this instant. The sprocket S is mounted on a shaft which is *separate* from a collinear shaft attached to AB at A. The pin at C is attached to one of the chain links.

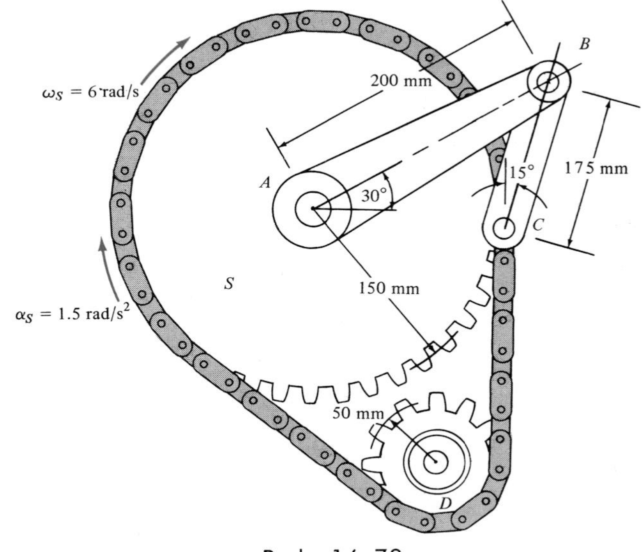

Prob. 16-72

16-73. At the instant shown, arm AB has an angular velocity of $\omega_{AB} = 0.5$ rad/s and an angular acceleration of $\alpha_{AB} = 0.8$ rad/s². Determine the angular velocity and angular acceleration of the dump bucket at this instant.

Prob. 16-73

16-74. The tied crank and gear mechanism gives rocking motion to crank AC, necessary for the operation of a printing press. If link DE has the instantaneous angular motion shown, determine the respective angular velocities and angular accelerations of gear F and crank AC at this instant.

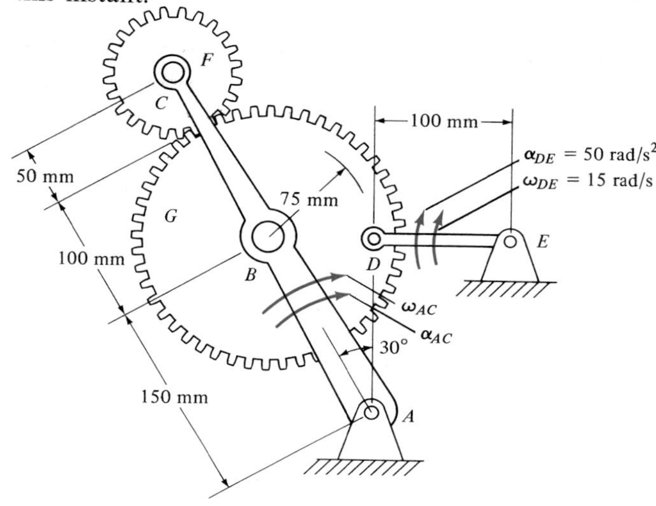

Prob. 16-74

16-75. Determine the angular velocity $\boldsymbol{\omega}_{FE}$ and the angular acceleration $\boldsymbol{\alpha}_{FE}$ of the plate FE of the stone-crushing mechanism at the instant AB and CD are both horizontal. At this instant $\theta = 30°$ and $\phi = 90°$. The driving link AB is turning with a constant angular velocity of $\omega_{AB} = 5$ rad/s. Set $b = 200$ mm.

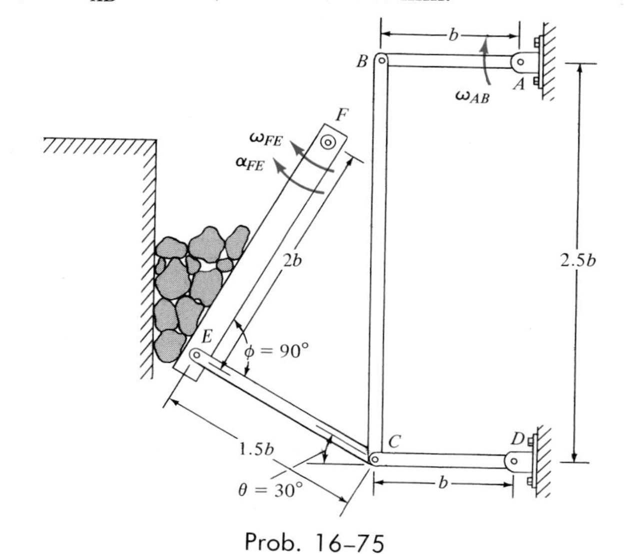

Prob. 16-75

16-75a. Solve Prob. 16-75 if $\omega_{AB} = 8$ rad/s and $b = 1.5$ ft.

16–10. Relative-General-Plane-Motion Analysis of a Particle or Rigid Body Using Translating and Rotating Axes

In the previous sections, the relative-motion analysis for velocity and acceleration was described using a translating coordinate system. This type of analysis is useful for determining the motion of points on the *same* rigid body, or the motion of points located on several pin-connected rigid bodies. In some problems, however, rigid bodies (mechanisms) are constructed such that *sliding* will occur at their connections. The kinematic analysis for such cases is best performed if the motion is analyzed using a coordinate system which both *translates* and *rotates*. Furthermore, this frame of reference is useful for analyzing the motions of two points on a mechanism which are *not* located on the *same* rigid body, and for specifying the kinematics of particle motion, when one of the particles is moving along a rotating path.

In the following analysis two equations will be developed which relate the velocity and acceleration of a point to the origin of a moving frame of reference, subjected to both a translation and a rotation in the plane.* Due to the generality in the derivation which follows, these two points may represent either two particles moving independently of one another or two points located on the same (or different) rigid bodies.

Position. Consider the two points A and B shown in Fig. 16–29*a*. Their location is specified by the position vectors $\mathbf{r}_A$ and $\mathbf{r}_B$, which are measured from the X, Y, Z coordinate system. As shown in the figure, the base point A represents the origin of the x, y, z coordinate system, which is assumed to be both translating and rotating with respect to the X, Y, Z system. The position of B with respect to A is determined by using the relative-position vector $\mathbf{r}_{B/A}$. The components of this vector may be expressed either in terms of unit vectors along the X, Y axes, i.e., $\mathbf{I}$ and $\mathbf{J}$, or by unit vectors along the x, y axes, i.e., $\mathbf{i}$ and $\mathbf{j}$. Although the magnitude of $\mathbf{r}_{B/A}$ is the same when measured in both coordinate systems, the direction of this vector will be different if the x, y axes are not parallel to the X, Y axes. For the proof which follows, $\mathbf{r}_{B/A}$ will be measured relative to the moving x, y frame of reference. Thus, if B has coordinates (x_B, y_B), Fig. 16–29*a*, $\mathbf{r}_{B/A}$ may be expressed as

$$\mathbf{r}_{B/A} = x_B\mathbf{i} + y_B\mathbf{j} \tag{16–23}$$

Using vector addition, the three position vectors in Fig. 16–29*a* may be related by the equation

$$\boxed{\mathbf{r}_B = \mathbf{r}_A + \mathbf{r}_{B/A}} \tag{16–24}$$

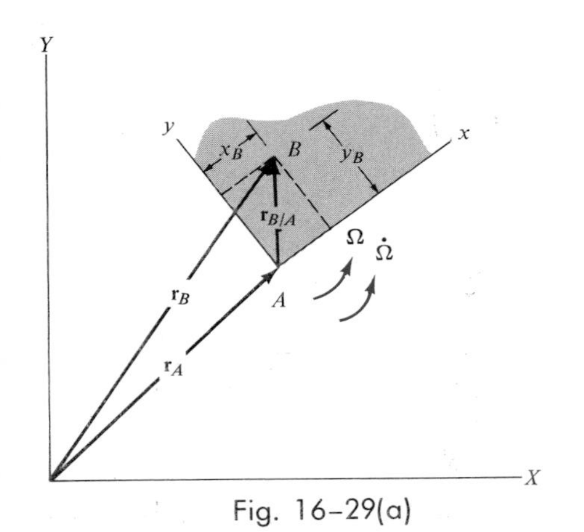

Fig. 16–29(a)

*The more general, spatial motion of the points is developed in Sec. 20-5.

At the instant considered, point A has a velocity of $\mathbf{v}_A$ and an acceleration of $\mathbf{a}_A$, while the angular velocity and angular acceleration of the axes are $\mathbf{\Omega}$ (omega) and $\dot{\mathbf{\Omega}} = d\mathbf{\Omega}/dt$, respectively. All these vectors are measured from the X, Y, Z frame of reference, although they may be expressed either in terms of $\mathbf{I}, \mathbf{J}, \mathbf{K}$ or $\mathbf{i}, \mathbf{j}, \mathbf{k}$ components. Since planar motion is specified, then by the right-hand rule, $\mathbf{\Omega}$ and $\dot{\mathbf{\Omega}}$ are always directed *perpendicular* to the reference plane of motion; where $\mathbf{v}_A$ and $\mathbf{a}_A$ lie in this plane.

Velocity. The velocity of point B is determined by taking the time derivative of Eq. 16–24, which yields

$$\mathbf{v}_B = \mathbf{v}_A + \frac{d\mathbf{r}_{B/A}}{dt} \tag{16–25}$$

The last term in this equation is evaluated using Eq. 16–23.

$$\begin{aligned} \frac{d\mathbf{r}_{B/A}}{dt} &= \frac{d}{dt}(x_B\mathbf{i} + y_B\mathbf{j}) \\ &= \frac{dx_B}{dt}\mathbf{i} + x_B\frac{d\mathbf{i}}{dt} + \frac{dy_B}{dt}\mathbf{j} + y_B\frac{d\mathbf{j}}{dt} \\ &= \left(\frac{dx_B}{dt}\mathbf{i} + \frac{dy_B}{dt}\mathbf{j}\right) + \left(x_B\frac{d\mathbf{i}}{dt} + y_B\frac{d\mathbf{j}}{dt}\right) \end{aligned} \tag{16–26}$$

The two terms in the first set of parentheses represent the components of velocity of point B as measured by an observer located at A and attached to the moving coordinate system. These terms will be denoted by vector $\mathbf{v}_{B/A}$. In the second set of parentheses the instantaneous time rate of change of the unit vectors $\mathbf{i}$ and $\mathbf{j}$ is measured by an observer located in the X, Y, Z coordinate system. These changes, $d\mathbf{i}$ and $d\mathbf{j}$, are due *only* to the instantaneous *rotation* $d\theta$ of the x, y, z reference system, Fig. 16–29*b*. As shown, the *magnitudes* of both $d\mathbf{i}$ and $d\mathbf{j}$ equal 1 $(d\theta)$, since $i = j = 1$. The *direction* of $d\mathbf{i}$ is defined by $+\mathbf{j}$, since $d\mathbf{i}$ is tangent to the path described by the tip of $\mathbf{i}$ in the limit as $\Delta t \to dt$. Likewise, $d\mathbf{j}$ acts in the $-\mathbf{i}$ direction. Hence,

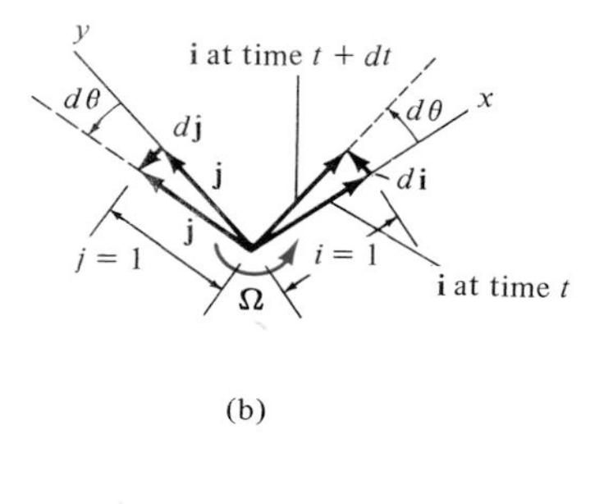

(b)

$$\frac{d\mathbf{i}}{dt} = \frac{d\theta}{dt}(\mathbf{j}) = \Omega\mathbf{j}, \qquad \frac{d\mathbf{j}}{dt} = \frac{d\theta}{dt}(-\mathbf{i}) = -\Omega\mathbf{i} \tag{16–27}$$

Viewing the axes in three dimensions, Fig. 16–29*c*, and noting that $\mathbf{\Omega} = \Omega\mathbf{k}$, we can express the above derivatives in terms of the cross product as

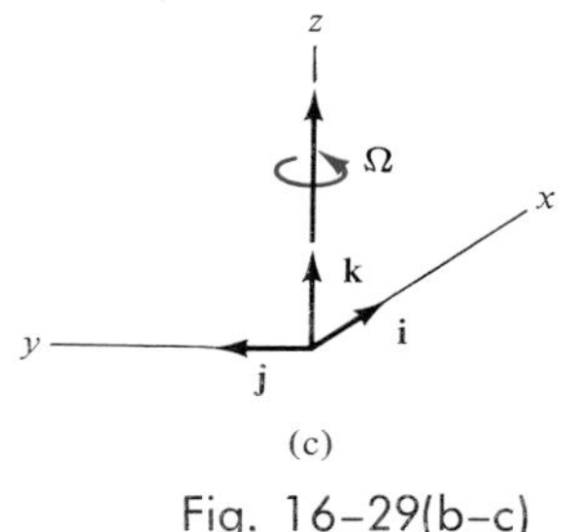

(c)

Fig. 16–29(b–c)

$$\frac{d\mathbf{i}}{dt} = \mathbf{\Omega} \times \mathbf{i}, \qquad \frac{d\mathbf{j}}{dt} = \mathbf{\Omega} \times \mathbf{j} \tag{16–28}$$

Substituting these results into Eq. 16–26 and using the distributive property of the vector cross product, we obtain

$$\frac{d\mathbf{r}_{B/A}}{dt} = \mathbf{v}_{B/A} + \mathbf{\Omega} \times (x_B\mathbf{i} + y_B\mathbf{j}) = \mathbf{v}_{B/A} + \mathbf{\Omega} \times \mathbf{r}_{B/A} \qquad (16\text{–}29)$$

Hence, Eq. 16–25 becomes

$$\mathbf{v}_B = \mathbf{v}_A + \mathbf{\Omega} \times \mathbf{r}_{B/A} + \mathbf{v}_{B/A} \qquad (16\text{–}30)$$

where

$\mathbf{v}_B$ = velocity of point B measured from the X, Y, Z reference

$\mathbf{v}_A$ = velocity of the origin A of the x, y, z reference, measured from the X, Y, Z reference

$\mathbf{v}_{B/A}$ = relative velocity of "B with respect to A," as measured by an observer attached to the *rotating* x, y, z reference

$\mathbf{\Omega}$ = instantaneous angular velocity of the x, y, z reference, measured from the X, Y, Z reference

$\mathbf{r}_{B/A}$ = relative-position vector drawn from point A to point B and measured with respect to the x, y, z reference

Acceleration. The acceleration of point B, observed from the X, Y, Z coordinate system, may be expressed in terms of its motion measured with respect to the rotating or moving system of coordinates by taking the time derivative of Eq. 16–30, i.e.,

$$\frac{d\mathbf{v}_B}{dt} = \frac{d\mathbf{v}_A}{dt} + \frac{d\mathbf{\Omega}}{dt} \times \mathbf{r}_{B/A} + \mathbf{\Omega} \times \frac{d\mathbf{r}_{B/A}}{dt} + \frac{d\mathbf{v}_{B/A}}{dt}$$

$$\mathbf{a}_B = \mathbf{a}_A + \dot{\mathbf{\Omega}} \times \mathbf{r}_{B/A} + \mathbf{\Omega} \times \frac{d\mathbf{r}_{B/A}}{dt} + \frac{d\mathbf{v}_{B/A}}{dt} \qquad (16\text{–}31)$$

Here $\dot{\mathbf{\Omega}} = d\mathbf{\Omega}/dt$ is the angular acceleration of the x, y, z coordinate system. For planar motion $\mathbf{\Omega}$ is always perpendicular to the plane of motion, and therefore $\dot{\mathbf{\Omega}}$ measures *only the change in magnitude* of $\mathbf{\Omega}$. The derivative $d\mathbf{r}_{B/A}/dt$ in Eq. 16–31 is defined by Eq. 16–29, so that

$$\mathbf{\Omega} \times \frac{d\mathbf{r}_{B/A}}{dt} = \mathbf{\Omega} \times \mathbf{v}_{B/A} + \mathbf{\Omega} \times (\mathbf{\Omega} \times \mathbf{r}_{B/A}) \qquad (16\text{–}32)$$

Computing the time derivative of $\mathbf{v}_{B/A} = (v_{B/A})_x\mathbf{i} + (v_{B/A})_y\mathbf{j}$, we have

$$\frac{d\mathbf{v}_{B/A}}{dt} = \left[\frac{d(v_{B/A})_x}{dt}\mathbf{i} + \frac{d(v_{B/A})_y}{dt}\mathbf{j}\right] + \left[(v_{B/A})_x\frac{d\mathbf{i}}{dt} + (v_{B/A})_y\frac{d\mathbf{j}}{dt}\right]$$

The two terms in the first set of brackets represent the components of acceleration of point B as measured by an observer located at A and attached to the moving coordinate system. These terms will be denoted by vector $\mathbf{a}_{B/A}$. The terms in the second set of brackets can be simplified using Eqs. 16–28. Hence,

$$\frac{d\mathbf{v}_{B/A}}{dt} = \mathbf{a}_{B/A} + \mathbf{\Omega} \times \mathbf{v}_{B/A} \qquad (16\text{–}33)$$

Substituting Eqs. 16–32 and 16–33 into Eq. 16–31 and rearranging terms yields

$$\mathbf{a}_B = \mathbf{a}_A + \dot{\mathbf{\Omega}} \times \mathbf{r}_{B/A} + \mathbf{\Omega} \times (\mathbf{\Omega} \times \mathbf{r}_{B/A}) + 2\mathbf{\Omega} \times \mathbf{v}_{B/A} + \mathbf{a}_{B/A} \qquad (16\text{–}34)$$

where

$\mathbf{a}_B$ = acceleration of point B measured from the X, Y, Z reference

$\mathbf{a}_A$ = acceleration of the origin A of the x, y, z reference, measured from the X, Y, Z reference

$\mathbf{a}_{B/A}$, $\mathbf{v}_{B/A}$ = relative acceleration and relative velocity of "B with respect to A," as measured by an observer attached to the *rotating* x, y, z reference

$\dot{\mathbf{\Omega}}$, $\mathbf{\Omega}$ = instantaneous angular acceleration and angular velocity of the x, y, z reference, measured from the X, Y, Z reference

$\mathbf{r}_{B/A}$ = relative-position vector drawn from point A to point B, and measured with respect to the x, y, z frame of reference

If the motions of points A and B are along *curved paths,* it is convenient to express the accelerations $\mathbf{a}_B$, $\mathbf{a}_A$, and $\mathbf{a}_{B/A}$ in Eq. 16–34 in terms of their normal and tangential components.

The third term in Eq. 16–34, $\dot{\mathbf{\Omega}} \times \mathbf{r}_{B/A}$, represents the effect of the angular acceleration caused by the rotation of the x, y, z frame. The fourth term, $\mathbf{\Omega} \times (\mathbf{\Omega} \times \mathbf{r}_{B/A})$, is called the *centripetal acceleration* and represents the acceleration component introduced by the angular velocity of the x, y, z frame. The fifth term, $2\mathbf{\Omega} \times \mathbf{v}_{B/A}$, is called the *Coriolis acceleration,* named after the French engineer G. C. Coriolis (1792–1843), who was the first to determine it. This term represents the *difference* in acceleration of B as measured from nonrotating and rotating x, y, z axes having an origin at A. The definition of the vector cross product insures that the Coriolis acceleration will always be perpendicular to both $\mathbf{\Omega}$ and $\mathbf{v}_{B/A}$.

PROCEDURE FOR ANALYSIS

The following four-step procedure should be used when applying Eqs. 16–30 and 16–34 to the solution of problems involving the planar motion of particles or rigid bodies.

Step 1: Choose an appropriate location for the origin and orientation of the axes for both the fixed X, Y, Z and moving x, y, z reference frames.

Most often solutions are easily obtained if at the instant considered: (1) the origins are coincident, (2) the axes are collinear, and/or (3) the axes are parallel. The moving frame should be selected fixed to the body or device where the relative motion occurs.
Step 2: After defining the origin A of the moving reference and specifying the moving point B, Eqs. 16–30 and 16–34 should be written in symbolic form

$$\mathbf{v}_B = \mathbf{v}_A + \boldsymbol{\Omega} \times \mathbf{r}_{B/A} + \mathbf{v}_{B/A}$$
$$\mathbf{a}_B = \mathbf{a}_A + \dot{\boldsymbol{\Omega}} \times \mathbf{r}_{B/A} + \boldsymbol{\Omega} \times (\boldsymbol{\Omega} \times \mathbf{r}_{B/A}) + 2\boldsymbol{\Omega} \times \mathbf{v}_{B/A} + \mathbf{a}_{B/A}$$

Step 3: Each of the vectors in these equations should be defined from the problem data and expressed in Cartesian vector form. This essentially requires a determination of (1) the motion of the moving reference, i.e., $\mathbf{v}_A$, $\mathbf{a}_A$, $\boldsymbol{\Omega}$, and $\dot{\boldsymbol{\Omega}}$, and (2) the motion of B measured with respect to the moving reference, i.e., $\mathbf{r}_{B/A}$, $\mathbf{v}_{B/A}$, and $\mathbf{a}_{B/A}$. The components of all these vectors may be selected along either the X, Y, Z axes or the x, y, z axes. The choice is arbitrary provided a consistent set of unit vectors is used.
Step 4: Substitute the data into the kinematic equations of *Step 2* and perform the vector operations.

The following examples numerically illustrate this four-step procedure.

Example 16–17

At the instant $\theta = 60°$, the rod in Fig. 16–30 has an angular velocity of 3 rad/s and an angular acceleration of 2 rad/s². At this same instant, the collar C is traveling outward along the rod such that when $x = 0.2$ m, the velocity is 2 m/s and the acceleration is 3 m/s², both measured relative to the rod. Determine the Coriolis acceleration and the velocity and acceleration of the collar at this instant.

Solution

Step 1: The origin of both coordinate systems is located at point O, Fig. 16–30. Since the motion of the collar is reported relative to the rod, the moving x, y, z frame of reference is *attached* to the rod.
Step 2: The motions of C and O are related by the equations

$$\mathbf{v}_C = \mathbf{v}_O + \boldsymbol{\Omega} \times \mathbf{r}_{C/O} + \mathbf{v}_{C/O} \qquad (1)$$
$$\mathbf{a}_C = \mathbf{a}_O + \dot{\boldsymbol{\Omega}} \times \mathbf{r}_{C/O} + \boldsymbol{\Omega} \times (\boldsymbol{\Omega} \times \mathbf{r}_{C/O}) + 2\boldsymbol{\Omega} \times \mathbf{v}_{C/O} + \mathbf{a}_{C/O} \qquad (2)$$

Step 3: It will be simpler to express the data in terms of **i, j, k** component vectors rather than **I, J, K** components. Hence,

Motion of moving reference	*Motion of C with respect to moving reference*
$\mathbf{v}_O = \mathbf{0}$	$\mathbf{r}_{C/O} = \{0.2\mathbf{i}\}$ m

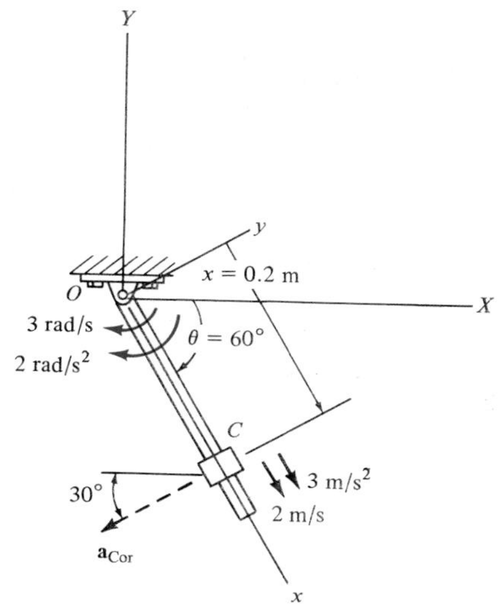

Fig. 16–30

$$\mathbf{a}_O = \mathbf{0} \qquad \mathbf{v}_{C/O} = \{2\mathbf{i}\}\ \text{m/s}$$
$$\boldsymbol{\Omega} = \{-3\mathbf{k}\}\ \text{rad/s} \qquad \mathbf{a}_{C/O} = \{3\mathbf{i}\}\ \text{m/s}^2$$
$$\dot{\boldsymbol{\Omega}} = \{-2\mathbf{k}\}\ \text{rad/s}^2$$

Step 4: From Eq. (2) the Coriolis acceleration is defined as

$$\mathbf{a}_{\text{Cor}} = 2\boldsymbol{\Omega} \times \mathbf{v}_{C/O} = 2(-3\mathbf{k}) \times (2\mathbf{i}) = \{-12\mathbf{j}\}\ \text{m/s}^2 \qquad \textit{Ans.}$$

This vector is shown dashed in Fig. 16–30. It may also be resolved into **I, J** components acting along the X and Y axes, which yields

$$\begin{aligned}\mathbf{a}_{\text{Cor}} &= -12\cos 30°\mathbf{I} - 12\sin 30°\mathbf{J}\\ &= \{-10.4\mathbf{I} - 6.0\mathbf{J}\}\ \text{m/s}^2 \qquad \textit{Ans.}\end{aligned}$$

The velocity and acceleration of the collar is determined by substituting the data into Eqs. (1) and (2) and evaluating the cross products, which yields

$$\begin{aligned}\mathbf{v}_C &= \mathbf{v}_O + \boldsymbol{\Omega} \times \mathbf{r}_{C/O} + \mathbf{v}_{C/O}\\ &= \mathbf{0} + (-3\mathbf{k}) \times (0.2\mathbf{i}) + 2\mathbf{i}\\ &= \{2\mathbf{i} - 0.6\mathbf{j}\}\ \text{m/s} \qquad \textit{Ans.}\end{aligned}$$

$$\begin{aligned}\mathbf{a}_C &= \mathbf{a}_O + \dot{\boldsymbol{\Omega}} \times \mathbf{r}_{C/O} + \boldsymbol{\Omega} \times (\boldsymbol{\Omega} \times \mathbf{r}_{C/O}) + 2\boldsymbol{\Omega} \times \mathbf{v}_{C/O} + \mathbf{a}_{C/O}\\ &= \mathbf{0} + (-2\mathbf{k}) \times (0.2\mathbf{i}) + (-3\mathbf{k}) \times [(-3\mathbf{k}) \times (0.2\mathbf{i})]\\ &\quad + 2(-3\mathbf{k}) \times (2\mathbf{i}) + 3\mathbf{i}\\ &= \mathbf{0} - 0.4\mathbf{j} - 1.8\mathbf{i} - 12\mathbf{j} + 3\mathbf{i}\\ &= \{1.2\mathbf{i} - 12.4\mathbf{j}\}\ \text{m/s}^2 \qquad \textit{Ans.}\end{aligned}$$

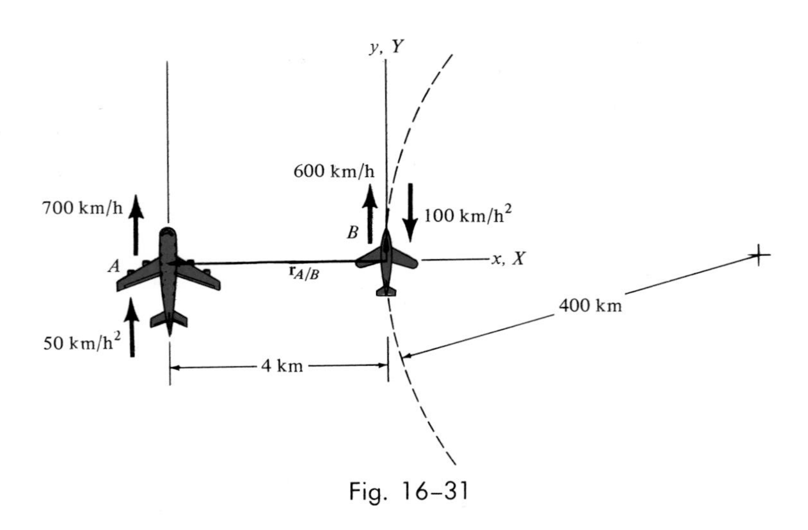

Fig. 16–31

Example 16–18

Two jet planes are flying at the same elevation as shown in Fig. 16–31. Plane A is flying along a straight-line path, and at the instant shown, it has a speed of 700 km/h and an acceleration of 50 km/h². Plane B is flying along a circular path at 600 km/h and decreasing its speed at the rate of 100 km/h². Determine the velocity and acceleration of A, as measured by the pilot of B.

Solution

Step 1: Since the relative motion of A with respect to B is being sought, the x, y, z axes are attached to the pilot at B, Fig. 16–31. At the *instant* considered, the origin B coincides with the origin of the fixed X, Y, Z, frame.

Step 2: The motions of A and B are related by the equations

$$\mathbf{v}_A = \mathbf{v}_B + \boldsymbol{\Omega} \times \mathbf{r}_{A/B} + \mathbf{v}_{A/B} \tag{1}$$

$$\mathbf{a}_A = \mathbf{a}_B + \dot{\boldsymbol{\Omega}} \times \mathbf{r}_{A/B} + \boldsymbol{\Omega} \times (\boldsymbol{\Omega} \times \mathbf{r}_{A/B}) + 2\boldsymbol{\Omega} \times \mathbf{v}_{A/B} + \mathbf{a}_{A/B} \tag{2}$$

Step 3: From the problem data, noting that B has both n and t components of acceleration since it is moving along a *curved* path, we have:

Motion of moving reference:

$$\mathbf{v}_B = \{600\mathbf{j}\}\ \text{km/h}$$

$$\mathbf{a}_B = (\mathbf{a}_B)_n + (\mathbf{a}_B)_t = \{900\mathbf{i} - 100\mathbf{j}\}\ \text{km/h}^2$$

$$\text{since, by Eq. 12–54, } (a_B)_n = \frac{v_B^2}{\rho} = \frac{(600)^2}{400} = 900\ \text{km/h}^2$$

$$\boldsymbol{\Omega} = \{-1.5\mathbf{k}\}\ \text{rad/h}$$

$$\text{since, by Eq. 16–12, } \Omega = \frac{v_B}{\rho} = \frac{600\ \text{km/h}}{400\ \text{km}} = 1.5\ \text{rad/h} \ \curvearrowright$$

$$\dot{\boldsymbol{\Omega}} = \{0.25\mathbf{k}\}\ \text{rad/h}^2$$

$$\text{since, by Eq. 16–14, } \dot{\Omega} = \frac{(a_B)_t}{\rho} = \frac{100\ \text{km/h}^2}{400\ \text{km}} = 0.25\ \text{rad/h}^2 \ \curvearrowleft$$

Motion of A with respect to moving reference:

$$\mathbf{r}_{A/B} = \{-4\mathbf{i}\}\ \text{km}, \qquad \mathbf{v}_{A/B} = ?, \qquad \mathbf{a}_{A/B} = ?$$

Step 4: Substituting this data into Eqs. (1) and (2), realizing that $\mathbf{v}_A = \{700\mathbf{j}\}$ km/h, and $\mathbf{a}_A = \{50\mathbf{j}\}$ km/h², we have

$$\mathbf{v}_A = \mathbf{v}_B + \boldsymbol{\Omega} \times \mathbf{r}_{A/B} + \mathbf{v}_{A/B}$$
$$700\mathbf{j} = 600\mathbf{j} + (-1.5\mathbf{k}) \times (-4\mathbf{i}) + \mathbf{v}_{A/B}$$
$$100\mathbf{j} = 6\mathbf{j} + \mathbf{v}_{A/B}$$
$$\mathbf{v}_{A/B} = \{94\mathbf{j}\}\ \text{km/h} \qquad \textit{Ans.}$$
$$v_{A/B} = 94\ \text{km/h}\ \uparrow$$

$$\mathbf{a}_A = \mathbf{a}_B + \dot{\boldsymbol{\Omega}} \times \mathbf{r}_{A/B} + \boldsymbol{\Omega} \times (\boldsymbol{\Omega} \times \mathbf{r}_{A/B}) + 2\boldsymbol{\Omega} \times \mathbf{v}_{A/B} + \mathbf{a}_{A/B}$$
$$50\mathbf{j} = (900\mathbf{i} - 100\mathbf{j}) + (0.25\mathbf{k}) \times (-4\mathbf{i}) + (-1.5\mathbf{k}) \times [(-1.5\mathbf{k}) \times (-4\mathbf{i})] + 2(-1.5\mathbf{k}) \times (94\mathbf{j}) + \mathbf{a}_{A/B}$$
$$= (900\mathbf{i} - 100\mathbf{j}) - 1\mathbf{j} + 9\mathbf{i} + 282\mathbf{i} + \mathbf{a}_{A/B}$$
$$\mathbf{a}_{A/B} = \{-1191\mathbf{i} + 151\mathbf{j}\}\ \text{km/h}^2 \qquad \textit{Ans.}$$

The motion of plane B as observed from plane A has been calculated in Example 12–21. It is suggested that the two methods of solution be compared.

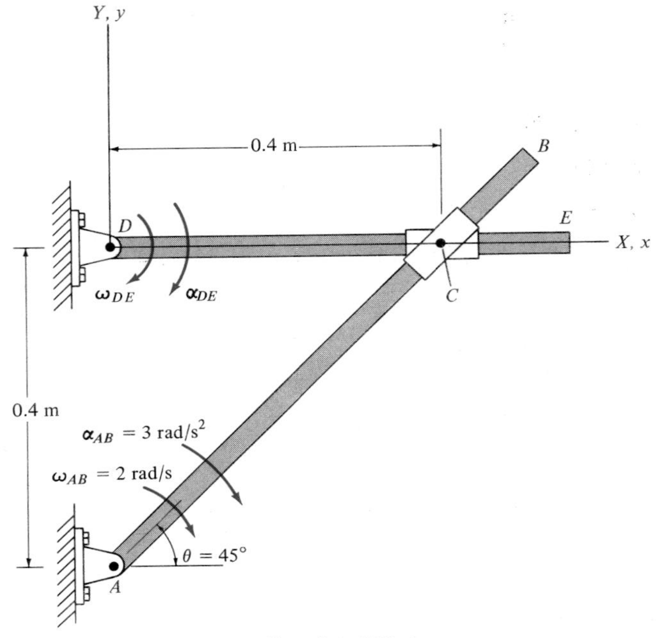

Fig. 16–32(a)

Example 16–19

The rod AB, shown in Fig. 16–32a, rotates clockwise such that it has an angular velocity of $\omega_{AB} = 2$ rad/s and an angular acceleration of $\alpha_{AB} = 3$ rad/s^2 when $\theta = 45°$. Determine the angular velocity and angular acceleration of rod DE at this instant. The collar at C is pin-connected to AB and slides over rod DE.

Solution

Step 1: The origins of both the fixed and moving frames of reference are located at point D, Fig. 16–32a. Furthermore, the x, y, z reference is attached to, and rotates with rod DE.

Step 2: The motions of points C and D are related by the equations

$$\mathbf{v}_C = \mathbf{v}_D + \mathbf{\Omega} \times \mathbf{r}_{C/D} + \mathbf{v}_{C/D} \quad (1)$$

$$\mathbf{a}_C = \mathbf{a}_D + \dot{\mathbf{\Omega}} \times \mathbf{r}_{C/D} + \mathbf{\Omega} \times (\mathbf{\Omega} \times \mathbf{r}_{C/D}) + 2\mathbf{\Omega} \times \mathbf{v}_{C/D} + \mathbf{a}_{C/D} \quad (2)$$

Step 3: All vector quantities will be expressed in terms of **i, j, k** unit vectors.

Motion of moving reference	*Motion of C with respect to moving reference*
$\mathbf{v}_D = \mathbf{0}$	$\mathbf{r}_{C/D} = \{0.4\mathbf{i}\}$ m

$$\mathbf{a}_D = \mathbf{0} \qquad \mathbf{v}_{C/D} = v_{C/D}\mathbf{i}$$
$$\boldsymbol{\Omega} = -\omega_{DE}\mathbf{k} \qquad \mathbf{a}_{C/D} = a_{C/D}\mathbf{i}$$
$$\dot{\boldsymbol{\Omega}} = -\alpha_{DE}\mathbf{k}$$

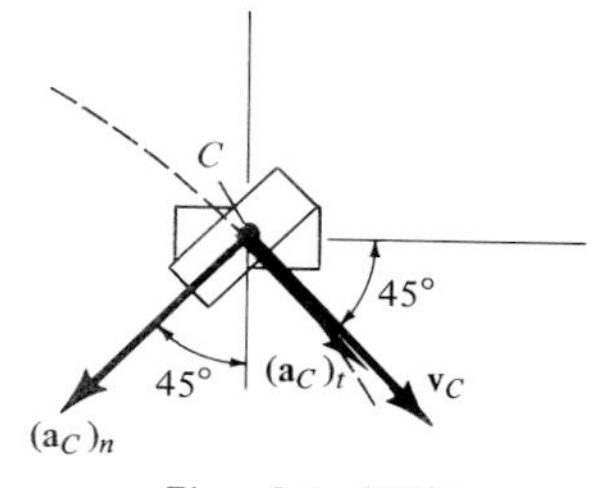

Fig. 16–32(b)

Since the collar moves along a *circular path,* it has the velocity and two components of acceleration shown in Fig. 16–32*b*.

Motion of point C:

$$v_C = \omega_{AB} r_{C/A} = 2(0.4\sqrt{2}) = 1.131 \text{ m/s}$$
$$\mathbf{v}_C = 1.131 \cos 45°\mathbf{i} - 1.131 \sin 45°\mathbf{j}$$
$$= \{0.8\mathbf{i} - 0.8\mathbf{j}\} \text{ m/s}$$
$$(a_C)_t = \alpha_{AB} r_{C/A} = 3(0.4\sqrt{2}) = 1.697 \text{ m/s}^2$$
$$(a_C)_n = \omega_{AB}^2 r_{C/A} = (2)^2(0.4\sqrt{2}) = 2.263 \text{ m/s}^2$$
$$\mathbf{a}_C = (1.697 \cos 45° - 2.263 \sin 45°)\mathbf{i} + (-1.697 \sin 45° - 2.263 \cos 45°)\mathbf{j}$$
$$= \{-0.4\mathbf{i} - 2.8\mathbf{j}\} \text{ m/s}^2$$

Step 4: Substituting the data into Eqs. (1) and (2), we have

$$\mathbf{v}_C = \mathbf{v}_D + \boldsymbol{\Omega} \times \mathbf{r}_{C/D} + \mathbf{v}_{C/D}$$
$$0.8\mathbf{i} - 0.8\mathbf{j} = \mathbf{0} + (-\omega_{DE}\mathbf{k}) \times (0.4\mathbf{i}) + v_{C/D}\mathbf{i}$$
$$0.8\mathbf{i} - 0.8\mathbf{j} = \mathbf{0} - 0.4\omega_{DE}\mathbf{j} + v_{C/D}\mathbf{i}$$

Equating the respective **i** and **j** components yields

$$v_{C/D} = 0.8 \text{ m/s}$$
$$\omega_{DE} = \frac{0.8}{0.4} = 2 \text{ rad/s}$$

Hence,

$$\boldsymbol{\omega}_{DE} = \boldsymbol{\Omega} = \{-2\mathbf{k}\} \text{ rad/s} \qquad \textit{Ans.}$$

$$\mathbf{a}_C = \mathbf{a}_D + \dot{\boldsymbol{\Omega}} \times \mathbf{r}_{C/D} + \boldsymbol{\Omega} \times (\boldsymbol{\Omega} \times \mathbf{r}_{C/D}) + 2\boldsymbol{\Omega} \times \mathbf{v}_{C/D} + \mathbf{a}_{C/D}$$
$$-0.4\mathbf{i} - 2.8\mathbf{j} = \mathbf{0} + (-\alpha_{DE}\mathbf{k}) \times (0.4\mathbf{i}) + (-2\mathbf{k}) \times [(-2\mathbf{k}) \times (0.4\mathbf{i})] + 2(-2\mathbf{k}) \times (0.8\mathbf{i}) + a_{C/D}\mathbf{i}$$
$$-0.4\mathbf{i} - 2.8\mathbf{j} = \mathbf{0} - 0.4\alpha_{DE}\mathbf{j} - 1.6\mathbf{i} - 3.2\mathbf{j} + a_{C/D}\mathbf{i}$$

Equating the respective **i** and **j** components and solving, yields

$$a_{C/D} = 1.2 \text{ m/s}^2$$
$$\alpha_{DE} = -1 \text{ rad/s}^2$$

Hence,

$$\boldsymbol{\alpha}_{DE} = \dot{\boldsymbol{\Omega}} = \{1\mathbf{k}\} \text{ rad/s}^2 \qquad \textit{Ans.}$$

***16–76.** The ball is moving along the slot of a rotating platform such that when it is at C it has a speed of 0.25 m/s, measured relative to the platform. What is the magnitude of the ball's Coriolis acceleration when it is at this point?

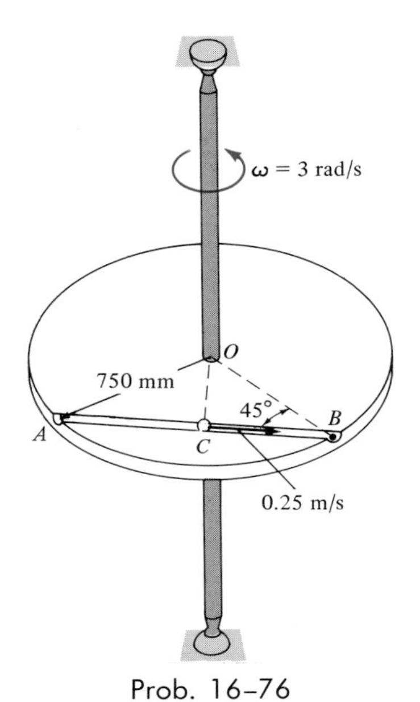

Prob. 16–76

16–77. At the instant shown, the ball B is rolling along the slot in the disk with a velocity of 600 mm/s and an acceleration of 150 mm/s^2, both measured with respect to the disk and directed away from O. If at the same instant, the disk has the angular velocity and angular acceleration shown, determine the velocity and acceleration of the ball, at this instant.

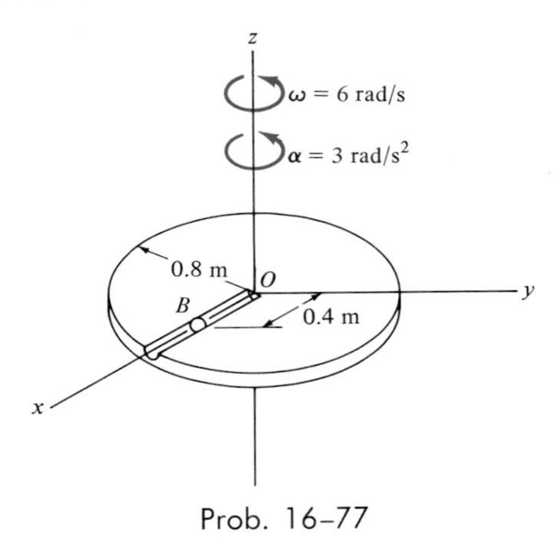

Prob. 16–77

16–78. The slider block B, which is attached to a cord, moves along the slot of a horizontal circular disk. If the cord is pulled down through the central hole A in the disk at a constant rate of $\dot{x} = -3$ m/s, determine the acceleration of the block at the instant $x = 0.1$ m. The disk has a constant angular velocity of $\omega_D = 2$ rad/s.

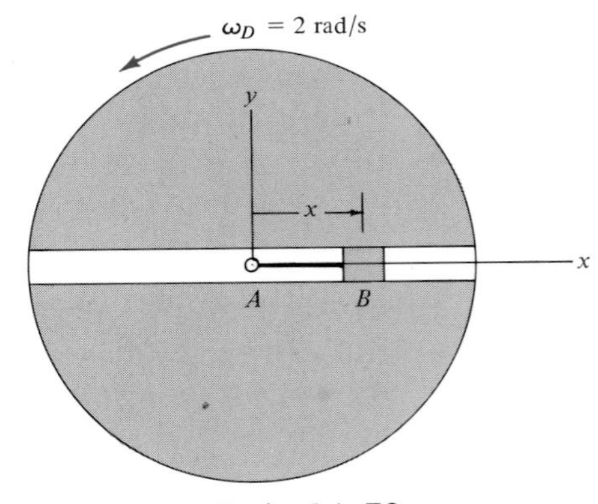

Prob. 16–78

16–79. Solve Prob. 16–78 assuming that at the instant $x = 0.1$ m, $\dot{x} = -3$ m/s, $\ddot{x} = 1.25$ m/s^2, $\omega_D = 2$ rad/s^2, and the disk has an *angular deceleration* of $\alpha_D = -4$ rad/s^2.

***16–80.** A girl stands at A on a platform which is rotating with a constant angular velocity of $\omega = 0.5$ rad/s. If she walks at a constant speed of $v = 0.75$ m/s on the platform, determine her acceleration (a) when she reaches point D in going along the path ADC, $d = 1$ m; and (b) when she reaches point B if she follows the path ABC, $r = 3$ m.

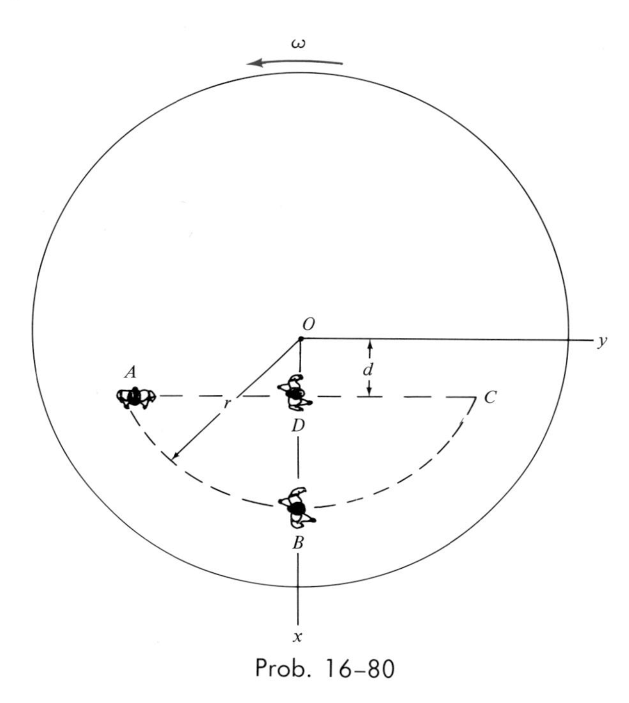

Prob. 16–80

16–81. A ride in an amusement park consists of a rotating platform P, having a constant angular velocity of $\omega_P = 2$ rad/s; and four cars, C, mounted on the platform, which have constant angular velocities of $\omega_{C/P} = 1.5$ rad/s, measured relative to the platform. At the instant shown, determine the velocity and acceleration of the passenger at A.

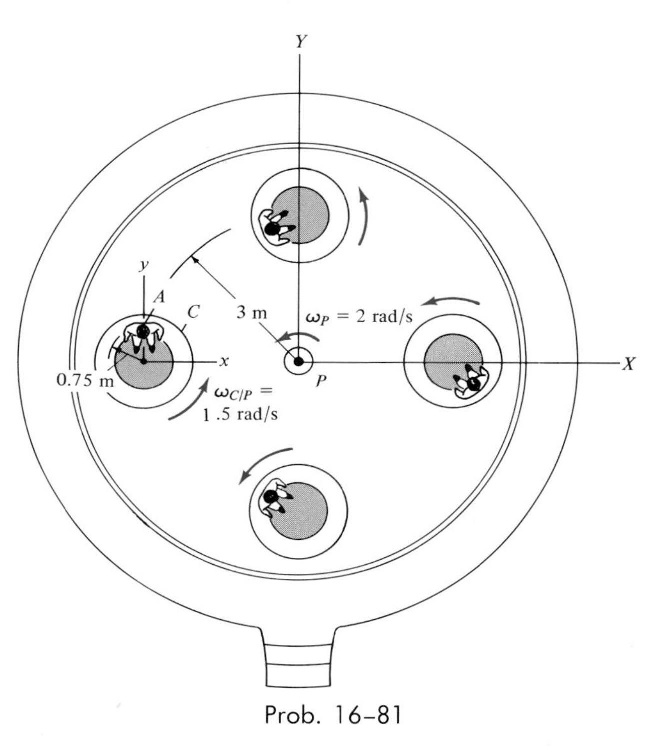

Prob. 16–81

***16–80a.** Solve Prob. 16–80 if $\omega = 0.25$ rad/s, $v = 2$ ft/s, $d = 3$ ft, and $r = 8$ ft.

16–82. Solve Prob. 16–81 assuming that at the instant shown, the car C has an *angular deceleration* of $\alpha_{C/P} = 0.5$ rad/s^2 measured relative to the platform. Use the data in Prob. 16–81.

16–83. If the slider block C is fixed to the disk that has a constant counterclockwise angular velocity of 4 rad/s, determine the angular velocity and angular acceleration of the slotted arm AB at the instant shown.

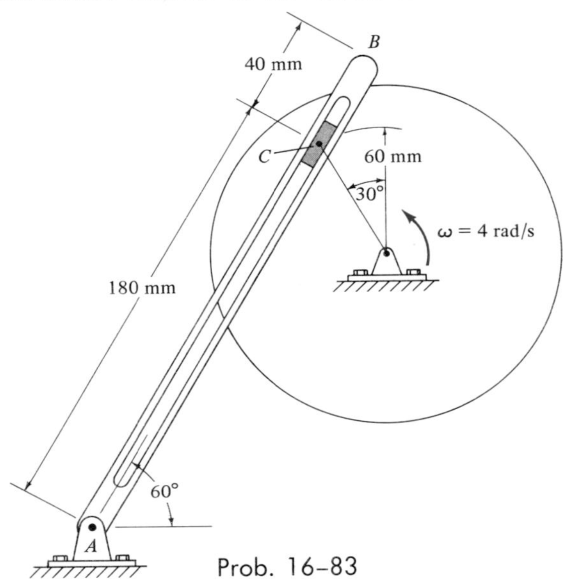

Prob. 16–83

16–85. If at the instant shown $a = 150$ mm, $b = 200$ mm, and link CD has an angular velocity of $\omega_{CD} = 4$ rad/s and an angular acceleration of $\alpha_{CD} = 2$ rad/s^2, determine the angular velocity and angular acceleration of rod AB at this instant. The collar at C is pin-connected to DC and slides over AB.

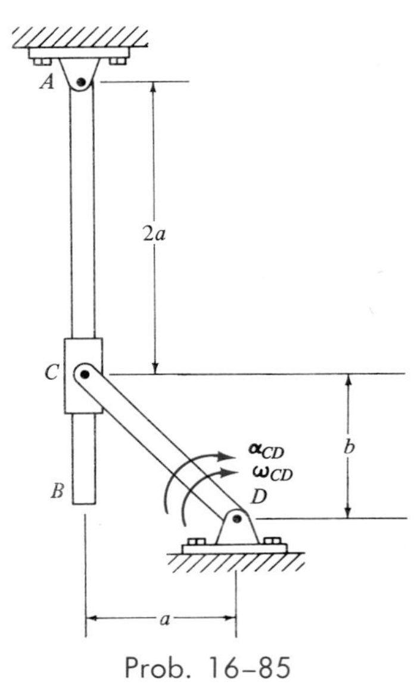

Prob. 16–85

* **16–84.** The two-link mechanism serves to amplify angular motion. Link AB has a pin at B which is confined to move within the slot of link CD. If at the instant shown, AB (input) has an angular velocity of $\omega_{AB} = 2.5$ rad/s and an angular acceleration of $\alpha_{AB} = 3$ rad/s^2, determine the angular velocity and angular acceleration of CD (output) at this instant.

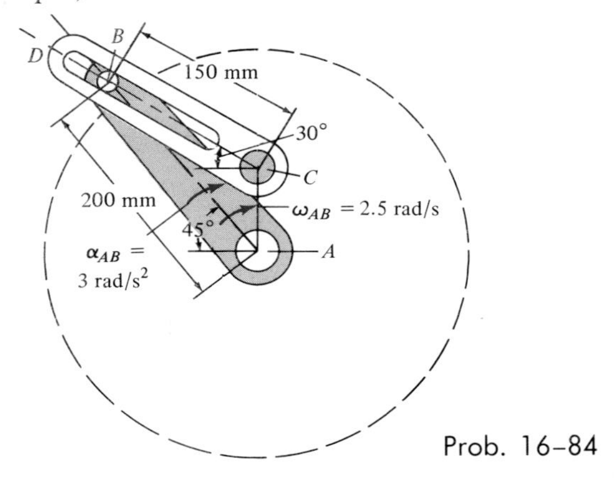

Prob. 16–84

16–85a. Solve Prob. 16–85 if $a = 1.5$ ft, $b = 2$ ft, $\omega_{CD} = 3$ rad/s, and $\alpha_{CD} = 4.5$ rad/s^2.

16–86. The block B of the quick-return mechanism is confined to move within the slot in member CD. If AB is rotating at a constant rate of $\omega_{AB} = 3$ rad/s, determine the angular velocity and angular acceleration of member CD at the instant shown.

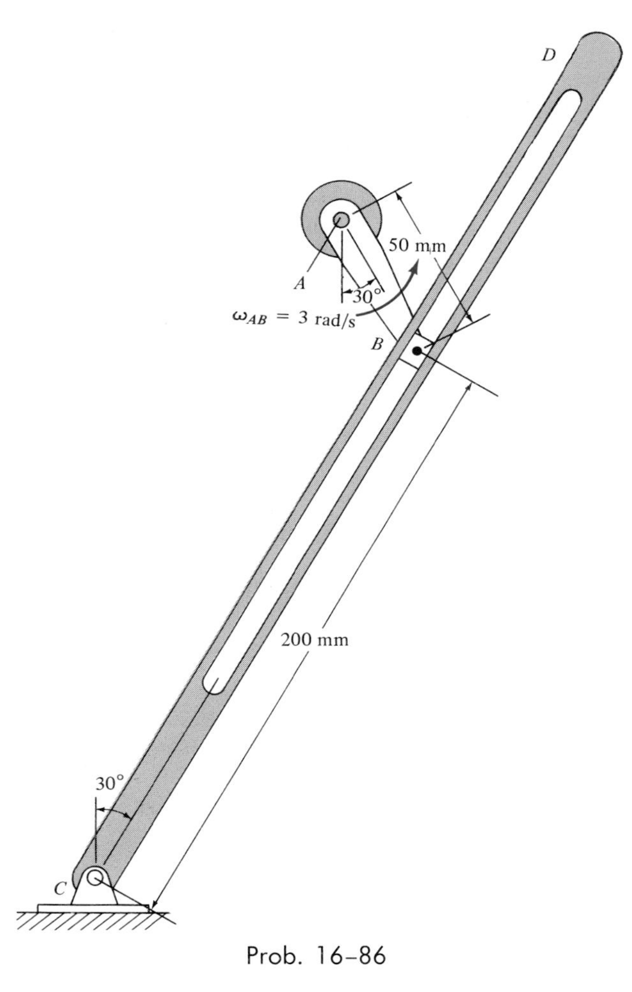

Prob. 16–86

16–87. The Geneva mechanism is used for a packaging system to convert constant angular motion into intermittent angular motion. The star wheel A makes one sixth of a revolution for each full revolution of the driving wheel B and the attached guide C. To do this, pin P, which is attached to B, slides into one of the radial slots of A, thereby turning wheel A, and then exits the slot. If B has a constant angular velocity of $\omega_B = 5$ rad/s, determine $\boldsymbol{\omega}_A$ and $\boldsymbol{\alpha}_A$ of wheel A at the instant $\theta = 30°$ as shown.

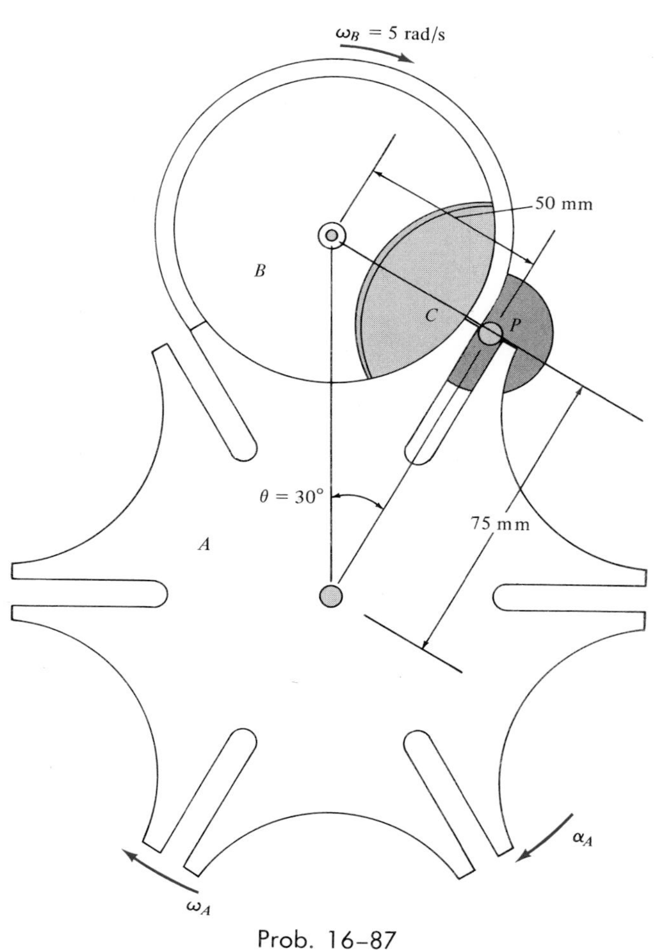

Prob. 16–87

17 Planar Kinetics of Rigid Bodies: Forces and Accelerations

17–1. Introduction

In the previous chapter the planar kinematics of rigid-body motion was presented in order of increasing complexity, that is, translation, rotation about a fixed axis, and general plane motion. The study of rigid-body kinetics in this chapter will be presented in somewhat the same order. The chapter begins by introducing a property of a body called the mass moment of inertia. Afterwards, a derivation of the equations of general plane motion for symmetrical rigid bodies is given, and then these equations are applied to specific problems of rigid-body translation, rotation about a fixed axis, and finally general plane motion. A rigid body subjected to any of these three types of motion may be analyzed in a fixed reference plane, because the path of motion of each particle of the body lies in a plane that is parallel to the reference plane. A kinetic study of these motions is referred to as the kinetics of planar motions or simply *planar kinetics*. The more complex study of the spatial kinetics of rigid-body motions, which includes planar motion of unsymmetrical rigid bodies, is presented in Chapter 21.

17–2. Mass Moment of Inertia

As shown in Chapter 13, the study of particle kinetics is based upon the equation of motion, $\mathbf{F} = m\mathbf{a}$, which relates the resultant force $\mathbf{F}$ acting on

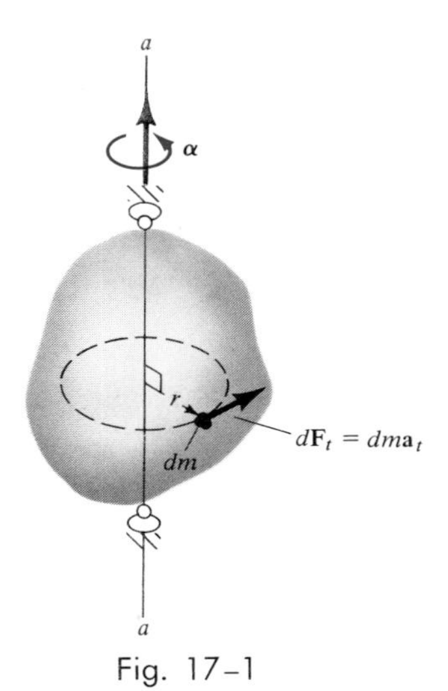

Fig. 17–1

the particle to its mass m and acceleration **a.** This equation applies as well to a rigid body. However, since a body has a definite size and shape, the force system applied to it will not necessarily be concurrent. As a result, the moments created by the forces give the body an angular acceleration. To show this, consider the body in Fig. 17–1 which is fixed to rotate about the *aa* axis. The body's angular motion can be studied by isolating an element of mass dm, located at a perpendicular distance r from the axis. Due to an external loading, any unbalanced *tangential-force component* $d\mathbf{F}_t$ acting on this element creates an angular motion of the element about the axis. Applying the equation of motion in the tangential direction $(\Sigma F_t = ma_t)$ yields $dF_t = dm\, a_t$. Since $a_t = r\alpha$, the moment of dF_t about the *aa* axis is therefore $dM = r\, dF_t = r^2\alpha\, dm$. The moment of the tangential forces acting on *all the elements* of the body is determined by integration, which gives $M = \int_m r^2\alpha\, dm$. However, since α is the same for all radial lines r extending from the axis to the element dm, α may be factored out of the integrand, leaving $M = I\alpha$, where

$$I = \int_m r^2\, dm \tag{17–1}$$

This integral is termed the *mass moment of inertia*. Since the formulation involves the distance r, Fig. 17–1, the value of I is *unique* for each axis *aa* about which it is computed. In the study of planar kinetics, however, the axis which is generally chosen for analysis passes through the body's mass center G and is *always* perpendicular to the plane of motion. The mass moment of inertia computed about this axis will be defined as I_G.

As a comparison, the *mass moment of inertia* is a measure of the resistance of a body to *angular acceleration* ($\mathbf{M} = I\boldsymbol{\alpha}$) in the same way that *mass* is a measure of the resistance to *acceleration* ($\mathbf{F} = m\mathbf{a}$). Since the moment of inertia is an important property, used throughout the study of rigid-body planar kinetics, methods used for its calculation will now be discussed.

PROCEDURE FOR INTEGRATION

In general, when integrating Eq. 17–1, it is best to choose a coordinate system which simplifies the equations that describe the boundary of the body. For example, cylindrical coordinates are generally appropriate when solving problems which involve bodies having circular boundaries.

If the body consists of material having a variable mass density, $\rho = \rho(x, y, z)$, the elemental mass dm of the body may be expressed in terms of its density and volume as $dm = \rho\, dV$. Consequently, the body's moment of inertia is computed using *volume elements* for integration, i.e.,

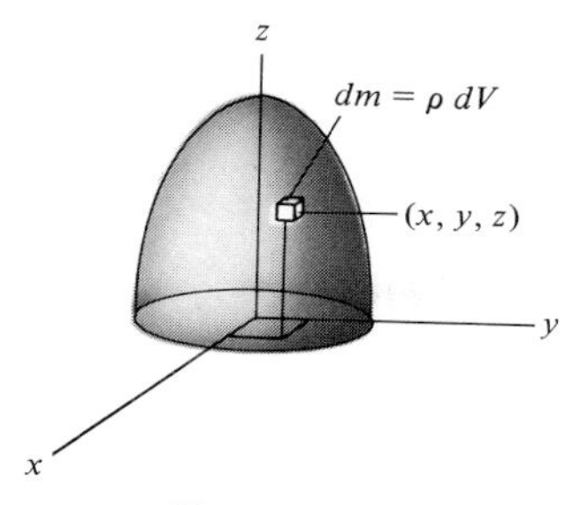

Fig. 17–2(a)

$$I = \int_V r^2 \rho \, dV \qquad (17\text{–}2)$$

In the special case of ρ being a *constant,* this term may be factored out of the integral and the integration is then purely a function of geometry,

$$I = \rho \int_V r^2 \, dV \qquad (17\text{–}3)$$

When the elemental volume, chosen for integration, has differential sizes in all three directions, e.g., $dV = dx\, dy\, dz$, Fig. 17–2*a,* the moment of inertia of the body must be computed using "triple integration." The integration process can, however, be simplified to a *single integration* provided the chosen elemental volume has a differential size or thickness in only *one direction.* Shell or disk elements are often used for this purpose.

Shell Element. If a *shell element* having a height z, radius $r = y$, and thickness dy is chosen for integration, Fig. 17–2*b,* then the volume is $dV = (2\pi y)(z)\, dy$. This element may be used in Eq. 17–2 or 17–3 for computing the moment of inertia I_z of the body about the z axis, since the *entire element,* due to its "thinness," lies at the *same* perpendicular distance $r = y$ from the z axis.* (See Example 17–1.)

Disk Element. If a disk element having a radius y and a thickness dz is chosen for integration, Fig. 17–2*c,* then the volume is $dV = (\pi y^2)\, dz$. In this case, however, the element is *finite* in the radial direction, and consequently parts of it *do not* all lie at the *same radial distance r* from the z axis. As a result, Eq. 17–2 or 17–3 cannot be used to determine I_z. Instead, to perform the integration, using this element, it is first necessary to determine the moment of inertia *of the element* about the z axis and then integrate this result. (See Example 17–3.)

*Note that this is also true for the element in Fig. 17–2*a.*

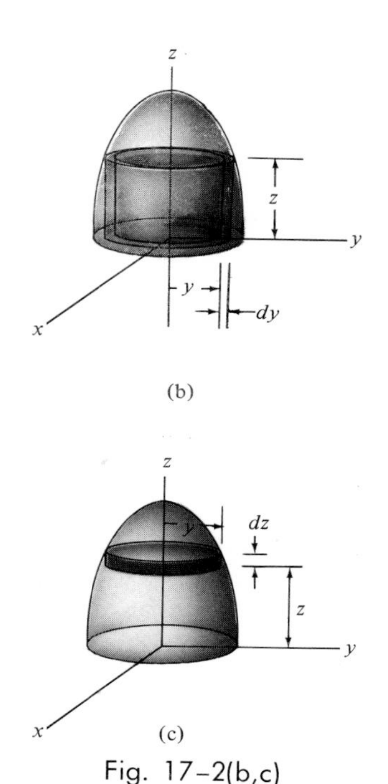

Fig. 17–2(b,c)

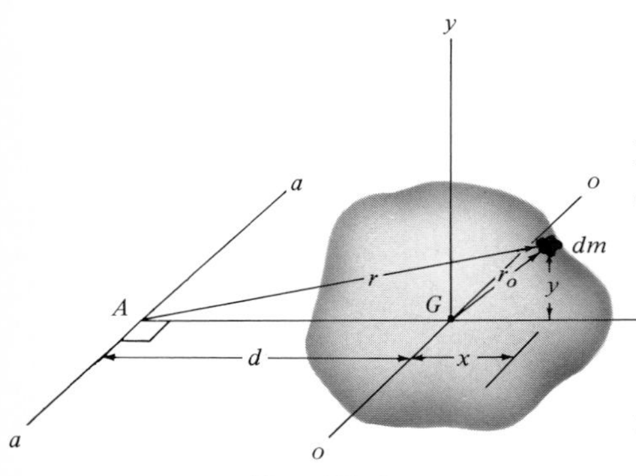

Fig. 17–3

Parallel-Axis Theorem. Provided the value of I_G, which is calculated about an axis passing through the body's mass center G, is known, the moment of inertia I of the body may be determined about any other *parallel axis* by using the *parallel-axis theorem.* This theorem can be derived by considering the body of mass m shown in Fig. 17–3. The x, y axes have their origin located on the oo axis which passes through the body's mass center G, whereas the corresponding *parallel axis aa,* passing through A, lies at a constant distance d away. Selecting the differential element of mass dm, which is located at point (x, y), and applying Eq. 17–1, we can express the moments of inertia of the body computed about the aa and oo axes as

$$I = \int_m r^2\, dm \qquad \text{and} \qquad I_G = \int_m r_o^2\, dm$$

respectively. Using the Pythagorean theorem, $r^2 = (d + x)^2 + y^2$, so that

$$I = \int_m [(d + x)^2 + y^2]\, dm$$

$$= \int_m (x^2 + y^2)\, dm + 2d \int_m x\, dm + d^2 \int_m dm$$

Since $r_o^2 = x^2 + y^2$, the first integral represents I_G. The second integral equals *zero,* since the oo axis passes through the body's mass center. Finally, the third integral represents the total mass m of the body. Hence, the moment of inertia about the aa axis can be written as

$$I = I_G + md^2 \tag{17–4}$$

where I_G is the moment of inertia about an axis passing through the mass center G, m is the mass of the body, and d is the perpendicular distance between the parallel axes.

Radius of Gyration. Occasionally, the moment of inertia of a body about a specified axis is reported in handbooks using the *radius of gyration.* When this length, k, and the body's mass m are known, the body's moment of inertia is determined from the equation

$$I = k^2 m \quad \text{or} \quad k = \sqrt{\frac{I}{m}} \tag{17–5}$$

Note the *similarity* between the definition of k in this formula and r in the equation $dI = r^2\,dm$, which defines the moment of inertia of an elemental mass dm of the body about an axis (see Eq. 17–1).

Composite Bodies. The parallel-axis theorem is often used to determine the moment of inertia of composite shapes when the moment of inertia I_G of each of the composite parts is either known (see Appendix D) or can be computed by integration. For example, if the body is constructed of a number of simple shapes such as disks, spheres, and rods, the moment of inertia of the body about any axis aa can be determined by adding together, algebraically, the moments of inertia of each of the composite shapes computed about the aa axis. The parallel-axis theorem is needed for the calculations if the center of gravity of each composite part does not lie on the aa axis. Example 17–2 numerically illustrates the procedure.

Example 17–1

Determine the moment of inertia of the right circular cylinder, shown in Fig. 17–4a, about the z axis. The mass density ρ of the material is constant.

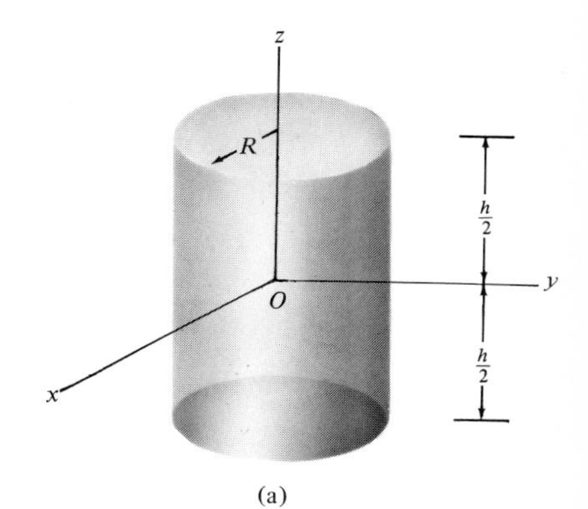

(a)

Solution

(***Thin-Shell Element***). This problem may be solved using the *thin-shell element* in Fig. 17–4b and single integration. The volume of the element is $dV = (2\pi r)(h)\,dr$, so that the mass is $dm = \rho\,dV = \rho(2\pi hr\,dr)$. Since the *entire element* lies at the same distance r from the z axis, the moment of inertia *of the element* is

$$dI_z = r^2\,dm = \rho 2\pi h r^3\,dr$$

Integrating over the entire region of the cylinder yields

$$I_z = \int_m r^2\,dm = \rho 2\pi h \int_0^R r^3\,dr = \frac{\rho\pi}{2} R^4 h$$

The mass of the cylinder is

$$m = \int_m dm = \rho 2\pi h \int_0^R r\,dr = \rho\pi h R^2$$

so that

$$I_z = \tfrac{1}{2} m R^2 \qquad \textit{Ans.}$$

Notice that the result is *independent* of the height h of the cylinder and is therefore applicable, as well, to *thin disks*.

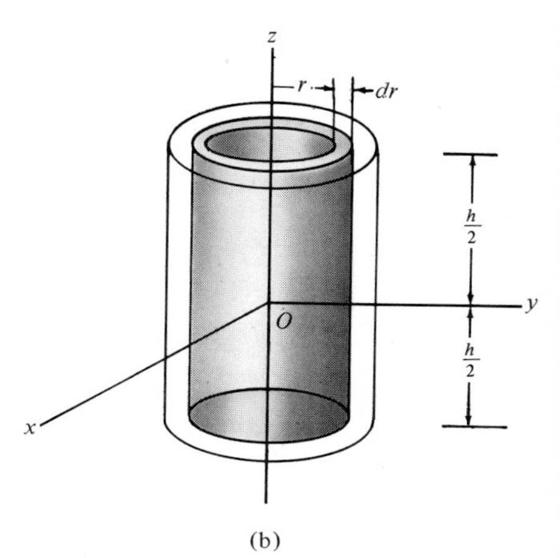

(b)

Fig. 17–4

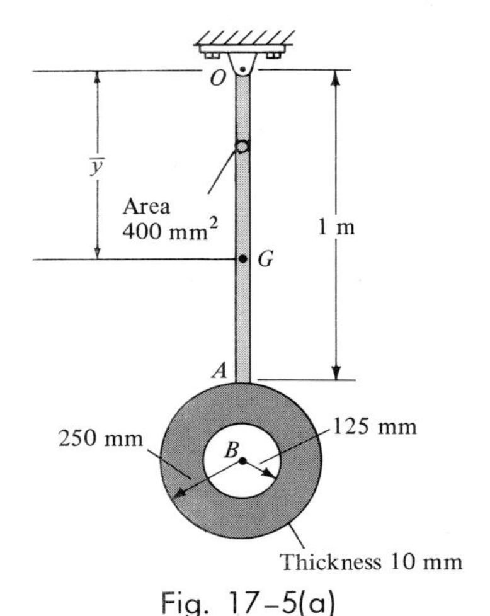

Fig. 17–5(a)

Example 17–2

The pendulum shown in Fig. 17–5*a* consists of a slender rod and a disk with a hole in it. The rod has a density of 7000 kg/m³ and a cross-sectional area of 400 mm², and the disk has a density of 8000 kg/m³ and a thickness of 10 mm. Compute the moment of inertia of the pendulum about an axis directed perpendicular to the page and passing through (a) the pin at *O*, and (b) the mass center *G* of the pendulum.

Solution

Part (a). The pendulum may be thought of as consisting of three composite parts, Fig. 17–5*b*: the rod *OA, plus* a 250-mm-radius disk, *minus* a 125-mm-radius disk. The moment of inertia about an axis passing through *O* can therefore be determined by computing the moment of inertia of each of these three composite parts about this axis and then *algebraically* adding the results. The computations are performed by using the parallel-axis theorem in conjunction with the data listed in Appendix D and Fig. 17–5*b*.

Rod. The moment of inertia of the slender rod *OA* about an axis perpendicular to the page and passing through the end point *O* of the rod, is $I_O = \frac{1}{3}ml^2$ (Appendix D). Hence, for the rod, we have

$$m_r = \rho_r V_r = 7000 \text{ kg/m}^3\ [4(10^{-4})\text{ m}^2(1\text{ m})] = 2.80\text{ kg}$$

$$\begin{aligned}(I_O)_r &= \tfrac{1}{3}m_r l^2\\ &= \tfrac{1}{3}(2.80\text{ kg})(1\text{ m})^2\\ &= 0.933\text{ kg}\cdot\text{m}^2\end{aligned}$$

This same value may be computed using $I_G = \frac{1}{12}ml^2$ (Appendix D) and the parallel-axis theorem, i.e.,

$$\begin{aligned}(I_O)_r &= \tfrac{1}{12}m_r l^2 + m_r d^2\\ &= \tfrac{1}{12}(2.80\text{ kg})(1\text{ m})^2 + 2.80\text{ kg}(0.5\text{ m})^2\\ &= 0.933\text{ kg}\cdot\text{m}^2\end{aligned}$$

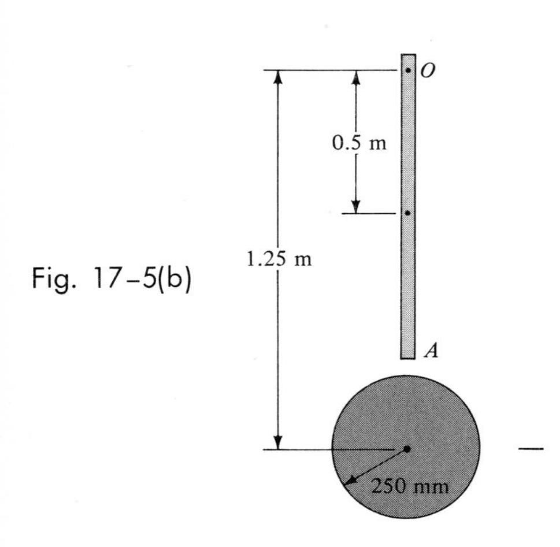

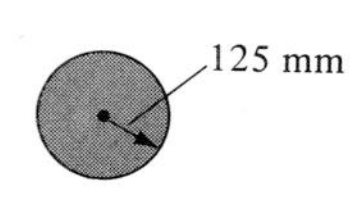

Fig. 17–5(b)

Disk. From Appendix D the moment of inertia of a thin disk about an axis perpendicular to the plane of the disk is $I_G = \frac{1}{2}mr^2$. The mass centers of *both* the 250-mm-radius disk and the 125-mm-radius disk (hole) are located at a distance of 1.25 m from point O. For the 250-mm-radius disk, we have

$$\begin{aligned} m_d &= \rho_d V_d = 8000\ \text{kg/m}^3[\pi(0.25\ \text{m})^2(0.01\ \text{m})] = 15.71\ \text{kg} \\ (I_O)_d &= \tfrac{1}{2}m_d r_d^2 + m_d d^2 \\ &= \tfrac{1}{2}(15.71\ \text{kg})(0.25\ \text{m})^2 + (15.71\ \text{kg})(1.25\ \text{m})^2 \\ &= 25.03\ \text{kg}\cdot\text{m}^2 \end{aligned}$$

Hole. For the 125-mm-radius disk (hole), we have

$$\begin{aligned} m_h &= \rho_h V_h = 8000\ \text{kg/m}^3[\pi(0.125\ \text{m})^2(0.01\ \text{m})] = 3.93\ \text{kg} \\ (I_O)_h &= \tfrac{1}{2}m_h r_h^2 + m_h d^2 \\ &= \tfrac{1}{2}(3.93\ \text{kg})(0.125\ \text{m})^2 + (3.93\ \text{kg})(1.25\ \text{m})^2 \\ &= 6.17\ \text{kg}\cdot\text{m}^2 \end{aligned}$$

The moment of inertia of the pendulum about the axis passing through point O is therefore the sum of the moments of inertia for the rod and the 250-mm-radius disk *less* the 125-mm-radius disk (hole),

$$\begin{aligned} I_O &= (I_O)_r + (I_O)_d - (I_O)_h \\ &= 0.93 + 25.03 - 6.17 \\ &= 19.79\ \text{kg}\cdot\text{m}^2 \end{aligned} \qquad \textit{Ans.}$$

Part (b). The mass center, G, will be located relative to the pin O. Assuming this distance to be $\bar{y}$, Fig. 17–5*a*, and using the formula for determining the mass center, we have

$$\begin{aligned} \bar{y} &= \frac{\Sigma \tilde{y} m}{\Sigma m} = \frac{\tilde{y}_r m_r + \tilde{y}_d m_d + \tilde{y}_h(-m_h)}{m_r + m_d - m_h} \\ &= \frac{0.5(2.80) + 1.25(15.71) + (1.25)(-3.93)}{2.80 + 15.71 - 3.93} = 1.11\ \text{m} \end{aligned}$$

The moment of inertia I_G may be computed in the same manner as I_O, which requires successive applications of the parallel-axis theorem in order to transfer the moment of inertia of each composite part to G. A more direct solution, however, involves applying the parallel-axis theorem using the result for I_O, i.e.,

$$\begin{aligned} I_O &= I_G + md^2 \\ 19.79 &= I_G + (2.80 + 15.71 - 3.93)(1.11)^2 \\ I_G &= 1.83\ \text{kg}\cdot\text{m}^2 \end{aligned} \qquad \textit{Ans.}$$

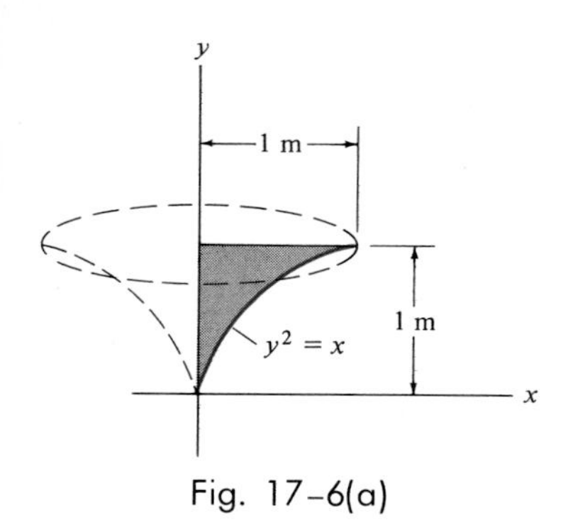

Fig. 17–6(a)

Example 17–3

A solid is formed by revolving the shaded area shown in Fig. 17–6*a* about the y axis. If the density of the material is ρ kg/m³, determine the radius of gyration k_y about the y axis.

Solution

(Thin-Disk Element). To obtain the radius of gyration k_y, it is first necessary to determine the mass m and the moment of inertia I_y. This will be done using a *thin-disk element,* as shown in Fig. 17–6*b*. Here the element intersects the curve at the arbitrary point (x, y) and has a mass

$$dm = \rho\, dV = \rho(\pi x^2)\, dy$$

Substituting $x = y^2$ and integrating with respect to y, from $y = 0$ to $y = 1$ m, yields the entire mass

$$m = \pi\rho \int_0^1 x^2\, dy = \pi\rho \int_0^1 y^4\, dy = 0.6283\rho \text{ kg}$$

Although all portions of the element are *not* located at the same distance from the y axis, it is still possible to determine the moment of inertia dI_y *of the element* about the y axis. In the first example it was shown that the moment of inertia of a cylinder about its longitudinal axis is $I = \frac{1}{2}mR^2$, where m and R are the mass and radius of the cylinder. Since the height h of the cylinder is not involved in this formula, the moment of inertia of the disk element having a mass dm and radius x in Fig. 17–6*b* is

$$dI_y = \tfrac{1}{2}(dm)x^2 = \tfrac{1}{2}[\rho(\pi x^2)dy]x^2$$

Integrating, the moment of inertia of the entire solid becomes

$$I_y = \frac{\pi\rho}{2}\int_0^1 x^4\, dy = \frac{\pi\rho}{2}\int_0^1 y^8\, dy = 0.1745\rho \text{ kg}\cdot\text{m}^2$$

Therefore,

$$k_y = \sqrt{\frac{I_y}{m}} = \sqrt{\frac{0.1745\rho}{0.6283\rho}} = 0.527 \text{ m} \qquad \textit{Ans.}$$

Fig. 17–6(b)

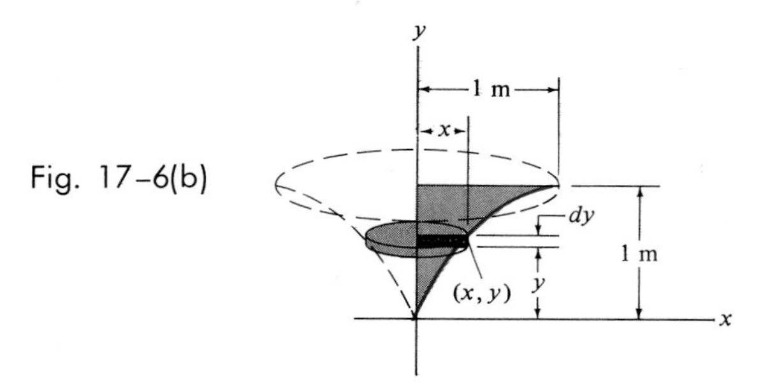

Problems

17–1. Determine the moment of inertia I_z for the slender rod which has a mass m. The rod's density ρ and cross-sectional area A are constant.

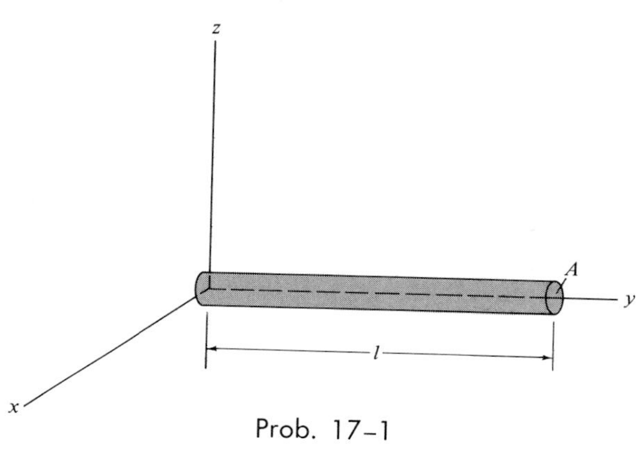

Prob. 17–1

17–2. Determine the moment of inertia of the homogeneous block of mass m with respect to its centroidal $\bar{x}$ axis. The block has a constant density ρ and the center of mass at its geometric center C.

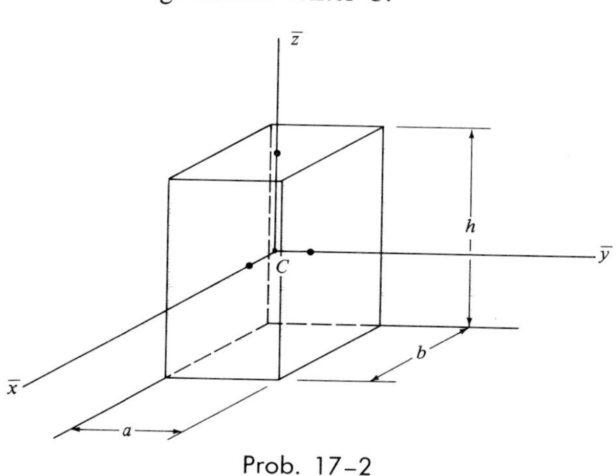

Prob. 17–2

17–3. Determine the moment of inertia of a homogeneous sphere of mass m and radius r with respect to the y axis. The density of the material, ρ, is constant.

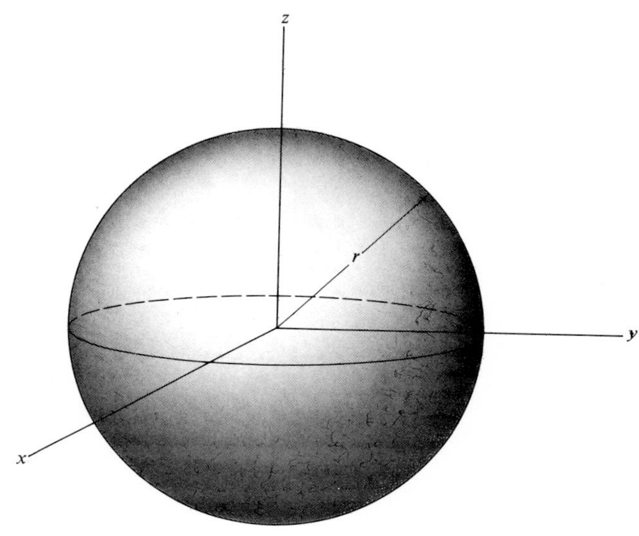

Prob. 17–3

*** 17–4.** The right circular cone is formed by revolving the shaded area around the x axis. Determine the moment of inertia I_x and express the result in terms of the total mass m of the cone. The cone has a constant density ρ.

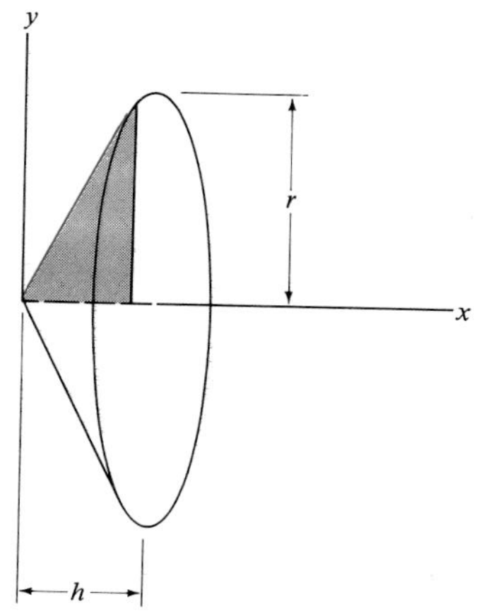

Prob. 17–4

17–5. The solid is formed by revolving the shaded area around the x axis, $a = 100$ mm. Determine the radius of gyration k_x. The mass density of the material is $\rho = 6\ \text{Mg/m}^3$.

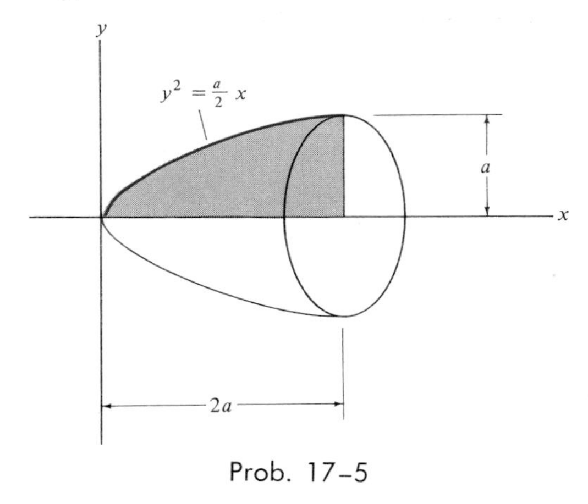

Prob. 17–5

17–5a. Solve Prob. 17–5 if $a = 2$ ft and $\rho = 300\ \text{lb/ft}^3$.

17–6. An ellipsoid is formed by rotating the shaded area about the x axis. Determine the moment of inertia of this body with respect to the x axis and express the result in terms of the mass m of the solid. Note that the boundary of the shaded area is defined by the equation $x^2/a^2 + y^2/b^2 = 1$. The density of the material, ρ, is constant.

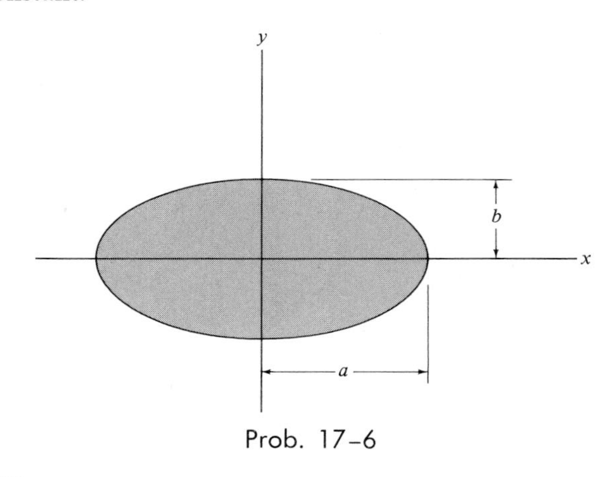

Prob. 17–6

17–7. Determine the moment of inertia of the hemispherical solid about the y axis. Express the result in terms of the mass m of the hemisphere. The density of the material, ρ, is constant.

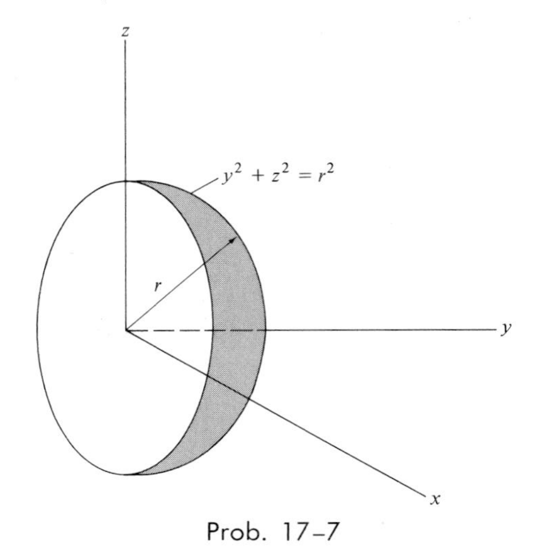

Prob. 17–7

*** 17–8.** The solid is formed by rotating the shaded area about the y axis. Determine the moment of inertia I_y. The density of the material is $\rho = 8\ \text{Mg/m}^3$.

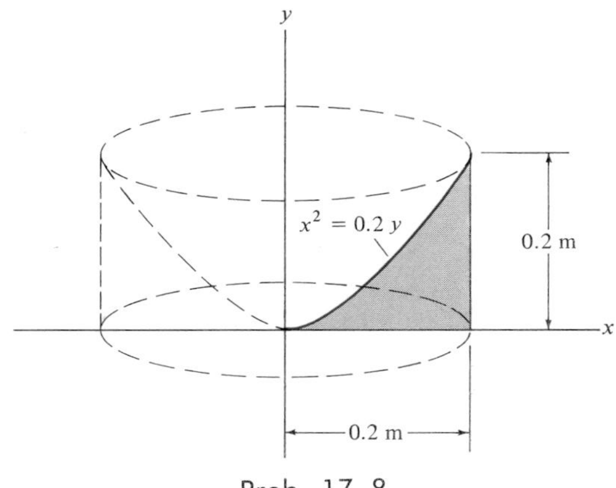

Prob. 17–8

17–9. Determine the moment of inertia I_z of the torus. The mass of the torus is m and the density ρ is constant.

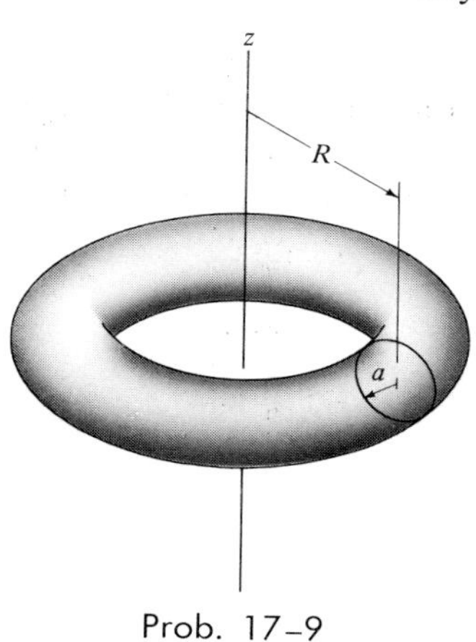

Prob. 17–9

17–10. Determine the moment of inertia of the wheel with respect to the z axis. The rim R and plate P both have a density of $\rho = 7\ \text{Mg/m}^3$. Set $r_i = 200$ mm, $r_o = 240$ mm, $t_i = 20$ mm, and $t_o = 80$ mm.

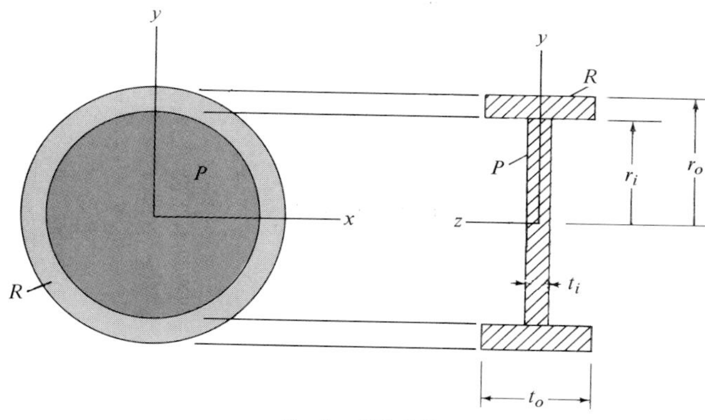

Prob. 17–10

17–10a. Solve Prob. 17–10 if $\rho = 600\ \text{lb/ft}^3$, $r_i = 0.7$ ft, $r_o = 0.8$ ft, $t_i = 0.1$ ft, and $t_o = 0.2$ ft.

17–11. Determine the moment of inertia I_z of the composite solid. Each semicircular cylinder has a density of $\rho = 8\ \text{Mg/m}^3$, and the rectangular block has a density of $\rho = 6\ \text{Mg/m}^3$. The z axis passes through the center of gravity G.

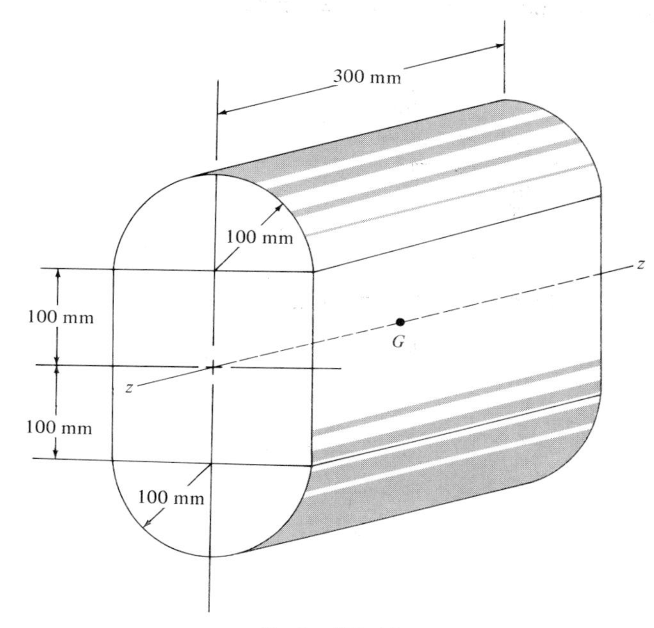

Prob. 17–11

*** 17–12.** Determine the moment of inertia I_x of the solid. The density of the material is $\rho = 6\ \text{Mg/m}^3$.

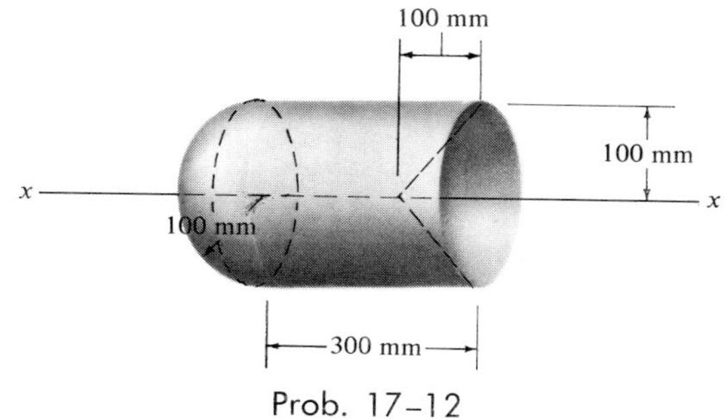

Prob. 17–12

17-13. Determine the moment of inertia I_z of the composite assembly. The z axis is directed perpendicular to the page and passes through the center of gravity G. Plate A has a thickness of 10 mm and the four slender rods each have a cross-sectional area of 800 mm^2. The material has a density of $\rho = 7.5\ Mg/m^3$.

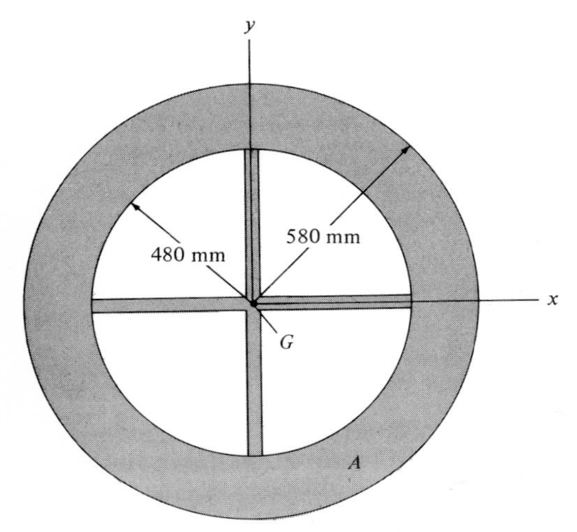

Prob. 17-13

17-14. Locate the center of mass G of the baseball bat by determining $\bar{y}$. Then, compute the mass moment of inertia of the bat about an axis which passes through G and is perpendicular to the plane of the page. For the calculation, consider the bat to be composed of a truncated cone and cylinder and neglect the size of the lip at A. The density of wood is $\rho_w = 700\ kg/m^3$.

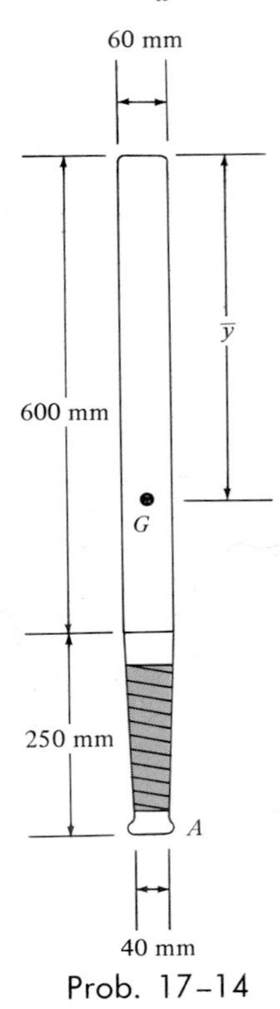

Prob. 17-14

17–15. The "mixer" on a blender consists of four equivalent thin rectangular blades, each having a density of $\rho_b = 4\ \text{Mg/m}^3$ and dimensions $a = 400$ mm, $b = 300$ mm, and $t = 10$ mm. If the blades are attached to a hollow cylinder having a density of $\rho_c = 5.5\ \text{Mg/m}^3$ and radii $r_i = 50$ mm, $r_o = 100$ mm, determine the moment of inertia of the assembly about the central z axis.

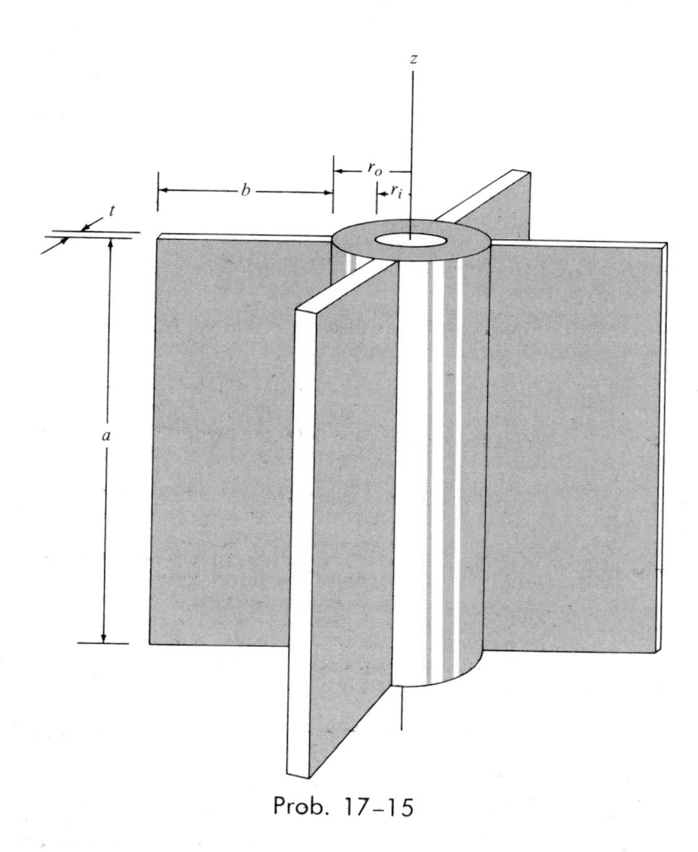

Prob. 17–15

17–15a. Solve Prob. 17–15 if $\rho_b = 525\ \text{lb/ft}^3$, $a = 2$ ft, $b = 1.50$ ft, $t = 0.1$ ft; and $\rho_c = 650\ \text{lb/ft}^3$, $r_i = 0.2$ ft, $r_o = 0.4$ ft.

17–3. Planar Kinetic Equations of Motion

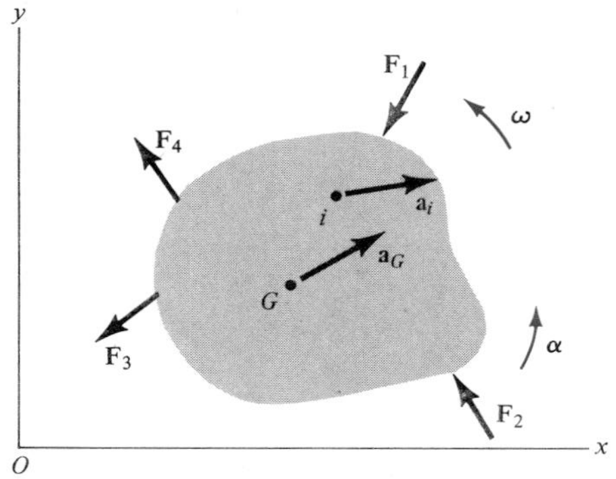

y, $\mathbf{F}_1$, ω, $\mathbf{F}_4$, i, $\mathbf{a}_i$, $\mathbf{a}_G$, G, α, $\mathbf{F}_3$, $\mathbf{F}_2$, x, O

Fig. 17–7(a)

Consider the rigid body (slab) shown in Fig. 17–7*a,* which has a mass m and is subjected to motion viewed in a reference plane. The *inertial frame of reference x, y* has its origin at point O. By definition, these axes do not rotate and are either fixed or translate with constant velocity. At the *instant* considered, the applied force system causes the slab to have an angular acceleration $\boldsymbol{\alpha}$ and angular velocity $\boldsymbol{\omega}$, while the center of mass of the slab has an acceleration $\mathbf{a}_G$.

Equation of Translational Motion. The free-body and kinetic diagrams for the arbitrary ith particle located in the body are shown in Fig. 17–7*b*. There are two types of forces which act on this particle. The *internal forces,* represented symbolically as $\Sigma\mathbf{f}_i$, are reactive forces which the other particles each exert on the ith particle. The resultant *external force* $\mathbf{F}_i$ represents the effect of gravitational, electrical, magnetic, or contact forces between adjacent bodies. Since this force system has been considered previously in Sec. 13–3, for the analysis of a system of n particles, the results may be used here, in which case the n particles are contained within the boundary of the body (or slab). Hence, if the equation of motion is applied to each of the n particles of the body, and the results added vectorially, it may be concluded that

$$\Sigma\mathbf{F} = m\mathbf{a}_G \tag{17–6}$$

This equation is referred to as the *equation of translational motion* for a rigid body. It states that *the resultant force of all the external forces acting on the body,* $\Sigma\mathbf{F}$, *is equal to the body's mass m times the acceleration of the mass center* $\mathbf{a}_G$. If $\mathbf{F}_R$ represents the resultant external force $\mathbf{F}_R = \Sigma\mathbf{F}$ which acts on the body, then from Eq. 17–6 it is necessary that $\mathbf{F}_R$ has a *magnitude* ma_G and a *line of action* which is *always* collinear with $m\mathbf{a}_G$.

For motion of the body (or slab) in the x-y plane, the external force system acting on the body may be expressed in terms of its x and y components. Also, the acceleration of the body's mass center may be written as $\mathbf{a}_G = (a_G)_x\mathbf{i} + (a_G)_y\mathbf{j}$. Vector equation 17–6 may then be written in the form of two independent scalar equations,

$$\begin{aligned} \Sigma F_x &= m(a_G)_x \\ \Sigma F_y &= m(a_G)_y \end{aligned} \tag{17–7}$$

Equation of Rotational Motion. Consider now the effects caused by the moments of the external force system, computed about the body's mass center G. As shown in Fig. 17–7*b,* vector $\mathbf{r}_{i/G}$ locates the ith particle of the body with respect to G. Hence, the moments of the external force resultant $\mathbf{F}_i$ and the internal forces $\Sigma\mathbf{f}_i$ about G (free-body diagram) must be equivalent to the moment created by $m_i\mathbf{a}_i$ about G (kinetic diagram), i.e.,

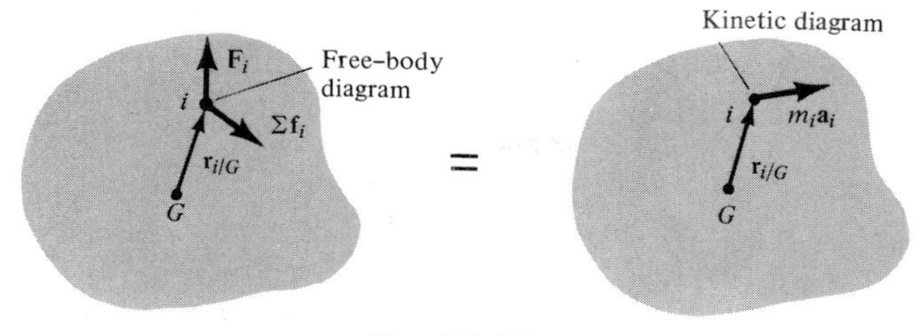

Fig. 17–7(b)

$$(\mathbf{r}_{i/G} \times \mathbf{F}_i) + (\mathbf{r}_{i/G} \times \Sigma\mathbf{f}_i) = (\mathbf{r}_{i/G} \times m_i\mathbf{a}_i) \qquad (17\text{–}8)$$

The acceleration $\mathbf{a}_i$ may be related to the acceleration of the body's mass center, $\mathbf{a}_G$, by Eq. 16–22, i.e.,

$$\mathbf{a}_i = \mathbf{a}_G + \boldsymbol{\alpha} \times \mathbf{r}_{i/G} - \omega^2\mathbf{r}_{i/G}$$

Substituting into Eq. 17–8, using the distributive property of the vector cross product, and noting that the last term $-m_i\omega^2(\mathbf{r}_{i/G} \times \mathbf{r}_{i/G}) = \mathbf{0}$, we get

$$(\mathbf{r}_{i/G} \times \mathbf{F}_i) + (\mathbf{r}_{i/G} \times \Sigma\mathbf{f}_i) = (m_i\mathbf{r}_{i/G} \times \mathbf{a}_G) + m_i\mathbf{r}_{i/G} \times (\boldsymbol{\alpha} \times \mathbf{r}_{i/G})$$

If equations like these are written for each of the n particles of the body, and the results are added together vectorially, the moments created by all the internal forces about G will cancel. This is because, by Newton's third law of motion, these forces occur in equal but opposite collinear pairs. Defining the moments created by the external forces about G by the sum $\Sigma\mathbf{M}_G$, we can therefore write the result as

$$\Sigma\mathbf{M}_G = (\Sigma m_i\mathbf{r}_{i/G}) \times \mathbf{a}_G + \Sigma[m_i\mathbf{r}_{i/G} \times (\boldsymbol{\alpha} \times \mathbf{r}_{i/G})]$$

The first term on the right side of this equation is zero, since $\Sigma m_i\mathbf{r}_{i/G} = (\Sigma m_i)\bar{\mathbf{r}}$, and the position vector $\bar{\mathbf{r}}$ relative to G is equal to zero by definition of the center of mass.* Furthermore, the second term on the right side may be simplified by noting that $\mathbf{r}_{i/G} \times (\boldsymbol{\alpha} \times \mathbf{r}_{i/G})$ is a vector having a *magnitude* of $r_{i/G}^2\,\alpha$ and a *direction* which is perpendicular to the plane of motion. Since $\Sigma\mathbf{M}_G$ also acts in this direction, the above equation can be written in the scalar form

$$\Sigma M_G = (\Sigma m_i r_{i/G}^2)\alpha$$

When the particle size $m_i \rightarrow dm$, the summation sign becomes an integral since the number of particles $n \rightarrow \infty$. Replacing $r_{i/G}$ by the generalized dimension r, we have

$$\Sigma M_G = \left(\int_m r^2\,dm\right)\alpha$$

*The center of mass of a body is defined in Sec. 9–2 of *Engineering Mechanics: Statics*.

The integral represents the *mass moment of inertia* of the body about an axis (the z' axis) passing through point G and directed perpendicular to the plane of motion. Indicating this term by I_G we can write the final result as

$$\Sigma M_G = I_G\alpha \tag{17-9}$$

This scalar equation is referred to as the *equation of rotational motion.* It states that *the sum of the moments of all the external forces computed about the body's mass center G is equal to the product of the mass moment of inertia of the body about G and the magnitude of the body's angular acceleration* $\boldsymbol{\alpha}$.

General Application of the Equations of Motion. From the above analysis, *three* scalar equations may be written to describe the general plane motion of a symmetrical rigid body,

$$\begin{aligned} \Sigma F_x &= m(a_G)_x \\ \Sigma F_y &= m(a_G)_y \\ \Sigma M_G &= I_G\alpha \end{aligned} \tag{17-10}$$

When applying these equations to the solution of problems, one should always draw the free-body and kinetic diagrams of the body at the instant considered. The *free-body diagram* graphically accounts for the terms involving ΣF_x, ΣF_y, and ΣM_G, Fig. 17–7*c*. The terms $m(a_G)_x$ and $m(a_G)_y$ are represented graphically on the *kinetic diagram,* Fig. 17–7*d*. Also shown on this diagram is $I_G\boldsymbol{\alpha}$, which has a *magnitude* of $I_G\alpha$ and a *direction* defined by $\boldsymbol{\alpha}$. The two diagrams are "equated" since the forces on the free-body diagram cause the accelerated motion indicated by the three vectors shown on the kinetic diagram.

Although the moment equation $\Sigma M_G = I_G\alpha$ applies *only* at point G, *other points* may be chosen for summing moments *provided* one accounts

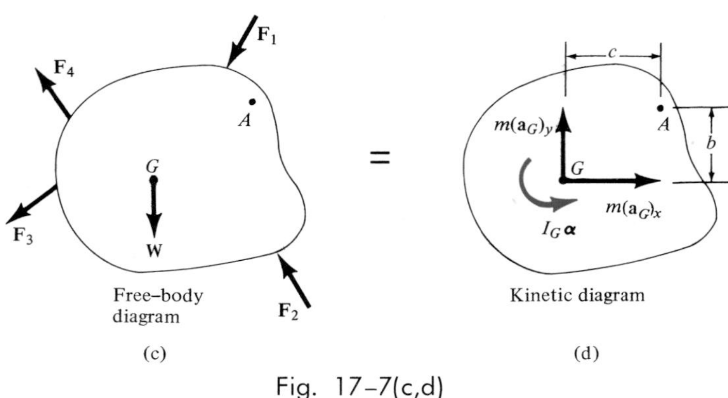

Fig. 17–7(c,d)

for the moments produced by $I_G\boldsymbol{\alpha}$ and the components of $m\mathbf{a}_G$ about the point. When these "kinetic" moments, ΣM_k, are computed, the vectors $m(\mathbf{a}_G)_x$ and $m(\mathbf{a}_G)_y$ are treated in the same manner as a force, that is, they can act at *any point along their line of action.* In a similar manner, $I_G\boldsymbol{\alpha}$ has the same properties as a couple and can therefore act at *any point* on the kinetic diagram. For example, if moments of the external forces are summed about the arbitrary point A on the free-body diagram, Fig. 17–7*c*, they are equivalent to the moment summation about point A on the kinetic diagram, Fig. 17–7*d*. Hence, if moments are assumed to be positive counterclockwise, the required moment equation applied at A becomes

$$\circlearrowleft +\Sigma M_A = \Sigma(M_k)_A; \quad \Sigma M_A = I_G\alpha + [m(a_G)_x]b - [m(a_G)_y]c$$

where the dimensions b and c are defined in Fig. 17–7*d*.

Although the equations of motion may be applied in vector form, throughout this chapter only a scalar approach will be used. A vector solution becomes advantageous only when the geometry of the problem becomes complicated, which occurs most often in three dimensions. A scalar solution of planar-motion problems is more direct and leads to greater insight regarding the physical aspects of the problem.

17–4. Equations of Motion: Translation of a Rigid Body

When a rigid body undergoes a *translation,* all the particles of the body have the *same acceleration,* so that $\mathbf{a}_G = \mathbf{a}$, Fig. 17–8*a*. Furthermore, $\boldsymbol{\alpha} = \mathbf{0}$, in which case the rotational equation of motion applied at point G reduces to a simplified form, $\Sigma M_G = 0$. Application of this and the translational equations of motion will now be discussed for each of the two types of translation presented in Chapter 16.

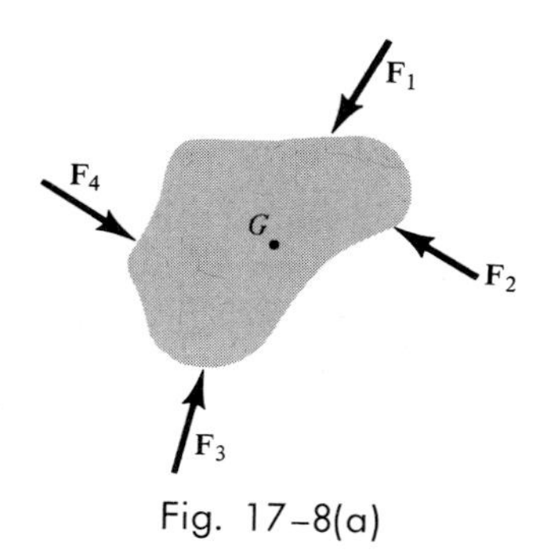

Fig. 17–8(a)

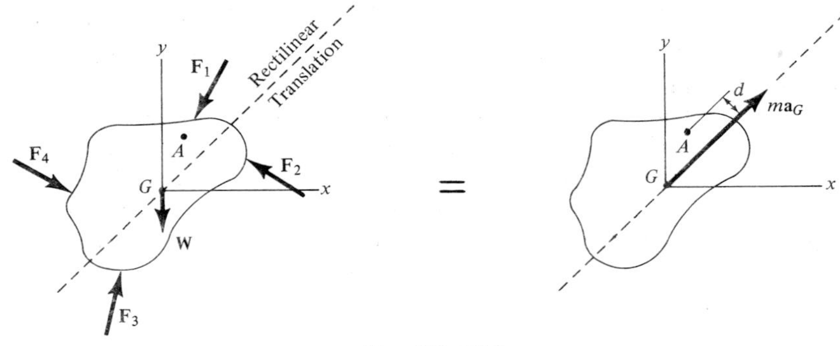

Fig. 17–8(b)

Rectilinear Translation. When a body is subjected to *rectilinear translation,* all the particles of the body (slab) travel along parallel straight-line paths. The free-body and kinetic diagrams, when the body in Fig. 17–8*a* is subjected to this motion, are shown in Fig. 17–8*b*. At the instant considered, the origin of the inertial reference is located at the center of mass G. Since $I_G\boldsymbol{\alpha} = \mathbf{0}$, only $m\mathbf{a}_G$ is shown on the kinetic diagram. Hence, the equations of motion which apply in this case become

$$\begin{aligned} \Sigma F_x &= m(a_G)_x \\ \Sigma F_y &= m(a_G)_y \\ \Sigma M_G &= 0 \end{aligned} \tag{17–11}$$

The last equation requires that the sum of the moments of all the external forces computed about the body's center of mass G be equal to zero. It is possible, of course, to sum moments about other points on the body, in which case the moment of $m\mathbf{a}_G$ must be taken into account. For example, if point A is chosen, which lies at a perpendicular distance d from the line of action of $m\mathbf{a}_G$, the following moment equation applies:

$$\circlearrowleft + \Sigma M_A = \Sigma(\mathcal{M}_k)_A; \qquad \Sigma M_A = (ma_G)d$$

Here the moment of the external forces about A (ΣM_A, free-body diagram) equals the moment of $m\mathbf{a}_G$ about A ($\Sigma(\mathcal{M}_k)_A$, kinetic diagram).

Curvilinear Translation. When a rigid body is subjected to *curvilinear translation,* all the particles of the body travel along *parallel curved paths.* For analysis, it is often convenient to use an inertial coordinate system having an origin at G and axes which are oriented in the normal and tangential directions to the path of motion, Fig. 17–8*c*. The three scalar equations of motion are then

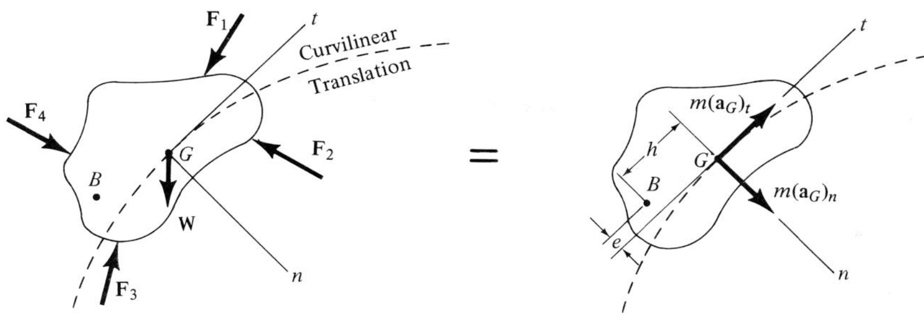

Fig. 17–8(c)

$$\Sigma F_n = m(a_G)_n$$
$$\Sigma F_t = m(a_G)_t \quad (17\text{–}12)$$
$$\Sigma M_G = 0$$

where $(a_G)_t$ and $(a_G)_n$ represent, respectively, the magnitudes of the tangential and normal components of acceleration of point G.

If the moment equation $\Sigma M_G = 0$ is replaced by a moment summation about the arbitrary point B, Fig. 17–8c, it is necessary to account for the moments, $\Sigma(M_k)_B$, of the two components $m(\mathbf{a}_G)_n$ and $m(\mathbf{a}_G)_t$ about this point. From the kinetic diagram, e and h represent the perpendicular distances (or "moment arms") from B to the line of action of the components. If positive moments are assumed to be clockwise, the required moment equation becomes

$$\curvearrowright + \Sigma M_B = \Sigma(M_k)_B; \quad \Sigma M_B = h[m(a_G)_n] - e[m(a_G)_t]$$

PROCEDURE FOR ANALYSIS

A three-step procedure should be followed when solving kinetic problems using the translational equations of motion.

Step 1: Draw the free-body and kinetic diagrams for the body. Recall that the *free-body diagram* is a graphical representation of all the external forces ($\Sigma\mathbf{F}$) which act on the body,* whereas the *kinetic diagram* graphically accounts for the components $m(\mathbf{a}_G)_x$, $m(\mathbf{a}_G)_y$ or $m(\mathbf{a}_G)_t$, $m(\mathbf{a}_G)_n$, Fig. 17–8*b* or 17–8*c*.

Step 2: Apply the three equations of motion, Eqs. 17–11 or Eqs. 17–12, with reference to an established x, y, or n, t inertial coordinate system. All the terms in these equations are computed directly from the data shown on the free-body and kinetic diagrams. To simplify the analysis, the

*See Chapter 5 of *Engineering Mechanics: Statics* for a discussion of free-body diagrams for rigid bodies.

moment equation $\Sigma M_G = 0$ can be replaced by the more general equation $\Sigma M_A = \Sigma(\mathcal{M}_k)_A$, where point A is usually located at the intersection of the lines of action of as many unknown forces as possible.

If the body is in contact with a *rough surface* and slipping occurs, use the frictional equation $F = \mu_k N$ to relate the normal force $\mathbf{N}$ to its associated frictional force $\mathbf{F}$.*

Step 3: Use kinematics if a complete solution cannot be obtained strictly from the equations of motion. For *rectilinear translation* with *variable acceleration,* use

$$a_G = \frac{dv_G}{dt}, \qquad a_G\, ds_G = v_G\, dv_G, \qquad v_G = \frac{ds_G}{dt}$$

For *rectilinear translation* with *constant acceleration,* use

$$v_G = (v_G)_1 + a_G t$$
$$v_G^2 = (v_G)_1^2 + 2a_G[s_G - (s_G)_1]$$
$$s_G = (s_G)_1 + (v_G)_1 t + \tfrac{1}{2}a_G t^2$$

For *curvilinear translation,* use

$$(a_G)_n = \frac{v_G^2}{\rho} = \omega^2 \rho$$

$$(a_G)_t = \frac{dv_G}{dt}, \qquad (a_G)_t\, ds_G = v_G\, dv_G, \qquad (a_G)_t = \alpha\rho$$

The following examples numerically illustrate application of this three-step procedure.

Example 17–4

The car shown in Fig. 17–9*a* has a mass of 2000 kg and a center of mass at G. If the "driving" wheels in the back are always slipping, whereas the front wheels freely rotate, determine the distance required for the car to reach a speed of 10 m/s starting from rest. Neglect the mass of the wheels. The coefficient of kinetic friction between the wheels and the road is $\mu_k = 0.25$.

Solution

Step 1: The free-body and kinetic diagrams of the car are shown in Fig. 17–9*b*. The rear-wheel frictional force $\mathbf{F}_B$ pushes the car forward and since *slipping occurs,* $\mathbf{F}_B$ is related to its associated normal $\mathbf{N}_B$ by

*Since many of the problems in rigid-body kinetics involve friction, it is suggested that one review the material on friction, covered in Secs. 8–1 and 8–2 of *Engineering Mechanics: Statics*.

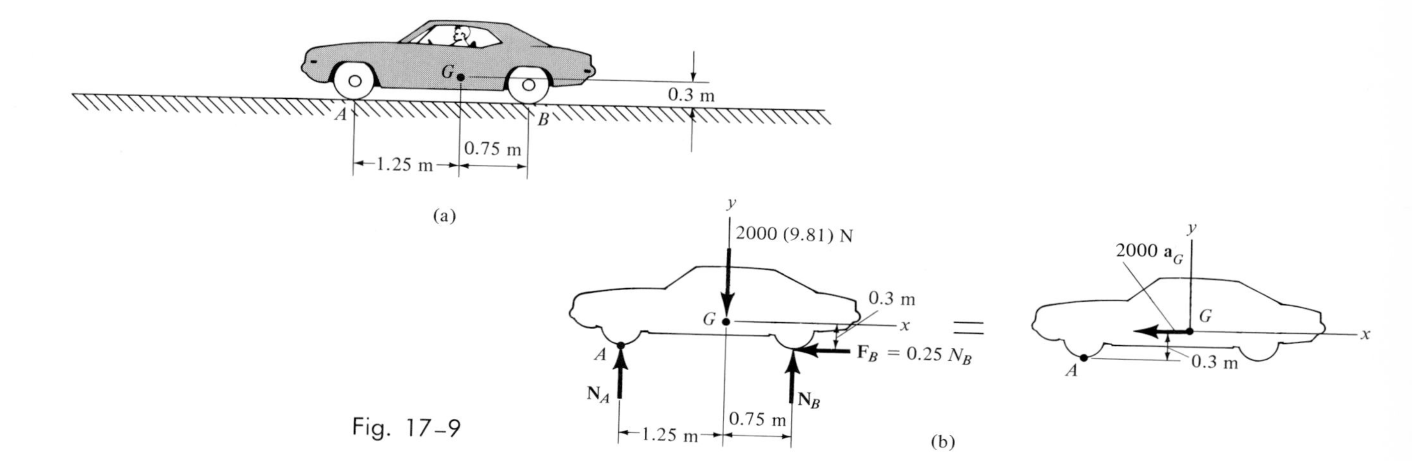

Fig. 17–9

$F_B = 0.25N_B$. The frictional force acting on the *front wheels* is *zero,* since these wheels have negligible mass.*

Step 2: Applying the equations of motion yields

$$\overset{+}{\leftarrow}\Sigma F_x = m(a_G)_x; \qquad 0.25N_B = 2000a_G \tag{1}$$

$$+\uparrow\Sigma F_y = m(a_G)_y; \qquad N_A + N_B - 2000(9.81) = 0 \tag{2}$$

$$\circlearrowleft + \Sigma M_G = 0; \qquad N_A(1.25) + 0.25N_B(0.3) - N_B(0.75) = 0 \tag{3}$$

The "moment" equation can also be applied at point A, which eliminates N_A from the equation. In this case the moment of the force system about A (free-body diagram) is equivalent to the moment of $m\mathbf{a}_G$ about A (kinetic diagram).

$$\circlearrowright + \Sigma M_A = \Sigma(M_k)_A; \quad N_B(2) - 2000(9.81)(1.25) = 2000(a_G)(0.3) \tag{4}$$

Solving Eqs. (1), (2), and (3) or the simpler set of Eqs. (1), (2), and (4) gives

$$a_G = 1.59 \text{ m/s}^2$$
$$N_A = 6.88 \text{ kN}$$
$$N_B = 12.74 \text{ kN}$$

Step 3: Since the acceleration is *constant,* we can obtain the car's displacement, using the equation

$(\overset{+}{\leftarrow})$

$$v_G^2 = (v_G)_1^2 + 2a_G[s_G - (s_G)_1]$$
$$10^2 = 0 + 2(1.59)(s_G - 0)$$
$$s_G = 31.4 \text{ m} \qquad \textit{Ans.}$$

*If the mass of the front wheels were to be included in the analysis, the frictional force acting at A would be *directed to the right* to create the counterclockwise rotation of the wheels. The problem solution for this case would be more involved since a general-plane-motion analysis of the wheels would have to be included (see Sec. 17–6).

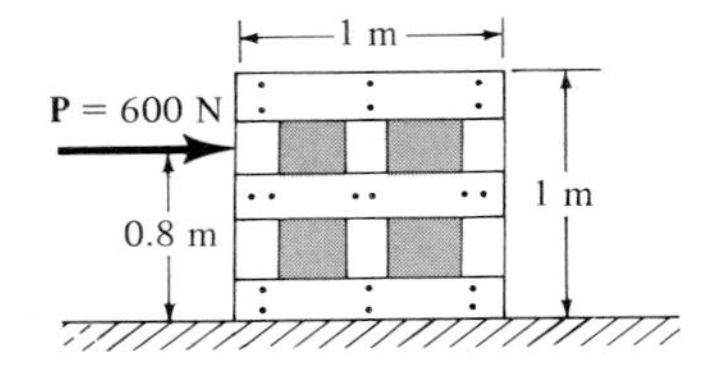

Fig. 17–10

Example 17–5

A uniform 50-kg crate rests on a horizontal surface for which the coefficient of kinetic friction is $\mu_k = 0.2$. If a force of $P = 600$ N is applied to the crate as shown in Fig. 17–10*a*, compute the velocity of the crate after it moves 3 m. Assume that the crate is originally at rest.

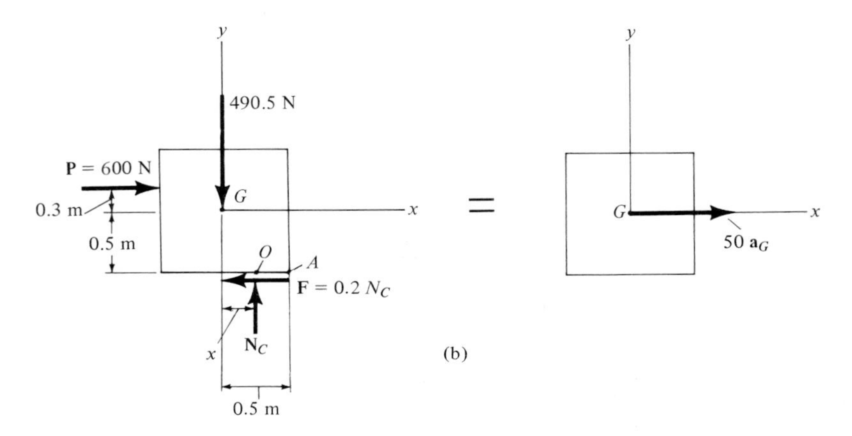

Solution

Step 1: The free-body and kinetic diagrams of the crate are shown in Fig. 17–10*b*. The force **P** can either cause the crate to slide or to tip over. Here it is assumed that the crate slides, so that $F = \mu_k N_C = 0.2N_C$. Also, the resultant normal force $\mathbf{N}_C$ acts at O, a distance of x (where $0 < x \leqslant 0.5$ m) from the crate's center line.*

Step 2: Applying the equations of motion yields

$$\overset{+}{\rightarrow}\Sigma F_x = m(a_G)_x; \qquad 600 - 0.2N_C = 50a_G \tag{1}$$

$$+\uparrow\Sigma F_y = m(a_G)_y; \qquad N_C - 490.5 = 0 \tag{2}$$

$$\circlearrowleft+\Sigma M_G = 0; \qquad -600(0.3) + N_C(x) - 0.2N_C(0.5) = 0 \tag{3}$$

Solving,

$$N_C = 490.5 \text{ N}$$
$$x = 0.47 \text{ m}$$
$$a_G = 10.0 \text{ m/s}^2$$

Since $x = 0.47$ m < 0.5 m, indeed the crate slides as originally assumed. If the solution had given a value of $x > 0.5$ m, the problem would have to be reworked with the assumption that tipping occurred. If this was the case, $\mathbf{N}_C$ would act at the *corner point A* and $F \leqslant 0.2N_C$, since in general the

*The line of action of $\mathbf{N}_C$ does not necessarily pass through the mass center G ($x = 0$), since $\mathbf{N}_C$ must counteract the effect of tipping caused by **P**. See Sec. 8–1 of *Engineering Mechanics: Statics.*

crate would *not* be on the verge of sliding at the instant it begins to tip.

Step 3: Since the acceleration is *constant*, the velocity in 3 m is

$$(\overset{+}{\rightarrow}) \qquad v_G^2 = (v_G)_1^2 + 2a_G[s_G - (s_G)_1]$$
$$v_G^2 = 0 + 2(10.0)(3 - 0)$$
$$v_G = 7.75 \text{ m/s} \qquad \textit{Ans.}$$

Example 17–6

The 100-kg beam BD shown in Fig. 17–11a is supported by two rods having negligible mass. At the instant $\theta = 30°$, the rods are both rotating with an angular velocity of $\omega = 6$ rad/s. Determine the force created in each rod and the angular acceleration of the rods at this instant.

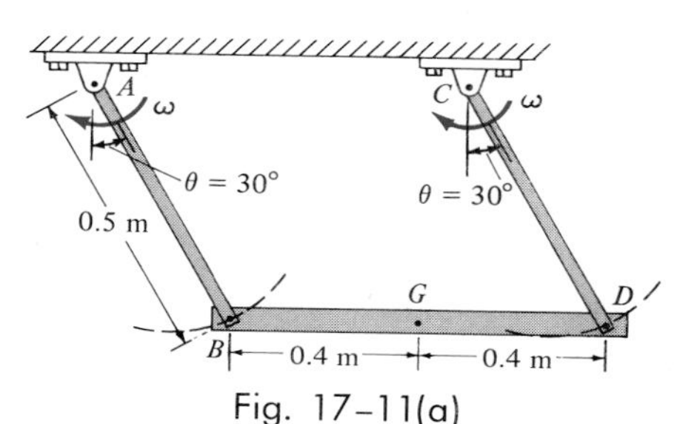

Fig. 17–11(a)

Solution I

Step 1: The beam moves with *curvilinear translation* since points B, D, and the center of mass G move along circular paths, each path having the same radius of 0.5 m. Using normal and tangential coordinates, the free-body and kinetic diagrams for the beam are shown in Fig. 17–11b. Because of the translation, G has the *same* acceleration and velocity as B, which is connected to both the rod and the beam. By studying the angular motion of rod AB, Fig. 17–11c, note that the tangential component of acceleration $\mathbf{a}_t$ acts downward to the left due to the clockwise direction of $\boldsymbol{\alpha}$. Furthermore, the normal component of acceleration $\mathbf{a}_n$ is *always* directed toward the center of curvature (toward point A for rod AB). Since the angular velocity of AB is 6 rad/s, then

$$a_n = \omega^2 r_{B/A} = (6)^2(0.5) = 18 \text{ m/s}^2$$

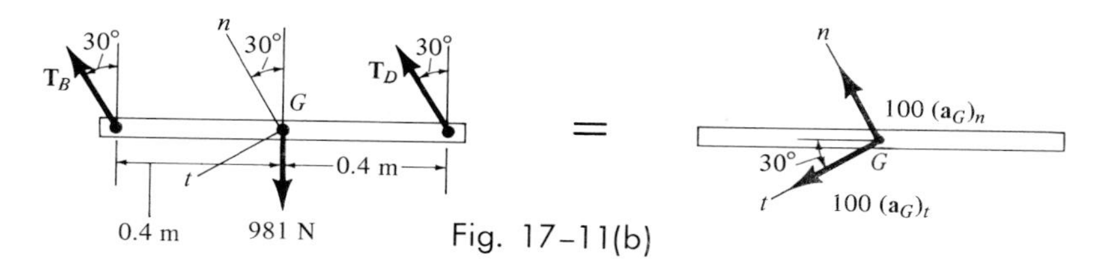

Fig. 17–11(b)

Step 2: Applying the equations of motion yields

$$+\nwarrow \Sigma F_n = m(a_G)_n; \quad T_B + T_D - 981 \cos 30° = 100(18) \qquad (1)$$
$$+\swarrow \Sigma F_t = m(a_G)_t; \quad 981 \sin 30° = 100(a_G)_t \qquad (2)$$
$$\circlearrowleft + \Sigma M_G = 0; \quad -(T_B \cos 30°)0.4 + (T_D \cos 30°)0.4 = 0 \qquad (3)$$

Simultaneous solution of these three equations gives

$$T_B = T_D = 1325 \text{ N} \qquad \textit{Ans.}$$
$$(a_G)_t = 4.905 \text{ m/s}^2$$

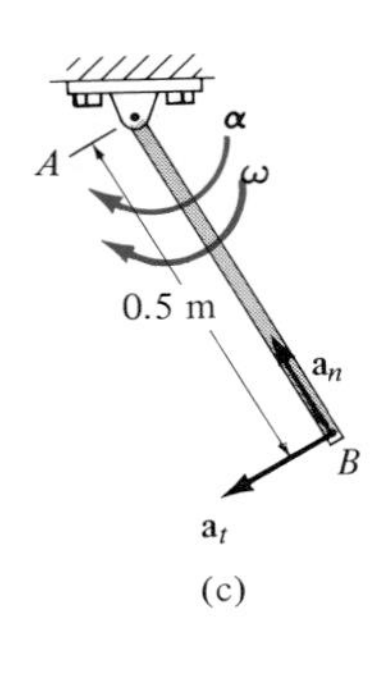

Step 3: The angular acceleration of each rod may now be found as shown in Fig. 17–11*c*. Using $a_t = (a_G)_t = r_{B/A}\alpha$ yields

$$\alpha = \frac{(a_G)_t}{r_{B/A}} = \frac{4.905\ \text{m/s}^2}{0.5\ \text{m}} = 9.81\ \text{rad/s}^2 \qquad \textit{Ans.}$$

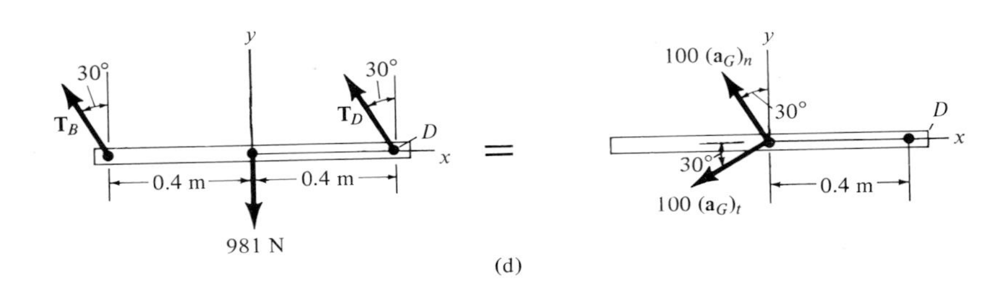

Fig. 17–11(c–d)

Solution II

Once the free-body and kinetic diagrams are established, Fig. 17–11*d*, the equations of translational motion can be applied in the x and y directions. This gives

$$\overset{+}{\rightarrow}\Sigma F_x = m(a_G)_x;$$
$$-T_B \sin 30° - T_D \sin 30° = -100(a_G)_n \sin 30° - 100(a_G)_t \cos 30° \qquad (4)$$
$$+\uparrow\Sigma F_y = m(a_G)_y; \quad T_B \cos 30° + T_D \cos 30° - 981$$
$$= 100(a_G)_n \cos 30° - 100(a_G)_t \sin 30° \qquad (5)$$

The rotational equation of motion will be applied at point D. Hence,

$$\circlearrowleft +\Sigma M_D = \Sigma(M_k)_D; \quad T_B \cos 30°(0.8) - 981(0.4)$$
$$= [100(a_G)_n \cos 30°](0.4) - [100(a_G)_t \sin 30°](0.4) \qquad (6)$$

Solving Eqs. (4), (5), and (6) simultaneously for T_B, T_D, and $(a_G)_t$ yields the same results previously obtained.

Example 17–7

As a result of the action of a 500-N horizontal force, the 10-kg block shown in Fig. 17–12*a* slides along the horizontal rail for which $\mu_k = 0.2$. If a 15-kg link is attached to the block at the pin connection A, determine the constant angle θ which the link makes with the vertical, the horizontal and vertical components of reaction at the pin A, and the acceleration of the system.

Solution

Step 1: Since the force acting on the block is constant, the link makes a constant angle θ with the vertical and therefore the block and the link have the *same* acceleration **a**. The free-body and kinetic diagrams for each

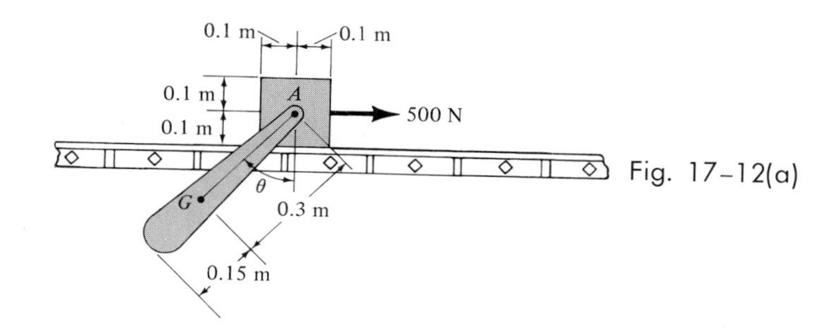

Fig. 17–12(a)

body are shown in Figs. 17–12*b* and 17–12*c*. The directions of the force components $\mathbf{A}_x$ and $\mathbf{A}_y$ have been assumed. In accordance with Newton's third law of motion, these components have equal magnitude and act in opposite directions on the free-body diagrams of the link and the block. Since the block is *moving* to the right, the frictional force has a magnitude of $F = \mu_k N_B = 0.2N_B$ and acts opposite to the motion of the block. Furthermore, assuming that the block does not tip, the normal force $\mathbf{N}_B$ acts at an unknown distance x from the center of the block such that $0 < x \leqslant 0.1$ m.

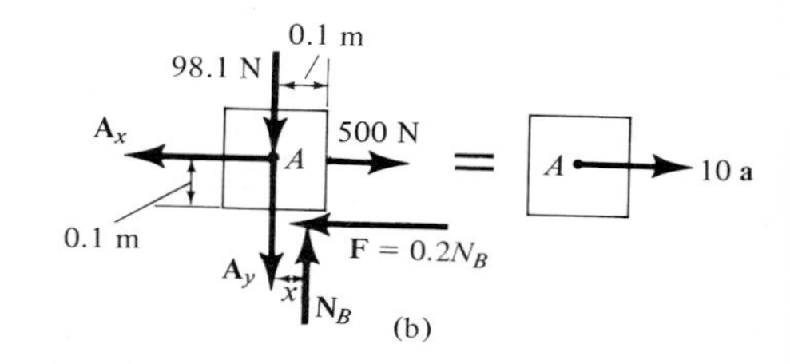

Step 2: Applying the three equations of motion to the block yields

$$\overset{+}{\rightarrow}\Sigma F_x = m(a_A)_x; \qquad 500 - A_x - 0.2N_B = 10a \tag{1}$$

$$+\uparrow\Sigma F_y = m(a_A)_y; \qquad N_B - 98.1 - A_y = 0 \tag{2}$$

$$\circlearrowleft+\Sigma M_A = 0; \qquad N_B(x) - 0.1(0.2N_B) = 0 \tag{3}$$

Applying the two translational equations of motion to the link, gives

$$\overset{+}{\rightarrow}\Sigma F_x = m(a_G)_x; \qquad A_x = 15a \tag{4}$$

$$+\uparrow\Sigma F_y = m(a_G)_y; \qquad A_y - 147.2 = 0 \tag{5}$$

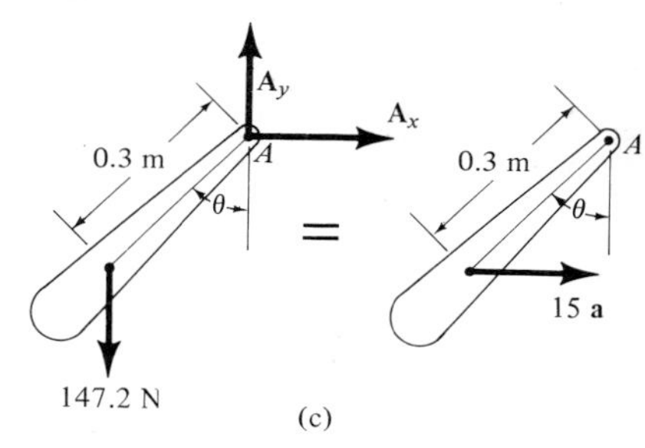

Fig. 17–12(b,c)

Moments will be summed about A in order to eliminate the two unknown force components $\mathbf{A}_x$ and $\mathbf{A}_y$ at this point. In reference to the free-body and kinetic diagrams, Fig. 17–12*c*, we have

$$\circlearrowright+\Sigma M_A = \Sigma(M_k)_A; \qquad 147.2(0.3\sin\theta) = 0.3\cos\theta(15a) \tag{6}$$

Solving Eqs. (1) through (6) for the six unknowns yields

$$N_B = 245.3 \text{ N}$$

$$x = 0.02 \text{ m} = 20 \text{ mm}$$

$$A_x = 270.6 \text{ N} \qquad \textit{Ans.}$$

$$A_y = 147.2 \text{ N} \qquad \textit{Ans.}$$

$$\theta = 61.4^\circ \qquad \textit{Ans.}$$

$$a = 18.0 \text{ m/s}^2 \qquad \textit{Ans.}$$

Since $x \leqslant 0.1$ m, the assumption that the block slides without tipping was correct.

Problems

Except when stated otherwise, throughout this chapter assume that the coefficients of static and kinetic friction are equal, i.e., $\mu = \mu_s = \mu_k$.

***17–16.** The car has a mass of 1.6 Mg and a mass center located at G. Determine the normal reactions of the wheels on the road and the acceleration of the car if it is rolling freely down the incline. Neglect the mass of the wheels.

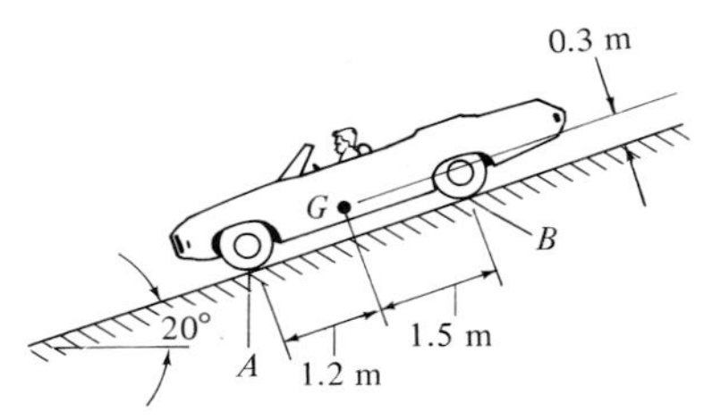

Prob. 17–16

17–17. In order to test its engine, the 500-kg missile is restrained from being fired by *four* short links such as AB, which are placed symmetrically around the nozzle, i.e., 90° apart. Determine the force developed in each link, knowing that without these restraints, the missile would accelerate upward at 25 m/s². Because of symmetry of geometry and loading, the force in each link is the same.

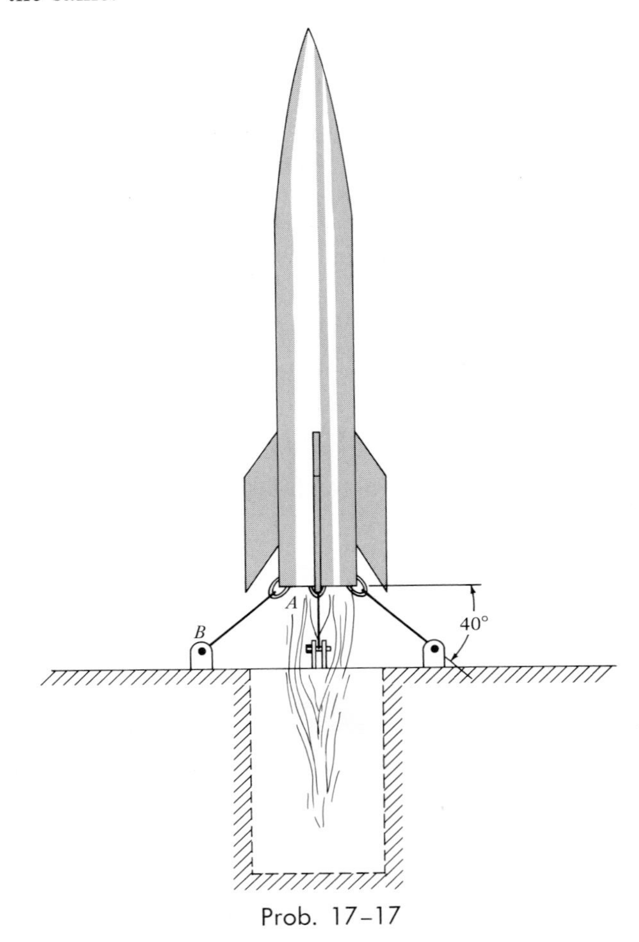

Prob. 17–17

17–18. The sports car has a mass of 1.4 Mg and a center of mass at G. Determine the shortest time it takes for it to reach a speed of 100 km/h starting from rest, if the motor only drives the rear wheels, whereas the front wheels are free rolling. The coefficient of friction between the wheels and the road is $\mu = 0.2$. Neglect the mass of the wheels for the calculation. If driving power could be supplied to all four wheels, what would be the shortest time for the car to reach a speed of 100 km/h?

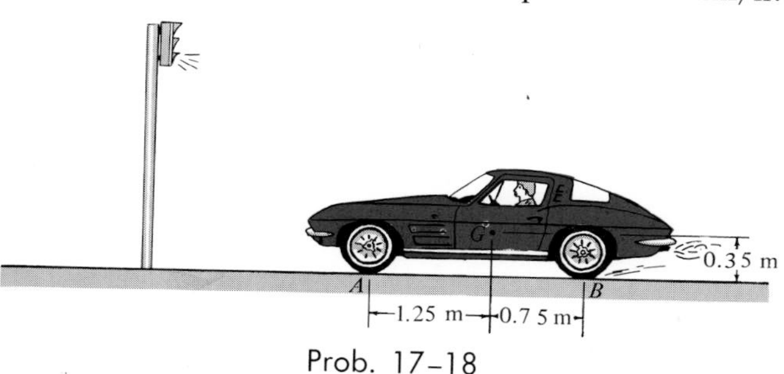

Prob. 17–18

17–19. The dragster has a mass of 1200 kg and a center of mass at G. If a braking parachute is attached at C and provides a horizontal braking force of $F = (1.6\,v^2)$ N, where v is in m/s, determine the critical speed the dragster can have upon releasing the parachute, such that the wheels at B are on the verge of leaving the ground; i.e., the normal reaction at B is zero. If such a condition occurs, determine the dragster's initial deceleration. Neglect the mass of the wheels and assume the engine is disengaged so that the wheels are freely rolling.

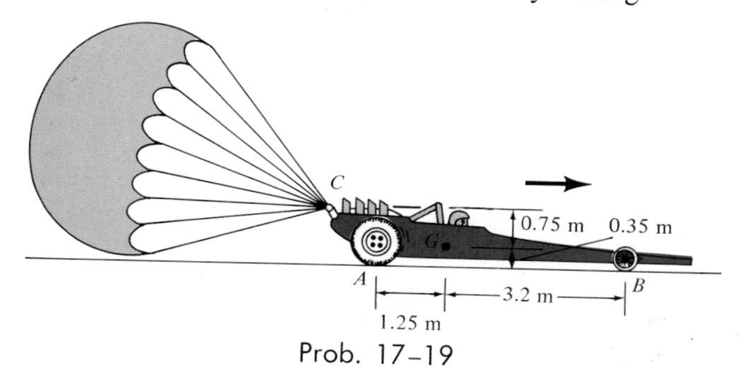

Prob. 17–19

***17–20.** A motorcyclist is traveling along the horizontally curved road which has a radius of $\rho = 40$ m. If the coefficient of friction between the tires and the road is $\mu = 0.5$, determine the maximum *constant* speed at which he may round the curve and the corresponding angle θ at which he must lean so as not to tip over or slip. The motorcycle and the rider have a total mass of $m = 200$ kg and a center of mass at G, $h = 1.2$ m.

Prob. 17–20

***17–20a.** Solve Prob. 17–20 if $\rho = 200$ ft, $\mu = 0.3$, and the weight of both the motorcycle and rider is $W = 400$ lb; $h = 3.5$ ft.

17–21. If the motorcyclist is rounding a 60° banked curve which has a radius of $\rho = 20$ m, determine the minimum constant speed at which he must travel, and the angle θ at which he must lean, so that he does not tip over or slip. The coefficient of friction between the tires and the road is $\mu = 0.5$. The motorcycle and rider have a total mass of 200 kg and a center of mass at G.

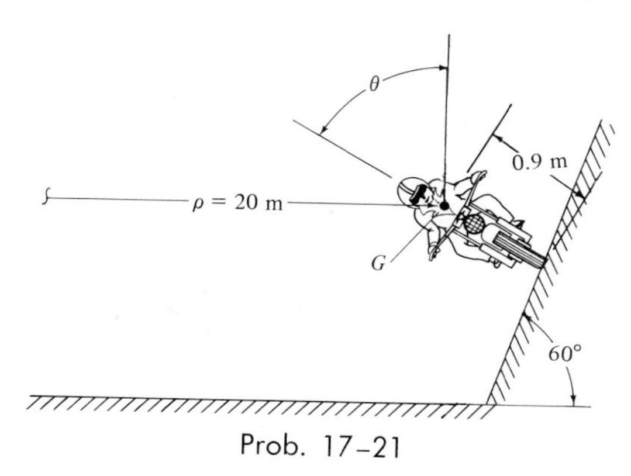

Prob. 17–21

17–22. The motorcycle has a mass of 125 kg and a center of mass at G_1, while the rider has a mass of 70 kg and a center of mass at G_2. If no slipping occurs, determine the minimum driving force $\mathbf{F}_B$ which must be supplied to the rear wheel B in order to create an acceleration of $a = 4 \text{ m/s}^2$. What are the normal reactions of the wheels on the ground? Neglect the mass of the wheels and assume that the front wheel is free to roll.

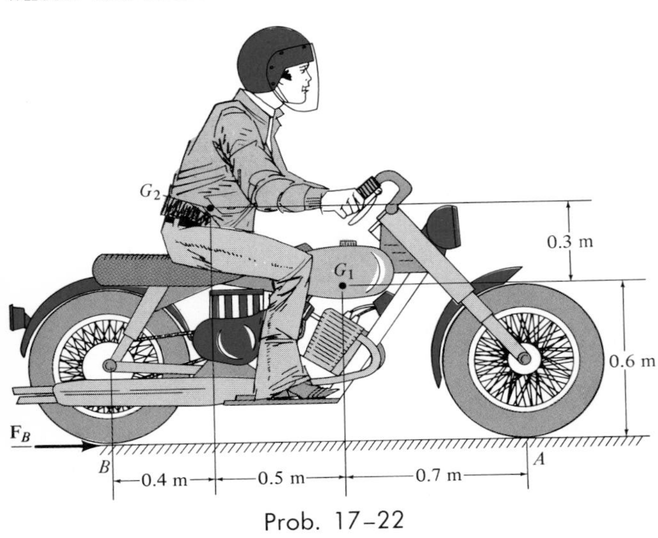

Prob. 17–22

17–23. A train, traveling at a constant speed of 15 m/s, is rounding a horizontal curve having a radius of 125 m, measured to the center of mass G. Determine the correct banking angle θ of the track so that the wheels of the train exert an equal force on both rails.

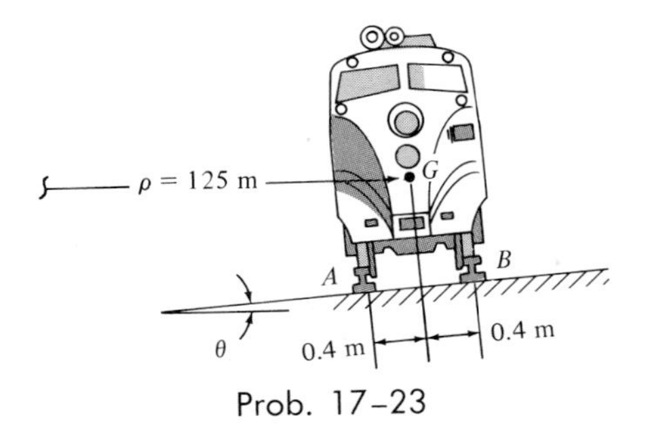

Prob. 17–23

***17–24.** The car has a mass of 1.5 Mg and a center of mass at G. If the coefficient of friction between the wheels and the road is $\mu = 0.30$, determine the minimum constant speed at which it may round the horizontally banked curve without sliding downward. The car travels in the same horizontal plane.

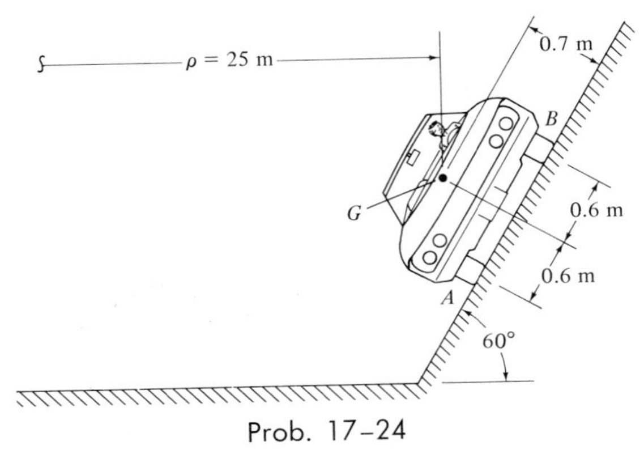

Prob. 17–24

17–25. The refrigerator has a mass of $m = 200$ kg and a center of mass at G. Determine the greatest constant force $\mathbf{F}$ which can be applied to the towing rope CD without causing the refrigerator to tip. The casters at A and B have negligible mass and are free to roll. Set $a = 0.6$ m, $b = 0.8$ m, and $d = 0.35$ m.

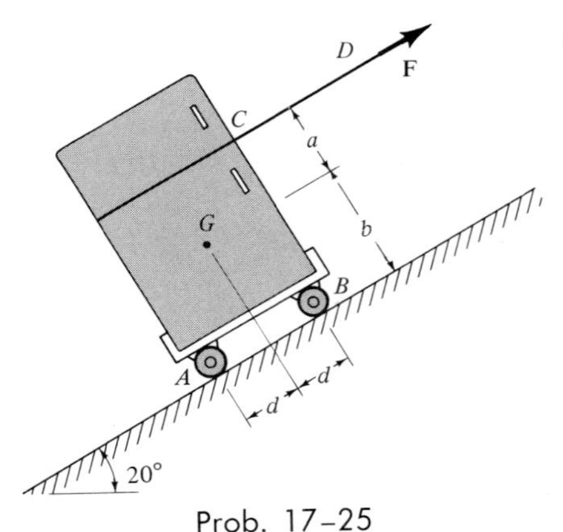

Prob. 17–25

17–25a. Solve Prob. 17–25 if the refrigerator has a weight of $W = 400$ lb, and $a = 2$ ft, $b = 2.8$ ft, $d = 1.4$ ft.

17–26. Solve Prob. 17–25, assuming that the casters at B stick while the casters at A are free to roll. The coefficient of kinetic friction between B and the floor is $\mu = 0.07$.

17–27. The rocket has a mass of 400 kg and a center of mass at G. It is supported on the monorail track at A and B by small rollers. If the engine provides a constant thrust **T** of 8 kN, determine the distance the rocket travels in 10 s, starting from rest. What are the normal reactions of the track on the rollers due to the motion? Neglect the loss of fuel in the calculation.

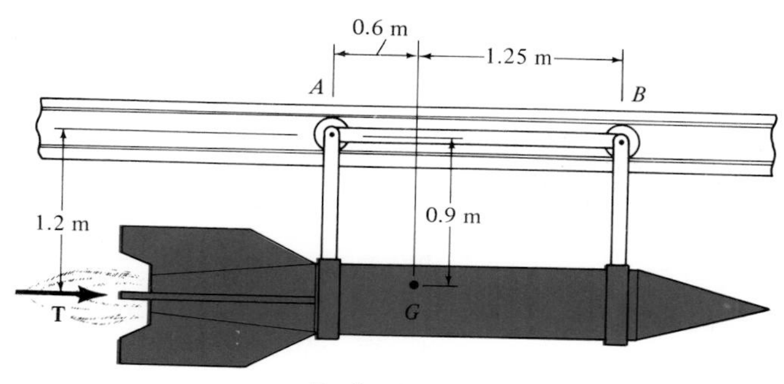

Prob. 17–27

***17–28.** Solve Prob. 17–27, assuming that the rollers at A and B are replaced by runners of negligible size and mass. The coefficient of friction between the runners and the track is $\mu = 0.2$.

17–29. The uniform crate C has a mass of 50 kg and it rests on the surface of a truck for which the coefficient of friction is $\mu = 0.30$. If the truck has an initial speed of $v_1 = 20$ m/s, determine the shortest distance s in which the truck can stop without causing the crate to slide. Assume that the crate does not tip.

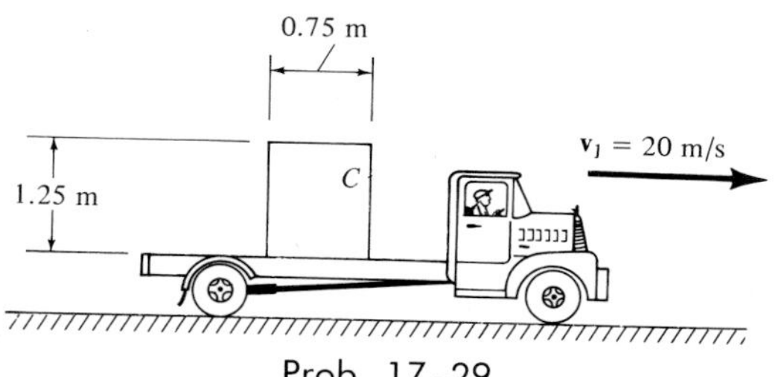

Prob. 17–29

17–30. The uniform connecting rod BC has a mass of $m = 3$ kg and is pin-connected at its end points. Determine the vertical forces which the pins exert on the ends B and C of the rod at the instant (a) $\theta = 0°$, and (b) $\theta = 90°$. The crank AB is turning with a constant angular velocity of $\omega_{AB} = 5$ rad/s. Set $l = 700$ mm and $r = 200$ mm.

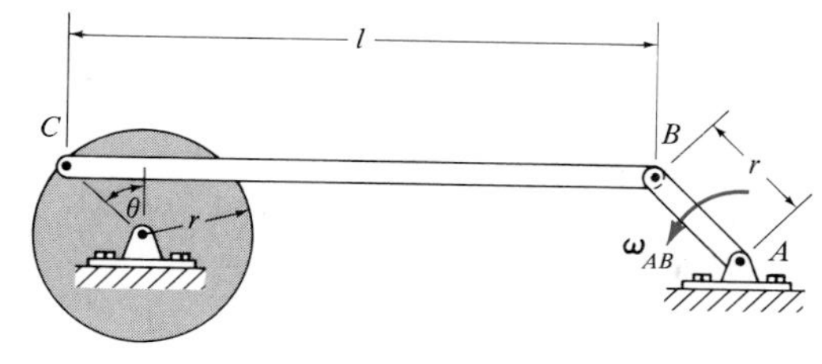

Prob. 17–30

17–30a. Solve Prob. 17–30 if rod BC has a weight of $W = 10$ lb, and $\omega_{AB} = 7$ rad/s, $l = 2.5$ ft, $r = 0.75$ ft.

17–31. The trailer portion of a truck has a mass of 3.5 Mg, with a center of mass at G. If a *uniform* crate, having a mass of 800 kg and a center of mass G_c, rests on the surface of the trailer, determine the horizontal and vertical components of reaction at the ball-and-socket joint (pin) A when the truck is decelerating at a constant rate of $a = 2.5$ m/s². Assume that the crate does not slip on the trailer and neglect the mass of the wheels. The wheels at B roll freely.

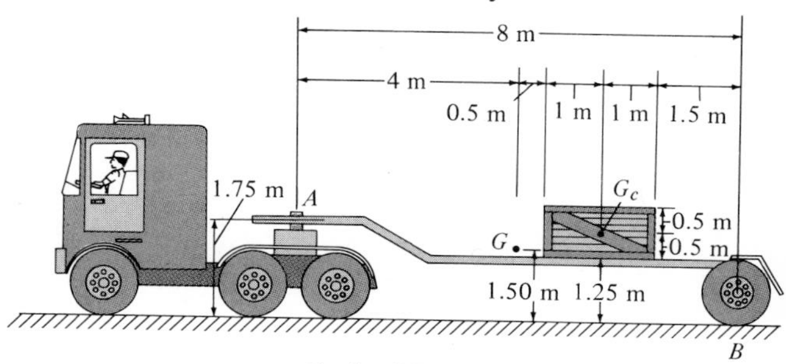

Prob. 17–31

***17–32.** If the coefficient of friction between the trailer and the crate in Prob. 17–31 is $\mu = 0.3$, determine the maximum allowable deceleration of the truck such that the crate does not slide on the trailer.

17–33. Solve Prob. 17–31 assuming that the coefficient of friction between the crate and the trailer is $\mu = 0.2$.

17–34. If the links AB and CD of the elevator frame in Prob. 17–35 are rotating counterclockwise at a *constant rate* of $\omega = 1.40$ rad/s ($\alpha = 0$), determine the normal and frictional forces which the 75-kg crate exerts on the elevator floor at the instant $\theta = 90°$. Use $\mu = 0.3$, $l = 1.25$ m.

17–35. The frame $BFED$ is used as an elevator to transport the crate that has a mass of $m = 75$ kg from one elevation ($\theta = 45°$) to another ($\theta = 180°$). If the coefficient of friction between the crate and the elevator floor is $\mu = 0.3$, determine the largest initial angular acceleration $\boldsymbol{\alpha}$ which the links AB and CD can have without causing the crate to slip. The frame is initially at rest when $\theta = 45°$. Assume that the crate does not tip, $l = 1.25$ m.

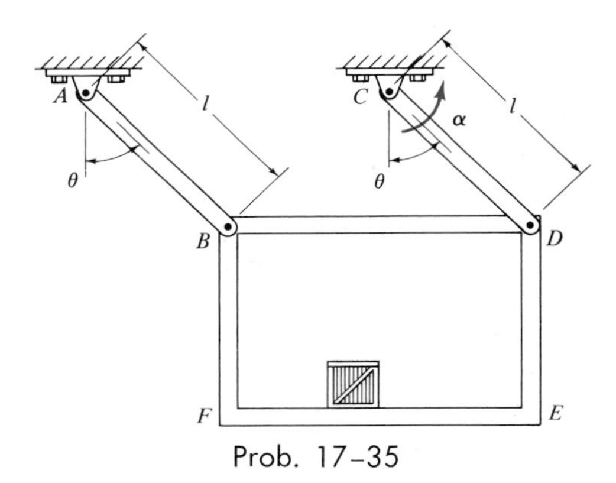

Prob. 17–35

17–35a. Solve Prob. 17–35 if the crate has a weight of $W = 125$ lb, and $\mu = 0.4$, $l = 3$ ft.

* **17–36.** The traction between two automobiles, A and B, is matched by connecting the rear bumpers with a cable CD. If A has a mass of 1800 kg, with center of mass at G_A, and B has a mass of 1500 kg, with center of mass at G_B, determine the tension developed in the cable and the acceleration of each vehicle. Slipping occurs only at the rear wheels F and H, where, in both cases, the coefficient of friction is $\mu = 0.4$. Neglect the mass of the wheels and assume that the front wheels at E and I are free to roll.

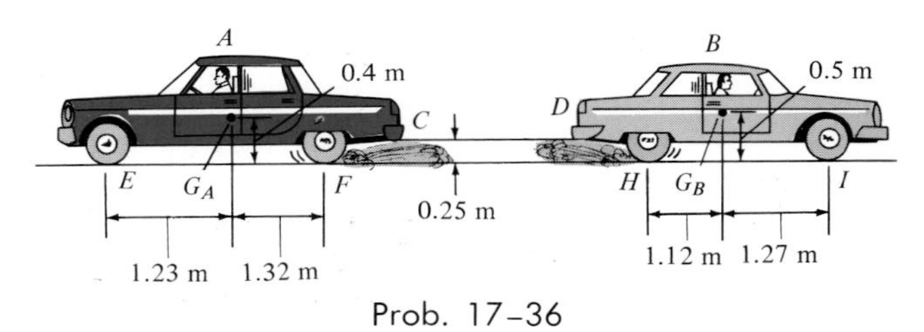

Prob. 17–36

17–5. Equations of Motion: Rotation of a Rigid Body About a Fixed Axis

Fig. 17–13(a)

Consider the rigid body (or slab) shown in Fig. 17–13*a*, which is constrained to rotate in the vertical plane about a fixed axis perpendicular to the page and passing through the pin at O. The angular velocity $\boldsymbol{\omega}$ and angular acceleration $\boldsymbol{\alpha}$ are caused by the external force system acting on the body. Because the body's center of mass G moves in a *circular path*, the acceleration of this point is represented by its tangential and normal components. The *tangential component of acceleration* has a *magnitude* of $(a_G)_t = \alpha r_G$ and acts in a *direction* which is consistent with the angular acceleration $\boldsymbol{\alpha}$. The *magnitude* of the *normal component of acceleration* is $(a_G)_n = \omega^2 r_G$. This component is *always directed* from point G to O regardless of the direction of $\boldsymbol{\omega}$.

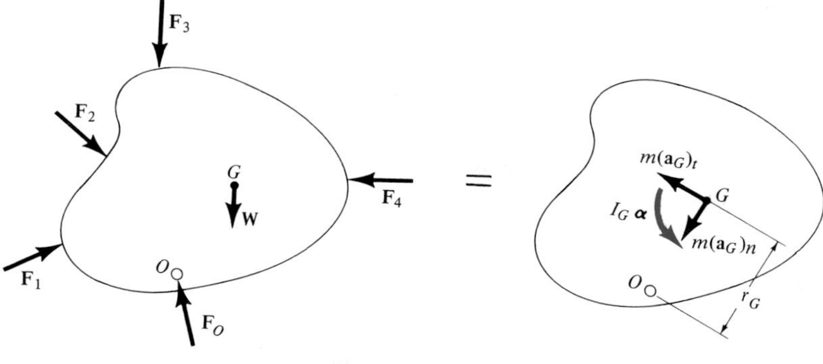

Fig. 17–13(b)

The free-body and kinetic diagrams for the body are shown in Fig. 17–13*b*. The weight of the body, $W = mg$, and the pin reaction $\mathbf{F}_O$ are included on the free-body diagram since they represent external forces acting on the body. The two components $m(\mathbf{a}_G)_t$ and $m(\mathbf{a}_G)_n$, shown on the kinetic diagram, are associated with the tangential and normal acceleration components of the body's mass center. These vectors act in the same *direction* as the acceleration components and have *magnitudes* of $m(a_G)_t$ and $m(a_G)_n$. The $I_G\boldsymbol{\alpha}$ vector acts in the same *direction* as $\boldsymbol{\alpha}$ and has a *magnitude* of $I_G\alpha$, where I_G is the body's moment of inertia, calculated about an axis which is perpendicular to the page and passing through G. From the derivation given in Sec. 17–3, the equations of motion which apply to the body may be written in the form

$$\begin{aligned} \Sigma F_n &= m(a_G)_n = m\omega^2 r_G \\ \Sigma F_t &= m(a_G)_t = m\alpha r_G \\ \Sigma M_G &= I_G\alpha \end{aligned} \tag{17–13}$$

The moment equation may be replaced by a moment summation about any arbitrary point on the body provided one accounts for the moments ΣM_k produced by $I_G\boldsymbol{\alpha}$, $m(\mathbf{a}_G)_t$, and $m(\mathbf{a}_G)_n$ about the point. In many problems it is convenient to sum moments about the pin at O in order to eliminate the *unknown* force $\mathbf{F}_O$ from the moment summation. From the kinetic diagram, Fig. 17–13*b*, this requires

$$\circlearrowleft{+}\Sigma M_O = \Sigma(M_k)_O; \qquad \Sigma M_O = r_G m(a_G)_t + I_G\alpha \tag{17–14}$$

Note that the moment of $m(\mathbf{a}_G)_n$ is not included in the summation since the line of action of this vector passes through O. Substituting $(a_G)_t = r_G\alpha$, one may rewrite the above equation as $\circlearrowleft{+}\Sigma M_O = (I_G + mr_G^2)\alpha$. From the parallel-axis theorem, $I = I_G + md^2$, the term in parentheses represents the *moment of inertia of the body about the fixed axis of rotation passing through O*. Denoting this term by I_O, we can write the three equations of motion for the body as

$$\begin{aligned} \Sigma F_n &= m(a_G)_n = m\omega^2 r_G \\ \Sigma F_t &= m(a_G)_t = m\alpha r_G \\ \Sigma M_O &= I_O\alpha \end{aligned} \tag{17-15}$$

For applications, one should remember that "$I_O\alpha$" accounts for the "moment" of *both* $(\mathbf{a}_G)_t$ *and* $I_G\boldsymbol{\alpha}$ about point O, Fig. 17–13*b*. In other words, $\Sigma M_O = \Sigma(M_k)_O = I_O\alpha$, as indicated by Eqs. 17–14 and 17–15.

PROCEDURE FOR ANALYSIS

A three-step procedure should be followed when solving problems which involve the rotation of a body about a fixed axis.

Step 1: Draw a free-body and kinetic diagram for the body, Fig. 17–13*b*, and compute the mass moment of inertia I_G or I_O.

Step 2: Apply the three equations of motion, Eqs. 17–13 or 17–15, with reference to an established x, y or n, t inertial coordinate system.

Step 3: Use kinematics if a complete solution cannot be obtained strictly from the equations of motion. In this regard, if the *angular acceleration is variable,* use

$$\alpha = \frac{d\omega}{dt}, \; \alpha\, d\theta = \omega\, d\omega, \; \omega = \frac{d\theta}{dt}$$

If the *angular acceleration is constant,* use

$$\begin{aligned} \omega &= \omega_1 + \alpha_c t \\ \theta &= \theta_1 + \omega_1 t + \tfrac{1}{2}\alpha_c t^2 \\ \omega^2 &= \omega_1^2 + 2\alpha_c(\theta - \theta_1) \end{aligned}$$

The following examples numerically illustrate application of this three-step procedure.

Example 17–8

The 30-kg disk, shown in Fig. 17–14*a,* is pin-supported at its center. Determine the number of revolutions it must make to attain an angular velocity of 20 rad/s starting from rest. It is acted upon by a constant force $F = 10$ N, which is applied to a cord wrapped around its periphery, and a constant couple $M = 5$ N · m. Neglect the mass of the cord in the calculation.

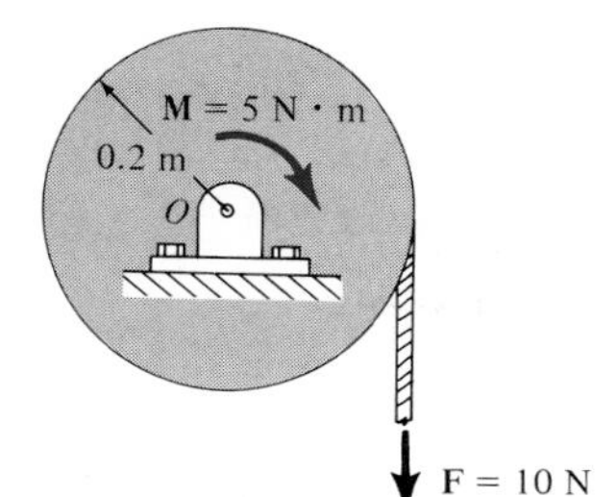

Fig. 17–14(a)

Solution

Step 1: The free-body and kinetic diagrams are shown in Fig. 17–14*b*. Since the mass center is not subjected to an acceleration, only $I_O\boldsymbol{\alpha}$ acts on the disk.

From Appendix D, the mass moment of inertia of the disk is

$$I_O = \tfrac{1}{2}mr^2 = \tfrac{1}{2}(30)(0.2)^2 = 0.6 \text{ kg} \cdot \text{m}^2$$

Step 2: Applying the equations of motion:

$$\xrightarrow{+} \Sigma F_x = m(a_G)_x; \qquad O_x = 0$$

$$+\uparrow \Sigma F_y = m(a_G)_y; \quad O_y - 294.3 - 10 = 0; \qquad O_y = 304.3 \text{ N}$$

$$\curvearrowright + \Sigma M_O = I_O\alpha; \qquad 10(0.2) + 5 = 0.6\alpha; \qquad \alpha = 11.7 \text{ rad/s}^2 \ \curvearrowright$$

Step 3: Since α is constant, the number of radians the disk must turn to obtain an angular velocity of 20 rad/s is

$$\curvearrowright + \qquad \omega^2 = \omega_1^2 + 2\alpha_c(\theta - \theta_1)$$

$$(20)^2 = 0 + 2(11.7)(\theta - 0)$$

$$\theta = 17.1 \text{ rad}$$

Hence,

$$\theta = 17.1 \text{ rad}\left(\frac{1 \text{ rev}}{2\pi \text{ rad}}\right) = 2.73 \text{ rev} \qquad \textit{Ans.}$$

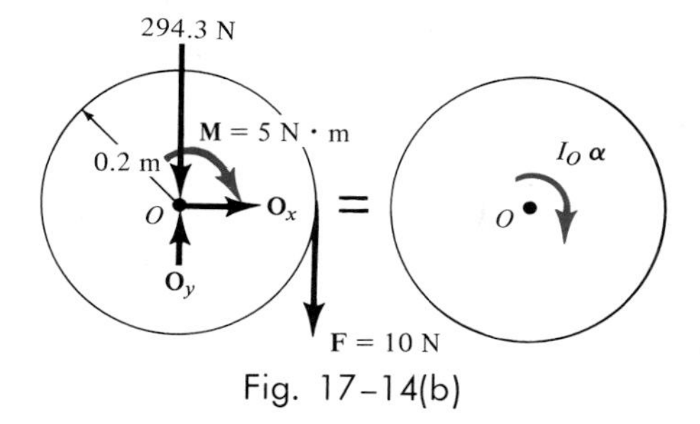

Fig. 17–14(b)

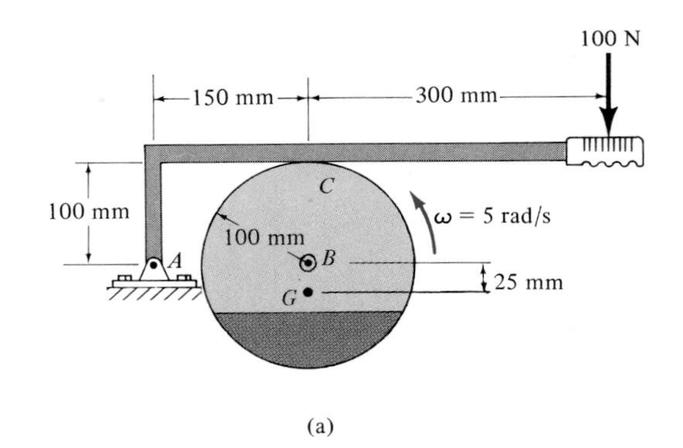

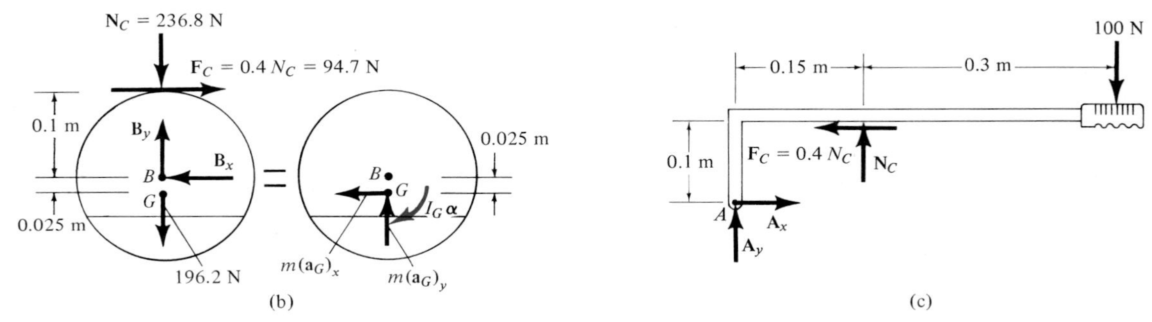

Fig. 17–15

Example 17–9

The unbalanced 20-kg flywheel shown in Fig. 17–15*a* has a radius of gyration of $k_G = 65$ mm about an axis passing through its mass center *G*. If a force of 100 N is applied to the hand brake, determine the horizontal and vertical components of reaction which the pin *B* exerts on the flywheel at the instant shown. The flywheel has a counterclockwise angular velocity of 5 rad/s at this instant, and the coefficient of kinetic friction between the brake and the flywheel is $\mu_k = 0.4$.

Solution I

Step 1: The free-body and kinetic diagrams for the flywheel are shown in Fig. 17–15*b*. The frictional force caused by the brake acts to the *right* in order to create an angular deceleration. Consequently, $I_G\boldsymbol{\alpha}$ acts in a *clockwise* sense and $m(\mathbf{a}_G)_x$ must act to the *left*, in accordance with the direction of $\boldsymbol{\alpha}$. Furthermore, in the normal or *y* direction, $m(\mathbf{a}_G)_y$ is directed upward, toward point *B*.

The moment of inertia of the flywheel about its mass center is determined from the radius of gyration and the mass, i.e.,

$$I_G = k_G^2 m = (0.065\text{ m})^2(20\text{ kg}) = 0.0845\text{ kg}\cdot\text{m}^2$$

Since the brake does not move, the laws of statics apply. The free-body diagram is shown in Fig. 17–15*c*.

Step 2: The normal force $\mathbf{N}_C$ acting on the brake can be obtained by summing moments about point A.

$$\circlearrowleft+\Sigma M_A = 0; \quad 0.4N_C(0.1) + N_C(0.15) - 100(0.45) = 0; \quad N_C = 236.8\text{ N}$$

Thus,

$$F_C = 0.4N_C = 0.4(236.8) = 94.7\text{ N}$$

Using these results and applying the equations of motion to the flywheel, noting that $m(a_G)_x = m\alpha r_G$ and $m(a_G)_y = m\omega^2 r_G$, Fig. 17–5*b*, we have

$$\overset{+}{\rightarrow}\Sigma F_x = m(a_G)_x; \quad -B_x + 94.7 = -20\alpha(0.025) \tag{1}$$

$$+\uparrow\Sigma F_y = m(a_G)_y; \quad -236.8 + B_y - 196.2 = 20(5)^2(0.025) \tag{2}$$

$$\circlearrowright+\Sigma M_G = I_G\alpha; \quad -B_x(0.025) + 94.7(0.125) = 0.0845\alpha \tag{3}$$

Solving yields

$$\alpha = 97.6\text{ rad/s}^2$$

$$B_x = 143.5\text{ N} \qquad \textit{Ans.}$$

$$B_y = 445.5\text{ N} \qquad \textit{Ans.}$$

Solution II

The rotational equation of motion may be applied at point B in order to eliminate $\mathbf{B}_x$ and $\mathbf{B}_y$ and thereby obtain a *direct solution* for α, Fig. 17–15*b*. This can be done in one of *two* ways, i.e., by using either $\Sigma M_B = \Sigma(\mathcal{M}_k)_B$ or $\Sigma M_B = I_B\alpha$. If the first of these equations is applied, then

$$\circlearrowright+\Sigma M_B = \Sigma(\mathcal{M}_k)_B; \qquad 94.7(0.1) = (0.0845)(\alpha) + [(20)\alpha(0.025)](0.025)$$

or

$$94.7(0.1) = 0.097\alpha \tag{4}$$

If $\Sigma M_B = I_B\alpha$ is applied, then by the parallel-axis theorem, $I = I_G + md^2$, the moment of inertia of the flywheel about B is

$$I_B = I_G + mr_G^2 = 0.0845 + 20(0.025)^2 = 0.097\text{ kg}\cdot\text{m}^2$$

Hence, from the free-body diagram, Fig. 17–15*b,* we require

$$\circlearrowright+\Sigma M_B = I_B\alpha; \qquad 94.7(0.1) = 0.097\alpha$$

which is the same as Eq. (4). Solving for α and substituting into Eq. (1) or (3) yields the answer for B_x obtained previously.

Example 17–10

The Charpy impact machine is used to determine the energy-absorption characteristics of a small specimen during impact. This test is performed by releasing the 50-kg pendulum shown in Fig. 17–16*a*, when $\theta = 0°$, and allowing it to fall freely to strike the specimen at *S*, $\theta = 90°$. If the pendulum has a mass center at *G* and a radius of gyration of $k_A = 0.75$ m about the pin at *A*, determine the reactive force at *A* (*a*) just before the pendulum strikes the specimen, and (b) when the pendulum has just been released.

Solution

Part (a). *Step 1:* The free-body and kinetic diagrams for the pendulum are shown in Fig. 17–16*b*, when the pendulum is in the *general position* θ. For convenience, the reaction components at *A* are shown acting in the *n* and *t* directions. Since the angular acceleration is clockwise, $I_G\boldsymbol{\alpha}$ acts clockwise and $m(\mathbf{a}_G)_t$ or $m\alpha r_G$ acts downward to the left in accordance with this rotation.

The moment of inertia of the pendulum about *A* can be computed since the radius of gyration about *A* and the mass of the pendulum are known. Using Eq. 17–5 gives

$$I_A = k_A^2 m = (0.75)^2 50 = 28.13 \text{ kg} \cdot \text{m}^2$$

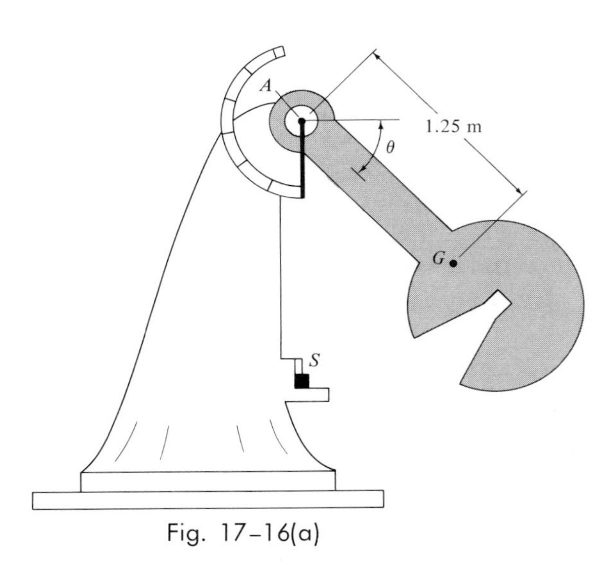

Fig. 17–16(a)

Step 2: Since the moment of inertia about the *pivot A* (fixed point) is known, moments will be summed about this point. This has an added advantage in that the unknown reactive force at *A* is *eliminated* from the moment equation.

$$+\nwarrow \Sigma F_n = m\omega^2 r_G \qquad A_n - 490.5 \sin\theta = 50\omega^2 1.25 \tag{1}$$

$$+\swarrow \Sigma F_t = m\alpha r_G \qquad A_t + 490.5 \cos\theta = 50\alpha 1.25 \tag{2}$$

$$\circlearrowleft + \Sigma M_A = I_A\alpha \qquad 490.5 \cos\theta(1.25) = 28.13\alpha \tag{3}$$

Note: Since the moment summation of the external forces about point A is equated to $I_A\alpha$, it is *not necessary* to account for the moments of $I_G\boldsymbol{\alpha}$ and $m(\mathbf{a}_G)_t$ about A. Why? (Refer to Eq. 17–15.)

Step 3: For a given angle θ, there are four unknowns in the above three equations: A_n, A_t, ω, and α. As shown by Eq. (3), α is *not constant;* rather it depends upon the position θ of the pendulum. By kinematics, the unknowns α and ω can be related to θ by the equation

$$\omega\, d\omega = \alpha\, d\theta \tag{4}$$

Equations (1) through (4) may be used to obtain the complete solution to the problem for any angle θ. Substituting Eq. (3) for α into Eq. (4) yields

$$\omega\, d\omega = \frac{490.5(1.25)}{(28.13)} \cos\theta\, d\theta$$

This equation can be integrated to obtain ω when $\theta = 90°$. Using the condition that $\omega_1 = 0$ at $\theta_1 = 0°$, gives

$$\int_0^{\omega} \omega\, d\omega = 21.80 \int_{0°}^{90°} \cos\theta\, d\theta$$

$$\frac{\omega^2}{2}\bigg|_0^{\omega} = 21.80(\sin\theta)\bigg|_{0°}^{90°} = 21.80(1 - 0) = 21.80$$

so that

$$\omega^2 = 43.60 \text{ (rad/s)}^2$$

Thus, for $\theta = 90°$, Eqs. (1) through (3) become

$$A_n - 490.5 = 50(43.60)1.25$$
$$A_t = 50(\alpha 1.25)$$
$$0 = 28.13\alpha$$

Solving yields

$$A_n = 3215.5 \text{ N} = 3.22 \text{ kN} \quad \uparrow \qquad \textit{Ans.}$$
$$A_t = 0$$
$$\alpha = 0$$

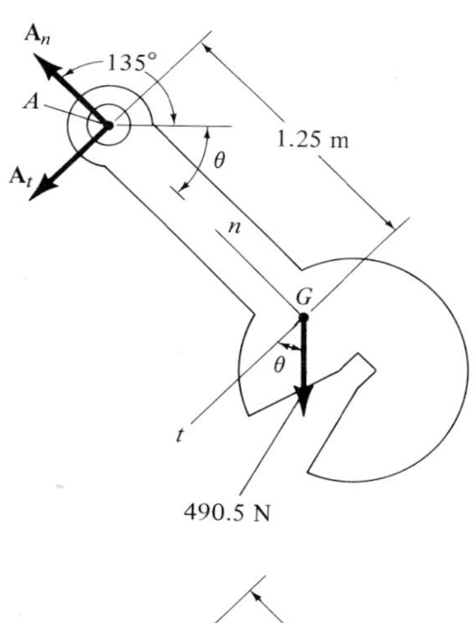

Part (b). When the pendulum is *just released* at $\theta = 0°$, $\omega = 0$, in which case solution of Eqs. (1) through (3) yields

$$A_n = 0$$
$$A_t = 871.8 \text{ N} \quad \uparrow \qquad \textit{Ans.}$$
$$\alpha = 21.8 \text{ rad/s}^2$$

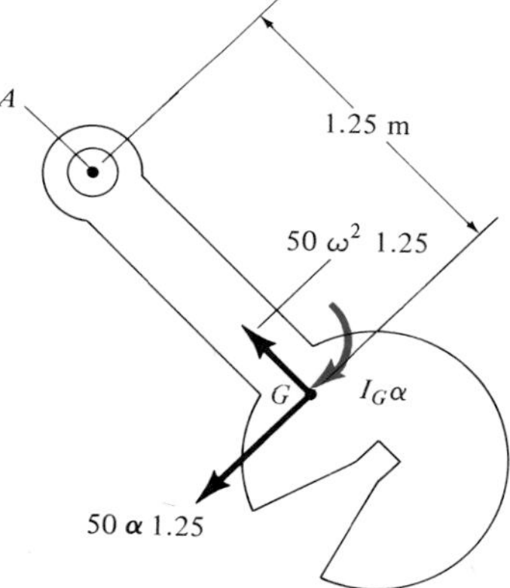

Fig. 17–16(b)

Example 17–11

The two smooth gears shown in Fig. 17–17a are used to lift the 100-kg crate. Gear B has a mass of 20 kg and a radius of gyration of $k_B = 0.315$ m. Gear A has a mass of 30 kg and a radius of gyration of $k_A = 0.260$ m. Determine the acceleration of the crate if a vertical force of 800 N is applied to a cable wrapped around the drum on gear B, as shown in the figure.

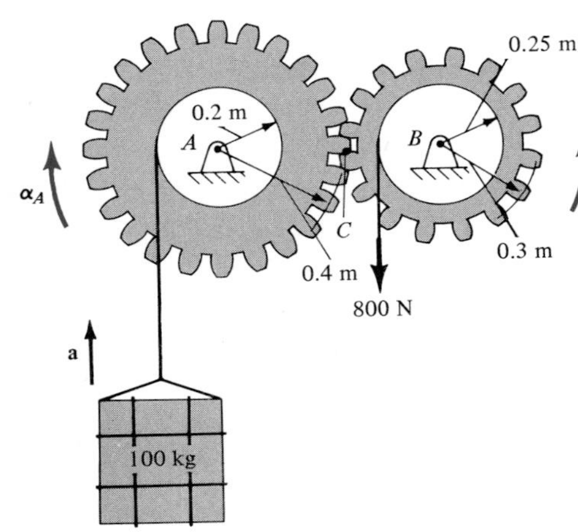

Fig. 17–17(a)

Solution

Step 1: The free-body and kinetic diagrams of each gear and the crate are shown in Fig. 17–17b.

The moments of inertia for gears A and B are

$$I_B = m_B k_B^2 = 20(0.315)^2 = 1.98 \text{ kg} \cdot \text{m}^2$$
$$I_A = m_A k_A^2 = 30(0.260)^2 = 2.03 \text{ kg} \cdot \text{m}^2$$

Step 2: Applying the rotational equation of motion about the mass center of each of the gears yields

$$\circlearrowright + \Sigma M_A = I_A \alpha_A; \qquad T(0.2) - P(0.4) = -2.03\alpha_A \tag{1}$$
$$\circlearrowright + \Sigma M_B = I_B \alpha_B; \qquad 800(0.25) - P(0.3) = 1.98\alpha_B \tag{2}$$

The crate is subjected to translation only; hence,

$$+\uparrow \Sigma F_y = m(a_G)_y; \qquad T - 981 = 100a \tag{3}$$

Equations (1) through (3) contain five unknowns, T, P, a, α_A, and α_B. (Applying the two translational equations of motion to each of the gears will not help in the solution, since these equations involve four other unknowns, A_x, A_y, B_x, and B_y.)

Step 3: Using *kinematics,* two more equations may be obtained to relate α_B, α_A, and a. Since the points of contact between the gears at C must have the same *tangential* components of acceleration, Fig. 17–17a, this requires that

$$(a_C)_t = r_B \alpha_B = r_A \alpha_A$$
$$= 0.3\alpha_B = 0.4\alpha_A$$

or

$$\alpha_B = 1.33\alpha_A \tag{4}$$

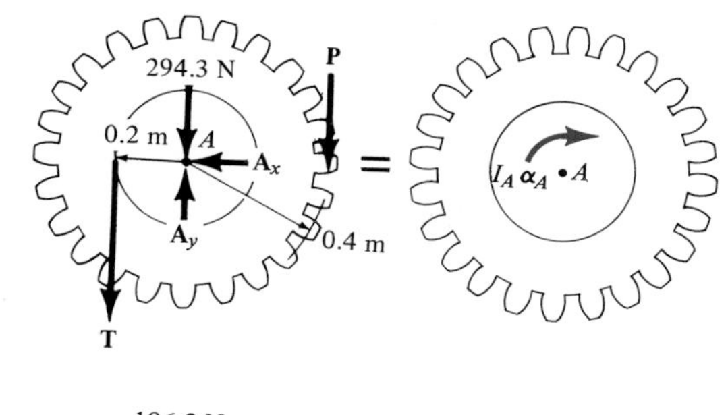

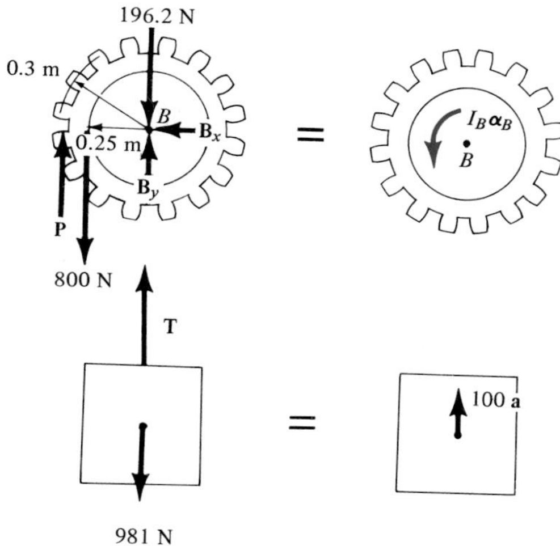

Fig. 17–17(b)

Also, the acceleration of the cable supporting the 100-kg crate may be related to the angular acceleration of gear A by the equation

$a = r\alpha;$ $$a = 0.2\alpha_A \tag{5}$$

It is important that the *directions* of $\boldsymbol{\alpha}_A$, $\boldsymbol{\alpha}_B$, and **a**, which were considered when Eqs. (1) through (3) were derived, Fig. 17–17*b*, be the *same* as the directions considered when Eqs. (4) and (5) are derived, Fig. 17–17*a*. This is necessary since the problem requires a simultaneous solution of equations. The results are:

$$P = 601.7 \text{ N}$$
$$T = 1128.4 \text{ N}$$
$$\alpha_A = 7.39 \text{ rad/s}^2$$
$$\alpha_B = 9.85 \text{ rad/s}^2$$
$$a = 1.48 \text{ m/s}^2 \qquad \textit{Ans.}$$

17–37. A 500-mm-diameter flywheel has a mass of 90 kg and it is free to rotate about its center. An inextensible cord is wrapped around the rim of the flywheel and when a 6-kg block is attached to the cord, the block attains a speed of 200 mm/s after moving downward a distance of 0.4 m starting from rest. Determine the radius of gyration of the flywheel about its central axis.

17–38. A 5-kg block A is attached to a cord which is wrapped around the rim of a 15-kg cylinder. Determine the speed of the block 2 s after it is released from rest. Neglect the weight of the cord.

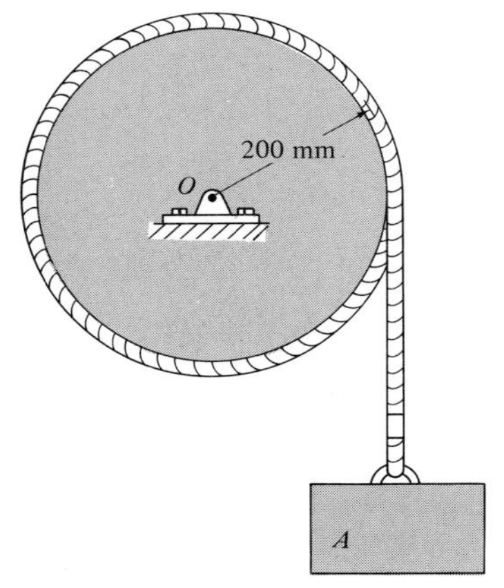

Prob. 17–38

17–39. Determine the force $\mathbf{T}_A$ which must be applied to the cable at A in order to give the 8-kg block B an upward acceleration of 400 mm/s². Assume that the cable does not slip over the surface of the 20-kg disk. Compute the tension in the vertical segment of the cord which supports the block and explain why this tension is different from that at A. The disk is pinned at its center C and is free to rotate.

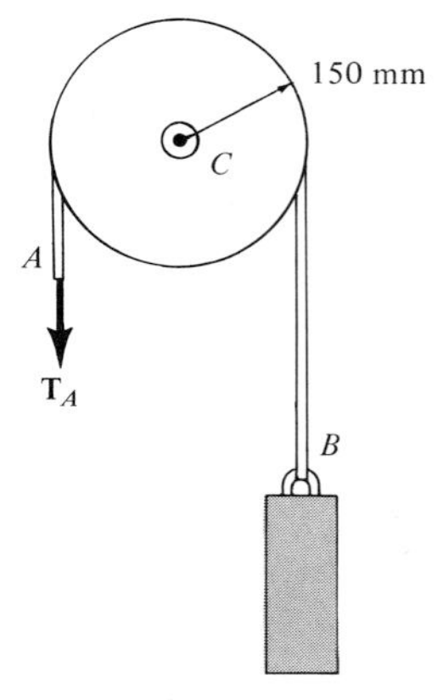

Prob. 17–39

*** 17–40.** The shaft consists of two cones, each having a mass of $m = 15$ kg. If the shaft is acted upon by a moment of $M = 3$ N · m as shown, determine the shaft's angular velocity in $t = 2$ s starting from rest. What are the reactions at A and B during the motion? $d = 0.2$ m and $l = 0.5$ m.

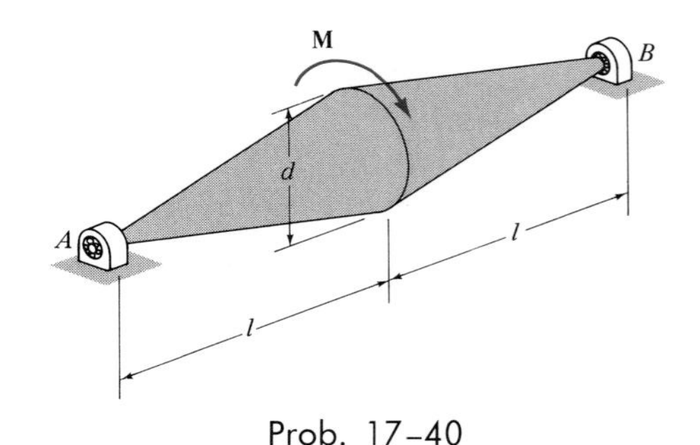

Prob. 17–40

*** 17–40a.** Solve Prob. 17–40 if each cone has a weight of $W = 20$ lb, and $M = 1.7$ lb · ft, $t = 1.5$ s, $d = 0.8$ ft, $l = 1.75$ ft.

17–41. The pendulum consists of a 20-kg sphere and a 5-kg slender rod. Compute the reaction at the pin O just after the cord AB is cut.

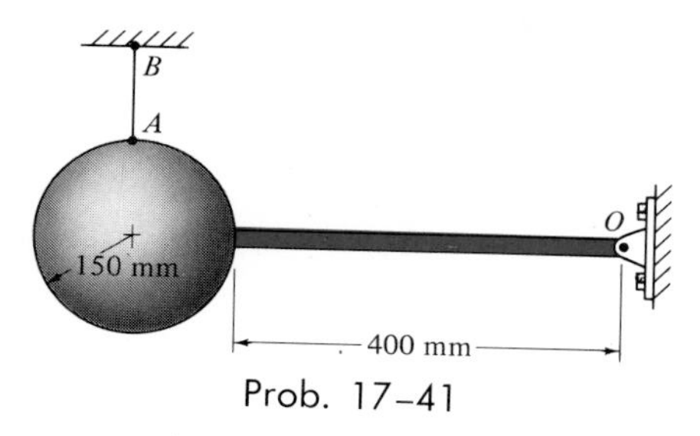

Prob. 17–41

17–42. The kinetic diagram representing the general rotational motion of a rigid body about a fixed axis is shown in the figure. Show that $I_G\boldsymbol{\alpha}$ may be eliminated by moving the vectors $m(\mathbf{a}_G)_t$ and $m(\mathbf{a}_G)_n$ to point P, located a distance $r_{GP} = k_G^2/r_{OG}$ from the center of mass G of the body. Here k_G represents the radius of gyration of the body about G. The point P is called the *center of percussion* of the body.

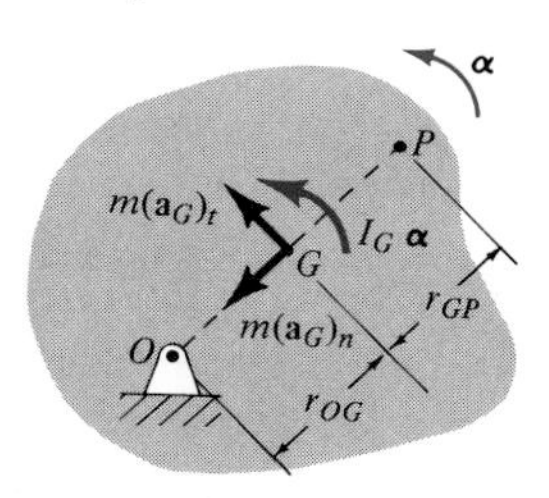

Prob. 17–42

17–43. Using the results of Prob. 17–14, i.e., $\bar{y} = 0.393$ m and $I_G = 0.0826$ kg $\cdot$ m^2, determine the distance h from the center of the grip, O, to the center of percussion P of the baseball bat. The bat has a mass of $m = 1.536$ kg. If the bat strikes a ball at the center of percussion, no stinging effect is felt in the hands of the batter at O. Explain why this is so. *Hint:* See Problem 17–42 and assume that the point of rotation is at O.

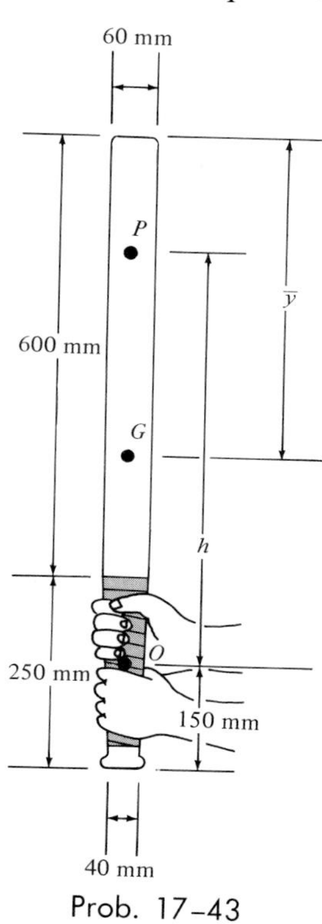

Prob. 17–43

*** 17–44.** A 17-kg roll of paper, originally at rest, is pin-supported at its ends to bracket AB. If the roll rests against a wall for which $\mu_C = 0.3$, and a force of 30 N is applied uniformly to the end of the sheet, determine the tension force in the bracket as the paper unwraps and the initial angular acceleration of the roll. For the calculation, treat the roll as a cylinder.

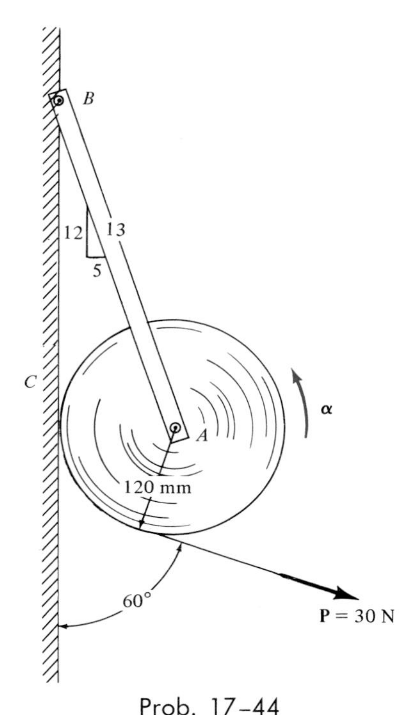

Prob. 17–44

17–45. The cylinder, having a mass of $m = 5$ kg, is initially at rest when it is placed into contact with the wall B and the rotor at A. If the rotor always maintains a constant clockwise angular velocity of $\omega = 6$ rad/s, determine the initial angular acceleration of the cylinder. The coefficient of friction at the contacting surfaces B and C is $\mu = 0.2$, and $r = 125$ mm.

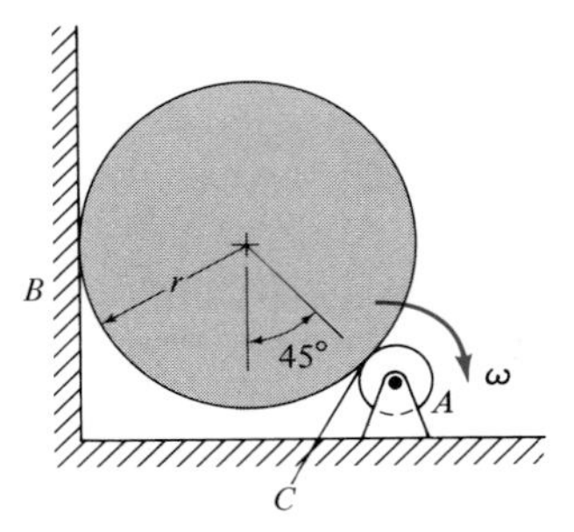

Prob. 17–45

17–45a. Solve Prob. 17–45 if the cylinder has a weight of $W = 15$ lb, and $\omega = 10$ rad/s, $\mu = 0.3$, $r = 0.4$ ft.

17–46. The lid of a pressure vessel has a mass of 60 kg and a radius of gyration about its mass center G of $k_G = 0.21$ m. In order to raise the lid, an operator applies a force of $F = 600$ N to the foot pad at E. Neglecting the mass of links CE and DB, determine the lid's initial angular acceleration and the horizontal and vertical components of reaction which the lid exerts on the hinge at A at the instant it begins to open. The hinges at A and C are connected to the pressure vessel.

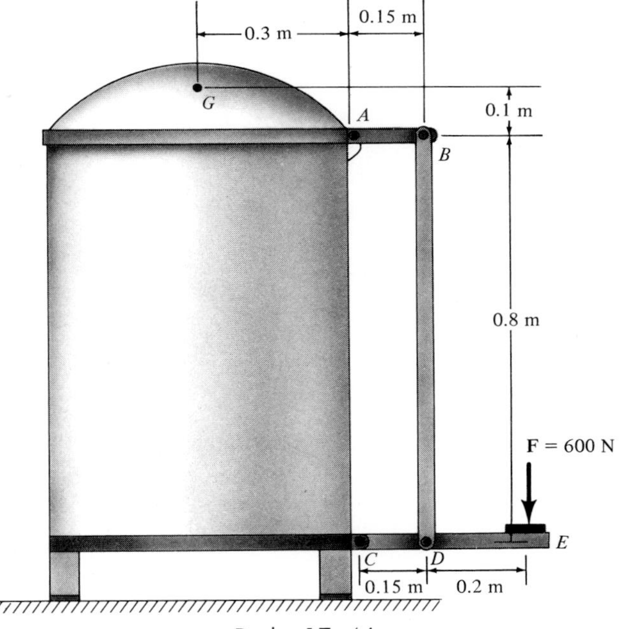

Prob. 17–46

17–47. As a man enters a room he pushes on the 30-kg smooth door with a constant force of $F = 60$ N. In doing so, his hand slides freely along the door from A to B at a constant rate of 0.5 m/s such that the force always remains perpendicular to the face of the door. If the door is originally at rest, when the force is applied at A, determine the door's angular velocity just at the instant the man's hand begins to slip off the edge B. The moment of inertia of the door about the z axis is $I_z = 14\ \text{kg}\cdot\text{m}^2$. How far has the door rotated during the time the force acts?

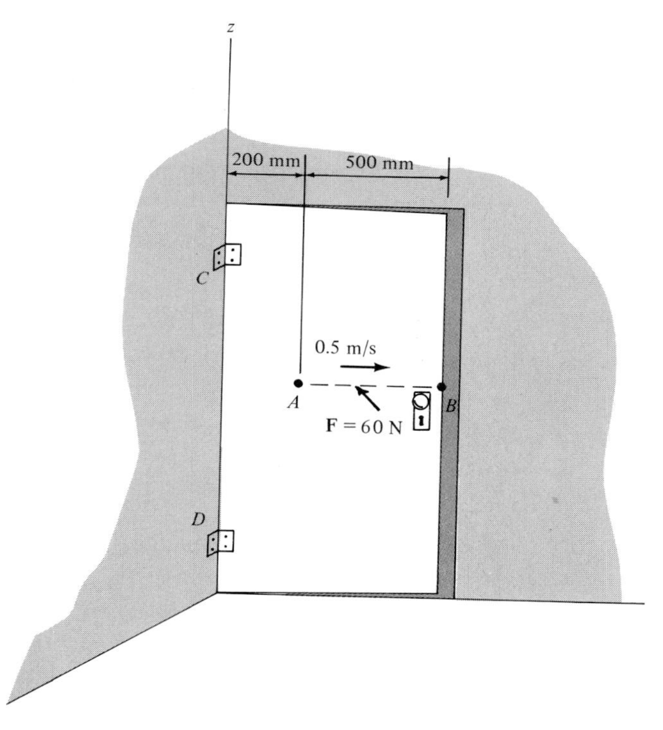

Prob. 17–47

***17–48.** The ring A has a mass of 60 kg. If it is placed over the rotor B, which always maintains a constant angular velocity of $\omega = 15$ rad/s, determine the angular acceleration of the ring when it is released from rest at the position shown. The wall at C is smooth, and the coefficient of friction between the ring and the rotor is $\mu = 0.2$. The radius of gyration of the ring is $k_G = 275$ mm.

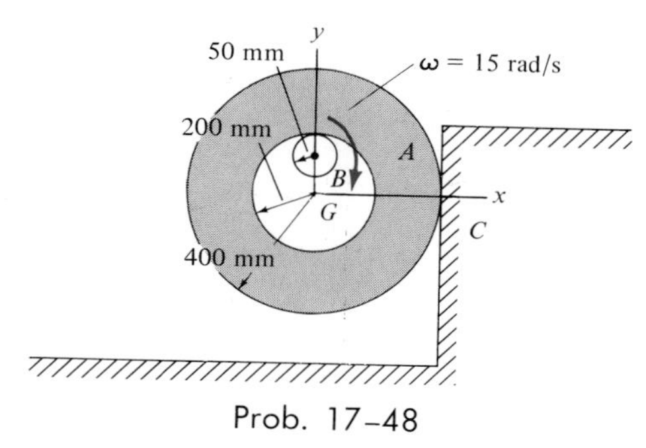

Prob. 17–48

17–49. Cable is unwound from a spool supported on small rollers at A and B by exerting a force of $T = 200$ N on the cable in the direction shown. Compute the time needed to unravel 8 m of cable from the spool if the spool and cable have a total mass of 600 kg and a centroidal radius of gyration of $k_O = 1.2$ m. For the calculation, neglect the mass of the cable being unwound and the mass of the rollers at A and B. The rollers turn with no friction.

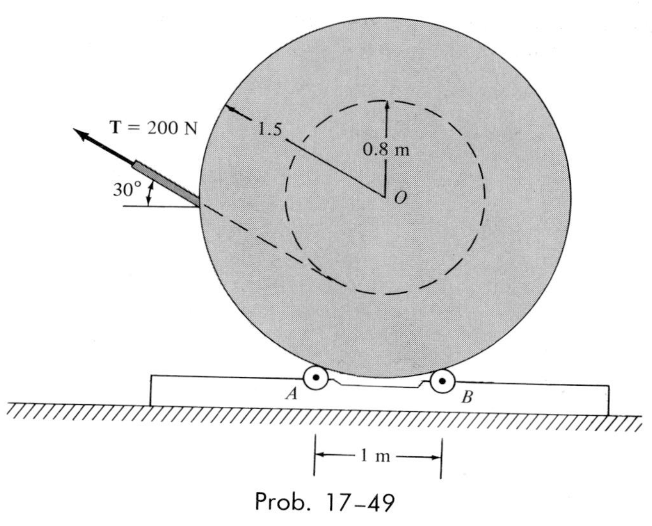

Prob. 17–49

17–50. At the instant shown, two forces, $F_1 = 600$ N and $F_2 = 400$ N, act on the slender rod, which has a mass of $m = 8$ kg. Determine the rod's initial angular acceleration and the horizontal and vertical components of reaction at the pin O. The rod is originally at rest and $a = 0.2$ m.

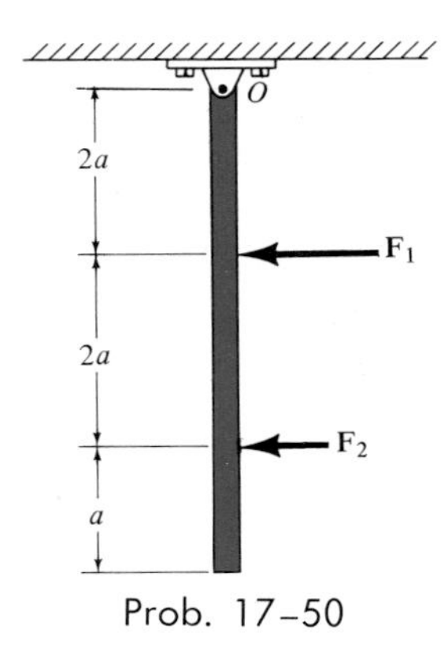

Prob. 17–50

17–50a. Solve Prob. 17–50 if $F_1 = 20$ lb, $F_2 = 15$ lb, the rod weighs $W = 10$ lb, and $a = 1$ ft.

17–51. The 15-kg cylinder is fixed to shaft AB, which is rotating with an angular velocity of $\omega = 40$ rad/s. If a force $F = 6$ N is applied to link CD, as shown, determine the time needed to stop the rotation. The coefficient of friction between CD and the cylinder is $\mu = 0.4$. Neglect friction at the bearings A and B.

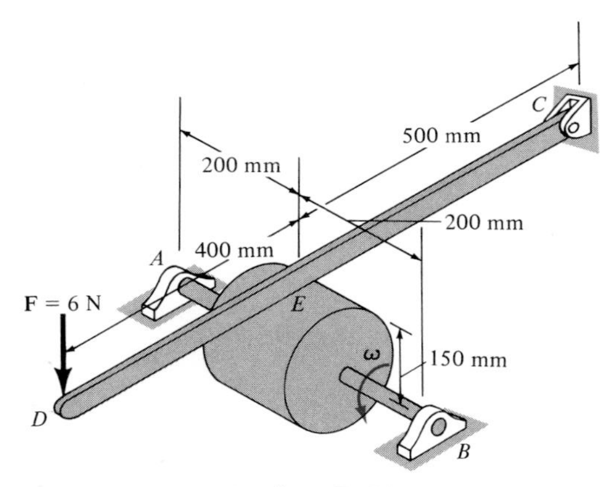

Prob. 17–51

***17–52.** The motor M supplies a constant torque or twist of 250 N · m to its connecting shaft. If the uniform boom OA has a mass of 60 kg, determine the horizontal and vertical reactive components at O at the instant the boom has an angular velocity of $\omega = 5$ rad/s and the cable is in the position shown. Neglect the mass of the pulley at B and of the cable, and assume that the boom can be approximated by a slender rod.

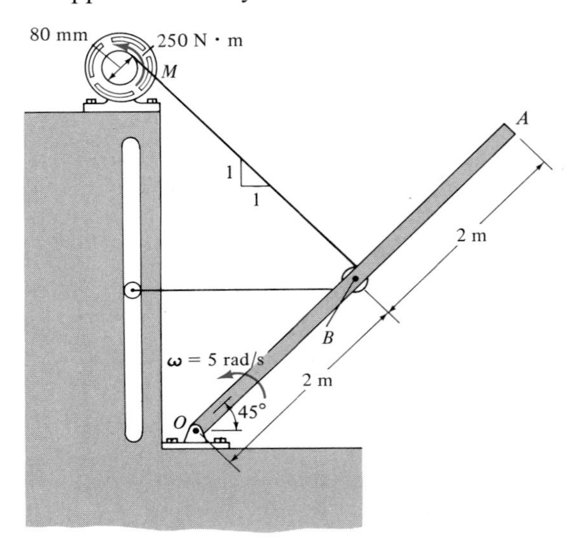

Prob. 17–52

17–53. The disk A maintains a constant clockwise angular velocity of 30 rad/s. If the 20-kg disk B is initially at rest when it is brought into contact with A, determine the time required for B to attain the same angular velocity as A. The coefficient of friction between the two disks is $\mu = 0.3$. Neglect the weight of bar BC.

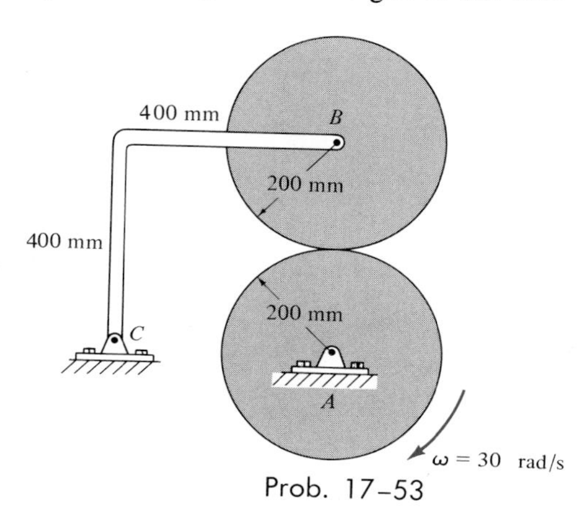

Prob. 17–53

17–54. Solve Prob. 17–53 assuming that disk A rotates with a constant *counterclockwise* angular velocity of 15 rad/s.

17–55. The uniform slender rod AB has a mass of $m = 4$ kg and a length of $l = 2$ m. If it is suspended horizontally by a spring at A and a cord at B, determine the angular acceleration of the rod and the acceleration of the rod's mass center at the instant the cord at B is cut. *Hint:* The stiffness of the spring is not needed for the calculation.

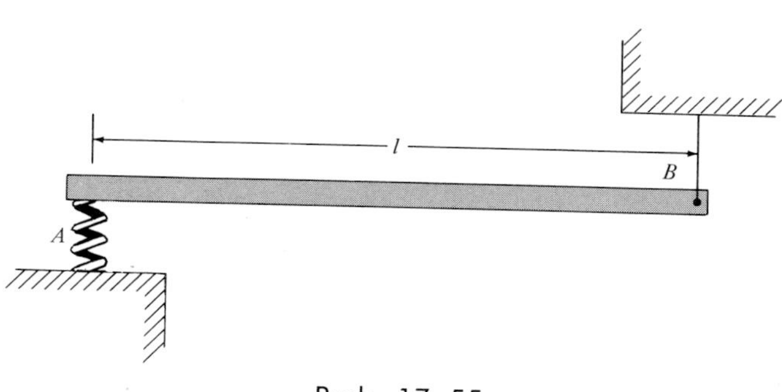

Prob. 17–55

17–55a. Solve Prob. 17–55 if the rod weighs 12 lb and $l = 3$ ft.

* **17–56.** The uniform slender rod has a mass of 5 kg. If the cord at A is cut, determine the reaction at the pin O, (a) when the rod is still in the horizontal position, and (b) when the rod swings to the vertical position.

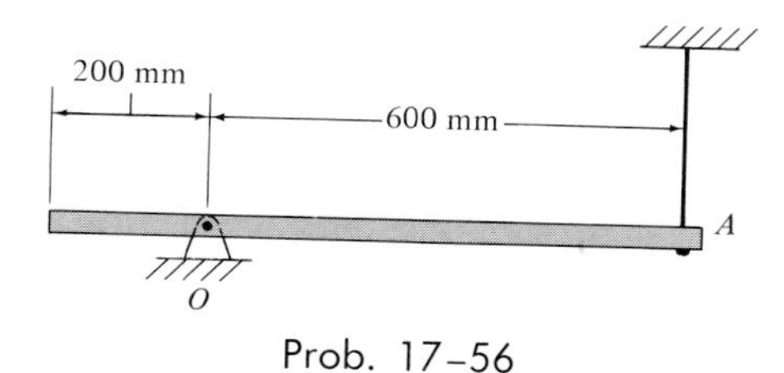

Prob. 17–56

17–57. In order to experimentally determine the moment of inertia I_G of a 4-kg connecting rod, the rod is suspended horizontally at A by a cord and at B by a piezoelectric sensor, an instrument used for measuring force. Under these equilibrium conditions, the force at B is measured as 14.6 N. If, at the instant the cord is cut, the reaction at B is measured as 9.3 N, determine the value of I_G. The support at B does not move when the measurement is taken. For the calculation, the horizontal location of G must be determined.

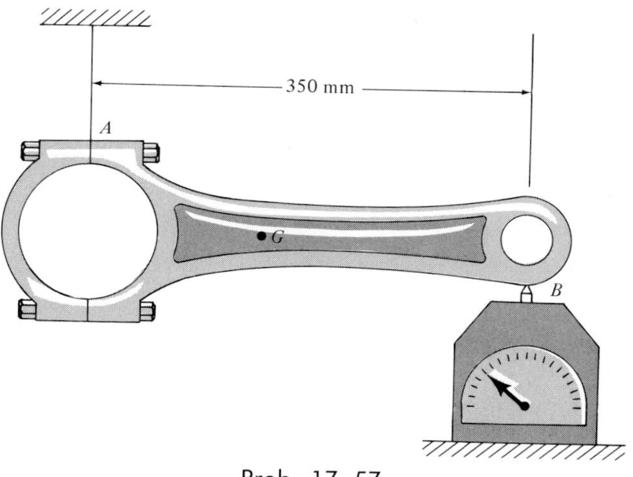

Prob. 17–57

17–58. The "Catherine wheel" is a firework that consists of a coiled tube of powder which is pinned at its center. If the powder burns at a constant rate of 100 g/s, such that the exhaust gases always exert a force having a constant magnitude of 0.3 N and directed tangent to the wheel, determine the angular velocity when 75% of the mass is burned off. Initially, the wheel is at rest, and has a mass of 500 g and a radius of $r = 75$ mm. For the calculation, always consider the wheel to be a thin disk.

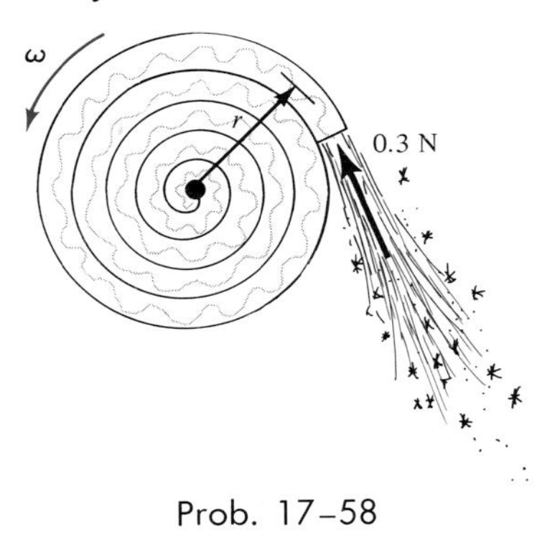

Prob. 17–58

17–59. If the 2-kg slender rod rotates about the z axis with a constant angular velocity of $\omega = 4$ rad/s, determine the axial force in the bar at a distance x from the axis of rotation.

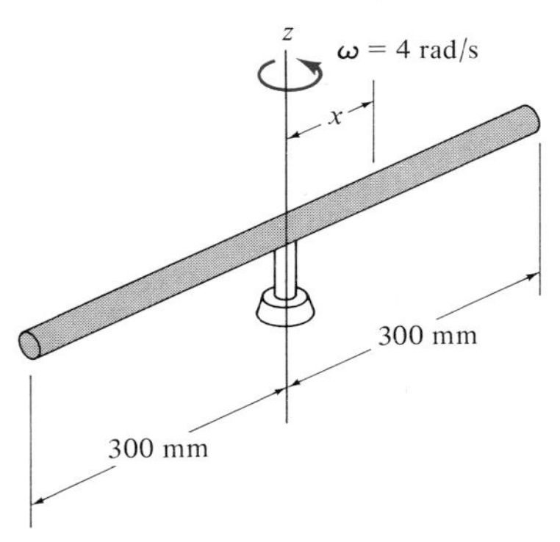

Prob. 17–59

*** 17–60.** The thin hoop has a mass of $m = 6$ kg and a radius of $r = 0.5$ m. If it is spinning on the smooth surface at a constant rate of $\omega = 8$ rad/s, determine the tension in the top of the hoop at A produced by the rotation. Assume that the hoop is "sleeping"; that is, its center O is not translating.

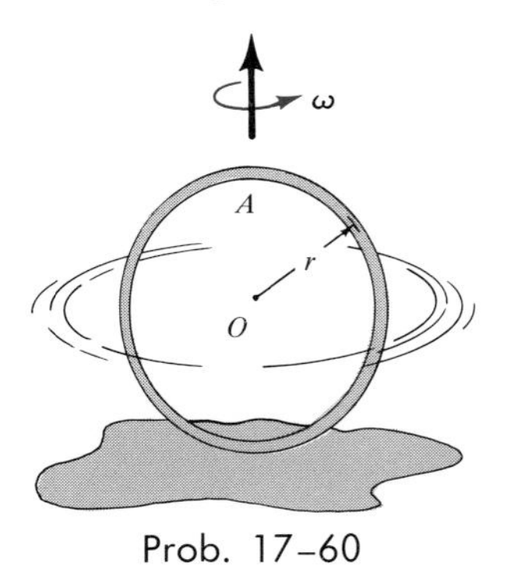

Prob. 17–60

*** 17–60a.** Solve Prob. 17–60 if the hoop has a weight of $W = 8$ lb, and $r = 1.25$ ft, $\omega = 10$ rad/s.

17–6. Equations of Motion: General Plane Motion of a Rigid Body

The rigid body (or slab) shown in Fig. 17–18*a* is subjected to general plane motion caused by the externally applied force system. The free-body and kinetic diagrams for the body are shown in Fig. 17–18*b*. The vector $m\mathbf{a}_G$ (shown dashed) has the same *direction* as the acceleration of the body's mass center, and $I_G\boldsymbol{\alpha}$ acts in the same *direction* as the angular acceleration $\boldsymbol{\alpha}$. If an x and y inertial coordinate system is chosen as shown, the three equations of motion may be written as

$$\begin{aligned} \Sigma F_x &= m(a_G)_x \\ \Sigma F_y &= m(a_G)_y \\ \Sigma M_G &= I_G\alpha \end{aligned} \qquad (17\text{–}16)$$

In some problems it may be convenient to sum moments about some point A, other than G. This is usually done in order to eliminate unknown forces from the moment summation. When used in this more general sense, the three equations of motion become

$$\begin{aligned} \Sigma F_x &= m(a_G)_x \\ \Sigma F_y &= m(a_G)_y \\ \Sigma M_A &= \Sigma(\mathcal{M}_k)_A \end{aligned} \qquad (17\text{–}17)$$

where $\Sigma(\mathcal{M}_k)_A$ represents the moment sum of $I_G\boldsymbol{\alpha}$ and $m\mathbf{a}_G$ (or its components) about A as determined by the data on the kinetic diagram.

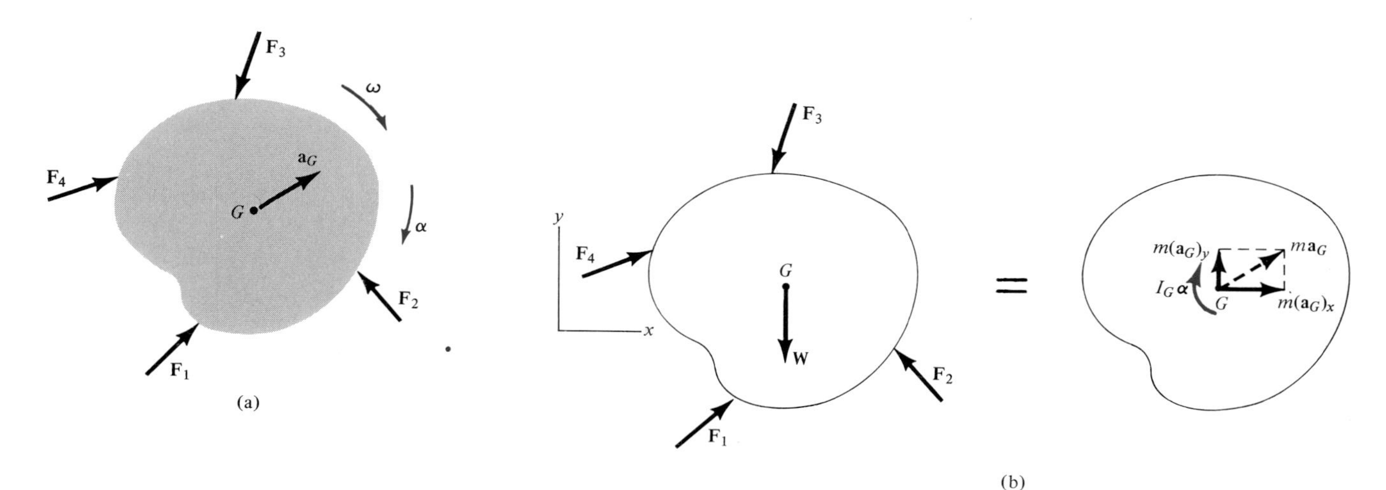

Fig. 17–18

PROCEDURE FOR ANALYSIS

A three-step procedure should be used to solve kinetics problems involving general plane motion.

Step 1: Draw the free-body and kinetic diagrams for the body, Fig. 17–18*b*; and if necessary, compute the mass moment of inertia I_G.

Step 2: Apply the three equations of motion, Eqs. 17–16 or 17–17, with reference to an established x, y inertial coordinate system.

Step 3: Use kinematics if a complete solution cannot be obtained strictly from the equations of motion. In this regard, additional equations may be obtained by using $\mathbf{v}_B = \mathbf{v}_A + \mathbf{v}_{B/A}$ and $\mathbf{a}_B = \mathbf{a}_A + \mathbf{a}_{B/A}$, which relate the motions of any two points A and B on a rigid body.

Frictional Rolling Problems. There is a class of planar kinetic problems which deserves special mentioning. These problems involve wheels, cylinders, or bodies of similar shape, which roll on a rough plane surface. Because of the applied loadings, it may not be known if the body *rolls without slipping,* or if it *slides as it rolls.* For example, consider the homogeneous disk shown in Fig. 17–19*a,* which has a mass m and is subjected to a known horizontal force **P.** Following the three-step procedure outlined above, the free-body and kinetic diagrams are shown in Fig. 17–19*b*. Applying the three equations of motion yields

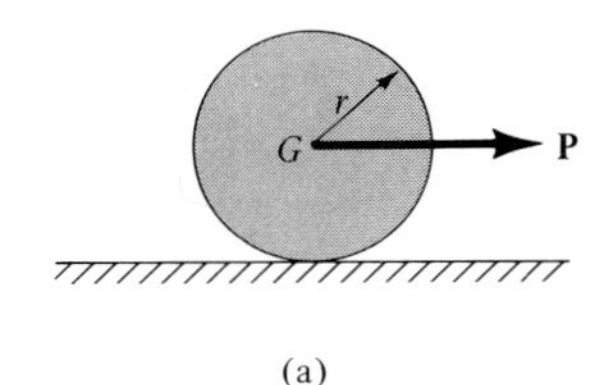

(a)

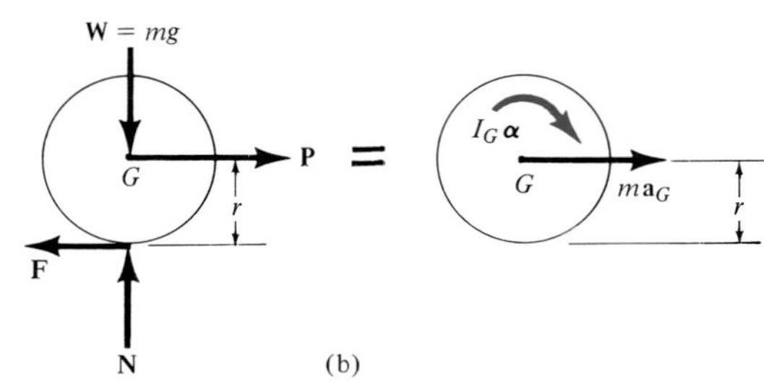

(b)

Fig. 17–19

$$\xrightarrow{+}\Sigma F_x = m(a_G)_x; \qquad P - F = ma_G \qquad (17\text{–}18)$$
$$+\uparrow\Sigma F_y = m(a_G)_y; \qquad N - mg = 0 \qquad (17\text{–}19)$$
$$\circlearrowleft +\Sigma M_G = I_G\alpha; \qquad Fr = I_G\alpha \qquad (17\text{–}20)$$

A fourth equation is needed since these *three equations* contain *four unknowns:* F, N, α, and a_G.

No Slipping. If the frictional force **F** is great enough to cause the disk to roll along the surface *without slipping*, then a_G may be related to α by the *kinematic equation*

$$a_G = \alpha r \qquad (17\text{–}21)$$

(See Example 16–3). The four unknowns are determined by *solving simultaneously* Eqs. 17–18 through 17–21. When the solution is obtained, the assumption of no slipping must be *checked*. In this regard, recall that no slipping occurs provided $F \leqslant \mu_s N$, where μ_s is the static coefficient of friction. If the inequality is satisfied, the problem is solved. However, if $F > \mu_s N$, the problem must be *reworked*, since then the disk slips as it rolls.

Slipping. In the case of slipping, α and a_G are *independent of one another* so that Eq. 17–21 does not apply. Instead, the magnitude of the frictional force **F** is related to the magnitude of the normal force **N**, using the coefficient of kinetic friction, μ_k, i.e.,

$$F = \mu_k N \qquad (17\text{–}22)$$

Hence, in this case Eqs. 17–18, 17–19, 17–20, and 17–22 are used for the solution. Examples 17–12 and 17–13 illustrate these concepts numerically.

Example 17–12

The 30-kg wheel shown in Fig. 17–20*a* has a mass center at G and a radius of gyration of $k_G = 0.15$ m. If a 200-N force is centrally applied to the wheel as shown, determine the acceleration of the center O. The wheel is initially at rest and both the static and kinetic coefficients of friction between the wheel and the horizontal plane at A are $\mu = 0.45$.

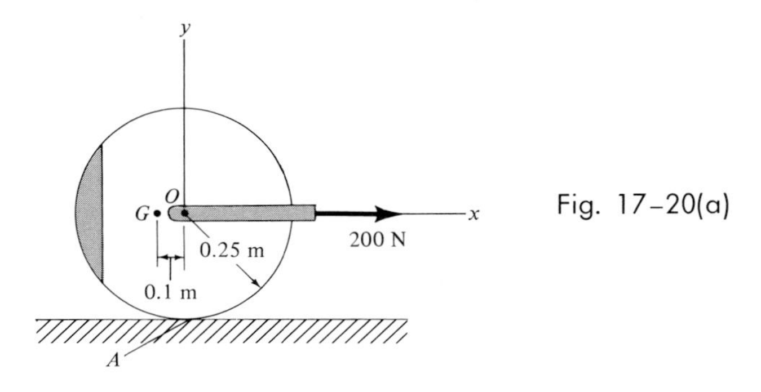

Fig. 17–20(a)

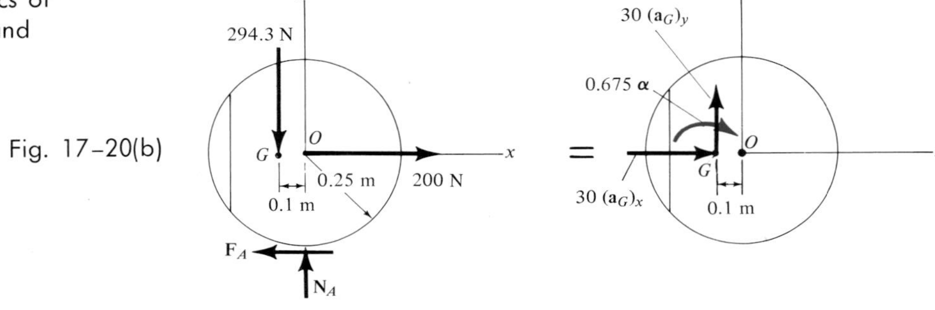

Fig. 17–20(b)

Solution

Step 1: The free-body and kinetic diagrams are shown in Fig. 17–20*b*. The mass center G of the wheel is not located at the geometric center O, and furthermore, the initial *path of motion* for this point is *unknown*. Therefore, the two components $m(\mathbf{a}_G)_x$ and $m(\mathbf{a}_G)_y$ are shown acting in the x and y directions on the kinetic diagram.

Using the radius of gyration and the mass, it is possible to determine the moment of inertia of the wheel about the mass center as

$$I_G = k_G^2 m = (0.15)^2 30 = 0.675 \text{ kg} \cdot \text{m}^2$$

Step 2: Applying the equations of motion gives

$$\overset{+}{\rightarrow}\Sigma F_x = m(a_G)_x; \qquad 200 - F_A = 30(a_G)_x \tag{1}$$

$$+\uparrow\Sigma F_y = m(a_G)_y; \qquad N_A - 294.3 = 30(a_G)_y \tag{2}$$

$$\circlearrowright +\Sigma M_G = I_G\alpha; \qquad F_A(0.25) - N_A(0.1) = 0.675\alpha \tag{3}$$

There are five unknowns in these three equations: F_A, N_A, $(a_G)_x$, $(a_G)_y$, and α.

No Slipping. *Step 3:* If it is assumed that the wheel rolls without slipping, the acceleration components $(\mathbf{a}_G)_x$ and $(\mathbf{a}_G)_y$ can be related to the wheel's angular acceleration $\boldsymbol{\alpha}$ using kinematics. In this case the acceleration of the *center* of the wheel has a magnitude of $a_O = \alpha r = \alpha(0.25)$ and is directed to the right. Furthermore, the wheel is originally at rest so that $\boldsymbol{\omega} = \mathbf{0}$. Using the data shown on the kinematic diagrams in Fig. 17–20*c*,

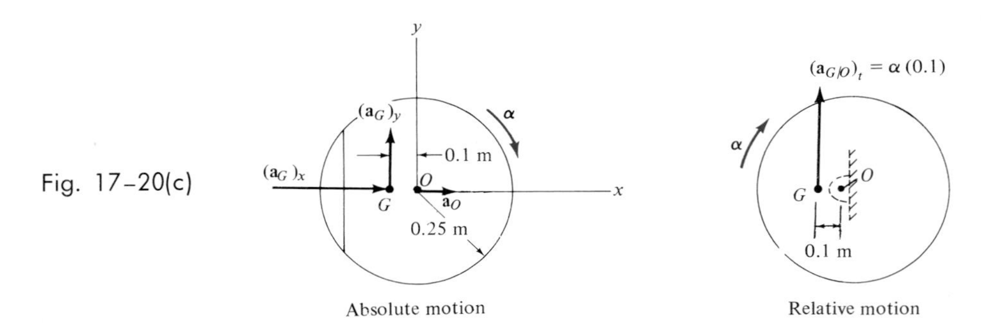

Fig. 17–20(c)

and choosing O as the base point, it is possible to relate the accelerations of points G and O by Eq. 16–21, i.e.,

$$\mathbf{a}_G = \mathbf{a}_O + (\mathbf{a}_{G/O})_t + (\mathbf{a}_{G/O})_n$$

$$\underset{\rightarrow}{(a_G)_x} + \underset{\uparrow}{(a_G)_y} = \underset{\rightarrow}{(0.25)\alpha} + \underset{\uparrow}{\alpha(0.1)} + 0$$

Equating the respective horizontal and vertical components, we have

$$(a_G)_x = 0.25\alpha \tag{4}$$

$$(a_G)_y = 0.1\alpha \tag{5}$$

Equations (1) through (5) may now be solved for the five unknowns, which yields

$$F_A = 146.0 \text{ N}$$
$$N_A = 316.0 \text{ N}$$
$$(a_G)_x = 1.80 \text{ m/s}^2$$
$$(a_G)_y = 0.722 \text{ m/s}^2$$
$$\alpha = 7.22 \text{ rad/s}^2$$

Our original assumption of no slipping requires $F_A \leqslant \mu N_A$. However, since $146.0 > 0.45(316.0) = 142.2$ N, the wheel will start to roll and slide at the same time.

Slipping. This requires $F_A = \mu N_A$, or

$$F_A = 0.45 N_A \tag{6}$$

The remaining equations are obtained using *kinematics*. If the center of the wheel has an acceleration of $\mathbf{a}_O$, Fig. 17–21*c*, Eq. 16–20, applied between G and O, becomes

$$\mathbf{a}_G = \mathbf{a}_O + (\mathbf{a}_{G/O})_t + (\mathbf{a}_{G/O})_n$$

$$\underset{\rightarrow}{(a_G)_x} + \underset{\uparrow}{(a_G)_y} = \underset{\rightarrow}{a_O} + \underset{\uparrow}{\alpha(0.1)} + 0$$

Equating the respective horizontal and vertical components,

$$(a_G)_x = a_O \tag{7}$$

$$(a_G)_y = 0.1\alpha \tag{8}$$

Solving Eqs. (1) through (3) and (6) through (8) gives

$$F_A = 140.2 \text{ N}$$
$$N_A = 311.6 \text{ N}$$
$$(a_G)_x = 1.99 \text{ m/s}^2$$
$$(a_G)_y = 0.577 \text{ m/s}^2$$
$$\alpha = 5.77 \text{ rad/s}^2$$
$$a_O = 1.99 \text{ m/s}^2 \qquad \textit{Ans.}$$

Example 17–13

The uniform slender beam shown in Fig. 17–21*a* has a mass of 100 kg. If the coefficients of static and kinetic friction between the end of the beam and the surface are both equal to $\mu_A = 0.25$, determine the beam's angular acceleration at the instant the 400-N horizontal force is applied. The beam is originally at rest.

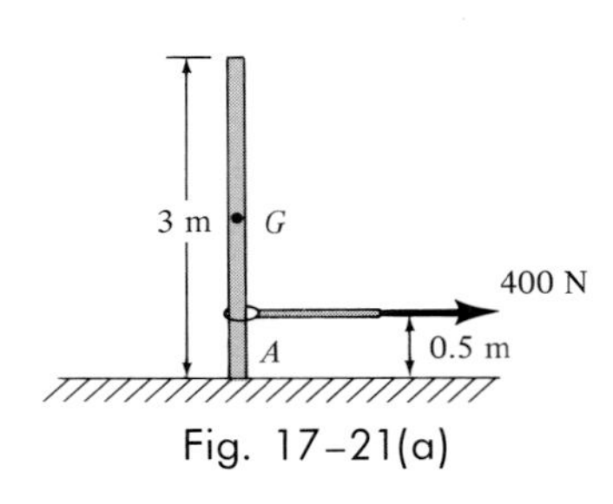

Fig. 17–21(a)

Solution

Step 1: The free-body and kinetic diagrams are shown in Fig. 17–21*b*. The path of motion of the mass center *G* will be along an unknown curved path having a radius of curvature which is initially parallel to the y axis. There is no normal or y component of acceleration since the beam is originally at rest, i.e., $\mathbf{v}_G = \mathbf{0}$.

If the beam is approximated by a uniform slender rod, the moment of inertia is

$$I_G = \tfrac{1}{12}ml^2 = \tfrac{1}{12}(100)(3)^2 = 75 \text{ kg} \cdot \text{m}^2$$

Step 2: Applying the equations of motion yields

$$\overset{+}{\rightarrow}\Sigma F_x = m(a_G)_x; \qquad 400 - F_A = 100a_G \qquad (1)$$

$$+\uparrow\Sigma F_y = m(a_G)_y; \qquad N_A - 981 = 0 \qquad (2)$$

$$\circlearrowleft + \Sigma M_G = I_G\alpha; \qquad F_A(1.5) - 400(1) = 75\alpha \qquad (3)$$

A fourth equation is needed to solve for the four unknowns N_A, F_A, a_G, and α.

No Slipping. *Step 3:* If it is assumed that no slipping occurs at *A*, then *A* acts as a "pivot" and the kinematic equation $a_G = \alpha r_{AG}$ may be applied, i.e.,

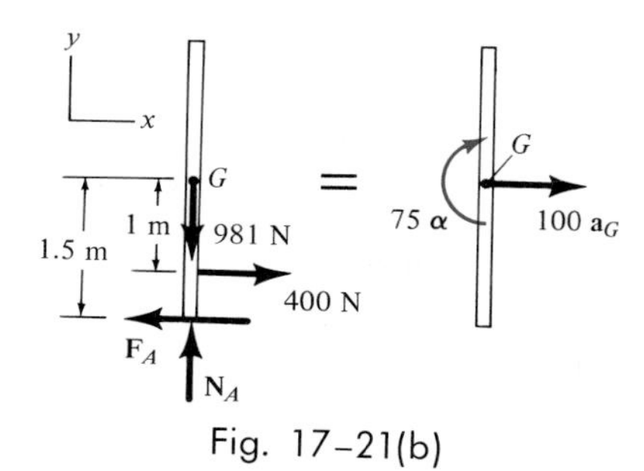

Fig. 17–21(b)

$$a_G = 1.5\alpha \tag{4}$$

Solving Eqs. (1) through (4) yields

$$N_A = 981 \text{ N}$$
$$F_A = 300 \text{ N}$$
$$a_G = 1.0 \text{ m/s}^2$$
$$\alpha = 0.667 \text{ rad/s}^2$$

Testing the original assumption of no slipping requires $F_A < \mu_A N_A$. Using the data, it is seen that this equation is not satisfied, since

$$300 > 0.25(981) = 245.3 \text{ N}$$

Slipping. The problem must be reworked with the assumption that the beam slips at A. For this case Eq. (4) does *not* apply. Instead, the frictional equation $F_A = \mu_A N_A$ is used, i.e.,

$$F_A = 0.25 N_A \tag{5}$$

Solving Eqs. (1) through (3) and (5) simultaneously yields

$$N_A = 981 \text{ N}$$
$$F_A = 245.3 \text{ N}$$
$$a_G = 1.55 \text{ m/s}^2$$
$$\alpha = -0.427 \text{ rad/s}^2 \qquad \textit{Ans.}$$

Because of the negative sign, the angular acceleration of the beam is counterclockwise.

Fig. 17–22(a)

Example 17–14

The 10-kg uniform ladder AB shown in Fig. 17–22*a* has its mass center at G. If it is released from rest in the position shown, determine the normal reactions at A and B at this instant. For the calculation, assume that the ladder is approximated by a uniform slender rod, and that the points of contact at A and B are smooth.

Solution

Step 1: The free-body and kinetic diagrams of the ladder are shown in Fig. 17–22*b*. The directions of $m(\mathbf{a}_G)_x$, $m(\mathbf{a}_G)_y$, and $I_G\boldsymbol{\alpha}$ have all been assumed.

From Appendix D, the moment of inertia of the ladder (rod) about its mass center is

$$I_G = \tfrac{1}{12}ml^2 = \tfrac{1}{12}(10)(3)^2 = 7.5\ \text{kg}\cdot\text{m}^2$$

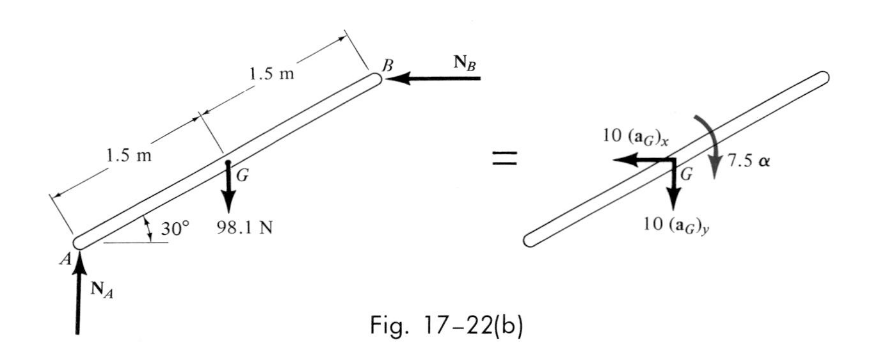

Fig. 17–22(b)

Step 2: Applying the equations of motion yields

$$\overset{+}{\leftarrow}\Sigma F_x = m(a_G)_x; \quad N_B = 10(a_G)_x \tag{1}$$

$$+\uparrow\Sigma F_y = m(a_G)_y; \quad N_A - 98.1 = -10(a_G)_y \tag{2}$$

$$\curvearrowleft+\Sigma M_G = I_G\alpha; \quad N_A(1.5\cos 30°) - N_B(1.5\sin 30°) = 7.5\alpha \tag{3}$$

There are five unknowns in these three equations: N_A, N_B, $(a_G)_x$, $(a_G)_y$, and α.

Step 3: The remaining two equations will be obtained by using kinematics to relate $(a_G)_x$ and $(a_G)_y$ to α. Since the ladder is released from rest, $\boldsymbol{\omega} = \mathbf{0}$. Furthermore, points A and B have *rectilinear motion* which is confined along the floor and wall, so that the *directions* of $\mathbf{a}_A$ and $\mathbf{a}_B$ are known. The kinematic diagrams shown in Fig. 17–22*c* indicate these motions, for which G is chosen as the base point. Note that the directions

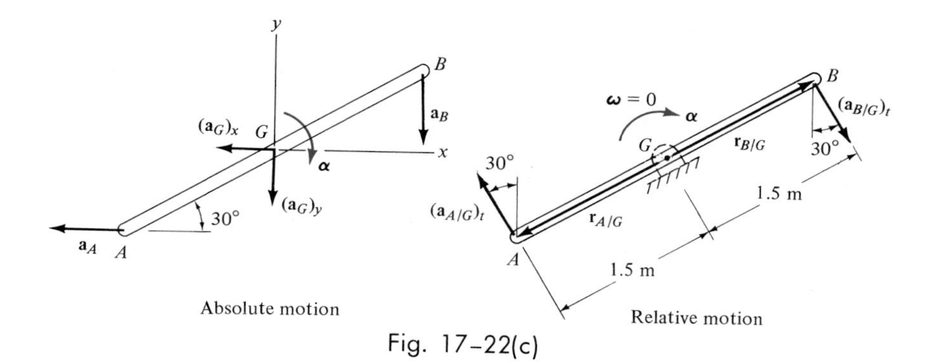

Fig. 17–22(c)

of $(\mathbf{a}_G)_x$, $(\mathbf{a}_G)_y$, and $\boldsymbol{\alpha}$ are consistent with those shown on the kinetic diagram. This is important since the equations of motion and kinematics must be solved simultaneously. Applying Eq. 16–21 between points B and G yields

$$\mathbf{a}_B = \mathbf{a}_G + (\mathbf{a}_{B/G})_t + (\mathbf{a}_{B/G})_n$$

$$\underset{\downarrow}{a_B} = \underset{\leftarrow}{(a_G)_x} + \underset{\downarrow}{(a_G)_y} + \underset{\searrow 30^\circ}{\alpha(1.5)} + 0$$

Equating the horizontal and vertical components,

$(\overset{+}{\leftarrow})$ $$0 = (a_G)_x + 0 - \alpha(1.5)\sin 30^\circ + 0 \tag{4}$$

$(+\downarrow)$ $$a_B = 0 + (a_G)_y + \alpha(1.5)\cos 30^\circ + 0$$

The second of these equations is of no use in the solution, since it contains the unknown a_B.

Applying Eq. 16–21 between points A and G, Fig. 17–22c, yields

$$\mathbf{a}_A = \mathbf{a}_G + (\mathbf{a}_{A/G})_t + (\mathbf{a}_{A/G})_n$$

$$\underset{\leftarrow}{a_A} = \underset{\leftarrow}{(a_G)_x} + \underset{\downarrow}{(a_G)_y} + \underset{\nwarrow 30^\circ}{\alpha(1.5)} + 0$$

Expanding and equating the horizontal and vertical components,

$(\overset{+}{\leftarrow})$ $$a_A = (a_G)_x + 0 + \alpha(1.5)\sin 30^\circ$$

$(+\downarrow)$ $$0 = 0 + (a_G)_y - \alpha(1.5)\cos 30^\circ \tag{5}$$

Solving Eqs. (1) through (5) simultaneously gives

$$\alpha = 4.24 \text{ rad/s}^2$$
$$(a_G)_x = 3.18 \text{ m/s}^2$$
$$(a_G)_y = 5.52 \text{ m/s}^2$$
$$N_A = 42.9 \text{ N} \qquad \textit{Ans.}$$
$$N_B = 31.8 \text{ N} \qquad \textit{Ans.}$$

17-61. If the disk in Fig. 17–19*a rolls without slipping,* show that when moments are summed about the instantaneous center of zero velocity, *IC,* it is possible to use the moment equation $\Sigma M_{IC} = I_{IC}\alpha$, where I_{IC} represents the moment of inertia of the disk calculated about the instantaneous axis of zero velocity.

17-62. The 3-kg ball rests on a horizontal surface and is being pushed forward by moving the piston P with a force of $F = 2$ N. If the ball rolls without slipping at C and the coefficient of friction between the piston and the ball at D is $\mu_D = 0.20$, compute the velocity of the ball's mass center G after the piston moves 50 mm. Neglect the mass of the piston and assume that the guides at A and B are smooth. The ball is originally at rest.

Prob. 17–62

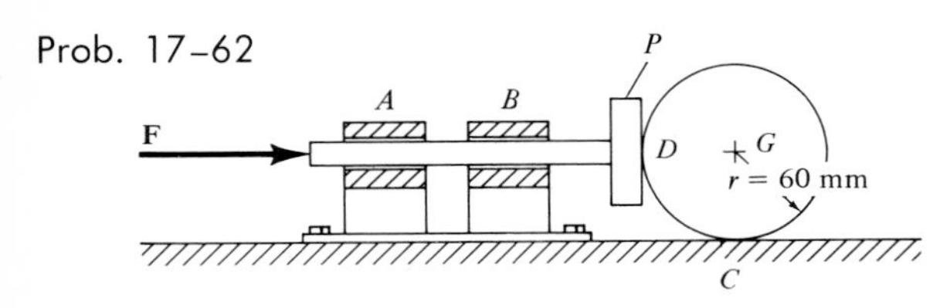

17-63. The spool and the wire wrapped around its core have a mass of 50 kg and a centroidal radius of gyration of $k_G = 235$ mm. If the coefficient of friction at the surface is $\mu_B = 0.15$, determine the angular acceleration of the spool after it is released from rest.

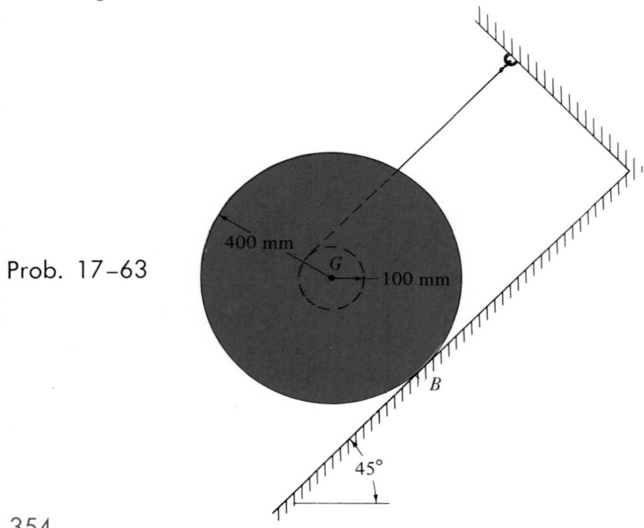

Prob. 17–63

***17-64.** The ring has a mass of 10 kg, a mass center at G, and a radius of gyration of $k_G = 135$ mm. If its angular velocity is $\omega_1 = 2$ rad/s when it is in the position shown, compute the angular acceleration at this instant and the normal force of the ring on the ground. Assume that slipping does not occur.

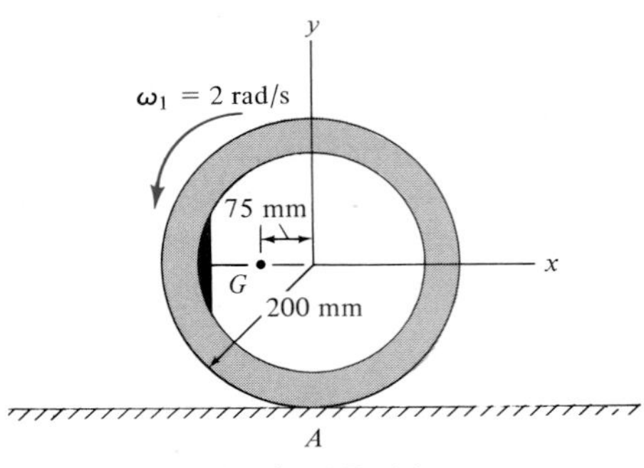

Prob. 17–64

17-65. The wheel has a mass of $m = 15$ kg, a radius of $r = 200$ mm, and a radius of gyration of $k_G = 130$ mm. If the coefficient of friction between the wheel and the plane is $\mu = 0.2$, determine the wheel's angular acceleration as it rolls down the incline. Set $\theta = 12°$.

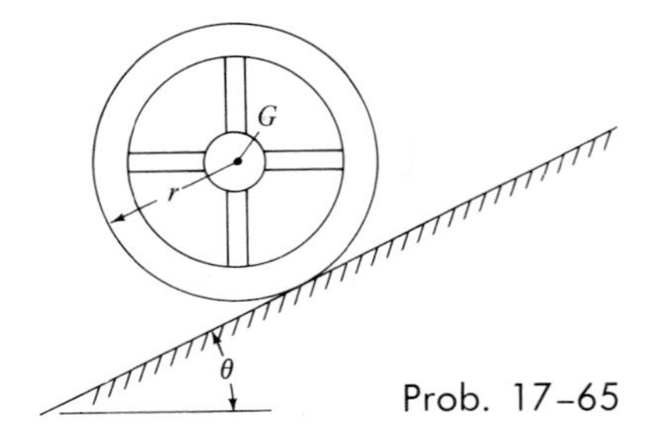

Prob. 17–65

17-65a. Solve Prob. 17–65 if the wheel weighs $W = 30$ lb, and $r = 1.25$ ft, $k_G = 0.92$ ft, $\mu = 0.3$, $\theta = 15°$.

17-66. Solve Prob. 17–65 if $\theta = 50°$.

17-67. Determine the maximum angle θ of the inclined plane in Prob. 17–65 so that the wheel rolls without slipping.

***17–68.** A uniform rod having a mass of 5 kg is pin-supported at A from a roller which rides on a horizontal track. At the instant shown the rod is swinging into the vertical position with an angular velocity of $\omega = 6$ rad/s. If a horizontal force of $F = 100$ N is applied to the roller at this instant, determine the acceleration of the roller. Neglect the mass of the roller and its size d in the computations.

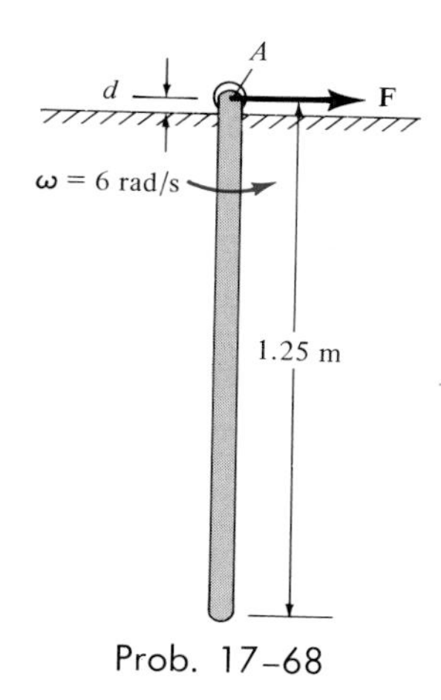

Prob. 17–68

17–69. Solve Prob. 17–68 assuming that the roller at A is replaced by a slider block having a negligible mass. The coefficient of friction between the block and the track is $\mu = 0.3$. Neglect the dimension d and the size of the block in the computations.

17–70. The bowling ball, having a mass of $m = 7$ kg and radius $r = 115$ mm, is cast horizontally onto an alley such that initially its center has a speed of $v = 2$ m/s and the ball has a backspin of $\omega = 20$ rad/s. If the coefficient of friction between the floor and the ball is $\mu = 0.12$, determine the distance the ball travels before it stops backspinning. For the calculation, neglect the finger holes in the ball and assume the ball has a uniform density.

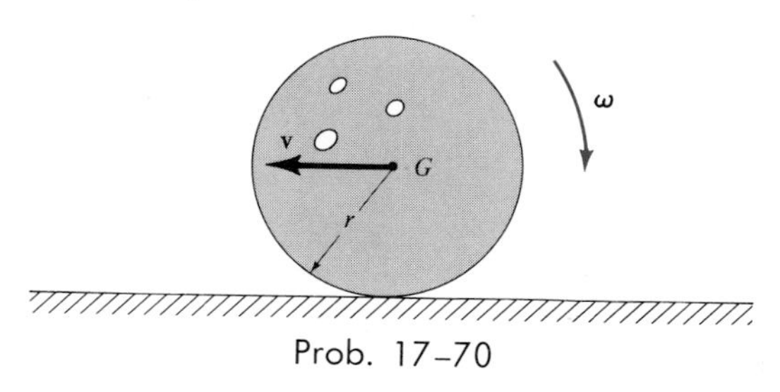

Prob. 17–70

17–70a. Solve Prob. 17–70 if the bowling ball weighs $W = 16$ lb, and $r = 0.375$ ft, $v = 8$ ft/s, $\omega = 25$ rad/s, $\mu = 0.2$.

17–71. A long strip of paper is wrapped into two rolls, each having a mass of 8 kg. Roll A is pin-supported about its center, whereas roll B is not centrally supported. If B is brought into contact with A and released from rest, determine the initial tension in the paper between the rolls and the angular acceleration of each roll. For the calculation, assume the rolls to be approximated by cylinders.

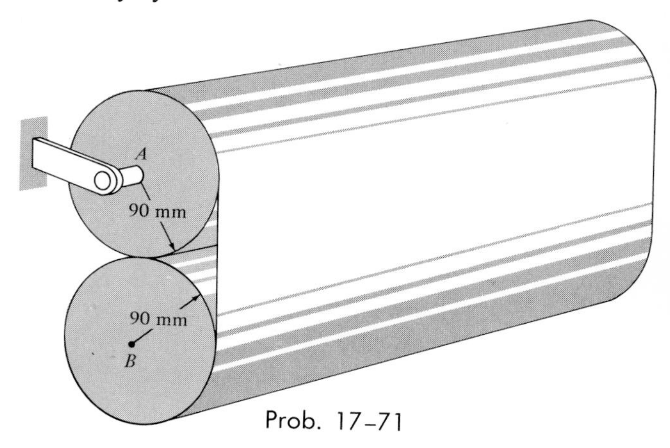

Prob. 17–71

*17-72. The rocket has a mass of 11 Mg and a center of mass at G. During operation, each of its two engines provides a thrust of $F = T = 120$ kN. If at a certain instant the thrust of engine A suddenly falls to $F = 60$ kN, while $T = 120$ kN, determine the angular acceleration of the rocket and the acceleration of its nose B while the rocket is still in the vertical position. The radius of gyration about an axis perpendicular to the plane of motion and passing through G is $k_G = 4.8$ m.

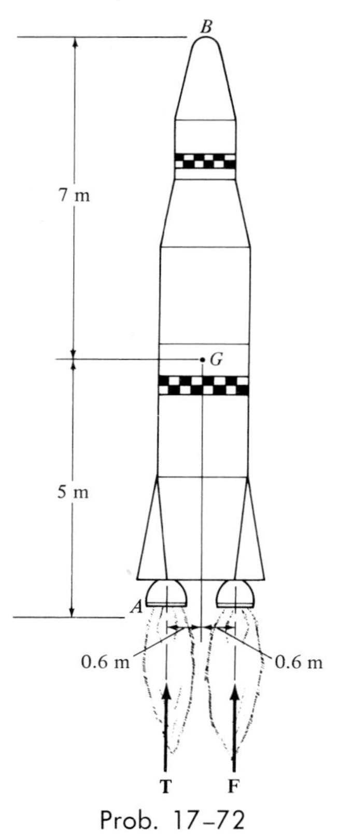

Prob. 17-72

17-73. The hoop, having a mass of 1.5 kg, is given an initial angular velocity of 12 rad/s when it is placed on a horizontal surface. Determine the velocity of its center O and its angular velocity in $t = 0.5$ s after it is released. Is the hoop still slipping on the surface at this time? The coefficient of friction between the hoop and the surface is $\mu_A = 0.3$. Neglect the thickness of the hoop for the calculation.

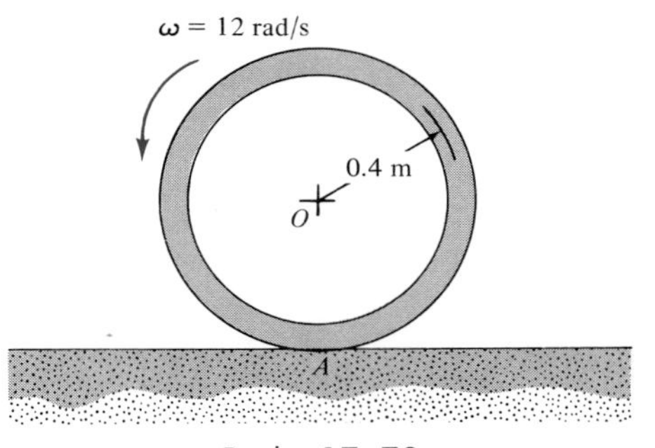

Prob. 17-73

17-74. If the cable CB is horizontal and the beam is at rest in the position shown, determine the tension in the cable at the instant the towing force $F = 1200$ N is applied. The coefficient of friction between the beam and the floor at A is $\mu_A = 0.3$. For the calculation, assume that the beam is a uniform slender rod having a mass of 100 kg.

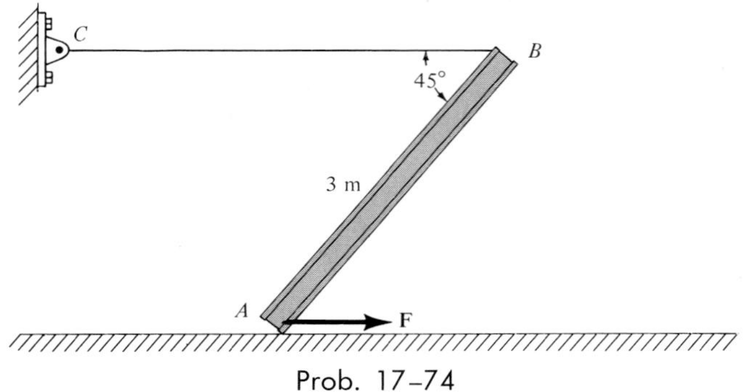

Prob. 17-74

17–75. The tube has a mass of $m = 200$ kg, an outer radius of $r = 400$ mm, and a centroidal radius of gyration of $k_G = 375$ mm. If it is originally at rest on the surface of a truck, determine its initial angular acceleration if the truck is given an acceleration of $a_t = 0.5\ \text{m/s}^2$. The tube rolls without slipping at B.

Prob. 17–75

17–75a. Solve Prob. 17–75 if the tube weighs $W = 50$ lb, and $r = 2.38$ ft, $k_G = 2.31$ ft, $a_t = 1.6\ \text{ft/s}^2$.

*** 17–76.** The cart and its contents have a mass of 30 kg and a mass center at G, excluding the wheels. Each of the two wheels has a mass of 3 kg and a radius of gyration of $k_O = 0.12$ m. If the cart is released from rest from the position shown, determine its speed after it travels 5 m down the incline. The coefficient of friction is $\mu_A = 0.3$ between the incline and A. The wheels roll without slipping at B.

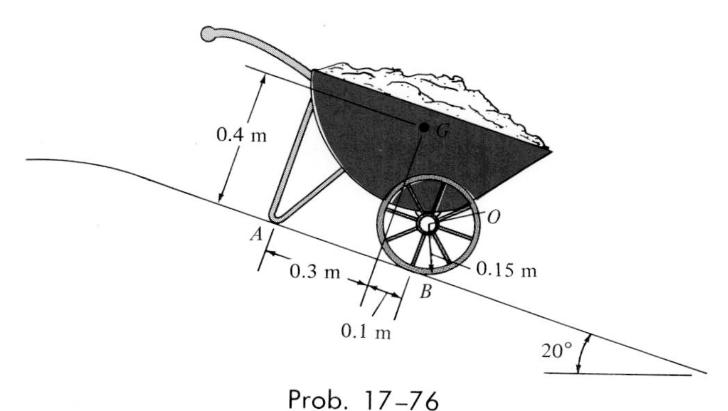

Prob. 17–76

17–77. A rocket CD, having a mass of 18 Mg with center of mass at G, is located in deep space so that the effect of gravitation (weight) can be neglected. The smaller rockets A and B each have a mass of 2 Mg and center of mass at G_A and G_B, respectively. If these rockets travel in a straight line so that initially they exert a constant thrust of $T = 8$ kN perpendicular to CD, determine the angular acceleration of CD and the acceleration of its center of mass G. Assume that CD is initially at rest and that its radius of gyration about an axis passing through G and perpendicular to the plane of motion is $k_G = 4.60$ m.

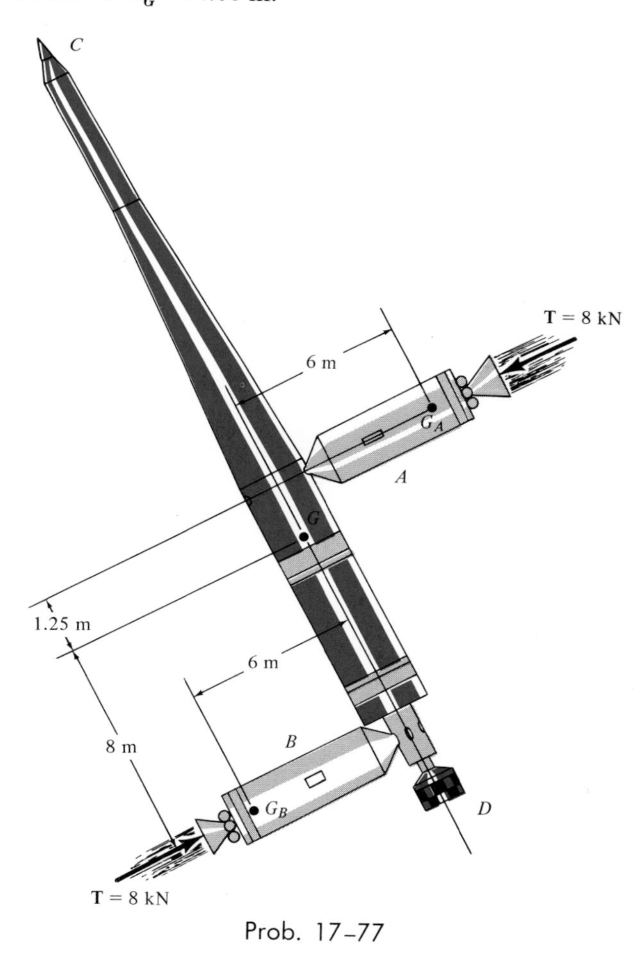

Prob. 17–77

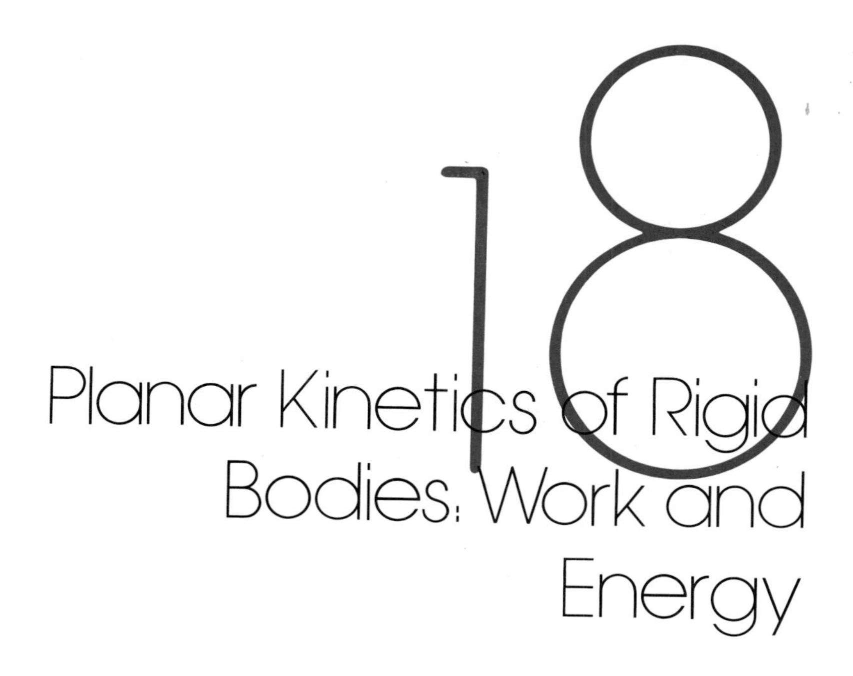

18 Planar Kinetics of Rigid Bodies: Work and Energy

18–1. Kinetic Energy of a Rigid Body

In this chapter work-and-energy methods for a rigid body subjected to plane motion will be discussed. The more general discussion of these methods, as applied to the spatial motion of a rigid body, is presented in Chapter 21.

Before discussing the principle of work and energy for a body, however, the methods for obtaining the body's kinetic energy when it is subjected to general plane motion, translation, or rotation about a fixed axis will be developed.

General Formulation. Consider the rigid body shown in Fig. 18–1, which is represented here by a *slab* moving in the *x*-*y* reference plane of motion. An arbitrary *i*th particle of the body, having a mass m_i, is located at $\mathbf{r}_{i/G}$ from the body's center of mass *G*. If, at the *instant* shown, this particle has a velocity $\mathbf{v}_i$, the particle's kinetic energy is

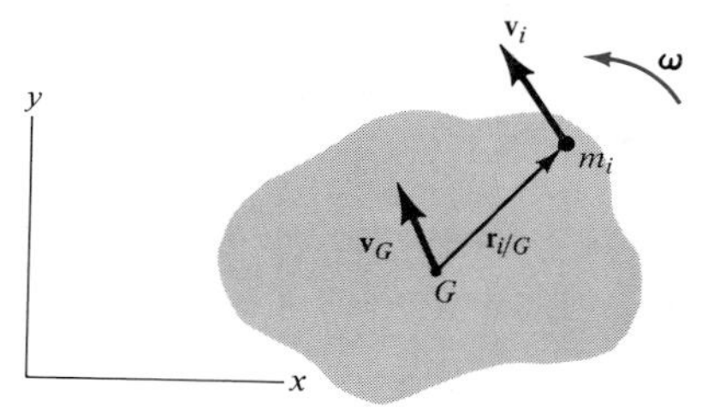

Fig. 18–1

$$T_i = \tfrac{1}{2}m_i(v_i)^2 = \tfrac{1}{2}m_i(\mathbf{v}_i \cdot \mathbf{v}_i)$$

The kinetic energy of the entire body is determined by writing similar expressions for each of the body's n particles and summing the results, i.e.,

$$T = \tfrac{1}{2}\Sigma m_i(\mathbf{v}_i \cdot \mathbf{v}_i)$$

This equation may be written in another manner by using kinematics to relate $\mathbf{v}_i$ to $\mathbf{v}_G$, the velocity of the body's mass center. In this case

$$\mathbf{v}_i = \mathbf{v}_G + \mathbf{v}_{i/G}$$

where $\mathbf{v}_{i/G}$ is the relative velocity of i with respect to G. Using this equation the kinetic energy can be expressed as

$$\begin{aligned} T &= \tfrac{1}{2}\Sigma m_i(\mathbf{v}_G + \mathbf{v}_{i/G}) \cdot (\mathbf{v}_G + \mathbf{v}_{i/G}) \\ &= \tfrac{1}{2}(\Sigma m_i)v_G^2 + v_G \cdot (\Sigma m_i v_{i/G}) + \tfrac{1}{2}\Sigma(m_i v_{i/G}^2) \end{aligned}$$

The first sum in parentheses represents the total mass m of the body. The second sum is *zero* because $\Sigma m_i v_{i/G} = d/dt(\Sigma m_i r_{i/G}) = d/dt(\Sigma m_i \bar{r}) = 0$; i.e., by definition of the mass center, the position vector $\bar{\mathbf{r}}$ relative to G is equal to zero. Furthermore, the relative velocity $v_{i/G} = \omega r_{i/G}$ (Eq. 16–12), which can be used to simplify the third term. Hence,

$$T = \tfrac{1}{2}mv_G^2 + \tfrac{1}{2}\omega^2\Sigma m_i r_{i/G}^2$$

When the particle mass $m_i \rightarrow dm$, the summation sign becomes an integral since the number of particles $n \rightarrow \infty$. Representing $r_{i/G}$ by the more "generalized" dimension r we have

$$T = \tfrac{1}{2}mv_G^2 + \tfrac{1}{2}\omega^2 \int_m r^2\, dm$$

The integral represents the mass moment of inertia I_G of the body about an axis passing through G and directed perpendicular to the plane of motion. The final result is therefore

$$T = \tfrac{1}{2}mv_G^2 + \tfrac{1}{2}I_G\omega^2 \qquad (18\text{–}1)$$

Both terms on the right side are always *positive,* since the velocities are squared. Furthermore, it may be verified that these terms have units of length times force, common units being m · N, which is expressed in joules (J).

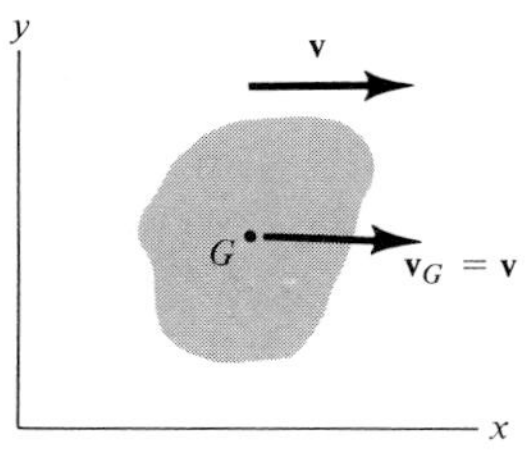

Fig. 18–2

Translation. When a rigid body of mass m is subjected to either rectilinear or curvilinear *translation,* the kinetic energy due to rotation is zero, since $\boldsymbol{\omega} = \mathbf{0}$. From Eq. 18–1, the kinetic energy of the body is, therefore,

$$T = \tfrac{1}{2}mv_G^2 \qquad (18\text{–}2)$$

where v_G is the magnitude of the body's translational velocity at the instant considered, Fig. 18–2.

Rotation About a Fixed Axis. When a rigid body is *rotating about a fixed axis* passing through point O, Fig. 18–3, its mass center has a velocity of $v_G = r_{G/O}\omega$. Hence, the body has both *translational* and *rotational* kinetic energy as defined by Eq. 18–1, i.e.,

$$T = \tfrac{1}{2}mv_G^2 + \tfrac{1}{2}I_G\omega^2$$

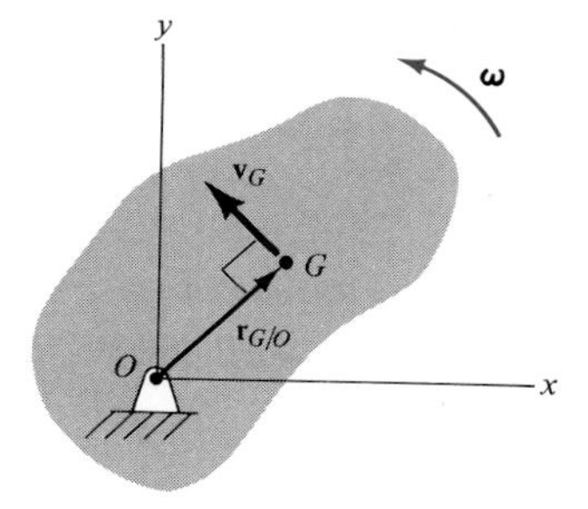

Fig. 18–3

The body's kinetic energy may be formulated in another manner by substituting $v_G = r_{G/O}\omega$ into the above equation, in which case $T = \tfrac{1}{2}(I_G + mr_{G/O}^2)\omega^2$, and, by the parallel-axis theorem, noting that the terms inside the parentheses represent the moment of inertia I_O of the body about an axis perpendicular to the plane of motion and passing through point O. Hence,*

$$T = \tfrac{1}{2}I_O\omega^2 \qquad (18\text{–}3)$$

From the derivation, this equation may be substituted for Eq. 18–1 since it accounts for *both* the translational kinetic energy of the body's mass center and the rotational kinetic energy of the body, computed about the mass center.

General Plane Motion. When a rigid body is subjected to general plane motion, Fig. 18–1, it has an angular velocity $\boldsymbol{\omega}$ and its mass center has a velocity $\mathbf{v}_G$. Hence, the kinetic energy is defined by Eq. 18–1, i.e.,

$$T = \tfrac{1}{2}mv_G^2 + \tfrac{1}{2}I_G\omega^2$$

Here it is seen that the total kinetic energy of the body consists of the *scalar* sum of the *translational* kinetic energy of the mass center, $\tfrac{1}{2}mv_G^2$, and the *rotational* kinetic energy of the body about its mass center, $\tfrac{1}{2}I_G\omega^2$.

Because energy is a scalar quantity, the total kinetic energy for a system of *connected* rigid bodies is the sum of the kinetic energies of all the moving parts. Depending upon the type of plane motion, the kinetic energy of *each body* is found by applying Eq. 18–1 or the alternative forms, Eq. 18–2 or 18–3.†

*The similarity between this derivation and that of $\Sigma M_O = I_O\alpha$ (Eq. 17–15) should be noted.

†A brief review of Secs. 16–5 through 16–7 may prove helpful in solving problems, since computations for kinetic energy require a kinematic analysis of velocity.

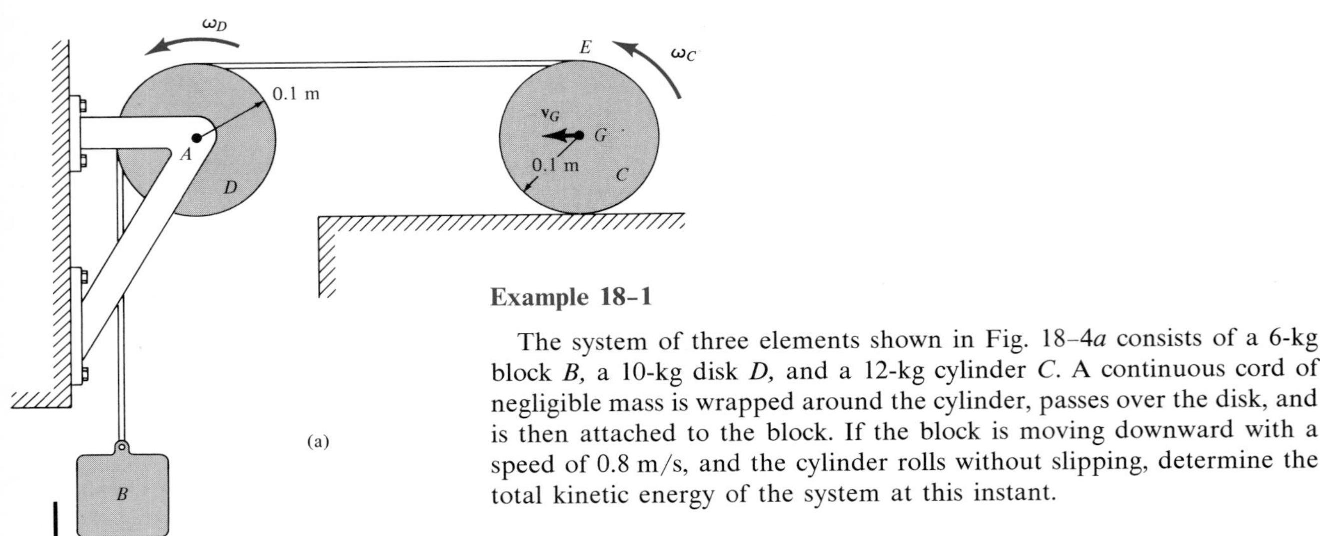

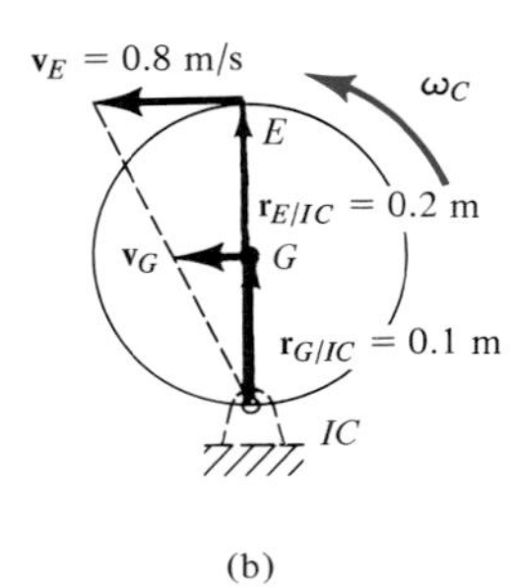

Fig. 18–4

Example 18–1

The system of three elements shown in Fig. 18–4*a* consists of a 6-kg block *B*, a 10-kg disk *D*, and a 12-kg cylinder *C*. A continuous cord of negligible mass is wrapped around the cylinder, passes over the disk, and is then attached to the block. If the block is moving downward with a speed of 0.8 m/s, and the cylinder rolls without slipping, determine the total kinetic energy of the system at this instant.

Solution

By inspection, the block is translating, the disk rotates about a fixed axis, and the cylinder has general plane motion. Hence, in order to compute the kinetic energy of the disk and cylinder, it is first necessary to determine ω_D, ω_C, and v_G, Fig. 18–4*a*. From the *kinematics* of the disk,

$$v_B = r_D\omega_D; \qquad 0.8 \text{ m/s} = (0.1 \text{ m})\omega_D; \qquad \omega_D = 8 \text{ rad/s}$$

Since the cylinder rolls without slipping, the instantaneous center of zero velocity is at the point of contact with the ground, Fig. 18–4*b*, hence,

$$v_E = r_{E/IC}\omega_C; \qquad 0.8 \text{ m/s} = (0.2 \text{ m})\omega_C; \qquad \omega_C = 4 \text{ rad/s}$$

$$v_G = r_{G/IC}\omega_C; \qquad v_G = (0.1 \text{ m})(4 \text{ rad/s}) = 0.4 \text{ m/s}$$

Using Eq. 18–2 to determine the kinetic energy of the block yields

$$T_B = \tfrac{1}{2}m_Bv_B^2; \qquad T_B = \tfrac{1}{2}(6 \text{ kg})(0.8 \text{ m/s})^2$$
$$= 1.92 \text{ J}$$

From Eq. 18–3 and Appendix D, the kinetic energy of the disk is

$$T_D = \tfrac{1}{2}I_D\omega_D^2; \qquad T_D = \tfrac{1}{2}(\tfrac{1}{2}m_Dr_D^2)\omega_D^2$$
$$= \tfrac{1}{2}[\tfrac{1}{2}(10 \text{ kg})(0.1 \text{ m})^2](8 \text{ rad/s})^2$$
$$= 1.60 \text{ J}$$

The kinetic energy of the cylinder can be computed using Eq. 18–1 and Appendix D.

$$T_C = \tfrac{1}{2}mv_G^2 + \tfrac{1}{2}I_G\omega_C^2; \quad T_C = \tfrac{1}{2}mv_G^2 + \tfrac{1}{2}(\tfrac{1}{2}m_Cr_C^2)\omega_C^2$$
$$= \tfrac{1}{2}(12 \text{ kg})(0.4 \text{ m/s})^2$$
$$+ \tfrac{1}{2}[\tfrac{1}{2}(12 \text{ kg})(0.1 \text{ m})^2](4 \text{ rad/s})^2$$
$$= 1.44 \text{ J}$$

The total kinetic energy of the system is therefore

$$T = T_B + T_D + T_C$$
$$= 1.92\text{ J} + 1.60\text{ J} + 1.44\text{ J}$$
$$= 4.96\text{ J} \qquad \textit{Ans.}$$

18–2. The Work of a Force

Work of a Variable Force. It was shown in Sec. 14–1 that if a *particle* is acted upon by a force **F**, the work done by this force in moving the particle along a path s is defined as

$$U = \int_s \mathbf{F} \cdot d\mathbf{s} = \int_s F \cos \theta \, ds \tag{18–4}$$

where θ is the angle between the "tails" of the force vector and the differential path displacement $d\mathbf{s}$. Consider now the work done by a system of three external forces acting on three particles of the rigid body shown in Fig. 18–5, as the body moves from position 1 to position 2. Since work is a scalar quantity, the total work done is simply the algebraic addition of the work of each force, i.e.,

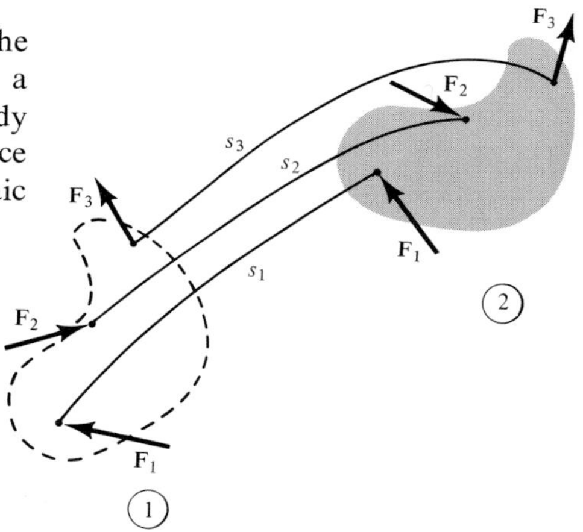

Fig. 18–5

$$U_F = \int_{s_1} \mathbf{F}_1 \cdot d\mathbf{s}_1 + \int_{s_2} \mathbf{F}_2 \cdot d\mathbf{s}_2 + \int_{s_3} \mathbf{F}_3 \cdot d\mathbf{s}_3 \tag{18–5}$$

Similar terms must be included in this equation if additional *external forces* are applied to the body. In general, the integration must account for the variation of the force's direction and magnitude, as the force moves along the path. Note that the work of the *internal forces* is not included in Eq. 18–5. These forces occur in equal but opposite collinear pairs, so as a result, when the body moves along the path, the work of one force cancels that of its counterpart. Furthermore, since the body is rigid, no relative movement between the forces occurs, so that no internal work is done.

The following types of forces are often encountered in planar kinetic problems involving a rigid body. The work of each force has been derived in Sec. 14–1 and is listed below as a summary.

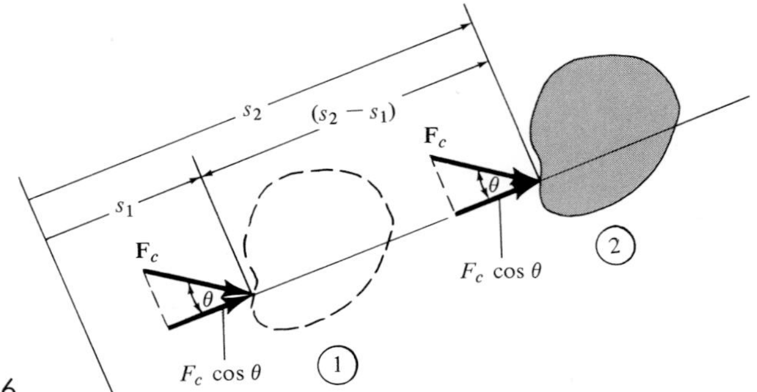

Fig. 18–6

Work of a Constant Force. If an external force $\mathbf{F}_c$ acts on a rigid body, Fig. 18–6, and maintains a constant magnitude F_c and constant direction θ, while the body undergoes a translation from s_1 to s_2, Eq. 18–4 can be integrated so that the work becomes

$$U_{F_c} = F_c \cos\theta\,(s_2 - s_1) \tag{18–6}$$

Here $F_c \cos\theta$ represents the magnitude of the component of force in the direction of displacement $(s_2 - s_1)$.

Work of a Weight. The weight of a body does work only when the body's center of mass G undergoes a *vertical displacement*. If this displacement is *downward* from y_1 to y_2, Fig. 18–7, the work is *positive,* since the weight and displacement are in the *same* direction.

$$U_W = W(y_2 - y_1) \tag{18–7}$$

Here the elevation change is considered to be small so that **W**, which is caused by gravitation, is constant.

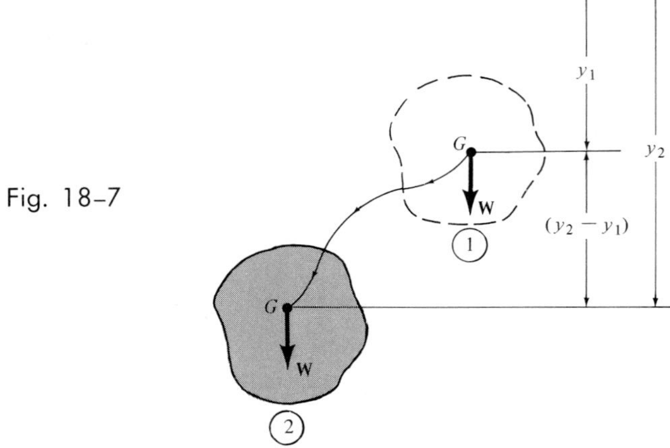

Fig. 18–7

Work of a Spring Force. If a linear elastic spring is attached to a body, the spring force, $F_s = kx$, *acting on the body* does work when the spring either stretches or compresses from x_1 to a *further* position x_2. In both cases the work will be *negative* since the *displacement of the body* is always in the opposite direction to the force, Fig. 18–8. The work done is

$$U_s = -(\tfrac{1}{2}kx_2^2 - \tfrac{1}{2}kx_1^2) \tag{18-8}$$

where $|x_2| > |x_1|$.

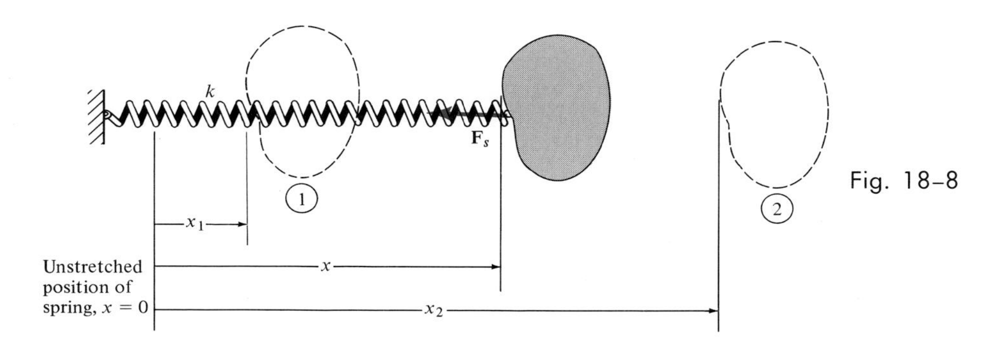

Fig. 18–8

Forces That Do No Work. There are some external forces that do no work when the body is displaced. These forces can act either at *fixed points* on the body or they can have a direction *perpendicular to their path of displacement*. Examples include the reactions at a pin support about which a body rotates, the normal reaction **N** acting on a body that moves along a surface, and the weight **W** of a body when the center of gravity of the body moves in a *horizontal plane,* Fig. 18–9. A frictional force $\mathbf{F}_r$ acting on a body as it *rolls without slipping* over a rough surface also does no work, Fig. 18–9. This is because, during any *instant of time dt,* $\mathbf{F}_r$ acts at a point having *zero velocity* (instantaneous center, *IC*.) Hence, for any differential rolling movement $d\mathbf{s}_{IC}$ of the body's *IC*, the work is $dU = \mathbf{F}_r \cdot d\mathbf{s}_{IC} = F_r(v_{IC}dt) = 0$, since $v_{IC} = 0$.

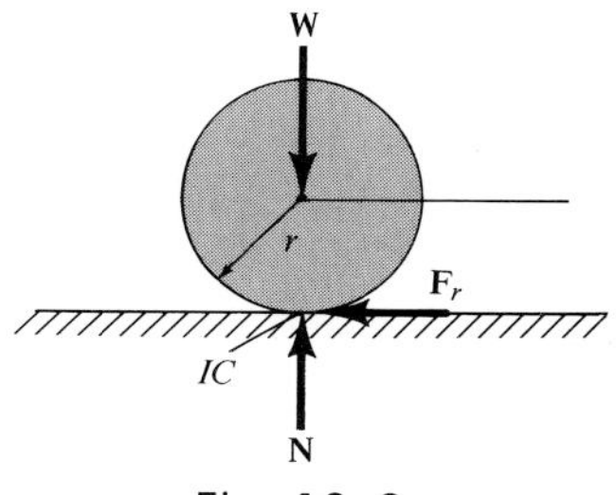

Fig. 18–9

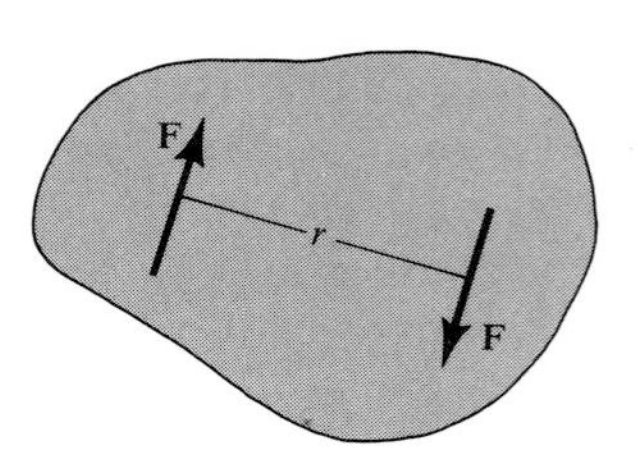

Fig. 18–10(a)

18–3. The Work of a Couple

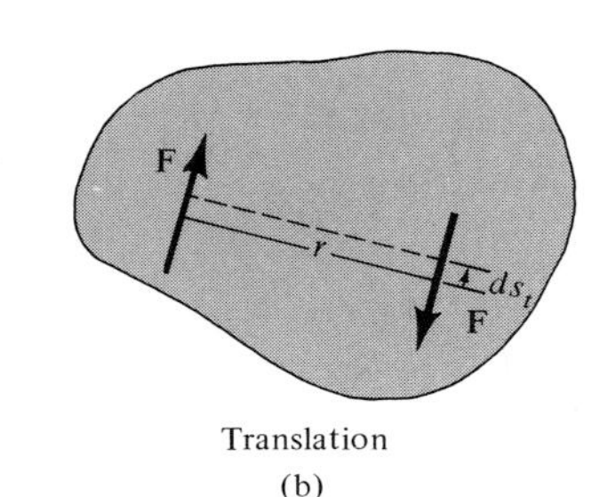

Translation
(b)

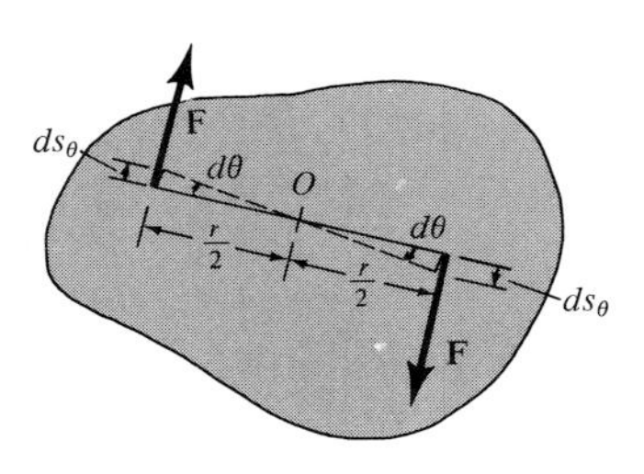

Rotation
(c)

Fig. 18–10(b,c)

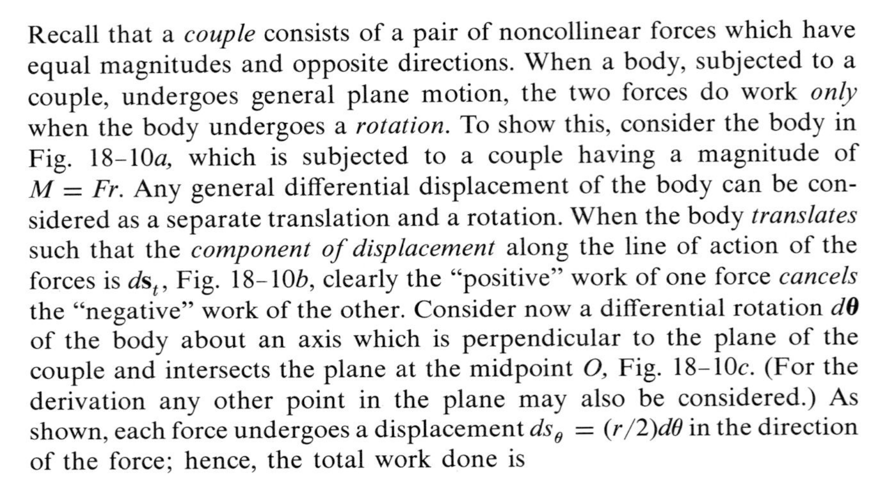

Recall that a *couple* consists of a pair of noncollinear forces which have equal magnitudes and opposite directions. When a body, subjected to a couple, undergoes general plane motion, the two forces do work *only* when the body undergoes a *rotation.* To show this, consider the body in Fig. 18–10*a,* which is subjected to a couple having a magnitude of $M = Fr$. Any general differential displacement of the body can be considered as a separate translation and a rotation. When the body *translates* such that the *component of displacement* along the line of action of the forces is $d\mathbf{s}_t$, Fig. 18–10*b*, clearly the "positive" work of one force *cancels* the "negative" work of the other. Consider now a differential rotation $d\boldsymbol{\theta}$ of the body about an axis which is perpendicular to the plane of the couple and intersects the plane at the midpoint *O,* Fig. 18–10*c*. (For the derivation any other point in the plane may also be considered.) As shown, each force undergoes a displacement $ds_\theta = (r/2)d\theta$ in the direction of the force; hence, the total work done is

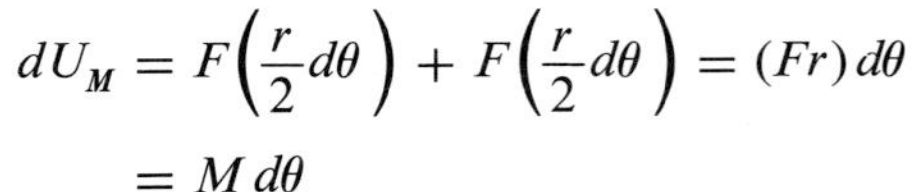

$$dU_M = F\left(\frac{r}{2}d\theta\right) + F\left(\frac{r}{2}d\theta\right) = (Fr)\,d\theta$$
$$= M\,d\theta$$

Here the line of action of $d\boldsymbol{\theta}$ is parallel to the line of action of **M.** This is *always the case for general plane motion,* since **M** and $d\boldsymbol{\theta}$ are perpendicular to the plane of motion. Furthermore, the resultant work is *positive* when **M** and $d\boldsymbol{\theta}$ are in the *same direction,* and *negative* if these vectors are in *opposite directions.*

When the body rotates in the plane through a finite angle θ, from θ_1 to θ_2, the work of a couple is

$$U_M = \int_{\theta_1}^{\theta_2} M\,d\theta \tag{18–9}$$

If **M** has a *constant magnitude,* then

$$U_M = M(\theta_2 - \theta_1) \tag{18–10}$$

where in all cases the angles θ_1 and θ_2 are measured in radians.

18–4. Principle of Work and Energy

In Sec. 14–2 the principle of work and energy was developed for a particle. By applying this principle to each of the particles of a rigid body and adding the results algebraically, since energy is a scalar, the principle of work and energy for a rigid body may be written as

$$T_1 + \Sigma U_{1-2} = T_2 \tag{18–11}$$

This equation states that the body's initial translational *and* rotational kinetic energy T_1, plus the work done by all the external forces and couples acting on the body, ΣU_{1-2}, as the body moves from its initial to its final position, is equal to the body's final translational *and* rotational kinetic energy T_2.

When several rigid bodies are pin-connected, connected by inextensible cables, or in mesh with one another, this equation may be applied to the entire system of connected bodies. In all these cases the internal forces, which hold the various members together, do no work, and hence are eliminated from the analysis.

PROCEDURE FOR ANALYSIS

The principle of work and energy is used to solve kinetic problems that involve *velocity, force,* and *displacement,* since these terms are involved in the formulation. For application the following three-step procedure should be used.

Step 1: Determine the kinetic-energy terms T_1 and T_2 by applying the equation $T = \frac{1}{2}mv_G^2 + \frac{1}{2}I_G\omega^2$ or an appropriate form of this equation developed in Sec. 18–1. In this regard, *kinematic diagrams* for velocity may be useful for determining v_G and ω, or for establishing a *relationship* between v_G and ω.

Step 2: Draw a free-body diagram of the body when it is located at an intermediate point along the path, in order to account for all the forces and couples which do work on the body. The work of each force and couple can be computed using the appropriate relations outlined in Secs. 18–2 and 18–3. Since *algebraic addition* of the work terms is required, it is important that the proper sign of each term be specified. Specifically, work is *positive* when the force (couple) is in the *same direction* as its displacement (rotation); otherwise, it is negative.

Step 3: Apply the principle of work and energy, $T_1 + \Sigma U_{1-2} = T_2$. Since this equation involves scalars, it can be used to solve for only one unknown when it is applied to a single rigid body. This is in contrast to the three scalar equations of motion (Eqs. 17–10) which may be written for the same body.

The following example problems numerically illustrate application of the above three-step procedure.

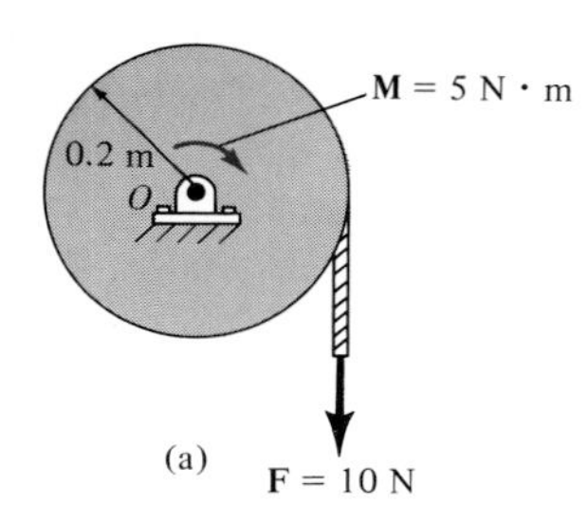

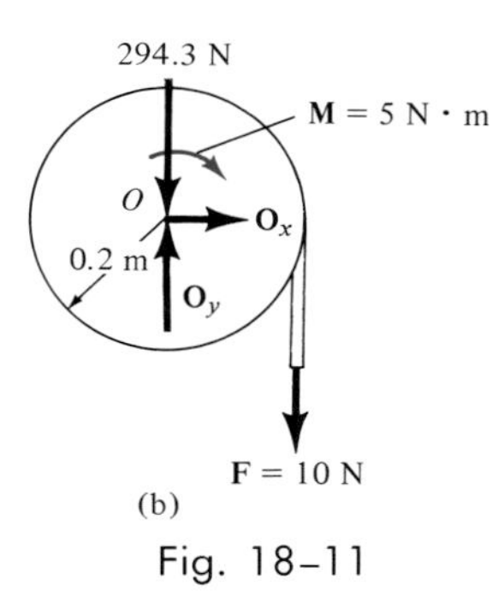

Fig. 18–11

Example 18–2

The 30-kg disk, shown in Fig. 18–11*a*, is pin-supported at its center. Determine the number of revolutions it must make to attain an angular velocity of 20 rad/s starting from rest. It is acted upon by a constant force $F = 10$ N, which is applied to a cord wrapped around its periphery, and a constant couple $M = 5$ N · m. Neglect the mass of the cord in the calculation.

Solution

Step 1: Since the disk rotates about a fixed axis, the kinetic energy can be computed using $T = \frac{1}{2}I_O\omega^2$, where the moment of inertia is $I_O = \frac{1}{2}mr^2$ (Appendix D). Initially, the disk is at rest, so that

$$T_1 = 0$$

$$T_2 = \tfrac{1}{2}I_O\omega_2^2 = \tfrac{1}{2}[\tfrac{1}{2}(30)(0.2)^2](20)^2 = 120 \text{ J}$$

Step 2: As shown on the free-body diagram, Fig. 18–11*b*, the pin reactions $\mathbf{O}_x$ and $\mathbf{O}_y$ and the weight (294.3 N) do no work, since they are not displaced. The *couple,* having a constant magnitude, does positive work $U_M = M\theta$ as the disk *rotates* through a clockwise angle of θ rad; and the *constant force* $\mathbf{F}$ does positive work $U_{F_c} = Fs$ as the cord *moves* downward $s = \theta r = \theta(0.2)$ m.

Step 3: Applying the equation of work and energy, we have

$$\{T_1\} + \{\Sigma U_{1-2}\} = \{T_2\}$$

$$\{T_1\} + \{M\theta + Fs\} = \{T_2\}$$

$$\{0\} + \{5\theta + (10)\theta(0.2)\} = \{120\}$$

$$\theta = 17.1 \text{ rad} = 17.1 \text{ rad}\left(\frac{1 \text{ rev}}{2\pi \text{ rad}}\right) = 2.73 \text{ rev} \qquad \textit{Ans.}$$

This problem has also been solved in Example 17–8. Compare the two methods of solution and note that since force, velocity, and displacement θ are involved, a work-energy approach yields the most direct solution.

Example 18–3

The 5-kg bar shown in Fig. 18–12*a* has a center of mass at *G*. If it is given an initial clockwise angular velocity of $\omega_1 = 10$ rad/s when $\theta = 90°$, compute the spring constant *k* so that it stops when $\theta = 0°$. What are the horizontal and vertical components of reaction at the pin *A* when $\theta = 0°$? The spring deforms 0.1 m when $\theta = 0°$.

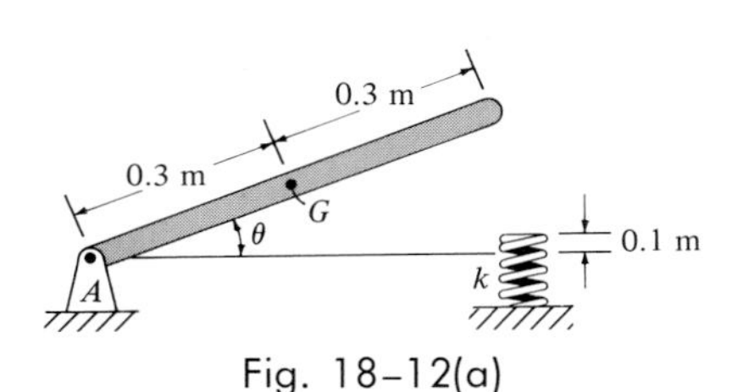

Fig. 18–12(a)

Solution

Step 1: Two kinematic diagrams for the bar when $\theta = 90°$ (position 1) and $\theta = 0°$ (position 2) are shown in Fig. 18–12*b*. The initial kinetic energy of the bar may be computed with reference to the point of rotation A or the center of mass G. If A is considered, it is necessary to apply $T_1 = \frac{1}{2}I_A\omega_1^2$. Using Appendix D to compute the moment of inertia of the bar about A, we get

$$I_A = \tfrac{1}{3}ml^2 = \tfrac{1}{3}(5)(0.6)^2 = 0.60 \text{ kg} \cdot \text{m}^2$$

Thus,

$$T_1 = \tfrac{1}{2}I_A\omega_1^2 = \tfrac{1}{2}(0.60)(10)^2 = 30 \text{ J}$$

The same result may be obtained by using $T = \frac{1}{2}mv_G^2 + \frac{1}{2}I_G\omega^2$, which applies to point G. To show this, note that

$$I_G = \tfrac{1}{12}ml^2 = \tfrac{1}{12}(5)(0.6)^2 = 0.15 \text{ kg} \cdot \text{m}^2$$

Then since $(v_G)_1 = \omega_1 r_{G/A} = 10(0.3) = 3$ m/s,

$$T_1 = \tfrac{1}{2}m(v_G)_1^2 + \tfrac{1}{2}I_G\omega_1^2 = \tfrac{1}{2}(5)(3)^2 + \tfrac{1}{2}(0.15)(10)^2 = 30 \text{ J}$$

Since in the final position $(v_G)_2 = \omega_2 = 0$, the final kinetic energy is

$$T_2 = 0$$

Step 2: The free-body diagram of the bar is shown in Fig. 18–12*c*. The reactions $\mathbf{A}_x$ and $\mathbf{A}_y$ do no work, since these forces do not move. The 49.05-N weight, centered at G, moves downward through a vertical height of $y = 0.3$ m. Since this displacement is in the *same* direction as the force, the work is *positive*. The spring force $\mathbf{F}_s$ does *negative work on the bar.* Why? This force acts while the spring is being compressed from zero to $x = 0.1$ m. Hence, the work is determined from $U_s = -\frac{1}{2}kx^2$.

Step 3: Applying the principle of work and energy, we have

$$\{T_1\} + \{\Sigma U_{1-2}\} = \{T_2\}$$
$$\{T_1\} + \{Wy - \tfrac{1}{2}kx^2\} = \{T_2\}$$
$$\{30\} + \{49.05(0.3) - \tfrac{1}{2}k(0.1)^2\} = \{0\}$$

Solving for k yields

$$k = 8943 \text{ N/m}$$
$$= 8.94 \text{ kN/m} \qquad \textit{Ans.}$$

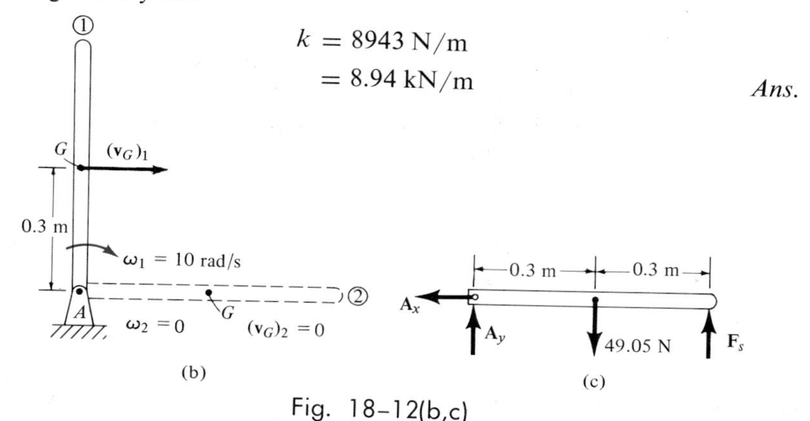

Fig. 18–12(b,c)

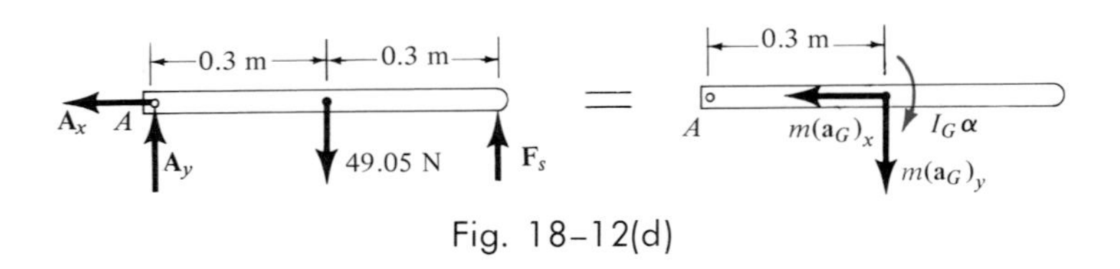

Fig. 18–12(d)

The pin reactions $\mathbf{A}_x$ and $\mathbf{A}_y$ must be determined by applying the equations of motion. (These forces cannot be obtained from the principle of work and energy since they do *no work*.) The free-body and kinetic diagrams are shown in Fig. 18–12*d*. When $\theta = 0°$, the magnitude of the spring force is $F_s = kx = (8943\ \text{N/m})(0.1\ \text{m}) = 894.3\ \text{N}$. Furthermore, $m(a_G)_x = m\omega^2 r_{G/A} = 0$, since $\boldsymbol{\omega} = \mathbf{0}$ at this instant; that is, the bar is motionless. Applying the equations of motion, we have

$$\overset{+}{\rightarrow}\Sigma F_x = m(a_G)_x; \qquad -A_x = 0$$

$$+\uparrow\Sigma F_y = m(a_G)_y; \quad A_y - 49.05 + 894.3 = -5(0.3)\alpha$$

$$\curvearrowleft +\Sigma M_A = I_A\alpha; \qquad 49.05(0.3) - 894.3(0.6) = 0.60\alpha$$

Solving yields

$$\alpha = -869.8\ \text{rad/s}^2$$

$$A_x = 0 \qquad \textit{Ans.}$$

$$A_y = 459.4\ \text{N} \qquad \textit{Ans.}$$

The negative sign for α indicates that the bar will begin to rotate counter-clockwise, just after coming to rest on the spring.

Example 18–4

The 10-kg rod shown in Fig. 18–13*a* is constrained so that its ends move along the grooved slots. The rod is initially at rest when $\theta = 0°$. If the slider block at B is acted upon by a horizontal force of $P = 50$ N, determine the angular velocity of the rod at the instant $\theta = 45°$. Neglect the mass of blocks A and B. (Why can the principle of work and energy be used to solve this problem?)

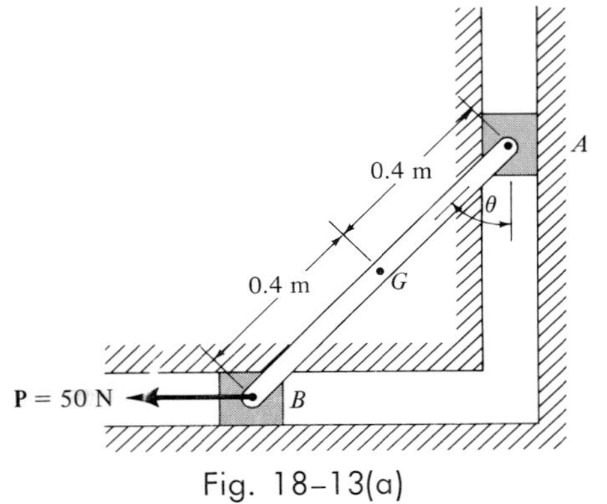

Fig. 18–13(a)

Solution

Step 1: Two kinematic diagrams of the rod, when it is in the initial position 1 and final position 2, are shown in Fig. 18–13*b*. When the rod is in position 1,

$$T_1 = 0$$

since $(\mathbf{v}_G)_1 = \boldsymbol{\omega}_1 = \mathbf{0}$. In position 2 the angular velocity is $\boldsymbol{\omega}_2$ and the velocity of the mass center is $(\mathbf{v}_G)_2$. Hence, the kinetic energy is

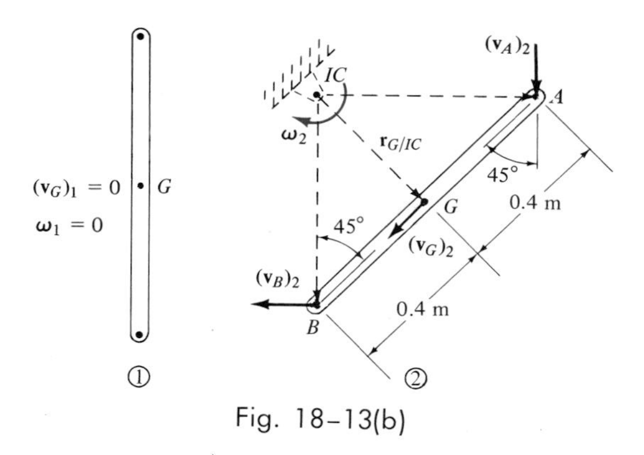

Fig. 18–13(b)

$$
\begin{aligned}
T_2 &= \tfrac{1}{2}m(v_G)_2^2 + \tfrac{1}{2}I_G(\omega_2)^2 \\
&= \tfrac{1}{2}(10)(v_G)_2^2 + \tfrac{1}{2}[\tfrac{1}{12}(10)(0.8)^2]\,(\omega_2)^2 \\
&= 5(v_G)_2^2 + 0.267(\omega_2)^2 \qquad (1)
\end{aligned}
$$

The two unknowns $(v_G)_2$ and ω_2 may be related via the instantaneous center of zero velocity of the rod, Fig. 18–13*b*. It is seen that as *A* moves downward with a velocity $(\mathbf{v}_A)_2$, *B* moves horizontally to the left with a velocity $(\mathbf{v}_B)_2$. Knowing these directions, the *IC* may be determined as shown in the figure. Hence,

$$
\begin{aligned}
(v_G)_2 &= r_{G/IC}\omega_2 \\
&= (0.4 \tan 45^\circ)\omega_2 \\
&= 0.4\omega_2
\end{aligned}
$$

Substituting into Eq. (1), we have

$$
\begin{aligned}
T_2 &= 5(0.4\omega_2)^2 + 0.267(\omega_2)^2 \\
&= 1.067(\omega_2)^2
\end{aligned}
$$

Step 2: The normal forces $\mathbf{N}_A$ and $\mathbf{N}_B$ shown on the free-body diagram, Fig. 18–13*c*, do no work as the rod is displaced. Why? The 98.1-N weight is displaced a vertical distance of $y = (0.4 - 0.4\cos 45^\circ)$ m; whereas the 50-N force moves a horizontal distance of $s = (0.8 \sin 45^\circ)$ m. The work done by both of these forces is *positive,* since the forces act in the same direction as their corresponding displacement.

Step 3: Applying the principle of work and energy gives

$$
\begin{aligned}
\{T_1\} + \{\Sigma U_{1-2}\} &= \{T_2\} \\
\{T_1\} + \{Wy + Ps\} &= \{T_2\} \\
\{0\} + \{98.1(0.4 - 0.4\cos 45^\circ) + 50(0.8 \sin 45^\circ)\} &= \{1.067(\omega_2)^2\}
\end{aligned}
$$

Solving for ω_2 gives

$$\omega_2 = 6.11 \text{ rad/s} \qquad \textit{Ans.}$$

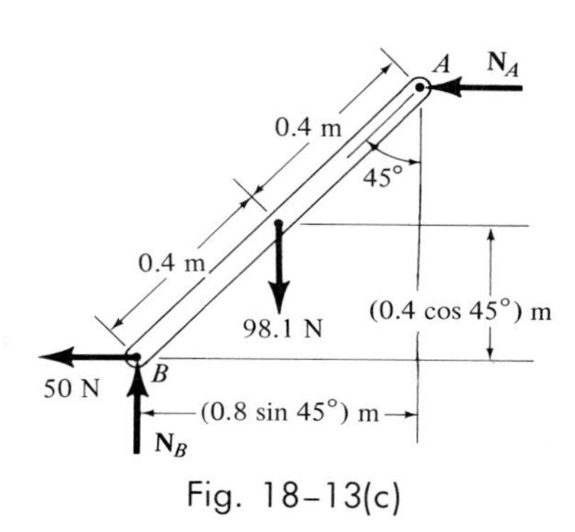

Fig. 18–13(c)

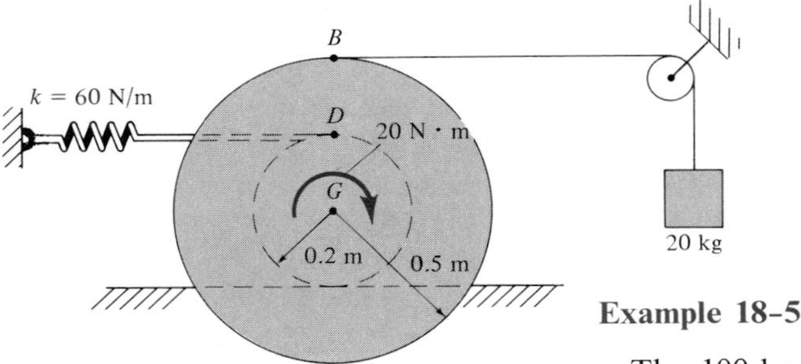

Fig. 18–14(a)

Example 18–5

The 100-kg wheel shown in Fig. 18–14*a* has a radius of gyration of $k_G = 0.25$ m about its center of mass G. If it is subjected to a clockwise couple of 20 N · m as it rolls on its inner hub without slipping, determine the wheel's angular velocity after the 20-kg block is released from rest and has fallen 0.4 m. The spring has a stiffness of $k = 60$ N/m and is initially unstretched when the block is released.

Solution

It is easier to analyze the system of both the wheel and block together, rather than treating each one separately. (A separate treatment involves two work equations and would introduce the work done by the unknown cable tension into the equations.)

Step 1: The kinematic diagrams of the system when it is in positions 1 and 2 are shown in Fig. 18–14*b*. Since the wheel and block are initially at rest,

$$T_1 = 0$$

When the wheel is in position 2, the unknown velocities of the wheel and block are ω_2, $(v_G)_2$, and $(v_B)_2$. The total kinetic energy of the system is, therefore,

$$\begin{aligned} T_2 &= \tfrac{1}{2}m_{wh}(v_G)_2^2 + \tfrac{1}{2}I_G(\omega_2)^2 + \tfrac{1}{2}m_{bl}(v_B)_2^2 \\ &= \tfrac{1}{2}100(v_G)_2^2 + \tfrac{1}{2}[100(0.25)^2](\omega_2)^2 + \tfrac{1}{2}20(v_B)_2^2 \\ &= 50(v_G)_2^2 + 3.125(\omega_2)^2 + 10(v_B)_2^2 \end{aligned} \qquad (1)$$

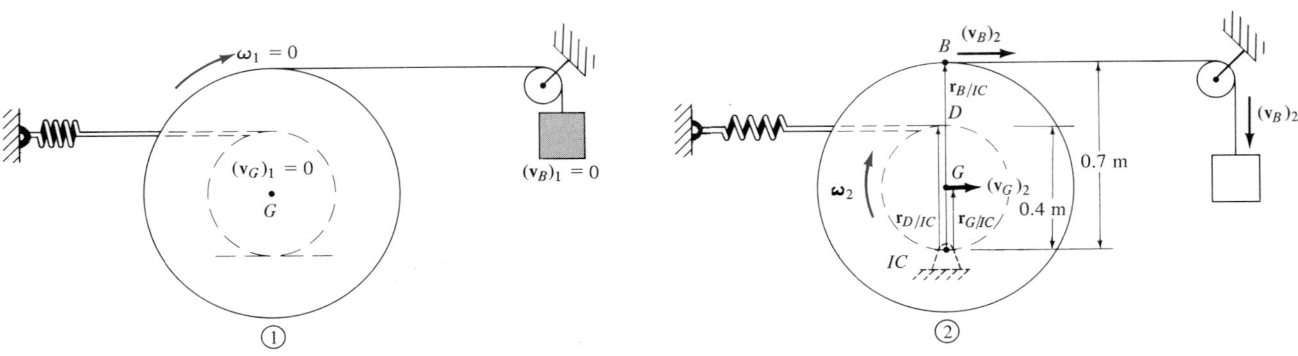

Fig. 18–14(b)

Using *kinematics*, the three unknowns ω_2, $(v_G)_2$, and $(v_B)_2$ may be related using the instantaneous center of zero velocity. Since the wheel does not slip as it rolls, the *IC* is located at the point of contact with the ground, Fig. 18–14*b*. Point *B* on the wheel moves with the same speed as the block, since the attached cable is inextensible. Therefore,

$$(v_B)_2 = r_{B/IC}\omega_2 = 0.7\omega_2$$
$$(v_G)_2 = r_{G/IC}\omega_2 = 0.2\omega_2$$

Substituting these relations into Eq. (1) we get

$$T_2 = 50(0.2\omega_2)^2 + 3.125(\omega_2)^2 + 10(0.7\omega_2)^2 = 10.025(\omega_2)^2$$

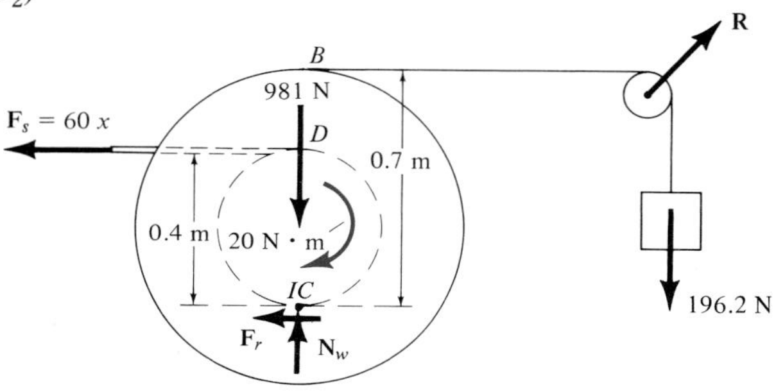

Fig. 18–14(c)

Step 2: Inspection of the free-body diagram, Fig. 18–14*c*, reveals that only the spring force $\mathbf{F}_s$, the 196.2-N weight of the block, and the 20-N · m couple do work while the system moves from position 1 to position 2. The reactions $\mathbf{N}_w$ and $\mathbf{R}$ do no work, since these forces do not move along their lines of action. Furthermore, the frictional force $\mathbf{F}_r$ does *no work*, since the *wheel does not slip as it rolls.*

The work of $\mathbf{F}_s$ may be computed by using $U_s = -\frac{1}{2}kx^2$. Since the wheel does not slip, as the block moves downward 0.4 m, the wheel rotates $\theta = s/r_{B/IC} = 0.4\text{ m}/0.7\text{ m} = 0.571$ rad, Fig. 18–14*b*. Hence, the spring stretches $x = \theta(r_{D/IC}) = 0.571\text{ rad}(0.4\text{ m}) = 0.229$ m. The couple does work because of the rotation, $\theta = 0.571$ rad. Since this rotation is in the same direction as the couple, the work done is positive and can be computed using $U_M = M\theta$.

Step 3: Applying the principle of work and energy to the system, we have

$$\{T_1\} + \{\Sigma U_{1-2}\} = \{T_2\}$$
$$\{0\} + \{M\theta + Ws - \tfrac{1}{2}kx^2\} = \{T_2\}$$
$$\{0\} + \{(20\text{ N}\cdot\text{m})(0.571\text{ rad}) + (196.2\text{ N})(0.4\text{ m}) - \tfrac{1}{2}(60\text{ N/m})(0.229\text{ m})^2\} = \{10.025(\omega_2)^2\}$$

or

$$88.33 = 10.025(\omega_2)^2$$

Thus,

$$\omega_2 = 2.97\text{ rad/s} \qquad \textit{Ans.}$$

Problems

Except when stated otherwise, throughout this chapter, assume that the coefficients of static and kinetic friction are equal, i.e., $\mu = \mu_s = \mu_k$.

18–1. The pendulum of the Charpy impact machine has a mass of 50 kg and a radius of gyration of $k_A = 0.75$ m. If it is released from rest when $\theta = 0°$, determine the angular velocity just before it strikes the specimen S, $\theta = 90°$. Compare the solution to that given in Example 17–10.

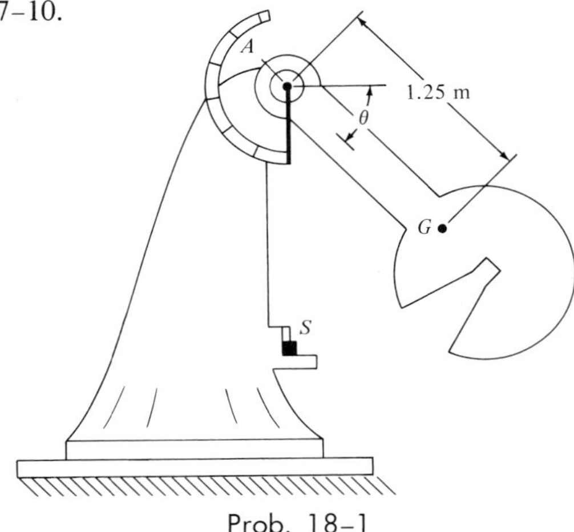

Prob. 18–1

18–2. The 12-kg plate is welded to a shaft which is supported horizontally by two smooth bearings at A and B. If the plate is released from rest from the position shown and falls downward, determine the angular velocity of the shaft at the instant it has rotated 180°.

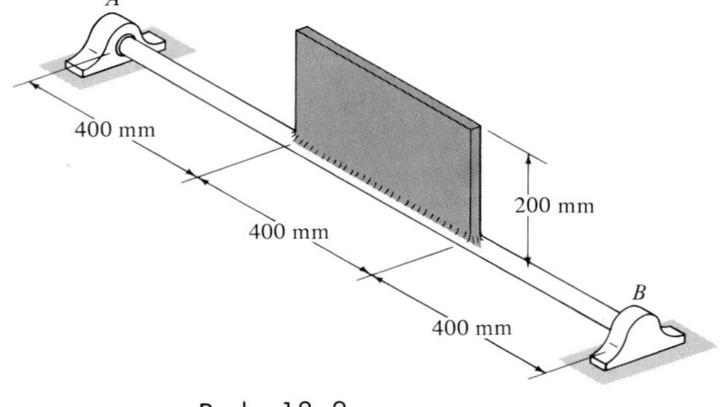

Prob. 18–2

18–3. At a given instant the body of mass m has an angular velocity $\boldsymbol{\omega}$ and its mass center has a velocity $\mathbf{v}_G$. Show that its kinetic energy can be represented as $T = \frac{1}{2} I_{IC} \omega^2$, where I_{IC} is the moment of inertia of the body computed about the instantaneous axis of zero velocity, located at a distance $r_{G/IC}$ from the mass center as shown.

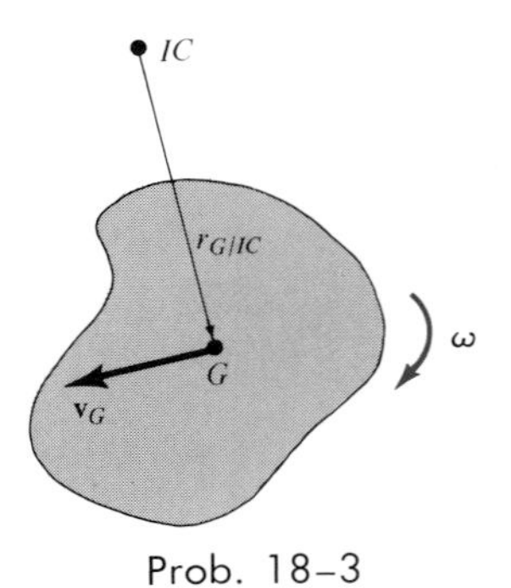

Prob. 18–3

* **18–4.** The 200-kg uniform door is closed by applying a constant handle force of 80 N, which is always directed perpendicular to the plane of the door. If the door is originally at rest and open at $\theta = 90°$, determine the door's angular velocity just before it closes ($\theta = 0°$). Neglect friction at the hinges. For the calculation, assume the door to be approximated by a thin plate.

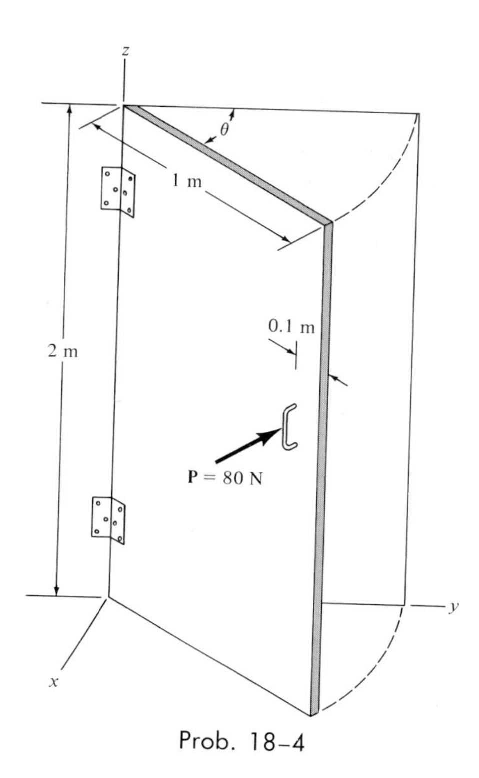

Prob. 18–4

18–5. A yo-yo has a mass of $m = 130$ g and a radius of gyration of $k_O = 20$ mm. If it is released from rest, determine how far it must descend in order to attain an angular velocity of $\omega = 60$ rad/s. Neglect the mass of the string and assume that the string is wound around the central peg such that the mean radius at which it unravels is $r = 8$ mm.

Prob. 18–5

18–5a. Solve Prob. 18–5 if the yo-yo weighs $W = 0.3$ lb, and $k_O = 0.06$ ft, $\omega = 50$ rad/s, $r = 0.02$ ft.

18–6. A spool of cable, originally at rest, has a mass of 150 kg and a radius of gyration of $k_G = 325$ mm. If the spool rests on two small rollers A and B, and a constant horizontal force $P = 300$ N is applied to the end of the cable, compute the angular velocity of the spool when 6 m of cable has been unraveled. Neglect friction and the mass of the rollers and unraveled cable.

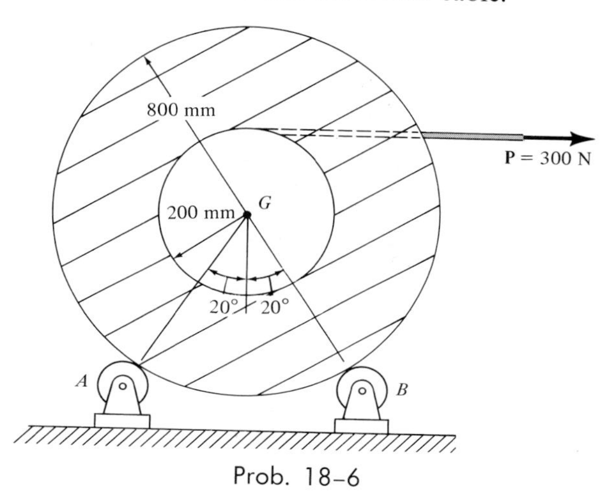

Prob. 18–6

18–7. The 60-kg wheel has a radius of gyration of $k_O = 125$ mm and is rotating at 20 rad/s when the vertical force of $P = 150$ N is applied to the brake handle. If the coefficient of friction between the brake at B and the wheel is $\mu = 0.35$, determine the number of revolutions the wheel makes before it stops. What are the horizontal and vertical components of reaction at A while the wheel is stopping?

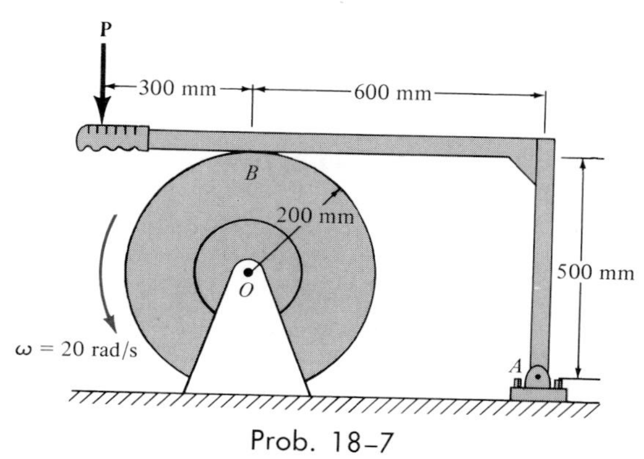

Prob. 18–7

*18–8. A 300-kg tree falls from the vertical position such that it pivots about its cut section at A. If the tree can be considered as a uniform rod, pin-supported at A, determine the speed of its top branch B just before it strikes the ground. What are the horizontal and vertical components of force at A at this instant?

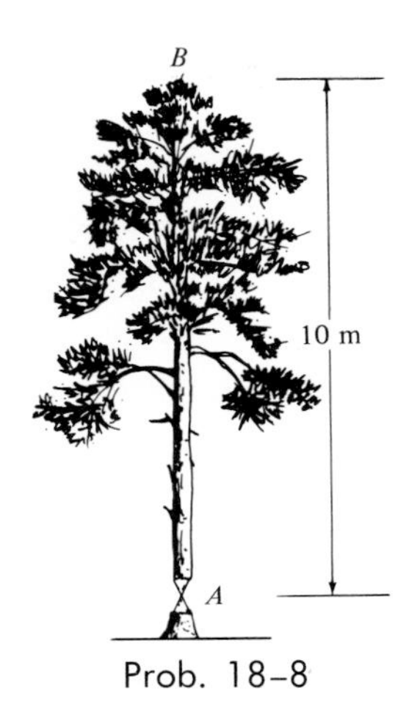

Prob. 18–8

18–10. The beam having a mass of $m = 225$ kg and length of $l = 3$ m is supported by two cables. If the cable at end B is cut so that the beam is released from rest when $\theta = 30°$, determine the speed at which the end A strikes the wall. Neglect friction at B, and set $d = 2$ m and $h = 3.4$ m. For the calculation of the moment of inertia, consider the beam to be a thin rod.

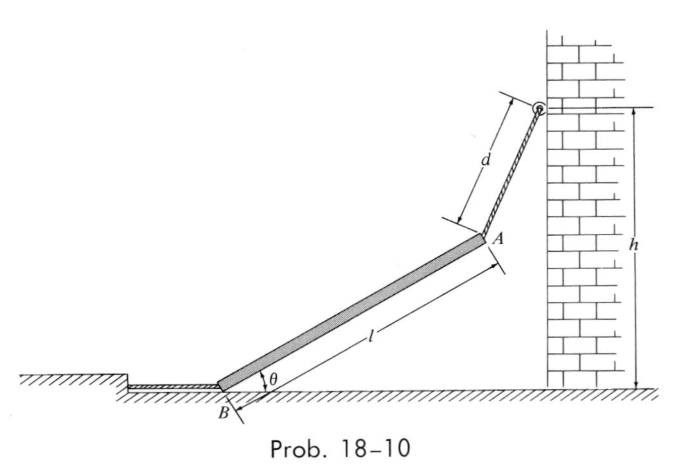

Prob. 18–10

18–9. The 12-kg rod AB is pin-connected at A and subjected to a couple of $M = 30$ N · m. If the rod is released from rest when the spring is unstretched, at $\theta = 30°$, determine the rod's angular velocity at the instant $\theta = 60°$. As the rod rotates, the spring always remains horizontal, because of the roller support at C.

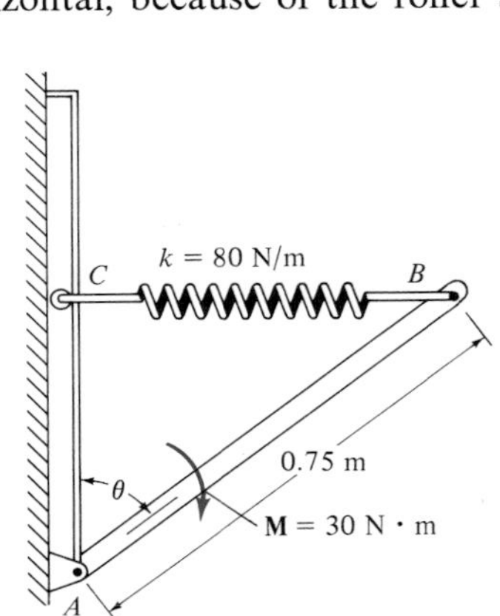

Prob. 18–9

18–10a. Solve Prob. 18–10 if the beam weighs $W = 150$ lb, and $\theta = 30°$, $l = 10$ ft, $d = 4$ ft, $h = 7.5$ ft.

18–11. The motor supplies a constant torque or twist of $M = 60$ N · m to an attached disk D of negligible mass. If the drum at A has a mass of 20 kg and a radius of gyration of $k_O = 125$ mm, determine the speed imparted to the 25-kg crate C after it rises 3 m starting from rest. Neglect the mass of the cables.

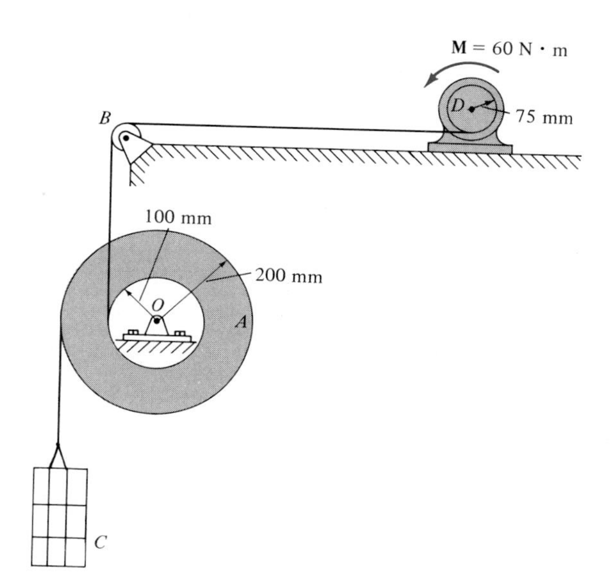

Prob. 18–11

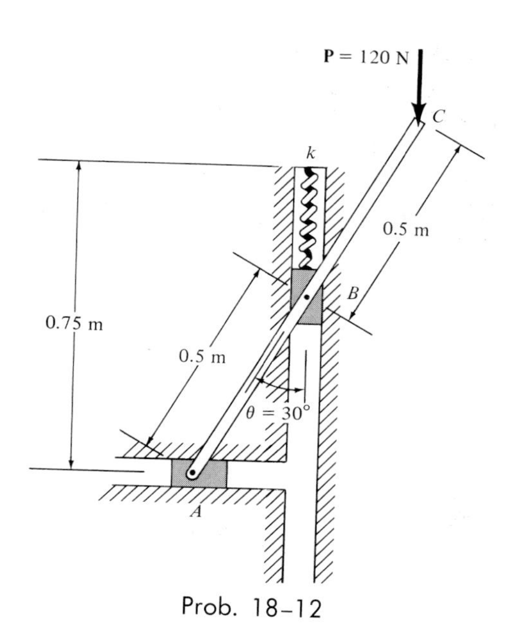

Prob. 18–12

18–13. A vertical force of $P = 1.2$ kN is used to lift the 20-kg wheel and the 40-kg crate B. Assuming that the rope does not slip over the wheel, determine the speed of B after the center O of the wheel rises 1.5 m starting from rest. The radius of gyration of the wheel is $k_O = 125$ mm.

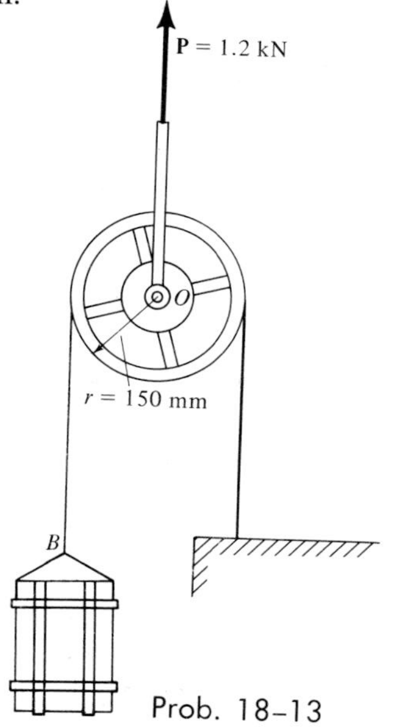

Prob. 18–13

***18–12.** The lever ABC has a mass of 4 kg and can be approximated by a slender rod. If a vertical force $P = 120$ N is applied to the lever from an at-rest position of $\theta = 30°$, determine the angular velocity of the lever at the instant $\theta = 90°$. The spring has a stiffness of $k = 500$ N/m and is unstretched when $\theta = 0°$. Neglect friction and the mass of the slider blocks at A and B.

18-14. The links AB and BC have a mass of 6 kg each. An elastic cable is attached to A and C and has an unstretched length of 0.2 m. If the links are extended out into the horizontal position, then released from rest ($\theta \approx 0°$) so that they move out of equilibrium, compute the velocity of C at the instant $\theta = 45°$. Neglect the mass of the roller.

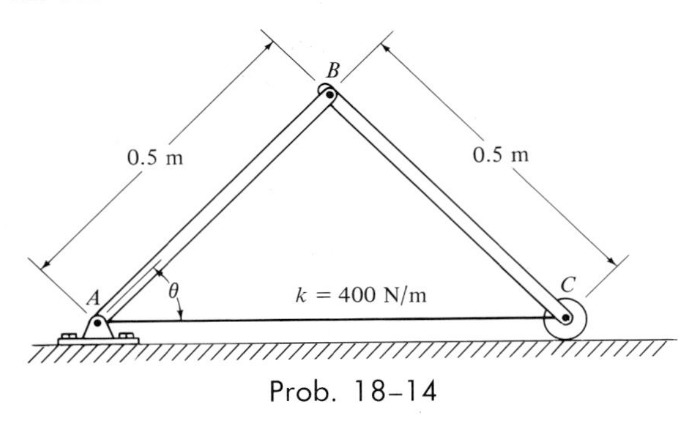

Prob. 18-14

18-15. A uniform ladder having a mass of $m = 20$ kg is released from rest when it is in the vertical position. If it is allowed to fall freely, determine the angle θ at which the bottom end A starts to lift off the ground. For the calculation, assume the ladder to be a slender rod of length $l = 3$ m and neglect friction at A.

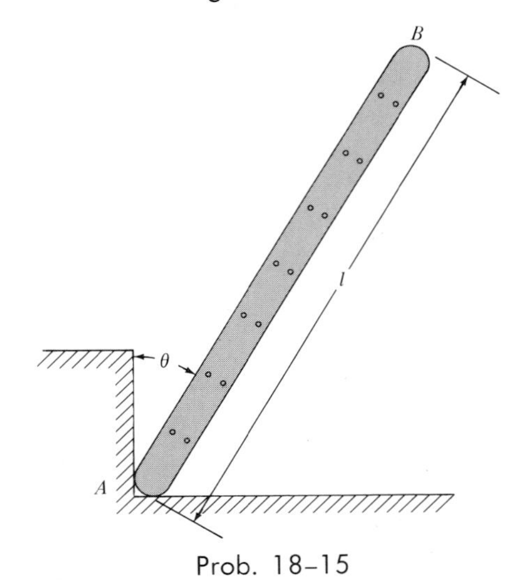

Prob. 18-15

18-15a. Solve Prob. 18-15 if the ladder weighs $W = 50$ lb and $l = 10$ ft.

* **18-16.** The motor supplies a constant torque or twist of $M = 50$ N · m to the gear at C. If gear B and the attached drum D have a total mass of 20 kg and a radius of gyration of $k_O = 120$ mm, determine the speed of the 15-kg crate A at the instant it rises 1.5 m, starting from rest. Neglect the mass of the cord and the pulley P.

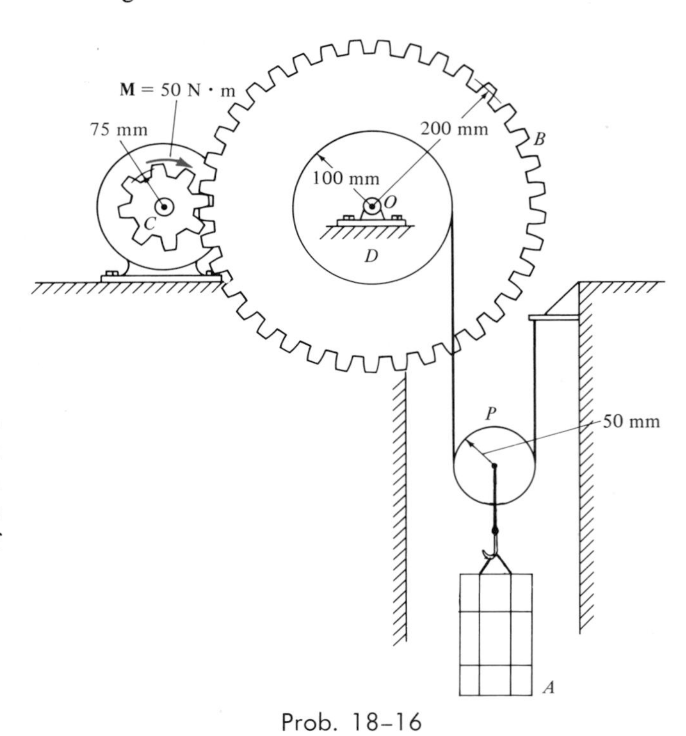

Prob. 18-16

18–5. Conservation of Energy

When a force system, acting on a rigid body, consists only of *conservative forces,* the conservation-of-energy theorem may be used to solve a problem which otherwise would be solved using the principle of work and energy. This theorem is often easier to apply since the work of a conservative force is *independent of the path* and only depends upon the initial and final positions of the body. It was shown in Sec. 14–5 that the work of a conservative force may be expressed as the difference in the body's potential energy measured from an arbitrarily selected horizontal reference or datum.

Gravitational Potential Energy. The weight of a body is a conservative force, and therefore, if a particle having a weight W_i is located y_i *above* an arbitrary datum plane, the gravitational potential energy for the particle is $V_g = W_i y_i$, Eq. 14–14. If it is assumed that the particle represents the ith particle of a rigid body, the gravitational potential energy for the body composed of n particles is determined by the scalar sum

$$V_g = \Sigma W_i y_i$$

The summation becomes an integral as $W_i \rightarrow dW$ and consequently $n \rightarrow \infty$, in which case we have

$$V_g = \int_W y\, dW = W y_G$$

or

$$V_g = W y_G \tag{18–12}$$

Thus, the *gravitational potential energy* of a body may be determined by knowing the height y_G of the body's *center of mass* from the datum plane and the body's weight **W**. In the above case, the potential energy is *positive,* since the weight has the ability to do *positive work* when the body is moved back to the datum plane, Fig. 18–15. If the body is located y_G *below* the datum, the gravitational potential energy is *negative,* i.e.,

$$V_g = -W y_G \tag{18–13}$$

since then the weight does *negative work* when the body is moved back to the datum.

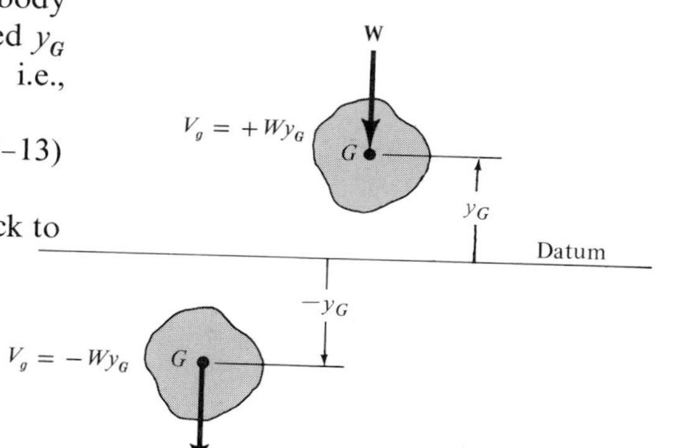

Fig. 18–15

Elastic Potential Energy. The force developed by an elastic spring is also defined as a conservative force. The *elastic potential energy* which a spring imparts to an attached body when the spring is elongated or compressed from an initial unstretched position ($x = 0$) to a final position x, Fig. 18–16, is

$$V_e = \tfrac{1}{2}kx^2 \tag{18–14}$$

In the deformed position, the spring force acting *on the body* has the capacity for doing positive work when the spring is returned back to its original undeformed position (see Sec. 14–5).

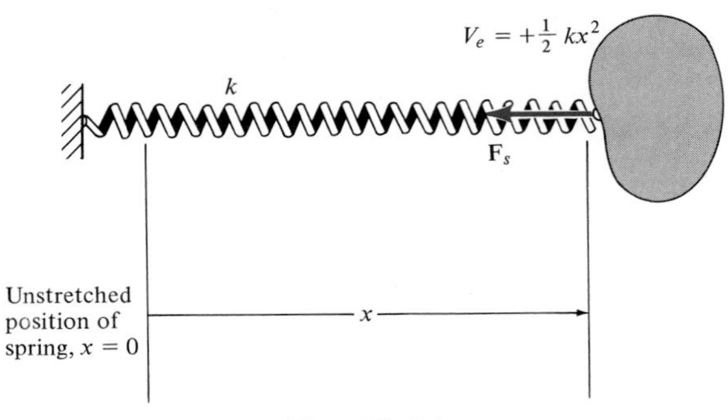

Fig. 18–16

In general, if a body is subjected to both gravitational and elastic forces, the total *potential energy* is expressed as the algebraic sum

$$V = V_g + V_e \tag{18–15}$$

Here measurement of V depends upon the location of the body with respect to a selected datum in accordance with Eqs. 18–12 through 18–14.

Conservation-of-Energy Theorem. In Sec. 14–6 the conservation-of-energy theorem was developed for a particle. By applying this theorem to each of the particles of a rigid body and adding the results algebraically, since energy is a scalar, the conservation-of-energy theorem for a rigid body may be written as

$$V_1 + T_1 = V_2 + T_2 \tag{18–16}$$

This theorem states that the *sum* of the potential and kinetic energy of the body remains *constant* when the body moves from one position to another. It also applies to a system of smooth, pin-connected rigid bodies, bodies connected by inextensible cords, and bodies in mesh with other bodies. In these cases the forces acting at all points of contact are *eliminated* from the analysis, since they occur in equal and opposite pairs and each pair of forces moves through an equal distance when the system undergoes a small displacement.

PROCEDURE FOR ANALYSIS

The conservation-of-energy theorem is used to solve problems involving *velocity, displacement,* and *conservative force systems*. For application, the following two-step procedure should be used.

Step 1: Draw two diagrams showing the body located at its initial and final positions along the path. If the center of mass, G, of the body is subjected to a *vertical displacement,* determine where to establish the fixed horizontal datum from which to measure the body's gravitational potential energy, V_g. Although the location of the datum is arbitrary, it is advantageous to place it through G when the body is either at the initial or final point of its path, since at the datum $V_g = 0$. Data pertaining to the elevation of the body from the datum, and the extension or compression of any connecting springs, can be determined from the geometry associated with the two diagrams. Also, for each body the velocity of G and the body's angular velocity should be indicated on the diagrams.

Step 2: Apply the conservation-of-energy theorem: $T_1 + V_1 = T_2 + V_2$. The calculation of the potential energy $V = V_g + V_e$ at the initial and final positions is formulated on the basis of applying $V_g = \pm W y_G$ (Eqs. 18–12 and 18–13) and $V_e = \frac{1}{2}kx^2$ (Eq. 18–14). The kinetic-energy terms T_1 and T_2 are determined from $T = \frac{1}{2}mv_G^2 + \frac{1}{2}I_G\omega^2$ or an appropriate form of this equation as developed in Sec. 18–1. In this regard, kinematic diagrams for velocity may be useful for determining v_G and ω, or for establishing a *relationship* between these quantities.

It is important to remember that *only problems involving conservative force systems may be solved by using the conservation-of-energy theorem.* As stated in Sec. 14–5, friction or other drag-resistant forces, which depend upon velocity or acceleration, are nonconservative. The work of such forces is transformed into thermal energy used to heat up the surfaces of contact, and consequently this energy is dissipated into the surroundings and may not be recovered. Therefore, problems involving frictional forces should either be solved by using the principle of work and energy, if it applies, or the equations of motion.

The following example problems numerically illustrate application of the above two-step procedure.

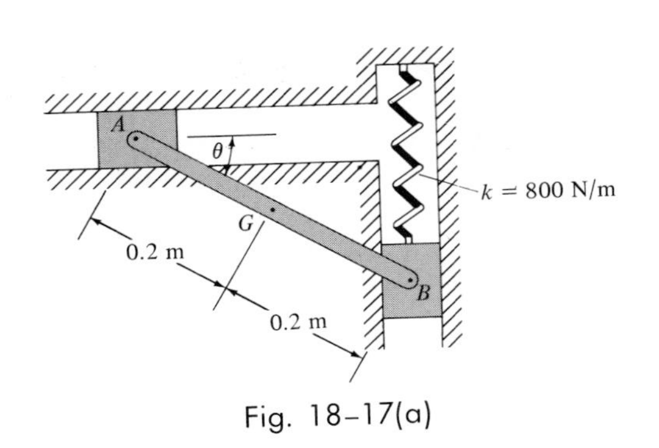

Fig. 18–17(a)

Example 18–6

The 10-kg rod AB shown in Fig. 18–17a is confined so that its ends move in the horizontal and vertical slots. The spring has a stiffness of $k = 800$ N/m and is unstretched when $\theta = 0°$. Determine the speed of the slider block at B when $\theta = 0°$, if AB is released from rest when $\theta = 30°$. Neglect the mass of the slider blocks.

Solution

Step 1: The two diagrams of the rod, when it is located at its initial and final positions, are shown in Fig. 18–17b. The datum plane, used to measure the gravitational potential energy of the system, is placed in line with the rod when $\theta = 0°$.

When the rod is in position 1, the center of mass G is located *below the datum,* so that the gravitational potential energy is *negative.* Furthermore, (positive) elastic potential energy is stored in the spring, since it is

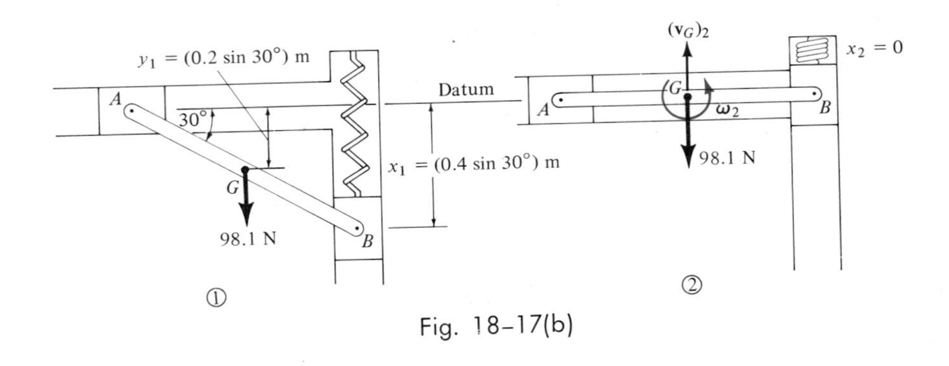

Fig. 18–17(b)

stretched a distance of $x_1 = (0.4 \sin 30°)$ m. Because the rod is released from rest, $(\mathbf{v}_G)_1 = \boldsymbol{\omega}_1 = \mathbf{0}$, so that $T_1 = 0$.

When the rod is in position 2, the potential energy of the rod is zero, since the spring is unstretched, $x_2 = 0$, and the center of mass G is located at the datum. In this position the angular velocity is $\boldsymbol{\omega}_2$ and the rod's mass center has a velocity of $(\mathbf{v}_G)_2$.

Step 2: Applying the conservation-of-energy theorem, we have

$$\{V_1\} + \{T_1\} = \{V_2\} + \{T_2\}$$

$$\{-Wy_1 + \tfrac{1}{2}k(x_1)^2\} + \{\tfrac{1}{2}m(v_G)_1^2 + \tfrac{1}{2}I_G(\omega_1)^2\}$$
$$= \{\tfrac{1}{2}k(x_2)^2\} + \{\tfrac{1}{2}m(v_G)_2^2 + \tfrac{1}{2}I_G(\omega_2)^2\}$$

$$\{-98.1(0.2 \sin 30°) + \tfrac{1}{2}(800)(0.4 \sin 30°)^2\} + \{0 + 0\}$$
$$= \{0\} + \{\tfrac{1}{2}(10)(v_G)_2^2 + \tfrac{1}{2}[\tfrac{1}{12}(10)(0.4)^2](\omega_2)^2\} \qquad (1)$$

Using *kinematics,* $(v_G)_2$ can be related to ω as shown in Fig. 18–17*c*. At the instant considered, the instantaneous center of zero velocity (IC) for the rod is at point A; hence,

$$(v_G)_2 = (r_{G/IC})\omega_2$$
$$(v_G)_2 = (0.2)\omega_2$$

Substituting this into Eq. (1) and solving for ω_2, we have

$$6.19 = 0.267(\omega_2)^2$$
$$\omega_2 = 4.82 \text{ rad/s}$$

With reference to Fig. 18–17*c*, the speed of the slider block at B is, therefore,

$$(v_B)_2 = (r_{B/IC})\omega_2 = 0.4(4.82)$$

or

$$(v_B)_2 = 1.93 \text{ m/s} \qquad \textit{Ans.}$$

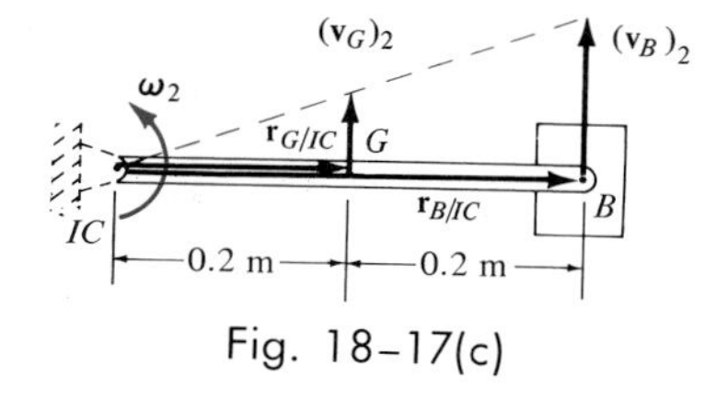

Fig. 18–17(c)

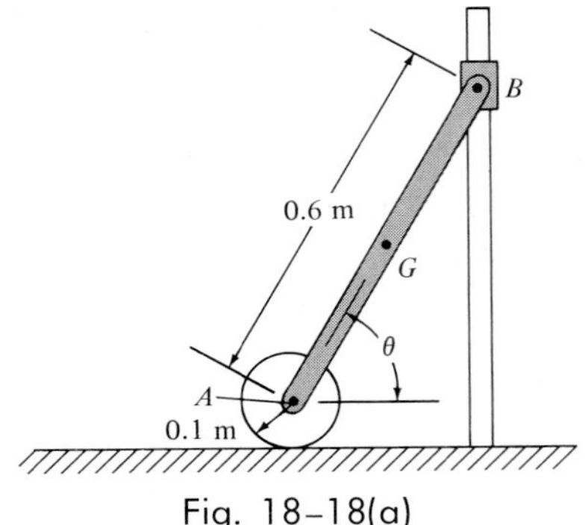

Fig. 18–18(a)

Example 18–7

The 10-kg homogeneous disk shown in Fig. 18–18*a* is attached to a uniform 5-kg rod *AB*. If the assembly is released from rest when $\theta = 60°$, determine the angular velocity of the rod when $\theta = 0°$. Assume that the disk rolls without slipping. Neglect friction along the guide and the mass of the slider block *B*.

Solution

Step 1: Two diagrams for the rod and disk when they are located at their initial and final positions are shown in Fig. 18–18*b*. For convenience the datum, which is horizontally fixed, passes through point *A*.

When the system is in position 1, the rod's weight has positive potential energy, since this force is above the datum and therefore has the capacity to do work in moving the rod and disk. Furthermore, the entire system is at rest in this position, so that $T_1 = 0$.

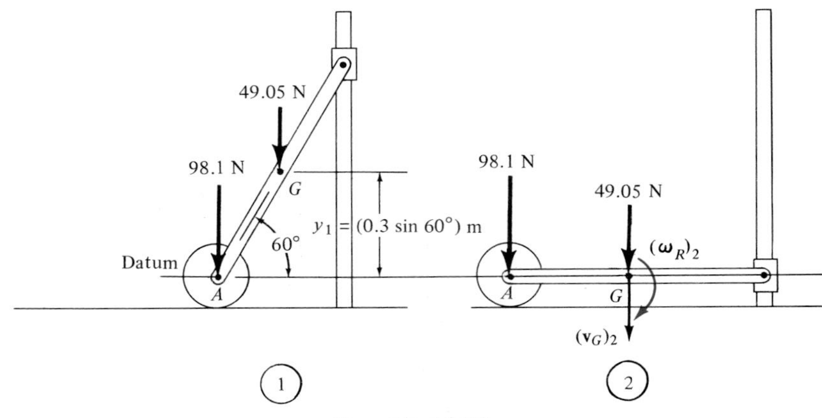

Fig. 18–18(b)

When the system is in position 2, both the weight of the rod and the weight of the disk have zero potential energy. Why? The rod has an angular velocity of $(\boldsymbol{\omega}_R)_2$ and its mass center has a velocity of $(\mathbf{v}_G)_2$. Since the rod is *fully extended* in this position, the disk is momentarily at rest. Therefore, $(\boldsymbol{\omega}_D)_2 = (\mathbf{v}_A)_2 = \mathbf{0}$.

Step 2: Applying the conservation-of-energy theorem, we have

$$\{V_1\} + \{T_1\} = \{V_2\} + \{T_2\}$$

$$\{W_R y_1\} + \{\tfrac{1}{2}m_R(v_G)_1^2 + \tfrac{1}{2}I_G(\omega_R)_1^2 + \tfrac{1}{2}m_D(v_A)_1^2 + \tfrac{1}{2}I_A(\omega_D)_1^2\}$$
$$= \{0\} + \{\tfrac{1}{2}m_R(v_G)_2^2 + \tfrac{1}{2}I_G(\omega_R)_2^2 + \tfrac{1}{2}m_D(v_A)_2^2 + \tfrac{1}{2}I_A(\omega_D)_2^2\}$$

$$\{49.05(0.3 \sin 60°)\} + \{0 + 0 + 0 + 0\}$$
$$= \{0\} + \{\tfrac{1}{2}(5)(v_G)_2^2 + \tfrac{1}{2}[\tfrac{1}{12}(5)(0.6)^2](\omega_R)_2^2 + 0 + 0\}$$

or

$$12.74 = 2.5(v_G)_2^2 + 0.075(\omega_R)_2^2 \qquad (1)$$

Using kinematics, $(v_G)_2$ can be related to $(\omega_R)_2$. When the rod is in position 2, the instantaneous center of zero velocity is at point A, Fig. 18–18*c*. Hence,

$$(v_G)_2 = r_{G/IC}(\omega_R)_2$$
$$(v_G)_2 = 0.3(\omega_R)_2$$

Substituting this equation into Eq. (1) yields

$$12.74 = 0.3(\omega_R)_2^2$$

Thus,

$$(\omega_R)_2 = 6.52 \text{ rad/s} \qquad \textit{Ans.}$$

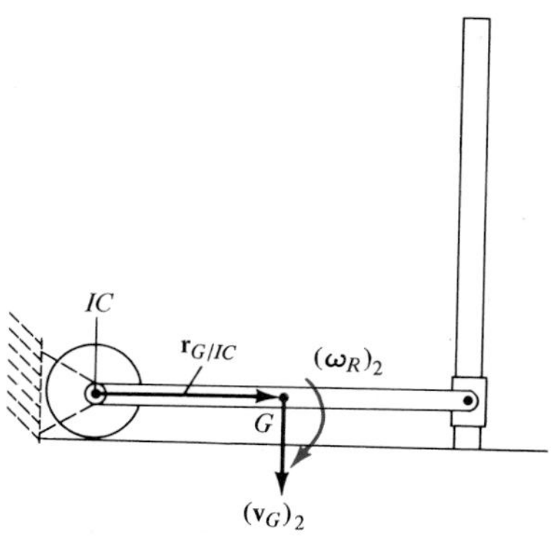

Fig. 18–18(c)

Problems

18–17. Solve Prob. 18–1 using the conservation-of-energy theorem.

18–18. Solve Prob. 18–2 using the conservation-of-energy theorem.

18–19. Solve Prob. 18–14 using the conservation-of-energy theorem.

***18–20.** Solve Prob. 18–5 using the conservation-of-energy theorem.

***18–20a.** Solve Prob. 18–5*a* using the conservation-of-energy theorem.

18–21. The 400-g rod *AB* rests along the smooth inner surface of a hemispherical bowl. If the rod is released from rest from the position shown, determine its angular velocity $\boldsymbol{\omega}$ at the instant it swings downward and becomes horizontal.

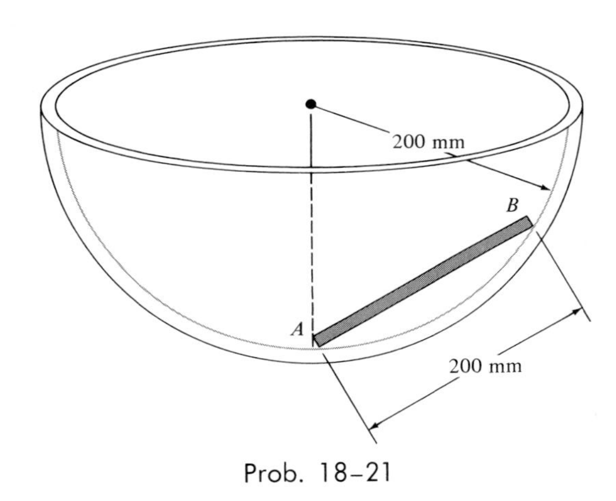

Prob. 18–21

18–22. A 7-kg automobile tire is released from rest at *A* on the incline and rolls without slipping to point *B*, from which it is launched into free-flight motion. Determine the maximum height *h* the tire attains after it is launched. The radius of gyration of the tire about its mass center is $k_G = 0.3$ m.

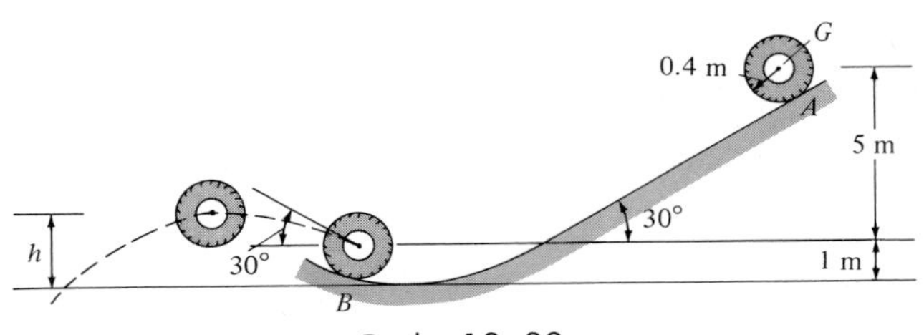

Prob. 18–22

18–23. The window *AB*, shown in side view, has a mass of 4 kg, a mass center at *G*, and a radius of gyration of $k_A = 0.2$ m about its hinged axis at *A*. If the spring is unstretched when $\theta \approx 0°$, and the window is released from rest from this position, determine the velocity with which the end *B* strikes the wall *D*, $\theta = 90°$. Neglect friction at the hinge *A* and at pulley *C*.

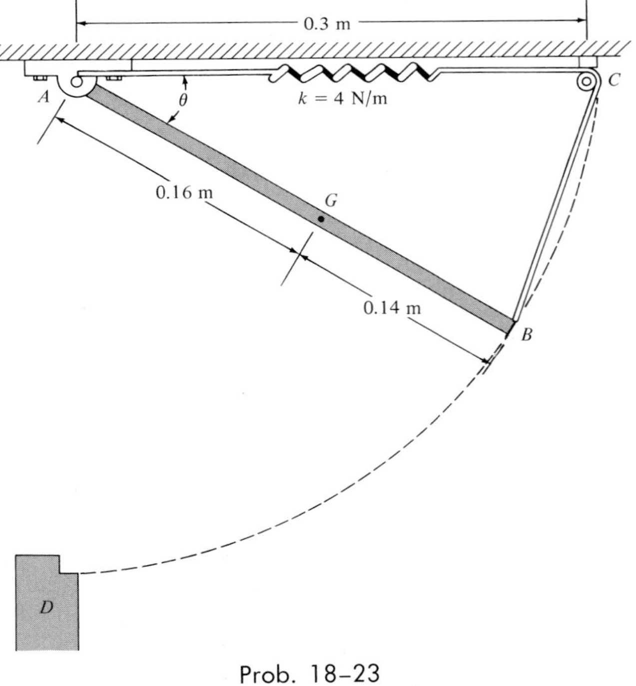

Prob. 18–23

*** 18–24.** A chain that has a negligible mass is draped over a sprocket which has a mass of 2 kg and a radius of gyration of $k_O = 55$ mm. If the 3-kg block A is released from rest in the position shown, $s = 1$ m, determine the angular velocity which the chain imparts to the sprocket when $s = 2$ m.

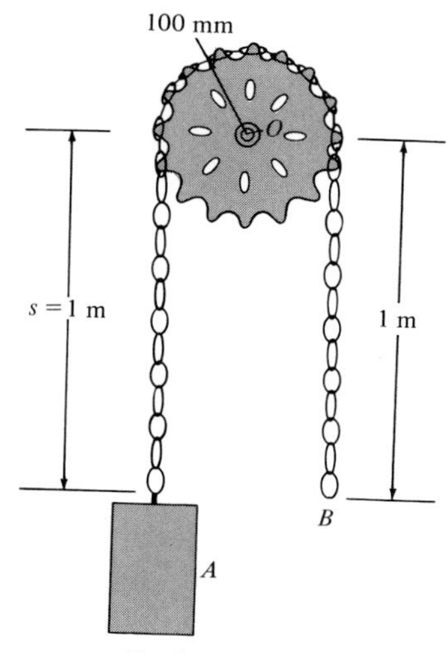

Prob. 18–24

18–25. The disk has a mass of $m_D = 5$ kg, a center of gravity at G, and a radius of gyration of $k_O = 80$ mm. A cord is placed over the rim of the disk and the attached blocks A and B are released from rest when the disk is in the position shown. If the cord does not slip on the rim, determine the angular velocity of the disk when it has rotated 180°. Blocks A and B have a mass of $m_A = 4$ kg and $m_B = 10$ kg, respectively. Set $\bar{r} = 50$ mm and $r = 125$ mm.

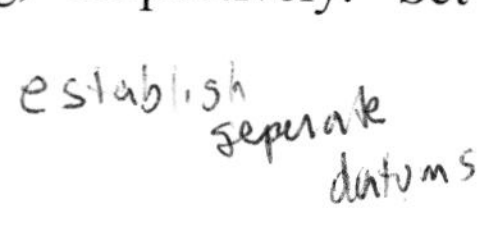

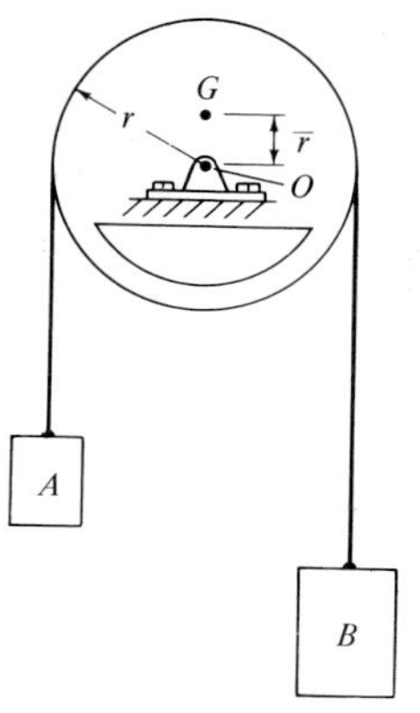

Prob. 18–25

18–25a. Solve Prob. 18–25 if the disk weighs $W_D = 25$ lb; $k_O = 0.55$ ft; blocks A and B weigh $W_A = 5$ lb and $W_B = 12$ lb, respectively; $\bar{r} = 0.2$ ft; and $r = 0.74$ ft.

18–26. The window AB, shown in side view, has a mass of 11 kg and a radius of gyration of $k_G = 0.2$ m. The attached elastic cable BC has a stiffness of $k = 40$ N/m and an unstretched length of 0.5 m. If the window is released from rest when $\theta \approx 0°$, determine its angular velocity just before it strikes the wall at D, $\theta = 90°$.

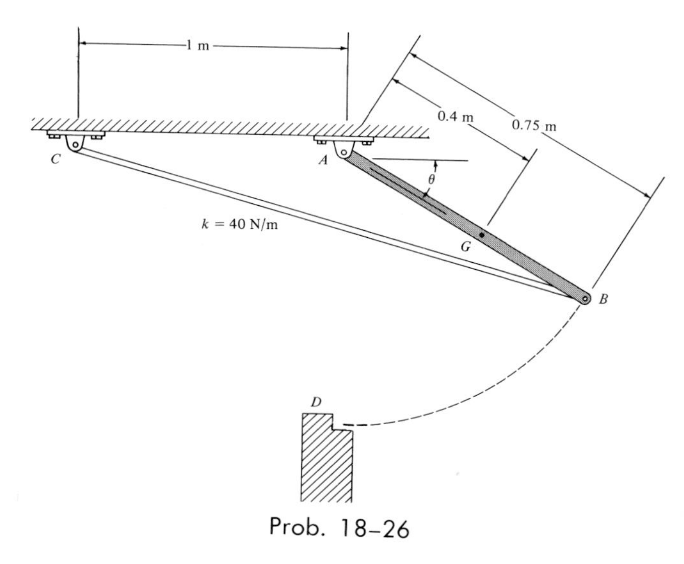

Prob. 18–26

18–27. The 14-kg disk has a counterclockwise angular velocity of 3 rad/s at the instant it is in the position shown. If the uniform rod AB has a mass of 4 kg and the slider block B has a mass of 2 kg, determine the angular velocity of the disk when AB is in its lowest vertical position. Neglect the effects of friction.

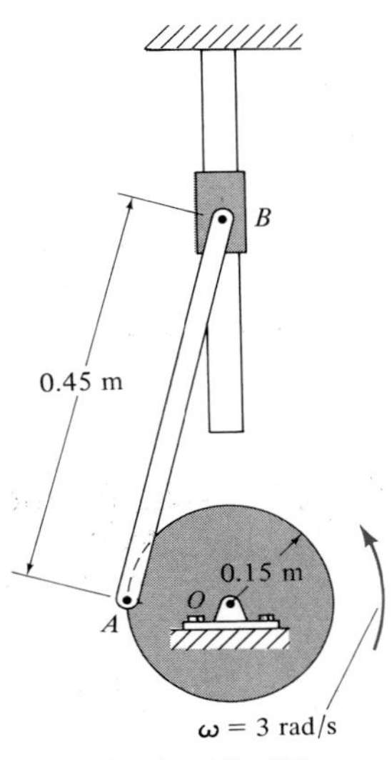

Prob. 18–27

* **18–28.** The uniform links AB and CD have a mass of 4 kg each, and the vertical member BC has a mass of 3 kg. If the spring is unstretched when $\theta = 0°$, determine the stiffness k such that when the links are released from rest at $\theta = 90°$, they reach a final *at rest position* at $\theta = 180°$. Neglect friction and the size of the pulley P.

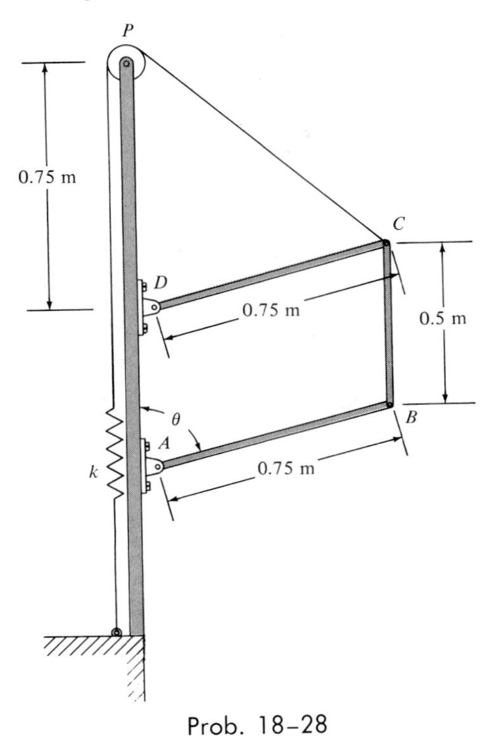

Prob. 18–28

18–29. If the stiffness of the spring in Prob. 18–28 is $k = 50$ N/m and the links are released from rest when $\theta = 90°$, determine the angular velocity of DC when $\theta = 180°$. Use the data in Prob. 18–28.

18–30. Pulley A and the attached drum have a mass of $m_A = 15$ kg and a centroidal radius of gyration of $k_B = 125$ mm. Determine the speed of the crate C, which has a mass of $m_C = 11$ kg, at the instant $(s)_2 = 2$ m. Initially, the crate is released from rest when $(s)_1 = 0.8$ m. The pulley at P "rolls" downward on the cord without slipping. For the calculation, neglect the mass of this pulley and the cord as it unravels from A. Set $r_o = 200$ mm, $r_i = 100$ mm, and $r = 50$ mm.

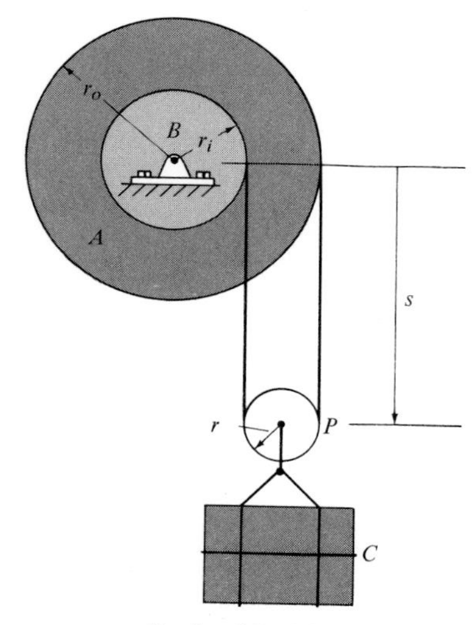

Prob. 18–30

18–30a. Solve Prob. 18–30 if pulley A and the attached drum have a weight of $W_A = 20$ lb, $k_B = 0.6$ ft, the crate has a weight of $W_C = 15$ lb, and $(s)_2 = 10$ ft, $(s)_1 = 4$ ft, $r_o = 0.8$ ft, $r_i = 0.4$ ft, $r = 0.2$ ft.

18–31. Its four wheels excluded, the mine car and its contents have a mass of 200 kg and a mass center at G. The wheels each have a mass of 15 kg and approximate thin disks. The car is ascending the 30° slope with a speed of $v_1 = 1.5$ m/s, when the cable C breaks at the instant shown. Determine the speed of the car after it rolls down the incline and moves along the horizontal track. The wheels roll without slipping.

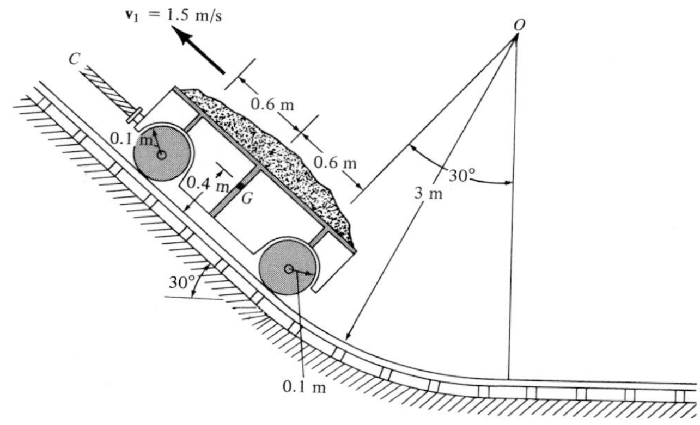

Prob. 18–31

*** 18–32.** The uniform bar AB has a mass of 12 kg and is pin-connected at A. If the support at B is removed ($\theta = 90°$), determine the velocity of the 5-kg block C at the instant the bar rotates downward to $\theta = 150°$. Neglect friction and the size of the pulley at D.

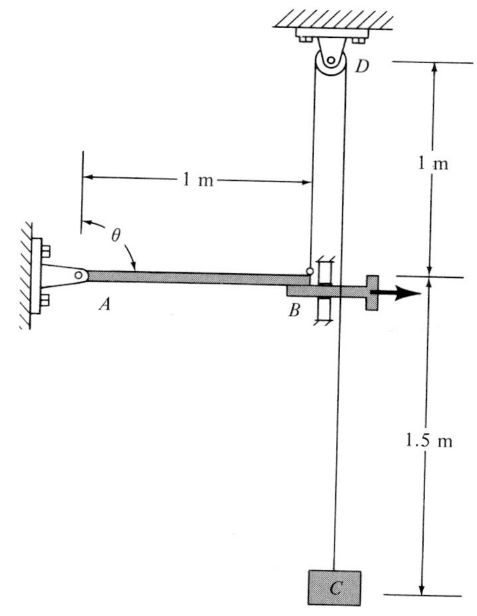

Prob. 18–32

18–33. A large roll of paper having a mass of $m = 20$ kg and radius $r = 150$ mm is resting over the edge of a table, such that the end of the paper on the roll is attached to the table's surface. If the roll is disturbed slightly from its equilibrium position, determine the angle θ at which it begins to leap off the table edge A as it falls. The centroidal radius of gyration of the roll is $k_G = 75$ mm.

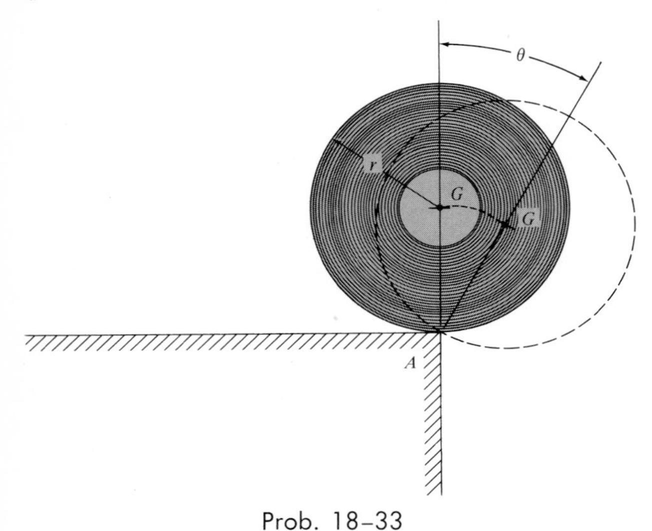

Prob. 18–33

19

Planar Kinetics of Rigid Bodies: Impulse and Momentum

19–1. Linear and Angular Momentum of a Rigid Body

The concepts of linear and angular momentum for a particle presented in Chapter 15 will be extended somewhat in order to determine momentum relationships which apply to a rigid body. The three planar motions that will be considered are translation, rotation about a fixed axis, and general plane motion. The more general discussion of momentum principles applied to the spatial motion of a rigid body is presented in Chapter 21.

Linear Momentum. The linear momentum $\mathbf{L}_i$ of a particle having a mass m_i and velocity $\mathbf{v}_i$ has been defined in Sec. 15–1 as $\mathbf{L}_i = m_i\mathbf{v}_i$. For the rigid body in Fig. 19–1*a*, the linear momentum may be obtained by summing (vectorially) the linear momenta of all *n* particles of the body. Thus,

$$\mathbf{L} = \Sigma m_i \mathbf{v}_i$$

This equation may be simplified by noting that the location $\mathbf{r}_G$ of the body's center of mass *G*, Fig. 19–1*a*, is determined from $m\mathbf{r}_G = \Sigma m_i \mathbf{r}_i$, where m is the total mass of the body. Taking the time derivative, i.e., $m\mathbf{v}_G = \Sigma m_i \mathbf{v}_i$, and substituting into the above equation, we have

$$\mathbf{L} = m\mathbf{v}_G \tag{19–1}$$

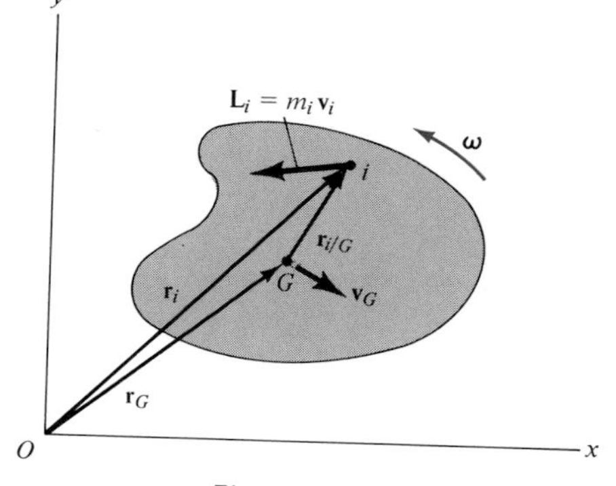

Fig. 19–1(a)

Hence, the *linear momentum* is a vector quantity, having a *magnitude* mv_G, which is commonly measured in units of kg · m/s or N · s; and a *direction* defined by $\mathbf{v}_G$, the instantaneous velocity of the mass center.

Angular Momentum. A detailed analysis for obtaining the angular momentum of a rigid body subjected to *general plane motion* will now be considered. The special cases of translation and rotation about a fixed axis will then be discussed. To simplify the analysis, the body is assumed to be symmetric with respect to the inertial x-y reference plane. Thus, the *general plane motion* of the body can be studied by considering a slab of the body lying in this plane, Fig. 19–1a. As indicated in Sec. 15–5, the angular momentum of a particle about a point has been defined as the moment of the linear-momentum vector about the point. Thus, the angular momentum of the body's ith particle, computed about the mass center G, is

$$(\mathbf{H}_i)_G = \mathbf{r}_{i/G} \times m_i \mathbf{v}_i$$

At the instant shown, the body has a known instantaneous angular velocity $\boldsymbol{\omega}$ and its mass center has an instantaneous velocity $\mathbf{v}_G$. Applying Eq. 16–19, we can express the velocity of the ith particle as $\mathbf{v}_i = \mathbf{v}_G + \boldsymbol{\omega} \times \mathbf{r}_{i/G}$. Hence, substituting into the above equation, and using the distributive property of the vector cross product, yields

$$\begin{aligned}(\mathbf{H}_i)_G &= m_i \mathbf{r}_{i/G} \times (\mathbf{v}_G + \boldsymbol{\omega} \times \mathbf{r}_{i/G}) \\ &= m_i(\mathbf{r}_{i/G} \times \mathbf{v}_G) + m_i \mathbf{r}_{i/G} \times (\boldsymbol{\omega} \times \mathbf{r}_{i/G})\end{aligned}$$

When similar expressions are written and summed for all n particles in the body, the total angular momentum becomes

$$\mathbf{H}_G = (\Sigma m_i \mathbf{r}_{i/G}) \times \mathbf{v}_G + \Sigma[m_i \mathbf{r}_{i/G} \times (\boldsymbol{\omega} \times \mathbf{r}_{i/G})]$$

By definition of the mass center the sum $\Sigma\, m_i \mathbf{r}_{i/G} = (\Sigma m_i)\bar{\mathbf{r}} = \mathbf{0}$, since the position vector $\bar{\mathbf{r}}$ relative to G is equal to zero. Hence, the first term on the right is equal to zero. The second term may be further simplified by noting that $\mathbf{r}_{i/G} \times (\boldsymbol{\omega} \times \mathbf{r}_{i/G})$ is a vector having a magnitude of $r_{i/G}^2 \omega$ and acts perpendicular to the plane of the slab. Since $\mathbf{H}_G$ also acts in this direction, the above equation can be written in the scalar form

$$H_G = (\Sigma m_i r_{i/G}^2)\omega$$

When the particle size $m_i \rightarrow dm$, the summation sign becomes an integral since the number of particles $n \rightarrow \infty$. Representing $r_{i/G}$ by a more "generalized" dimension r, yields

$$H_G = \left(\int_m r^2\, dm\right)\omega$$

The integral represents the moment of inertia I_G of the body about an axis which is perpendicular to the slab and passes through point G. The final result, written in scalar form, is therefore

$$H_G = I_G \omega \qquad (19\text{–}2)$$

Thus, in the case of general plane motion of a symmetrical rigid body (represented here as a slab) the *angular momentum* $\mathbf{H}_G$ of the body is a vector quantity having a *magnitude* $I_G\omega$, which is commonly measured in units of $\text{kg}\cdot\text{m}^2/\text{s}$ or $\text{N}\cdot\text{m}\cdot\text{s}$, and a *direction* defined by $\boldsymbol{\omega}$, which is always perpendicular to the slab. Since $\boldsymbol{\omega}$ is a free vector, $\mathbf{H}_G$ *can act at any point on the body* (or slab) provided it preserves its same magnitude and direction. Furthermore, since angular momentum is equal to the moment of the linear momentum, the *line of action of* $\mathbf{L}$ *must pass through the body's mass center G* in order to preserve the correct magnitude of $\mathbf{H}_G$, see Fig. 19–1*b*. In other words, $\mathbf{L}$ creates *zero* angular momentum about G, so that simply $\mathbf{H}_G = I_G\boldsymbol{\omega}$.

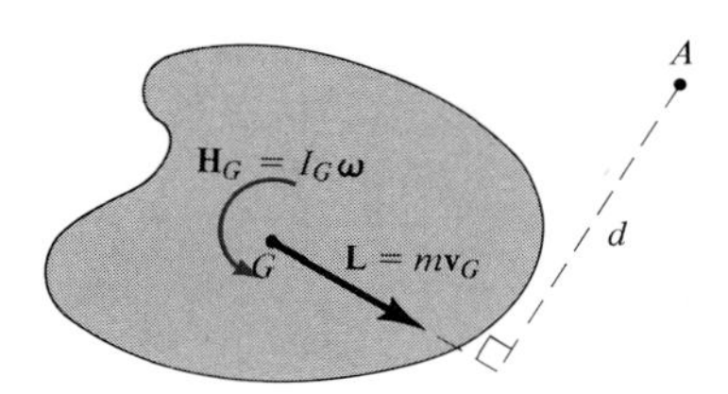

Fig. 19–1(b)

Translation. When a rigid body of mass m is subjected to either rectilinear or curvilinear *translation,* Fig. 19–2*a,* its mass center has a velocity of $\mathbf{v}_G = \mathbf{v}$ and $\boldsymbol{\omega} = \mathbf{0}$. Hence, the linear momentum and the angular momentum, computed about G, become

$$\mathbf{L} = m\mathbf{v}_G \qquad (19\text{–}3)$$
$$\mathbf{H}_G = \mathbf{0}$$

If the angular momentum is computed about any other point A, on or off the body, Fig. 19–2*a,* the "moment" of the linear momentum $\mathbf{L}$ must be computed about the point. Since d is the "moment arm" as shown in the figure, then

$$\circlearrowleft + H_A = (d)(mv_G)$$

Rotation About a Fixed Axis. When a rigid body is *rotating about a fixed axis* passing through point O, Fig. 19–2*b,* the linear momentum and the angular momentum computed about G are

$$L = mv_G \qquad (19\text{–}4)$$
$$H_G = I_G\omega$$

It is sometimes convenient to compute the angular momentum of the body about point O. In this case it is necessary to account for the "moments" of *both* $\mathbf{L}$ and $\mathbf{H}_G$ about O. Noting that $\mathbf{L}$ (or $\mathbf{v}_G$) is always *perpendicular* to $\mathbf{r}_{G/O}$, we have

$$\circlearrowleft + H_O = I_G\omega + r_{G/O}(mv_G) \qquad (19\text{–}5)$$

This equation may be *reduced* by first substituting $v_G = r_{G/O}\omega$, in which case $H_O = (I_G + mr^2_{G/O})\omega$, and, by the parallel-axis theorem, noting that the terms inside the parentheses represent the moment of inertia I_O of the body about an axis perpendicular to the plane of motion and passing through point O. Hence,*

$$H_O = I_O\omega \qquad (19\text{–}6)$$

*The similarity between this derivation and that of Eq. 17–15 ($\Sigma M_O = I_O\alpha$) and Eq. 18–3 ($T = \frac{1}{2}I_O\omega^2$) should be noted.

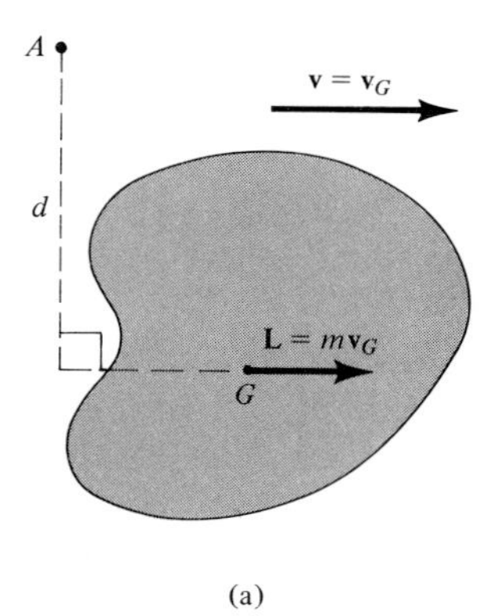

(a)

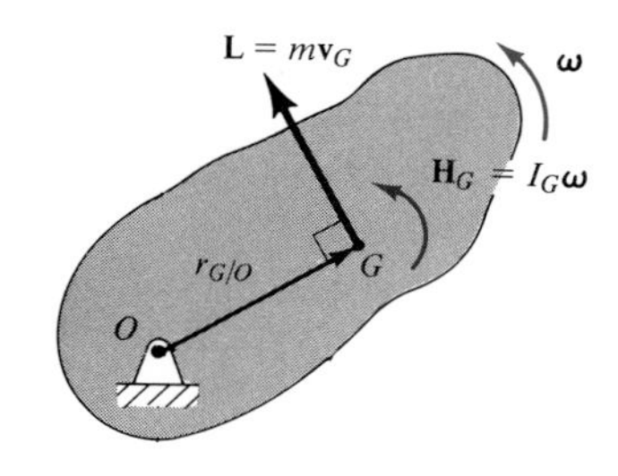

(b)

Fig. 19–2(a–b)

For the computation, then, either Eq. 19–5 or 19–6 can be used.

General Plane Motion. When a rigid body is subjected to general plane motion, Fig. 19–2*c*, the linear momentum, and the angular momentum computed about *G*, become

$$\begin{aligned} L &= mv_G \\ H_G &= I_G\omega \end{aligned} \qquad (19\text{–}7)$$

If the angular momentum is computed about a point, *A*, located either on or off the body, Fig. 19–2*c*, it is necessary to compute the moments of *both* **L** and $\mathbf{H}_G$ about this point. Hence,

$$\circlearrowleft + H_A = I_G\omega + (d)(mv_G)$$

Here *d* is the moment arm, as shown in the figure.

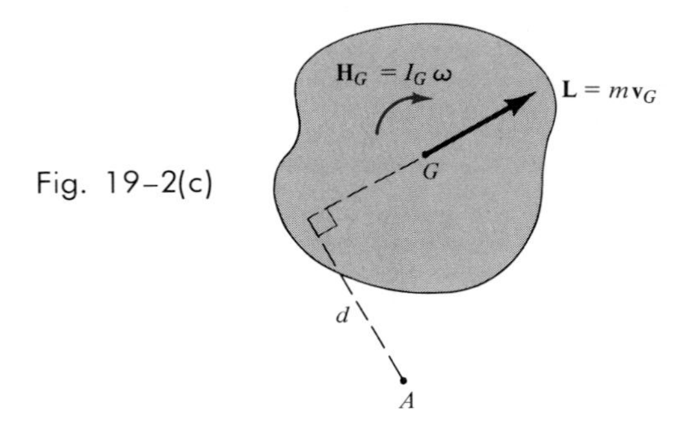

Fig. 19–2(c)

19–2. Principle of Impulse and Momentum for a Rigid Body

Principle of Linear Impulse and Momentum. The equation of motion for a rigid body, as stated by Eq. 17–6, can be written as

$$\Sigma\mathbf{F} = m\mathbf{a}_G = m\frac{d\mathbf{v}_G}{dt}$$

Since the mass *m* of the body is constant,

$$\Sigma\mathbf{F} = \frac{d}{dt}(m\mathbf{v}_G)$$

Multiplying both sides by *dt* and integrating between the limits $t = t_1$, $\mathbf{v}_G = (\mathbf{v}_G)_1$ and $t = t_2$, $\mathbf{v}_G = (\mathbf{v}_G)_2$, yields

$$\Sigma\int_{t_1}^{t_2} \mathbf{F}\,dt = m(\mathbf{v}_G)_2 - m(\mathbf{v}_G)_1 \qquad (19\text{–}8)$$

This equation is referred to as the *principle of linear impulse and momentum*. It states that the sum of all the impulses created by the *external force system* which acts on the body during the time interval t_1 to t_2 is equal to the change in the linear momentum of the body during the time interval.

Principle of Angular Impulse and Momentum. If the body is subjected to *general plane motion,* using Eq. 17–9 yields

$$\Sigma M_G = I_G\alpha = I_G\frac{d\omega}{dt}$$

Since the moment of inertia, I_G, is constant,

$$\Sigma M_G = \frac{d}{dt}(I_G\omega)$$

Multiplying both sides by dt and integrating from $t = t_1, \omega = \omega_1$ to $t = t_2$, $\omega = \omega_2$, gives

$$\Sigma\int_{t_1}^{t_2} M_G\,dt = I_G\omega_2 - I_G\omega_1 \tag{19–9}$$

In a similar manner, for *rotation about a fixed axis* passing through point O, Eq. 17–15 ($\Sigma M_O = I_O\alpha$) when integrated becomes

$$\Sigma\int_{t_1}^{t_2} M_O\,dt = I_O\omega_2 - I_O\omega_1 \tag{19–10}$$

Equations 19–9 and 19–10 are referred to as the *principle of angular impulse and momentum*. As stated, the sum of the angular impulses acting on the body during the time interval t_1 to t_2 is equal to the change in the angular momentum of the body during this time interval. In particular, the angular impulse considered is determined by integrating the moments about point G or O of all the external forces and couples applied to the body.

To summarize the preceding concepts, if motion is occurring in the x-y plane, using impulse and momentum principles the following *three scalar equations* may be written which describe the *planar motion* of the body:

$$\begin{aligned} m(v_{Gx})_1 + \Sigma\int_{t_1}^{t_2} F_x dt &= m(v_{Gx})_2 \\ m(v_{Gy})_1 + \Sigma\int_{t_1}^{t_2} F_y dt &= m(v_{Gy})_2 \\ I_G\omega_1 + \Sigma\int_{t_1}^{t_2} M_G dt &= I_G\omega_2 \end{aligned} \tag{19–11}$$

The first two of these equations represent the principle of linear impulse and momentum in the x-y plane (Eq. 19–8), and the third equation represents the principle of angular impulse and momentum about the z axis (Eq. 19–9).

Equations 19–11 may also be applied to an entire system of connected bodies rather than to each body separately. Doing this eliminates the need to include reactive impulses which occur at the connections since they are *internal* to the system. The resultant equations may be written in symbolic form as

$$\left(\sum \begin{matrix}\text{syst. linear}\\ \text{momentum}\end{matrix}\right)_{x1} + \left(\sum \begin{matrix}\text{syst. linear}\\ \text{impulse}\end{matrix}\right)_{x(1-2)} = \left(\sum \begin{matrix}\text{syst. linear}\\ \text{momentum}\end{matrix}\right)_{x2}$$

$$\left(\sum \begin{matrix}\text{syst. linear}\\ \text{momentum}\end{matrix}\right)_{y1} + \left(\sum \begin{matrix}\text{syst. linear}\\ \text{impulse}\end{matrix}\right)_{y(1-2)} = \left(\sum \begin{matrix}\text{syst. linear}\\ \text{momentum}\end{matrix}\right)_{y2}$$

$$\left(\sum \begin{matrix}\text{syst. angular}\\ \text{momentum}\end{matrix}\right)_{O1} + \left(\sum \begin{matrix}\text{syst. angular}\\ \text{impulse}\end{matrix}\right)_{O(1-2)} = \left(\sum \begin{matrix}\text{syst. angular}\\ \text{momentum}\end{matrix}\right)_{O2}$$

(19–12)

As indicated, the system's angular momentum and angular impulse must be computed with respect to the *same fixed reference point O* for all the bodies of the system.

PROCEDURE FOR ANALYSIS

Impulse-and-momentum principles are used to solve kinetic problems that involve *velocity, force,* and *time* since these terms are involved in the formulation. For application a two-step procedure should be followed.

Step 1: Draw the momentum- and impulse-vector diagrams for the body or system of bodies. Each of these diagrams represents an outlined shape of the body, which graphically accounts for the data required for each of the three terms in Eqs. 19–11 or 19–12.

An appropriate set of momentum- and impulse-vector diagrams for a rigid body subjected to general plane motion is shown in Fig. 19–3. Note that the linear-momentum vectors $m\mathbf{v}_G$ are applied at the body's mass center, Figs. 19–3*a* and 19–3*c;* whereas the angular-momentum vectors, $I_G\boldsymbol{\omega}$, are free vectors, and therefore, like a couple, they may be applied at any point on the body.

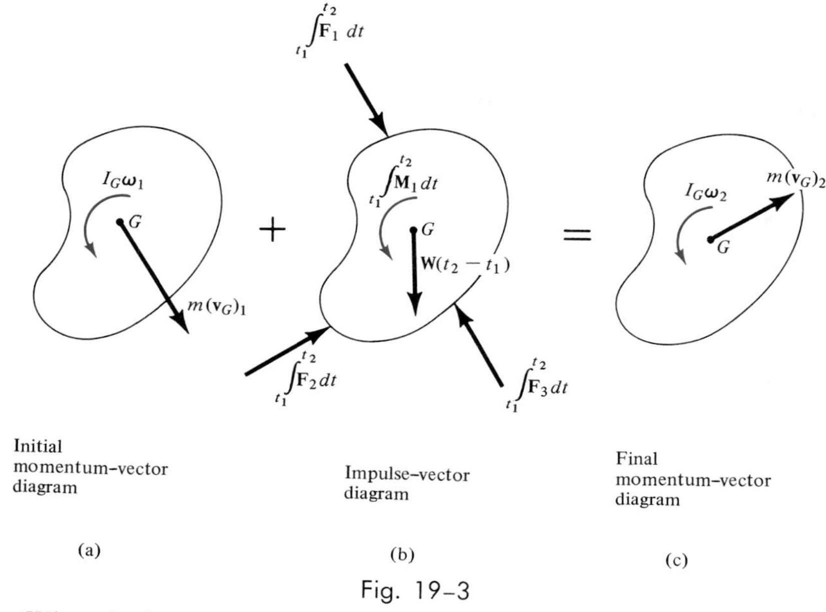

Fig. 19–3

When the impulse-vector diagram is constructed, Fig. 19–3*b*, vectors **F** and **M** which vary with time are indicated by $\int_{t_1}^{t_2} \mathbf{F}(t)dt$ and $\int_{t_1}^{t_2} \mathbf{M}(t)\,dt$. However, if **F** and **M** are *constant* during the time interval t_1 to t_2, the integration of the impulses yields $\mathbf{F}(t_2 - t_1)$ and $\mathbf{M}(t_2 - t_1)$, respectively. Such is the case for the body's weight **W**, shown in Fig. 19–3*b*.

Step 2: Apply the three scalar Equations 19–11 (or 19–12) by determining the vector components for each of the terms in these equations directly from the momentum- and impulse-vector diagrams. In cases where the body is rotating about a fixed axis, Eq. 19–10 may be substituted for the third of Eqs. 19–11.

If more than three equations are needed for a complete solution, it may be possible to relate the velocity of the body's mass center to the body's angular velocity, using *kinematics*. If these motions appear to be complicated, kinematic (velocity) diagrams may be helpful in obtaining the necessary relation.

In general, a method to be used for the solution of a particular type of problem should be decided upon *before* attempting to solve the problem. As indicated by the terms in Eqs. 19–11, the *principle of impulse and momentum* is most suitable for solving problems which involve *velocity, force,* and *time*. For some problems, however, a combination of the equations of motion, and its two integrated forms, the principle of work and energy and the principle of impulse and momentum, will yield the most direct solution to the problem.

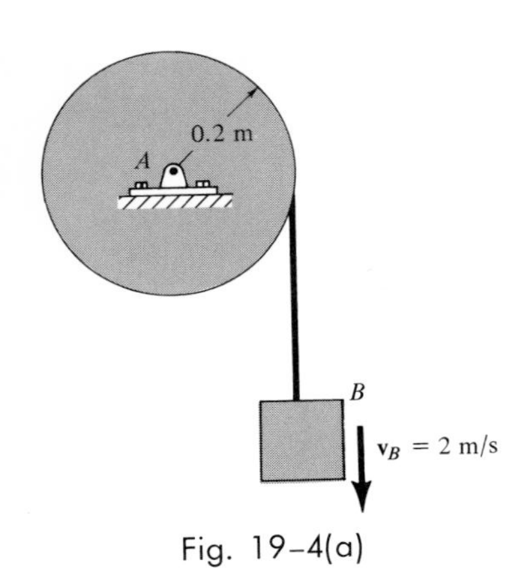

Fig. 19–4(a)

Example 19–1

The 6-kg block, shown in Fig. 19–4*a,* is attached to a cord which is wrapped around the periphery of a 20-kg disk. If the block is initially moving downward with a speed of 2 m/s, determine its speed in 3 s. Neglect the mass of the cord in the calculation.

Solution I

Step 1: The momentum- and impulse-vector diagrams for the block and disk are shown in Fig. 19–4*b*. Since the linear momentum for the block acts downward, the angular momentum of the disk is clockwise. All the impulsive forces shown on the impulse-vector diagram are *constant* throughout the 3-s time interval, since gravity is the cause of motion.

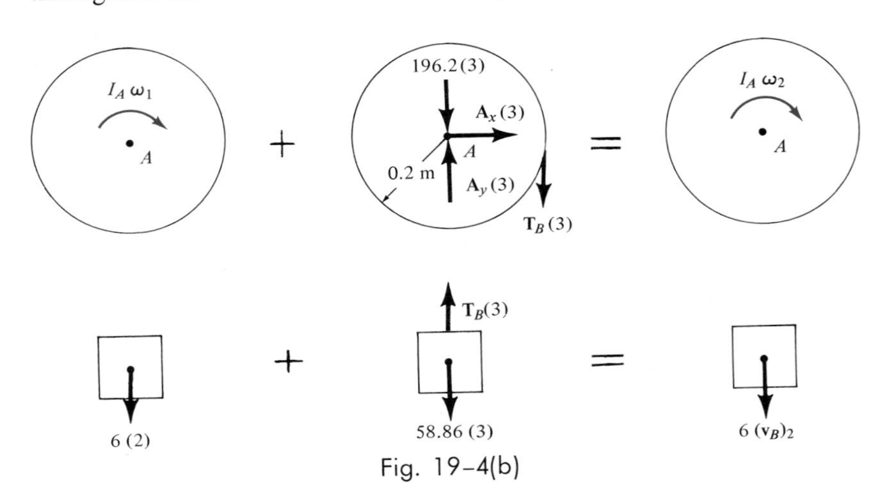

Fig. 19–4(b)

Step 2: The moment of inertia of the disk about its fixed axis of rotation is

$$I_A = \tfrac{1}{2}mr^2 = \tfrac{1}{2}(20)(0.2)^2 = 0.40 \text{ kg} \cdot \text{m}^2$$

At any instant of time, $v_B = \omega r = \omega(0.2)$. Therefore, initially at $t_1 = 0$, $(v_B)_1 = 2$ m/s, $\omega_1 = 10$ rad/s; and at $t_2 = 3$ s, $\omega_2 = (v_B)_2/0.2 = 5(v_B)_2$.

Applying the *principle of linear impulse and momentum* to the block, Fig. 19–4*b,* gives

$$(+\downarrow) \qquad m_B(v_B)_1 + \Sigma \int_{t_1}^{t_2} F_y\, dt = m_B(v_B)_2$$

$$(6)(2) + 58.86(3) - T_B(3) = (6)(v_B)_2 \tag{1}$$

If the *principle of angular impulse and momentum* is applied to the disk about *A*, the unknown reactive impulses at the pin will be eliminated since they create zero "moment" about *A*. Noting that $\omega_1 = 10$ rad/s and $\omega_2 = 5(v_B)_2$, we have

$$(\circlearrowleft +) \qquad I_A\omega_1 + \Sigma \int_{t_1}^{t_2} M_A\, dt = I_A\omega_2$$

$$0.40(10) + T_B(3)(0.2) = 0.40[5(v_B)_2] \qquad (2)$$

Solving Eqs. (1) and (2) simultaneously, yields

$$(v_B)_2 = 13.0 \text{ m/s} \qquad \textit{Ans.}$$

$$T_B = 36.8 \text{ N}$$

Solution II

Step 1: A more *direct solution* to this problem may be obtained by applying the principle of impulse and momentum to the *entire system,* consisting of the block, the cable, and the disk. The cable tension $\mathbf{T}_B$ is thereby eliminated from the analysis, since it acts as an internal force. The momentum- and impulse-vector diagrams for the system are shown in Fig. 19–4*c*.

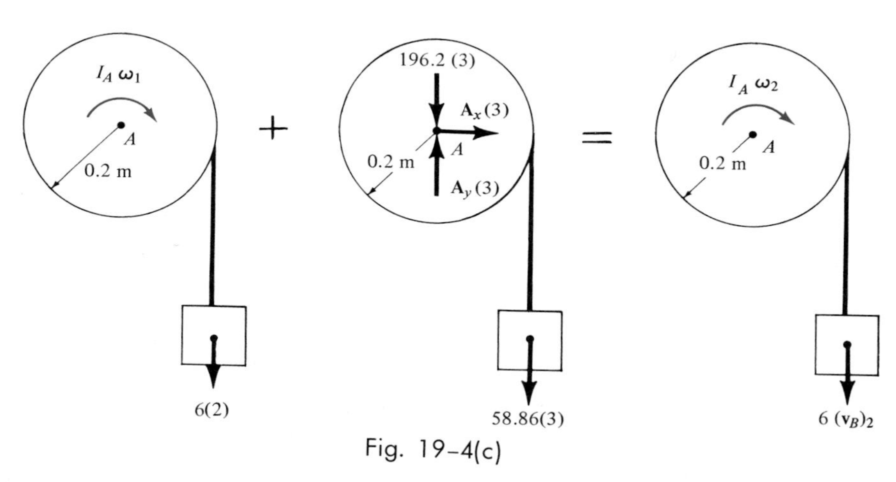

Fig. 19–4(c)

Step 2: When the principle of angular impulse and momentum is applied about point *A*, it is necessary to account for the "moments" of both the linear momentum and linear impulses about this point. From Fig. 19–4*c*, we have

$$\left(\Sigma \begin{matrix}\text{syst. angular}\\ \text{momentum}\end{matrix}\right)_{A1} + \left(\Sigma \begin{matrix}\text{syst. angular}\\ \text{impulse}\end{matrix}\right)_{A(1-2)} = \left(\Sigma \begin{matrix}\text{syst. angular}\\ \text{momentum}\end{matrix}\right)_{A2}$$

$$(\circlearrowleft +) \qquad [m_B(v_B)_1(r) + I_A\omega_1] + [(W_Bt)r] = [m_B(v_B)_2 r + I_A\omega_2]$$

Since $I_A = 0.40 \text{ kg} \cdot \text{m}^2$, $t = 3$ s, $\omega_1 = 10$ rad/s, and $\omega_2 = 5(v_B)_2$, we have

$$(6)(2)(0.2) + 0.40(10) + 58.86(3)(0.2) = (6)(v_B)_2(0.2) + 0.40[5(v_B)_2]$$

Solving this equation for $(v_B)_2$ yields

$$(v_B)_2 = 13.0 \text{ m/s} \qquad \textit{Ans.}$$

Example 19–2

The 100-kg spool shown in Fig. 19–5*a* has a radius of gyration of $k_G = 0.35$ m. A cable is wrapped around the central hub of the spool and a variable horizontal force having a magnitude of $P = (t + 10)$ N is applied, where t is measured in seconds. If the spool is initially at rest, determine its angular velocity in 5 s. Assume that the spool rolls without slipping.

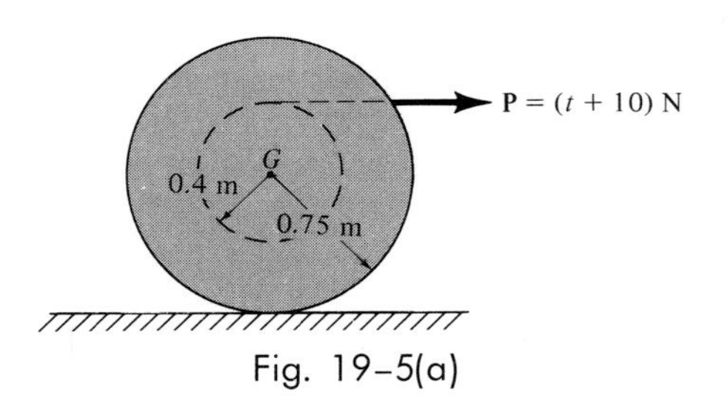

Fig. 19–5(a)

Solution I

Step 1: The momentum- and impulse-vector diagrams for the spool are shown in Fig. 19–5*b*. Note that the impulse created by the frictional force **F** must be represented by an integral, since both **F** and the cable force **P** are *variable*.

Step 2: The moment of inertia of the spool about its mass center is

$$I_G = mk_G^2 = (100)(0.35)^2 = 12.25 \text{ kg} \cdot \text{m}^2$$

Since the spool does not slip, the instantaneous center of zero velocity is at the point of contact with the ground. From the *kinematics* of rotation, the velocity of G can be expressed in terms of the spool's angular velocity $\boldsymbol{\omega}$ as $(v_G)_2 = 0.75\omega_2$, Fig 19–5*b*.

Applying Eqs. 19–11, in reference to Fig. 19–5*b,* we have

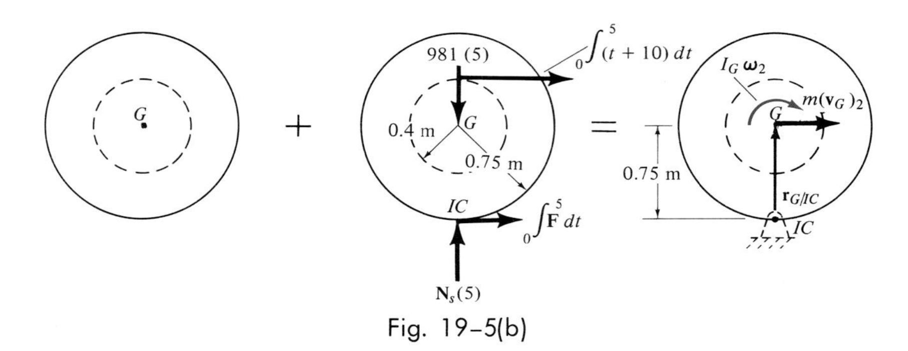

Fig. 19–5(b)

$(\overset{+}{\rightarrow})$

$$m(v_{Gx})_1 + \Sigma \int_{t_1}^{t_2} F_x\, dt = m(v_{Gx})_2$$

$$0 + \int_0^5 (t + 10)\, dt + \int_0^5 F\, dt = 100(0.75\omega_2)$$

$(+\uparrow)$

$$m(v_{Gy})_1 + \Sigma \int_{t_1}^{t_2} F_y\, dt = m(v_{Gy})_2$$

$$0 + N_s(5) - 981(5) = 0$$

$(\circlearrowright +)$

$$I_G\omega_1 + \Sigma \int_{t_1}^{t_2} M_G\, dt = I_G\omega_2$$

$$0 + \int_0^5 (t + 10)(0.4)\, dt - \int_0^5 F(0.75)\, dt = 12.25\omega_2$$

Simplifying these equations gives

$$62.5 + \int_0^5 F\, dt = 75\omega_2 \qquad (1)$$

$$5N_s - 4905 = 0 \qquad (2)$$

$$25 - 0.75 \int_0^5 F\, dt = 12.25\omega_2 \qquad (3)$$

Eliminating the unknown $\int_0^5 F\, dt$ between Eqs. (1) and (3) and solving for ω_2 yields

$$\omega_2 = 1.05 \text{ rad/s} \qquad \textit{Ans.}$$

Although the normal force is not needed for the solution, it can be determined by solving Eq. (2), which gives

$$N_s = 981.0 \text{ N}$$

Solution II

One can eliminate the unknown normal and frictional impulses and thereby obtain a direct solution for ω_2 by computing the angular momentum and angular impulses about the *IC*. Using the data on the momentum- and impulse-vector diagrams, Fig. 19–5*b*, we have

$(\circlearrowright +)$

$$I_G\omega_1 + r_{G/IC}[m_G(v_G)_1] + \Sigma \int_{t_1}^{t_2} M_{IC} dt = I_G\omega_2 + r_{G/IC}[m_G(v_G)_2]$$

$$0 + 0 + \int_0^5 (t + 10)(1.15)\, dt = 12.25\omega_2 + 0.75[100(v_G)_2]$$

Since $(v_G)_2 = 0.75\omega_2$, then

$$(1.15) \int_0^5 (t + 10)\, dt = 68.5\omega_2$$

which yields

$$\omega_2 = 1.05 \text{ rad/s} \qquad \textit{Ans.}$$

Example 19–3

The structure for a ride in an amusement park consists of two symmetrical arms AB and AC which are mounted on a vertical shaft, as shown in Fig. 19–6*a*. The cars at B and C, including their passengers, each have a mass of 125 kg and a radius of gyration about a vertical axis through their mass center G_c of $(k_{Gc})_z = 0.5$ m. Neglecting the mass of the framework, determine the speed of the cars in 10 s starting from rest if a motor supplies a torque of $M = 600(1 - e^{-0.2t})$ N · m about the axis of the shaft.

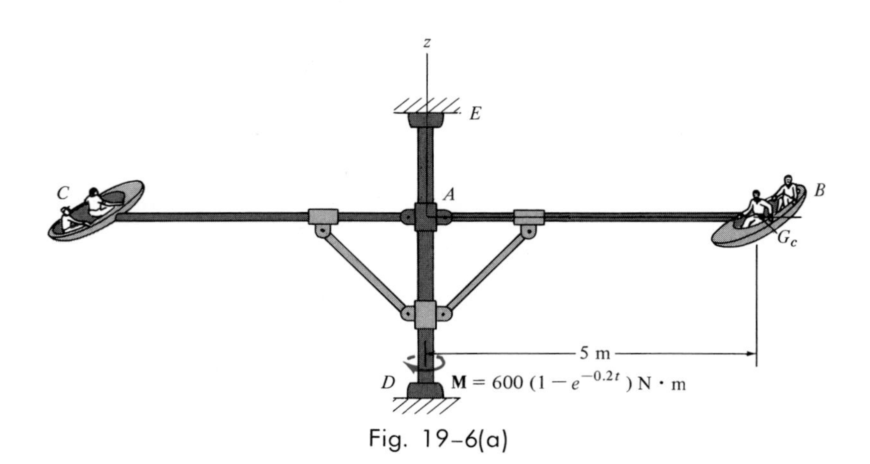

Fig. 19–6(a)

Solution

Step 1: Using a top view, the momentum- and impulse-vector diagrams for the ride are shown in Fig. 19–6*b*. The impulses created by the bearings at D and E have an unknown magnitude and direction, although they do pass through the z axis of the shaft. The cars have angular and linear momentum since they are subjected to rotational motion and their mass centers have a velocity.

Step 2: Computing the moment of inertia of each car, we have

$$(I_{Gc})_z = m_c(k_{Gc})_z^2 = 125(0.5)^2 = 31.3 \text{ kg} \cdot \text{m}^2$$

Applying the principle of angular impulse and momentum about point A (z axis) to eliminate the unknown impulses at the bearings D and E, yields

$$\left(\Sigma \begin{matrix}\text{syst. angular}\\ \text{momentum}\end{matrix}\right)_{A1} + \left(\Sigma \begin{matrix}\text{syst. angular}\\ \text{impulse}\end{matrix}\right)_{A1} = \left(\Sigma \begin{matrix}\text{syst. angular}\\ \text{momentum}\end{matrix}\right)_{A2}$$

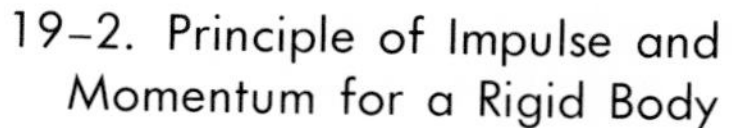

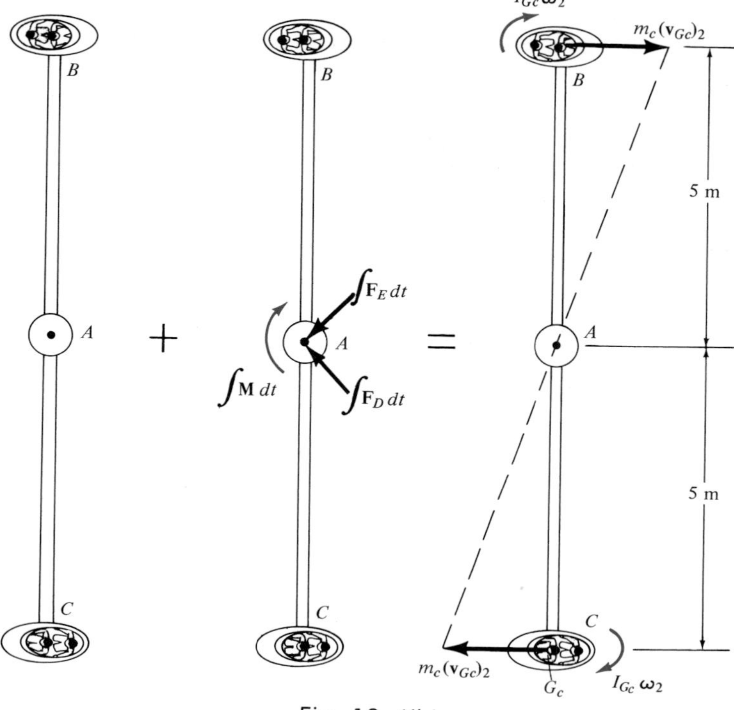

Fig. 19–6(b)

From the kinematics of rotation about a fixed axis, $(v_{Gc})_2 = 5\omega_2$. Thus,

$$(\circlearrowleft +) \quad [0] + \left[\int_0^{10} 600(1 - e^{-0.2t})\, dt \right] = 2(31.3)\omega_2 + 2(125)(5\omega_2)5$$

$$3406.0 = 6312.6\omega_2$$

Solving for ω_2, we have

$$\omega_2 = 0.540 \text{ rad/s}$$

so that

$$(v_{Gc})_2 = 5(0.540) = 2.70 \text{ m/s} \qquad \textit{Ans.}$$

Example 19-4

A monkey having a mass of 2 kg steps off a platform, Fig. 19–7*a*, and begins to climb up a rope with a constant speed of 1.2 m/s, measured relative to the rope. If the rope is wound around a 100-kg spool, determine the time t required for the spool to rotate at $\omega_2 = 2$ rad/s. What are the components of reaction at the pin O at time t? Neglect the mass of the rope in the calculation. The spool has a radius of gyration of $k_O = 0.3$ m.

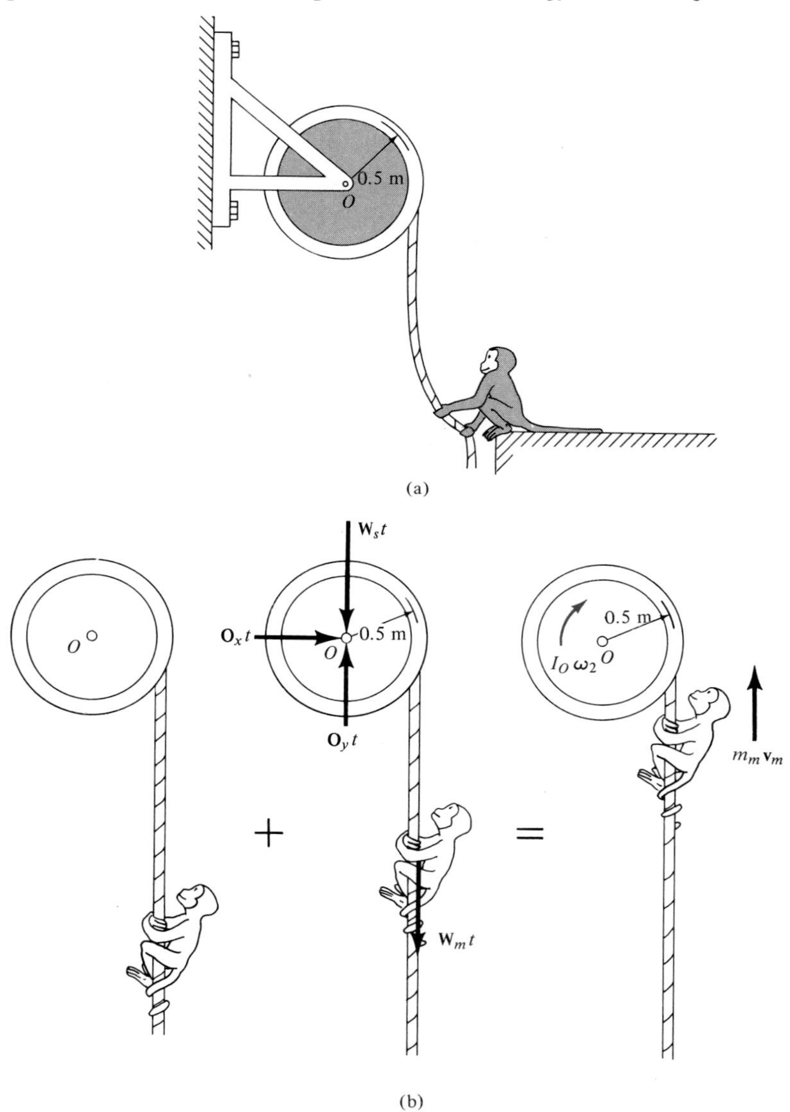

Fig. 19–7

Solution

Step 1: If the monkey and spool are considered as a *single system,* the impulses of the monkey on the rope are *internal* and therefore eliminated from the analysis. The momentum- and impulse-vector diagrams for the monkey-spool system are shown in Fig. 19–7*b*. The reaction at the pin O and the weights of the spool, $\mathbf{W}_s$, and monkey, $\mathbf{W}_m$, create impulses on the system. When the spool has an angular velocity $\boldsymbol{\omega}_2$, the rope is unwinding (downward) at a rate of $v_R = \omega_2 r = 2(0.5) = 1$ m/s. Consequently, if the positive direction is upward, the monkey has a velocity of $v_m = v_R + v_{m/R} = (-1 + 1.2) = 0.2$ m/s (upward), as seen by an observer on the ground. Hence, the final momentum of the monkey is $m_m v_m = 2(0.2) = 0.4$ kg · m/s, Fig. 19–7*b*.

Step 2: The mass moment of inertia of the spool about O is

$$I_O = mk_O^2 = 100(0.3)^2 = 9 \text{ kg} \cdot \text{m}^2$$

The principle of angular impulse and momentum may be applied about point O to eliminate the unknown impulsive reaction components at this point. It is required that

$$\left(\sum \begin{matrix}\text{syst. angular}\\ \text{momentum}\end{matrix}\right)_{O1} + \left(\sum \begin{matrix}\text{syst. angular}\\ \text{impulse}\end{matrix}\right)_{O(1-2)} = \left(\sum \begin{matrix}\text{syst. angular}\\ \text{momentum}\end{matrix}\right)_{O2}$$

$(\curvearrowleft +)$
$$[0] + [W_m t(r)] = [I_O\omega_2 + (m_m v_m)r]$$
$$[0] + [2(9.81)(t)(0.5)] = [9(2) - (0.4)(0.5)]$$
$$t = 1.81 \text{ s} \qquad \textit{Ans.}$$

Using this result, and applying the principle of linear impulse and momentum to the system in reference to Fig. 19–7*b*, we have

$$\left(\sum \begin{matrix}\text{syst. linear}\\ \text{momentum}\end{matrix}\right)_{x1} + \left(\sum \begin{matrix}\text{syst. linear}\\ \text{impulse}\end{matrix}\right)_{x(1-2)} = \left(\sum \begin{matrix}\text{syst. linear}\\ \text{momentum}\end{matrix}\right)_{x2}$$

$(\overset{+}{\rightarrow})$
$$[0] + [O_x(1.81)] = [0]$$
$$O_x = 0 \qquad \textit{Ans.}$$

$$\left(\sum \begin{matrix}\text{syst. linear}\\ \text{momentum}\end{matrix}\right)_{y1} + \left(\sum \begin{matrix}\text{syst. linear}\\ \text{impulse}\end{matrix}\right)_{y(1-2)} = \left(\sum \begin{matrix}\text{syst. linear}\\ \text{momentum}\end{matrix}\right)_{y2}$$

$(+\uparrow)$
$$[0] + [-W_s t + O_y t - W_m t] = [m_m v_m]$$
$$[0] + [-100(9.81)(1.81) + O_y(1.81) - 2(9.81)(1.81)] = [0.4]$$
$$O_y = 1000.8 \text{ N} \qquad \textit{Ans.}$$

Problems

Except when stated otherwise, throughout this chapter, **assume** that the coefficients of static and kinetic friction **are** equal, i.e., $\mu = \mu_s = \mu_k$.

19–1. Gear A is pinned at B and rotates along the periphery of the gear rack R. If A has a mass of 4 kg and a radius of gyration of $k_B = 80$ mm, determine the angular momentum of gear A about point C when $\omega_{CB} = 30$ rad/s and (a) $\omega_R = 0$, (b) $\omega_R = 20$ rad/s.

Prob. 19–1

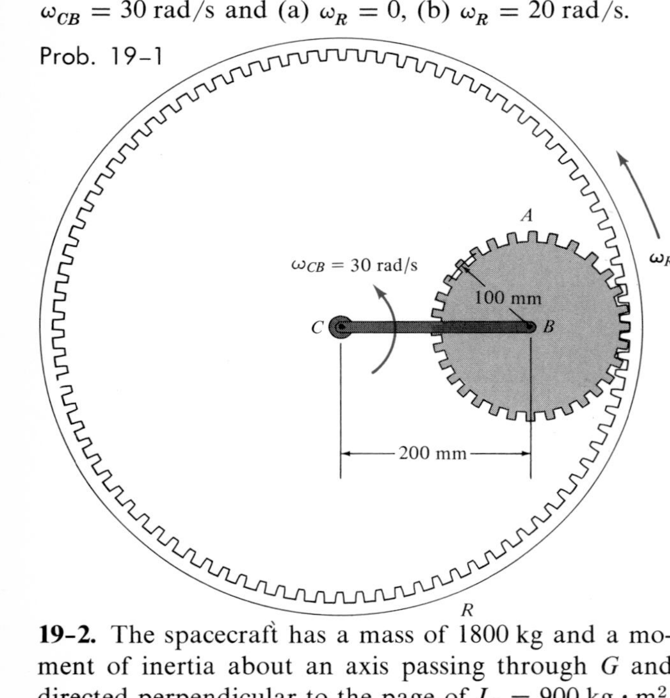

19–2. The spacecraft has a mass of 1800 kg and a moment of inertia about an axis passing through G and directed perpendicular to the page of $I_G = 900$ kg · m². If it is traveling forward with a speed of $v_G = 1500$ m/s and executes a turn by means of two jets, which provide a constant thrust of 600 N for 0.3 s, determine the spacecraft's angular velocity just after the jets are turned off.

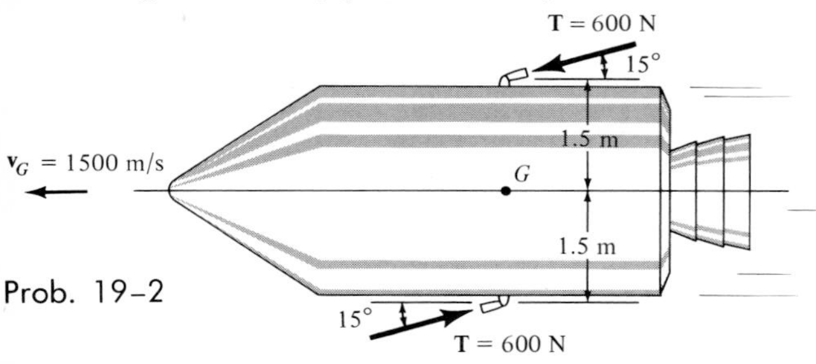

Prob. 19–2

19–3. The device shown is used to test the effectiveness of a tire to resist skidding. The coefficient of friction between the 15-kg tire and the surface of the conveyor belt at C is $\mu = 0.3$. If the belt is moving with a constant speed of 12 m/s when the tire, having zero rotation, is placed in direct contact with it, determine the approximate length of skid mark the tire makes on the belt during the time the tire is slipping. What is the force in link AB while slipping occurs? Neglect the weight of the link. The radius of gyration of the tire is $k_A = 0.115$ m.

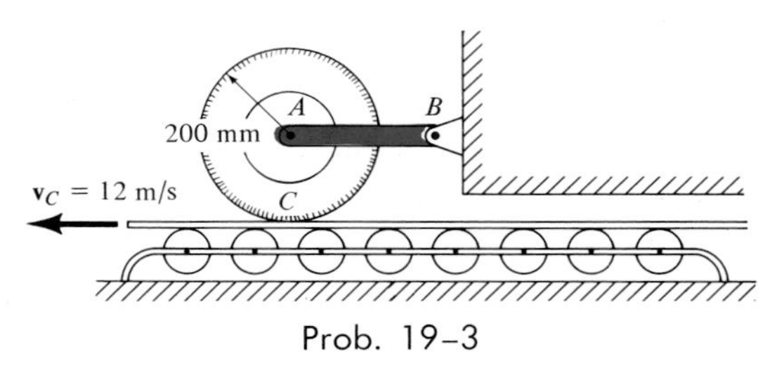

Prob. 19–3

*** 19–4.** The 50-kg cylinder has an angular velocity of 30 rad/s when it is brought into contact with the horizontal surface at C. If the coefficient of friction is $\mu = 0.2$, determine how long it takes for the cylinder to stop spinning. What force is developed at the pin A during this time? The axis of the cylinder is connected to *two* symmetrical links. (Only AB is shown.) For the computation, neglect the weight of the links.

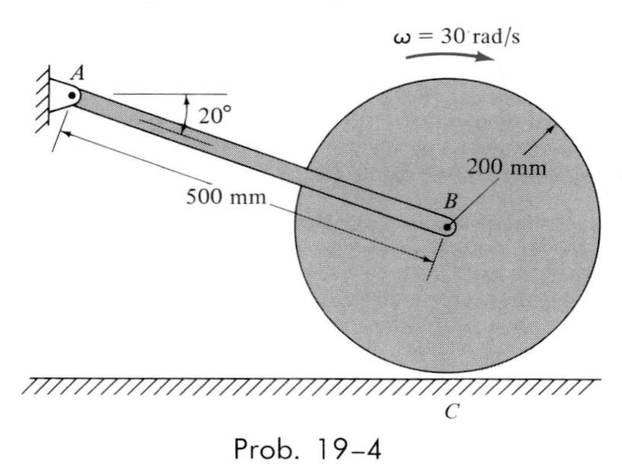

Prob. 19–4

19-5. The wheel has a radius of $r = 300$ mm, a mass of $m = 9$ kg, and a radius of gyration of $k_O = 225$ mm. If it is released from rest and rolls down the plane without slipping, determine the speed of its center O in $t = 3$ s.

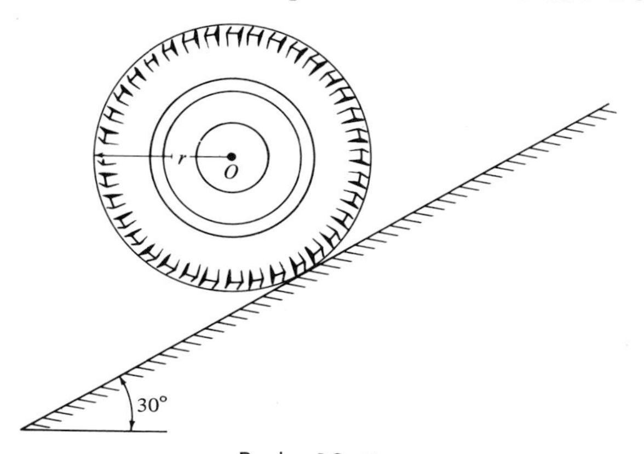

Prob. 19-5

19-5a. Solve Prob. 19-5 if $r = 1.5$ ft, the weight of the wheel is $W = 23$ lb, $k_O = 0.8$ ft, and $t = 2$ s.

19-6. Show that if a slab is rotating about a fixed axis perpendicular to the slab and passing through its mass center G, the angular momentum is the same when computed about any other point P on the slab.

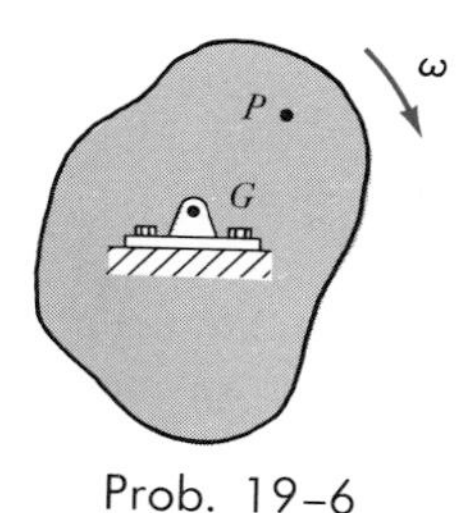

Prob. 19-6

19-7. At a given instant, the body has a linear momentum of $\mathbf{L} = m\mathbf{v}_G$ and an angular momentum, computed about its mass center, of $\mathbf{H}_G = I_G\boldsymbol{\omega}$. Show that the angular momentum of the body, computed about the instantaneous center of zero velocity, IC, can be expressed as $\mathbf{H}_{IC} = I_{IC}\boldsymbol{\omega}$, where I_{IC} represents the body's moment of inertia computed about the instantaneous axis of zero velocity. As shown, the IC is located at a distance of $r_{G/IC}$ away from the mass center G.

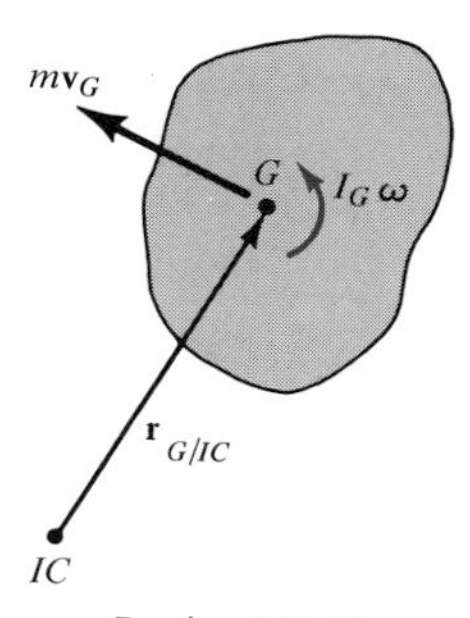

Prob. 19-7

* **19-8.** The rigid body (slab) has a mass m and is rotating with an angular velocity $\boldsymbol{\omega}$ about an axis passing through the fixed point O. Show that the momentum of all the particles composing the body can be represented by a single vector, having a magnitude of mv_G and acting through point P, called the *center of percussion*, which lies at a distance $r_{P/G} = k_G^2/r_{G/O}$ from the mass center G. Here k_G is the radius of gyration of the body, computed about an axis perpendicular to the plane of motion and passing through G.

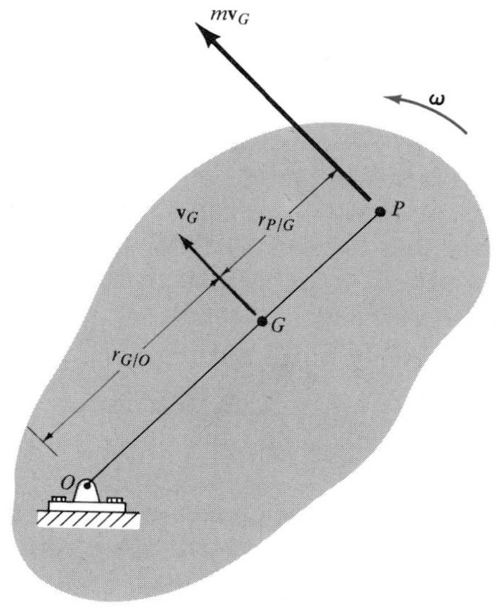

Prob. 19-8

19-9. A 5-kg disk A is mounted on arm BC, which has a negligible mass. If a constant torque or twist of $M = (3e^{0.5t})$ N · m, where t is measured in seconds, is applied to the arm at C, determine the angular velocity of BC in 1.5 s starting from rest. Solve the problem assuming that (a) the disk is set in a smooth bearing at B so that it rotates with curvilinear translation, (b) the disk is fixed to the shaft BC, and (c) the disk is given an initial freely spinning angular velocity of $\boldsymbol{\omega}_D = \{-10\mathbf{k}\}$ rad/s prior to application of the torque.

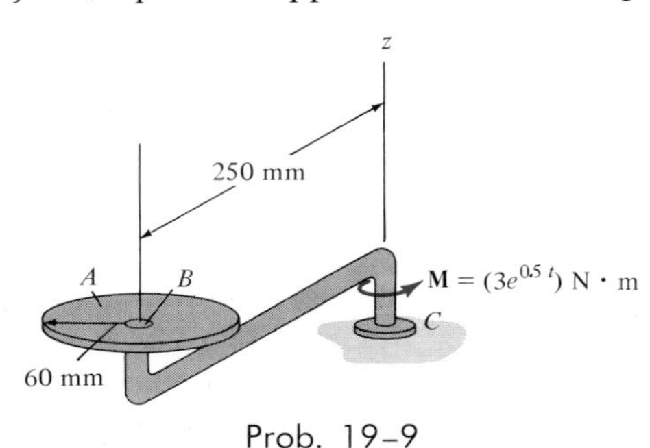

Prob. 19-9

19-10. The spool has a mass of $m = 20$ kg and a radius of gyration of $k_O = 0.3$ m. A cord is wrapped around its inner hub and is subjected to a variable force of $P = (20t + 50)$ N, where t is measured in seconds. If the spool rolls without slipping at A, determine the speed of the center O, $t = 4$ s after the force is applied. The spool is initially at rest. Set $r_o = 0.45$ m and $r_i = 0.2$ m.

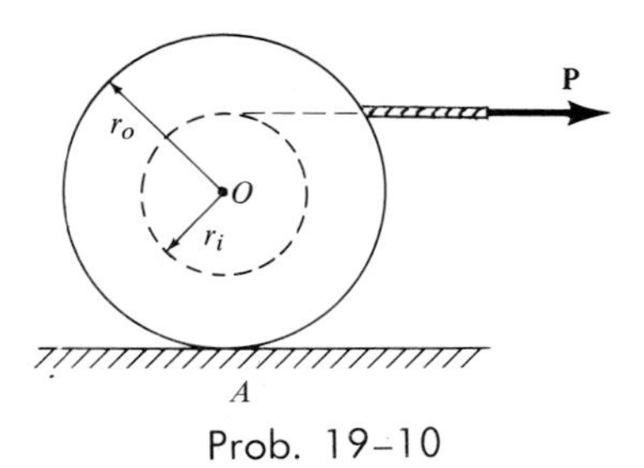

Prob. 19-10

19-10a. Solve Prob. 19-10 if the spool has a weight of $W = 50$ lb; $k_O = 1.25$ ft; $P = (10t + 4)$ lb, where t is measured in seconds; $t = 2$ s, $r_o = 1.75$ ft, and $r_i = 1$ ft.

19-11. If the ball has a mass of 1.5 kg and is thrown onto a *rough surface* with a velocity of 750 mm/s parallel to the surface, determine the amount of backspin, $\boldsymbol{\omega}$, it must be given so that it stops spinning at the same instant that its forward velocity is zero. It is not necessary to know the coefficient of friction at A for the calculation.

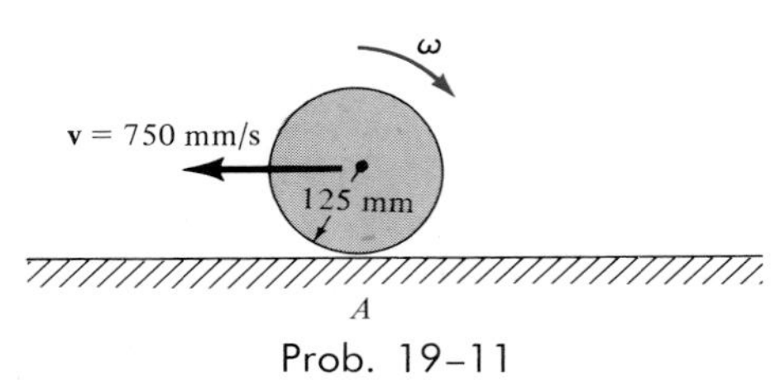

Prob. 19-11

*** 19-12.** The 50-kg drum, having a radius of gyration of $k_O = 222$ mm, rolls along an inclined plane for which the coefficient of friction is $\mu = 0.2$. If the drum is released from rest, determine the maximum angle θ for the incline so that it rolls without slipping at A.

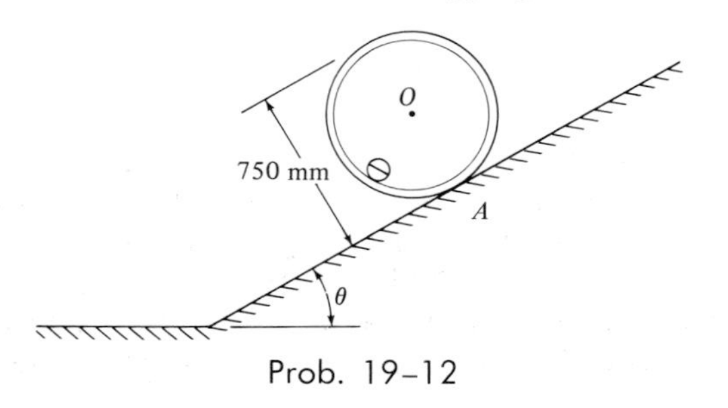

Prob. 19-12

19–13. The double pulley consists of two wheels which are attached to one another and turn at the same rate. The pulley has a mass of 15 kg and a radius of gyration of $k_O = 110$ mm. If the block at A has a mass of 40 kg and the container at B has a mass of 85 kg, including its contents, determine the speed of the container in 3 s after it is released from rest.

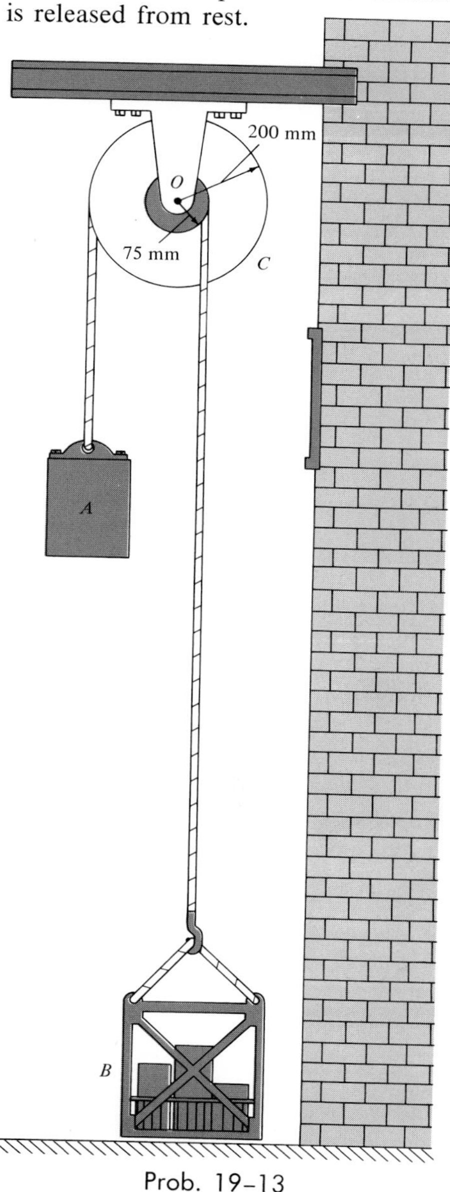

Prob. 19–13

19–14. A constant torque or twist of $M = 0.05$ N · m is applied to the center gear A. If the system starts from rest, determine the angular velocity of each of the three (equal) smaller gears in 2 s. The smaller gears (B) are pinned at their centers, and the mass and centroidal radii of gyration of the gears are given in the figure.

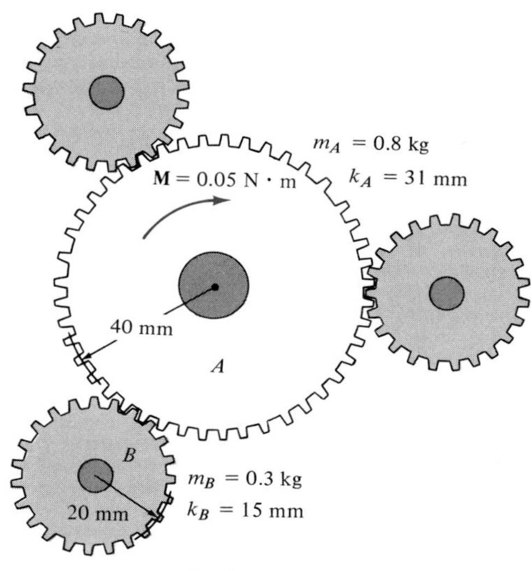

Prob. 19–14

19–15. Gear A has a mass of $m_A = 50$ g, a radius of $r = 30$ mm, and a radius of gyration of $k_O = 18$ mm. The coefficient of friction between the gear rack B and the horizontal surface is $\mu = 0.3$. If the rack has a mass of $m_B = 20$ g, and it is initially sliding to the left with a velocity of $(v_B)_1 = 80$ mm/s, determine the constant moment **M** which must be applied to the gear to increase the motion of the rack so that in $t = 2.5$ s it will have a velocity of $(v_B)_2 = 120$ mm/s to the left. Neglect friction between the rack and the gear and assume that the gear exerts *only* a horizontal force on the rack.

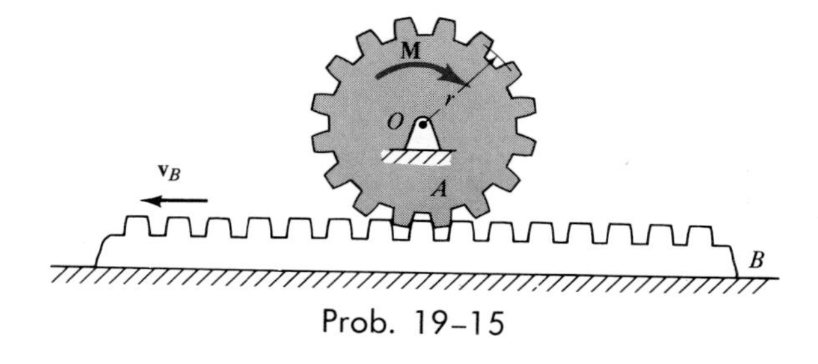

Prob. 19–15

19–15a. Solve Prob. 19–15 if the gear has a weight of $W_A = 1.3$ lb, and $r = 0.2$ ft, $k_O = 0.13$ ft, $\mu = 0.2$; and if the rack has a weight of $W_B = 0.7$ lb, and $(v_B)_1 = 0.6$ ft/s, $t = 1.6$ s, $(v_B)_2 = 1.9$ ft/s.

***19–16.** The 15-kg disk has an angular velocity of $\omega = 30$ rad/s. If the brake ABC is applied such that the magnitude of force **P** varies with time as shown, determine the time needed to stop the disk. The coefficient of friction at B is $\mu = 0.4$.

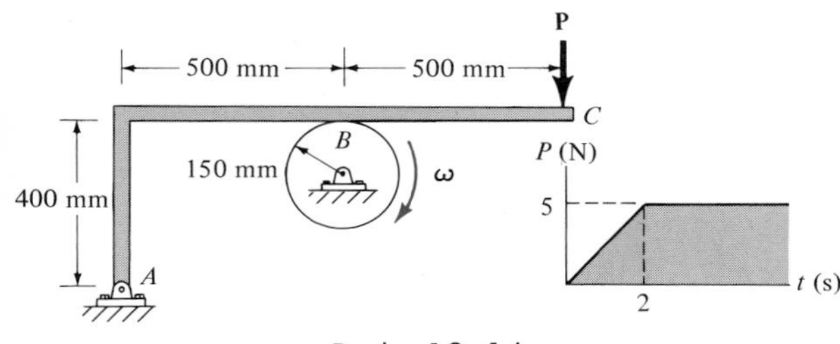

Prob. 19–16

19–17. The flywheel A has a mass of 40 kg and a radius of gyration of $k_C = 110$ mm. Disk B has a mass of 15 kg, is pinned at D, and is coupled to the flywheel using a belt which is subjected to a tension such that it does not slip at its contacting surfaces. If a motor supplies a counterclockwise torque or twist to the flywheel having a magnitude of $M = (5t)$ N · m, where t is measured in seconds, determine the angular velocity of the disk 4 s after the motor is turned on. Initially, the flywheel is at rest.

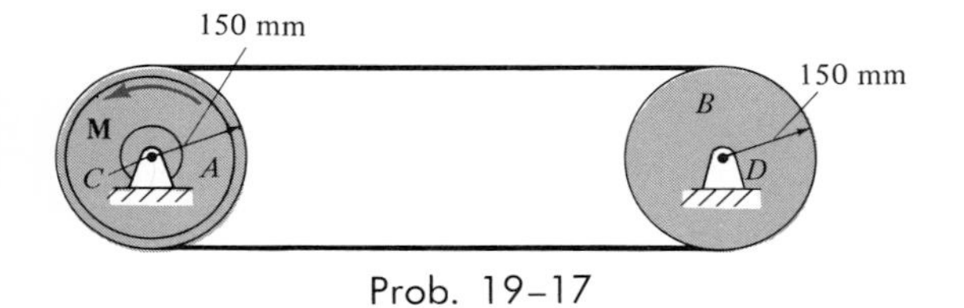

Prob. 19–17

19–18. The 20-kg flywheel A has a radius of gyration of $k_C = 200$ mm. Disk B has a mass of 35 kg and is coupled to the flywheel using a belt which is subjected to a tension, such that it does not slip at its contacting surfaces. If a motor supplies a clockwise torque or twist to the flywheel having a magnitude of $M = 4(1 - e^{-t})$ N · m, where t is in seconds, determine the angular velocity of the disk in $t = 2$ s starting from rest. Neglect friction at the pins C and D.

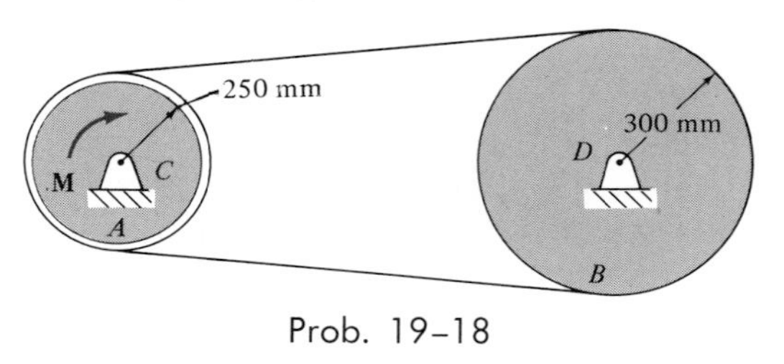

Prob. 19–18

19–19. The drum A has a mass of 120 kg and a centroidal radius of gyration of $k_O = 210$ mm. If a torque or twist of $M = (50t)$ N · m, where t is in seconds, is applied to the drum, determine the speed of the 450-kg mine car C in 10 s. The car is originally resting along the stop s and the cables are loose. Neglect friction and the mass of both the pulley B and the car wheels for the calculation. *Hint:* First determine the time needed to set the car in motion.

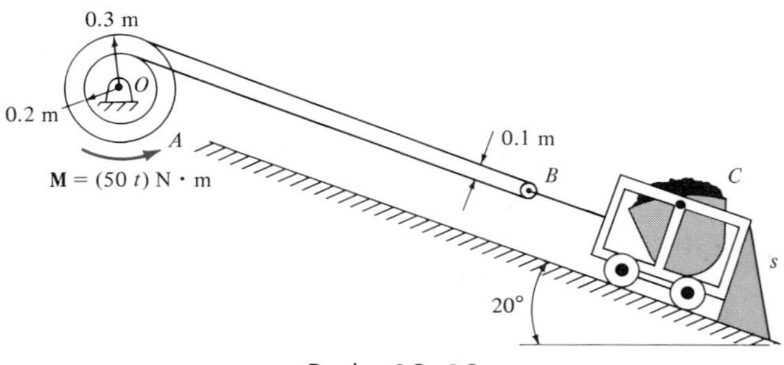

Prob. 19–19

19–3. Conservation of Momentum

Conservation of Linear Momentum. If the sum of all the *linear impulses* acting on a system of connected rigid bodies is *zero,* the linear momentum of the system is constant or conserved. Consequently, the first two of Eqs. 19–12 reduce to the form

$$\left(\sum \begin{matrix}\text{syst. linear}\\ \text{momentum}\end{matrix}\right)_1 = \left(\sum \begin{matrix}\text{syst. linear}\\ \text{momentum}\end{matrix}\right)_2 \qquad (19\text{–}13)$$

This equation is referred to as the *conservation of linear momentum.*

Without inducing appreciable errors in the computations, it may be possible to apply Eq. 19–13 in a specified direction, for which the linear impulses are small or *nonimpulsive*. Specifically, nonimpulsive forces occur when small forces act over very short periods of time. For example, the impulse created by the force of a tennis racket hitting a ball, during a very short time interval Δt, is large, whereas the impulse of the weight of the ball during this time is small by comparison and may therefore be neglected in the motion analysis of the ball during Δt.

Conservation of Angular Momentum. The angular momentum of a system of connected rigid bodies is conserved about the system's center of mass, G, or a point O lying on an axis of rotation of the system, when the sum of all the angular impulses created by the external forces acting on the system is zero or appreciably small (nonimpulsive) when computed about these points. The third of Eqs. 19–12 then becomes

$$\left(\sum \begin{matrix}\text{syst. angular}\\ \text{momentum}\end{matrix}\right)_{O1} = \left(\sum \begin{matrix}\text{syst. angular}\\ \text{momentum}\end{matrix}\right)_{O2} \qquad (19\text{–}14)$$

This equation is referred to as the *conservation of angular momentum.* In the case of a single rigid body, Eq. 19–14 applied to point G becomes $(I_G\omega)_1 = (I_G\omega)_2$. To illustrate an application of this equation, consider a swimmer who executes a somersault after jumping off a diving board. By tucking his arms and legs in close to his chest, he *decreases* his body's moment of inertia and thus *increases* his angular velocity. If he straightens out just before entering the water, his body's moment of inertia is *increased* and his angular velocity *decreases* ($I_G\omega$ must be constant). Since the weight of his body creates a linear impulse during the time of motion, this example also illustrates that the angular momentum of a body is conserved and yet the linear momentum is *not*. Such cases occur whenever the external forces creating the linear impulse pass through either the center of mass of the body or a fixed axis of rotation.

PROCEDURE FOR ANALYSIS

Generally, the equations defining the conservation of linear or angular momentum are used to determine the final velocity of a body *just after* the time period considered. Furthermore, by applying these equations to a *system* of bodies, the internal impulses acting within the system, which may be unknown, are eliminated from the analysis, since they occur in equal but opposite collinear pairs. For applications, the following two-step procedure should be used.

Step 1: Draw the momentum- and impulse-vector diagrams for the body or system of bodies. From the impulse-vector diagram, it is possible to classify each of the applied forces as being either "impulsive" or "nonimpulsive." In general, for short times, "nonimpulsive forces" consist of the weight of a body, the force of a slightly deformed spring, or any force that is *known to be small* when compared to the impulsive force.

By inspection of the impulse-vector diagram, it will be possible to tell if the conservation of linear or angular momentum applies. Specifically, the *conservation of linear momentum* applies in a given direction when *no* external impulsive forces act on the body or system in that direction; whereas the *conservation of angular momentum* applies about a fixed point O or at the mass center G of a body or system of bodies when all the external impulsive forces acting on the body or system create zero moment (or zero angular impulse) about O or G.

Step 2: Apply the conservation of linear or angular momentum in the appropriate directions. Most often, the scalar component equations can be formulated by resolving the vector components or summing "moments" *directly* from the momentum- and impulse-vector diagrams. Use *kinematics* if further equations are necessary for the solution of a problem. If the motion appears to be complicated, kinematic (velocity) diagrams may be helpful in obtaining the necessary kinematic relations.

In general, if it is necessary to determine an *internal impulsive force* acting on only one body of a system, the body must be *isolated* and the principle of linear or angular impulse and momentum must be applied *to the body*. After the impulse $\int F\,dt$ is calculated, then, provided the time Δt for which the impulse acts is known, the *average impulsive force* F_{avg} can be determined from $F_{avg} = \int F\,dt/\Delta t$.

The following examples numerically illustrate application of the above procedure.

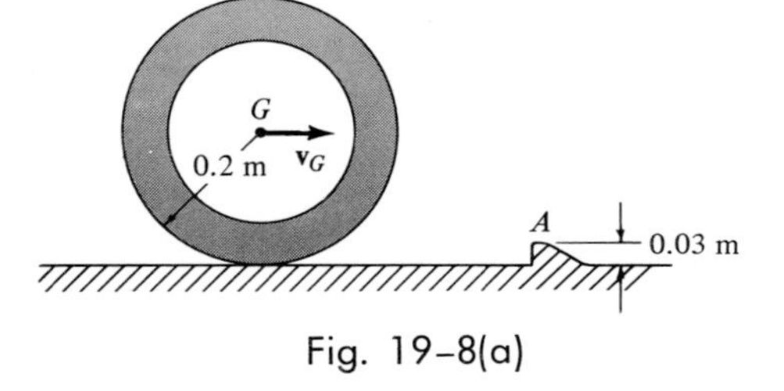

Fig. 19–8(a)

Example 19–5

The 10-kg wheel shown in Fig. 19–8*a* has a centroidal radius of gyration of $k_G = 0.125$ m. Assuming that the wheel does not slip or rebound, determine the minimum velocity $\mathbf{v}_G$ it must have to just roll over the obstruction at A.

Solution

Step 1: Since no slipping or rebounding occurs, the wheel essentially *pivots* about point A when contact occurs. The momentum- and impulse-vector diagrams are shown in Fig. 19–8*b*. These diagrams indicate, respectively, the momentum of the wheel *just before impact,* the impulses given to the wheel *during impact,* and the momentum of the wheel *just after impact*. Note that only two impulses (forces) act on the wheel. (The normal force created by the horizontal plane is zero. Why?) By comparison, the impulse at A is much greater than that caused by the weight ($W = 10(9.81) = 98.1$ N), and since the time of impact is very short, the weight can be considered nonimpulsive. The impulsive force $\mathbf{F}$ at A has both an unknown magnitude and an unknown direction θ. To eliminate this force from the analysis, during impact *only* the nonimpulsive weight force creates an *angular impulse about point A.* Furthermore, since $(98.1\Delta t)d \approx 0$, angular momentum about A is essentially *conserved.*

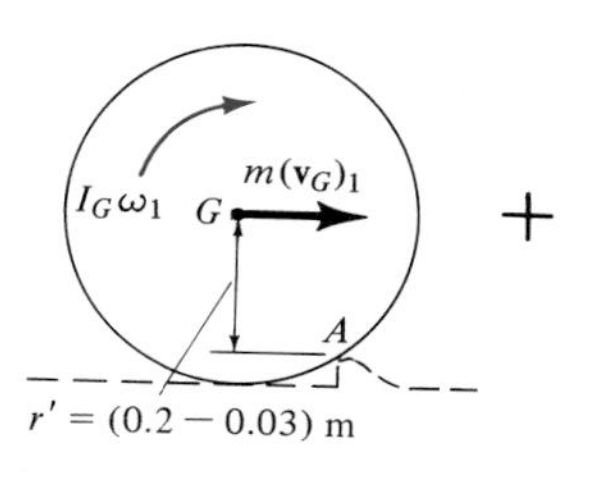

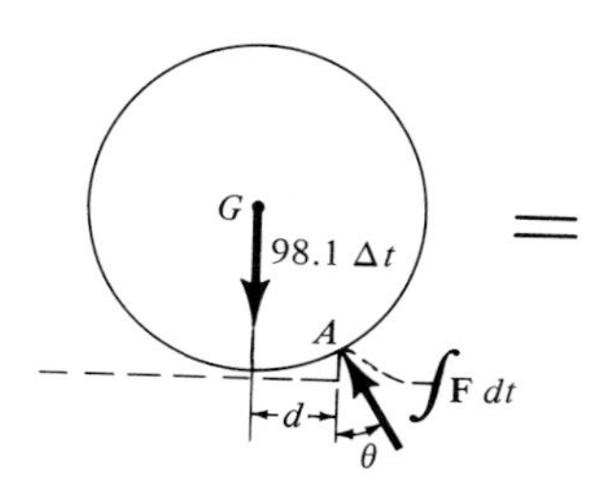

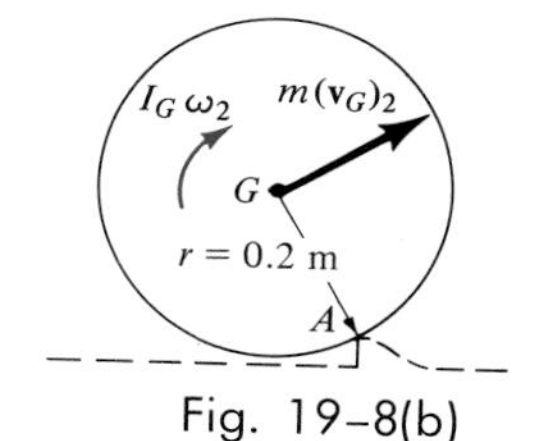

Fig. 19–8(b)

Step 2: The moment of inertia of the wheel about G is determined from the radius of gyration and the mass,

$$I_G = mk_G^2 = (10)(0.125)^2 = 0.156 \text{ kg} \cdot \text{m}^2$$

The angular momentum when computed about A must account for both the moment of the linear momentum, $m\mathbf{v}_G$, about A *and* the angular momentum $\mathbf{H}_G = I_G\boldsymbol{\omega}$, which is due to the rotational motion of the wheel about G. Hence, applying the conservation of angular momentum about A in reference to Fig. 19–8*b* yields

$$(\mathbf{H}_A)_1 = (\mathbf{H}_A)_2$$
$$r'm(v_G)_1 + I_G\omega_1 = rm(v_G)_2 + I_G\omega_2$$
$$(0.2 - 0.03)(10)(v_G)_1 + (0.156)(\omega_1) = (0.2)(10)(v_G)_2 + (0.156)(\omega_2)$$

From *kinematics,* since no slipping occurs the velocity of the mass center $\mathbf{v}_G$ is related to the angular velocity $\boldsymbol{\omega}$ by the equation $\omega = v_G/r = v_G/0.2 = 5v_G$. Substituting this into the above equation and simplifying yields

$$(v_G)_2 = 0.892(v_G)_1 \tag{1}$$

In order to roll over the obstruction, the wheel must pass the dashed position 3 shown in Fig. 19–8*c*. Hence, if $(v_G)_2$ (or $(v_G)_1$) is to be a minimum, it is necessary that the kinetic energy of the wheel at position 2 be equal to the potential energy at position 3. Constructing the datum through the center of mass, as shown in the figure, and applying the conservation-of-energy theorem, we have

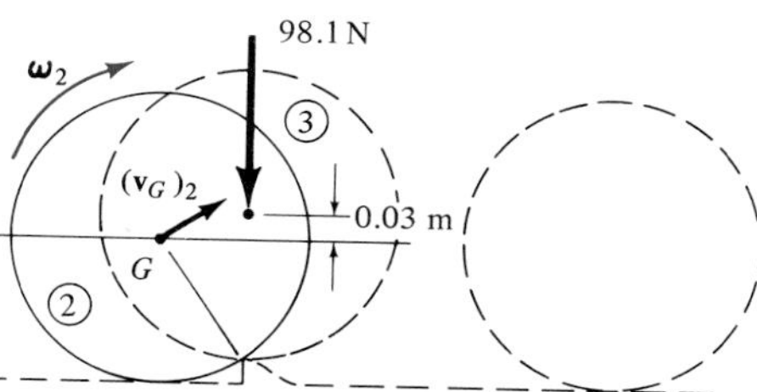

Fig. 19–8(c)

$$\{T_2\} + \{V_2\} = \{T_3\} + \{V_3\}$$
$$\{\tfrac{1}{2}(10)(v_G)_2^2 + \tfrac{1}{2}(0.156)(\omega_2)^2\} + \{0\} = \{0\} + \{(98.1)(0.03)\}$$

Substituting $\omega_2 = 5(v_G)_2$ and Eq. (1) into this equation, and simplifying, yields

$$3.98(v_G)_1^2 + 1.55(v_G)_1^2 = 2.94$$

Thus,

$$(v_G)_1 = 0.729 \text{ m/s} \qquad \textit{Ans.}$$

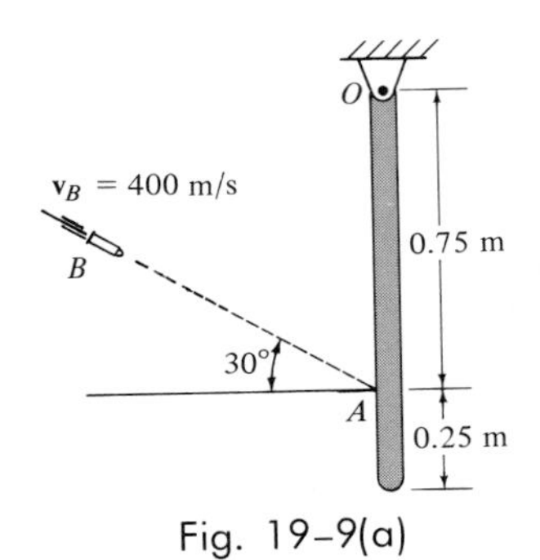

Fig. 19–9(a)

Example 19–6

The 5-kg slender rod shown in Fig. 19–9*a* is pinned at O and is initially at rest. If a 4-g bullet is fired into the rod with a velocity of 400 m/s, as shown in the figure, determine the angle θ_{max} through which the rod swings. After impact the bullet becomes embedded in the rod.

Solution

Step 1: The impulse which the bullet exerts on the rod can be eliminated from the analysis and the angular velocity of the rod just after impact can be determined by considering the bullet and rod as a single system. The momentum- and impulse-vector diagrams for this system are shown in Fig. 19–9*b*. The momentum diagrams are drawn *just before and just after impact*. During impact, the bullet and rod exchange equal but opposite *internal impulses* at A. As shown on the impulse-vector diagram, the impulses that are external to the system are due to the reactions at O and the weights of the bullet and rod. Since Δt is very short, the "moments" of these impulses about point O are essentially zero and therefore angular momentum is conserved about this point.

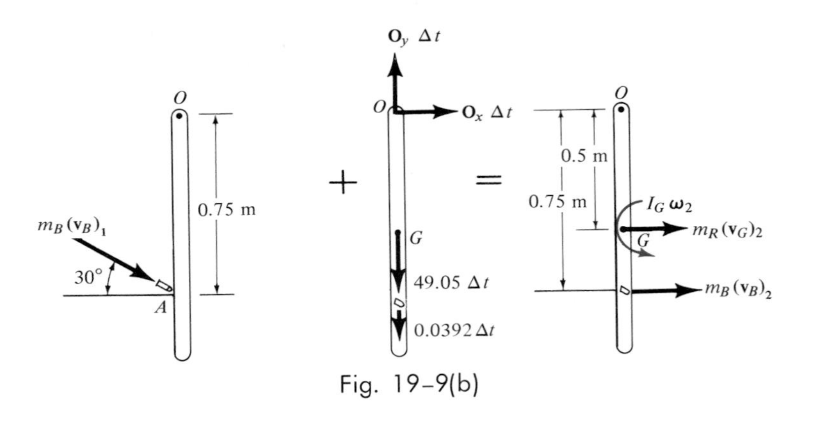

Fig. 19–9(b)

Step 2: From Fig. 19–9*b*, we have

$$(\circlearrowright +) \qquad \Sigma(H_O)_1 = \Sigma(H_O)_2$$

$$m_B(v_B)_1 \cos 30°(0.75\text{ m}) = m_B(v_B)_2(0.75\text{ m}) + m_R(v_G)_2(0.5\text{ m}) + I_G\omega_2$$

$$(0.004)(400 \cos 30°)(0.75) = (0.004)(v_B)_2(0.75) + (5)(v_G)_2(0.5) + [\tfrac{1}{12}(5)(1)^2]\omega_2$$

or

$$1.039 = 0.003(v_B)_2 + 2.50(v_G)_2 + 0.417\omega_2 \qquad (1)$$

Since the rod is pinned at O, using kinematics the magnitudes of $(\mathbf{v}_G)_2$, $(\mathbf{v}_B)_2$, and $\boldsymbol{\omega}_2$ may be related to one another, Fig. 19–9c. Hence,

$$(v_G)_2 = 0.5\omega_2 \quad (2)$$
$$(v_B)_2 = 0.75\omega_2 \quad (3)$$

Solving Eqs. (1) through (3) simultaneously yields

$$\omega_2 = 0.622 \text{ rad/s}$$
$$(v_G)_2 = 0.311 \text{ m/s}$$
$$(v_B)_2 = 0.467 \text{ m/s}$$

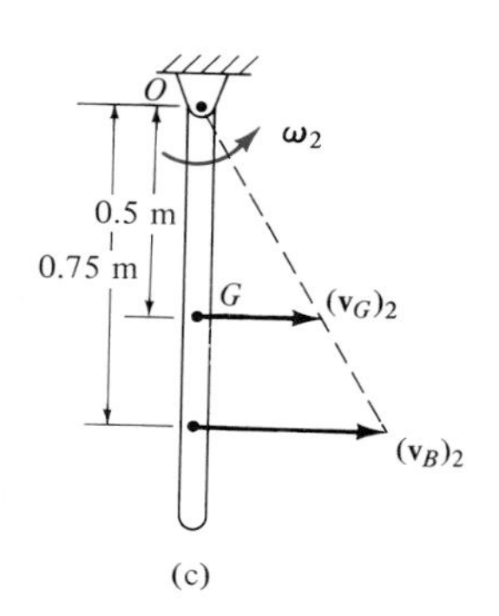

The angle θ_{max} through which the rod swings may be computed by applying the conservation-of-energy theorem to the rod just after the impact. The datum is established through the initial position of the rod's center of mass, Fig. 19–9d. It is required that the initial kinetic energy of the rod and bullet be converted entirely into potential energy. Thus,

$$\{T_2\} + \{V_2\} = \{T_3\} + \{V_3\}$$
$$\{\tfrac{1}{2}m_R(v_G)_2^2 + \tfrac{1}{2}I_G\omega_2^2 + \tfrac{1}{2}m_B(v_B)_2^2\} + \{0\} = \{0\} + \{W_R y_R + W_B y_B\}$$
$$\{\tfrac{1}{2}(5)(0.311)^2 + \tfrac{1}{2}[\tfrac{1}{12}(5)(1)^2](0.622)^2 + \tfrac{1}{2}(0.004)(0.467)^2\} + \{0\}$$
$$= \{0\} + \{5(9.81)(0.5)(1 - \cos\theta_{max}) + (0.004)(9.81)(0.75)(1 - \cos\theta_{max})\}$$

Solving for θ_{max},

$$\cos\theta_{max} = 0.987$$
$$\theta_{max} = 9.25° \qquad \textit{Ans.}$$

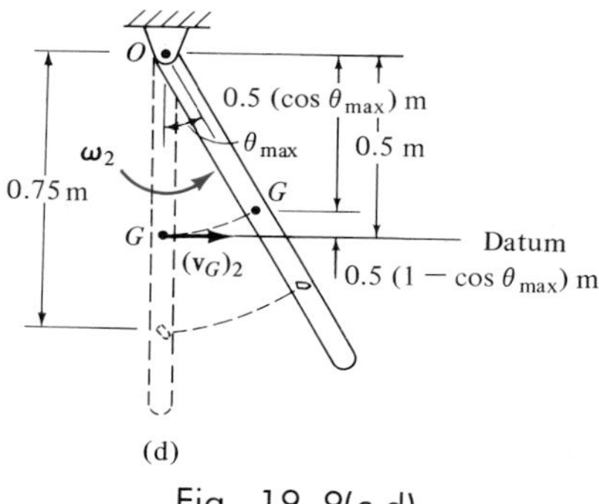

Fig. 19–9(c,d)

Problems

*** 19–20.** Disk A has a mass of $m_A = 5$ kg and a radius of $r_A = 125$ mm. It is fixed to a vertical shaft and is freely rotating with it at $\omega_1 = 11$ rad/s. Disk B has a mass of $m_B = 3$ kg, a radius of $r_B = 75$ mm, and it is not rotating. Neglecting the mass of the axle, determine the angular velocity of both disks after B slides freely down the shaft, comes in contact with A, and slipping between the disks has stopped. What average frictional moment does B exert on A if slipping between the two disks occurs for $t = 0.8$ s?

*** 19–20a.** Solve Prob. 19–20 if disks A and B weigh $W_A = 12$ lb and $W_B = 8$ lb, respectively; $r_A = 0.5$ ft; $r_B = 0.35$ ft; $\omega_1 = 6$ rad/s; and $t = 0.65$ s.

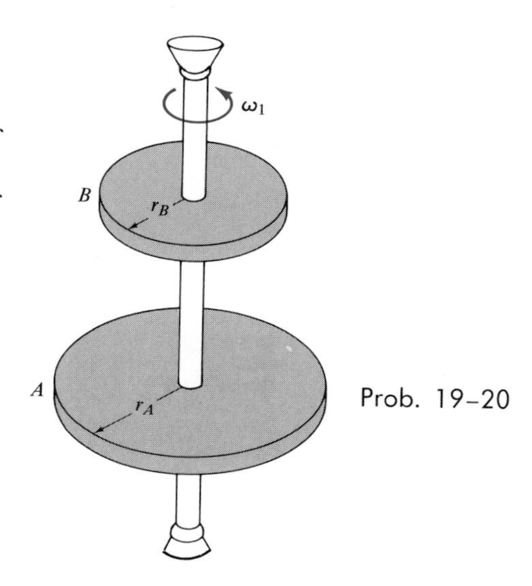

Prob. 19–20

19-21. The space satellite has a mass of 93 kg and a moment of inertia $I_z = 0.830 \text{ kg} \cdot \text{m}^2$, excluding the four solar panels A, B, C, and D. Each solar panel has a mass of 15 kg and can be approximated as a thin plate. If the satellite is originally spinning about the z axis at a constant rate of $\omega_z = 0.5$ rad/s when $\theta = 90°$, determine the rate of spin if all the panels are raised and reach the upward position, $\theta = 0°$, at the same instant.

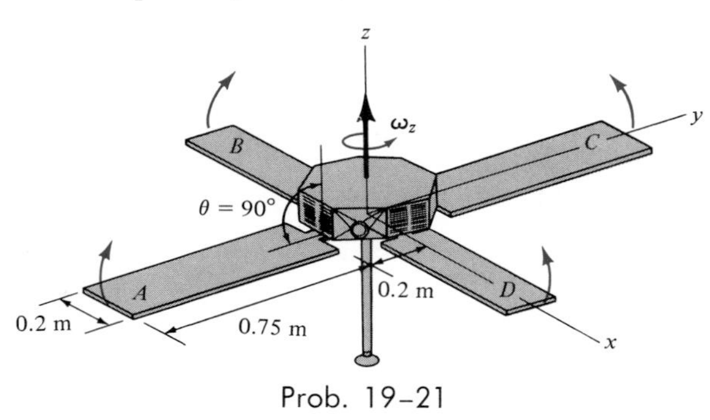

Prob. 19–21

19-22. A horizontal circular platform has a mass of 125 kg and a radius of gyration about the z axis, passing through its center O, of $k_O = 3.3$ m. The platform is free to rotate about the z axis and is initially at rest. A man, having a mass of 70 kg, begins to run along the edge in a circular path of radius 4 m. If he has a speed of 1.5 m/s, and maintains this speed relative to the platform, compute the angular velocity of the platform.

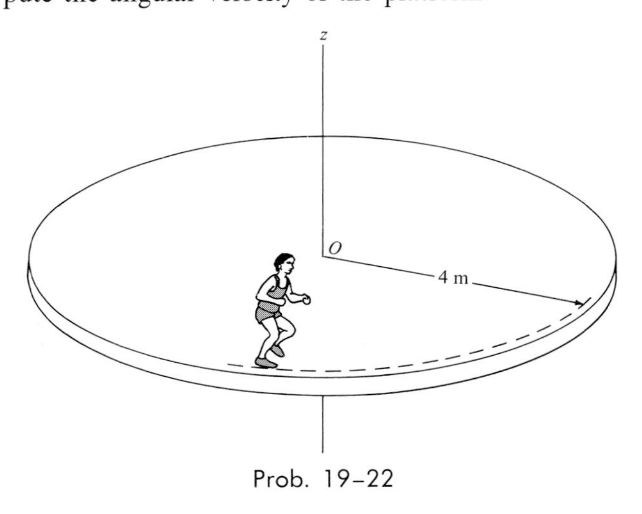

Prob. 19–22

19-23. Two hamsters are placed on the surface of a platform that has a mass of 3 kg. The platform has a radius of gyration of $k_O = 430$ mm about the vertical axis and is free to rotate in the horizontal plane. The hamsters each have a mass of 200 g and can run at a speed of 300 mm/s relative to the platform. Determine the angular velocity of the platform if they run in circular paths in opposite directions, as shown.

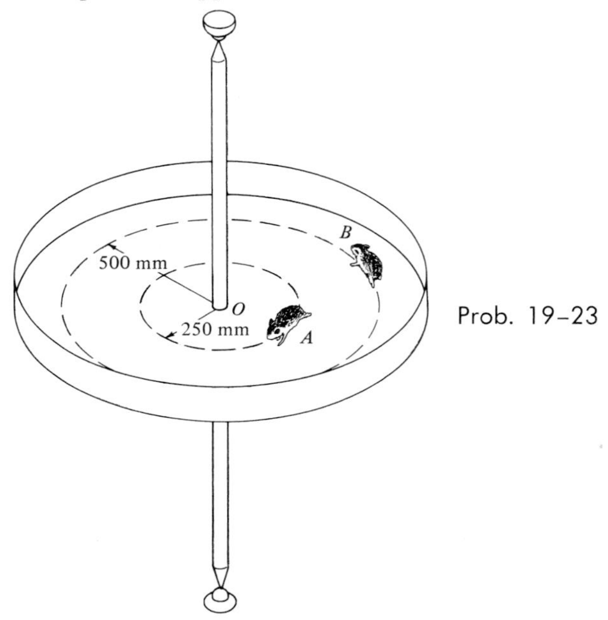

Prob. 19–23

* **19-24.** Solve Prob. 19–23 assuming that both hamsters run in the same direction along their paths.

19-25. The square plate, where $a = 200$ mm, has a mass of $m = 4$ kg and is rotating on the smooth surface with a constant angular velocity of $\omega_1 = 15$ rad/s. Determine the new angular velocity of the plate just after its corner strikes the peg P and the plate starts to rotate about P without rebounding.

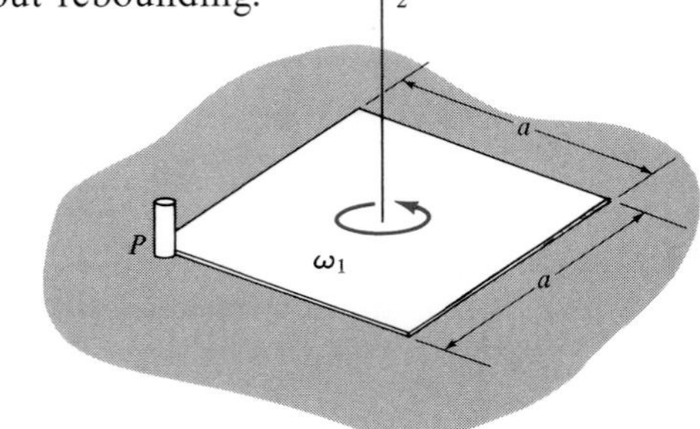

Prob. 19–25

19–25a. Solve Prob. 19–25 if $a = 0.75$ ft, the plate weighs $W = 6$ lb, and $\omega_1 = 2$ rad/s.

19–26. Two acrobats, each having a mass of 70 kg, ride in the cages of a space wheel. Each 40-kg cage can rotate about a horizontal axis passing through its center (A or B) so that the acrobats remain upright by running within the cage as the entire frame rotates about its horizontal axis at O. If the frame has a mass of 190 kg, *excluding* the cages, and the radii of gyration of the cages and frame are $k_A = k_B = 0.9$ m and $k_O = 1.83$ m, respectively, determine the angular velocity ω_O of the frame if the acrobats run so that the floor of the cage has a relative speed of 2 m/s. For the calculation, assume that the center of mass of each acrobat coincides with the axis of rotation (A or B) of each cage.

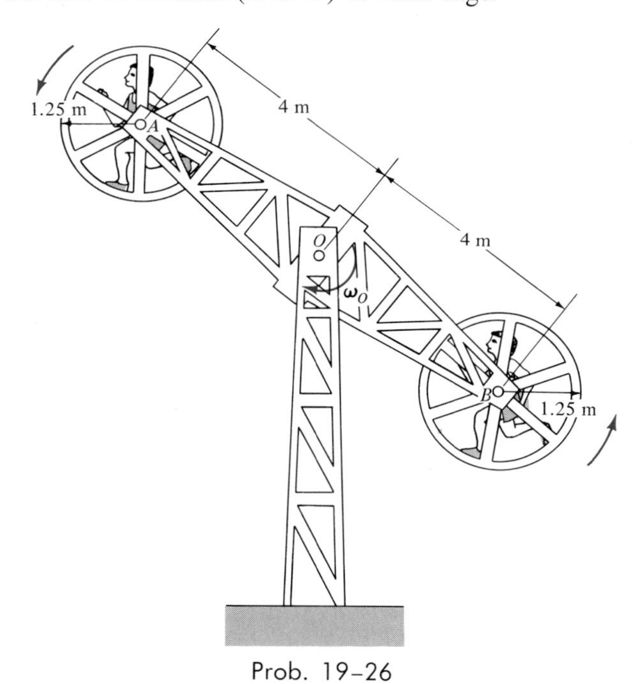

Prob. 19–26

19–27. For safety reasons, the 20-kg supporting leg of a sign is designed to break away with negligible resistance at B when the leg is subjected to the impact of a car. Assuming that the leg is pin-supported at A, and can be approximated by a thin rod, determine the impulse the car bumper exerts on it, if after the impact the leg appears to rotate upward to an angle $\theta_{max} = 135°$.

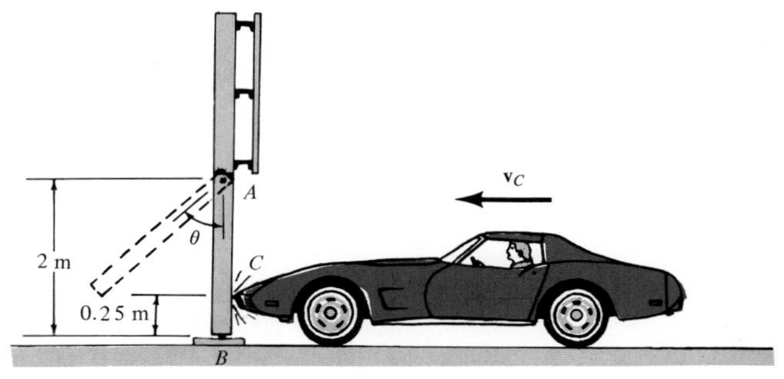

Prob. 19–27

***19–28.** The uniform rod AB has a mass of 3 kg and is released from rest without rotating from the position shown. As it falls, the end A strikes a hook S, which provides a permanent connection. Determine the speed at which the other end B strikes the wall at C.

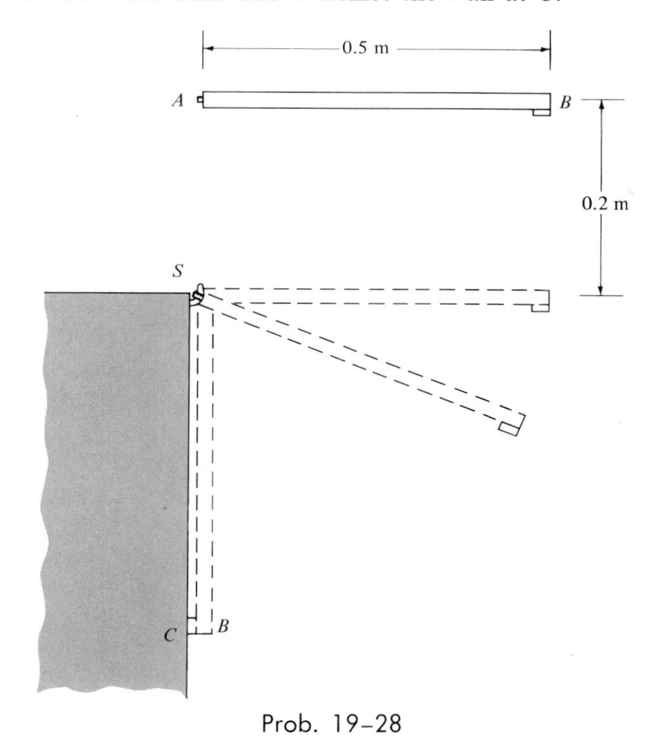

Prob. 19–28

19–29. The plank has a mass of 30 kg, center of mass at G, and it rests on the two sawhorses at A and B. If the end D is raised 200 mm above the top of the sawhorses and is released from rest, determine how high end C will rise from the top of the sawhorses after the plank falls so that it rotates clockwise about A, strikes and pivots on the sawhorse at B, and rotates clockwise off the sawhorse at A.

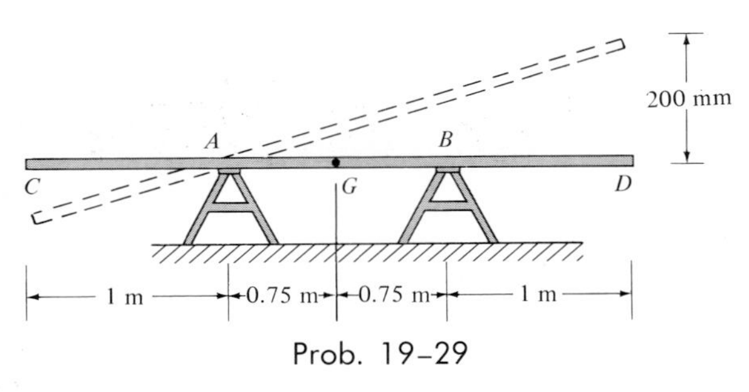

Prob. 19–29

19–30. Determine the height h at which a bullet can strike the disk that has a mass of $m = 1.5$ kg and cause it to roll without slipping at A. Set $r = 50$ mm. For the calculation this requires that the frictional force at A be essentially zero.

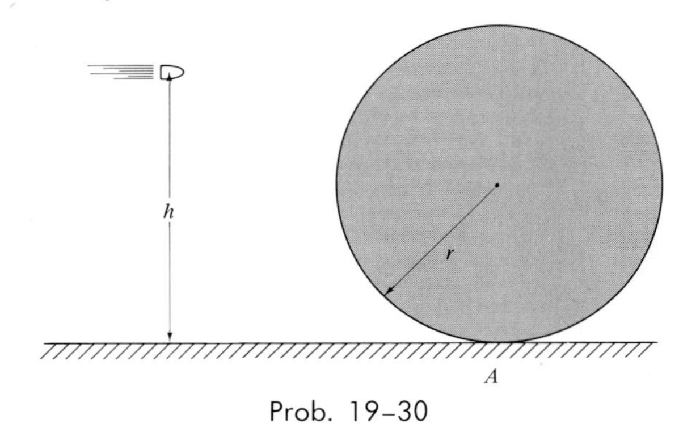

Prob. 19–30

19–30a. Solve Prob. 19–30 if the disk weighs $W = 14$ lb, and $r = 0.25$ ft.

19–31. A thin rod having a mass m and length l is balanced vertically as shown. Determine the height h at which it can be struck with a horizontal force **F** and not slip at the floor. For the solution this requires that the frictional force at A be essentially zero.

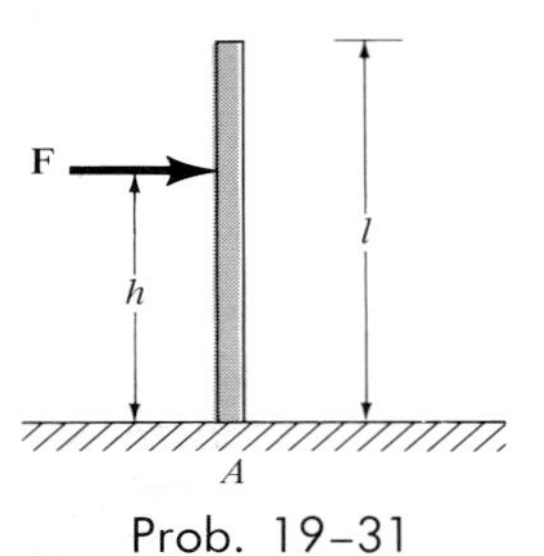

Prob. 19–31

* **19–32.** Determine the height h at which a billiard ball of mass m must be struck so that no frictional force develops between it and the table at A. Assume that the cue C only exerts a horizontal force **P** on the ball.

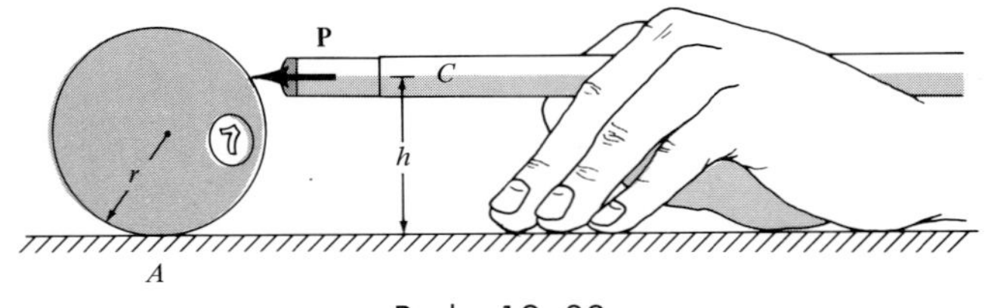

Prob. 19–32

19–33. *Eccentric impact* of two bodies occurs when the line connecting the mass centers of the bodies *does not* coincide with the line of impact. This situation is shown in the figure, where just before impact at C, body B is rotating with an angular velocity of $\boldsymbol{\omega}_1$, such that $(v_B)_1 = \omega_1 r$; and the velocity of the contact point on body A is $(\mathbf{u}_A)_1$, such that the component of velocity along the line of impact is $(\mathbf{v}_A)_1$, where $(v_A)_1 > (v_B)_1$. Just after collision, B has an angular velocity of $\boldsymbol{\omega}_2$, such that $(v_B)_2 = \omega_2 r$; and the contact point at A has a velocity of $(\mathbf{u}_A)_2$, with a component of velocity along the line of impact of $(\mathbf{v}_A)_2$. Assuming that body B has a moment of inertia I_O at O and mass m_B, and body A has a mass m_A, apply the conservation of angular momentum for *both* bodies about point O and the principle of angular impulse and momentum to *each* body about O, and show that the above velocities are related by $e = [(v_B)_2 - (v_A)_2]/[(v_A)_1 - (v_B)_1]$, where e is the coefficient of restitution between the bodies as defined by Eq. 15–17. *Suggestion:* Follow the procedure outlined in Sec. 15–4, which was used to derive Eq. 15–18.

Prob. 19–33

Just before collision

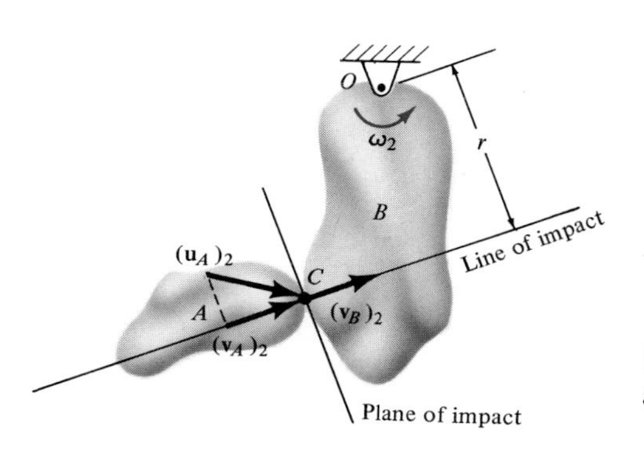

Just afer collision

19–34. The disk has a mass of 20 kg. If it is released from rest when $\theta = 30°$, determine the maximum angle θ of rebound after it collides with the wall. The coefficient of restitution between the disk and wall is $e = 0.75$. When $\theta = 0°$, the disk hangs such that it just touches the wall. Neglect friction at the pin C. *Hint:* Use the equation for e defined in Prob. 19–33.

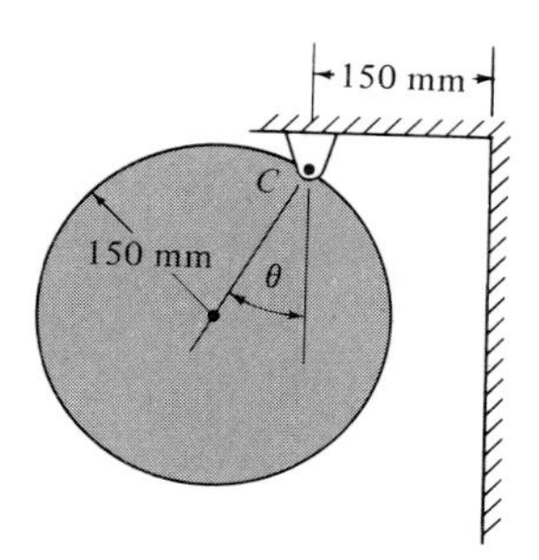

Prob. 19–34

19–35. The ball has a mass of $m = 6$ kg and rolls without slipping along a horizontal plane with a velocity of $v_O = 100$ mm/s. Provided it does not slip or rebound, determine the velocity of its center O as it just starts to roll up the inclined plane. Set $r = 75$ mm.

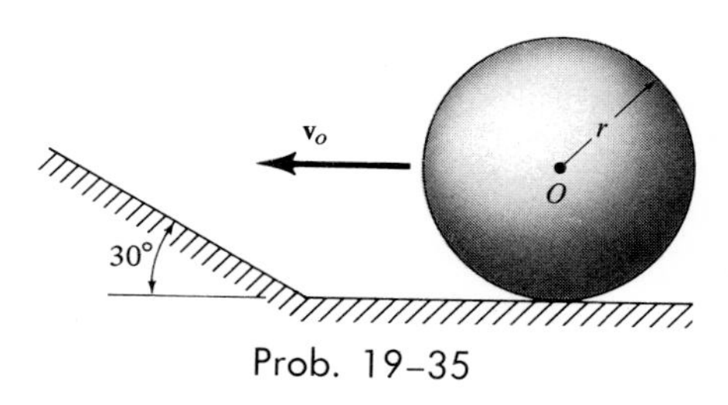

Prob. 19–35

19–35a. Solve Prob. 19–35 if the ball weighs $W = 3$ lb, and $v_O = 0.5$ ft/s, $r = 0.12$ ft.

*** 19-36.** The uniform pole has a mass of 20 kg and falls from rest when $\theta = 90°$, until it strikes the edge at A, $\theta = 60°$. If the pole then begins to pivot about this point after contact, determine the pole's angular velocity just after the impact. Assume that the pole does not slip at B as it falls until it strikes A.

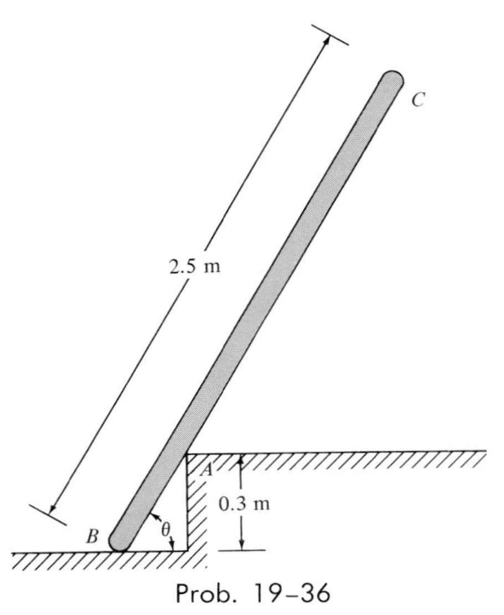

Prob. 19–36

19-37. A ball having a mass of 7 kg and initial speed of $v_1 = 2$ m/s rolls over a 30-mm-long depression. Assuming that the ball does not rebound off the edges of contact, first A, then B, determine its final velocity when it reaches the other side.

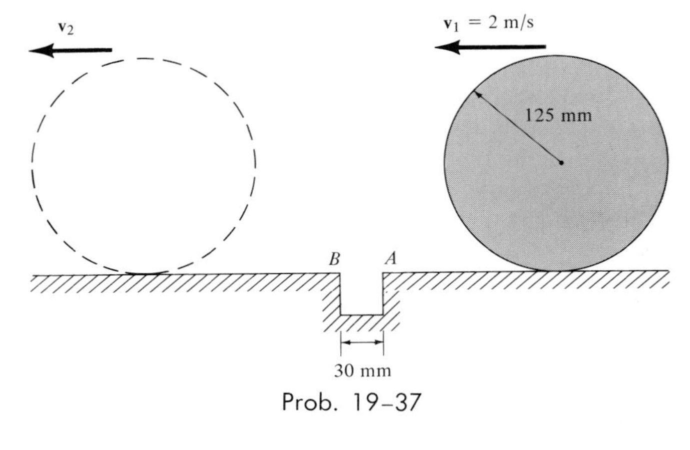

Prob. 19–37

20 Spatial Kinematics of a Rigid Body

20–1. Introduction

Three types of planar rigid-body motion have been discussed in Chapter 16: translation, rotation about a fixed axis, and general plane motion. In this chapter the spatial motion of a rigid body, which consists of rotation about a fixed point and general motion, will be discussed. This is followed by a more general study of spatial motion of particles and rigid bodies using coordinate systems which both translate and rotate. The analysis of these motions is more complex than planar motion analysis, since the *direction* of the body's angular velocity $\boldsymbol{\omega}$ is *not* constant. Hence, the body's angular acceleration $\boldsymbol{\alpha}$ will measure both a change in the *magnitude* and *direction* of $\boldsymbol{\omega}$. In order to simplify the motion's three-dimensional aspects, throughout the chapter we will make complete use of vector analysis.*

20–2. Rotation of a Rigid Body About a Fixed Point

When a rigid body rotates about a fixed point, the distance r from the point to a particle P located in the body is the *same* for *any position* of the body. Thus, the path of motion for the particle lies on the *surface of a sphere* having a radius r and centered at the fixed point. Since motion along this path occurs only from a series of rotations made during a finite time interval, it is perhaps wise to first develop a familiarity with some of the properties of rotational displacements.

*A brief review of vector analysis is given in Appendix B.

Euler's Theorem. This theorem states that two "component" rotations about different axes passing through a point are equivalent to a single resultant rotation about an axis passing through the point. If more than two rotations are applied, they can be combined into pairs, and each pair further reduced to combine into one rotation.

Finite Rotations. If the component rotations used in Euler's theorem are *finite,* it is important that the *order* in which they are applied be maintained. This is because finite rotations do *not* obey the law of vector addition, and hence they cannot be classified as vector quantities. To show this, consider the two finite rotations $\boldsymbol{\theta}_1 + \boldsymbol{\theta}_2$, applied to the block in Fig. 20–1*a*. Each rotation has a magnitude of 90° and a direction defined by the right-hand rule, as indicated by the arrowhead. The resultant orientation of the block is shown at the right. When these two rotations are applied in the order $\boldsymbol{\theta}_2 + \boldsymbol{\theta}_1$, as shown in Fig. 20–1*b*, the resultant position of the block is *not* the same as it is in Fig. 20–1*a*. Consequently, *finite rotations* do not obey the commutative law of addition ($\boldsymbol{\theta}_1 + \boldsymbol{\theta}_2 \neq \boldsymbol{\theta}_2 + \boldsymbol{\theta}_1$), and therefore *they cannot be classified as vectors.* If smaller, yet finite, rotations had been used to illustrate this point, e.g., 10° instead of 90°, the *resultant* orientation of the block, after each combination of rotations, would also be different; however, in this case, only by a small amount.

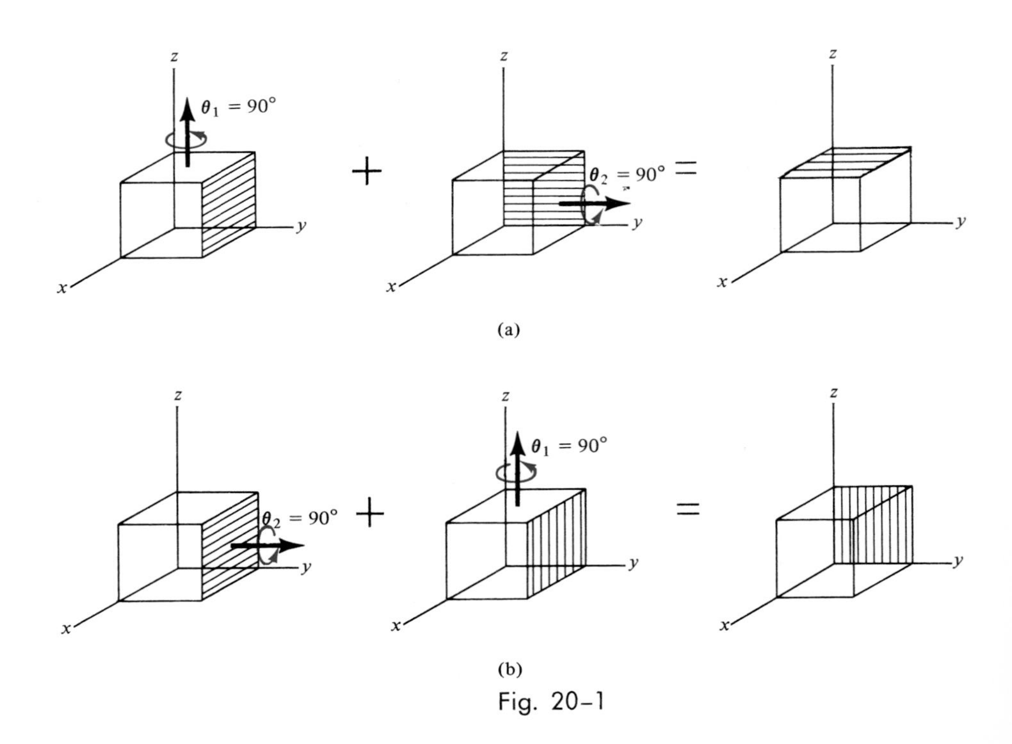

Fig. 20–1

Infinitesimal Rotations. When defining the angular motions of a body subjected to spatial motion, only rotations which are *infinitesimally small* will be considered. *Such rotations may be classified as vectors, since they can be added vectorially in any manner.* To prove this, let us, for purposes of simplicity, consider the rigid body itself to be a sphere which is allowed to rotate about its central fixed point O, Fig. 20–2a. Two infinitesimal rotations, $d\boldsymbol{\theta}_1$ and $d\boldsymbol{\theta}_2$, will be imposed on the body and the motion of particle P will be studied. If the initial position of P is defined by the vector $\mathbf{r}$, Fig. 20–20a, then after the first rotation $d\boldsymbol{\theta}_1$, P is displaced by an amount $r\,d\boldsymbol{\theta}_1$, which can be expressed by the cross product $d\boldsymbol{\theta}_1 \times \mathbf{r}$. By vector addition, the new position of P is then

$$\mathbf{r}_1 = \mathbf{r} + d\boldsymbol{\theta}_1 \times \mathbf{r}$$

The second rotation, $d\boldsymbol{\theta}_2$, gives a displacement $d\boldsymbol{\theta}_2 \times \mathbf{r}_1$, so that the final position of P is

$$\begin{aligned}\mathbf{r}_2 = \mathbf{r}_1 + d\boldsymbol{\theta}_2 \times \mathbf{r}_1 &= (\mathbf{r} + d\boldsymbol{\theta}_1 \times \mathbf{r}) + d\boldsymbol{\theta}_2 \times (\mathbf{r} + d\boldsymbol{\theta}_1 \times \mathbf{r})\\ &= \mathbf{r} + (d\boldsymbol{\theta}_1 + d\boldsymbol{\theta}_2) \times \mathbf{r} + d\boldsymbol{\theta}_2 \times (d\boldsymbol{\theta}_1 \times \mathbf{r})\end{aligned}$$

The last term of this equation may be neglected, since it represents a product of two differentials. Therefore,

$$\mathbf{r}_2 = \mathbf{r} + (d\boldsymbol{\theta}_1 + d\boldsymbol{\theta}_2) \times \mathbf{r} \tag{20–1}$$

Had the two successive rotations occurred in the order $d\boldsymbol{\theta}_2$ then $d\boldsymbol{\theta}_1$, Fig. 20–2b, the resultant positions would have been

$$\mathbf{r}_1' = \mathbf{r} + d\boldsymbol{\theta}_2 \times \mathbf{r}$$

and

$$\begin{aligned}\mathbf{r}_2 = \mathbf{r}_1' + d\boldsymbol{\theta}_1 \times \mathbf{r}_1' &= (\mathbf{r} + d\boldsymbol{\theta}_2 \times \mathbf{r}) + d\boldsymbol{\theta}_1 \times (\mathbf{r} + d\boldsymbol{\theta}_2 \times \mathbf{r})\\ &= \mathbf{r} + (d\boldsymbol{\theta}_2 + d\boldsymbol{\theta}_1) \times \mathbf{r} + d\boldsymbol{\theta}_1 \times (d\boldsymbol{\theta}_2 \times \mathbf{r})\end{aligned}$$

Again neglecting the last term, since it is a second-order differential, we have

$$\mathbf{r}_2 = \mathbf{r} + (d\boldsymbol{\theta}_2 + d\boldsymbol{\theta}_1) \times \mathbf{r} \tag{20–2}$$

Since the vector cross product obeys the distributive law, comparison of Eqs. 20–1 and 20–2 proves that the final position is the same regardless of the order of the applied rotations. It is therefore possible to conclude that infinitesimal rotations $d\boldsymbol{\theta}$ are indeed vectors, since these quantities have both a magnitude and direction for which the order of (vector) addition is not important, i.e., $d\boldsymbol{\theta}_1 + d\boldsymbol{\theta}_2 = d\boldsymbol{\theta}_2 + d\boldsymbol{\theta}_1$. Furthermore, as shown in Figs. 20–2a and 20–2b, the two "component" rotations $d\boldsymbol{\theta}_1$ and $d\boldsymbol{\theta}_2$ are equivalent to the single resultant rotation $d\boldsymbol{\theta} = d\boldsymbol{\theta}_1 + d\boldsymbol{\theta}_2$, Fig. 20–2$c$, a consequence of Euler's theorem.

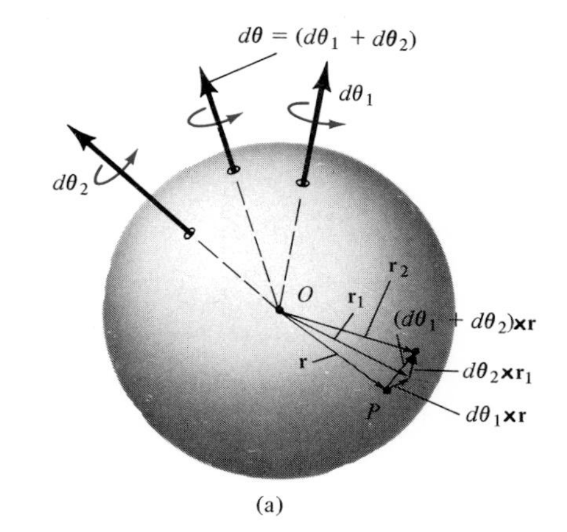

(a)

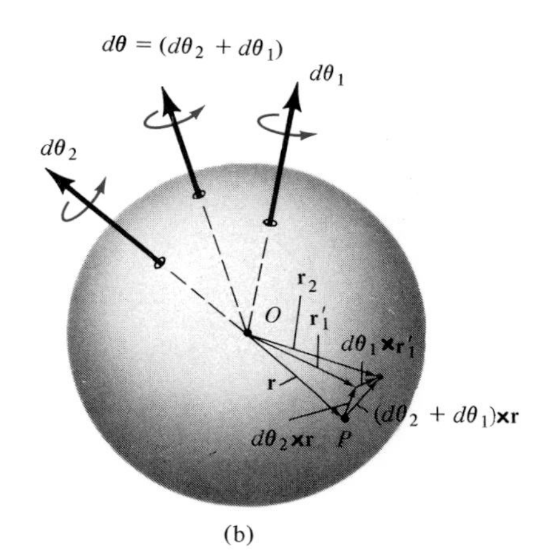

(b)

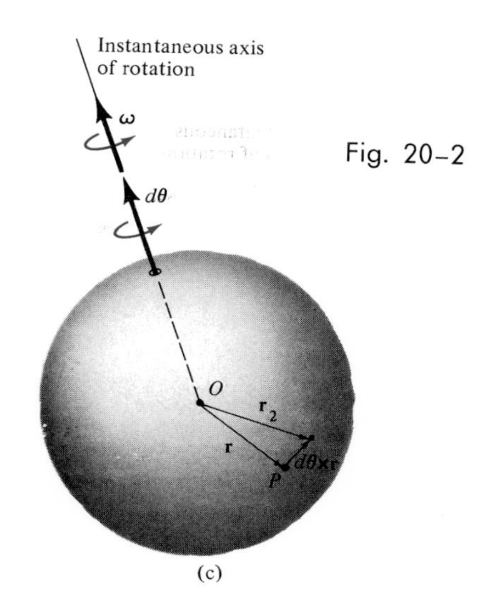

(c)

Fig. 20–2

Angular Velocity. If the body is subjected to an angular rotation $d\boldsymbol{\theta}$ about a fixed point, the instantaneous angular velocity of the body is defined by the time derivative,

$$\boldsymbol{\omega} = \dot{\boldsymbol{\theta}} \tag{20–3}$$

The line specifying the direction of $\boldsymbol{\omega}$, which is collinear with $d\boldsymbol{\theta}$, is referred to as the *instantaneous axis of rotation,* Fig. 20–2*c*. In general, this axis changes direction during each instant of time. Since $\boldsymbol{\omega}$ is a vector quantity, it follows from vector addition that if the body is subjected to two component angular motions, $\boldsymbol{\omega}_1 = \dot{\boldsymbol{\theta}}_1$ and $\boldsymbol{\omega}_2 = \dot{\boldsymbol{\theta}}_2$, the resultant angular velocity is $\boldsymbol{\omega} = \boldsymbol{\omega}_1 + \boldsymbol{\omega}_2$.

Velocity. Once $\boldsymbol{\omega}$ is specified, the instantaneous velocity of any point P on the body can be determined by the cross product,

$$\mathbf{v}_P = \boldsymbol{\omega} \times \mathbf{r}_P \tag{20–4}$$

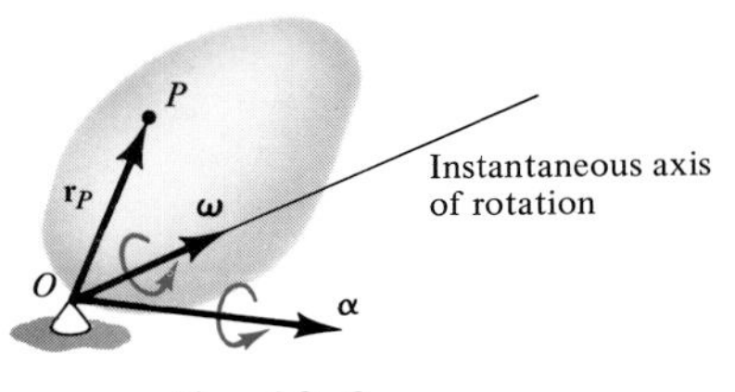

Fig. 20–3

Here $\mathbf{r}_P$ defines the position of P measured from the fixed point O, Fig. 20–3. The form of this equation is the same as Eq. 16–13, which defines the velocity of a particle located on a body subjected to rotation about a fixed axis.

Angular Acceleration. The body's angular acceleration is determined from the time derivative of the angular velocity, i.e.,

$$\boldsymbol{\alpha} = \dot{\boldsymbol{\omega}} \tag{20–5}$$

For motion about a fixed point, $\boldsymbol{\alpha}$ must account for a change in *both* the magnitude and direction of $\boldsymbol{\omega}$, so that, in general, $\boldsymbol{\alpha}$ is not directed along the instantaneous axis of rotation, Fig. 20–3.

As the direction of the instantaneous axis of rotation (or the line of action of $\boldsymbol{\omega}$) changes in space, the locus of points defined by the axis generates a fixed *space cone*. If the change in this axis is viewed with respect to the rotating body, the locus of the axis generates a *body cone,* Fig. 20–4. At any given instant, these cones are tangent along the instantaneous axis of rotation, and when the body is in motion, the body cone appears to roll either on the inside or the outside surface of the fixed space cone. Provided the paths defined by the open ends of the cones are described by the tip of the $\boldsymbol{\omega}$ vector, $\boldsymbol{\alpha}$ must act tangent to these paths at any given instant, since the time rate of change of $\boldsymbol{\omega}$ is equal to $\boldsymbol{\alpha}$, Fig. 20–4.

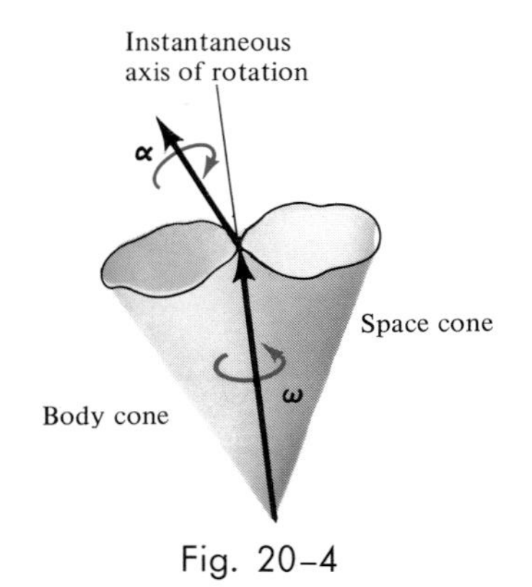

Fig. 20–4

Acceleration. If $\boldsymbol{\omega}$ and $\boldsymbol{\alpha}$ are known at a given instant, the acceleration of any point P on the body can be obtained by time differentiation of Eq. 20–4, which yields

$$\mathbf{a}_P = \boldsymbol{\alpha} \times \mathbf{r}_P + \boldsymbol{\omega} \times (\boldsymbol{\omega} \times \mathbf{r}_P) \tag{20–6}$$

The form of this equation is the same as Eq. 16–16, which defines the acceleration of a point located on a body subjected to rotation about a fixed axis.

20–3. The Time Derivative of a Vector Measured from a Fixed and Translating-Rotating System

In many types of problems involving the motion of a body about a fixed point, the angular velocity $\boldsymbol{\omega}$ is specified in terms of its component angular motions. For example, the disk in Fig. 20–5 spins about the horizontal y axis at $\boldsymbol{\omega}_s$ while it rotates or precesses about the vertical z axis at $\boldsymbol{\omega}_p$. Therefore, its resultant angular velocity is $\boldsymbol{\omega} = \boldsymbol{\omega}_s + \boldsymbol{\omega}_p$. If the angular acceleration $\boldsymbol{\alpha}$ of such a body is to be determined, it is sometimes easier to compute the time derivative of $\boldsymbol{\omega}$, Eq. 20–5, by using a coordinate system which has a *rotation* defined by one or more of the components of $\boldsymbol{\omega}$.* For this reason, and for other uses later, an equation will presently be derived that relates the time derivative of any vector **A**, defined from a rotating and translating reference, to its derivative defined from a fixed reference.

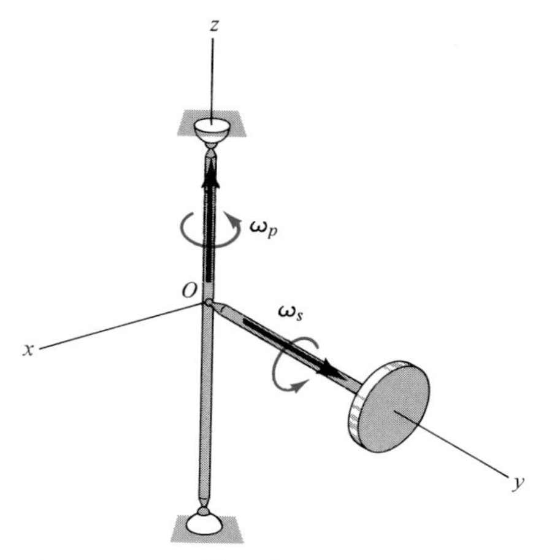

Fig. 20–5

Consider the x, y, z axes of the moving frame of reference to have an angular velocity $\boldsymbol{\Omega}$ which is measured from the fixed X, Y, Z axes, Fig. 20–6a. In the following discussion, it will be convenient to express vector **A** in terms of its **i, j, k** components which define the directions of the moving axes. Hence,

$$\mathbf{A} = A_x\mathbf{i} + A_y\mathbf{j} + A_z\mathbf{k}$$

In general, the time derivative of **A** must account for the change in both the vector's magnitude and direction. However, if this derivative is taken *with respect to the moving frame of reference,* only a change in the magnitudes of the components of **A** must be accounted for, since the directions of **i, j,** and **k** do not change with respect to the moving reference. Hence,

$$(\dot{\mathbf{A}})_{xyz} = \dot{A}_x\mathbf{i} + \dot{A}_y\mathbf{j} + \dot{A}_z\mathbf{k} \qquad (20\text{–}7)$$

When the time derivative of **A** is taken *with respect to the fixed frame of reference,* the *directions* of **i, j,** and **k** change only on account of the rotation $\boldsymbol{\Omega}$ of the axes and not their translation. Hence, in general,

$$(\dot{\mathbf{A}})_{XYZ} = (\dot{A}_x\mathbf{i} + \dot{A}_y\mathbf{j} + \dot{A}_z\mathbf{k} + A_x\dot{\mathbf{i}} + A_y\dot{\mathbf{j}} + A_z\dot{\mathbf{k}}) \qquad (20\text{–}8)$$

The time derivatives of the unit vectors will now be considered. For example, $\dot{\mathbf{i}} = d\mathbf{i}/dt$ reflects only a change in the direction of **i** with respect to time since **i** has a fixed magnitude of 1 unit. As shown in Fig. 20–6b, the change in **i**, $d\mathbf{i}$, is *tangent to the path* described by the tip of **i** as **i** moves because of the rotation $\boldsymbol{\Omega}$. Accounting for both the magnitude and direction of $d\mathbf{i}$, we can therefore express $\dot{\mathbf{i}}$ using the cross product, $\dot{\mathbf{i}} = \boldsymbol{\Omega} \times \mathbf{i}$. In general,

$$\dot{\mathbf{i}} = \boldsymbol{\Omega} \times \mathbf{i}, \qquad \dot{\mathbf{j}} = \boldsymbol{\Omega} \times \mathbf{j}, \qquad \dot{\mathbf{k}} = \boldsymbol{\Omega} \times \mathbf{k} \qquad (20\text{–}9)$$

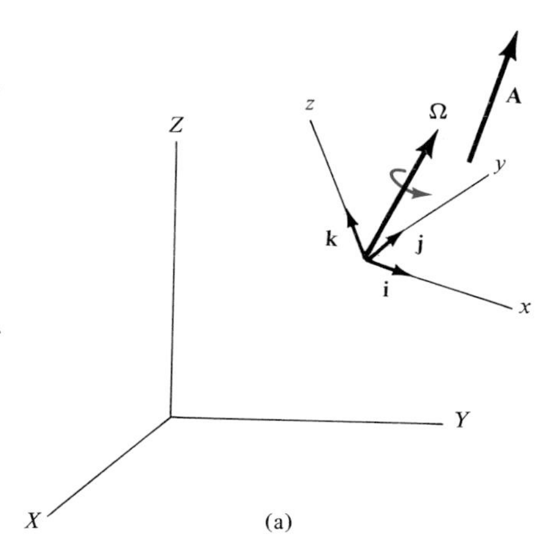

(a)

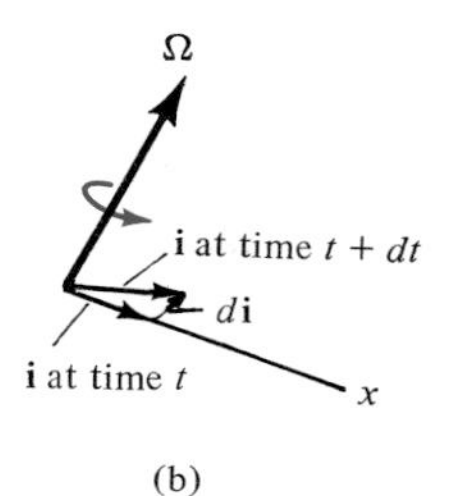

(b)

Fig. 20–6

*In the case of the spinning disk, Fig. 20–5, the x, y, z axes may be given an angular velocity of $\boldsymbol{\omega}_p$.

These formulations were also developed in Sec. 16–10, regarding planar motion of the axes. Substituting the results into Eq. 20–8 and using Eq. 20–7, yields

$$(\dot{\mathbf{A}})_{XYZ} = (\dot{\mathbf{A}})_{xyz} + \mathbf{\Omega} \times \mathbf{A} \quad (20\text{–}10)$$

This result is rather important and it will be used throughout Sec. 20–6 and Chapter 21. In words, it states that the time derivative of **A**, as observed from the fixed X, Y, Z frame of reference, is equal to the time rate of change of **A** as observed from the x, y, z rotating and translating frame of reference (Eq. 20–7) plus $\mathbf{\Omega} \times \mathbf{A}$, the change of **A** caused by the rotation of the x, y, z frame.

The following two example problems numerically illustrate the use of this equation for obtaining the angular acceleration of a body rotating about a fixed point.

Example 20–1

The disk shown in Fig. 20–7*a* is spinning about its horizontal axis with a constant angular velocity of $\omega_s = 3$ rad/s, while the horizontal platform upon which the disk is mounted is rotating about the vertical axis at a constant rate of $\omega_p = 1$ rad/s. Determine the angular acceleration of the disk and the velocity and acceleration of point A on the disk when it is in the position shown.

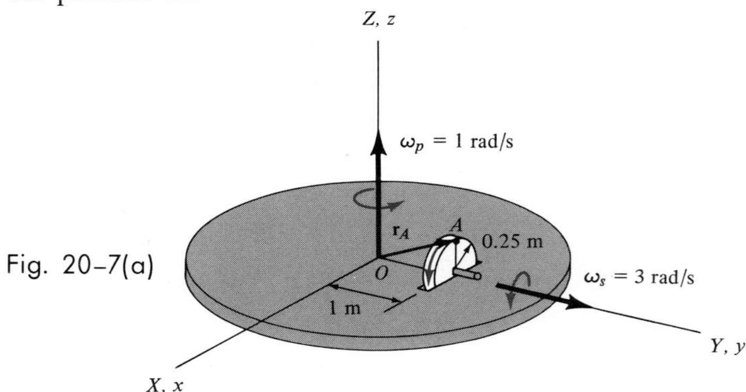

Fig. 20–7(a)

Solution

Point O represents a fixed point of rotation for the disk if one considers a hypothetical extension of the disk to this point. To determine the velocity and acceleration of point A, it is first necessary to determine the resultant angular velocity $\boldsymbol{\omega}$ and angular acceleration $\boldsymbol{\alpha}$ of the disk, since these vectors are used in Eqs. 20–4 and 20–6. The angular velocity is simply the vector addition of the two component motions. Thus,

$$\boldsymbol{\omega} = \boldsymbol{\omega}_s + \boldsymbol{\omega}_p = \{3\mathbf{j} + 1\mathbf{k}\} \text{ rad/s}$$

At first glance it may appear that the disk is not actually rotating with this angular velocity, since it is generally more difficult to imagine the resultant of angular motions in comparison with linear motions. To further understand the angular motion, the appearance of the problem may be changed without changing its characteristics. This is done by replacing the disk with a body cone rolling over an inverted space cone, Fig. 20–7*b*. The instantaneous axis of rotation is along the line of contact of the cones. As shown, this axis defines the direction of the resultant angular velocity $\boldsymbol{\omega}$ having components $\boldsymbol{\omega}_s$ and $\boldsymbol{\omega}_p$.

Since the magnitude of $\boldsymbol{\omega}$ is constant, only a change in its direction, as seen from a fixed reference, creates the angular acceleration $\boldsymbol{\alpha}$ of the disk.

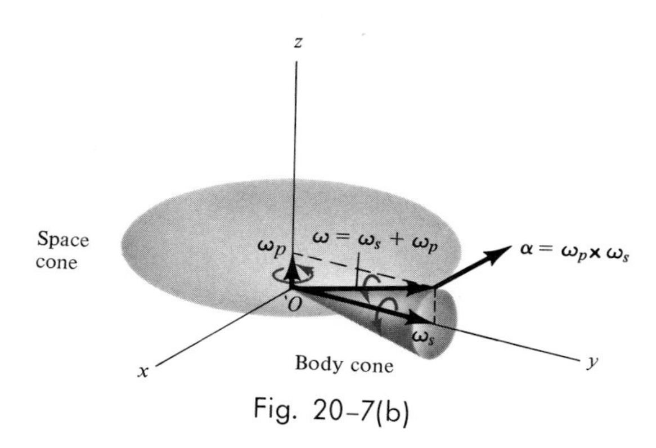

Fig. 20–7(b)

Solution I

One way to obtain $\boldsymbol{\alpha}$ is to compute the time derivative of *each of the two components* of $\boldsymbol{\omega}$ using Eq. 20–10. At the instant shown in Fig. 20–7*a*, imagine the fixed X, Y, Z and rotating x, y, z frames to be in coincidence. If the rotating x, y, z frame is chosen to have an angular velocity of $\boldsymbol{\Omega} = \boldsymbol{\omega}_p = \{1\mathbf{k}\}$ rad/s, then $\boldsymbol{\omega}_s$ will *always* be directed along the y (not Y) axis, and the time rate of change of $\boldsymbol{\omega}_s$ as seen from x, y, z is *zero,* i.e., $(\dot{\boldsymbol{\omega}}_s)_{xyz} = \mathbf{0}$ (the magnitude of $\boldsymbol{\omega}_s$ is constant). Consequently,

$$(\dot{\boldsymbol{\omega}}_s)_{XYZ} = (\dot{\boldsymbol{\omega}}_s)_{xyz} + \boldsymbol{\omega}_p \times \boldsymbol{\omega}_s = \mathbf{0} + (1\mathbf{k}) \times (3\mathbf{j}) = \{-3\mathbf{i}\} \text{ rad/s}^2$$

By the same choice of axes rotation, $\boldsymbol{\Omega} = \boldsymbol{\omega}_p$; or with $\boldsymbol{\Omega} = \mathbf{0}$, the time derivative $(\dot{\boldsymbol{\omega}}_p)_{xyz} = \mathbf{0}$, since $\boldsymbol{\omega}_p$ is *always* directed along the z (or Z) axis and has a constant magnitude. Hence,

$$(\dot{\boldsymbol{\omega}}_p)_{XYZ} = (\dot{\boldsymbol{\omega}}_p)_{xyz} + \boldsymbol{\omega}_p \times \boldsymbol{\omega}_p = \mathbf{0} + \mathbf{0} = \mathbf{0}$$

The angular acceleration is therefore

$$\boldsymbol{\alpha} = (\dot{\boldsymbol{\omega}})_{XYZ} = (\dot{\boldsymbol{\omega}}_s)_{XYZ} + (\dot{\boldsymbol{\omega}}_p)_{XYZ} = \{-3\mathbf{i}\} \text{ rad/s}^2 \qquad \textit{Ans.}$$

Solution II

The angular acceleration may also be obtained by attaching the rotating x, y, z frame of reference to the disk so that $\mathbf{\Omega} = \boldsymbol{\omega} = \boldsymbol{\omega}_s + \boldsymbol{\omega}_p$. In this case, as the axes rotate, $\boldsymbol{\omega}_p$, because of its fixed vertical position, will have *variable projections* along the x and z axes. To obtain the time rate of change of these projections, it is necessary to place the x, y, z axes in an arbitrary position, ϕ (phi) and ψ (psi), Fig. 20–7*c*, determine $(\dot{\boldsymbol{\omega}})_{xyz}$, then evaluate the result at $\phi = 0°$, $\psi = 0°$. Hence, in this more general position, when the X, Y, Z and x, y, z axes are not coincident, the angular velocity is seen to be independent of ϕ and can be expressed as

$$\boldsymbol{\omega} = -\omega_p \sin\psi\mathbf{i} + \omega_s\mathbf{j} + \omega_p \cos\psi\mathbf{k}$$

The time derivative of $\boldsymbol{\omega}$ with respect to the rotating x, y, z frame is, therefore,

$$(\dot{\boldsymbol{\omega}})_{xyz} = -\omega_p \cos\psi\dot{\psi}\mathbf{i} + 0\mathbf{j} - \omega_p \sin\psi\dot{\psi}\mathbf{k}$$

Since $\omega_s = \dot{\psi}$ and $\psi = 0°$ for the instant considered, we have

$$(\dot{\boldsymbol{\omega}})_{xyz} = -\omega_p\omega_s\mathbf{i}$$

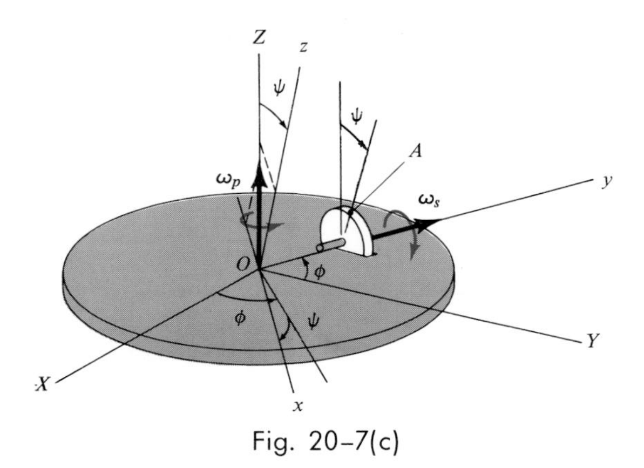

Fig. 20–7(c)

Applying Eq. 20–10 to obtain the time derivative with respect to the fixed X, Y, Z axes, realizing that $\mathbf{\Omega} = \boldsymbol{\omega}$, we have

$$\boldsymbol{\alpha} = (\dot{\boldsymbol{\omega}})_{XYZ} = (\dot{\boldsymbol{\omega}})_{xyz} + \boldsymbol{\omega} \times \boldsymbol{\omega} \tag{1}$$

$$= -\omega_p\omega_s\mathbf{i} + \mathbf{0}$$

$$= -(1)(3)\mathbf{i} = \{-3\mathbf{i}\}\ \text{rad/s}^2 \qquad \textit{Ans.}$$

By comparison, solution I, where $\boldsymbol{\Omega} = \boldsymbol{\omega}_p$, yields an easier computation for $\boldsymbol{\alpha}$ because both components, $\boldsymbol{\omega}_p$ and $\boldsymbol{\omega}_s$, as viewed from the rotating reference, have a constant direction and magnitude. Consequently, for the calculation, $(\dot{\boldsymbol{\omega}}_p)_{xyz} = (\dot{\boldsymbol{\omega}}_s)_{xyz} = \mathbf{0}$.

Once $\boldsymbol{\omega}$ and $\boldsymbol{\alpha}$ are determined, the velocity and acceleration of point A can be computed using Eqs. 20–4 and 20–6. Realizing that $\mathbf{r}_A = \{1\mathbf{j} + 0.25\mathbf{k}\}$ m, Fig. 20–7*a*, we have

$$\mathbf{v}_A = \boldsymbol{\omega} \times \mathbf{r}_A = \begin{vmatrix} \mathbf{i} & \mathbf{j} & \mathbf{k} \\ 0 & 3 & 1 \\ 0 & 1 & 0.25 \end{vmatrix} = \{-0.25\mathbf{i}\} \text{ m/s} \qquad \textit{Ans.}$$

$$\begin{aligned} \mathbf{a}_A &= \boldsymbol{\alpha} \times \mathbf{r}_A + \boldsymbol{\omega} \times (\boldsymbol{\omega} \times \mathbf{r}_A) \\ &= (-3\mathbf{i}) \times (1\mathbf{j} + 0.25\mathbf{k}) + (3\mathbf{j} + 1\mathbf{k}) \times \begin{vmatrix} \mathbf{i} & \mathbf{j} & \mathbf{k} \\ 0 & 3 & 1 \\ 0 & 1 & 0.25 \end{vmatrix} \\ &= -3\mathbf{k} + 0.75\mathbf{j} + (3\mathbf{j} + 1\mathbf{k}) \times (-0.25\mathbf{i}) \\ &= \{0.50\mathbf{j} - 2.25\mathbf{k}\} \text{ m/s}^2 \end{aligned} \qquad \textit{Ans.}$$

Example 20–2

At the instant when $\theta = 60°$, because of mechanical action the gyrotop in Fig. 20–8 has three components of angular motion directed as shown and having magnitudes which are defined as follows:

spin: $\omega_s = 10$ rad/s, increasing at the rate of 6 rad/s²
nutation: $\omega_n = 3$ rad/s, increasing at the rate of 2 rad/s²
precession: $\omega_p = 5$ rad/s, increasing at the rate of 4 rad/s²

Determine the angular velocity and angular acceleration of the top at this instant.

Solution

The top is rotating about the fixed point O. If the fixed and rotating frames are considered to be coincident at the instant shown, then the angular velocity can be expressed in terms of **i, j, k** components, appropriate to the x, y, z frame, i.e.,

$$\boldsymbol{\omega} = -\omega_n \mathbf{i} + \omega_s \sin\theta \mathbf{j} + (\omega_p + \omega_s \cos\theta)\mathbf{k} \qquad (1)$$

Substituting the numerical data, gives

$$\begin{aligned} \boldsymbol{\omega} &= -3\mathbf{i} + 10 \sin 60°\mathbf{j} + (5 + 10 \cos 60°)\mathbf{k} \\ &= \{-3\mathbf{i} + 8.66\mathbf{j} + 10\mathbf{k}\} \text{ rad/s} \end{aligned} \qquad \textit{Ans.}$$

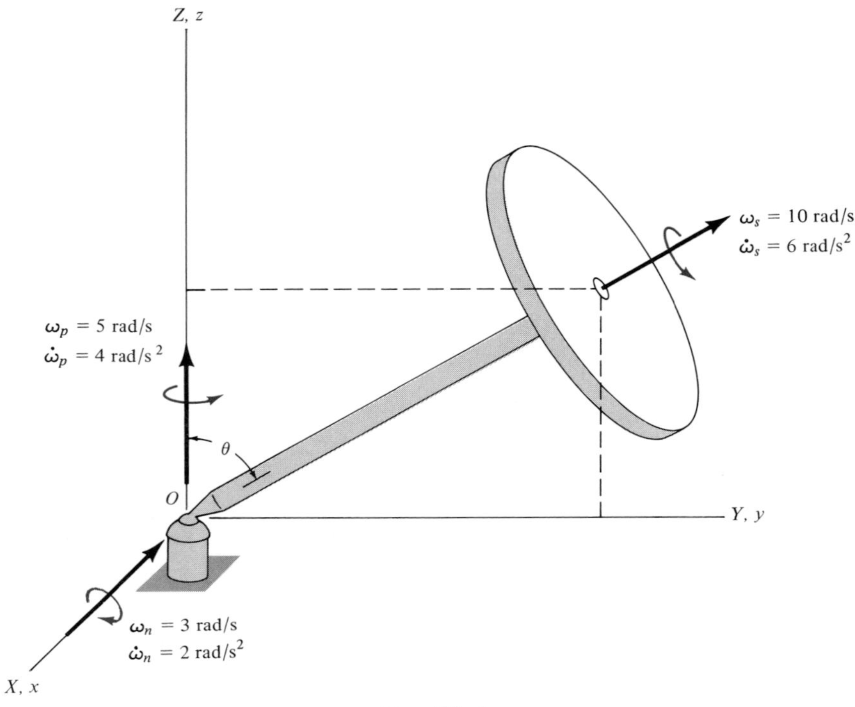

Fig. 20–8

As in solution I of Example 20–1, the angular acceleration $\boldsymbol{\alpha}$ will be determined by investigating separately the time rate of change of *each of the angular velocity components* as observed from the fixed reference X, Y, Z. *This is done by using Eq. 20-10 and choosing an* $\boldsymbol{\Omega}$ *for the* x, y, z *reference so that the component of* $\boldsymbol{\omega}$ *which is being considered is viewed as having a constant direction when observed from* x, y, z.

Careful examination of the motion of the top reveals that the motion of $\boldsymbol{\omega}_s$ is viewed as having a *constant direction* relative to x, y, z if these axes rotate at $\boldsymbol{\Omega} = \boldsymbol{\omega}_p + \boldsymbol{\omega}_n$. Since $\boldsymbol{\omega}_n$ *always* lies in the fixed X-Y plane, this vector has a *constant direction* if the motion is viewed from axes x, y, z having a rotation of $\boldsymbol{\Omega} = \boldsymbol{\omega}_p$ (not $\boldsymbol{\Omega} = \boldsymbol{\omega}_s + \boldsymbol{\omega}_p$). Finally, the component $\boldsymbol{\omega}_p$ is *always directed* along the Z axis so that here it is not necessary to think of x, y, z as rotating, i.e., $\boldsymbol{\Omega} = \mathbf{0}$. Expressing the data in terms of the $\mathbf{i}, \mathbf{j}, \mathbf{k}$ components, we therefore have

For $\boldsymbol{\omega}_s$, $\boldsymbol{\Omega} = \boldsymbol{\omega}_n + \boldsymbol{\omega}_p$;

$$\begin{aligned}(\dot{\boldsymbol{\omega}}_s)_{XYZ} &= (\dot{\boldsymbol{\omega}}_s)_{xyz} + (\boldsymbol{\omega}_n + \boldsymbol{\omega}_p) \times \boldsymbol{\omega}_s \\ &= (6 \sin 60°\mathbf{j} + 6 \cos 60°\mathbf{k}) + (-3\mathbf{i} + 5\mathbf{k}) \\ &\quad \times (10 \sin 60°\mathbf{j} + 10 \cos 60°\mathbf{k}) \\ &= \{-43.30\mathbf{i} + 20.20\mathbf{j} - 22.98\mathbf{k}\} \text{ rad/s}^2\end{aligned}$$

For $\boldsymbol{\omega}_n$, $\boldsymbol{\Omega} = \boldsymbol{\omega}_p$;

$$
\begin{aligned}
(\dot{\boldsymbol{\omega}}_n)_{XYZ} &= (\dot{\boldsymbol{\omega}}_n)_{xyz} + \boldsymbol{\omega}_p \times \boldsymbol{\omega}_n \\
&= -2\mathbf{i} + (5\mathbf{k}) \times (-3\mathbf{i}) \\
&= \{-2\mathbf{i} - 15\mathbf{j}\} \text{ rad/s}^2
\end{aligned}
$$

For $\boldsymbol{\omega}_p$, $\boldsymbol{\Omega} = \mathbf{0}$;

$$
\begin{aligned}
(\dot{\boldsymbol{\omega}}_p)_{XYZ} &= (\dot{\boldsymbol{\omega}}_p)_{xyz} + \mathbf{0} \times \boldsymbol{\omega}_p \\
&= \{4\mathbf{k}\} \text{ rad/s}^2
\end{aligned}
$$

Thus, the angular acceleration of the top can be written as

$$
\begin{aligned}
\boldsymbol{\alpha} &= (\dot{\boldsymbol{\omega}}_s)_{XYZ} + (\dot{\boldsymbol{\omega}}_n)_{XYZ} + (\dot{\boldsymbol{\omega}}_p)_{XYZ} \\
&= \{-45.30\mathbf{i} + 5.20\mathbf{j} - 18.98\mathbf{k}\} \text{ rad/s}^2 \qquad \textit{Ans.}
\end{aligned}
$$

This problem may also be solved by attaching the x, y, z axes to the top such that $\boldsymbol{\Omega} = \boldsymbol{\omega} = \boldsymbol{\omega}_s + \boldsymbol{\omega}_n + \boldsymbol{\omega}_p$, as in solution II of Example 20–1. However, in this case the solution for $\boldsymbol{\alpha}$ is more difficult to obtain, since the time derivative of $\boldsymbol{\omega}$'s components relative to the x, y, z axes must be determined when the top is in a general position and then evaluated at the instant considered. Since this method of analysis will be useful for studying the kinetics of tops and gyroscopes in Chapter 21, a detailed discussion of the kinematics is presented in the next section and the analysis is applied to the solution of this problem in Example 20–3.

20–4. Euler Angles

One method for specifying the position of a rigid body at any given instant is to define three components of a position vector which locates a point in the body, and three finite angular displacements to describe the orientation of the body about this point. Unfortunately, the three finite angular displacements cannot be represented as vectors, since it has been shown in Sec. 20–2 that they do not obey the commutative law of vector addition. Provided, however, these displacements are made in a certain *orderly fashion,* the use of this method becomes practical for analytical work. Furthermore, when the angles are defined as the *Euler angles* ϕ, θ, ψ (phi, theta, psi), this method is particularly well suited for defining the rotational motion of bodies about a fixed point. A method will be established in this section for defining the three Euler angles; and later in Chapter 21 this method will be used for the kinetic analysis of symmetrical tops and gyroscopes and bodies subjected to torque-free motions.

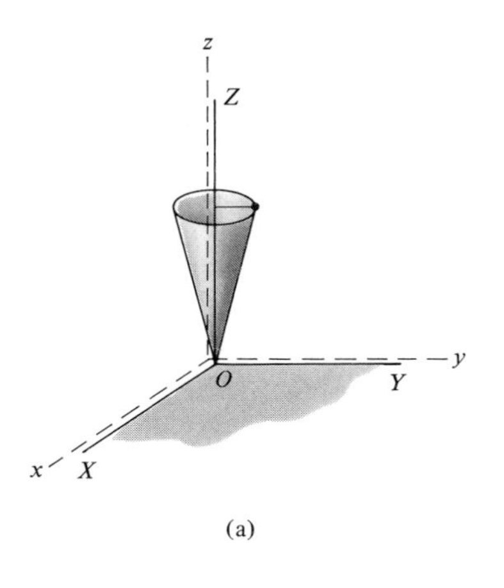

To illustrate how the Euler angles are formed, reference is made to the conical top shown in Fig. 20–9*a*. The top is attached to point O and has an orientation relative to the fixed X, Y, Z axes at some instant of time, as shown in Fig. 20–9*d*. To define this final position, a second set of x, y, z axes will be needed. For purposes of discussion, assume that this reference is fixed in the top. Starting with the X, Y, Z and x, y, z axes in coincidence, Fig. 20–9*a*, the final position of the top is determined using the following three orderly steps:

1. Rotate the top about the Z (or z) axis through an angle ϕ $(0 \leqslant \phi < 2\pi)$, Fig. 20–9*b*.
2. Rotate the top about the x axis through an angle θ $(0 \leqslant \theta \leqslant \pi)$, Fig. 20–9*c*.
3. Rotate the top about the z axis through an angle ψ $(0 \leqslant \psi < 2\pi)$ to obtain the final position, Fig. 20–9*d*.

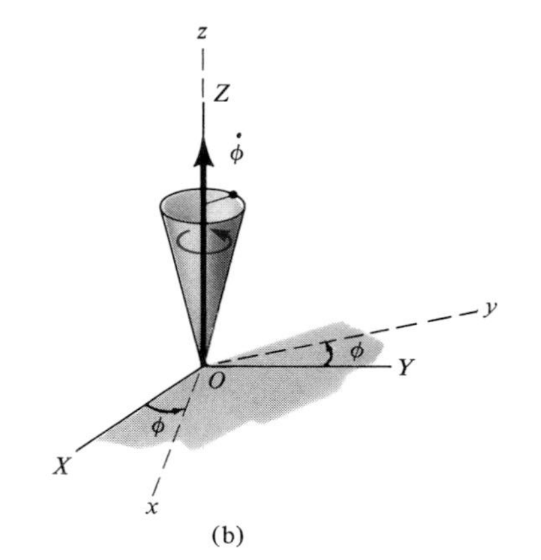

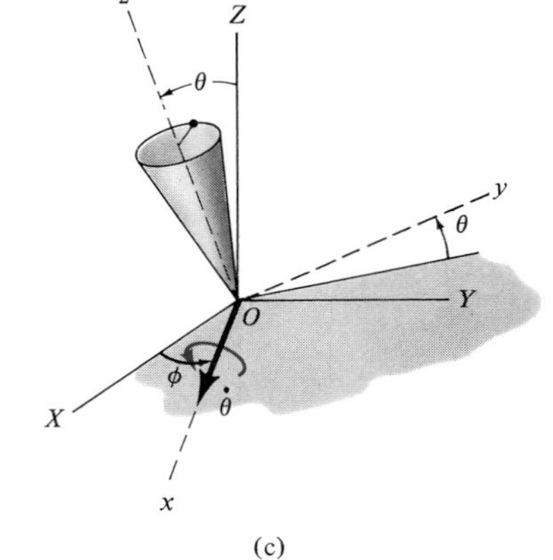

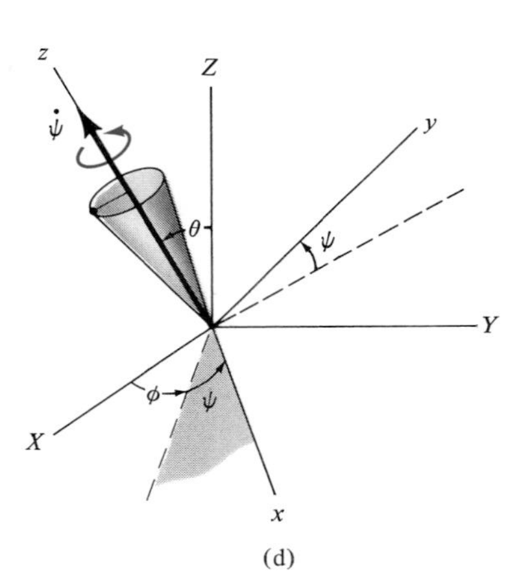

Fig. 20–9

Angular Motion. When the top is in motion, in general, each of the Euler angles is changing with time. Although finite rotations are not vectors, infinitesimal rotations $d\boldsymbol{\phi}$, $d\boldsymbol{\theta}$, and $d\boldsymbol{\psi}$ are vectors, and thus the angular velocity $\boldsymbol{\omega}$ of the top can be expressed in terms of the time derivatives of the Euler angles. The angular-velocity components, $\dot{\phi}$, $\dot{\theta}$, and $\dot{\psi}$, are known as the *precession, nutation,* and *spin,* respectively. Their positive directions are shown in Fig. 20–9. Although these vectors are not all perpendicular (orthogonal) to one another, $\boldsymbol{\omega}$ of the top can still be expressed in terms of these three components along any orthogonal set of coordinate axes.

Angular-Motion Components Along the X, Y, Z Axes. If the projections along the fixed X, Y, Z axes are considered, Fig. 20–10, the spin $\boldsymbol{\omega}_s$ and nutation $\boldsymbol{\omega}_n$ will have to be resolved using the Euler angles θ and ϕ. In particular, the component of spin in the X-Y plane is $\dot{\psi}$ sin θ. From Fig. 20–10, this component makes an angle ϕ with the negative Y axis so that the components of $\dot{\psi}\sin\theta$ along the X and Y axes are $(\dot{\psi}\sin\theta)\sin\phi$ and $(-\dot{\psi}\sin\theta)\cos\phi$, respectively. As a result, the angular velocity is

$$\begin{aligned}\boldsymbol{\omega} &= \omega_X\mathbf{I} + \omega_Y\mathbf{J} + \omega_Z\mathbf{K}\\ &= (\dot{\theta}\cos\phi + \dot{\psi}\sin\theta\sin\phi)\mathbf{I} + (\dot{\theta}\sin\phi\\ &\quad -\dot{\psi}\sin\theta\cos\phi)\mathbf{J} + (\dot{\phi} + \dot{\psi}\cos\theta)\mathbf{K} \qquad (20\text{–}11)\end{aligned}$$

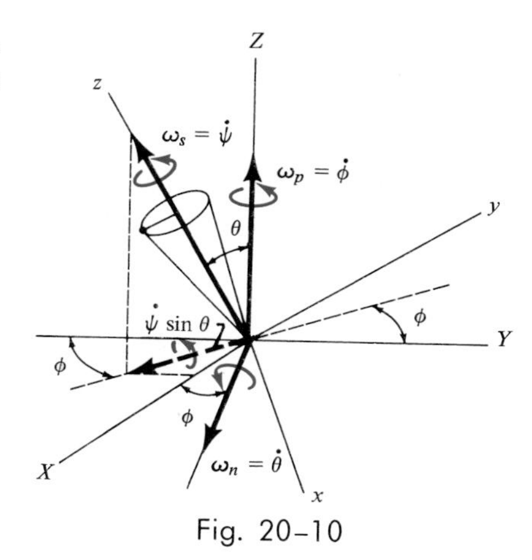

Fig. 20–10

Since this equation provides a general expression for the angular velocity at any instant of time, the angular acceleration is obtained by time differentiation, i.e.,

$$\begin{aligned}\boldsymbol{\alpha} = (\dot{\boldsymbol{\omega}})_{XYZ} &= (\ddot{\theta}\cos\phi - \dot{\theta}\dot{\phi}\sin\phi + \ddot{\psi}\sin\theta\sin\phi + \dot{\psi}\dot{\theta}\cos\theta\sin\phi\\ &\quad + \dot{\psi}\dot{\phi}\sin\theta\cos\phi)\mathbf{I}\\ &\quad + (\ddot{\theta}\sin\phi + \dot{\theta}\dot{\phi}\cos\phi - \ddot{\psi}\sin\theta\cos\phi - \dot{\psi}\dot{\theta}\cos\theta\cos\phi\\ &\quad + \dot{\psi}\dot{\phi}\sin\theta\sin\phi)\mathbf{J} + (\ddot{\phi} + \ddot{\psi}\cos\theta - \dot{\psi}\dot{\theta}\sin\theta)\mathbf{K} \qquad (20\text{–}12)\end{aligned}$$

Angular-Motion Components Along the x, y, z Axes, $\boldsymbol{\Omega} \neq \boldsymbol{\omega}$. The components of $\boldsymbol{\omega}$ may also be resolved along the rotating x, y, z axes. For example, consider the top in Fig. 20–11, for which the x axis is oriented such that at the instant considered, the spin angle $\psi = 0$ and *the x, y, z axes follow the motion of the body only in nutation and precession,* i.e., $\boldsymbol{\Omega} = \boldsymbol{\omega}_p + \boldsymbol{\omega}_n$. This being the case, the spin and components of nutation are always directed along the x and z axes. Hence, the angular velocity of the body is specified only in terms of the Euler angle θ, i.e.,

$$\begin{aligned}\boldsymbol{\omega} &= \omega_x\mathbf{i} + \omega_y\mathbf{j} + \omega_z\mathbf{k}\\ &= \dot{\theta}\mathbf{i} + (\dot{\phi}\sin\theta)\mathbf{j} + (\dot{\phi}\cos\theta + \dot{\psi})\mathbf{k} \qquad (20\text{–}13)\end{aligned}$$

Since motion of the axes is not affected by the spin component,

$$\begin{aligned}\boldsymbol{\Omega} &= \Omega_x\mathbf{i} + \Omega_y\mathbf{j} + \Omega_z\mathbf{k}\\ &= \dot{\theta}\mathbf{i} + (\dot{\phi}\sin\theta)\mathbf{j} + (\dot{\phi}\cos\theta)\mathbf{k} \qquad (20\text{–}14)\end{aligned}$$

Using these results, the angular acceleration of the top is obtained by applying Eq. 20–10. Hence,

$$\boldsymbol{\alpha} = (\dot{\boldsymbol{\omega}})_{XYZ} = (\dot{\boldsymbol{\omega}})_{xyz} + \boldsymbol{\Omega} \times \boldsymbol{\omega}$$

The term $(\dot{\boldsymbol{\omega}})_{xyz}$ is simply the time derivative of the **i, j, k** components of $\boldsymbol{\omega}$ (Eq. 20–13). Computing these derivatives, substituting Eqs. 20–13 and 20–14 into the above equation, and carrying out the cross product, we get

$$\begin{aligned}\boldsymbol{\alpha} &= (\ddot{\theta} + \dot{\psi}\dot{\phi}\sin\theta)\mathbf{i} + (\ddot{\phi}\sin\theta + \dot{\phi}\dot{\theta}\cos\theta - \dot{\theta}\dot{\psi})\mathbf{j}\\ &\quad + (\ddot{\psi} + \ddot{\phi}\cos\theta - \dot{\phi}\dot{\theta}\sin\theta)\mathbf{k} \qquad (20\text{–}15)\end{aligned}$$

As an exercise show that the **i, j, k** components in this equation can be resolved into the corresponding **I, J, K** components, defined by Eq. 20–12 for the fixed axes.

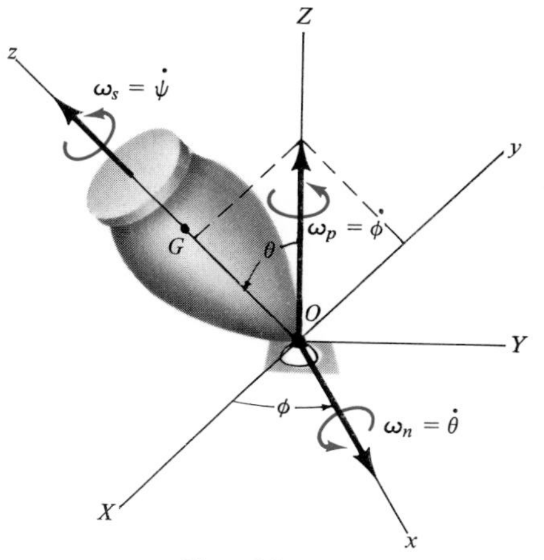

Fig. 20–11

Angular-Motion Components Along the x, y, z Axes, $\mathbf{\Omega} = \boldsymbol{\omega}$. The computation for $\boldsymbol{\alpha}$ can also be performed by choosing the x, y, z axes fixed in the body, in which case

$$\mathbf{\Omega} = \boldsymbol{\omega} = \boldsymbol{\omega}_s + \boldsymbol{\omega}_n + \boldsymbol{\omega}_p \tag{20–16}$$

Applying Eq. 20–10, yields

$$\boldsymbol{\alpha} = (\dot{\boldsymbol{\omega}})_{XYZ} = (\dot{\boldsymbol{\omega}})_{xyz} + \boldsymbol{\omega} \times \boldsymbol{\omega}$$

or

$$\boldsymbol{\alpha} = (\dot{\boldsymbol{\omega}})_{XYZ} = (\dot{\boldsymbol{\omega}})_{xyz} \tag{20–17}$$

As indicated, for this special case of $\mathbf{\Omega} = \boldsymbol{\omega}$, the time derivative for $\boldsymbol{\omega}$ is *equivalent* when computed with respect to either the *fixed X, Y, Z or rotating x, y, z axes*. Consequently, the components for $\boldsymbol{\alpha} = (\dot{\boldsymbol{\omega}})_{XYZ}$ expressed by Eq. 20–12 are the *same* as those for a set of rotating axes attached to the body, provided the x, y, z axes are *coincident* with the fixed X, Y, Z axes at the instant considered, i.e., $\theta = \phi = 0°$ in Fig. 20–11.

Example 20–3

Determine the angular acceleration of the gyrotop in Example 20–2 assuming that $\mathbf{\Omega} = \boldsymbol{\omega} = \boldsymbol{\omega}_s + \boldsymbol{\omega}_n + \boldsymbol{\omega}_p$ for the x, y, z axes.

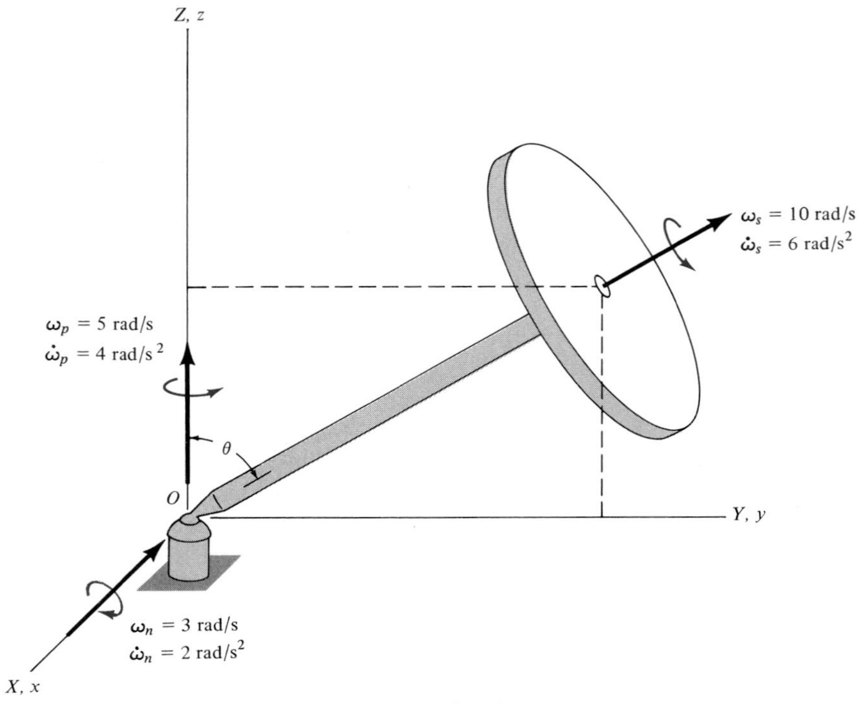

Fig. 20–12

Solution

As shown in Fig. 20–12, the x, y, z axes are fixed to the top and *coincident* with the X, Y, Z axes at the instant considered. Comparing the position of the gyrotop in relation to the x, y, z and X, Y, Z frames defined in Fig. 20–9*d*, it is seen that $\phi = \psi = 0$. Furthermore, according to the right-hand rule, since the gyrotop has been rotated in the *negative direction* about the x axis, $\theta = -60°$. Listing the problem data, gives

$$\phi = 0° \qquad \theta = -60° \qquad \psi = 0°$$

$$\omega_p = \dot{\phi} = 5 \text{ rad/s} \qquad \omega_n = \dot{\theta} = -3 \text{ rad/s} \qquad \omega_s = \dot{\psi} = 10 \text{ rad/s}$$

$$\dot{\omega}_p = \ddot{\phi} = 4 \text{ rad/s}^2 \qquad \dot{\omega}_n = \ddot{\theta} = -2 \text{ rad/s}^2 \qquad \dot{\omega}_s = \ddot{\psi} = 6 \text{ rad/s}^2$$

If Eq. 20–17 is applied, $\boldsymbol{\alpha}$ can be expressed either in terms of **I, J, K** or **i, j, k** components, using Eq. 20–12. Substituting the above data into this equation, we have

$$\begin{aligned}\boldsymbol{\alpha} = (\dot{\boldsymbol{\omega}})_{XYZ} = (\dot{\boldsymbol{\omega}})_{xyz} \\ &= [-2\cos 0° - (-3)(5)\sin 0° + 6(\sin -60°)\sin 0° \\ &\quad + 10(-3)(\cos -60°)\sin 0° + (10)(5)(\sin -60°)\cos 0°]\mathbf{i} \\ &\quad + [-2\sin 0° + (-3)(5)\cos 0° - 6(\sin -60°)\cos 0° \\ &\quad - 10(-3)(\cos -60°)\cos 0° + 10(5)(\sin -60°)\sin 0°]\mathbf{j} \\ &\quad + [4 + 6(\cos -60°) - 10(-3)(\sin -60°)]\mathbf{k}\end{aligned}$$

or

$$\boldsymbol{\alpha} = \{-45.30\mathbf{i} + 5.20\mathbf{j} - 18.98\mathbf{k}\} \text{ rad/s}^2 \qquad \textit{Ans.}$$

This is the same result as obtained in Example 20–2.

20–5. General Motion of a Rigid Body

The most general spatial motion of a rigid body occurs when any point on the body has a specified velocity $\mathbf{v}$ and acceleration $\mathbf{a}$, and the body is rotating with an angular velocity $\boldsymbol{\omega}$ and angular acceleration $\boldsymbol{\alpha}$. In general, none of these vectors will be collinear, since $\mathbf{a}$ and $\boldsymbol{\alpha}$ measure the change in both the magnitude and direction of $\mathbf{v}$ and $\boldsymbol{\omega}$, respectively.

Shown in Fig. 20–13 is a rigid body subjected to general spatial motion for which the instantaneous angular velocity is $\boldsymbol{\omega}$ and the angular acceleration is $\boldsymbol{\alpha}$. If point A has a known motion of $\mathbf{v}_A$ and $\mathbf{a}_A$, the motion of any other point, B, may be determined by using a relative-motion analysis. In this section a translating coordinate system will be used to define the relative motion, and in the next section a reference that is both rotating and translating will be considered.

If the origin of the translating coordinate system x, y, z ($\boldsymbol{\Omega} = \mathbf{0}$) is located at the base point A, then, at the instant shown, the spatial motion of the body may be regarded as the sum of an instantaneous translation of the body having a motion of $\mathbf{v}_A$ and $\mathbf{a}_A$ and a rotation of the body about an

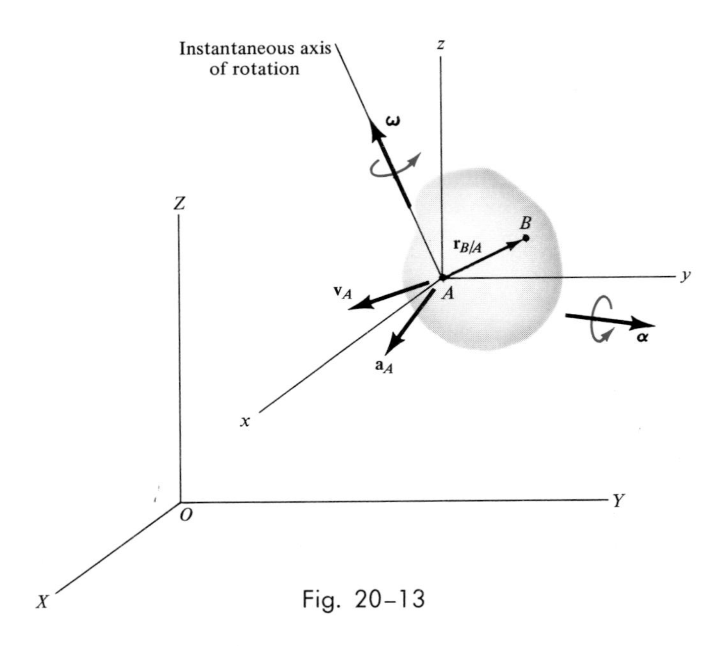

Fig. 20–13

instantaneous axis passing through the base point. Since the body is rigid, the motion of point B measured by an observer located at A is the same as *motion of the body about a fixed point*. This relative motion occurs about the instantaneous axis of rotation and is defined by $\mathbf{v}_{B/A} = \boldsymbol{\omega} \times \mathbf{r}_{B/A}$ (Eq. 20–4) and $\mathbf{a}_{B/A} = \boldsymbol{\alpha} \times \mathbf{r}_{B/A} + \boldsymbol{\omega} \times (\boldsymbol{\omega} \times \mathbf{r}_{B/A})$ (Eq. 20–6). For a translating axis the relative motions are related to absolute motions by $\mathbf{v}_B = \mathbf{v}_A + \mathbf{v}_{B/A}$ and $\mathbf{a}_B = \mathbf{a}_A + \mathbf{a}_{B/A}$ (Eqs. 16–18 and 16–21), so that the absolute velocity and acceleration of point B can be determined from the equations

$$\mathbf{v}_B = \mathbf{v}_A + (\boldsymbol{\omega} \times \mathbf{r}_{B/A}) \tag{20–18}$$

and

$$\mathbf{a}_B = \mathbf{a}_A + (\boldsymbol{\alpha} \times \mathbf{r}_{B/A}) + \boldsymbol{\omega} \times (\boldsymbol{\omega} \times \mathbf{r}_{B/A}) \tag{20–19}$$

These two equations are identical to those describing the general plane motion of a rigid body (Eqs. 16–19 and 16–22). However, difficulty in application arises for general spatial motion, because $\boldsymbol{\alpha}$ measures the change in *both* the magnitude and direction of $\boldsymbol{\omega}$. (Recall that for general plane motion $\boldsymbol{\alpha}$ and $\boldsymbol{\omega}$ are always parallel or perpendicular to the plane of motion, and therefore $\boldsymbol{\alpha}$ measures only a change in the magnitude of $\boldsymbol{\omega}$.) In some problems the constraints or connections of a body will require that the directions of the angular motions or displacement paths of points in the body be defined. As illustrated in the following example, this information is useful for obtaining some of the terms in the above equations.

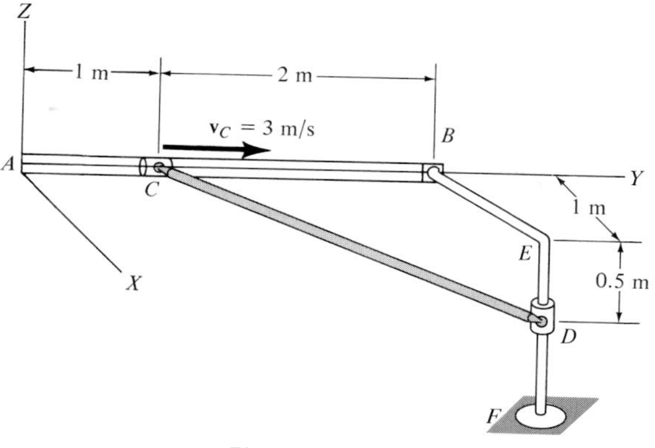

Fig. 20–14(a)

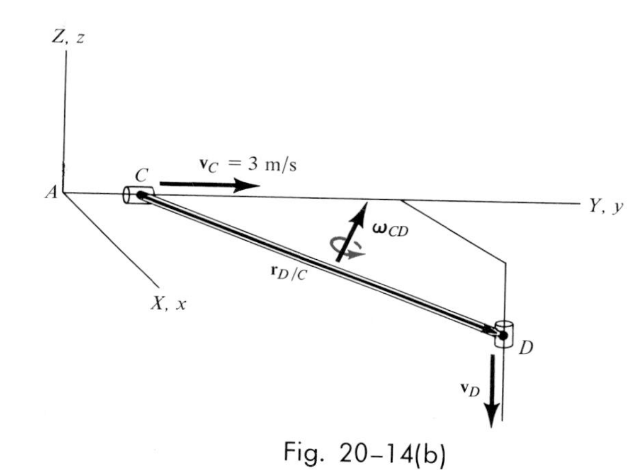

Fig. 20–14(b)

Example 20–4

One end of the rigid bar CD shown in Fig. 20–14a slides along the grooved wall slot AB, and the other end slides along the vertical member EF. If the collar at C is moving towards B at a speed of 3 m/s, determine the velocity of the collar at D and the angular velocity of the bar. The bar is connected to the collars at its end points by ball-and-socket joints.

Solution

Bar CD is subjected to general motion. Why? The velocity of point D on the bar may be related to the velocity of point C, the "base point," by the equation

$$\mathbf{v}_D = \mathbf{v}_C + (\boldsymbol{\omega}_{CD} \times \mathbf{r}_{D/C}) \tag{1}$$

Referring to Fig. 20–14b, the fixed and translating frames of reference are assumed to coincide at the instant considered. Expressing $\mathbf{v}_D$, $\mathbf{v}_C$, $\mathbf{r}_{D/C}$, and $\boldsymbol{\omega}_{CD}$ in Cartesian vector form, we have

$$\begin{aligned} \mathbf{v}_D &= -v_D\mathbf{k} \\ \mathbf{v}_C &= \{3\mathbf{j}\} \text{ m/s} \\ \mathbf{r}_{D/C} &= \{1\mathbf{i} + 2\mathbf{j} - 0.5\mathbf{k}\} \text{ m} \\ \boldsymbol{\omega}_{CD} &= \omega_x\mathbf{i} + \omega_y\mathbf{j} + \omega_z\mathbf{k} \end{aligned}$$

Substituting these quantities into Eq. (1) gives

$$-v_D\mathbf{k} = 3\mathbf{j} + \begin{vmatrix} \mathbf{i} & \mathbf{j} & \mathbf{k} \\ \omega_x & \omega_y & \omega_z \\ 1 & 2 & -0.5 \end{vmatrix}$$

Expanding this expression and equating the respective **i, j, k** components yields

$$-0.5\omega_y - 2\omega_z = 0 \tag{2}$$

$$0.5\omega_x + 1\omega_z + 3 = 0 \tag{3}$$

$$2\omega_x - 1\omega_y + v_D = 0 \tag{4}$$

These equations contain four unknowns*: ω_x, ω_y, ω_z, and v_D. To determine a fourth equation it is necessary to specify the direction of the bar's angular velocity $\boldsymbol{\omega}_{CD}$. Since the bar is connected at its end points using ball-and-socket joints, any component of $\boldsymbol{\omega}_{CD}$ acting along the axis of the bar has no effect on moving the collars. This is because the bar is *free to rotate* about its axis. Therefore, *any* arbitrary magnitude of the component of $\boldsymbol{\omega}_{CD}$ can be assumed in this direction without changing the solution of Eqs. (2), (3), and (4). However, if $\boldsymbol{\omega}_{CD}$ is specified as acting *perpendicular* to the axis of the bar, then $\boldsymbol{\omega}_{CD}$ must have a unique magnitude to satisfy the equations. Perpendicularity is guaranteed provided the dot product of $\boldsymbol{\omega}_{CD}$ and $\mathbf{r}_{D/C}$ is zero.† Hence,

$$\boldsymbol{\omega}_{CD} \cdot \mathbf{r}_{D/C} = (\omega_x\mathbf{i} + \omega_y\mathbf{j} + \omega_z\mathbf{k}) \cdot (1\mathbf{i} + 2\mathbf{j} - 0.5\mathbf{k}) = 0$$

$$1\omega_x + 2\omega_y - 0.5\omega_z = 0 \tag{5}$$

Solving Eqs. (2) through (5) simultaneously yields

$$\omega_x = -4.86 \text{ rad/s}$$

$$\omega_y = 2.29 \text{ rad/s}$$

$$\omega_z = -0.571 \text{ rad/s}$$

$$v_D = 12.00 \text{ m/s}$$

so that

$$\boldsymbol{\omega}_{CD} = \{-4.86\mathbf{i} + 2.29\mathbf{j} - 0.571\mathbf{k}\} \text{ rad/s} \qquad \textit{Ans.}$$

and

$$\mathbf{v}_D = \{12.00\mathbf{k}\} \text{ m/s} \qquad \textit{Ans.}$$

*Although this is the case the magnitude of $\mathbf{v}_D$ can be obtained. For example, solve Eqs. (2) and (3) for ω_y and ω_x in terms of ω_z and substitute into Eq. (4). It will be noted that ω_z will *cancel out,* which will allow a solution for v_D.

†See Eq. B–14 Appendix B.

Problems

20-1. The electric fan is mounted on a swivel support such that the fan rotates about the vertical z axis at a constant rate of $\omega_1 = 2$ rad/s, while the blade is spinning at $\omega_2 = 50$ rad/s. If $\phi = 45°$ throughout the motion, determine the angular velocity and the angular acceleration of the blade.

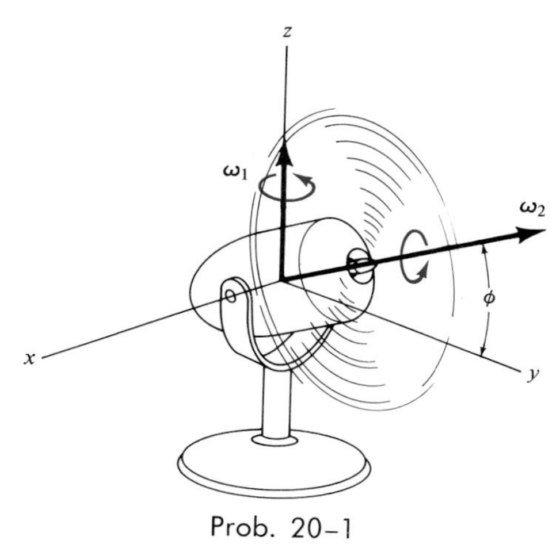

Prob. 20-1

20-2. The propeller of an airplane is rotating at a constant speed of $\omega_s\mathbf{i}$, while the plane is undergoing a turn at a constant rate of ω_t. Compute the angular acceleration of the propeller if (a) the turn is horizontal, i.e., $\omega_t\mathbf{k}$, and (b) the turn is vertical, downward, i.e., $\omega_t\mathbf{j}$.

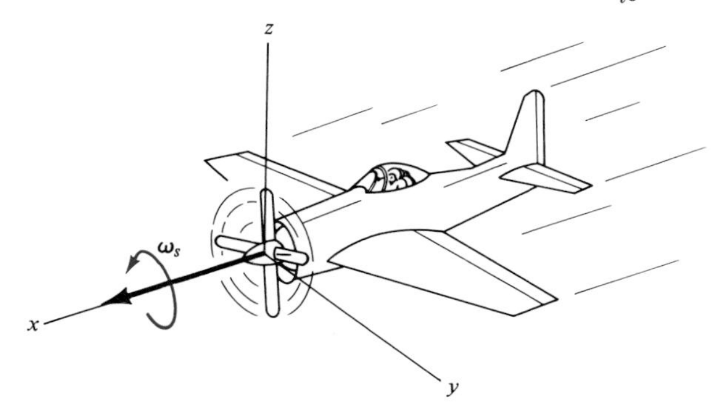

Prob. 20-2

20-3. The anemometer located on the ship at A is spinning about its own axis at a rate $\boldsymbol{\omega}_s$, while the ship is rolling about the x axis at the rate $\boldsymbol{\omega}_x$ and about the y axis at the rate $\boldsymbol{\omega}_y$. Compute the angular velocity and angular acceleration of the anemometer at the instant the ship is level as shown. Assume that the magnitudes of all components of angular velocity are constant and that the rolling motion caused by the sea is independent in the x and y directions.

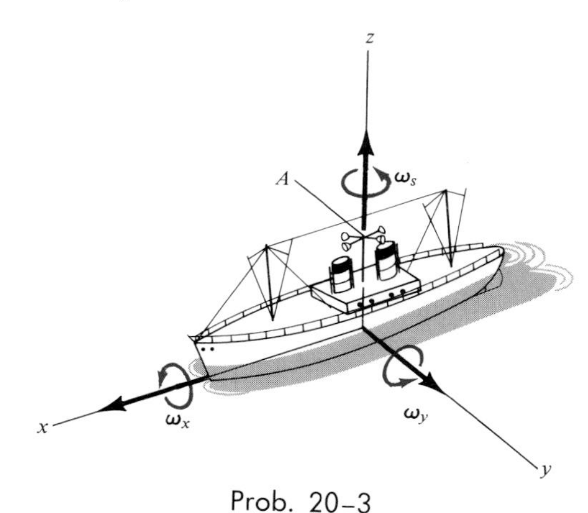

Prob. 20-3

***20-4.** Gears A and B are fixed while gears C and D are free to rotate on the shaft S. If the shaft is turning about the z axis at a constant rate of $\omega_1 = 3$ rad/s, determine the magnitudes of the angular velocity and angular acceleration of gear C.

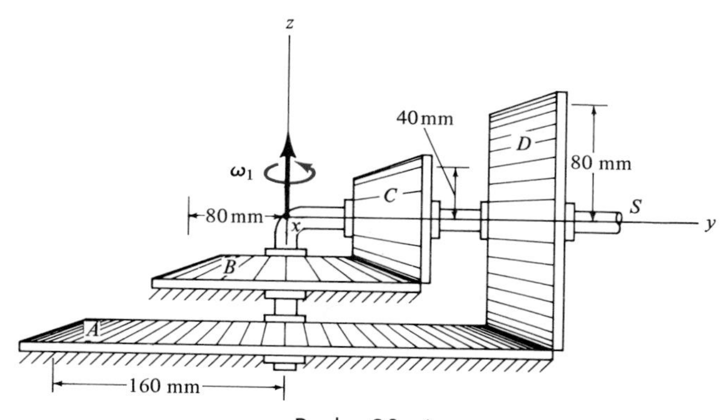

Prob. 20-4

20-5. The construction boom OA is rotating about the vertical z axis with a constant angular velocity of $\omega_1 = 0.08$ rad/s, while it is rotating downward with a constant angular velocity of $\omega_2 = 0.06$ rad/s. Compute the velocity and acceleration of point A located at the tip of the boom at the instant shown. $l = 35$ m and $h = 15$ m.

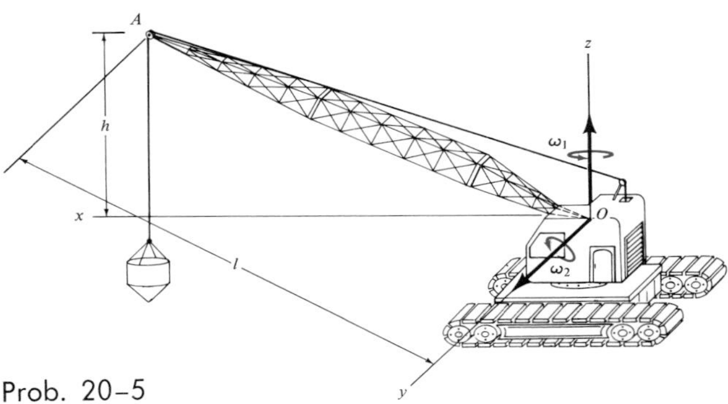

Prob. 20-5

20-5a. Solve Prob. 20-5 if $\omega_1 = 0.7$ rad/s, $\omega_2 = 0.4$ rad/s, $l = 120$ ft, and $h = 50$ ft.

20-6. The three bevel gears are in mesh. If gears A and B are rotating with the angular velocities shown, determine the angular velocity of gear C about the shaft DE. What is the angular velocity of DE about the y axis?

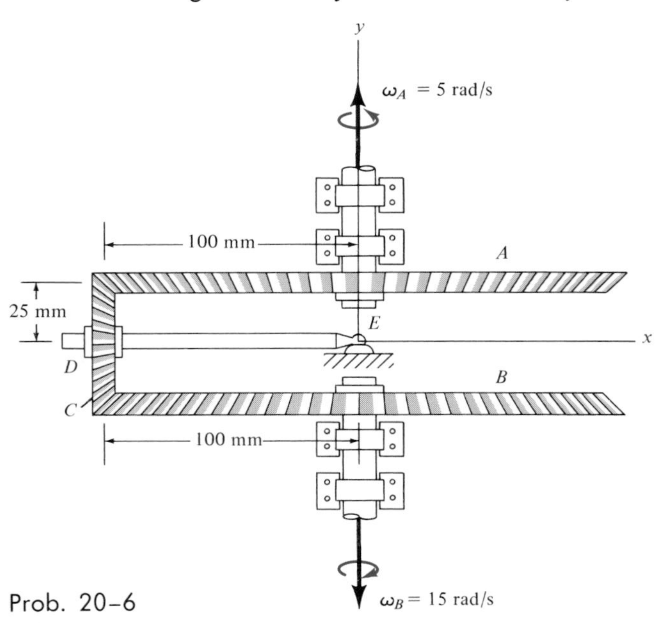

Prob. 20-6

20-7. The radar-tracking antenna is following a jet plane. At the instant $\theta = 30°$ and $\phi = 60°$, the angular rates of change are $\dot{\theta} = 0.2$ rad/s and $\dot{\phi} = 0.5$ rad/s. Compute the velocity and acceleration of the signal horn A at this instant. The distance OA is 0.8 m.

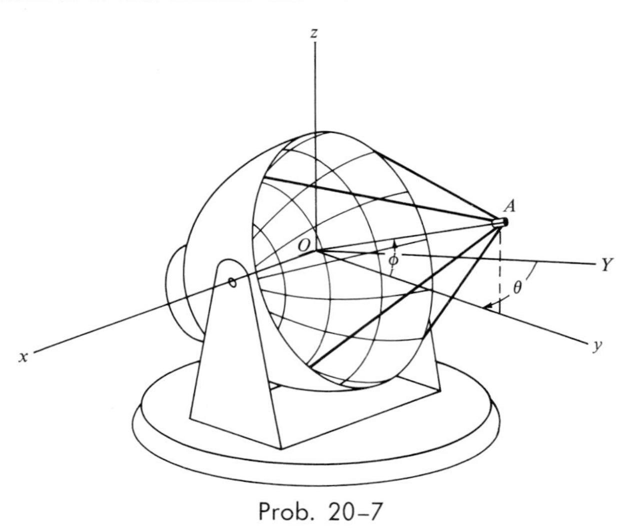

Prob. 20-7

***20-8.** The right circular cone rotates about the vertical at a constant rate of $\omega_1 = 2$ rad/s, without slipping on the horizontal plane. Determine the magnitudes of the velocity and acceleration of points B and C.

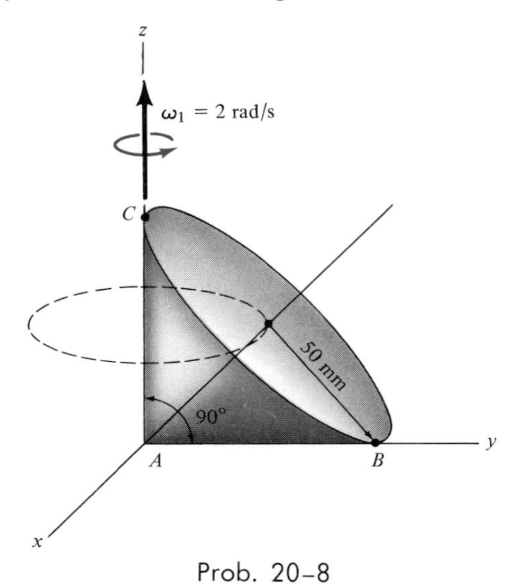

Prob. 20-8

20–9. The shaft BD is connected to a ball-and-socket joint at B and a beveled gear A is attached to its other end. The gear is in mesh with a fixed gear C. If the shaft and gear A are *spinning* with a constant angular velocity of $\omega_1 = 6$ rad/s in the direction shown, determine the angular velocity and angular acceleration of gear A.

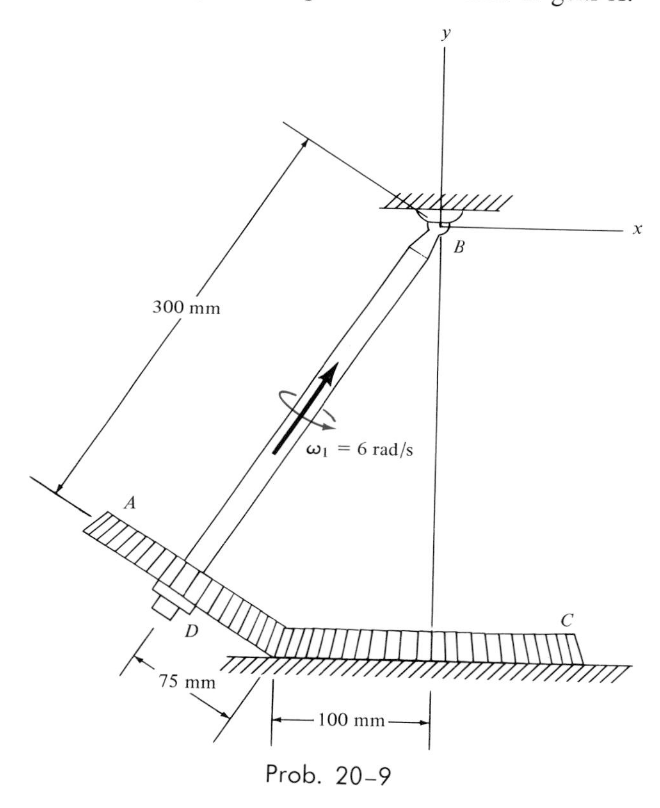

Prob. 20–9

20–10. Gear A is fixed to the crankshaft S, while gear C is fixed and gear B is free to rotate. If the crankshaft is turning at $\omega_1 = 50$ rad/s about its axis, determine the magnitudes of the angular velocity of the propeller and the angular acceleration of gear B. $r_A = 150$ mm and $r_B = 25$ mm.

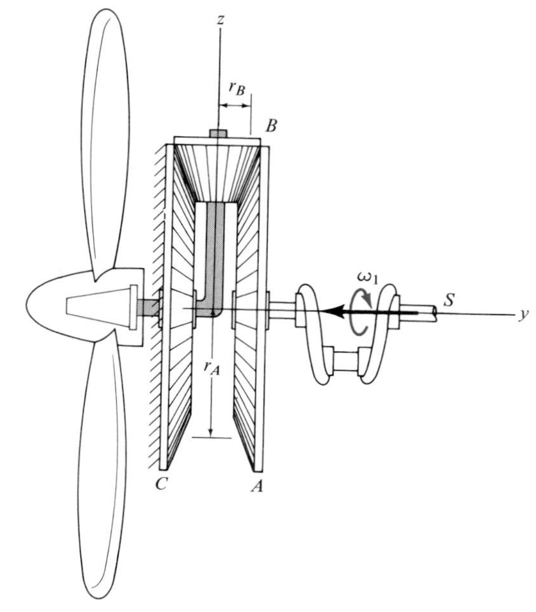

Prob. 20–10

20–10a. Solve Prob. 20–10 if $\omega_1 = 80$ rad/s, $r_A = 0.3$ ft, $r_B = 0.1$ ft.

20–11. The conical spool rolls on the surface of the plate without slipping. If the axle AB has an angular velocity of $\omega_1 = 3$ rad/s and an angular acceleration of $\alpha_1 = 2$ rad/s^2 at the instant shown, determine the angular velocity and angular acceleration of the spool at this instant.

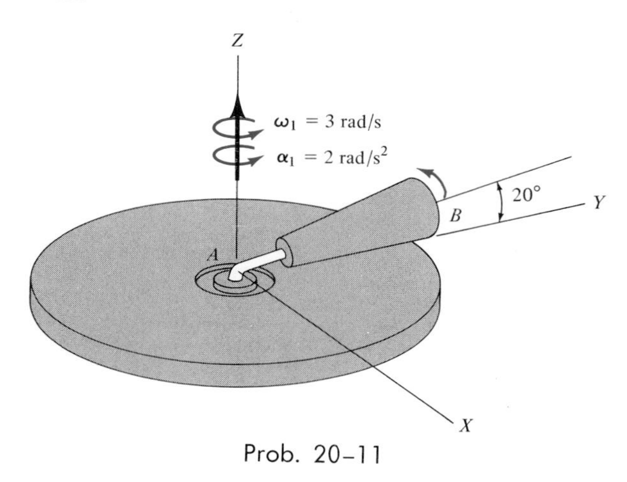

Prob. 20–11

***20–12.** The differential of an automobile allows the two rear wheels to rotate at different speeds when the automobile travels along a curve. For operation, the rear axles are attached to the wheels at one end and have beveled gears A and B on their other ends. The differential case D is placed over the left axle, but can rotate about C, independently of the axle. The case supports a pinion gear E on a shaft, which meshes with gears A and B. Finally, a ring gear G is *fixed* to the differential case, so that the case rotates with the ring gear when the latter is driven by the drive pinion H. This gear, like the differential case, is free to rotate about the left wheel axle. If the drive pinion is turning at $\omega_H = 120$ rad/s and the pinion gear E is spinning about its shaft at $\omega_E = 20$ rad/s, compute the angular velocity, ω_A and ω_B, of each axle.

Prob. 20–12

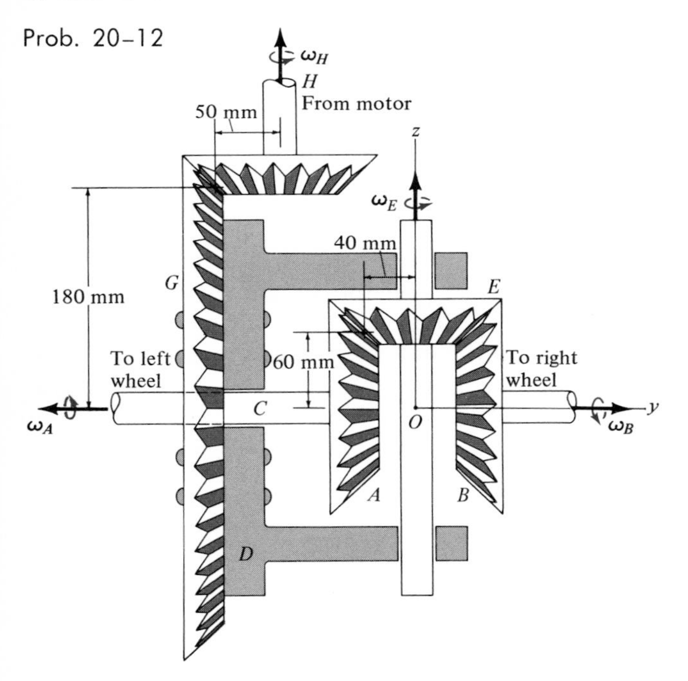

20–13. A thin rod is initially coincident with the Z axis when it is given three rotations defined by the Euler angles $\phi = 60°$, $\theta = 15°$, and $\psi = 30°$. If these rotations are given in the order stated, determine the final coordinate direction angles α, β, and γ of the rod with respect to the X, Y, and Z coordinate axes. Are these angles the same for any order of the rotations? Why?

20–14. Show that the angular velocity of a body, in terms of Euler angles ϕ, θ, and ψ, may be expressed as $\boldsymbol{\omega} = (\dot{\phi} \sin\theta \sin\psi + \dot{\theta}\cos\psi)\mathbf{i} + (\dot{\phi}\sin\theta\cos\psi - \dot{\theta}\sin\psi)\mathbf{j} + (\dot{\phi}\cos\theta + \dot{\psi})\mathbf{k}$, where **i**, **j**, and **k** are unit vectors directed along the x, y, and z axes shown in Fig. 21–9d.

20–15. The rod AB is attached to collars at its ends by ball-and-socket joints. If, at a given instant, collar A has a velocity of $v_A = 1.5$ m/s as shown, determine the velocity of collar B. $a = 0.4$ m, $b = 1.2$ m, $c = 0.6$ m.

Prob. 20–15

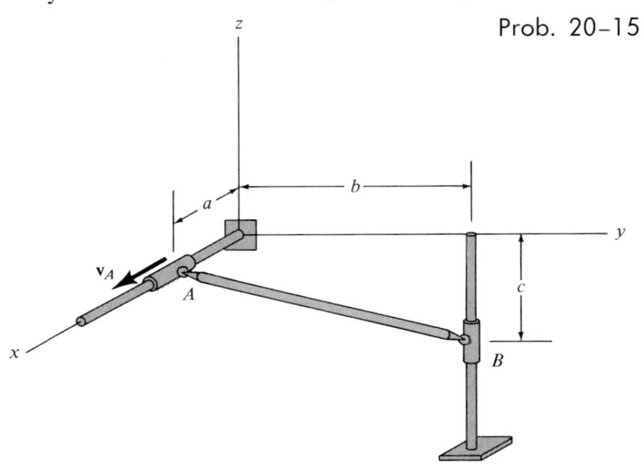

20–15a. Solve Prob. 20–15 if $v_A = 4$ ft/s, $a = 1$ ft, $b = 3$ ft, $c = 1.5$ ft.

***20–16.** Rod AB is attached to a disk and a collar by ball-and-socket joints. If the disk is rotating at a constant angular velocity of 4 rad/s, determine the velocity and acceleration of the collar at A at the instant shown. Assume that the angular velocity of the rod is directed perpendicular to the axis of the rod.

Prob. 20–16

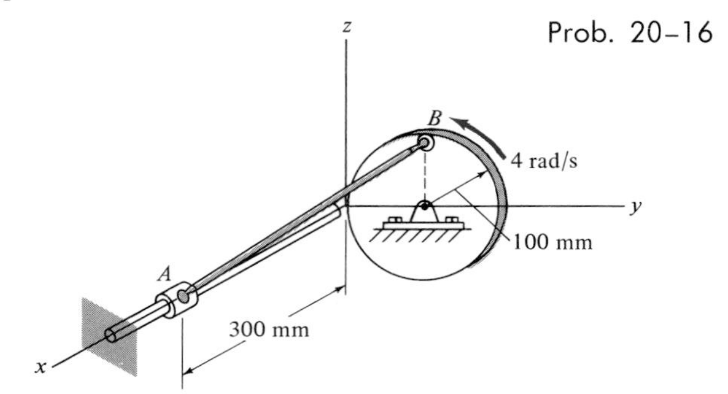

20–17. If the rod is attached to smooth collars at its end points by ball-and-socket joints, determine the speed of B if A is moving downward at a constant speed of 0.5 m/s. Also, determine the angular velocity of the rod if it is directed perpendicular to the axis of the rod.

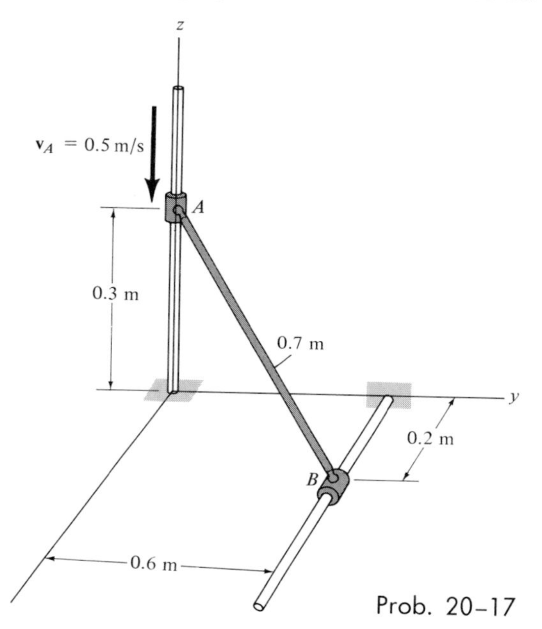

Prob. 20–17

20–18. Disk A is rotating at a constant angular velocity of 10 rad/s. If rod BC is joined to the disk and a collar by ball-and-socket joints, determine the velocity of collar B and the component of angular velocity normal to the rod at the instant shown.

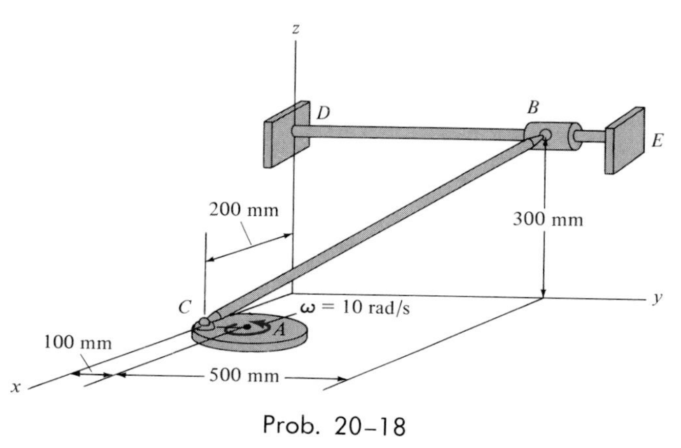

Prob. 20–18

20–19. Solve Prob. 20–18 if the connection at B consists of a pin as shown in the figure below, rather than a ball-and-socket joint. *Hint:* The constraint allows rotation of the rod both along bar DE (**j** direction) and along the axis of the pin (**n** direction). Since there is no rotational component in the **u** direction, i.e., perpendicular to **n** and **j** where $\mathbf{u} = \mathbf{j} \times \mathbf{n}$, an additional equation for solution can be obtained from $\boldsymbol{\omega} \cdot \mathbf{u} = 0$. The vector **n** is in the same direction as $\mathbf{r}_{B/C} \times \mathbf{r}_{D/C}$.

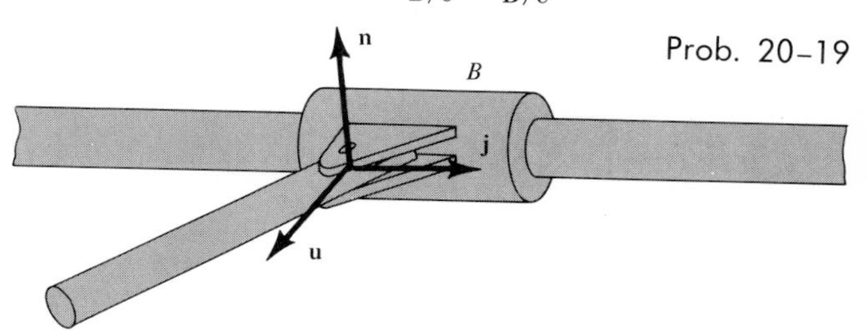

Prob. 20–19

***20–20.** The end C of the bent rod rests on the horizontal plane, while end points A and B are restricted to move along the grooved slots. If at the instant shown, A is moving downward with a constant speed of $v_A = 0.5$ m/s, determine the angular velocity of the rod and the velocities of points B and C. $a = 0.4$ m, $b = 0.3$ m, and $c = 1$ m.

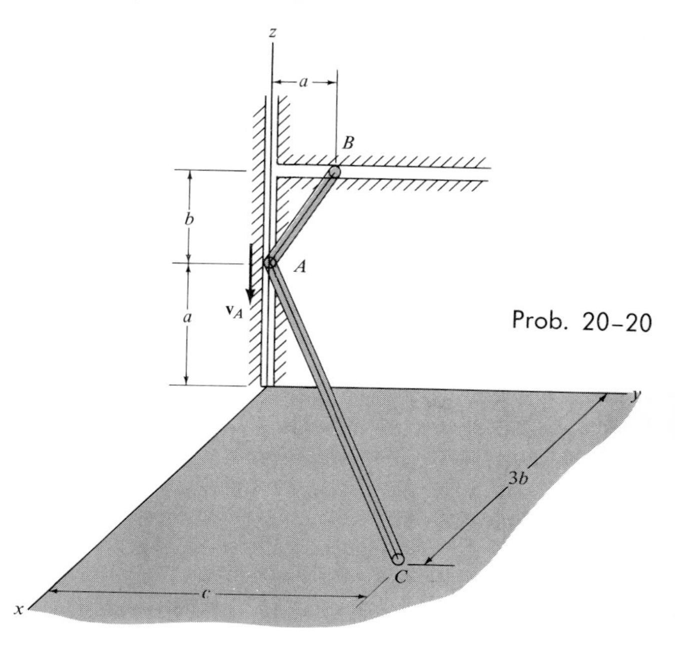

Prob. 20–20

***20–20a.** Solve Prob. 20–20 if $v_A = 1$ ft/s, $a = 0.8$ ft, $b = 0.6$ ft, and $c = 2$ ft.

20–6. Relative-Motion Analysis Using Translating and Rotating Axes

The most general way to analyze the spatial motion of a rigid body requires the use of a system of x, y, z axes which both translate and rotate relative to a second frame X, Y, Z. This analysis also provides a means for determining the motions of two points on a mechanism which are not located in the same rigid body, and for determining the relative motion of one particle with respect to another when one or both particles are moving along *rotating paths*. In this section two equations will be developed which relate the velocities and accelerations of two points A and B, of which one point moves relative to a frame of reference subjected to both translation and rotation. Because of the generality in the derivation, A and B may represent either two particles moving independently of one another or two points located in the same (or different) rigid bodies.

As shown in Fig. 20–15, the locations of points A and B are specified relative to the X, Y, Z frame of reference by position vectors $\mathbf{r}_A$ and $\mathbf{r}_B$. The base point A represents the origin of the x, y, z coordinate system, which is translating and rotating with respect to X, Y, Z. At the instant considered, the velocity and acceleration of point A are $\mathbf{v}_A$ and $\mathbf{a}_A$, respectively; and the angular velocity and angular acceleration of the x, y, z axes are $\mathbf{\Omega}$ and $\dot{\mathbf{\Omega}} = d\mathbf{\Omega}/dt$, respectively. All these vectors are *measured* with respect to the X, Y, Z frame of reference, although they may be expressed in Cartesian component form along either set of axes.

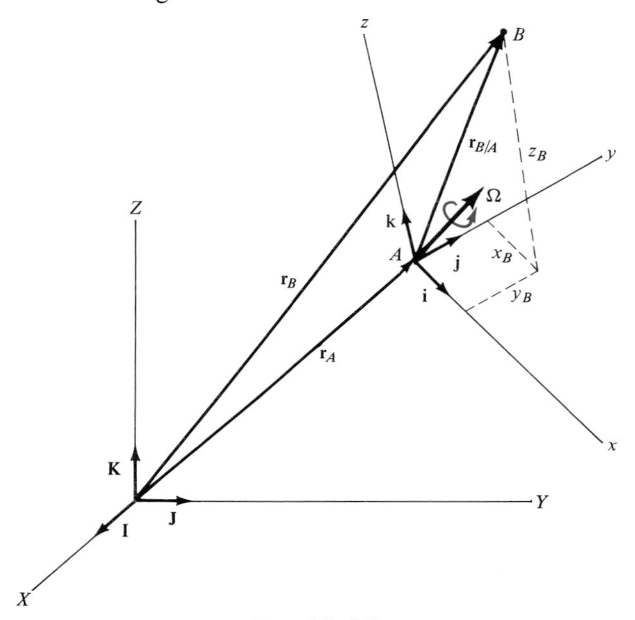

Fig. 20–15

Position. If the position of "B with respect to A" is specified by the *relative-position vector* $\mathbf{r}_{B/A}$, Fig. 20–15, then, by vector addition,

$$\mathbf{r}_B = \mathbf{r}_A + \mathbf{r}_{B/A} \tag{20–20}$$

Velocity. The velocity of point B, measured from X, Y, Z, is determined by taking the time derivative of Eq. 20–20, which yields

$$(\dot{\mathbf{r}}_B)_{XYZ} = (\dot{\mathbf{r}}_A)_{XYZ} + (\dot{\mathbf{r}}_{B/A})_{XYZ} \tag{20–21}$$

The first two terms represent $\mathbf{v}_B$ and $\mathbf{v}_A$. The last term is evaluated by applying Eq. 20–10, since $\mathbf{r}_{B/A}$ is measured between two points in a rotating reference. Hence,

$$(\dot{\mathbf{r}}_{B/A})_{XYZ} = (\dot{\mathbf{r}}_{B/A})_{xyz} + \mathbf{\Omega} \times \mathbf{r}_{B/A} = \mathbf{v}_{B/A} + \mathbf{\Omega} \times \mathbf{r}_{B/A} \tag{20–22}$$

Here $\mathbf{v}_{B/A}$ is the relative velocity of B with respect to A measured from x, y, z. Substituting into Eq. 20–21, we have

$$\mathbf{v}_B = \mathbf{v}_A + \mathbf{\Omega} \times \mathbf{r}_{B/A} + \mathbf{v}_{B/A} \tag{20–23}$$

where

$\mathbf{v}_B$ = velocity of B

$\mathbf{v}_A$ = velocity of the origin A of the x, y, z frame of reference

$\mathbf{v}_{B/A}$ = relative velocity of "B with respect to A" as measured by an observer attached to the rotating x, y, z frame of reference

$\mathbf{\Omega}$ = instantaneous angular velocity of the x, y, z frame of reference

$\mathbf{r}_{A/B}$ = relative position of "B with respect to A" as measured from the x, y, z frame of reference

Acceleration. The acceleration of point B, measured from X, Y, Z, is determined by taking the time derivative of Eq. 20–23, which yields

$$(\dot{\mathbf{v}}_B)_{XYZ} = (\dot{\mathbf{v}}_A)_{XYZ} + (\dot{\mathbf{\Omega}})_{XYZ} \times \mathbf{r}_{B/A} + \mathbf{\Omega} \times (\dot{\mathbf{r}}_{B/A})_{XYZ} + (\dot{\mathbf{v}}_{B/A})_{XYZ} \tag{20–24}$$

The time derivatives defined in the first and second terms represent $\mathbf{a}_B$ and $\mathbf{a}_A$ respectively. The fourth term is evaluated using Eq. 20–22, and the last term is evaluated by applying Eq. 20–10, which yields

$$(\dot{\mathbf{v}}_{B/A})_{XYZ} = (\dot{\mathbf{v}}_{B/A})_{xyz} + \mathbf{\Omega} \times \mathbf{v}_{B/A} = \mathbf{a}_{B/A} + \mathbf{\Omega} \times \mathbf{v}_{B/A}$$

Here $\mathbf{a}_{B/A}$ is the relative acceleration of B with respect to A, measured from x, y, z. Substituting this result, and Eq. 20–22, into Eq. 20–24 and simplifying, we have

$$\mathbf{a}_B = \mathbf{a}_A + \dot{\mathbf{\Omega}} \times \mathbf{r}_{B/A} + \mathbf{\Omega} \times (\mathbf{\Omega} \times \mathbf{r}_{B/A}) + 2\mathbf{\Omega} \times \mathbf{v}_{B/A} + \mathbf{a}_{B/A} \qquad (20\text{–}25)$$

where

$\mathbf{a}_B$ = acceleration of B

$\mathbf{a}_A$ = acceleration of the origin A of the x, y, z frame of reference

$\mathbf{a}_{B/A}$, $\mathbf{v}_{B/A}$ = relative acceleration and relative velocity of "B with respect to A" as measured by an observer attached to the rotating x, y, z frame of reference

$\dot{\mathbf{\Omega}}$, $\mathbf{\Omega}$ = instantaneous angular acceleration and instantaneous angular velocity of the x, y, z frame of reference

$\mathbf{r}_{B/A}$ = relative position of "B with respect to A" as measured from the x, y, z frame of reference

Equations 20–23 and 20–25 are identical to those used in Sec. 16–10 for analyzing relative plane motion. In that case, however, application of these equations is simplified since $\mathbf{\Omega}$ and $\dot{\mathbf{\Omega}}$ have a constant direction which is always perpendicular to the plane of motion. For spatial motion, $\dot{\mathbf{\Omega}}$ must be computed by using Eq. 20–10, since it depends upon the change in both the magnitude and direction of $\mathbf{\Omega}$. Furthermore, in some problems, calculation of $\mathbf{v}_A$, $\mathbf{a}_A$ and $\mathbf{v}_{B/A}$, $\mathbf{a}_{B/A}$ must also be performed by using Eq. 20–10, since these quantities depend upon the angular rate at which $\mathbf{r}_A$ and $\mathbf{r}_{B/A}$ are "swinging" as measured from their respective frames of reference.

PROCEDURE FOR ANALYSIS

The following four-step procedure should be used when applying Eqs. 20–23 and 20–25 to solve problems involving the spatial motion of particles or rigid bodies.

Step 1: Define the location and orientation of the X, Y, Z and x, y, z coordinate axes. Most often solutions are easily obtained if at the instant considered: (1) the origins are *coincident,* (2) the axes are collinear, and/or (3) the axes are parallel. Since several components of angular velocity may be involved in a problem, the calculations will be reduced if the x, y, z axes are selected such that only one component of angular velocity is observed in this frame ($\mathbf{\Omega}_{B/A}$) and the frame rotates with $\mathbf{\Omega}$ defined by the other components of angular velocity.

Step 2: After the origin of the moving reference, A, is defined and the moving point B is specified, Eqs. 20–23 and 20–25 should be written in symbolic form as

$$\mathbf{v}_B = \mathbf{v}_A + \mathbf{\Omega} \times \mathbf{r}_{B/A} + \mathbf{v}_{B/A}$$
$$\mathbf{a}_B = \mathbf{a}_A + \dot{\mathbf{\Omega}} \times \mathbf{r}_{B/A} + \mathbf{\Omega} \times (\mathbf{\Omega} \times \mathbf{r}_{B/A}) + 2\mathbf{\Omega} \times \mathbf{v}_{B/A} + \mathbf{a}_{B/A}$$

Step 3: Vectors $\mathbf{\Omega}$, $\mathbf{r}_A$, and $\mathbf{r}_{B/A}$ should be defined from the problem data and represented in Cartesian form. *Motion of the moving reference* ($\mathbf{v}_A$, $\mathbf{a}_A$, and $\dot{\mathbf{\Omega}}$) is determined by applying Eq. 20–10 to compute the time derivatives of $\mathbf{r}_A$ and $\mathbf{\Omega}$. In a similar manner, if $\mathbf{r}_{B/A}$ has an angular motion $\mathbf{\Omega}_{B/A}$ when observed from the moving reference, then the *motion of B with respect to the moving reference* ($\mathbf{v}_{B/A}$ and $\mathbf{a}_{B/A}$) must be determined by applying Eq. 20–10 to compute the time derivatives of $\mathbf{r}_{B/A}$. In all cases, the kinematic quantities which are involved in computing the time derivatives of $\mathbf{r}_A$, $\mathbf{r}_{B/A}$, and $\mathbf{\Omega}$ should be treated as *variables.* After the final forms of $\mathbf{v}_A$, $\mathbf{a}_A$, $\mathbf{v}_{B/A}$, $\mathbf{a}_{B/A}$, and $\dot{\mathbf{\Omega}}$ are obtained, numerical problem data may be substituted and the kinematic terms evaluated. The components of all these vectors may be selected either along the X, Y, Z axes or along x, y, z. The choice is arbitrary, provided a consistent set of unit vectors is used.

Step 4: Substitute the data into the kinematic equations of *Step 2* and perform the vector operations.

The following two examples numerically illustrate this four-step procedure.

Example 20–5

A motor M is fixed to the surface of a platform that has an angular motion as shown in Fig. 20–16. A rod AB is attached to the motor, rotates about the axis of the motor at an angular speed of $\omega_M = 3$ rad/s, and has an angular acceleration of $\alpha_M = 1$ rad/s^2, as shown. A smooth collar C passes over the rod, and at the instant the rod is in the vertical position, the collar, located 0.25 m from A, is moving downward along the rod with a velocity of 3 m/s and an acceleration of 2 m/s^2. Determine the absolute velocity and acceleration of the collar at this instant.

Solution

Step 1: The origin of the fixed X, Y, Z reference is chosen at the center of the platform, and the origin of the moving x, y, z frame at point A. At the instant considered, the axes are oriented as shown in Fig. 20–16. Since the collar is subjected to two components of angular motion, $\boldsymbol{\omega}_p$ and $\boldsymbol{\omega}_M$, it is viewed as having an angular velocity of $\mathbf{\Omega}_{C/A} = \boldsymbol{\omega}_M$ in x, y, z and the x, y, z axes are attached to the platform so that $\mathbf{\Omega} = \boldsymbol{\omega}_p$.

Step 2: Equations 20–23 and 20–25, applied to points C and A, become

$$\mathbf{v}_C = \mathbf{v}_A + \mathbf{\Omega} \times \mathbf{r}_{C/A} + \mathbf{v}_{C/A} \tag{1}$$

$$\mathbf{a}_C = \mathbf{a}_A + \dot{\mathbf{\Omega}} \times \mathbf{r}_{C/A} + \mathbf{\Omega} \times (\mathbf{\Omega} \times \mathbf{r}_{C/A}) + 2\mathbf{\Omega} \times \mathbf{v}_{C/A} + \mathbf{a}_{C/A} \tag{2}$$

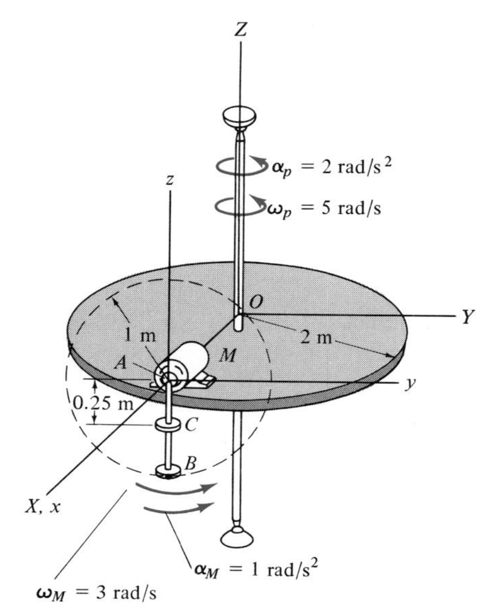

Fig. 20–16

Step 3: Motion of moving reference:

$$\begin{aligned}
\boldsymbol{\Omega} &= \boldsymbol{\omega}_p = 5\mathbf{k} \\
\dot{\boldsymbol{\Omega}} &= (\dot{\boldsymbol{\omega}}_p)_{xyz} + \boldsymbol{\omega}_p \times \boldsymbol{\omega}_p \\
&= 2\mathbf{k} + 5\mathbf{k} \times 5\mathbf{k} = 2\mathbf{k} \\
\mathbf{r}_A &= 2\mathbf{i} \\
\mathbf{v}_A &= \dot{\mathbf{r}}_A = (\dot{\mathbf{r}}_A)_{xyz} + \boldsymbol{\omega}_p \times \mathbf{r}_A \\
&= \mathbf{0} + (5\mathbf{k} \times 2\mathbf{i}) = 10\mathbf{j} \\
\mathbf{a}_A &= \ddot{\mathbf{r}}_A = (\ddot{\mathbf{r}}_A)_{xyz} + \dot{\boldsymbol{\omega}}_p \times \mathbf{r}_A + \boldsymbol{\omega}_p \times \dot{\mathbf{r}}_A \\
&= \mathbf{0} + (2\mathbf{k} \times 2\mathbf{i}) + (5\mathbf{k} \times 10\mathbf{j}) = -50\mathbf{i} + 4\mathbf{j}
\end{aligned}$$

Motion of C with respect to moving reference:

$$\begin{aligned}
\boldsymbol{\Omega}_{C/A} &= \boldsymbol{\omega}_M = 3\mathbf{i} \\
\dot{\boldsymbol{\Omega}}_{C/A} &= \dot{\boldsymbol{\omega}}_M = 1\mathbf{i} \\
\mathbf{r}_{C/A} &= -0.25\mathbf{k} \\
\mathbf{v}_{C/A} &= \dot{\mathbf{r}}_{C/A} = (\dot{\mathbf{r}}_{C/A})_{xyz} + \boldsymbol{\omega}_M \times \mathbf{r}_{C/A} \\
&= -3\mathbf{k} + [3\mathbf{i} \times (-0.25\mathbf{k})] = 0.75\mathbf{j} - 3\mathbf{k} \\
\mathbf{a}_{C/A} &= \ddot{\mathbf{r}}_{C/A} = (\ddot{\mathbf{r}}_{C/A})_{xyz} + \dot{\boldsymbol{\omega}}_M \times \mathbf{r}_{C/A} + \boldsymbol{\omega}_M \times \dot{\mathbf{r}}_{C/A} \\
&= -2\mathbf{k} + [(1\mathbf{i}) \times (-0.25\mathbf{k})] + [(3\mathbf{i}) \times (0.75\mathbf{j} - 3\mathbf{k})] \\
&= 9.25\mathbf{j} + 0.25\mathbf{k}
\end{aligned}$$

Step 4: Substituting the data into Eqs. (1) and (2) yields

$$\begin{aligned}
\mathbf{v}_C &= \mathbf{v}_A + \boldsymbol{\Omega} \times \mathbf{r}_{C/A} + \mathbf{v}_{C/A} \\
&= 10\mathbf{j} + [5\mathbf{k} \times (-0.25\mathbf{k})] + (0.75\mathbf{j} - 3\mathbf{k}) \\
&= \{10.75\mathbf{j} - 3\mathbf{k}\}\ \text{m/s} \qquad \textit{Ans.}
\end{aligned}$$

$$\begin{aligned}
\mathbf{a}_C &= \mathbf{a}_A + \dot{\boldsymbol{\Omega}} \times \mathbf{r}_{C/A} + \boldsymbol{\Omega} \times (\boldsymbol{\Omega} \times \mathbf{r}_{C/A}) + 2\boldsymbol{\Omega} \times \mathbf{v}_{C/A} + \mathbf{a}_{C/A} \\
&= (-50\mathbf{i} + 4\mathbf{j}) + [2\mathbf{k} \times (-0.25\mathbf{k})] + 5\mathbf{k} \times [5\mathbf{k} \times (-0.25\mathbf{k})] \\
&\quad + 2[5\mathbf{k} \times (0.75\mathbf{j} - 3\mathbf{k})] + (9.25\mathbf{j} + 0.25\mathbf{k}) \\
&= \{-57.5\mathbf{i} + 13.25\mathbf{j} + 0.25\mathbf{k}\}\ \text{m/s}^2 \qquad \textit{Ans.}
\end{aligned}$$

Example 20–6

The pendulum shown in Fig. 20–17 consists of two rods. AB is pin-supported at A and swings only in the Y-Z plane, whereas a bearing at B allows the attached rod BD to spin about rod AB. At a given instant, the rods have the angular motions shown. If a collar C, located 0.2 m from B, has a velocity of 3 m/s and an acceleration of 2 m/s^2 along the rod, determine the absolute velocity and acceleration of the collar at this instant.

Solution

Step 1: Since the rod is rotating about the fixed point A, the origin of the X, Y, Z frame will be chosen at A. Motion of the collar is conveniently observed from B, so the origin of the x, y, z frame is located at this point. Three choices can be made as to the "fixity" of the x, y, z axes. They can be fixed to rod BD so that $\mathbf{\Omega} = \boldsymbol{\omega}_1 + \boldsymbol{\omega}_2$ and the collar appears only to move radially outward along BD, hence $\mathbf{\Omega}_{C/B} = \mathbf{0}$; the axes can have a spin $\mathbf{\Omega} = \boldsymbol{\omega}_1$, in which case the collar appears not only to have a radial motion, but also a rotation $\mathbf{\Omega}_{C/B} = \boldsymbol{\omega}_2$; or the axes can move with curvilinear translation, $\mathbf{\Omega} = \mathbf{0}$, in which case the collar appears to have both an angular velocity of $\mathbf{\Omega}_{C/B} = \boldsymbol{\omega}_1 + \boldsymbol{\omega}_2$ and radial motion. The solution of each of these three cases will be presented to better illustrate the concepts involved.

Step 2: From Eqs. 20–23 and 20–25 it is required that

$$\mathbf{v}_C = \mathbf{v}_B + \mathbf{\Omega} \times \mathbf{r}_{C/B} + \mathbf{v}_{C/B} \qquad (1)$$

$$\mathbf{a}_C = \mathbf{a}_B + \dot{\mathbf{\Omega}} \times \mathbf{r}_{C/B} + \mathbf{\Omega} \times (\mathbf{\Omega} \times \mathbf{r}_{C/B}) + 2\mathbf{\Omega} \times \mathbf{v}_{C/B} + \mathbf{a}_{C/B} \qquad (2)$$

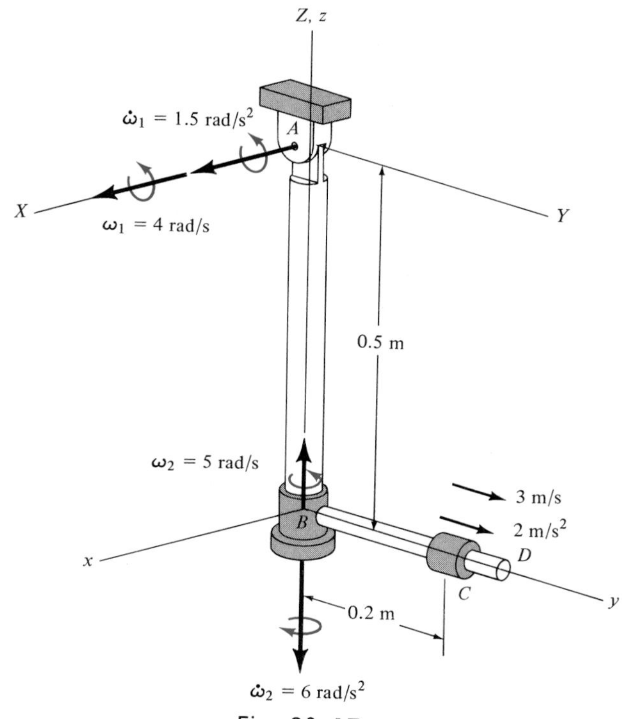

Fig. 20–17

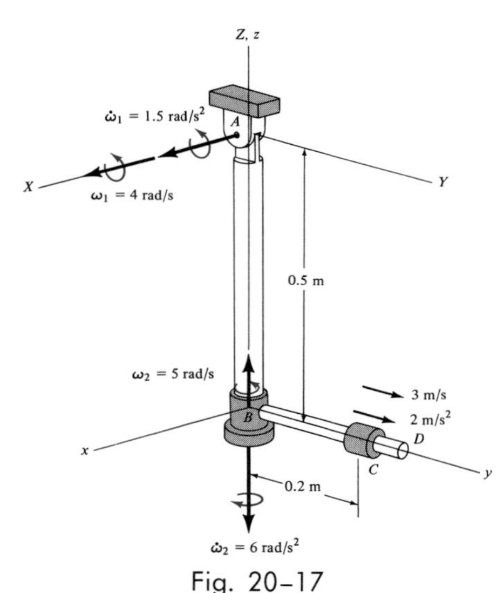

Fig. 20–17

Solution I $(\boldsymbol{\Omega} = \boldsymbol{\omega}_1 + \boldsymbol{\omega}_2,\ \boldsymbol{\Omega}_{C/B} = \mathbf{0})$.

Step 3: Motion of moving reference:

$$\boldsymbol{\Omega} = \boldsymbol{\omega}_1 + \boldsymbol{\omega}_2 = 4\mathbf{i} + 5\mathbf{k}$$

From the constraints of the problem $\boldsymbol{\omega}_1$ does not change direction; however, the direction of $\boldsymbol{\omega}_2$ is changed by $\boldsymbol{\omega}_1$. Thus, a simple way of obtaining $\dot{\boldsymbol{\Omega}}$ is to consider x', y', z' axes coincident with the X, Y, Z axes at A, such that the primed axes have an angular velocity $\boldsymbol{\omega}_1$. Then

$$\begin{aligned}\dot{\boldsymbol{\Omega}} &= \dot{\boldsymbol{\omega}}_1 + \dot{\boldsymbol{\omega}}_2 = [(\dot{\boldsymbol{\omega}}_1)_{x'y'z'} + \boldsymbol{\omega}_1 \times \boldsymbol{\omega}_1] + [(\dot{\boldsymbol{\omega}}_2)_{x'y'z'} + \boldsymbol{\omega}_1 \times \boldsymbol{\omega}_2] \\ &= [1.5\mathbf{i} + \mathbf{0}] + [-6\mathbf{k} + (4\mathbf{i} \times 5\mathbf{k})] = 1.5\mathbf{i} - 20\mathbf{j} - 6\mathbf{k}\end{aligned}$$

Also,

$$\mathbf{r}_B = -0.5\mathbf{k}$$

Here again only $\boldsymbol{\omega}_1$ changes the direction of $\mathbf{r}_B$ so that the time derivatives of $\mathbf{r}_B$ can be computed from the primed axes used above, which are rotating at $\boldsymbol{\omega}_1$. Hence

$$\begin{aligned}\mathbf{v}_B = \dot{\mathbf{r}}_B &= (\dot{\mathbf{r}}_B)_{x'y'z'} + \boldsymbol{\omega}_1 \times \mathbf{r}_B \\ &= \mathbf{0} + 4\mathbf{i} \times (-0.5\mathbf{k}) = 2\mathbf{j} \\ \mathbf{a}_B = \ddot{\mathbf{r}}_B &= [(\ddot{\mathbf{r}}_B)_{x'y'z'} + \boldsymbol{\omega}_1 \times (\dot{\mathbf{r}}_B)_{x'y'z'}] + \dot{\boldsymbol{\omega}}_1 \times \mathbf{r}_B + \boldsymbol{\omega}_1 \times \dot{\mathbf{r}}_B \\ &= [\mathbf{0} + \mathbf{0}] + [1.5\mathbf{i} \times (-0.5\mathbf{k})] + (4\mathbf{i} \times 2\mathbf{j}) = 0.75\mathbf{j} + 8\mathbf{k}\end{aligned}$$

Motion of C with respect to moving reference:

$$\begin{aligned}\boldsymbol{\Omega}_{C/B} &= \mathbf{0} \\ \dot{\boldsymbol{\Omega}}_{C/B} &= \mathbf{0} \\ \mathbf{r}_{C/B} &= 0.2\mathbf{j} \\ \mathbf{v}_{C/B} = \dot{\mathbf{r}}_{C/B} &= (\dot{\mathbf{r}}_{C/B})_{xyz} + \boldsymbol{\Omega}_{C/B} \times \mathbf{r}_{C/B} \\ &= 3\mathbf{j} + \mathbf{0} = 3\mathbf{j} \\ \mathbf{a}_{C/B} = \ddot{\mathbf{r}}_{C/B} &= [(\ddot{\mathbf{r}}_{C/B})_{xyz} + \boldsymbol{\Omega}_{C/B} \times (\dot{\mathbf{r}}_{C/B})_{xyz}] + \dot{\boldsymbol{\Omega}}_{C/B} \times \mathbf{r}_{C/B} + \boldsymbol{\Omega}_{C/B} \times \dot{\mathbf{r}}_{C/B} \\ &= [2\mathbf{j} + \mathbf{0}] + \mathbf{0} + \mathbf{0} = 2\mathbf{j}\end{aligned}$$

Step 4: Substituting the data into Eqs. (1) and (2) yields

$$\begin{aligned}\mathbf{v}_C &= \mathbf{v}_B + \boldsymbol{\Omega} \times \mathbf{r}_{C/B} + \mathbf{v}_{C/B} \\ &= 2\mathbf{j} + [(4\mathbf{i} + 5\mathbf{k}) \times (0.2\mathbf{j})] + 3\mathbf{j} \\ &= \{-1\mathbf{i} + 5\mathbf{j} + 0.8\mathbf{k}\}\ \text{m/s} \qquad \textit{Ans.}\end{aligned}$$

$$\begin{aligned}\mathbf{a}_C &= \mathbf{a}_B + \dot{\boldsymbol{\Omega}} \times \mathbf{r}_{C/B} + \boldsymbol{\Omega} \times (\boldsymbol{\Omega} \times \mathbf{r}_{C/B}) + 2\boldsymbol{\Omega} \times \mathbf{v}_{C/B} + \mathbf{a}_{C/B} \\ &= (0.75\mathbf{j} + 8\mathbf{k}) + [(1.5\mathbf{i} - 20\mathbf{j} - 6\mathbf{k}) \times (0.2\mathbf{j})] \\ &\quad + (4\mathbf{i} + 5\mathbf{k}) \times [(4\mathbf{i} + 5\mathbf{k}) \times 0.2\mathbf{j}] + 2[(4\mathbf{i} + 5\mathbf{k}) \times 3\mathbf{j}] + 2\mathbf{j} \\ &= \{-28.8\mathbf{i} - 5.45\mathbf{j} + 32.3\mathbf{k}\}\ \text{m/s}^2 \qquad \textit{Ans.}\end{aligned}$$

Solution II $(\boldsymbol{\Omega} = \boldsymbol{\omega}_1, \boldsymbol{\Omega}_{C/B} = \boldsymbol{\omega}_2)$.
Step 3: Motion of moving reference:

$$\boldsymbol{\Omega} = \boldsymbol{\omega}_1 = 4\mathbf{i}$$
$$\dot{\boldsymbol{\Omega}} = (\dot{\boldsymbol{\omega}}_1)_{xyz} + \boldsymbol{\omega}_1 \times \boldsymbol{\omega}_1 = 1.5\mathbf{i} + \mathbf{0} = 1.5\mathbf{i}$$
$$\mathbf{r}_B = -0.5\mathbf{k}$$
$$\mathbf{v}_B = \dot{\mathbf{r}}_B = (\mathbf{r}_B)_{xyz} + \boldsymbol{\omega}_1 \times \mathbf{r}_B$$
$$= \mathbf{0} + 4\mathbf{i} \times (-0.5\mathbf{k}) = 2\mathbf{j}$$
$$\mathbf{a}_B = \ddot{\mathbf{r}}_B = [(\ddot{\mathbf{r}}_B)_{xyz} + \boldsymbol{\omega}_1 \times (\dot{\mathbf{r}}_B)_{xyz}] + \dot{\boldsymbol{\omega}}_1 \times \mathbf{r}_B + \boldsymbol{\omega}_1 \times \dot{\mathbf{r}}_B$$
$$= [\mathbf{0} + \mathbf{0}] + [1.5\mathbf{i} \times (-0.5\mathbf{k})] + (4\mathbf{i} \times 2\mathbf{j}) = 0.75\mathbf{j} + 8\mathbf{k}$$

Motion of C with respect to moving reference:

$$\boldsymbol{\Omega}_{C/B} = \boldsymbol{\omega}_2 = 5\mathbf{k}$$
$$\dot{\boldsymbol{\Omega}}_{C/B} = \dot{\boldsymbol{\omega}}_2 = -6\mathbf{k}$$
$$\mathbf{r}_{C/B} = 0.2\mathbf{j}$$
$$\mathbf{v}_{C/B} = \dot{\mathbf{r}}_{C/B} = (\dot{\mathbf{r}}_{C/B})_{xyz} + \boldsymbol{\omega}_2 \times \mathbf{r}_{C/B}$$
$$= 3\mathbf{j} + (5\mathbf{k} \times 0.2\mathbf{j}) = -1\mathbf{i} + 3\mathbf{j}$$
$$\mathbf{a}_{C/B} = \ddot{\mathbf{r}}_{C/B} = [(\ddot{\mathbf{r}}_{C/B})_{xyz} + \boldsymbol{\omega}_2 \times (\dot{\mathbf{r}}_{C/B})_{xyz}] + \dot{\boldsymbol{\omega}}_2 \times \mathbf{r}_{C/B} + \boldsymbol{\omega}_2 \times \dot{\mathbf{r}}_{C/B}$$
$$= [2\mathbf{j} + (5\mathbf{k} \times 3\mathbf{j})] + (-6\mathbf{k} \times 0.2\mathbf{j}) + [5\mathbf{k} \times (-1\mathbf{i} + 3\mathbf{j})]$$
$$= -28.8\mathbf{i} - 3\mathbf{j}$$

Step 4: Substituting the data into Eqs. (1) and (2) yields

$$\mathbf{v}_C = \mathbf{v}_B + \boldsymbol{\Omega} \times \mathbf{r}_{C/B} + \mathbf{v}_{C/B}$$
$$= 2\mathbf{j} + (4\mathbf{i} \times 0.2\mathbf{j}) + (-1\mathbf{i} + 3\mathbf{j})$$
$$= \{-1\mathbf{i} + 5\mathbf{j} + 0.8\mathbf{k}\}\ \text{m/s} \qquad \textit{Ans.}$$

$$\mathbf{a}_C = \mathbf{a}_B + \dot{\boldsymbol{\Omega}} \times \mathbf{r}_{C/B} + \boldsymbol{\Omega} \times (\boldsymbol{\Omega} \times \mathbf{r}_{C/B}) + 2\boldsymbol{\Omega} \times \mathbf{v}_{C/B} + \mathbf{a}_{C/B}$$
$$= (0.75\mathbf{j} + 8\mathbf{k}) + (1.5\mathbf{i} \times 0.2\mathbf{j}) + [4\mathbf{i} \times (4\mathbf{i} \times 0.2\mathbf{j})]$$
$$+ 2[4\mathbf{i} \times (-1\mathbf{i} + 3\mathbf{j})] + (-28.8\mathbf{i} - 3\mathbf{j})$$
$$= \{-28.8\mathbf{i} - 5.45\mathbf{j} + 32.3\mathbf{k}\}\ \text{m/s}^2 \qquad \textit{Ans.}$$

Solution III $(\boldsymbol{\Omega} = \mathbf{0}, \boldsymbol{\Omega}_{C/B} = \boldsymbol{\omega}_1 + \boldsymbol{\omega}_2)$.
Step 3: Motion of moving reference:

$$\boldsymbol{\Omega} = \mathbf{0}$$
$$\dot{\boldsymbol{\Omega}} = \mathbf{0}$$

From solution I:

$$\mathbf{r}_B = -0.5\mathbf{k}$$
$$\mathbf{v}_B = 2\mathbf{j}$$
$$\mathbf{a}_B = 0.75\mathbf{j} + 8\mathbf{k}$$

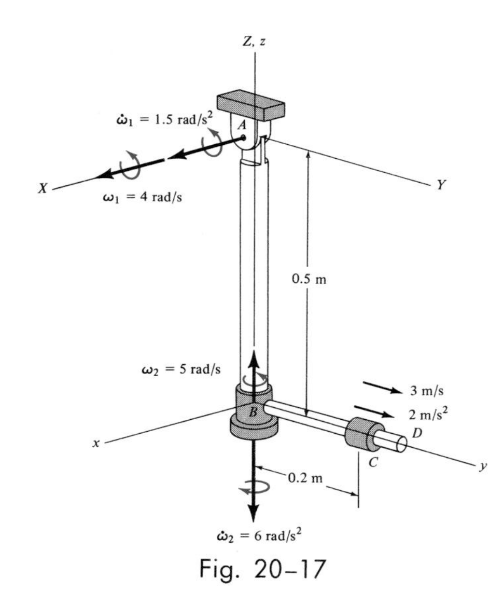

Fig. 20–17

Motion of C with respect to moving reference:

$$\mathbf{\Omega}_{C/B} = \boldsymbol{\omega}_1 + \boldsymbol{\omega}_2 = 4\mathbf{i} + 5\mathbf{k}$$
$$\dot{\mathbf{\Omega}}_{C/B} = \dot{\boldsymbol{\omega}}_1 + \dot{\boldsymbol{\omega}}_2 = 1.5\mathbf{i} - 6\mathbf{k}$$
$$\mathbf{r}_{C/B} = 0.2\mathbf{j}$$
$$\mathbf{v}_{C/B} = \dot{\mathbf{r}}_{C/B} = (\dot{\mathbf{r}}_{C/B})_{xyz} + (\boldsymbol{\omega}_1 + \boldsymbol{\omega}_2) \times \mathbf{r}_{C/B}$$
$$= 3\mathbf{j} + (4\mathbf{i} + 5\mathbf{k}) \times 0.2\mathbf{j}$$
$$= -1\mathbf{j} + 3\mathbf{j} + 0.8\mathbf{k}$$
$$\mathbf{a}_{C/B} = \ddot{\mathbf{r}}_{C/B} = [(\ddot{\mathbf{r}}_{C/B})_{xyz} + (\boldsymbol{\omega}_1 + \boldsymbol{\omega}_2) \times (\dot{\mathbf{r}}_{C/B})_{xyz}] + [(\dot{\boldsymbol{\omega}}_1 + \dot{\boldsymbol{\omega}}_2) \times \mathbf{r}_{C/B}]$$
$$+ [(\boldsymbol{\omega}_1 + \boldsymbol{\omega}_2) \times \dot{\mathbf{r}}_{C/B}]$$
$$= [2\mathbf{j} + (4\mathbf{i} + 5\mathbf{k}) \times 3\mathbf{j}] + [(1.5\mathbf{i} - 6\mathbf{k}) \times 0.2\mathbf{j}]$$
$$+ [(4\mathbf{i} + 5\mathbf{k}) \times (-1\mathbf{i} + 3\mathbf{j} + 0.8\mathbf{k})]$$
$$= -28.8\mathbf{i} - 6.2\mathbf{j} + 24.3\mathbf{k}$$

Step 4: Substituting the data into Eqs. (1) and (2) yields

$$\mathbf{v}_C = \mathbf{v}_B + \mathbf{\Omega} \times \mathbf{r}_{C/B} + \mathbf{v}_{C/B}$$
$$= 2\mathbf{j} + \mathbf{0} + (-1\mathbf{i} + 3\mathbf{j} + 0.8\mathbf{k})$$
$$= \{-1\mathbf{i} + 5\mathbf{j} + 0.8\mathbf{k}\}\ \text{m/s} \qquad \textit{Ans.}$$

$$\mathbf{a}_C = \mathbf{a}_B + \dot{\mathbf{\Omega}} \times \mathbf{r}_{C/B} + \mathbf{\Omega} \times (\mathbf{\Omega} \times \mathbf{r}_{C/B}) + 2\mathbf{\Omega} \times \mathbf{v}_{C/B} + \mathbf{a}_{C/B}$$
$$= (0.75\mathbf{j} + 8\mathbf{k}) + \mathbf{0} + \mathbf{0} + \mathbf{0} + (-28.8\mathbf{i} - 6.2\mathbf{j} + 24.3\mathbf{k})$$
$$= \{-28.8\mathbf{i} - 5.45\mathbf{j} + 32.3\mathbf{k}\}\ \text{m/s}^2 \qquad \textit{Ans.}$$

Problems

20–21. The particle P slides around the circular hoop with a constant angular velocity of $\dot{\theta} = 5$ rad/s, while the hoop rotates about the X axis at a constant rate of $\omega = 3$ rad/s. If at the instant shown, the hoop is in the x-y plane and the angle $\theta = 45°$, determine the velocity and acceleration of the particle at this instant.

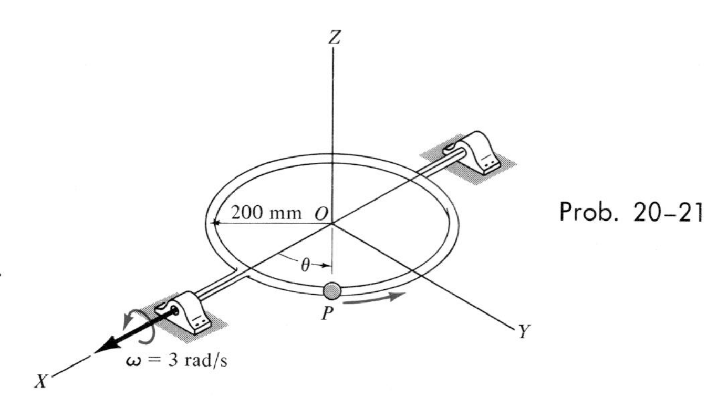

Prob. 20–21

20–22. At a given instant, rod BD is rotating about the vertical axis BD with an angular velocity of $\omega_{BD} = 4$ rad/s and an angular acceleration of $\alpha_{BD} = 3$ rad/s². If at this same instant $\theta = 60°$, link AC is rotating downward such that $\dot{\theta} = 2$ rad/s, and $\ddot{\theta} = 1.5$ rad/s², determine the velocity and acceleration of point A on the link at this instant.

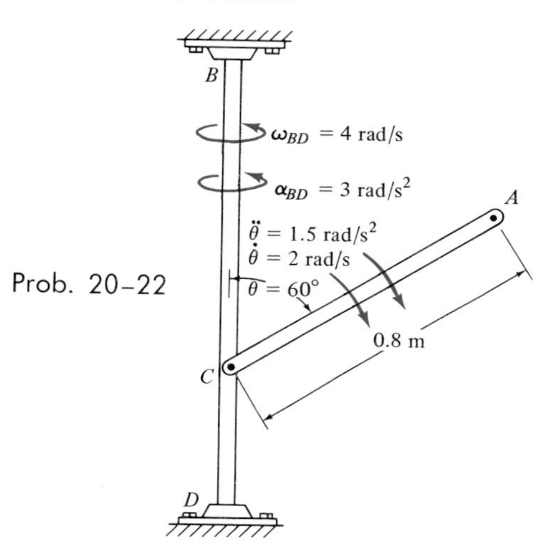

Prob. 20–22

20–23. The boom AB of the locomotive crane is rotating about the Z axis with an angular velocity of $\omega_1 = 6$ rad/s and has a rate of increase of $\dot{\omega}_1 = 2$ rad/s². At the same instant, $\theta = 30°$ and the boom is rotating upward at a constant rate of $\dot{\theta} = 3$ rad/s. Determine the velocity and acceleration of the tip B of the boom at this instant.

Prob. 20–23

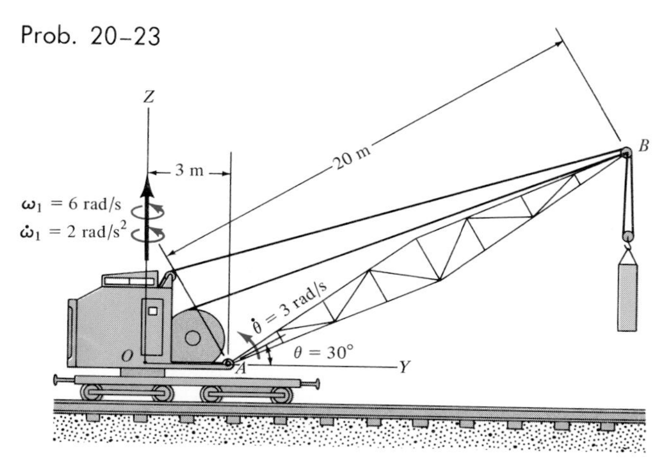

***20–24.** At the instant shown, the frame $ABCD$ is rotating about the X axis with an angular velocity of $\omega_1 = 6$ rad/s and an angular acceleration of $\alpha_1 = 2$ rad/s². At the same instant, the rotating rod EGH has an angular motion *relative to the frame,* as shown in the figure. If a collar P is moving along the rod from G to H with a velocity of 0.5 m/s and has an acceleration of 0.7 m/s², both measured relative to rod EGH, determine the collar's velocity and acceleration at this instant.

Prob. 20–24

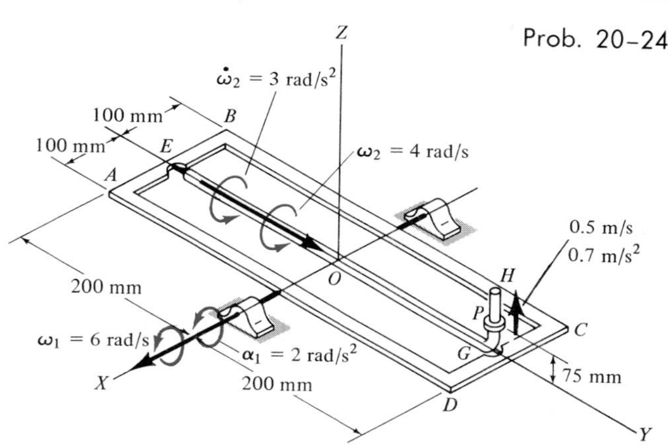

20–25. At a given instant, the radar antenna has an angular motion about the Z axis of $\omega_1 = 3$ rad/s, and $\dot{\omega}_1 = 2$ rad/s². At this same instant $\theta = 30°$, the angular motion about the X axis is $\omega_2 = 1.5$ rad/s, and $\dot{\omega}_2 = 4$ rad/s². Determine the velocity and acceleration of the signal horn A at this instant. The distance from O to A is $d = 1.5$ m.

Prob. 20–25

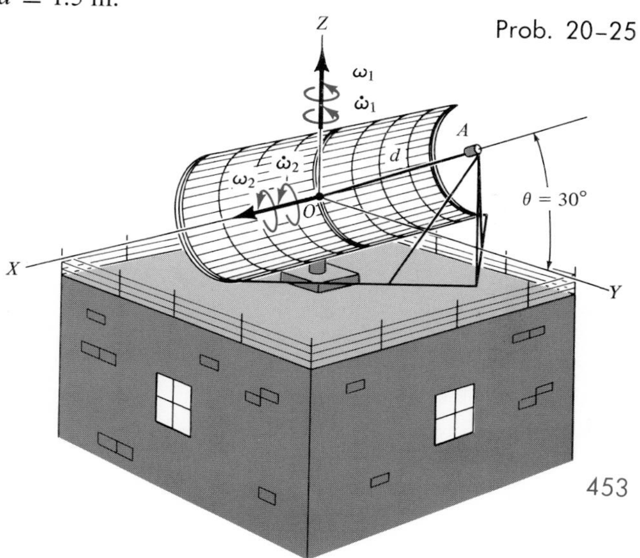

20-25a. Solve Prob. 20-25 if $\omega_1 = 2$ rad/s, $\dot{\omega}_1 = 0.5$ rad/s^2, $\theta = 30°$, $\omega_2 = 0.75$ rad/s, $\dot{\omega}_2 = 1$ rad/s^2, and $d = 3$ ft.

20-26. A ride at an amusement park consists of cars suspended from a frame that is rotating about both the horizontal and vertical axes with the angular motion shown. Compute the velocity and acceleration of the tip B of the boom AB if, at the instant shown, $\theta = 30°$ and the hydraulic cylinder CD is extending such that it causes θ to increase at a constant rate of $\dot{\theta} = 0.6$ rad/s.

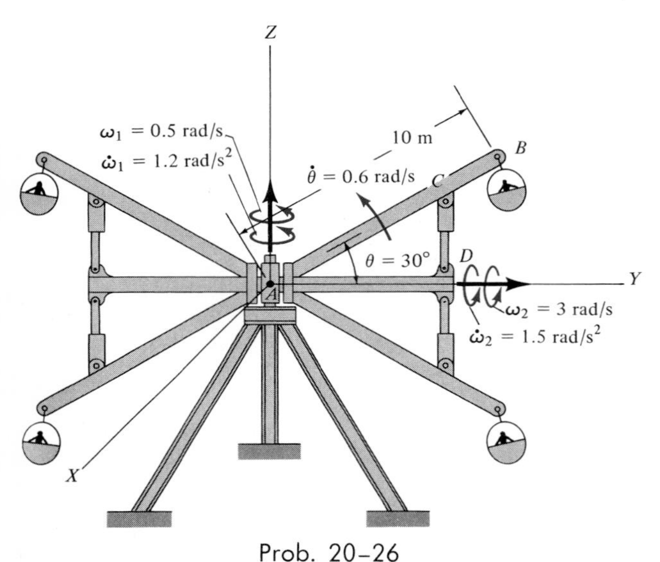

Prob. 20-26

20-27. The load is being lifted upward at a constant rate of 2 m/s relative to the crane boom AB. At the instant shown, the boom is rotating about the vertical at a constant angular rate of $\omega_1 = 3$ rad/s and the trolly T is moving outward along the boom at a constant rate of $v_t = 1.5$ m/s. Furthermore, at this same instant the cables supporting the load are vertical and they are swinging in the Y-Z plane at an angular rate of 5 rad/s, with an increase in the rate of swing of 1 rad/s^2. Determine the velocity and acceleration of the center G of the load at this instant, i.e., when $s = 4$ m and $h = 3$ m.

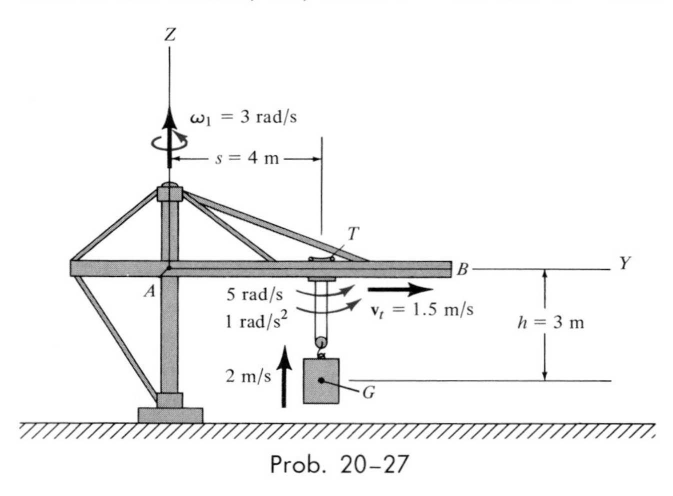

Prob. 20-27

***20–28.** Work Prob. 20–27 assuming that at the instant shown the speed of the trolly T is decreasing at the rate of $2\ \text{m/s}^2$ and ω_1 is increasing at the rate of $\dot{\omega}_1 = 2\ \text{rad/s}^2$. Use the data in Prob. 20–27.

20–29. At a given instant the boom AB of the tower crane is rotating about the Z axis with the motion shown. At this same instant, $\theta = 60°$ and the boom is rotating downward such that $\dot{\theta} = -0.3\ \text{rad/s}$ and $\ddot{\theta} = -0.5\ \text{rad/s}^2$. Determine the velocity and acceleration of the tip of the boom, A, at this instant. The boom has a length of $l_{AB} = 80\ \text{m}$.

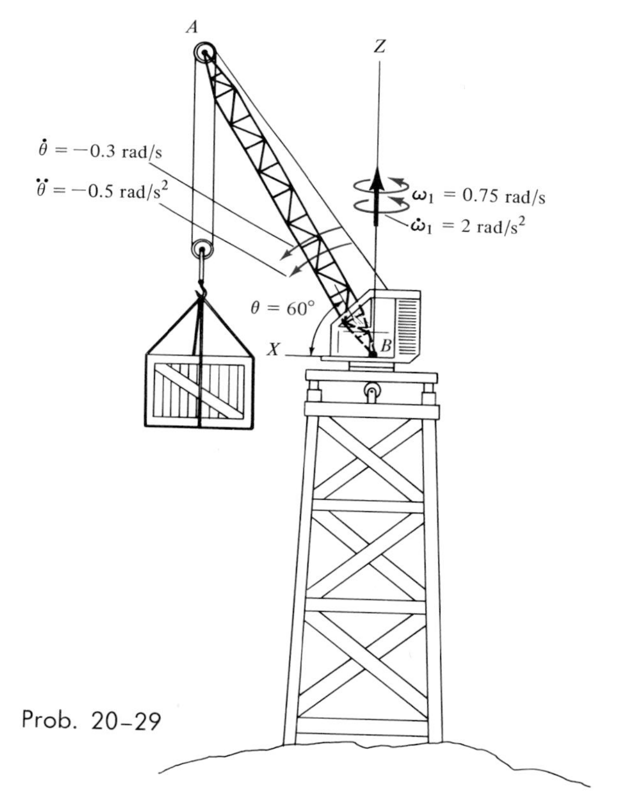

Prob. 20–29

20–30. At the instant shown, the helicopter is moving upward with a velocity of $v_H = 10\ \text{m/s}$ and has an acceleration of $a_H = 2\ \text{m/s}^2$. At the same instant the helicopter *frame* H, *not* the horizontal blade, is rotating about a vertical axis with an angular velocity of $\omega_H = 1.5\ \text{rad/s}$. If the tail blade B is rotating with an angular velocity of $\omega_{B/H} = 200\ \text{rad/s}$, measured relative to H, determine the velocity and acceleration of point P located on the tip of the blade at the instant the blade is in the vertical position. $l = 7\ \text{m}$ and $r = 0.75\ \text{m}$.

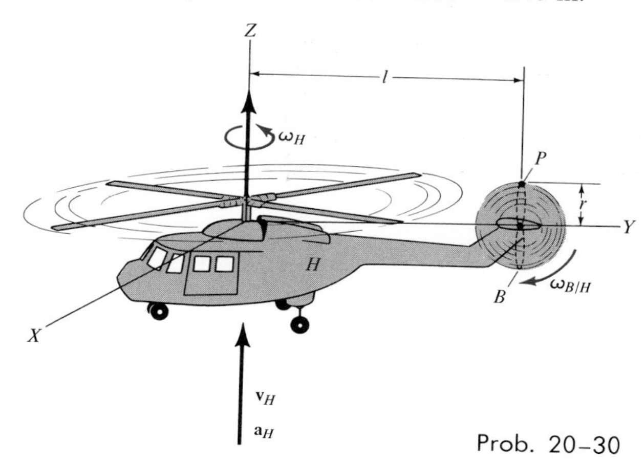

Prob. 20–30

20–30a. Solve Prob. 20–30 if $v_H = 4\ \text{ft/s}$, $a_H = 3\ \text{ft/s}^2$, $\omega_H = 0.8\ \text{rad/s}$, $\omega_{B/H} = 150\ \text{rad/s}$, $l = 20\ \text{ft}$, and $r = 2.5\ \text{ft}$.

20–31. At a given instant, the pillar crane has angular motion about the Z axis as shown. At this same instant the cable AC is being drawn in such that the constant rate of increase of θ is $\dot{\theta} = 0.8\ \text{rad/s}$, while the motor at D is drawing in its attached cable at a constant rate of $3\ \text{m/s}$. Assuming that the load L is very great, so that its suspending cable remains essentially vertical, determine the velocity and acceleration of the crate's mass center G at the instant shown, when $\theta = 30°$.

Prob. 20–31

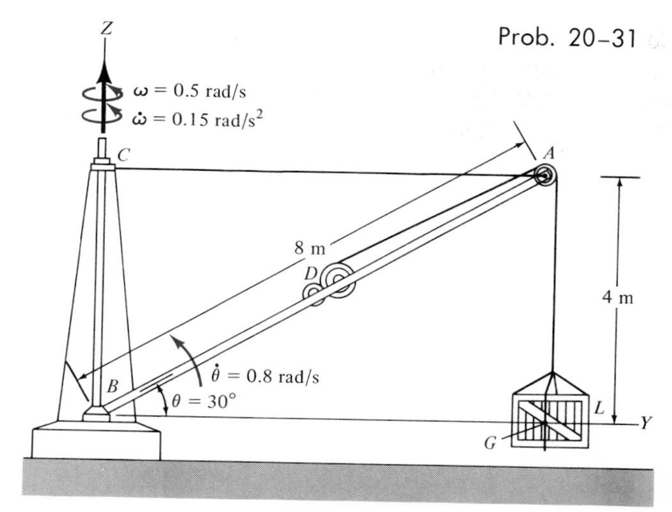

***20–32.** The arm AB is rotating about the fixed pin A at a constant rate of $\omega_1 = 4$ rad/s, while the rod BD is rotating about the Z axis at a constant rate of $\omega_2 = 5$ rad/s. At the instant the mechanism is in the position shown, collar C is moving *along the rod* with a velocity of 3 m/s and an acceleration of 2 m/s², *both measured relative to the rod.* Determine the absolute velocity and acceleration of the collar at this instant.

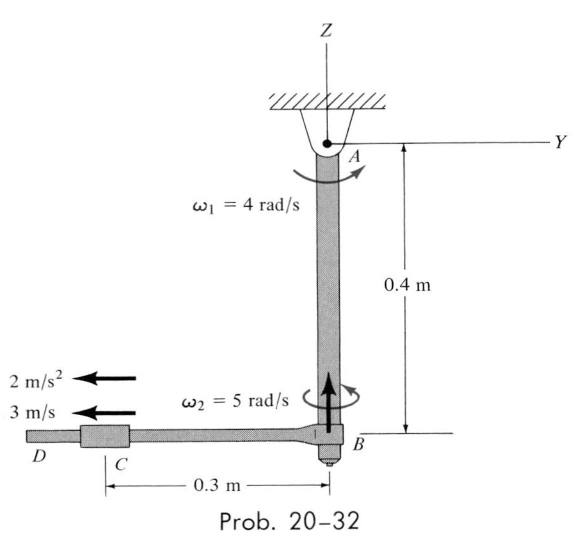

Prob. 20–32

20–33. Solve Prob. 20–32 assuming that arm AB has an increase in its angular rate of rotation of $\dot{\omega}_1 = 1.5$ rad/s² and rod BD has an angular *deceleration* of $\dot{\omega}_2 = 0.8$ rad/s² (opposite to $\boldsymbol{\omega}_2$). Use the data specified in Prob. 20–32.

20–34. At a given instant, the tower crane is turning, while the trolley T is moving outward along the boom with the motion shown. At this same instant, the concrete bucket B is swinging toward the vertical such that $\dot{\theta} = -6$ rad/s and $\ddot{\theta} = -2$ rad/s², both measured with respect to the trolley. If the cable AB is being shortened at a constant rate of 0.5 m/s, compute the velocity and acceleration of the tip C of the bucket at this instant.

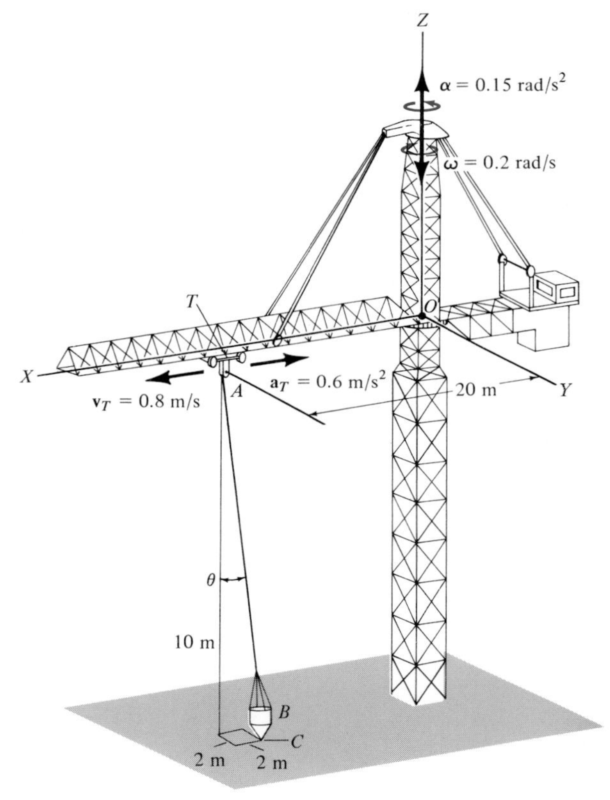

Prob. 20–34

21
Spatial Kinetics of a Rigid Body

*21–1. Introduction

In general, as a body moves through space, it has a simultaneous translation and rotation at a given instant. The kinematic aspects of the *translation* have been discussed in Chapter 17, where it was shown that a system of external forces acting on the body may be related to the acceleration of the body's mass center by the equation $\Sigma \mathbf{F} = m\mathbf{a}_G$. In this chapter emphasis is placed primarily upon the *rotational* aspects of rigid-body motion, since motion of the body's mass center, defined by $\Sigma \mathbf{F} = m\mathbf{a}_G$, is treated in the same manner as particle motion.

The rotational equations of motion relate the body's components of angular motion to the moment components created by the external forces about some point located either on or off the body. To apply these equations, it is first necessary to formulate the moments and products of inertia of the body, and to compute the body's angular momentum. For this reason these topics will be presented first. Afterward, the principles of impulse and momentum and work and energy will be at our disposal for solving spatial motion rigid-body problems. The rotational equations of rigid-body motion are developed in Sec. 21–5. Of special interest are problems involving the motion of an unsymmetrical body about a fixed axis; motion of a gyroscope, Sec. 21–6; and torque-free motion, Sec. 21–7.

*21–2. Mass Moments and Products of Inertia

When studying the planar kinetics of a body, it was necessary to introduce the mass moment of inertia I_G, which was computed about an axis perpendicular to the plane of motion and passing through the mass center G. For the kinetic analysis of spatial motions, it will be necessary to calculate six inertial quantities. These terms, called the mass moments and products of inertia, describe in a particular way the distribution of mass for a body relative to a given coordinate system that has a specified orientation and point of origin.

Mass Moment of Inertia. Consider the rigid body shown in Fig. 21–1. The *mass moment of inertia* for a differential element dm of the body about any one of the three coordinate axes is defined as the product of the mass of the element and the square of the shortest distance from the axis to the element. For example, as noted in the figure, $r_x = \sqrt{y^2 + z^2}$, so that the mass moment of inertia of dm about the x axis is

$$dI_{xx} = r_x^2\, dm = (y^2 + z^2)\, dm$$

The moment of inertia I_{xx} for the entire body is determined by integrating this expression over the entire mass of the body. Hence, for each of the axes, we may write

$$\begin{aligned} I_{xx} &= \int_m r_x^2\, dm = \int_m (y^2 + z^2)\, dm \\ I_{yy} &= \int_m r_y^2\, dm = \int_m (z^2 + x^2)\, dm \\ I_{zz} &= \int_m r_z^2\, dm = \int_m (x^2 + y^2)\, dm \end{aligned} \qquad (21\text{–}1)$$

Here it is seen that the mass moment of inertia is *always a positive quantity,* since it is the summation of the product of the mass $dm,$ which is always positive, and distances squared.

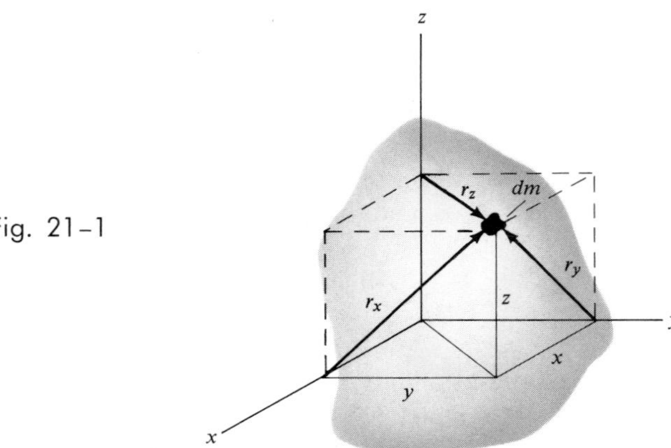

Fig. 21–1

Mass Product of Inertia. The *mass product of inertia* for a differential element dm is defined with respect to a set of *two orthogonal planes* as the product of the mass of the element and the perpendicular (or shortest) distances from the planes to the element. For example, with respect to the x-z and y-z planes, the mass product of inertia dI_{xy} for the element dm, shown in Fig. 21–1, is

$$dI_{xy} = xy\,dm$$

Note also that $dI_{yx} = dI_{xy}$. By integrating over the entire mass, the product of inertia of the body, for each combination of planes, may be expressed as

$$I_{xy} = I_{yx} = \int_m xy\,dm$$

$$I_{yz} = I_{zy} = \int_m yz\,dm \qquad (21\text{–}2)$$

$$I_{xz} = I_{zx} = \int_m xz\,dm$$

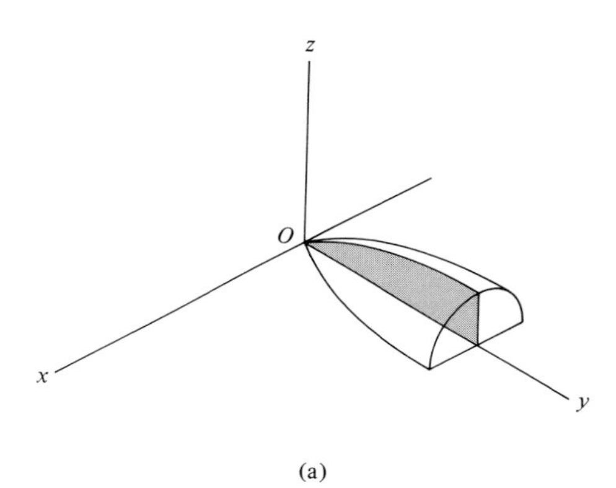

(a)

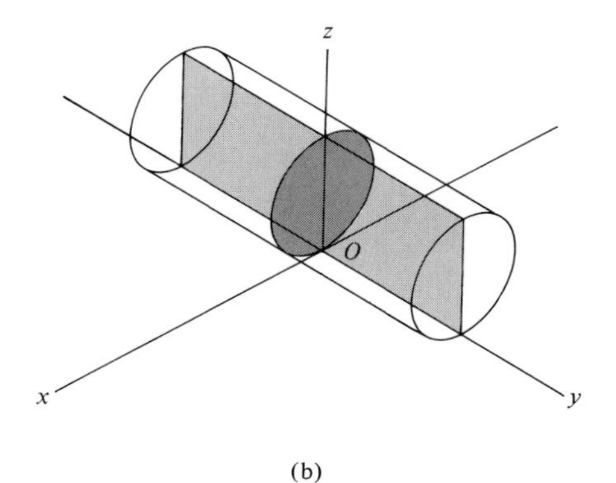

(b)

Fig. 21–2

Unlike the mass moment of inertia, which is always positive, the mass product of inertia may be positive, negative, or zero. The result depends upon the signs of the two defining coordinates, which vary independently from one another. In particular, if either one or both of the orthogonal planes are *planes of symmetry* for the mass, the *product of inertia* with respect to these planes will be *zero*. In such cases, elements of mass will occur in *pairs*, located on each side of the plane of symmetry. On one side of the plane the product of inertia for the element will be positive, while on the other side the product of inertia for the corresponding element will be negative, the sum therefore yielding zero. An example of this is shown in Fig. 21–2. In the first case, Fig. 21–2*a*, the y-z plane is a plane of symmetry, and hence for point O, $I_{xz} = I_{xy} = 0$. Computation for I_{yz} will yield a *positive* result, since all elements of mass are located using only positive y and z coordinates. For the cylinder, with the coordinate axes located as shown in Fig. 21–2*b*, the x-z and y-z planes are both planes of symmetry. Thus, for point O, $I_{zx} = I_{yz} = I_{xy} = 0$.

Parallel-Axis and Parallel-Plane Theorems. The techniques of integration which are used to determine the moment of inertia of a body were described in Sec. 17–2. Also discussed were methods to determine the moment of inertia of a composite body, i.e., a body that is composed of simpler segments, as tabulated in Appendix D. In both of these cases, the *parallel-axis theorem* is necessary for the calculations. This theorem, which was proven in Sec. 17–2, is used to transfer the moment of inertia of a body from an axis passing through its mass center G to a parallel axis passing through some other point. In this regard, if G has coordinates x_G, y_G, and z_G defined from the x, y, z axes, Fig. 21–3, then the parallel-axis theorems, used to calculate the moments of inertia about the x, y, z axes, are

$$\begin{aligned} I_{xx} &= (I_{x'x'})_G + m(y_G^2 + z_G^2) \\ I_{yy} &= (I_{y'y'})_G + m(x_G^2 + z_G^2) \\ I_{zz} &= (I_{z'z'})_G + m(x_G^2 + y_G^2) \end{aligned} \tag{21-3}$$

The products of inertia of a body or a composite are computed in the same manner as the body's moments of inertia. Here, however, the *parallel-plane theorems* are important. These theorems are used to transfer the products of inertia of the body from a set of three orthogonal planes passing through the body's mass center to a corresponding set of three parallel planes passing through some other point. Defining the perpendicular distances between the planes as x_G, y_G, and z_G, Fig. 21–3, the parallel-plane theorems can be written as

$$\begin{aligned} I_{xy} &= (I_{x'y'})_G + m x_G y_G \\ I_{yz} &= (I_{y'z'})_G + m y_G z_G \\ I_{zx} &= (I_{z'x'})_G + m z_G x_G \end{aligned} \tag{21-4}$$

The derivation of these formulas is similar to that given for the parallel-axis theorem, Sec. 17–2.

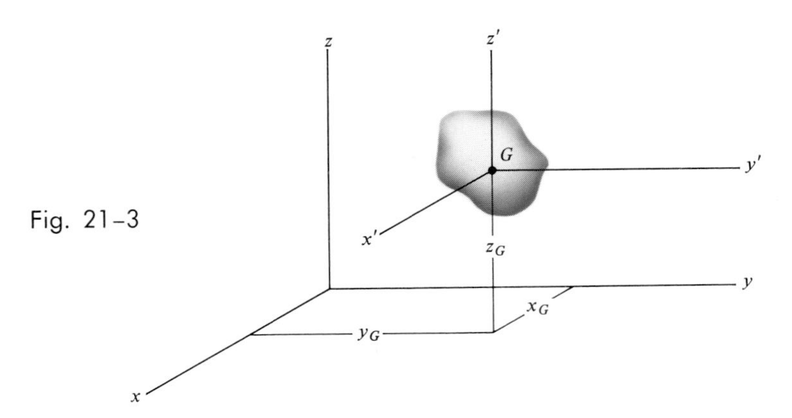

Fig. 21–3

Inertia Tensor. The inertial properties of a body are completely characterized by nine terms, six of which are independent of one another. Once this set of terms is defined using Eqs. 21–1 and 21–2, it may be grouped into an array which has the following form:

$$\begin{pmatrix} I_{xx} & -I_{xy} & -I_{xz} \\ -I_{yx} & I_{yy} & -I_{yz} \\ -I_{zx} & -I_{zy} & I_{zz} \end{pmatrix} \qquad (21\text{–}5)$$

This array is called an *inertia tensor*. It has a unique set of values for a body when it is computed for each location of the origin and inclination of the coordinate axes.

Moment of Inertia About an Arbitrary Axis. Consider the body shown in Fig. 21–4, where the nine elements of the inertia tensor have been computed for the x, y, z axes having an origin at O. Here we wish to compute the moment of inertia of the body about the Oa axis, for which the direction is defined by the unit vector $\mathbf{u}_a$. By definition $I_{aa} = \int b^2\, dm$, where b is the *perpendicular distance* from dm to Oa. If the position of dm is located using $\mathbf{r}$, then $b = r \sin\theta$, which represents the *magnitude* of the cross product $\mathbf{u}_a \times \mathbf{r}$. Hence, the moment of inertia can be expressed as

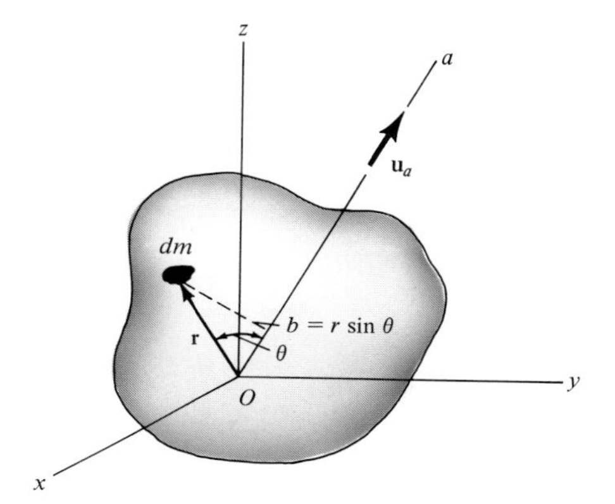

Fig. 21–4

$$I_{aa} = \int_m |(\mathbf{u}_A \times \mathbf{r})|^2 dm = \int_m (\mathbf{u}_a \times \mathbf{r}) \cdot (\mathbf{u}_a \times \mathbf{r})\, dm$$

Provided $\mathbf{u}_a = u_x\mathbf{i} + u_y\mathbf{j} + u_z\mathbf{k}$ and $\mathbf{r} = x\mathbf{i} + y\mathbf{j} + z\mathbf{k}$, so that $\mathbf{u}_a \times \mathbf{r} = (u_y z - u_z y)\mathbf{i} + (u_z x - u_x z)\mathbf{j} + (u_x y - u_y x)\mathbf{k}$, then, after substituting and performing the dot-product operation, the moment of inertia becomes

$$I_{aa} = \int_m [(u_y z - u_z y)^2 + (u_z x - u_x z)^2 + (u_x y - u_y x)^2]\, dm$$

$$= u_x^2 \int_m (y^2 + z^2)\, dm + u_y^2 \int_m (z^2 + x^2)\, dm + u_z^2 \int_m (x^2 + y^2)\, dm$$

$$-2u_x u_y \int_m xy\, dm - 2u_y u_z \int_m yz\, dm - 2u_z u_x \int_m zx\, dm$$

Recognizing the integrals to be the moments and products of inertia of the body, Eqs. 21–1 and 21–2, we have

$$I_{aa} = I_{xx}u_x^2 + I_{yy}u_y^2 + I_{zz}u_z^2 - 2I_{xy}u_x u_y - 2I_{yz}u_y u_z - 2I_{zx}u_z u_x \qquad (21\text{–}6)$$

Thus, if the inertia tensor is specified for the x, y, z axes, the moment of inertia of the body about the inclined Oa axis can be computed by using Eq. 21–6. For the calculation the direction cosines u_x, u_y, and u_z of the axes must be determined. These numbers specify the cosines of the coordinate direction angles α, β, and γ made between the Oa axis and the x, y, z axes, respectively. (See Appendix B.)

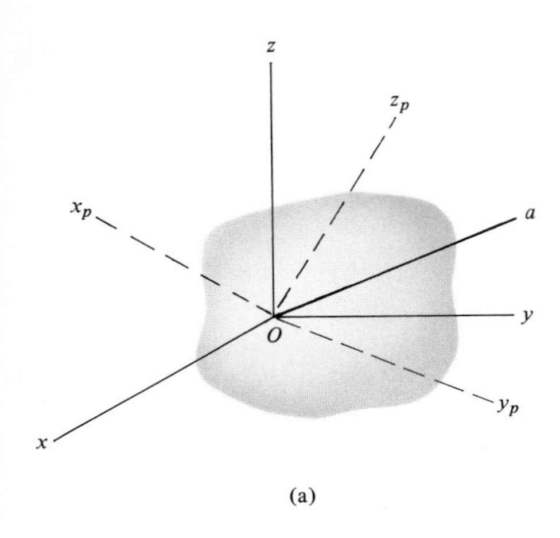

(a)

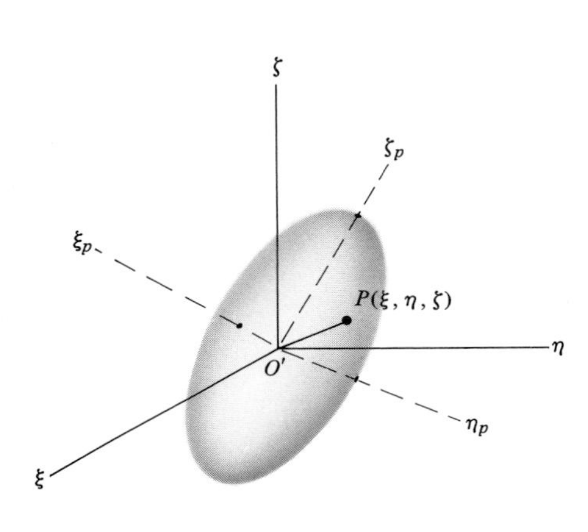

(b)

Fig. 21–5

Ellipsoid of Inertia. Using Eq. 21–6, consider now the *variation* of I_{aa} as the direction of the Oa axis changes, Fig. 21–5*a*. The results can be described graphically if a *plot* is made of the locus of point P, which represents the end point of a line segment having a magnitude of $O'P = 1/\sqrt{I_{aa}}$ and direction defined by $\mathbf{u}_a$. Confusion is avoided if the plot is made on the ξ, η, ζ (xi, eta, zeta) axes, which are parallel to the x, y, z axes, Fig. 21–5*b*. It is seen that the locus, so created, forms a three-dimensional surface defined as an ellipsoid. To show this, note that point P on the surface, Fig. 21–5*b*, has coordinates (ξ, η, ζ), and the direction cosines of $O'P$ are the same as those for the Oa axis, since $O'P$ and Oa are parallel. Thus,

$$u_x = \frac{\xi}{O'P} = \xi\sqrt{I_{aa}}$$

$$u_y = \frac{\eta}{O'P} = \eta\sqrt{I_{aa}}$$

$$u_z = \frac{\zeta}{O'P} = \zeta\sqrt{I_{aa}}$$

Substituting these values into Eq. 21–6 and canceling the common factor I_{aa} yields

$$I_{xx}\xi^2 + I_{yy}\eta^2 + I_{zz}\zeta^2 - 2I_{xy}\xi\eta - 2I_{yz}\eta\zeta - 2I_{zx}\zeta\xi = 1 \qquad (21\text{–}7)$$

This equation defines the surface of the ellipsoid shown in Fig. 21–5*b*. Since the constants I_{xx}, I_{yy}, I_{zz}, I_{xy}, I_{yz}, and I_{zx} are included in the terms of this equation, the surface is called the *ellipsoid of inertia*. Representing the inertia tensor graphically at a point using the ellipsoid of inertia is similar to using an arrow to represent a vector. Specifically, for each point O in space, the size, shape, and orientation of the ellipsoid of inertia for the body will be unique.

Equation 21–7 can be simplified if the ξ, η, and ζ coordinate axes are oriented along the ellipsoid's three axes of symmetry. This special set of axes, called ξ_p, η_p, and ζ_p, is shown in Fig. 21–5*b*. The corresponding parallel set of x_p, y_p, and z_p axes for the body is shown in Fig. 21–5*a*. These axes are called the *principal axes of inertia*. Using the ξ_p, η_p, and ζ_p axes, the ellipsoid of inertia may be represented as

$$I_x\xi_p^2 + I_y\eta_p^2 + I_z\zeta_p^2 = 1$$

Here $I_x = I_{xx}$, $I_y = I_{yy}$, and $I_z = I_{zz}$ are termed the *principal moments of inertia* for the body which are computed from the *principal axes of inertia*, x_p, y_p, and z_p, shown in Fig. 21–5*a*. Comparing the above equation with Eq. 21–7, it is seen that *the products of inertia for the body are zero when computed with respect to the principal axes of inertia.*

From this analysis it may therefore be concluded that regardless of the shape of the body and the point of origin O considered, a set of principal axes may always be established at the point so that the products of inertia for the body are zero when computed with respect to these axes. When this is done, the inertia tensor for the body is said to be "diagonalized" and may be written in the simplified form

$$\begin{pmatrix} I_x & 0 & 0 \\ 0 & I_y & 0 \\ 0 & 0 & I_z \end{pmatrix}$$

Of these three principal moments of inertia, one will be a maximum and another a minimum of the body's moments of inertia.

Mathematical determination of the direction of principal axes of inertia will not be discussed in this section. (See Prob. 21–13.) There are many cases, however, in which the principal axes may be determined by inspection. From the previous discussion, it was noted that if the coordinate axes are oriented such that *two* of the three orthogonal planes containing the axes are planes of *symmetry* for the body, then all the products of inertia for the body are zero with respect to the coordinate planes, and hence the coordinate axes are principal axes of inertia. For example, the x, y, z axes shown in Fig. 21–2*b* represent the principal axes of inertia for the cylinder at point O. This concept may be extended to include bodies which have partial rotational symmetry, such as the equilateral triangular plate shown in Fig. 21–6*a*. If the plate is rotated 120° about the x axis, it will coincide with both its original geometry and mass distribution. It follows that the ellipsoid of inertia for the plate remains the *same* for each 120° rotation. Hence, $O'P = O'Q = O'R$, Fig. 21–6*b*, so that the ellipsoid has an axis of symmetry ξ_p, which is parallel to the x axis (a principal axis). The y and z axes, which are perpendicular to the x axis, will *also* represent principal axes, since these axes correspond to η_p and ζ_p for the ellipsoid. This argument may be extended to include bodies having higher than three degrees of symmetry as illustrated here.

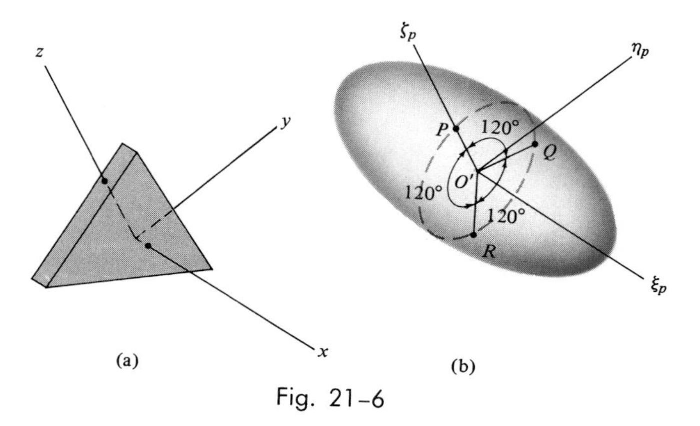

Fig. 21–6

Example 21–1

Determine the moment of inertia of the bent rod shown in Fig. 21–7*a* about the *Aa* axis. The mass of each of the three segments is shown in the figure.

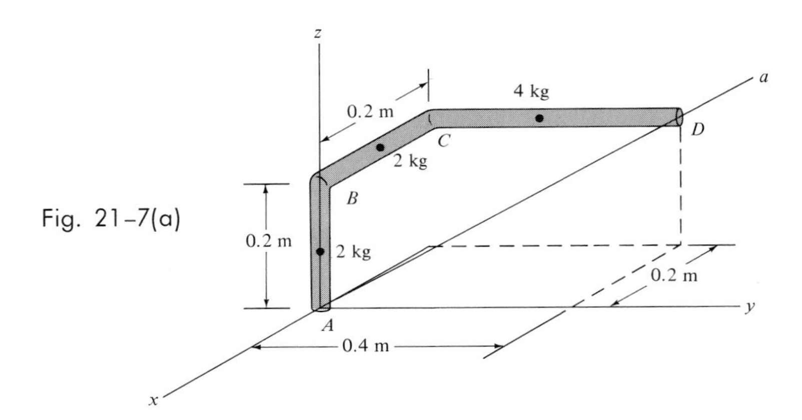

Fig. 21–7(a)

Solution

The moment of inertia I_{Aa} can be computed by using Eq. 21–6. It is first necessary, however, to determine the moments and products of inertia of the rod about the *x, y, z* axes. This is done using the data in Appendix D and the parallel-axis and parallel-plane theorems, Eqs. 21–3 and 21–4. Dividing the rod into three parts and locating the mass centers of each of the three segments of the rod, Fig. 21–7*b*, the calculations are ordered such that they apply to segments *AB*, *BC*, and *CD*, respectively.

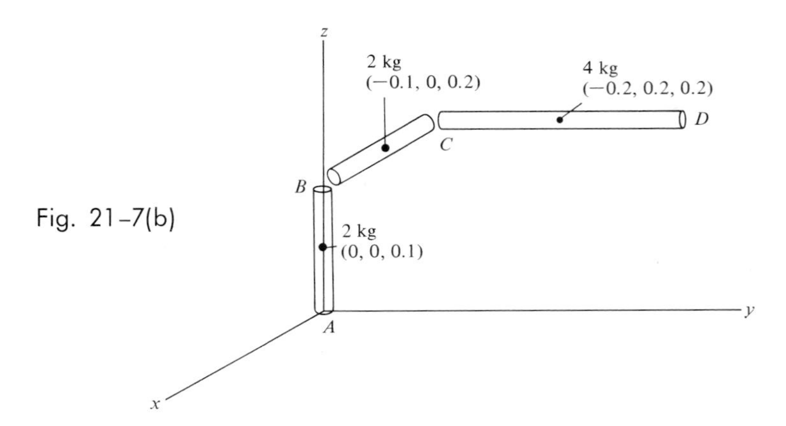

Fig. 21–7(b)

$$I_{xx} = [\tfrac{1}{12}(2)(0.2)^2 + 2(0.1)^2] + [0 + 2(0.2)^2] + [\tfrac{1}{12}(4)(0.4)^2 + 4((0.2)^2 + (0.2)^2)] = 0.480 \text{ kg} \cdot \text{m}^2$$

$$I_{yy} = [\tfrac{1}{12}(2)(0.2)^2 + 2(0.1)^2] + [\tfrac{1}{12}(2)(0.2)^2 + 2((-0.1)^2 + (0.2)^2)] + [0 + 4((-0.2)^2 + (0.2)^2)] = 0.453 \text{ kg} \cdot \text{m}^2$$

$$I_{zz} = [0 + 0] + [\tfrac{1}{12}(2)(0.2)^2 + 2(0.1)^2] + [\tfrac{1}{12}(4)(0.4)^2 + 4((-0.2)^2 + (0.2)^2)] = 0.400 \text{ kg} \cdot \text{m}^2$$

$$I_{xy} = [0 + 0] + [0 + 0] + [0 + 4(-0.2)(0.2)] = -0.160 \text{ kg} \cdot \text{m}^2$$

$$I_{yz} = [0 + 0] + [0 + 0] + [0 + 4(0.2)(0.2)] = 0.160 \text{ kg} \cdot \text{m}^2$$

$$I_{zx} = [0 + 0] + [0 + 2(0.2)(-0.1)] + [0 + 4(0.2)(-0.2)] = -0.200 \text{ kg} \cdot \text{m}^2$$

The Aa axis is defined by

$$\mathbf{u}_{Aa} = \frac{\mathbf{r}_D}{r_D} = \frac{-0.2\mathbf{i} + 0.4\mathbf{j} + 0.2\mathbf{k}}{\sqrt{(-0.2)^2 + (0.4)^2 + (0.2)^2}} = -0.408\mathbf{i} + 0.816\mathbf{j} + 0.408\mathbf{k}$$

Thus,

$$u_x = -0.408, \qquad u_y = 0.816, \qquad u_z = 0.408$$

Substituting the computed data into Eq. 21–6, we have

$$\begin{aligned} I_{Aa} &= I_{xx}u_x^2 + I_{yy}u_y^2 + I_{zz}u_z^2 - 2I_{xy}u_xu_y - 2I_{yz}u_yu_z - 2I_{zx}u_zu_x \\ &= 0.480(-0.408)^2 + (0.453)(0.816)^2 + 0.400(0.408)^2 \\ &\quad - 2(-0.160)(-0.408)(0.816) \\ &\quad - 2(0.160)(0.816)(0.408) - 2(-0.200)(0.408)(-0.408) \\ &= 0.168 \text{ kg} \cdot \text{m}^2 \end{aligned}$$

Ans.

Problems

21–1. Determine by direct integration the mass product of inertia I_{yz} for the homogeneous prism. The density of the material is ρ. Express the result in terms of the total mass m of the prism.

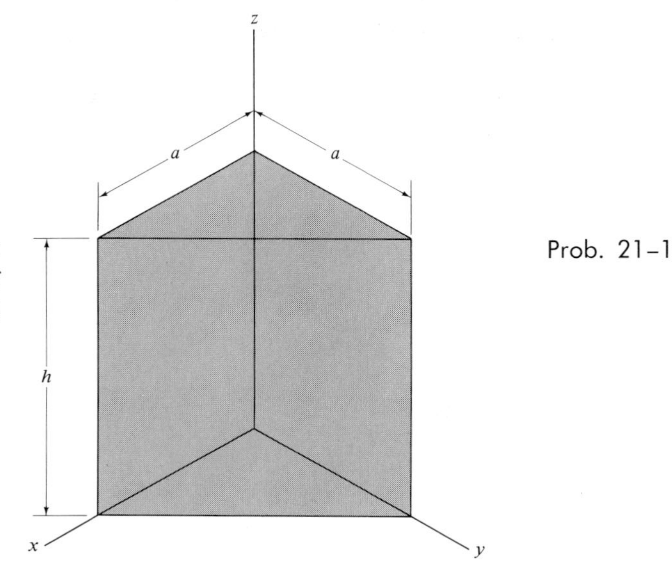

Prob. 21–1

21–2. Determine by direct integration the mass product of inertia I_{xy} of the homogeneous prism in Prob. 21–1. The density of the material is ρ. Express the result in terms of the total mass m of the prism.

21–3. Determine by direct integration the mass product of inertia I_{yz} for the homogeneous tetrahedron. The mass density of the material is ρ. Express the result in terms of the total mass m of the solid.

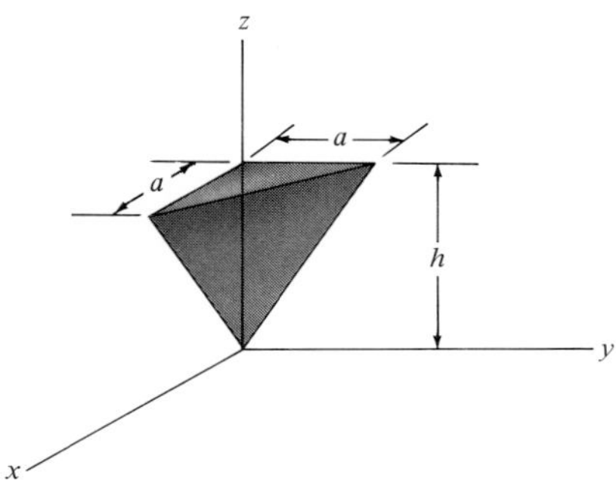

Prob. 21–3

***21–4.** Determine by direct integration the mass product of inertia I_{xy} for the homogeneous tetrahedron. The density of the material is ρ. Express the result in terms of the total mass m of the solid. *Suggestion:* Use a triangular element of thickness dz and then express dI_{xy} in terms of the size and mass of the element using the result of Prob. 21–2.

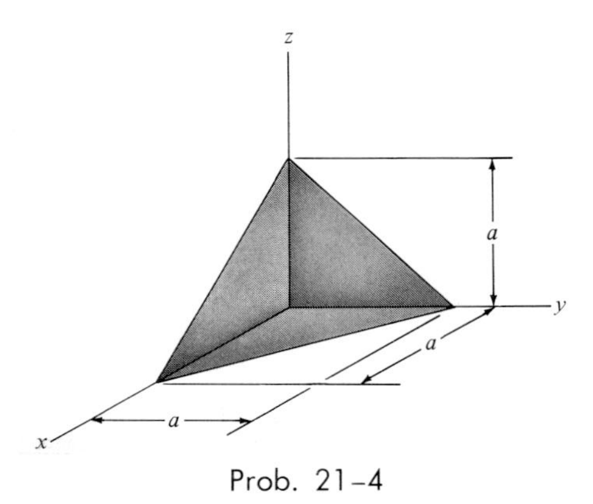

Prob. 21–4

21–5. Determine the moment of inertia of the cone about the z' axis. The mass of the cone is $m = 4$ kg, the *height* is $h = 300$ mm, and the radius is $r = 75$ mm.

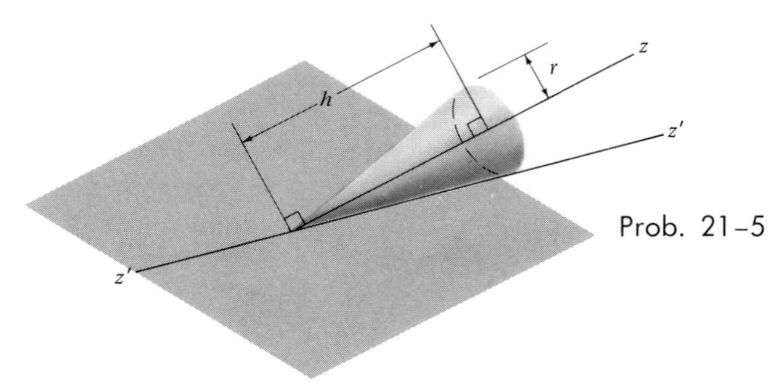

Prob. 21–5

21–5a. Solve Prob. 21–5 if the cone has a weight of $W = 15$ lb, $h = 2$ ft, and $r = 1$ ft.

21–6. Compute the mass moment of inertia of the disk about the axis of shaft AB. The disk has a total mass of 8 kg.

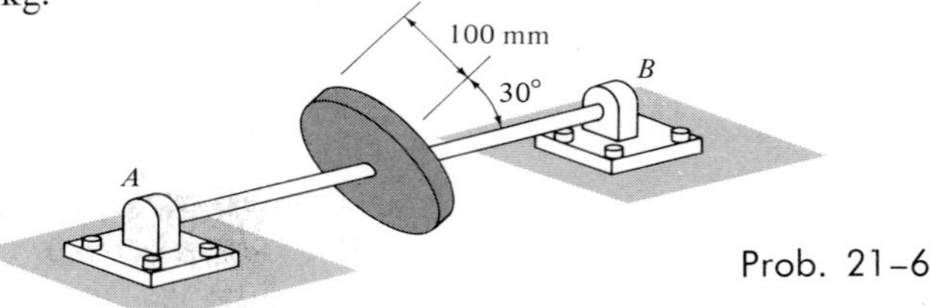

Prob. 21–6

21–7. Determine the moment of inertia of both the 2-kg rod CD and the 5-kg disk about the z axis, which is collinear with the shaft AB.

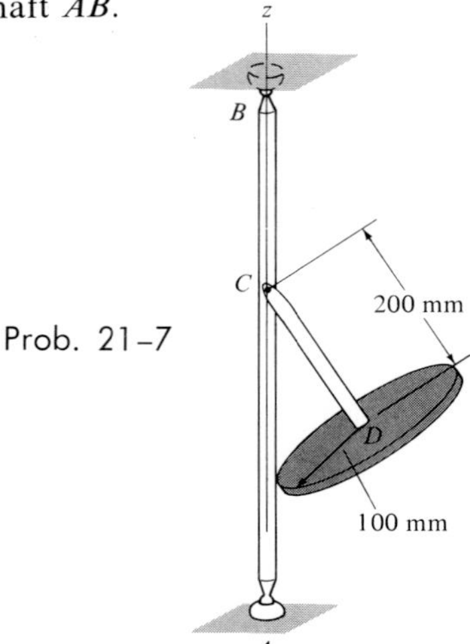

Prob. 21–7

*21-8. Determine the moment of inertia of the 5-kg circular plate about the axis of rod OA.

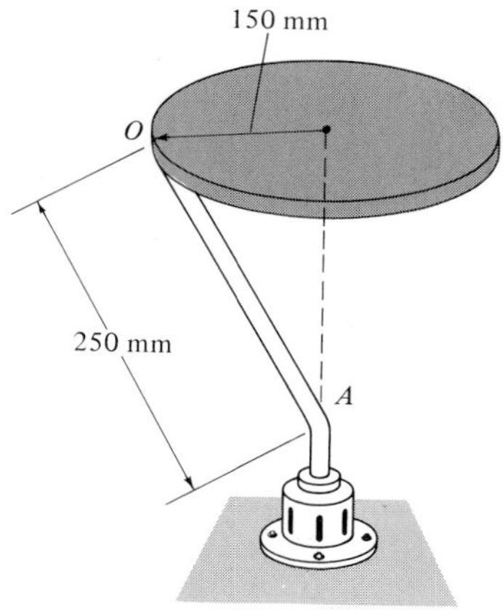

Prob. 21-8

21-9. The top consists of a cone having a mass of 0.7 kg and a hemisphere of mass 0.2 kg. Determine the moment of inertia I_z when the top is in the position shown.

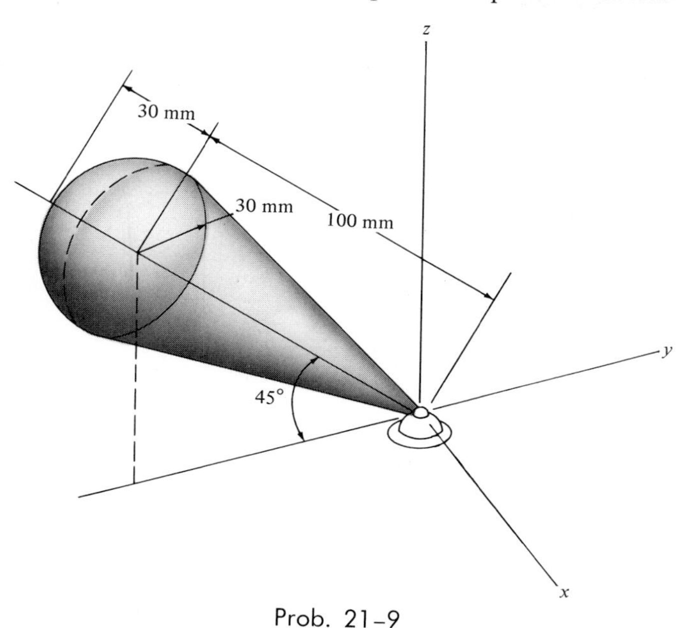

Prob. 21-9

21-10. Compute the moment of inertia of the rod-and-disk assembly about the xx axis. The disks each have a mass of $m_D = 4$ kg and a radius of $r = 75$ mm. Each of the two rods has a mass of $m_r = 1.5$ kg. $l = 300$ mm.

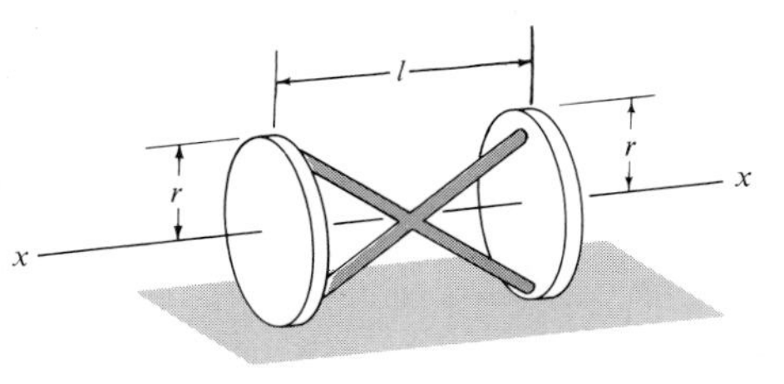

Prob. 21-10

21-10a. Solve Prob. 21-10 if each disk weighs $W_D = 2$ lb, $r = 1$ ft, each rod weighs $W_r = 2$ lb, and $l = 2$ ft.

21-11. Compute the moment of inertia of the rod-and-thin-ring assembly about the z axis. The rods and ring have a mass of 3 kg/m.

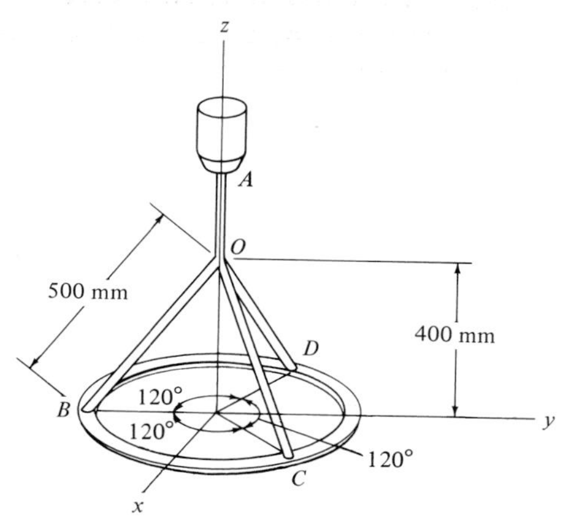

Prob. 21-11

21-12. Show that the sum of the moments of inertia of a body, $I_{xx} + I_{yy} + I_{zz}$, is independent of the orientation of the x, y, z axes and thus depends only on the location of the origin.

*21–3. Angular Momentum of a Rigid Body

Consider the rigid body in Fig. 21–8, which has a total mass m and center of mass located at G. The X, Y, Z coordinate system represents an inertial frame of reference, and hence, its axes are fixed or translate with a constant velocity. The angular momentum as measured from this reference will be computed relative to the arbitrary point A located in the body. The position vectors $\mathbf{r}_A$ and $\boldsymbol{\rho}_A$ are drawn from the origin of coordinates to point A, and from A to the ith particle P of the body. If the mass of P is Δm_i, the angular momentum about point A is

$$\Delta \mathbf{H}_A = \boldsymbol{\rho}_A \times \Delta m_i \mathbf{v}_i$$

where $\mathbf{v}_i$ represents the particle's velocity as measured from the X, Y, Z coordinate system. If the body has an angular velocity $\boldsymbol{\omega}$ at the instant considered, $\mathbf{v}_i$ may be related to the velocity of A by the kinematic equation

$$\mathbf{v}_i = \mathbf{v}_A + \boldsymbol{\omega} \times \boldsymbol{\rho}_A$$

Thus,

$$\begin{aligned} \Delta \mathbf{H}_A &= \boldsymbol{\rho}_A \times (\mathbf{v}_A + \boldsymbol{\omega} \times \boldsymbol{\rho}_A)\,\Delta m_i \\ &= (\boldsymbol{\rho}_A\,\Delta m_i) \times \mathbf{v}_A + \boldsymbol{\rho}_A \times (\boldsymbol{\omega} \times \boldsymbol{\rho}_A)\,\Delta m_i \end{aligned}$$

For the entire body, summing all n particles,

$$\mathbf{H}_A = (\Sigma \boldsymbol{\rho}_A\,\Delta m_i) \times \mathbf{v}_A + \Sigma \boldsymbol{\rho}_A \times (\boldsymbol{\omega} \times \boldsymbol{\rho}_A)\,\Delta m_i$$

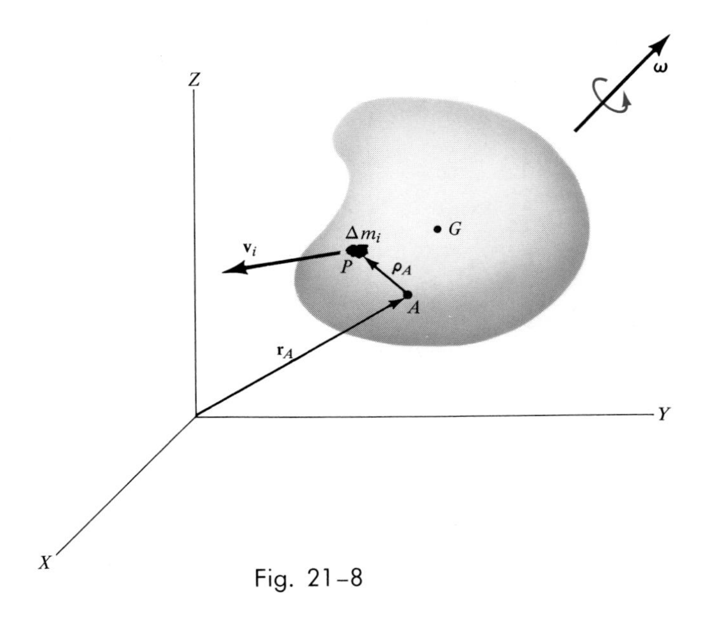

Fig. 21–8

As $n \to \infty$, then $\Delta m_i \to dm$ and the summations become integrals:

$$\mathbf{H}_A = \left(\int_m \boldsymbol{\rho}_A \, dm\right) \times \mathbf{v}_A + \int_m \boldsymbol{\rho}_A \times (\boldsymbol{\omega} \times \boldsymbol{\rho}_A) \, dm \qquad (21\text{–}8)$$

Fixed Point *O*. If A becomes a *fixed point* O on the body, Fig. 21–9*a*, then $\mathbf{v}_A = \mathbf{0}$ and Eq. 21–8 reduces to

$$\mathbf{H}_O = \int_m \boldsymbol{\rho}_O \times (\boldsymbol{\omega} \times \boldsymbol{\rho}_O) \, dm \qquad (21\text{–}9)$$

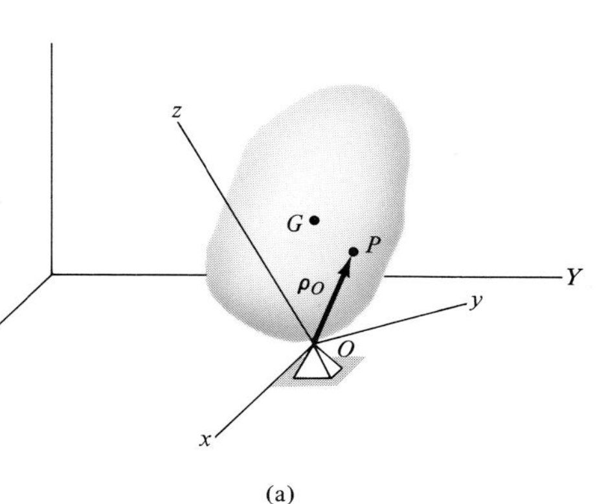

(a)

Center of Mass *G*. If A is located at the *center of mass* G of the body, Fig. 21–9*b*, then $\int_m \boldsymbol{\rho}_A \, dm = \mathbf{0}$ and

$$\mathbf{H}_G = \int_m \boldsymbol{\rho}_G \times (\boldsymbol{\omega} \times \boldsymbol{\rho}_G) \, dm \qquad (21\text{–}10)$$

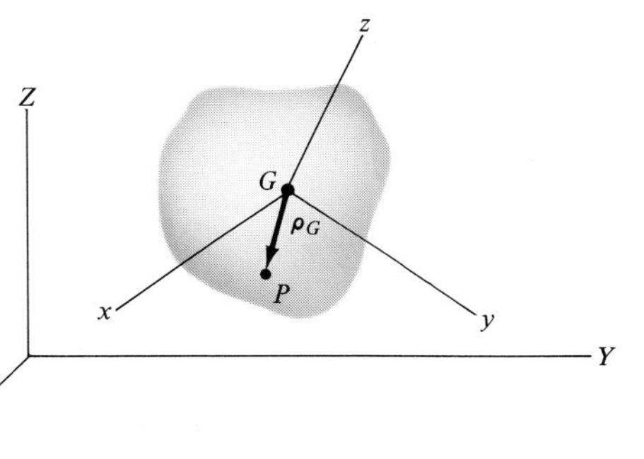

(b)

Arbitrary Point *A*. In general, A may be some point other than O or G, in which case Eq. 21–8 may nevertheless be simplified by noting that $\boldsymbol{\rho}_A = \boldsymbol{\rho}_G + \boldsymbol{\rho}_{G/A}$, Fig. 21–9*c*. Substituting into Eq. 21–8, we have

$$\mathbf{H}_A = \int_m (\boldsymbol{\rho}_G + \boldsymbol{\rho}_{G/A}) \times \mathbf{v}_A \, dm + \int_m (\boldsymbol{\rho}_G + \boldsymbol{\rho}_{G/A}) \times [\boldsymbol{\omega} \times (\boldsymbol{\rho}_G + \boldsymbol{\rho}_{G/A})] \, dm$$

$$= \left(\int_m \boldsymbol{\rho}_G \, dm\right) \times \mathbf{v}_A + (\boldsymbol{\rho}_{G/A} \times \mathbf{v}_A) \int_m dm + \int_m \boldsymbol{\rho}_G \times (\boldsymbol{\omega} \times \boldsymbol{\rho}_G) \, dm$$

$$+ \left(\int_m \boldsymbol{\rho}_G \, dm\right) \times (\boldsymbol{\omega} \times \boldsymbol{\rho}_{G/A}) + \boldsymbol{\rho}_{G/A} \times \left(\boldsymbol{\omega} \times \int_m \boldsymbol{\rho}_G \, dm\right)$$

$$+ \boldsymbol{\rho}_{G/A} \times (\boldsymbol{\omega} \times \boldsymbol{\rho}_{G/A}) \int_m dm$$

The first, fourth, and fifth terms on the right side are zero, since they contain the form $\int_m \boldsymbol{\rho}_G \, dm$, which is zero by definition of the mass center. Also, the third term on the right side is defined by Eq. 21–10. Thus,

$$\mathbf{H}_A = (\boldsymbol{\rho}_{G/A} \times \mathbf{v}_A)m + \mathbf{H}_G + \boldsymbol{\rho}_{G/A} \times (\boldsymbol{\omega} \times \boldsymbol{\rho}_{G/A})m$$
$$= \boldsymbol{\rho}_{G/A} \times [\mathbf{v}_A + \boldsymbol{\omega} \times \boldsymbol{\rho}_{G/A}]m + \mathbf{H}_G$$

Since $\mathbf{v}_G = \mathbf{v}_A + \boldsymbol{\omega} \times \boldsymbol{\rho}_{G/A}$, then

$$\mathbf{H}_A = (\boldsymbol{\rho}_{G/A} \times m\mathbf{v}_G) + \mathbf{H}_G \qquad (21\text{–}11)$$

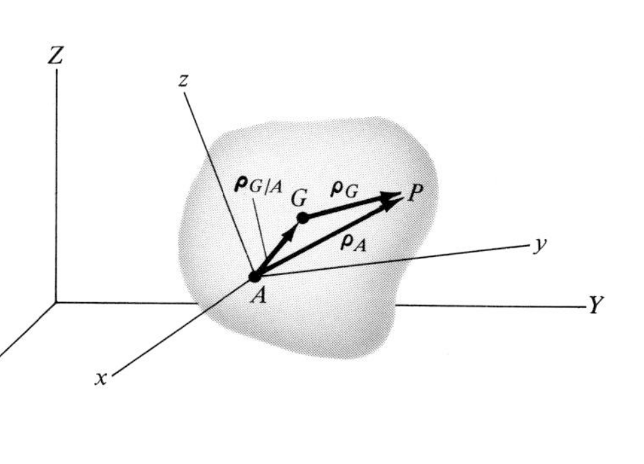

(c)

Fig. 21–9

It is seen from this equation that the angular momentum of the body about the arbitrary point A consists of two parts—the moment of the linear momentum $m\mathbf{v}_G$ of the body* about point A added (vectorially) to the angular momentum $\mathbf{H}_G$ of the body about the mass center. Equation 21–11 may also be used for computing the angular momentum of the body about a fixed point O in the body; the results, of course, will be the same as those computed using the more convenient Eq. 21–9.

Rectangular Components of H. To make practical use of Eqs. 21–9 through 21–11, the angular momentum must be expressed in terms of its scalar components. For this purpose, it is convenient to choose a second set of x, y, z axes having an arbitrary orientation relative to the X, Y, Z axes, Fig. 21–9. For a general explanation, note that Eqs. 21–9 through 21–11 all contain the form

$$\mathbf{H} = \int_m \boldsymbol{\rho} \times (\boldsymbol{\omega} \times \boldsymbol{\rho})\, dm$$

Expressing $\mathbf{H}$, $\boldsymbol{\rho}$, and $\boldsymbol{\omega}$ in terms of x, y, and z components, we have

$$H_x\mathbf{i} + H_y\mathbf{j} + H_z\mathbf{k} = \int_m (x\mathbf{i} + y\mathbf{j} + z\mathbf{k}) \times [(\omega_x\mathbf{i} + \omega_y\mathbf{j} + \omega_z\mathbf{k}) \times (x\mathbf{i} + y\mathbf{j} + z\mathbf{k})]\, dm$$

Expanding the cross products and combining terms yields

$$\begin{aligned} H_x\mathbf{i} + H_y\mathbf{j} + H_z\mathbf{k} = &\left[\omega_x \int_m (y^2 + z^2)\, dm - \omega_y \int_m xy\, dm - \omega_z \int_m xz\, dm\right]\mathbf{i} \\ &+ \left[-\omega_x \int_m xy\, dm + \omega_y \int_m (x^2 + z^2)\, dm - \omega_z \int_m yz\, dm\right]\mathbf{j} \\ &+ \left[-\omega_x \int_m zx\, dm - \omega_y \int_m yz\, dm + \omega_z \int_m (x^2 + y^2)\, dm\right]\mathbf{k} \end{aligned}$$

Equating the respective $\mathbf{i}, \mathbf{j}, \mathbf{k}$ components and recognizing that the integrals represent the mass moments and products of inertia, we obtain

$$\begin{aligned} H_x &= I_{xx}\omega_x - I_{xy}\omega_y - I_{xz}\omega_z \\ H_y &= -I_{yx}\omega_x + I_{yy}\omega_y - I_{yz}\omega_z \\ H_z &= -I_{zx}\omega_x - I_{zy}\omega_y + I_{zz}\omega_z \end{aligned} \qquad (21\text{–}12)$$

These three equations represent the scalar form of the $\mathbf{i}$, $\mathbf{j}$, and $\mathbf{k}$ components of $\mathbf{H}_O$ or $\mathbf{H}_G$ (given in vector form by Eqs. 21–9 and 21–10).

*The linear momentum $\mathbf{L} = m\mathbf{v}_G$ was defined in Sec. 19–1 for any general motion.

The angular momentum of the body about the arbitrary point A, other than the fixed point O or the center of mass G, may also be expressed in scalar form. Here it is necessary to use Eq. 21–11 and to represent $\boldsymbol{\rho}_{G/A}$ and $\mathbf{v}_G$ in Cartesian component form, carry out the cross-product operation, and substitute the components, Eqs. 21–12, for $\mathbf{H}_G$.

Equations 21–12 may be simplified further if the x, y, z coordinate axes are oriented such that they become *principal axes of inertia* for the body at the point. It was shown in Sec. 21–2 that when these axes are used, computation of the products of inertia $I_{xy} = I_{yz} = I_{zx} = 0$. If the principal moments of inertia about the x, y, z axes are represented as $I_x = I_{xx}$, $I_y = I_{yy}$, and $I_z = I_{zz}$, the three components of angular momentum become

$$\begin{aligned} H_x &= I_x\omega_x \\ H_y &= I_y\omega_y \\ H_z &= I_z\omega_z \end{aligned} \tag{21–13}$$

Principle of Impulse and Momentum. Now that the means for computing the angular momentum for a body have been presented, the *principle of impulse and momentum,* as discussed in Sec. 19–2, may be used to solve kinetic problems which involve *force, velocity, and time.* For this case, the following two vector equations are available:

$$m(\mathbf{v}_G)_1 + \Sigma \int_{t_1}^{t_2} \mathbf{F}\, dt = m(\mathbf{v}_G)_2 \tag{21–14}$$

$$\mathbf{H}_1 + \Sigma \int_{t_1}^{t_2} \mathbf{M}\, dt = \mathbf{H}_2 \tag{21–15}$$

In three dimensions each vector term can be represented by three scalar components, and therefore a total of *six scalar equations* can be written. Three equations relate the linear impulse and momentum in the x, y, and z directions; and three equations relate the body's angular impulse and momentum about these axes. Before applying Eqs. 21–14 and 21–15 to the solution of problems, the material in Sec. 19–2 and 19–3 should be reviewed.

*21–4. Kinetic Energy of a Rigid Body

Consider the rigid body in Fig. 21–10, which has a total mass m and center of mass located at G. The kinetic energy of the ith particle P of the body having a mass Δm_i and velocity $\mathbf{v}_i$, measured relative to the inertial X, Y, Z frame of reference, is

$$\Delta T_i = \tfrac{1}{2}\Delta m_i v_i^2 = \tfrac{1}{2}\Delta m_i(\mathbf{v}_i \cdot \mathbf{v}_i)$$

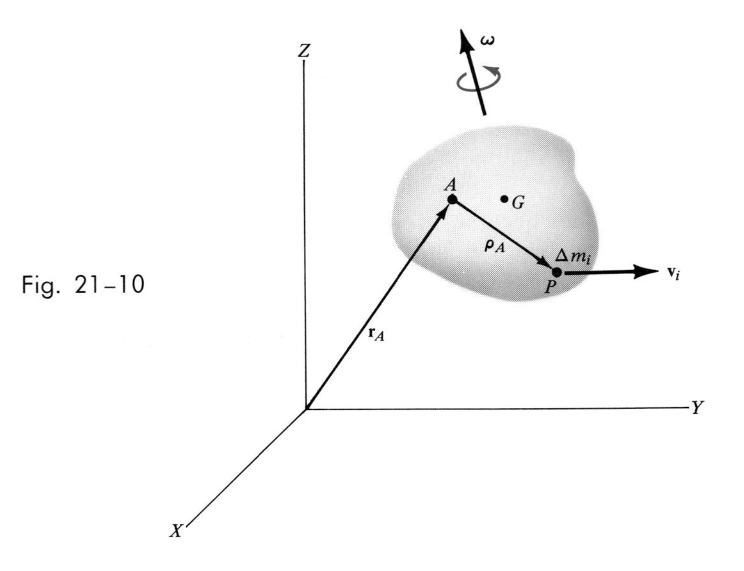

Fig. 21–10

Provided the velocity $\mathbf{v}_A$ of an arbitrary point A on the body is known, $\mathbf{v}_i$ may be related to $\mathbf{v}_A$ by the equation

$$\mathbf{v}_i = \mathbf{v}_A + \boldsymbol{\omega} \times \boldsymbol{\rho}_A$$

where $\boldsymbol{\omega}$ is the instantaneous angular velocity of the body measured from the X, Y, Z coordinate system, and $\boldsymbol{\rho}_A$ is a position vector drawn from A to P. Using this expression for $\mathbf{v}_i$, the kinetic energy for the particle may be written as

$$\begin{aligned}\Delta T_i &= \tfrac{1}{2}\Delta m_i\,(\mathbf{v}_A + \boldsymbol{\omega} \times \boldsymbol{\rho}_A)\cdot(\mathbf{v}_A + \boldsymbol{\omega} \times \boldsymbol{\rho}_A)\\ &= \tfrac{1}{2}(\mathbf{v}_A \cdot \mathbf{v}_A)\,\Delta m_i + \mathbf{v}_A \cdot (\boldsymbol{\omega} \times \boldsymbol{\rho}_A)\,\Delta m_i + \tfrac{1}{2}(\boldsymbol{\omega} \times \boldsymbol{\rho}_A)\cdot(\boldsymbol{\omega} \times \boldsymbol{\rho}_A)\,\Delta m_i\end{aligned}$$

The kinetic energy for the entire body is obtained by summing the kinetic energies of all n particles of the body. Thus,

$$T = \tfrac{1}{2}(\mathbf{v}_A \cdot \mathbf{v}_A)\Sigma\,\Delta m_i + \mathbf{v}_A \cdot (\boldsymbol{\omega} \times \Sigma\boldsymbol{\rho}_A\,\Delta m_i) + \tfrac{1}{2}\Sigma(\boldsymbol{\omega} \times \boldsymbol{\rho}_A)\cdot(\boldsymbol{\omega} \times \boldsymbol{\rho}_A)\,\Delta m_i$$

If $n \to \infty$, then $\Delta m_i \to dm$, and the summations become integrals, so that

$$T = \tfrac{1}{2}m(\mathbf{v}_A \cdot \mathbf{v}_A) + \mathbf{v}_A \cdot \left(\boldsymbol{\omega} \times \int_m \boldsymbol{\rho}_A\,dm\right) + \tfrac{1}{2}\int_m (\boldsymbol{\omega} \times \boldsymbol{\rho}_A)\cdot(\boldsymbol{\omega} \times \boldsymbol{\rho}_A)\,dm$$

The last term on the right may be rewritten using the vector identity $\mathbf{a} \times \mathbf{b} \cdot \mathbf{c} = \mathbf{a} \cdot \mathbf{b} \times \mathbf{c}$, where $\mathbf{a} = \boldsymbol{\omega}$, $\mathbf{b} = \boldsymbol{\rho}_A$, and $\mathbf{c} = \boldsymbol{\omega} \times \boldsymbol{\rho}_A$. The final result is, therefore,

$$T = \tfrac{1}{2}m(\mathbf{v}_A \cdot \mathbf{v}_A) + \mathbf{v}_A \cdot \left(\boldsymbol{\omega} \times \int_m \boldsymbol{\rho}_A\,dm\right) + \tfrac{1}{2}\boldsymbol{\omega} \cdot \int_m \boldsymbol{\rho}_A \times (\boldsymbol{\omega} \times \boldsymbol{\rho}_A)\,dm \qquad (21\text{–}16)$$

This equation is rarely used for application because of computation involving the integrals. Simplification occurs, however, if the reference point A is either a fixed point O or the center of mass G.

Fixed Point O. If A is a *fixed point* O on the body, Fig. 21–9a, then $\mathbf{v}_A = \mathbf{0}$, and using Eq. 21–9, we can express Eq. 21–16 as

$$T = \tfrac{1}{2}\boldsymbol{\omega} \cdot \mathbf{H}_O \tag{21–17}$$

Given that $\boldsymbol{\omega} = \omega_x\mathbf{i} + \omega_y\mathbf{j} + \omega_z\mathbf{k}$ and $\mathbf{H}_O = (H_x)_O\mathbf{i} + (H_y)_O\mathbf{j} + (H_z)_O\mathbf{k}$, where the three scalar components for $\mathbf{H}_O$ are defined by Eqs. 21–12, the kinetic energy may be expressed in an alternative form by substituting these expressions into Eq. 21–17 and carrying out the vector operations. The result is

$$T = \tfrac{1}{2}I_{xx}\omega_x^2 + \tfrac{1}{2}I_{yy}\omega_y^2 + \tfrac{1}{2}I_{zz}\omega_z^2 - I_{xy}\omega_x\omega_y - I_{yz}\omega_y\omega_z - I_{zx}\omega_z\omega_x \tag{21–18}$$

If the x, y, z coordinate axes, having origin at O, are oriented in the body such that these axes represent *principal axes of inertia,* then $I_{xy} = I_{yz} = I_{zx} = 0$, and

$$T = \tfrac{1}{2}I_x\omega_x^2 + \tfrac{1}{2}I_y\omega_y^2 + \tfrac{1}{2}I_z\omega_z^2 \tag{21–19}$$

Center of Mass G. If A is located at the *center of mass G* of the body, Fig. 21–9b, then $\int_m \boldsymbol{\rho}_A\, dm = \mathbf{0}$ and, using Eq. 21–10, we can write Eq. 21–16 as

$$T = \tfrac{1}{2}mv_G^2 + \tfrac{1}{2}\boldsymbol{\omega} \cdot \mathbf{H}_G \tag{21–20}$$

In a manner similar to that for a fixed point, the last term on the right side may be represented in scalar form, in which case

$$T = \tfrac{1}{2}mv_G^2 + \tfrac{1}{2}I_{xx}\omega_x^2 + \tfrac{1}{2}I_{yy}\omega_y^2 + \tfrac{1}{2}I_{zz}\omega_z^2 - I_{xy}\omega_x\omega_y - I_{yz}\omega_y\omega_z - I_{zx}\omega_z\omega_x \tag{21–21}$$

If the x, y, z axes are *principal axes of inertia* for the body at point G, then

$$T = \tfrac{1}{2}mv_G^2 + \tfrac{1}{2}I_x\omega_x^2 + \tfrac{1}{2}I_y\omega_y^2 + \tfrac{1}{2}I_z\omega_z^2 \tag{21–22}$$

From the above equations it may be seen that the kinetic energy consists of two parts, namely, the translational kinetic energy of the mass center, $\tfrac{1}{2}mv_G^2$, and the body's rotational kinetic energy.

Principle of Work and Energy. Using one of the above expressions for computing the kinetic energy of a body, the *principle of work and energy* may be applied to solve kinetic problems which involve *force, velocity, and displacement*. For this case only one scalar equation can be written for each body, i.e.,

$$T_1 + \Sigma U_{1-2} = T_2 \tag{21-23}$$

Before applying this equation to the solution of problems the material in Chapter 18 should be reviewed.

Example 21-2

The bent rod in Fig. 21-11*a* has a mass of 1 kg/m. Initially, the rod lies in the same horizontal plane and is released from rest when the end A is 0.5 m above the hook at E. Assuming that the rod falls uniformly without rotation, determine the angular velocity of the rod just after A falls onto E. The hook at E provides a permanent connection for the rod because of the spring-lock mechanism S.

Fig. 21-11(a)

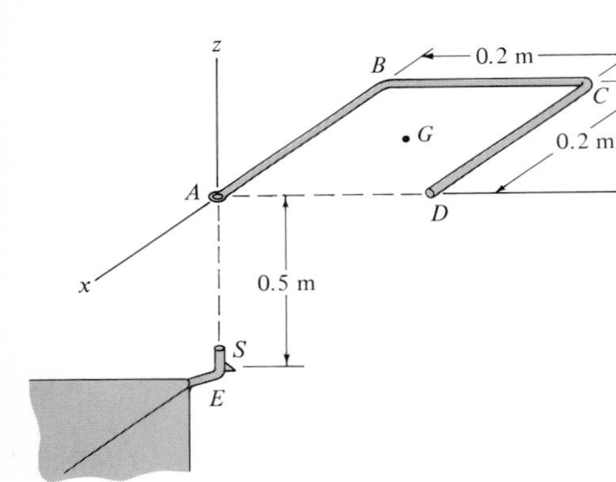

Solution

The velocity of the rod just *before* end A strikes the hook can be determined by applying the principle of work and energy (or the conservation-of-energy theorem, Eq. 18-16). Since the rod does not rotate while it is falling, the kinetic energy equation 21-21 reduces to the simplified form $T = \frac{1}{2}mv_G^2$, which involves only translation.

Since the total mass of the rod is (1 kg/m)(0.6 m) = 0.6 kg, we have

$$\{T_1\} + \{\Sigma U_{1-2}\} = \{T_2\}$$

$$\{0\} + \{0.6(9.81)(0.5)\} = \tfrac{1}{2}(0.6)(v_G)_1^2$$

$$(v_G)_1 = 3.13 \text{ m/s}$$

The momentum- and impulse-vector diagrams for the rod are shown in Fig. 21-11*b*. During the short time Δt, the impulsive force $\mathbf{F}$ acting at A changes the momentum of the rod. (The impulse created by the rod's weight $\mathbf{W}$ during this time is small compared to $\int \mathbf{F}\,dt$, so that it is neglected, i.e., the weight is a non-impulsive force.) Hence, the angular momentum of the rod is *conserved* about point A since the moment of $\int \mathbf{F}\,dt$ about point A is zero. (The moment of $\mathbf{W}\,\Delta t$ is neglected.) Equation 21–11 must be used for computing the angular momentum of the rod, since A does not become a *fixed point* until *after* the impulsive interaction with the hook. Thus, applying the conservation of angular momentum about A, Fig. 21–11*b*, gives

$$(\mathbf{H}_A)_1 = (\mathbf{H}_A)_2$$

$$\mathbf{r}_{G/A} \times m(\mathbf{v}_G)_1 = \mathbf{r}_{G/A} \times m(\mathbf{v}_G)_2 + (\mathbf{H}_G)_2 \qquad (1)$$

The scalar components of this equation may be determined once the position $(\bar{x}, \bar{y})$ of the mass center G is known and the moments of inertia of the rod relative to a set of coordinate axes x', y', z', having an origin at the mass center, have been computed. With reference to Fig. 21–11*c*, $\bar{y} = 0.1$ m, because of symmetry, and

$$\bar{x} = \frac{\Sigma \tilde{x} m}{\Sigma m}$$

$$= \frac{(-0.1 \text{ m})(0.2 \text{ kg}) + (-0.2 \text{ m})(0.2 \text{ kg}) + (-0.1 \text{ m})(0.2 \text{ kg})}{0.6 \text{ kg}}$$

$$= -0.133 \text{ m}$$

Thus,

$$\mathbf{r}_{G/A} = \{-0.133\mathbf{i} + 0.1\mathbf{j}\} \text{ m}$$

The primed axes are principal axes of inertia for the rod because $I_{x'y'} = I_{x'z'} = I_{z'y'} = 0$. Hence, from Eqs. 21–13, $(\mathbf{H}_G)_2 = I_{x'}\omega_x\mathbf{i} + I_{y'}\omega_y\mathbf{j} + I_{z'}\omega_z\mathbf{k}$. Substituting into Eq. (1), we have

$$(-0.133\mathbf{i} + 0.1\mathbf{j}) \times [(0.6)(-3.13\mathbf{k})] = (-0.133\mathbf{i} + 0.1\mathbf{j}) \times [(0.6)(-v_G)_2\mathbf{k}] + I_{x'}\omega_x\mathbf{i} + I_{y'}\omega_y\mathbf{j} + I_{z'}\omega_z\mathbf{k}$$

Expanding and equating the respective **i**, **j**, and **k** components yields

$$I_{x'}\omega_x - 0.060(v_G)_2 = -0.188 \qquad (2)$$

$$I_{y'}\omega_y - 0.080(v_G)_2 = -0.250 \qquad (3)$$

$$I_{z'}\omega_z = 0 \qquad (4)$$

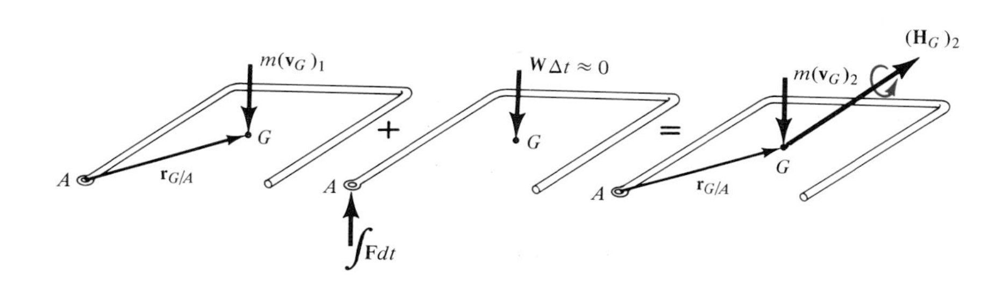

Fig. 21–11(b)

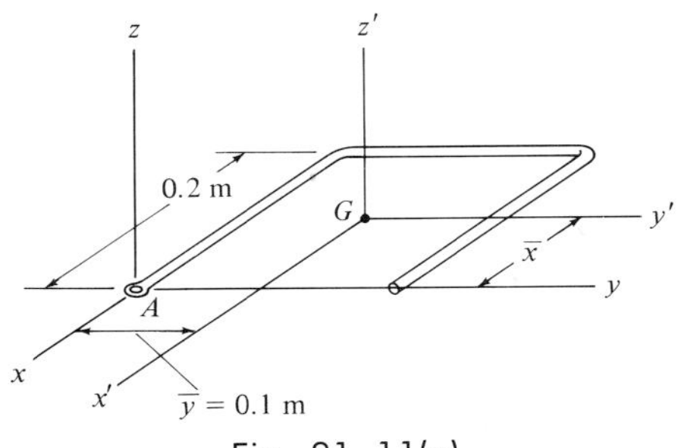

Fig. 21–11(c)

Only the moments of inertia $I_{x'}$ and $I_{y'}$ must be calculated, since it is seen that $\omega_z = 0$. Thus, from Fig. 21–11*c*, using Appendix D and the parallel-axis theorem, we have

$$I_{x'} = 2[0 + (0.2 \text{ kg})(0.1 \text{ m})^2] + [\tfrac{1}{12}(0.2 \text{ kg})(0.2 \text{ m})^2] = 4.67(10^{-3}) \text{ kg} \cdot \text{m}^2$$
$$I_{y'} = 2[\tfrac{1}{12}(0.2 \text{ kg})(0.2 \text{ m})^2 + (0.2 \text{ kg})(0.133 \text{ m} - 0.1 \text{ m})^2] + [0 + (0.2 \text{ kg})(0.2 \text{ m} - 0.133 \text{ m})^2] = 2.67(10^{-3}) \text{ kg} \cdot \text{m}^2$$

Another equation may be obtained by relating $\boldsymbol{\omega}$ to $(\mathbf{v}_G)_2$ using *kinematics*. Since $\omega_z = 0$ and after impact the rod rotates about the fixed point A, Eq. 20–4 may be applied, in which case

$$\mathbf{v}_G = \boldsymbol{\omega} \times \mathbf{r}_{G/A}$$
$$-(v_G)_2\mathbf{k} = (\omega_x\mathbf{i} + \omega_y\mathbf{j}) \times (-0.133\mathbf{i} + 0.1\mathbf{j})$$
$$-(v_G)_2 = 0.1\omega_x + 0.133\omega_y \tag{5}$$

Using the computed values for $I_{x'}$ and $I_{y'}$ and solving Eqs. (2) through (5) simultaneously gives

$$\omega_x = -6.59 \text{ rad/s}$$
$$\omega_y = -14.9 \text{ rad/s}$$
$$\omega_z = 0$$
$$(v_G)_2 = 2.62 \text{ m/s}$$

Thus,

$$\boldsymbol{\omega} = \{-6.59\mathbf{i} - 14.9\mathbf{j}\} \text{ rad/s} \qquad \textit{Ans.}$$
$$(\mathbf{v}_G)_2 = \{-2.62\mathbf{k}\} \text{ m/s}$$

It should be noted that $\boldsymbol{\omega}$ acts along the instantaneous axis of rotation for the rod just after impact (see Sec. 20–2). The direction cosines for this axis, which passes through point A, are the components of the unit vector for $\boldsymbol{\omega}$, i.e., $\mathbf{u} = \boldsymbol{\omega}/\omega$.

Example 21–3

A 5-N · m torque is applied to the vertical shaft CD shown in Fig. 21–12*a*, which allows the 10-kg gear A to turn freely about CE. Assuming that gear A starts from rest, determine the angular velocity of CD after it has turned two revolutions. Neglect the mass of shaft CD and axle CE and assume that gear A is approximated by a thin disk.

Solution

The principle of work and energy may be applied to solve this problem. Why? If shaft CD, the axle CE, and gear A are considered as a system of connected bodies, only the applied torque **M** does work. For two revolutions of CD, this work is

$$\Sigma U_{1-2} = (5 \text{ N} \cdot \text{m})(4\pi \text{ rad}) = 62.83 \text{ J}$$

Since the gear is initially at rest, the initial kinetic energy of the system is zero. A kinematic diagram for the gear is shown in Fig. 21–12*b*. If the angular velocity of CD is taken as $\boldsymbol{\omega}_{CD}$, then the angular velocity of gear A is $\boldsymbol{\omega}_A = \boldsymbol{\omega}_{CD} + \boldsymbol{\omega}_{CE}$. The gear may be imagined as a portion of a hypothetical massless extended body which is rotating about the *fixed point C*. The instantaneous axis of rotation for this body is along line CH, because both points C and H on the body (gear) have zero velocity and must therefore lie on this axis. This requires that the components $\boldsymbol{\omega}_{CD}$ and $\boldsymbol{\omega}_{CE}$ be related by the equation

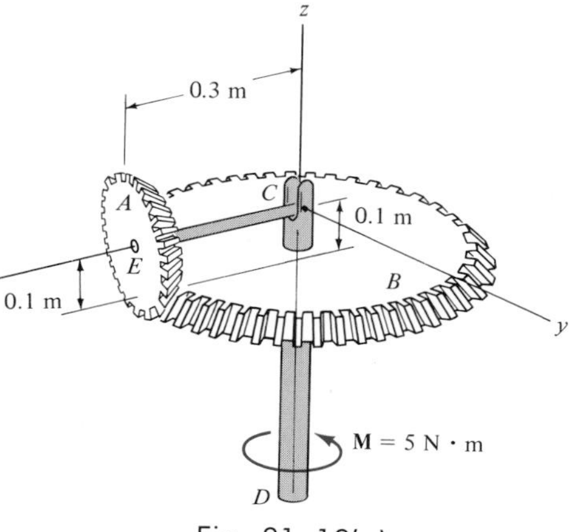

Fig. 21–12(a)

$$\frac{\omega_{CD}}{0.1\text{ m}} = \frac{\omega_{CE}}{0.3\text{ m}} \quad \text{or} \quad \omega_{CE} = 3\omega_{CD}$$

The angular velocity of the gear at the instant considered is, therefore,

$$\begin{aligned} \boldsymbol{\omega}_A &= -\omega_{CE}\mathbf{i} + \omega_{CD}\mathbf{k} \\ &= -3\omega_{CD}\mathbf{i} + \omega_{CD}\mathbf{k} \end{aligned} \tag{1}$$

The x, y, z axes in Fig. 21–12*a* represent *principal axes of inertia* at C for the gear. Since point C is a fixed point of rotation, Eq. 21–19 may be applied to determine the kinetic energy.

$$T = \tfrac{1}{2}I_x\omega_x^2 + \tfrac{1}{2}I_y\omega_y^2 + \tfrac{1}{2}I_z\omega_z^2 \tag{2}$$

From Appendix D and the parallel-axis theorem, the mass moments of inertia of the gear about point C are as follows:

$$I_x = \tfrac{1}{2}(10)(0.1)^2 = 0.05\text{ kg}\cdot\text{m}^2$$
$$I_y = I_z = \tfrac{1}{4}(10)(0.1)^2 + 10(0.3)^2 = 0.925\text{ kg}\cdot\text{m}^2$$

Since $\omega_x = -3\omega_{CD}$, $\omega_y = 0$, $\omega_z = \omega_{CD}$, Eq. (2) becomes

$$T_A = \tfrac{1}{2}(0.05)(-3\omega_{CD})^2 + 0 + \tfrac{1}{2}(0.925)(\omega_{CD})^2 = 0.6875\omega_{CD}^2$$

Applying the principle of work and energy, we obtain

$$\{T_1\} + \{\Sigma U_{1-2}\} = \{T_2\}$$
$$\{0\} + \{62.83\} = \{0.6875\omega_{CD}^2\}$$

Therefore,

$$\omega_{CD} = 9.56\text{ rad/s} \qquad \textit{Ans.}$$

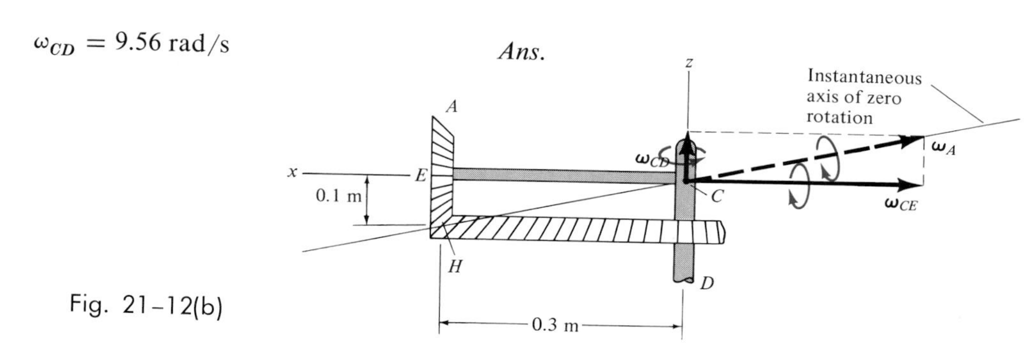

Fig. 21–12(b)

21-13. If a body contains *no planes of symmetry,* the principal moments of inertia can be computed mathematically. To show how this is done, consider the rigid body which is spinning with an angular velocity $\boldsymbol{\omega}$, directed along one of its principal axes of inertia. If the principal moment of inertia about this axis is I, the angular momentum can be expressed as $\mathbf{H} = I\boldsymbol{\omega} = I\omega_x\mathbf{i} + I\omega_y\mathbf{j} + I\omega_z\mathbf{k}$. The components of $\mathbf{H}$ may also be expressed by Eqs. 21-12, where the inertia tensor is assumed to be known. Equate the **i, j,** and **k** components of both expressions for **H** and consider ω_x, ω_y, and ω_z to be unknown. The solution of these three equations is obtained provided the determinant of the coefficients is zero. Show that this determinant, when expanded, yields the cubic equation

$$I^3 - (I_{xx} + I_{yy} + I_{zz})I^2 + (I_{xx}I_{yy} + I_{yy}I_{zz} + I_{zz}I_{xx} - I_{xy}^2 - I_{yz}^2 - I_{zx}^2)I - (I_{xx}I_{yy}I_{zz} - 2I_{xy}I_{yz}I_{zx} - I_{xx}I_{yz}^2 - I_{yy}I_{zx}^2 - I_{zz}I_{xy}^2) = 0$$

The three positive roots of I, obtained from the solution of this equation, represent the principal moments of inertia I_x, I_y, and I_z.

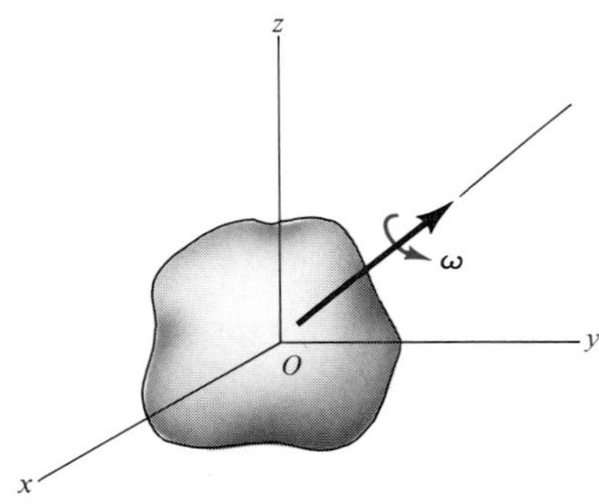

Prob. 21-13

21-14. Gear A has a mass of 3 kg and a radius of gyration of $k_z = 75$ mm. Gears B and C each have a mass of 200 g and a radius of gyration about the axis of their connecting shaft of 15 mm. If the gears are in mesh and C has an angular velocity of $\boldsymbol{\omega}_C = \{20\mathbf{j}\}$ rad/s, determine the total angular momentum for the system of three gears.

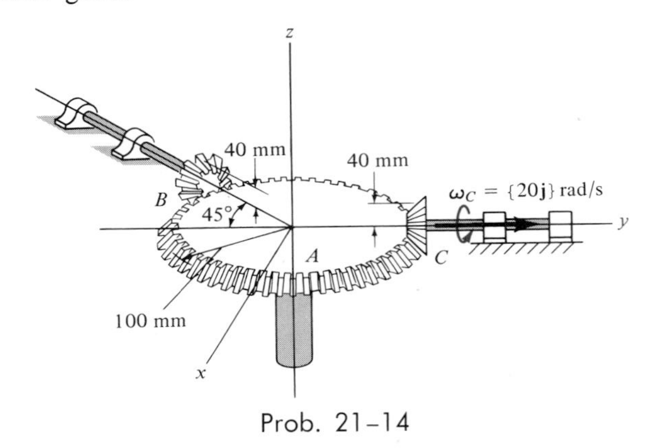

Prob. 21-14

21-15. The circular disk has a mass of $m = 5$ kg and a radius of $r = 200$ mm. If it is mounted on the shaft AB at an angle of $\theta = 40°$ with the vertical, determine the kinetic energy of the disk if the shaft is rotating with an angular velocity of $\omega = 20$ rad/s.

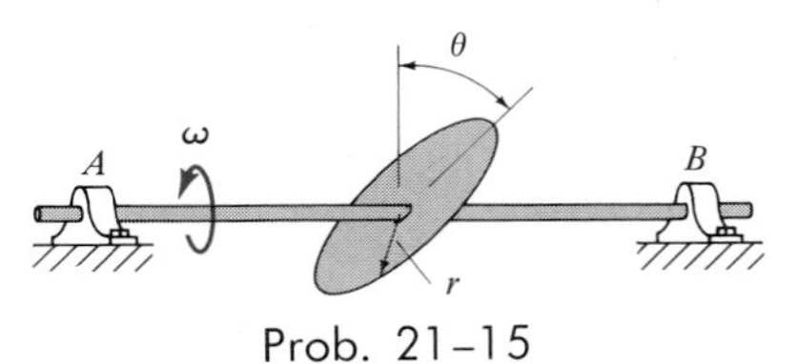

Prob. 21-15

21-15a. Solve Prob. 21-15 if the disk weighs $W = 20$ lb, and $r = 0.5$ ft, $\theta = 45°$, $\omega = 6$ rad/s.

***21-16.** Compute the magnitude of the angular momentum of the disk in Prob. 21-15 at the instant shown.

21–17. The rod AB has a mass of 2 kg and is attached to two smooth collars at its end points by ball-and-socket joints. If collar A is moving downward at a speed of 0.5 m/s, determine the kinetic energy of the rod at the instant shown. Assume that at this instant the angular velocity of the rod is directed perpendicular to the rod's axis. *Note:* The motion has been specified in Prob. 20–17.

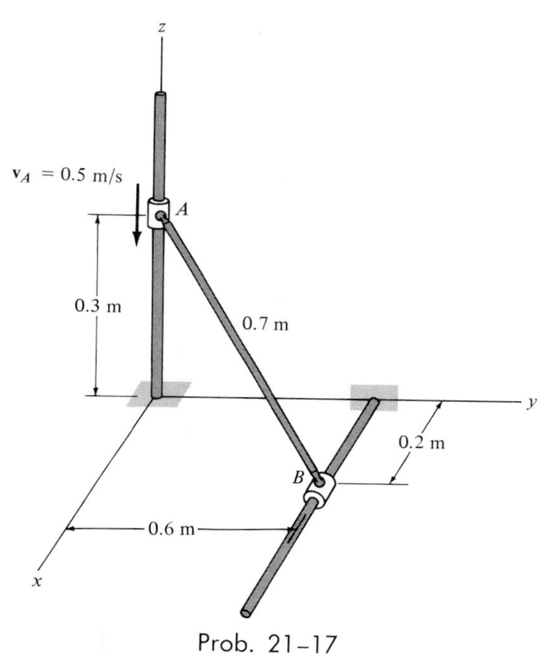

Prob. 21–17

21–18. Compute the magnitude of the angular momentum of rod AB in Prob. 21–17 about the mass center G at the instant shown. *Note:* The motion has been specified in the solution of Prob. 20–17.

21–19. The 5-kg rectangular plate is free to rotate about the y axis because of the bearing supports at A and B. When the plate is balanced in the vertical plane, a 3-g bullet is fired into it, perpendicular to its surface, with a velocity of $\mathbf{v} = \{-1500\mathbf{i}\}$ m/s. Compute the angular velocity of the plate when it rotates through an angle of 180°. If the bullet strikes corner D with this same velocity, instead of at C, does the angular velocity remain the same? Why or why not?

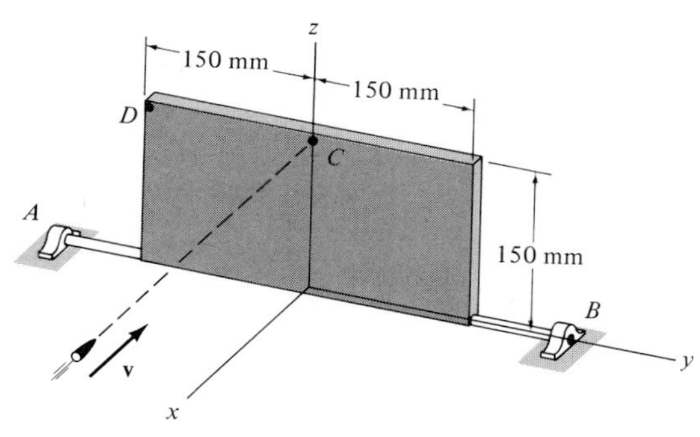

Prob. 21–19

***21–20.** The space capsule has a mass of $m_s = 2.5$ Mg and the radii of gyration are $k_x = k_z = 0.8$ m and $k_y = 0.5$ m. If it is initially coasting with a velocity of $\mathbf{v}_G = \{800\mathbf{j}\}$ m/s, compute its angular velocity just after it is struck by a meteoroid having a mass of $m_m = 0.75$ kg and a speed of $v_m = 1.5$ km/s, directed as shown. Assume that the meteoroid embeds itself into the capsule at $A(x_A = 0, y_A = 0.5 \text{ m}, z_A = 0.75 \text{ m})$ and that the capsule initially has no angular velocity.

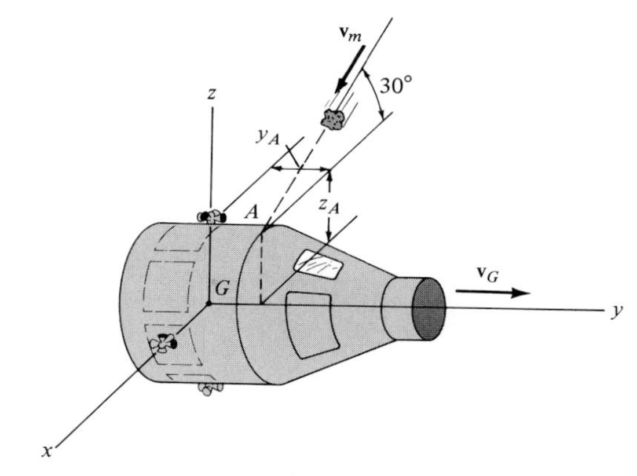

Prob. 21–20

***21–20a.** Solve Prob. 21–20 if the space capsule weighs $W_s = 6000$ lb (on earth); $k_x = k_z = 6$ ft; $k_y = 2$ ft; $\mathbf{v}_G = \{400\mathbf{j}\}$ ft/s; the meteoroid weighs $W_m = 1$ lb (on earth); $v_m = 6000$ ft/s; $(x_A = 0, y_A = 2 \text{ ft}, z_A = 1.5 \text{ ft})$.

21–21. The rod has a mass of 4 kg/m and is suspended from parallel cables at A and B. If the rod has an angular velocity of 2 rad/s about the z axis when it is in its lowest position, determine how high the center of mass G rises at the instant the rod stops swinging.

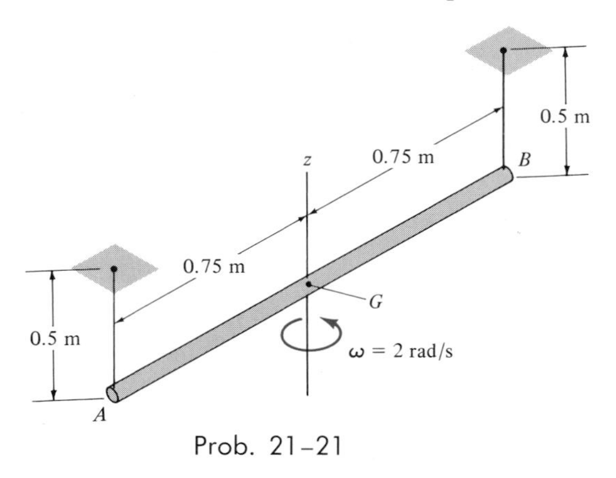

Prob. 21–21

21–22. The assembly consists of the 2-kg uniform link OA, the 3-kg rod AB, and the 1.5-kg slider block B that slides along the horizontal rod CD. If OA is released from rest when $\theta = 0°$, determine its angular velocity at the instant it swings downward to $\theta = 90°$. The connections at A and B consist of ball-and-socket joints and there is a pin at O.

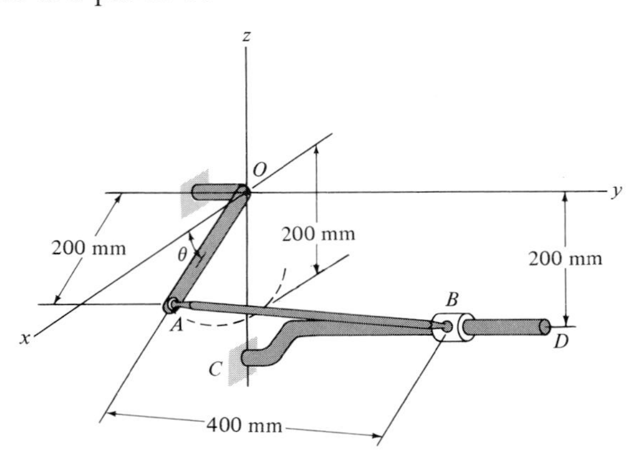

Prob. 21–22

21–23. The thin plate has a mass of 10 kg and is suspended from a ball-and-socket joint at O. A 3-g bullet is fired with a speed of 250 m/s into the plate at its corner and becomes embedded in the plate. If the bullet travels in a plane parallel to the x-z plane at an angle of 45° with the horizontal as shown, determine the angular momentum and the instantaneous axis of rotation of the plate just after impact.

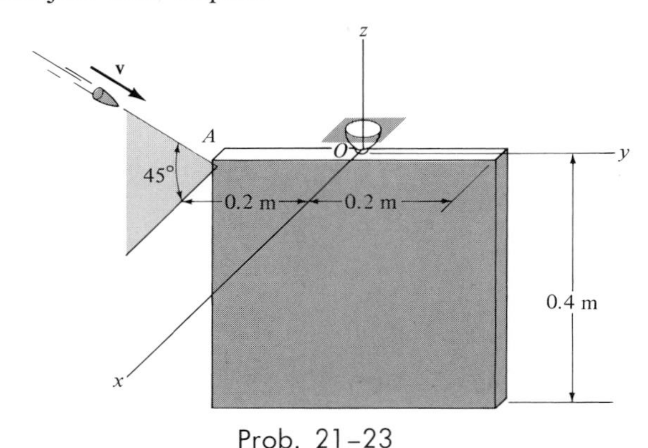

Prob. 21–23

***21–24.** The rod assembly has a mass of 3 kg/m and is rotating with a constant angular velocity of $\boldsymbol{\omega} = \{2\mathbf{k}\}$ rad/s when the looped end at C encounters a hook at S, which provides a permanent connection. Determine the angular velocity of the assembly immediately after impact.

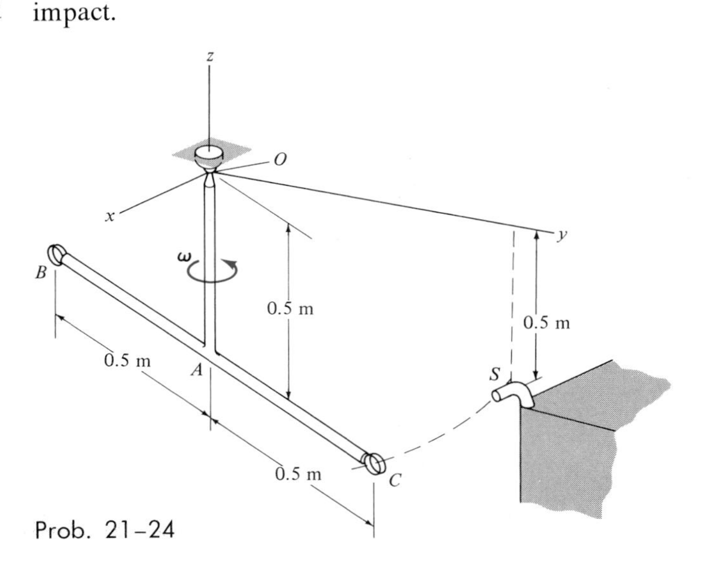

Prob. 21–24

21–25. The circular plate has a mass of $m = 3$ kg and a diameter of $d = 200$ mm. If it is released from rest and falls horizontally, $h = 500$ mm, onto the hook at S, which provides a permanent connection, determine the velocity of the mass center of the plate just after the connection with the hook is made.

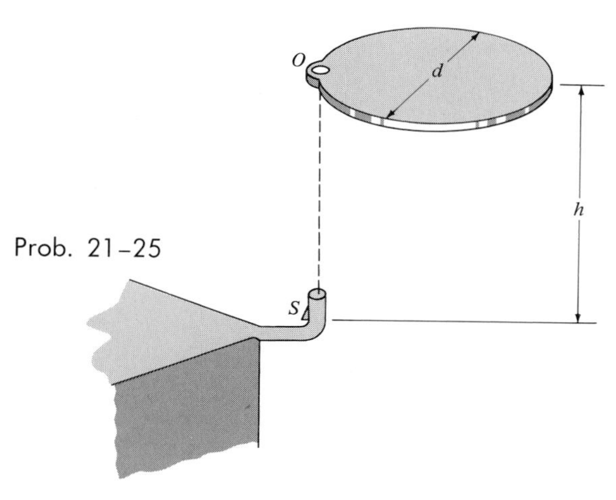

Prob. 21–25

21–25a. Solve Prob. 21–25 if the plate weighs $W = 10$ lb, and $d = 1$ ft, $h = 2$ ft.

21–26. Gear A has a mass of 0.5 kg. If it is released from rest from its highest position shown and rolls along the fixed inclined gear C, determine its spin about shaft OB at the instant it reaches its lowest point. The shaft is connected to a ball-and-socket joint at O and fixed to the gear at B. Neglect the mass of shaft OB and assume that gear A is a uniform disk having a radius of 50 mm.

Prob. 21–26

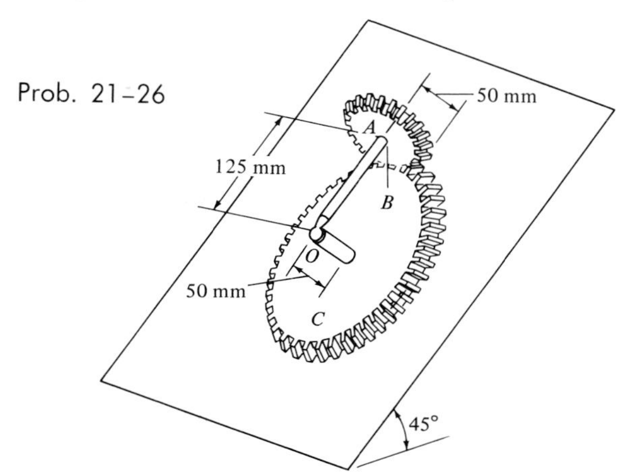

21–27. Solve Prob. 21–26 assuming that shaft OB has a mass of 120 g.

***21–28.** A thin plate, having a mass of 5 kg, is suspended from one of its corners by a ball-and-socket joint O. If a stone strikes the plate perpendicular to its surface, at an adjacent corner A, with an impulse of $\mathbf{I}_s = \{-4\mathbf{i}\}$ N · s, determine the instantaneous axis of rotation for the plate and the impulse created at O.

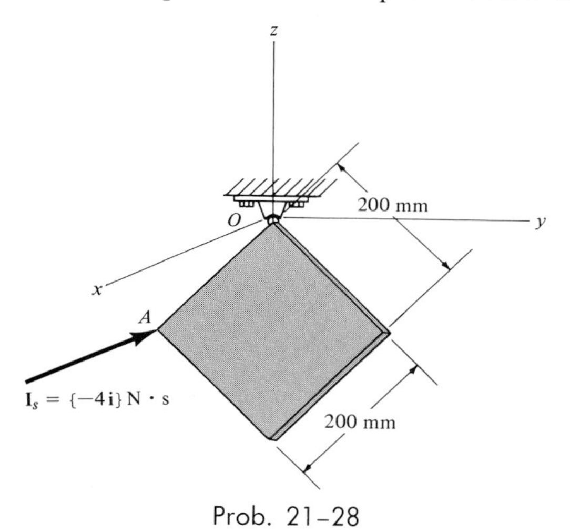

Prob. 21–28

*21–5. Equations of Motion of a Rigid Body

Having become familiar with the techniques used to describe both the inertia properties and the angular momentum of a body, one can write the equations which describe the motion of the body in their most useful forms.

Equations of Translational Motion. The *translational motion* of a body is defined in terms of the acceleration of the body's mass center, $\mathbf{a}_G$, which is measured from an inertial X, Y, Z reference. The equation of translational motion for the body can be written in vector form as

$$\Sigma \mathbf{F} = m\mathbf{a}_G \tag{21–24}$$

or by the three scalar equations

$$\begin{aligned} \Sigma F_x &= m(a_G)_x \\ \Sigma F_y &= m(a_G)_y \\ \Sigma F_z &= m(a_G)_z \end{aligned} \tag{21–25}$$

Here, $\Sigma \mathbf{F} = \Sigma F_x \mathbf{i} + \Sigma F_y \mathbf{j} + \Sigma F_z \mathbf{k}$ represents the sum of all the external forces acting on the body.

Equations of Rotational Motion. The *rotational motion* of a body will be considered with respect to one of two points: either point O, which is fixed to an inertial reference and therefore does not accelerate, or the body's center of mass G. As developed in Sec. 15–6, the equations

$$\Sigma \mathbf{M}_O = \frac{d\mathbf{H}_O}{dt} \tag{21–26}$$

$$\Sigma \mathbf{M}_G = \frac{d\mathbf{H}_G}{dt} \tag{21–27}$$

relate the moments of the external forces applied to a system of particles, to the angular momentum of the particles measured with respect to either point O or the center of mass G for the system. Since a rigid body is composed of many particles, these same equations may also be used to relate the moments of the external forces acting on the body to the angular momentum of the body about point O or G, respectively.

The scalar components of the angular momentum $\mathbf{H}_O$ or $\mathbf{H}_G$ are defined by Eqs. 21–12 or, if principal axes of inertia are used either at point O or G, by Eqs. 21–13. If these components are computed about x, y, z axes that are *rotating* with an angular velocity $\mathbf{\Omega}$, which may be *different* from the body's angular velocity $\boldsymbol{\omega}$, then the time derivative, $\dot{\mathbf{H}} = d\mathbf{H}/dt$, as used in Eqs. 21–26 and 21–27, must account for the rotation of the x, y, z axes as measured from the inertial X, Y, Z axes. Hence, the time derivative of $\mathbf{H}$ must be determined from Eq. 20–10, in which case Eqs. 21–26 and 21–27 become

$$\begin{aligned}\Sigma\mathbf{M}_O &= (\dot{\mathbf{H}}_O)_{xyz} + \mathbf{\Omega} \times \mathbf{H}_O \\ \Sigma\mathbf{M}_G &= (\dot{\mathbf{H}}_G)_{xyz} + \mathbf{\Omega} \times \mathbf{H}_G\end{aligned} \tag{21–28}$$

Here $(\dot{\mathbf{H}})_{xyz}$ is the time rate of change of $\mathbf{H}$ measured from the x, y, z reference.

There are three ways in which one can define the motion of the x, y, z axes. Obviously, motion of this reference should be chosen to yield the simplest set of moment equations for the solution of a particular problem.

***x, y, z* Axes Having Motion $\mathbf{\Omega} = \mathbf{0}$.** If the body has general motion, Fig. 21–13*a*, the x, y, z axes may be chosen with origin at G, such that the axes only *translate* relative to the inertial X, Y, Z frame of reference. Doing this would certainly simplify Eq. 21–28, since $\mathbf{\Omega} = \mathbf{0}$. However, the body may have a rotation $\boldsymbol{\omega}$ about these axes, and therefore the moments and products of inertia of the body would have to be expressed as *functions of time*. In most cases this would be a difficult task, so that such a choice of axes has restricted value.

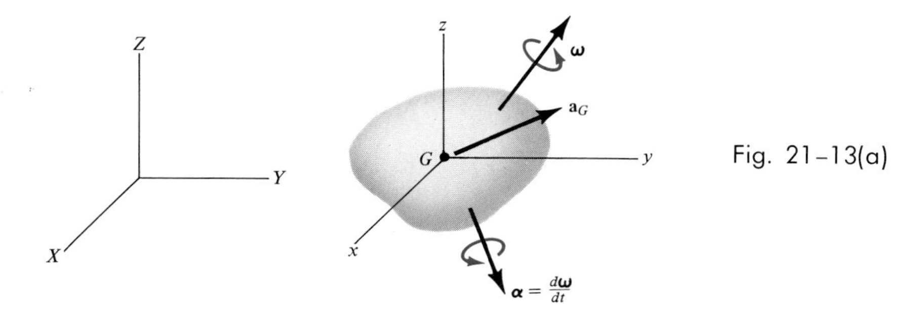

Fig. 21–13(a)

***x, y, z* Axes Having Motion $\mathbf{\Omega} = \boldsymbol{\omega}$.** The x, y, z axes may be chosen such that they are *fixed in and move with the body*. This case is shown in Fig. 21–13*b* for a body having a fixed point O. The moments and products of inertia of the body relative to these axes will be *constant* during the motion. Since $\mathbf{\Omega} = \boldsymbol{\omega}$, Eqs. 21–28 become

$$\begin{aligned}\Sigma\mathbf{M}_O &= (\dot{\mathbf{H}}_O)_{xyz} + \boldsymbol{\omega} \times \mathbf{H}_O \\ \Sigma\mathbf{M}_G &= (\dot{\mathbf{H}}_G)_{xyz} + \boldsymbol{\omega} \times \mathbf{H}_G\end{aligned} \tag{21–29}$$

These vector equations may also be expressed as three scalar equations, using Eqs. 21–12. Neglecting the subscripts O and G, we have

$$\begin{aligned}\Sigma M_x\mathbf{i} + \Sigma M_y\mathbf{j} + \Sigma M_z\mathbf{k} = {} & (I_{xx}\dot{\omega}_x - I_{xy}\dot{\omega}_y - I_{xz}\dot{\omega}_z)\mathbf{i} \\ & + (-I_{yx}\dot{\omega}_x + I_{yy}\dot{\omega}_y - I_{yz}\dot{\omega}_z)\mathbf{j} \\ & + (-I_{zx}\dot{\omega}_x - I_{zy}\dot{\omega}_y + I_{zz}\dot{\omega}_z)\mathbf{k} \\ & + (\omega_x\mathbf{i} + \omega_y\mathbf{j} + \omega_z\mathbf{k}) \times (I_{xx}\omega_x - I_{xy}\omega_y - I_{xz}\omega_z)\mathbf{i} \\ & + (\omega_x\mathbf{i} + \omega_y\mathbf{j} + \omega_z\mathbf{k}) \times (-I_{yx}\omega_x + I_{yy}\omega_y - I_{yz}\omega_z)\mathbf{j} \\ & + (\omega_x\mathbf{i} + \omega_y\mathbf{j} + \omega_z\mathbf{k}) \times (-I_{zx}\omega_x - I_{zy}\omega_y + I_{zz}\omega_z)\mathbf{k}\end{aligned}$$

Computing the cross products and equating the respective **i**, **j**, and **k** components yields

$$\begin{aligned}\Sigma M_x &= I_{xx}\dot{\omega}_x - (I_{yy} - I_{zz})\omega_y\omega_z - I_{xy}(\dot{\omega}_y - \omega_z\omega_x) - I_{yz}(\omega_y^2 - \omega_z^2) \\ &\qquad - I_{zx}(\dot{\omega}_z + \omega_x\omega_y) \\ \Sigma M_y &= I_{yy}\dot{\omega}_y - (I_{zz} - I_{xx})\omega_z\omega_x - I_{yz}(\dot{\omega}_z - \omega_x\omega_y) - I_{zx}(\omega_z^2 - \omega_x^2) \\ &\qquad - I_{xy}(\dot{\omega}_x + \omega_y\omega_z) \\ \Sigma M_z &= I_{zz}\dot{\omega}_z - (I_{xx} - I_{yy})\omega_x\omega_y - I_{zx}(\dot{\omega}_x - \omega_y\omega_z) - I_{xy}(\omega_x^2 - \omega_y^2) \\ &\qquad - I_{yz}(\dot{\omega}_y + \omega_z\omega_x)\end{aligned} \tag{21–30}$$

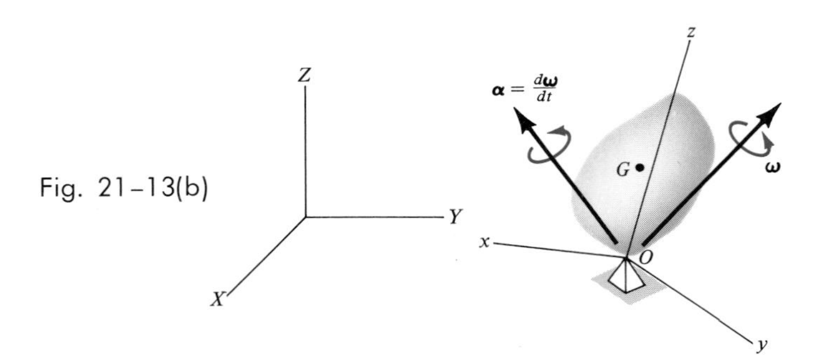

Fig. 21–13(b)

If the x, y, and z axes are chosen as *principal axes of inertia,* the products of inertia are zero, $I_{xx} = I_x$, etc., and the above equations reduce to the form

$$\begin{aligned}\Sigma M_x &= I_x\dot{\omega}_x - (I_y - I_z)\omega_y\omega_z \\ \Sigma M_y &= I_y\dot{\omega}_y - (I_z - I_x)\omega_z\omega_x \\ \Sigma M_z &= I_z\dot{\omega}_z - (I_x - I_y)\omega_x\omega_y\end{aligned} \tag{21–31}$$

This set of equations is known historically as the *Euler equations of motion,* named after the Swiss mathematician Leonhard Euler (1707–1783).

Both Eqs. 21–30 and 21–31 apply *only* for moments summed either about point O or G.

When applying these equations it should be realized that $\dot{\omega}_x$, $\dot{\omega}_y$, and $\dot{\omega}_z$ represent the time derivatives of the magnitudes of the x, y, z components of $\boldsymbol{\omega}$. Since the x, y, z axes are rotating at $\boldsymbol{\Omega} = \boldsymbol{\omega}$, then, from Eq. 20–10, it may be noted that $(\dot{\boldsymbol{\omega}})_{XYZ} = (\dot{\boldsymbol{\omega}})_{xyz} + \boldsymbol{\omega} \times \boldsymbol{\omega}$. Since $\boldsymbol{\omega} \times \boldsymbol{\omega} = \mathbf{0}$, $(\dot{\boldsymbol{\omega}})_{XYZ} = (\dot{\boldsymbol{\omega}})_{xyz}$. This important result indicates that the required time derivative of $\boldsymbol{\omega}$ can be obtained either by first finding the components of $\boldsymbol{\omega}$ along the x, y, z axes, and then taking the time derivative of the magnitudes of these components, i.e., $(\dot{\boldsymbol{\omega}})_{xyz}$, or by first finding the time derivative of $\boldsymbol{\omega}$ with respect to the X, Y, Z axes, i.e., $(\dot{\boldsymbol{\omega}})_{XYZ}$ and then determining the components $\dot{\omega}_x$, $\dot{\omega}_y$, and $\dot{\omega}_z$. In practice, it is generally easier to compute $\dot{\omega}_x$, $\dot{\omega}_y$, and $\dot{\omega}_z$ on the basis of finding $(\dot{\boldsymbol{\omega}})_{XYZ}$. (Compare solutions I and II of Example 20–1.)

***x, y, z* Axes Having Motion $\boldsymbol{\Omega} \neq \boldsymbol{\omega}$.** To simplify the calculations for the time derivative of $\boldsymbol{\omega}$, it is often convenient to choose the x, y, z axes having an angular velocity $\boldsymbol{\Omega}$ which is different from the angular velocity $\boldsymbol{\omega}$ of the body. This is particularly suitable for the analysis of spinning tops and gyroscopes which are *symmetrical* about the spinning axis.* When this is the case, the moments and products of inertia remain constant during the motion. An example is given in Fig. 21–13*c*. Even though the disk is spinning with a constant rate $\boldsymbol{\omega}$, the moments of inertia of the disk about the (principal) x, y, z axes are the same for *any* angular velocity $\boldsymbol{\Omega}$ of the x, y, z axes, *provided* $\boldsymbol{\Omega}$ is collinear with $\boldsymbol{\omega}$.

Fig. 21–13(c)

Equations 21–28 are applicable for such a set of chosen axes. Each of these two vector equations may be reduced to a set of three scalar equations which are derived in a manner similar to Eqs. 21–30. Most often, the x, y, z axes chosen are principal axes for the body, so that the *Euler equations of motion,* analogous to Eqs. 21–31, become

$$\begin{aligned}\Sigma M_x &= I_x\dot{\omega}_x - I_y\Omega_z\omega_y + I_z\Omega_y\omega_z\\ \Sigma M_y &= I_y\dot{\omega}_y - I_z\Omega_x\omega_z + I_x\Omega_z\omega_x\\ \Sigma M_z &= I_z\dot{\omega}_z - I_x\Omega_y\omega_x + I_y\Omega_x\omega_y\end{aligned} \qquad (21\text{–}32)$$

where $\Omega_x, \Omega_y, \Omega_z$ represent the x, y, z components of $\boldsymbol{\Omega}$, measured from the inertial frame of reference.

Any one of these sets of moment equations (Eqs. 21–30, 21–31, or 21–32) represents a series of three first-order nonlinear differential equations. These equations are "coupled," since the angular velocity components are present in all the terms. Success in determining the solution for a particular problem therefore depends upon what is unknown in these

*A detailed discussion of such devices is given in Sec. 21–6.

equations. Difficulty certainly arises when one attempts to solve for the unknown components of $\boldsymbol{\omega}$, given the external moments as functions of time. Further complications can arise if the moment equations are coupled to the three scalar equations of translation (Eqs. 21–25). This can happen because of the existence of kinematic constraints which relate the rotation of the body to the translation of its mass center, as in the case of a hoop which rolls without slipping. Problems necessitating the simultaneous solution of differential equations generally require application of numerical methods with the aid of a computer. In many engineering problems, however, one is required to determine the applied moments acting on the body, given information about the motion of the body. Fortunately, many of these types of problems have direct solutions, so that there is no need to resort to computer techniques.

PROCEDURE FOR ANALYSIS

It is recommended that the following four-step procedure be used for solving problems involving the spatial motion of a rigid body:

Step 1: Draw a *free-body diagram* of the body at the instant considered and specify the x, y, z coordinate system. The origin of this reference must be located either at the body's mass center G, or at point O, considered fixed in an inertial reference frame and located either on the body or on a hypothetical massless extension of the body. Depending upon the nature of the problem, a decision should be made as to what type of rotational motion $\boldsymbol{\Omega}$ this coordinate system should have, i.e., $\boldsymbol{\Omega} = \mathbf{0}$, $\boldsymbol{\Omega} = \boldsymbol{\omega}$, or $\boldsymbol{\Omega} \neq \boldsymbol{\omega}$. When deciding, one should keep in mind that the moment equations are simplified when the axes move in such a manner that they represent principal axes of inertia for the body at all times.

Step 2: Compute the necessary moments and products of inertia, and the components of the body's angular velocity and angular acceleration. In some cases a *kinematic diagram* might be helpful, since it provides a graphical aid for determining the acceleration of the body's mass center and for computing the angular acceleration. The angular velocity can usually be determined from the constraints of the body or from the given problem data. If this vector is changing direction, Eq. 20–10 $[(\dot{\boldsymbol{\omega}})_{XYZ} = (\dot{\boldsymbol{\omega}})_{xyz} + \boldsymbol{\Omega} \times \boldsymbol{\omega}]$ must be used to determine the components of angular acceleration. In particular, note that if $\boldsymbol{\Omega} = \boldsymbol{\omega}$ then $(\dot{\boldsymbol{\omega}})_{XYZ} = (\dot{\boldsymbol{\omega}})_{xyz}$.

Step 3: Apply either the two vector equations 21–24 and 21–28, or the six scalar component equations appropriate to the x, y, z coordinate axes chosen for the problem.

Step 4: If necessary, use *kinematics* to relate some of the unknowns to obtain further equations necessary for a complete solution to the problem.

The following example problems numerically illustrate application of this four-step procedure.

Example 21-4

The gear shown in Fig. 21–14*a* has a mass of 10 kg and is mounted at an angle of 10° with a rotating shaft having negligible mass. The moment of inertia of the gear about the z axis is $I_z = 0.1 \text{ kg} \cdot \text{m}^2$, and the moments of inertia about the x and y axes are $I_x = I_y = 0.05 \text{ kg} \cdot \text{m}^2$. If the shaft is rotating with a constant angular velocity of $\omega_{AB} = 30 \text{ rad/s}$, determine the reactions of the bearing supports A and B at the instant the gear is in the position shown.

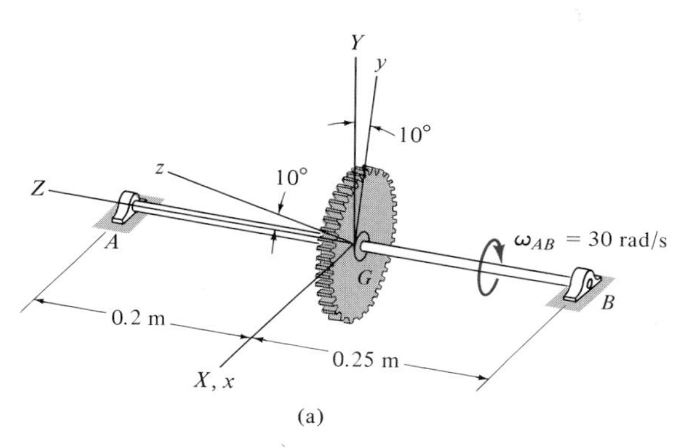

Fig. 21–14

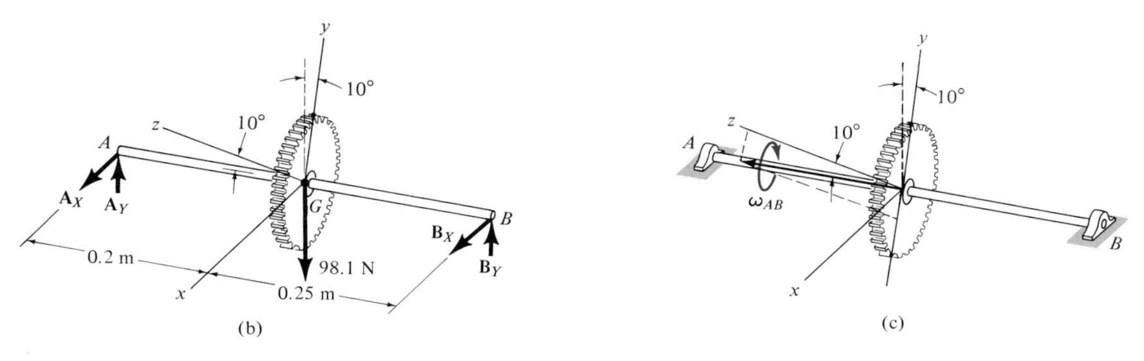

Solution I

(*Scalar Analysis*). *Step 1:* The free-body diagram of the gear and shaft is shown in Fig. 21–14*b*. The origin of the x, y, z coordinate system is located at the gear's center of mass G, which is also a fixed point. The axes are fixed in and rotate with the gear; hence $\boldsymbol{\Omega} = \boldsymbol{\omega}$.

Step 2: As shown in Fig. 21–14*c*, the angular velocity $\boldsymbol{\omega}$ of the gear is constant in magnitude and is always directed along the axis of the shaft AB. Since this vector is measured from the X, Y, Z inertial frame of reference, for any position of the x, y, z axes $\boldsymbol{\omega}$ can be expressed in terms of the x, y, z coordinates as

$$\boldsymbol{\omega} = \{-30 \sin 10°\mathbf{j} + 30 \cos 10°\mathbf{k}\} \text{ rad/s} \qquad (1)$$

Hence,

$$\omega_x = 0, \qquad \omega_y = -30 \sin 10°, \qquad \omega_z = 30 \cos 10°$$

Since $\boldsymbol{\omega}$ has a constant direction and magnitude when observed from X, Y, Z, its time rate of change $(\dot{\boldsymbol{\omega}})_{XYZ} = \mathbf{0}$. Hence, since $\boldsymbol{\Omega} = \boldsymbol{\omega}$, $(\dot{\boldsymbol{\omega}})_{xyz} = \mathbf{0}$ or

$$\dot{\omega}_x = \dot{\omega}_y = \dot{\omega}_z = 0$$

Also, since G is a fixed point,

$$(a_x)_G = (a_y)_G = (a_z)_G = 0$$

Step 3: Applying the Euler equations of motion, Eqs. 21–31, yields

$$\Sigma M_x = I_x \dot{\omega}_x - (I_y - I_z)\omega_y \omega_z$$

$$-(A_Y)(0.2) + (B_Y)(0.25) = 0 - (0.05 - 0.1)(-30 \sin 10°)(30 \cos 10°)$$

$$-0.2A_Y + 0.25B_Y = -7.70 \qquad (2)$$

$$\Sigma M_y = I_y \dot{\omega}_y - (I_z - I_x)\omega_z \omega_x$$

$$A_X(0.2) \cos 10° - B_X(0.25) \cos 10° = 0 + 0$$

$$A_X = 1.25B_X \qquad (3)$$

$$\Sigma M_z = I_z \dot{\omega}_z - (I_x - I_y)\omega_x \omega_y$$

$$A_X(0.2) \sin 10° - B_X(0.25) \sin 10° = 0$$

Again,

$$A_X = 1.25B_X$$

Applying Eqs. 21–25, we have

$$\Sigma F_X = m(a_G)_X; \qquad A_X + B_X = 0 \qquad (4)$$

$$\Sigma F_Y = m(a_G)_Y; \qquad A_Y + B_Y - 98.1 = 0 \qquad (5)$$

$$\Sigma F_Z = m(a_G)_Z; \qquad 0 = 0$$

Solving Eqs. (2) through (5) simultaneously gives

$$A_X = B_X = 0 \qquad \textit{Ans.}$$

$$A_Y = 71.6 \text{ N} \qquad \textit{Ans.}$$

$$B_Y = 26.4 \text{ N} \qquad \textit{Ans.}$$

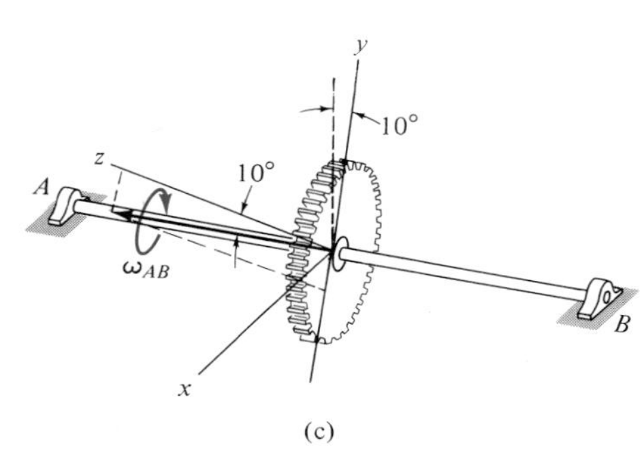

Fig. 21–14(b,c)

Solution II

(*Vector Analysis*). Since the x, y, z axes are chosen fixed to the flywheel, this problem may be solved by using Eq. 21–29. It is first necessary to determine the components of the angular velocity $\boldsymbol{\omega}$ and the angular momentum $\mathbf{H}_G$ of the flywheel with respect to the x, y, z axes. Using Eqs. 21–13 and Eq. (1), we have

$$H_x = I_x \omega_x = 0$$

$$H_y = I_y \omega_y = 0.05(-30 \sin 10°) = -0.260 \text{ kg} \cdot \text{m}^2/\text{s}$$

$$H_z = I_z \omega_z = 0.1(30 \cos 10°) = 2.95 \text{ kg} \cdot \text{m}^2/\text{s}$$

Thus,

$$\mathbf{H}_G = \{-0.260\mathbf{j} + 2.95\mathbf{k}\} \text{ kg} \cdot \text{m}^2/\text{s}$$

Since the x, y, z frame of reference rotates with the same constant angular velocity as the flywheel, the angular momentum of the flywheel, $\mathbf{H}_G$, remains *constant* with respect to an observer located in this rotating frame; hence, $(\dot{\mathbf{H}}_G)_{xyz} = \mathbf{0}$, and Eq. 21–29 becomes

$$\Sigma \mathbf{M}_G = \boldsymbol{\omega} \times \mathbf{H}_G$$

$$\mathbf{r}_A \times \mathbf{F}_A + \mathbf{r}_B \times \mathbf{F}_B = \boldsymbol{\omega} \times \mathbf{H}_G \tag{6}$$

$$\begin{aligned}(-0.2 \sin 10°\mathbf{j} + 0.2 \cos 10°\mathbf{k}) \times (A_X\mathbf{i} + A_Y \cos 10°\mathbf{j} + A_Y \sin 10°\mathbf{k})\\ + (0.25 \sin 10°\mathbf{j} - 0.25 \cos 10°\mathbf{k})\\ \times (B_X\mathbf{i} + B_Y \cos 10°\mathbf{j} + B_Y \sin 10°\mathbf{k})\\ = (-30 \sin 10°\mathbf{j} + 30 \cos 10°\mathbf{k}) \times (-0.260\mathbf{j} + 2.95\mathbf{k})\end{aligned} \tag{7}$$

This equation is rather cumbersome to expand. It appears more suitable to apply Eq. (6) in terms of components along the axes of the inertial X, Y, Z frame of reference simply because all the vectors, except $\mathbf{H}_G$, are easily expressed in these directions. Representing the directions of the X, Y, and Z axes by the unit vectors **I, J,** and **K,** respectively, we may write

$$\boldsymbol{\omega} = \{30\mathbf{K}\} \text{ rad/s}$$

$$\mathbf{H}_G = H_y(\cos 10°\mathbf{J} - \sin 10°\mathbf{K}) + H_z(\sin 10°\mathbf{J} + \cos 10°\mathbf{K})$$

Hence, substituting into Eq. (6),

$$\begin{aligned}0.2\mathbf{K} \times (A_X\mathbf{I} + A_Y\mathbf{J}) + (-0.25\mathbf{K}) \times (B_X\mathbf{I} + B_Y\mathbf{J})\\ = (30\mathbf{K}) \times [-0.260(\cos 10°\mathbf{J} - \sin 10°\mathbf{K})\\ + 2.95(\sin 10°\mathbf{J} + \cos 10°\mathbf{K})]\end{aligned} \tag{8}$$

Expanding either Eq. (7) or (8) and equating the respective unit-vector components yields

$$-0.2A_Y + 0.25B_Y = -7.70$$

and

$$A_X = 1.25B_X$$

which are the same as Eqs. (2) and (3) of the previous solution. Also, application of vector equation 21–24 ($\Sigma\mathbf{F} = m\mathbf{a}_G$) yields the scalar equations (4) and (5), previously obtained.

Example 21–5

The airplane shown in Fig. 21–15*a* is in the process of making a steady *horizontal* turn at the rate of ω_p. During this motion, the airplane's propeller is spinning at the rate of ω_s. For (a) a two-bladed propeller, Fig. 21–15*b*, and (b) a four-bladed propeller, Fig. 21–15*d*, determine the moments which the propeller-shaft exerts on the propeller when two of the blades are in the vertical position. For simplicity, assume *two* blades to be

Fig. 21–15(a)

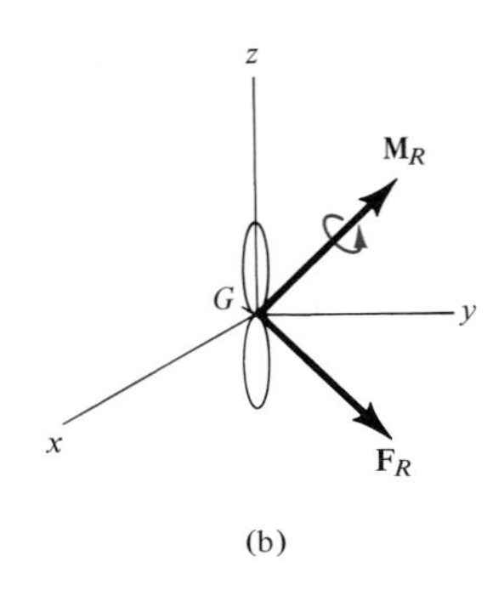

(b)

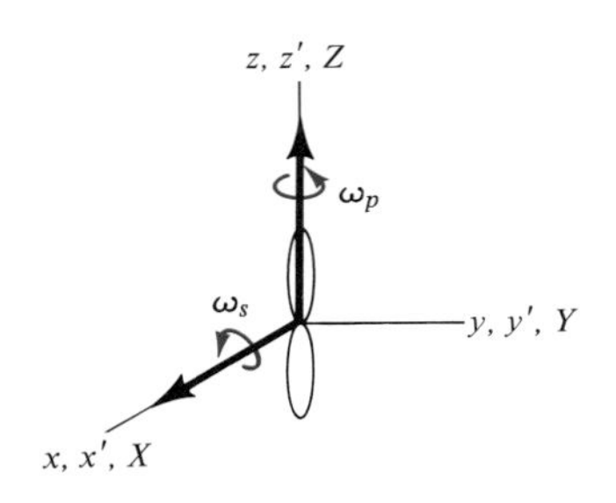

(c)

Fig. 21–15(b–c)

a uniform slender bar having a moment of inertia I about an axis perpendicular to the blades and passing through their center, and zero moment of inertia about a longitudinal axis.

Solution

Part (a). *Step 1:* The free-body diagram of the propeller is shown in Fig. 21–15*b*. The effect of the connecting shaft on the propeller is indicated by the resultants $\mathbf{F}_R$ and $\mathbf{M}_R$. If it is assumed that the x, y, z axes are *fixed to the propeller,* then $\mathbf{\Omega} = \boldsymbol{\omega}$. Furthermore, the axes so chosen represent the principal axes of inertia for the propeller.

Step 2: The moments of inertia I_x and I_y are equal ($I_x = I_y = I$) and $I_z = 0$. The angular velocity at the instant shown, as measured from a fixed X, Y, Z frame coincident with x, y, z, Fig. 21–15*c*, is

$$\boldsymbol{\omega} = \boldsymbol{\omega}_s + \boldsymbol{\omega}_p = \omega_s\mathbf{i} + \omega_p\mathbf{k}$$

so that the components of $\boldsymbol{\omega}$ are

$$\omega_x = \omega_s, \qquad \omega_y = 0, \qquad \omega_z = \omega_p$$

Since $\boldsymbol{\omega}$ is *changing direction,* the time derivative of $\boldsymbol{\omega}$ will be computed with respect to the fixed X, Y, Z axes and *then* $(\dot{\boldsymbol{\omega}})_{XYZ}$ will be resolved into components along the moving x, y, z axes. To simplify this calculation, a third coordinate system x', y', z' will be chosen which has an angular velocity $\mathbf{\Omega}' = \boldsymbol{\omega}_p$ and is coincident with the X, Y, Z axes at the instant shown. Thus

$$\begin{aligned}(\dot{\boldsymbol{\omega}})_{XYZ} &= (\dot{\boldsymbol{\omega}})_{x'y'z'} + \mathbf{\Omega}' \times \boldsymbol{\omega} \\ &= (\dot{\boldsymbol{\omega}}_s)_{x'y'z'} + (\dot{\boldsymbol{\omega}}_p)_{x'y'z'} + \boldsymbol{\omega}_p \times (\boldsymbol{\omega}_s + \boldsymbol{\omega}_p) \\ &= \mathbf{0} + \mathbf{0} + \boldsymbol{\omega}_p \times \boldsymbol{\omega}_s + \boldsymbol{\omega}_p \times \boldsymbol{\omega}_p\end{aligned} \tag{1}$$

Since $\boldsymbol{\omega}_p \times \boldsymbol{\omega}_p = \mathbf{0}$,

$$(\dot{\boldsymbol{\omega}})_{XYZ} = \omega_p\mathbf{k} \times \omega_s\mathbf{i} = \omega_p\omega_s\mathbf{j}$$

Therefore, since $\mathbf{\Omega} = \boldsymbol{\omega}$, $(\dot{\boldsymbol{\omega}})_{XYZ} = (\dot{\boldsymbol{\omega}})_{xyz}$ and

$$\dot{\omega}_x = 0, \qquad \dot{\omega}_y = \omega_p\omega_s, \qquad \dot{\omega}_z = 0$$

Step 3: Using Eqs. 21–31, since $\mathbf{\Omega} = \boldsymbol{\omega}$, we have

$$\begin{aligned}\Sigma M_x &= I_x\dot{\omega}_x - (I_y - I_z)\omega_y\omega_z \\ M_x &= I(0) - (I - 0)(0)\omega_p \\ M_x &= 0\end{aligned} \qquad \textit{Ans.}$$

$$\begin{aligned}\Sigma M_y &= I_y\dot{\omega}_y - (I_z - I_x)\omega_z\omega_x \\ M_y &= I(\omega_p\omega_s) - (0 - I)\omega_p\omega_s \\ M_y &= 2I\omega_p\omega_s\end{aligned} \qquad \textit{Ans.}$$

$$\Sigma M_z = I_z\dot{\omega}_z - (I_x - I_y)\omega_x\omega_y$$
$$M_z = 0(0) - (I - I)(\omega_s)(0)$$
$$M_z = 0 \qquad \textit{Ans.}$$

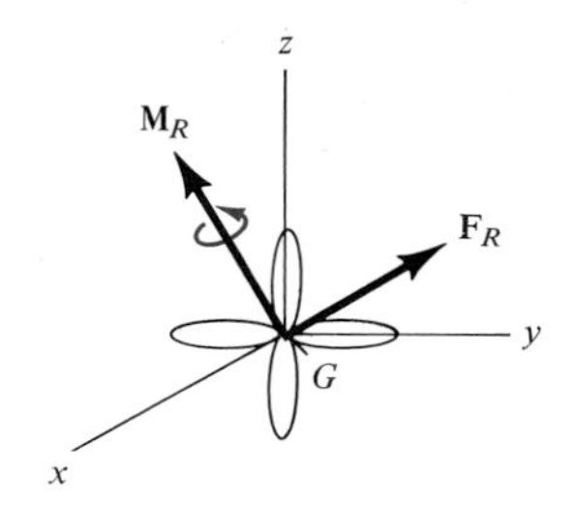

(d)

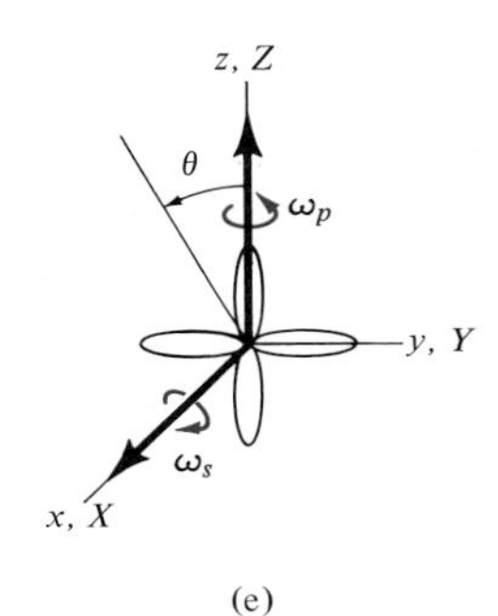

(e)

Fig. 21–15(d–e)

Part (b). *Step 1:* The free-body diagram of the four-bladed propeller is shown in Fig. 21–15*d*. In this case the x, y, z axes will not be fixed to the propeller blade; instead, they will be fixed to the airplane, and thus turn at a constant rate of $\mathbf{\Omega} = \boldsymbol{\omega}_p$. This choice is suitable since the axes so chosen always represent principal axes of inertia for any angle of orientation, θ, of the propeller,* Fig. 21–15*e*. Furthermore, the values of I_x, I_y, and I_z are constant with respect to time as the propeller rotates with respect to these axes. (This is not the case for a two-bladed propeller.)

Step 2: For any angle θ, the moments of inertia are

$$I_x = 2I$$
$$I_y = I_z = I$$

The angular velocities of the x, y, z axes and the propeller are observed from the fixed X, Y, Z axes, Fig. 21–15*e*. Hence the components are

$$\Omega_x = 0 \qquad \Omega_y = 0 \qquad \Omega_z = \omega_p$$
$$\omega_x = \omega_s \qquad \omega_y = 0 \qquad \omega_z = \omega_p$$

Since the angular velocity of the x, y, z axes is $\mathbf{\Omega} = \boldsymbol{\omega}_p$, the components of $\boldsymbol{\omega}$ do *not* appear to change direction when observed in x, y, z. Furthermore, since the magnitude of $\boldsymbol{\omega}$ is also constant when viewed from the x, y, z axes, the components of $(\dot{\boldsymbol{\omega}})_{xyz}$ are

$$\dot{\omega}_x = 0, \qquad \dot{\omega}_y = 0, \qquad \dot{\omega}_z = 0$$

Step 3: Using Eqs. 21–31, since $\mathbf{\Omega} = \boldsymbol{\omega}$, we have

$$\Sigma M_x = I_x\dot{\omega}_x - I_y\Omega_z\omega_y + I_z\Omega_y\omega_z$$
$$M_x = 2I(0) - I\omega_p(0) + I(0)\omega_p$$
$$M_x = 0 \qquad \textit{Ans.}$$

$$\Sigma M_y = I_y\dot{\omega}_y - I_z\Omega_x\omega_z + I_x\Omega_z\omega_x$$
$$M_y = I(0) - I(0)\omega_p + 2I\omega_p\omega_s$$
$$M_y = 2I\omega_p\omega_s \qquad \textit{Ans.}$$

$$\Sigma M_z = I_z\dot{\omega}_z - I_x\Omega_y\omega_x + I_y\Omega_x\omega_y$$
$$M_z = I(0) - 2I(0)\omega_s + I(0)(0)$$
$$M_z = 0 \qquad \textit{Ans.}$$

This problem can be solved by fixing the axes to the propeller as in part (a); however, this will require computation of $\dot{\boldsymbol{\omega}}$ by means of Eq. (1).

*Refer to the discussion in Sec. 21–2 that pertains to Fig. 21–6.

Example 21-6

The two rods and a shaft shown in Fig. 21–16*a* have a mass of 1.5 kg/m. If a resultant frictional moment **M**, developed in bearings *A* and *B*, causes the shaft to decelerate at 1 rad/s^2 when it has an initial angular velocity of $\omega = 6$ rad/s, determine the reactions on the bearings at the instant shown. What is the frictional moment **M** causing the deceleration?

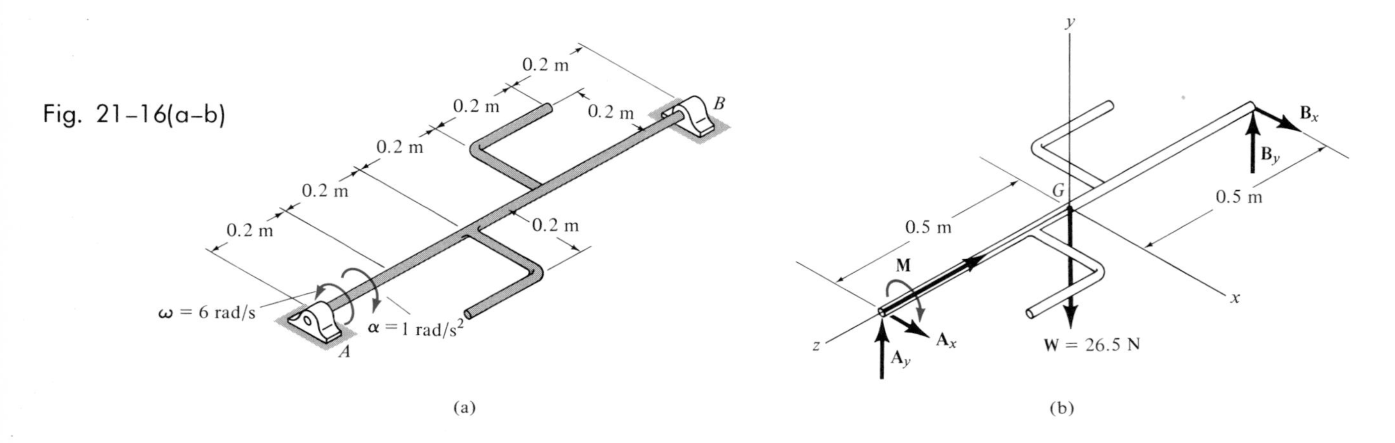

Fig. 21–16(a–b)

Solution I

Step 1: Since the assembly has a total rod length of 1.8 m, the total weight is $W = (1.5 \text{ kg/m})(1.8 \text{ m})(9.81 \text{ m/s}^2) = 26.5$ N. This force is applied at the mass center *G* on the free-body diagram, Fig. 21–16*b*. The resultant bearing frictional moment is represented by **M**, which is considered as a free vector in determining the external reactions and therefore can be applied at any point on the shaft. The *x*, *y*, *z* axes are located at the mass center *G* and are fixed in and rotate with the shaft. Hence, $\mathbf{\Omega} = \boldsymbol{\omega}$.

Step 2: The moments and products of inertia about the *x*, *y*, *z* axes are computed using Appendix D and the parallel-axis and parallel-plane theorems. The inertia terms which will be required for the solution are

$$I_{zz} = 2[\tfrac{1}{12}(1.5)(0.2)(0.2)^2 + 1.5(0.2)(0.1)^2] + 2[0 + (1.5)(0.2)(0.2)^2]$$
$$= 0.032 \text{ kg} \cdot \text{m}^2$$
$$I_{xy} = I_{yx} = I_{yz} = I_{zy} = 0$$
$$I_{xz} = I_{zx} = 2[0 + 1.5(0.2)(0.1)(0.1)] + 2[0 + 1.5(0.2)(0.2)(0.2)]$$
$$= 0.030 \text{ kg} \cdot \text{m}^2$$

Since points *A* and *B* are fixed at all times, the angular velocity $\boldsymbol{\omega}$ is always directed along the *z* axis of the shaft, Fig. 21–16*c*. The time derivative $\dot{\boldsymbol{\omega}}$ therefore measures only a change in magnitude of $\boldsymbol{\omega}$. (Recall that $(\dot{\boldsymbol{\omega}})_{XYZ} = (\dot{\boldsymbol{\omega}})_{xyz}$ since $\mathbf{\Omega} = \boldsymbol{\omega}$.) From the problem data,

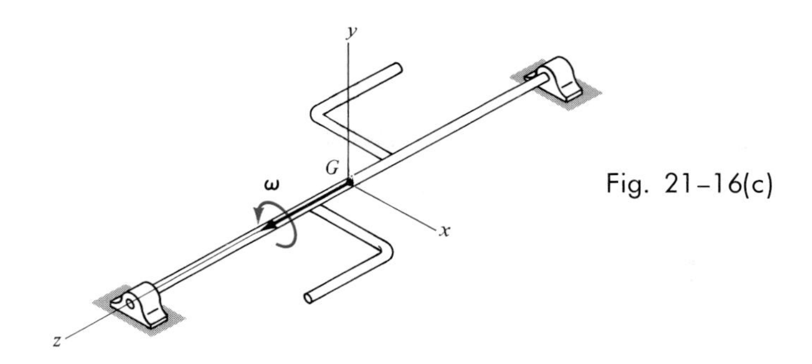

Fig. 21–16(c)

$$\omega_x = 0, \qquad \omega_y = 0, \qquad \omega_z = 6 \text{ rad/s}$$
$$\dot{\omega}_x = 0, \qquad \dot{\omega}_y = 0, \qquad \dot{\omega}_z = -1 \text{ rad/s}^2$$

Because G is a fixed point, $\mathbf{a}_G = \mathbf{0}$.
Step 3: Applying Eqs. 21–25, we have

$$\Sigma F_x = m(a_G)_x; \qquad A_x + B_x = 0 \qquad (1)$$
$$\Sigma F_y = m(a_G)_y; \qquad A_y + B_y - 26.5 = 0 \qquad (2)$$

Equations 21–30 must be used, since $\mathbf{\Omega} = \boldsymbol{\omega}$ and the x, y, z axes are *not* principal axes of inertia. Using the computed data, these equations reduce to the form

$$\Sigma M_x = -I_{zx}\dot{\omega}_z$$
$$-A_y(0.5) + B_y(0.5) = -0.03(-1)$$
$$-A_y + B_y = 0.06 \qquad (3)$$
$$\Sigma M_y = -I_{zx}\omega_z^2$$
$$A_x(0.5) - B_x(0.5) = -0.030(6)^2$$
$$A_x - B_x = -2.16 \qquad (4)$$
$$\Sigma M_z = I_{zz}\dot{\omega}_z$$
$$-M = 0.032(-1)$$
$$M = 0.032 \text{ N} \cdot \text{m} \qquad \textit{Ans.}$$

Solving Eqs. (1) through (4) simultaneously gives

$$A_x = -1.08 \text{ N} \qquad \textit{Ans.}$$
$$A_y = 13.22 \text{ N} \qquad \textit{Ans.}$$
$$B_x = 1.08 \text{ N} \qquad \textit{Ans.}$$
$$B_y = 13.28 \text{ N} \qquad \textit{Ans.}$$

Since A_x is negative, $\mathbf{A}_x$ acts in a direction opposite to that shown in Fig. 21–16*b*.

Solution II

The moment equations may *also* be established by direct application of vector equation 21–29. It is first necessary to compute the angular momentum $\mathbf{H}_G$ using Eqs. 21–12. Since

$$\boldsymbol{\omega} = \{6\mathbf{k}\}\ \text{rad/s}$$

then

$$H_x = -I_{xz}\omega_z = -0.030\omega_z$$
$$H_y = 0$$
$$H_z = I_{zz}\omega_z = 0.032\omega_z$$

Therefore,

$$\mathbf{H}_G = H_x\mathbf{i} + H_y\mathbf{j} + H_z\mathbf{k} = (-0.030\mathbf{i} + 0.032\mathbf{k})\omega_z$$

and

$$(\dot{\mathbf{H}}_G)_{xyz} = (-0.030\mathbf{i} + 0.032\mathbf{k})\dot{\omega}_z$$

Substituting $\omega_z = 6$ rad/s and $\dot{\omega}_z = -1$ rad/s^2 into these equations and applying Eq. 21–29 yields

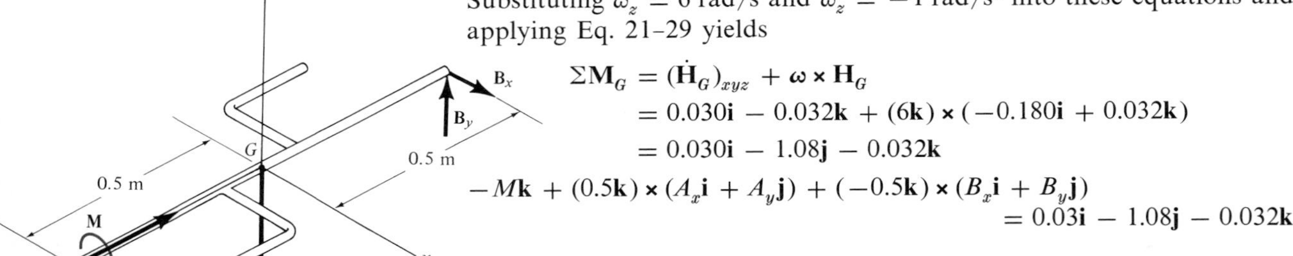

$$\Sigma\mathbf{M}_G = (\dot{\mathbf{H}}_G)_{xyz} + \boldsymbol{\omega} \times \mathbf{H}_G$$
$$= 0.030\mathbf{i} - 0.032\mathbf{k} + (6\mathbf{k}) \times (-0.180\mathbf{i} + 0.032\mathbf{k})$$
$$= 0.030\mathbf{i} - 1.08\mathbf{j} - 0.032\mathbf{k}$$

$$-M\mathbf{k} + (0.5\mathbf{k}) \times (A_x\mathbf{i} + A_y\mathbf{j}) + (-0.5\mathbf{k}) \times (B_x\mathbf{i} + B_y\mathbf{j}) = 0.03\mathbf{i} - 1.08\mathbf{j} - 0.032\mathbf{k}$$

Fig. 21–16(b)

Expanding and equating the respective **i, j,** and **k** components yields Eqs. (3) and (4) and the solution for M obtained previously.

From this analysis it is seen that the bearing reactions will *rotate* with the shaft, since they are directed along the x, y axes. Consequently, from a fixed reference these reactions are applied *periodically,* thereby causing the shaft to vibrate. From the calculations it is seen that the magnitudes of the reactions are, in part, proportional to the square of the angular speed, since $\Sigma M_y = -I_{zx}\omega_z^2$. As a result, for cases of high rotational velocities, the periodically applied reactions may cause severe damage to the supports and the shaft. As noted from the moment equations (3) and (4), the adverse effects of rotation may be eliminated by *balancing* the shaft such that I_{zx} and all the other *products of inertia* for the shaft are *equal to zero.* This may be done either by removing the forked rods on the shaft, or by adding two equivalent forked rods such that all the rods are symmetrically located about the mass center. Under these conditions, the shaft is said to be *dynamically balanced.*

Problems

21–29. Derive the scalar form of the rotational equation of motion along the x axis, when $\mathbf{\Omega} \neq \boldsymbol{\omega}$ and the moments and the products of inertia of the body are constant with respect to time.

21–30. Derive the scalar form of the rotational equation of motion along the x axis, when $\mathbf{\Omega} \neq \boldsymbol{\omega}$ and the moments and products of inertia of the body are not constant with respect to time.

21–31. Derive the Euler equations of motion for $\mathbf{\Omega} \neq \boldsymbol{\omega}$, i.e., Eqs. 21–32.

***21–32.** The 4-kg circular disk is mounted off-center on a shaft which is supported by bearings at A and B. If the shaft is rotating at a speed of $\omega = 7$ rad/s, determine the reactions at the bearings when the disk is in the position shown.

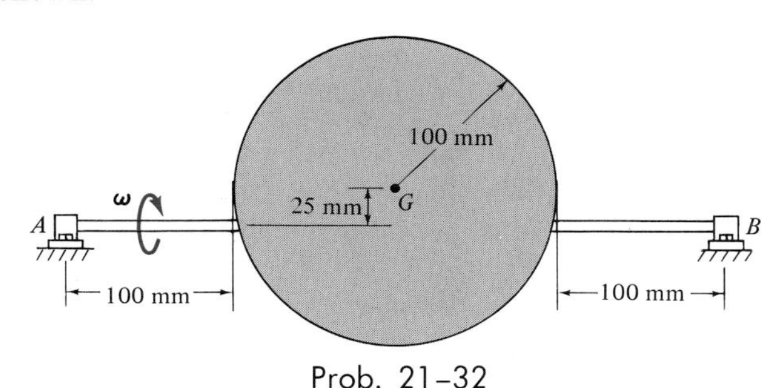

Prob. 21–32

21–33. The disk, having a mass of 5 kg, is mounted eccentrically on the axis of shaft AB. If the shaft is rotating at a speed of 12 rad/s, compute the reactions at the bearing supports when the disk is in the position shown.

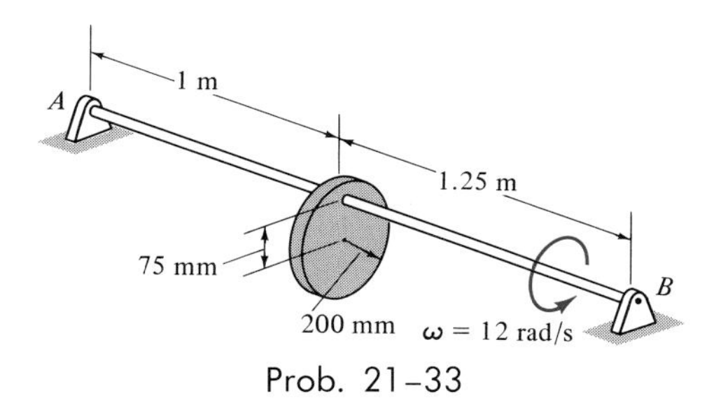

Prob. 21–33

21–34. The square plate, having a mass of 10 kg, is mounted on the shaft AB so that the plane of the plate makes an angle of $\theta = 30°$ with the vertical. If the shaft is turning with an angular velocity of 25 rad/s, determine the reactions at the bearing supports A and B at the instant the plate is in the position shown.

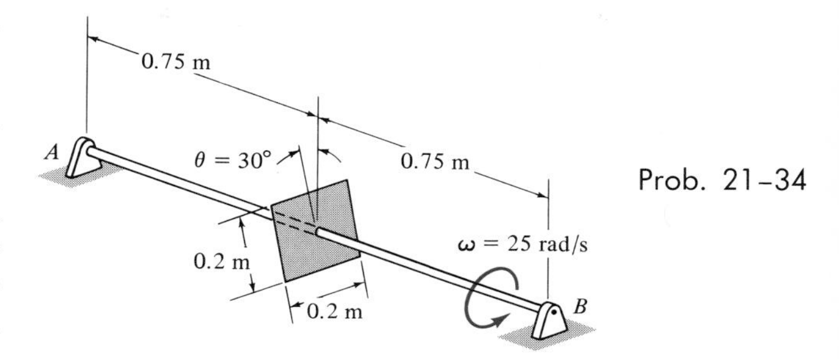

Prob. 21–34

21–35. The bar has a mass of $m = 0.5$ kg and rests against the smooth corner walls of a box at A and B. At the instant shown, the box has an upward velocity of $v = 2$ m/s and an acceleration of $a = 1.5$ m/s^2. If the box is open at the top, determine the components of force which the corners exert on the bar. Set $b = 200$ mm.

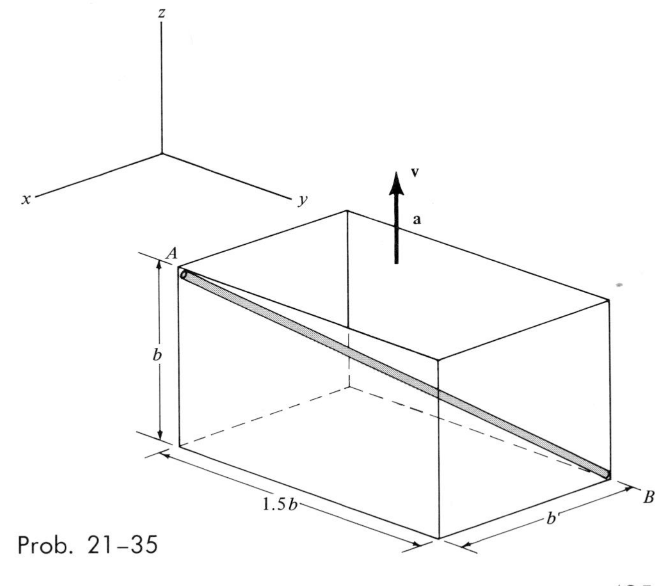

Prob. 21–35

21–35a. Solve Prob. 21–35 if the bar weighs $W = 4$ lb, and $v = 5$ ft/s, $a = 2$ ft/s², $b = 2$ ft.

***21–36.** The 18-kg rectangular block rotates with a constant angular velocity of 6 rad/s about the AB axis. The support at A is a smooth journal bearing which develops reactions normal to the shaft. The support at B is a smooth thrust bearing which develops reactions both normal and along the axis of the shaft (z axis). Determine the reaction components at A and B when the block is in the position shown.

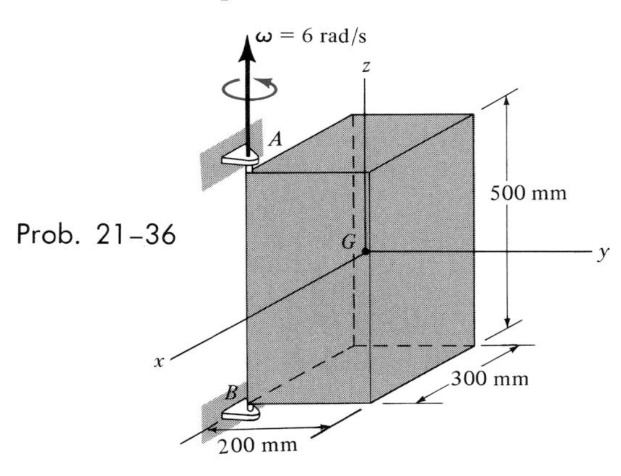

Prob. 21–36

21–37. Determine the magnitude of the torque $\mathbf{T} = T\mathbf{k}$ which must be applied to the block in Prob. 21–36 to give it an angular acceleration of $\boldsymbol{\alpha} = \{12\mathbf{k}\}$ rad/s².

21–38. The 4-kg shaft AB is supported by a rotating arm. The support at A is a journal bearing, which develops reactions normal to the shaft. The support at B is a thrust bearing, which develops reactions both normal to the shaft and along the axis of the shaft. Neglecting friction, determine the reactions at these supports when the frame rotates with a constant angular velocity of $\omega = 15$ rad/s.

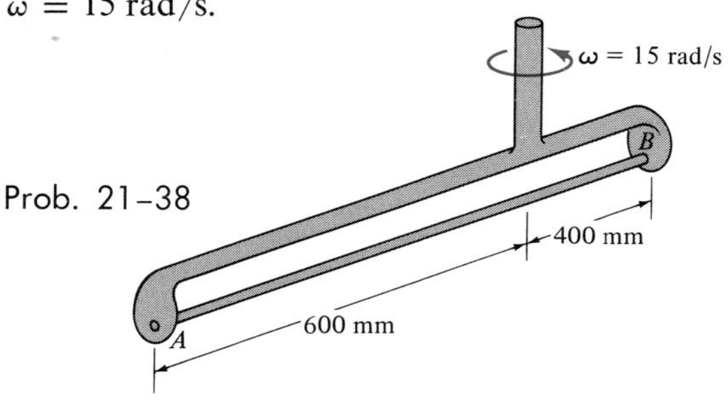

Prob. 21–38

21–39. A stone crusher consists of a large thin disk which is pin-connected to a horizontal axle. If the axle is turning at a constant rate of 10 rad/s, determine the normal force which the disk exerts on the stones. Assume that the disk rolls without slipping and has a uniform mass of 50 kg.

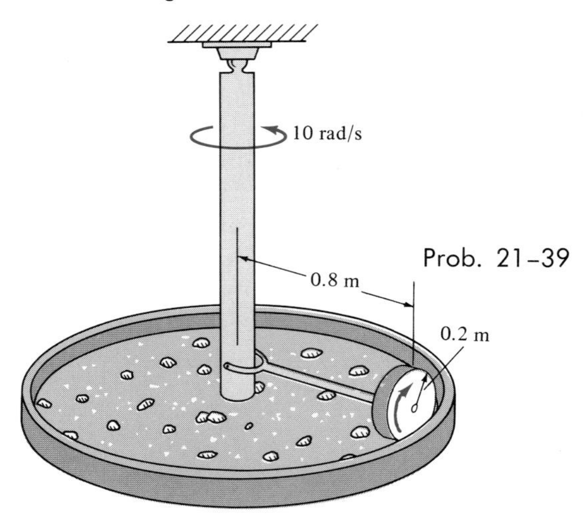

Prob. 21–39

***21–40.** The uniform hatch door, having a mass of $m = 14$ kg and a mass center at G, is supported in the horizontal plane by bearings at A and B. If a vertical force of $F = 250$ N is applied to the door as shown, determine the components of reaction at the bearings and the angular acceleration of the door. The bearing at A will resist a component of force in the y direction, whereas the bearing at B will not. For the calculation, assume the door to be a thin plate. Set $a = 100$ mm, $b = 200$ mm, $c = 150$ mm, and $d = 30$ mm.

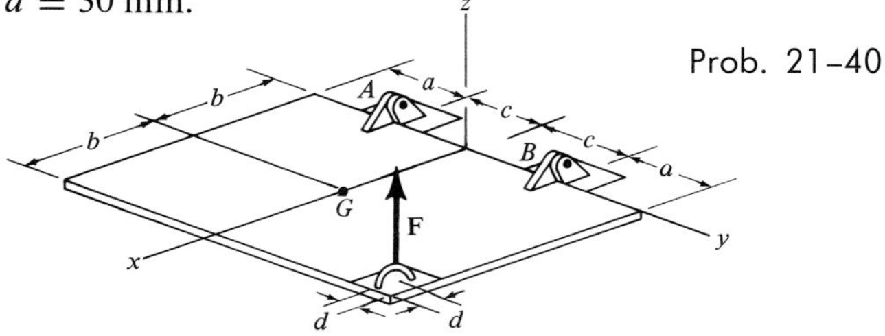

Prob. 21–40

***21–40a.** Solve Prob. 21–40 if the door has a weight of $W = 20$ lb, and $F = 60$ lb, $a = 1$ ft, $b = 1.5$ ft, $c = 1$ ft, $d = 0.25$ ft.

21–41. The 8-kg cylinder is rotating about shaft AB with a constant angular speed of $\omega = 7$ rad/s. If the supporting shaft at C, initially at rest, is given an angular acceleration of $\alpha_C = 20$ rad/s², determine the components of reaction at the bearings A and B. The bearing at A cannot support a force component along the x axis, whereas the bearing at B does.

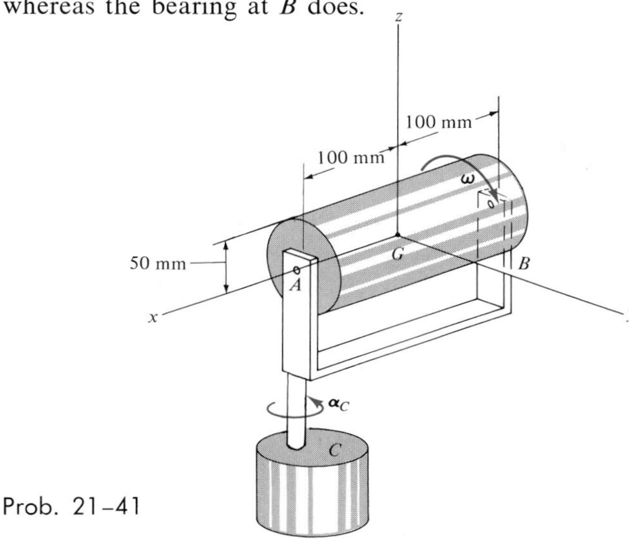

Prob. 21–41

21–42. Solve Prob. 21–41 assuming that the cylinder is rotating about the z axis with an angular velocity of $\boldsymbol{\omega}_C = \{3\mathbf{k}\}$ rad/s when $\alpha_C = 20$ rad/s².

21–43. The 8-kg slender rod AB is pinned at A and held at B by a cord. The axle CD is supported at the ends by ball-and-socket joints and is rotating with a constant angular velocity of 5 rad/s. Determine the tension developed in the cord, and the magnitude of force developed at the pin A.

Prob. 21–43

***21–44.** Determine the tension in the cord and the magnitude of force at the pin A in Prob. 21–43 if the supporting axle has an angular acceleration of $\alpha = 1.8$ rad/s² at the instant $\omega = 2$ rad/s. Assume that $\boldsymbol{\alpha}$ acts in the same direction as $\boldsymbol{\omega}$ shown in Prob. 21–43.

21–45. The bent uniform rod ACD has a mass of $m = 2$ kg/m and is supported at A by a pin and at B by a cord. If the vertical shaft rotates with a constant angular velocity of $\omega = 20$ rad/s, determine the components of force developed at A and the tension in the cord. Set $a = 300$ mm and $b = 125$ mm.

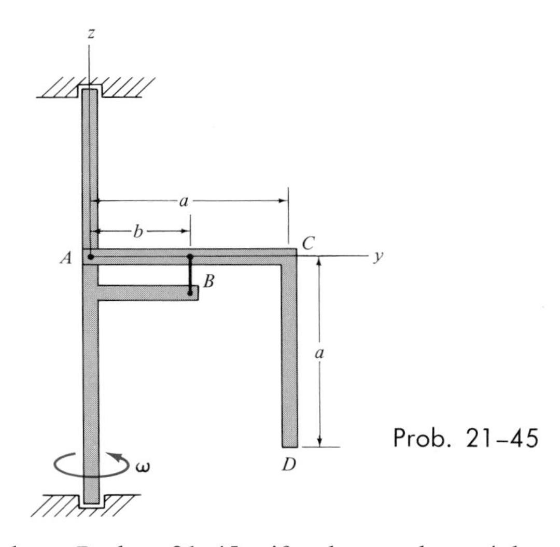

Prob. 21–45

21–45a. Solve Prob. 21–45 if the rod weighs $W = 5$ lb/ft, and $\omega = 20$ rad/s, $a = 1$ ft, $b = 0.5$ ft.

21–46. Two uniform rods, each having a mass of 10 kg, are pin-connected to the edge of a rotating disk. If the disk has a constant angular velocity of $\omega_D = 4$ rad/s, determine the angle θ made by each rod during the motion and the components of the force and moment developed at the pin A. *Suggestion:* Use the x, y, z axes oriented as shown.

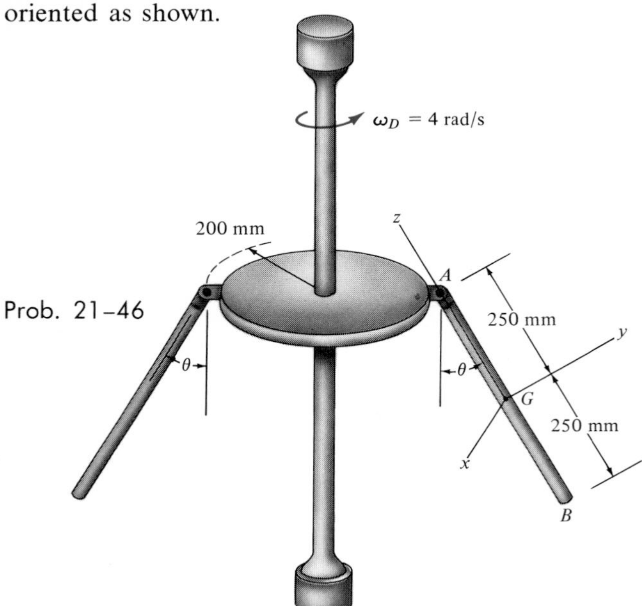

Prob. 21–46

21–47. Four spheres are connected to shaft AB. If $m_C = 1$ kg and $m_E = 2$ kg, determine the mass of D and F and the angles of the rods, θ_D and θ_F, so that the shaft is dynamically balanced, that is, so that the bearings at A and B exert only vertical reactions on the shaft as it rotates. Neglect the mass of the rods.

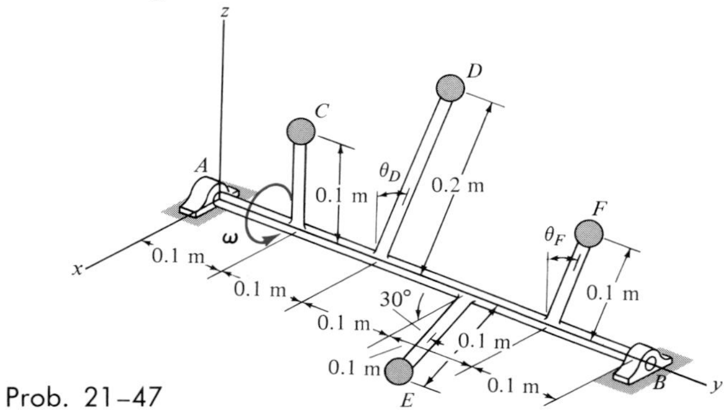

Prob. 21–47

***21–48.** The shaft is constructed from a rod which has a mass of 2 kg/m. Determine the components of reaction at bearings A and B if at the instant shown, it is freely spinning and has an angular velocity of $\omega = 30$ rad/s. What is the angular acceleration of the shaft at this instant? The bearing at A can support a component of force in the y direction, whereas the bearing at B cannot.

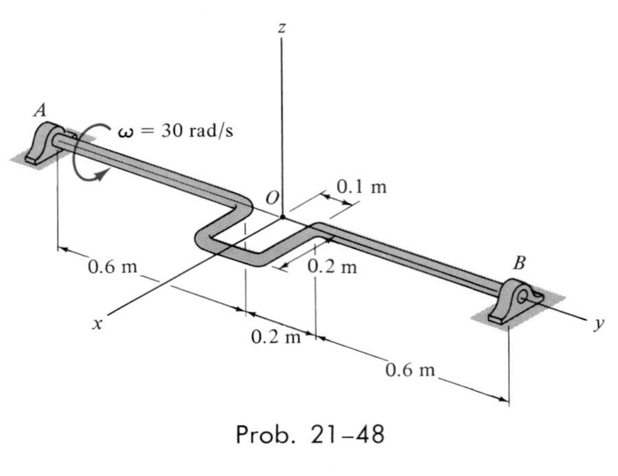

Prob. 21–48

*21–6. Gyroscopic Motion

In this section the equations used for analyzing the motion of a body which is symmetrical with respect to an axis and moving about a fixed point lying on the axis will be developed. These equations will then be applied to study the motion of spinning tops and, a particularly interesting device, the gyroscope.

The body's motion will be analyzed using Euler angles ϕ, θ, and ψ, which were discussed in Sec. 20–4. Here, by convention, the x, y, z coordinate axes are subjected to precession $\dot{\phi}$ and nutation $\dot{\theta}$, but, since they have no spin, $\dot{\psi} = 0$. When the z axis represents the axis of symmetry for the body, as in the case of the top shown in Fig. 21–17, the axes represent *principal axes of inertia* of the body for any spinning rotation of the body about these axes. Hence, the moments of inertia are constant and will be represented as $I_{xx} = I_{yy} = I$ and $I_{zz} = I_z$. The components of the angular velocity of the body along the axes so chosen are defined by Eqs. 20–13 and 20–14, respectively. Since $\mathbf{\Omega} \neq \boldsymbol{\omega}$, the Euler equations 21–32 must be used to establish the rotational equations of motion. Substituting into these equations the respective angular velocities, their corresponding time derivatives, and the moment of inertia components yields

$$\begin{aligned}
\Sigma M_x &= I(\ddot{\theta} - \dot{\phi}^2 \sin\theta\cos\theta) + I_z\dot{\phi}\sin\theta(\dot{\phi}\cos\theta + \dot{\psi}) \\
\Sigma M_y &= I(\ddot{\phi}\sin\theta + 2\dot{\phi}\dot{\theta}\cos\theta) - I_z\dot{\theta}(\dot{\phi}\cos\theta + \dot{\psi}) \\
\Sigma M_z &= I_z(\ddot{\psi} + \ddot{\phi}\cos\theta - \dot{\phi}\dot{\theta}\sin\theta)
\end{aligned} \qquad (21\text{–}33)$$

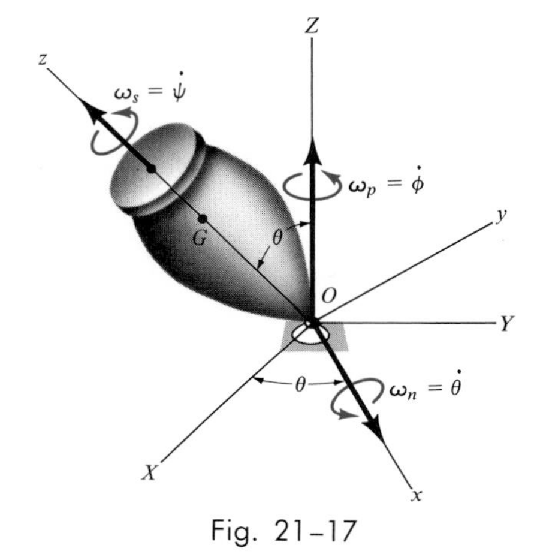

Fig. 21–17

Each moment summation applies only at the fixed point O or the center of mass G of the body. Since the equations represent a coupled set of nonlinear second-order differential equations, in general a closed-form solution may not be obtained. Instead, the Euler angles ϕ, θ, and ψ may be obtained graphically as functions of time using numerical analysis and computer techniques.

A special case, however, does exist for which simplification of Eqs. 21–33 is possible. Commonly referred to as *steady precession,* it occurs when the nutation angle θ, precession $\dot{\phi}$, and spin $\dot{\psi}$ all remain *constant.* Equations 21–33 then reduce to the form

$$\Sigma M_x = -I\dot{\phi}^2\sin\theta\cos\theta + I_z\dot{\phi}\sin\theta\,(\dot{\phi}\cos\theta + \dot{\psi}) \qquad (21\text{–}34)$$

$$\Sigma M_y = 0$$

$$\Sigma M_z = 0$$

Equation 21–34 may be further simplified by noting that, from Eq. 20–13, $\omega_z = \dot{\phi}\cos\theta + \dot{\psi}$, so that

$$\Sigma M_x = -I\dot{\phi}^2\sin\theta\cos\theta + I_z\dot{\phi}(\sin\theta)\omega_z$$

or

$$\Sigma M_x = \dot{\phi}\sin\theta(I_z\omega_z - I\dot{\phi}\cos\theta) \qquad (21\text{–}35)$$

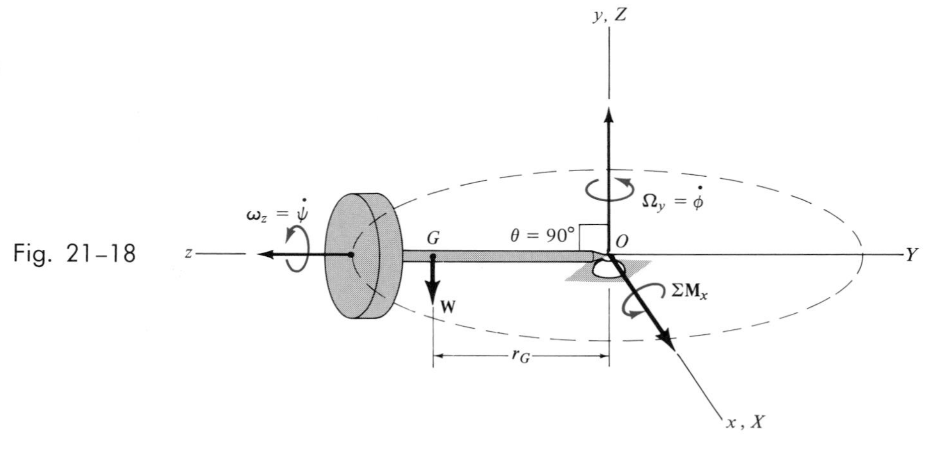

Fig. 21–18

It is interesting to note what effects the spin $\dot{\psi}$ has on the moment about the x axis. In this regard consider the spinning rotor shown in Fig. 21–18. Here $\theta = 90°$, in which case Eq. 21–34 reduces to the form

$$\Sigma M_x = I_z \dot{\phi} \dot{\psi}$$

or

$$\Sigma M_x = I_z \Omega_y \omega_z \tag{21–36}$$

From the figure it is seen that the vectors $\Sigma \mathbf{M}_x$, $\mathbf{\Omega}_y$, and $\boldsymbol{\omega}_z$ all act along their respective *positive axes* and therefore are mutually perpendicular. Instinctively, one would expect the rotor to fall down under the influence of gravity! However, this is not the case at all provided the product $I_z \Omega_y \omega_z$ is correctly chosen to counterbalance the moment $W r_G$ of the rotor's weight about O. This unusual phenomenon of rigid-body motion is often referred to as the *gyroscopic effect*.

Perhaps a more intriguing demonstration of the gyroscopic effect comes from studying the action of a *gyroscope*, frequently referred to as a *gyro*. A gyro is a rotor which spins at a very high rate about its axis of symmetry. This rate of spin is considerably greater than its precessional rate of rotation about the vertical axis. Hence, for all practical purposes, the angular momentum of the gyro can be assumed directed along its axis of spin. Thus, for the gyro rotor shown in Fig. 21–19, $\omega_z \gg \Omega_y$, and the magnitude of the angular momentum about point O, as computed by Eqs. 21–13, reduces to the form $H_O = I_z \omega_z$. Since both the magnitude and direction of $\mathbf{H}_O$ are constant as observed from x, y, z, direct application of Eq. 21–28 yields

Fig. 21–19

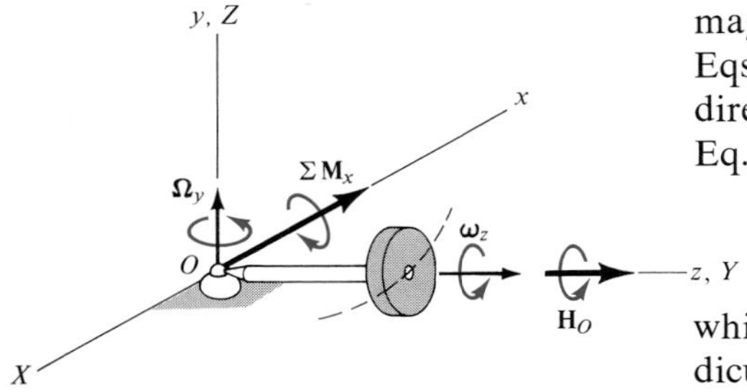

$$\Sigma \mathbf{M}_x = \mathbf{\Omega}_y \times \mathbf{H}_O \tag{21–37}$$

which reduces to Eq. 21–36 since $\Sigma \mathbf{M}_x$, $\mathbf{\Omega}_y$, and $\mathbf{H}_O$ are mutually perpendicular and $H_O = I_z \omega_z$.

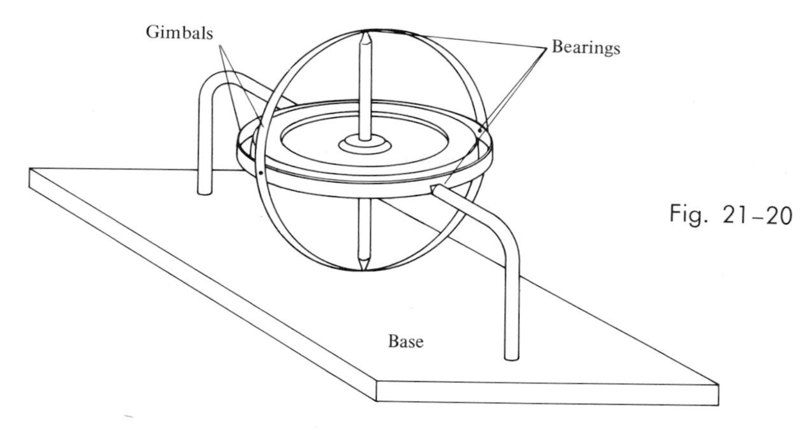

Fig. 21–20

When a gyro is mounted in gimbal rings, Fig. 21–20, it becomes free of external moments applied to its base. Thus, in theory, its angular momentum **H** will never precess but, instead, maintain its same fixed orientation along the axis of spin when the base is rotated. This type of gyroscope is called a *free gyro* and is useful as a gyrocompass when the spin axis of the gyro is directed north. In reality, the gimbal mechanism is never completely free of friction, so that such a device is useful only for the local navigation of ships and aircraft. The gyroscopic effect is also useful as a means of stabilizing both the rolling motion of ships at sea and the trajectories of missiles and projectiles. As noted in Example 21–5, this effect is of significant importance in the design of shafts and bearings for rotors which are subjected to forced precessions.

Example 21–7

The top shown in Fig. 21–21*a* has a mass of 0.5 kg and is precessing about the vertical axis at a constant angle of $\theta = 60°$. If it spins with an angular velocity of $\boldsymbol{\omega}_s = 100$ rad/s, determine the precessional velocity $\boldsymbol{\omega}_p$. Assume that the axial and transverse moments of inertia of the top are $4.5(10^{-4})\ \text{kg}\cdot\text{m}^2$ and $12.0(10^{-4})\ \text{kg}\cdot\text{m}^2$, respectively, measured with respect to the fixed point O.

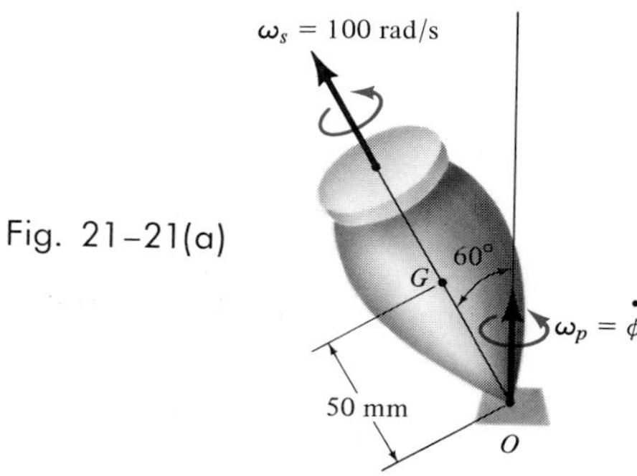

Fig. 21–21(a)

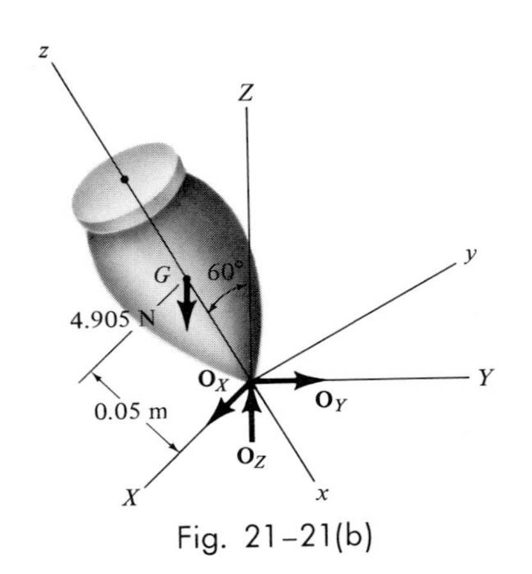

Fig. 21–21(b)

Solution I

(*Scalar Analysis*). Equation 21–34 will be used for the solution since the motion is a *steady precession.* As shown on the free-body diagram, Fig. 21–21*b*, the coordinate axes are established in the usual manner, that is, with the positive *z* axis in the direction of spin, the positive *Z* axis in the direction of precession, and the positive *x* axis in the direction of the moment $\Sigma \mathbf{M}_x$ (refer to Fig. 21–17). The origin is located at the fixed point *O*. Thus,

$$\Sigma M_x = -I\dot{\phi}^2 \sin\theta \cos\theta + I_z\dot{\phi}\sin\theta\,(\dot{\phi}\cos\theta + \dot{\psi})$$

$$4.905(0.05)\sin 60^\circ = -12.0(10^{-4})\dot{\phi}^2 \sin 60^\circ \cos 60^\circ + 4.5(10^{-4})\dot{\phi}\sin 60^\circ\,(\dot{\phi}\cos 60^\circ + 100)$$

or

$$\dot{\phi}^2 - 120.0\dot{\phi} + 654.0 = 0 \qquad (1)$$

Solving this quadratic equation for the precession gives

$$\dot{\phi} = 114.3 \text{ rad/s} \qquad \text{(high precession)} \qquad \textit{Ans.}$$

and

$$\dot{\phi} = 5.72 \text{ rad/s} \qquad \text{(low precession)} \qquad \textit{Ans.}$$

Solution II

(*Vector Analysis*). Using a vector approach, the angular velocity of the top and the axis can be obtained from Eqs. 20–13 and 20–14 or directly from Fig. 21–21*a*. In any case,

$$\boldsymbol{\omega} = \dot{\phi}\sin 60^\circ\mathbf{j} + (\dot{\phi}\cos 60^\circ + 100)\mathbf{k}$$

$$\boldsymbol{\Omega} = \dot{\phi}\sin 60^\circ\mathbf{j} + \dot{\phi}\cos 60^\circ\mathbf{k}$$

Using Eq. 21–13 yields

$$\mathbf{H}_O = (I_x\omega_x)\mathbf{i} + (I_y\omega_y)\mathbf{j} + (I_z\omega_z)\mathbf{k}$$

$$= [12.0(10^{-4})](\dot{\phi}\sin 60^\circ)\mathbf{j} + [4.5(10^{-4})](\dot{\phi}\cos 60^\circ + 100)\mathbf{k}$$

Since $\mathbf{H}_O$ is constant, relative to the *x*, *y*, *z* coordinate system, Eq. 21–28 reduces to the form

$$\Sigma\mathbf{M}_O = \boldsymbol{\Omega} \times \mathbf{H}_O$$

$$4.905(0.05\sin 60^\circ)\mathbf{i} = (\dot{\phi}\sin 60^\circ\mathbf{j} + \dot{\phi}\cos 60^\circ\mathbf{k}) \times [12.0(10^{-4})(\dot{\phi}\sin 60^\circ)\mathbf{j} + 4.5(10^{-4})(\dot{\phi}\cos 60^\circ + 100)\mathbf{k}]$$

When expanded, each of the terms in this equation gives components only in the **i** (or *x*) direction. After simplification,

$$\dot{\phi}^2 - 120.0\dot{\phi} + 654.0 = 0$$

which is the same as Eq. (1) obtained previously.

Example 21-8

The 1-kg disk shown in Fig. 21-22*a* is spinning about its axis with a constant angular velocity of $\omega_D = 70$ rad/s. The block at *B* has a mass of 2 kg, and by adjusting its position *s* one can change the angular velocity of precession of the disk about its supporting pivot at *O*. Compute the position *s* which will enable the disk to have a constant precessional velocity of $\omega_p = 0.5$ rad/s about the pivot. Neglect the weight of the shaft.

Solution

The free-body diagram of the disk is shown in Fig. 21-22*b*, where **F** represents the force reaction of the shaft on the disk. The origin for both the *x, y, z* and *X, Y, Z* coordinate systems is located at point *O*, which represents a *fixed point* for the disk. (Although point *O* does not lie on the disk, imagine a massless extension of the disk to this point.) In the conventional sense, the *Z* axis is chosen along the axis of precession, and the *z* axis is along the axis of spin, so that $\theta = 90°$. Since the precession is *steady,* Eq. 21-35 may be used for the solution. This equation reduces to

$$\Sigma M_x = \dot{\phi}\, I_z \omega_z$$

which is essentially Eq. 21-36. Substituting the required data gives

$$9.81(0.2) - F(0.2) = 0.5[\tfrac{1}{2}(1)(0.05)^2](-70)$$
$$F = 10.0 \text{ N}$$

As shown on the free-body diagram of the shaft and block *B*, Fig. 21-22*c*, summing moments about the *x* axis requires

$$(19.62)\, s = (10.0)(0.20)$$
$$= 0.101\,9 \text{ m} = 101.9 \text{ mm} \qquad \textit{Ans.}$$

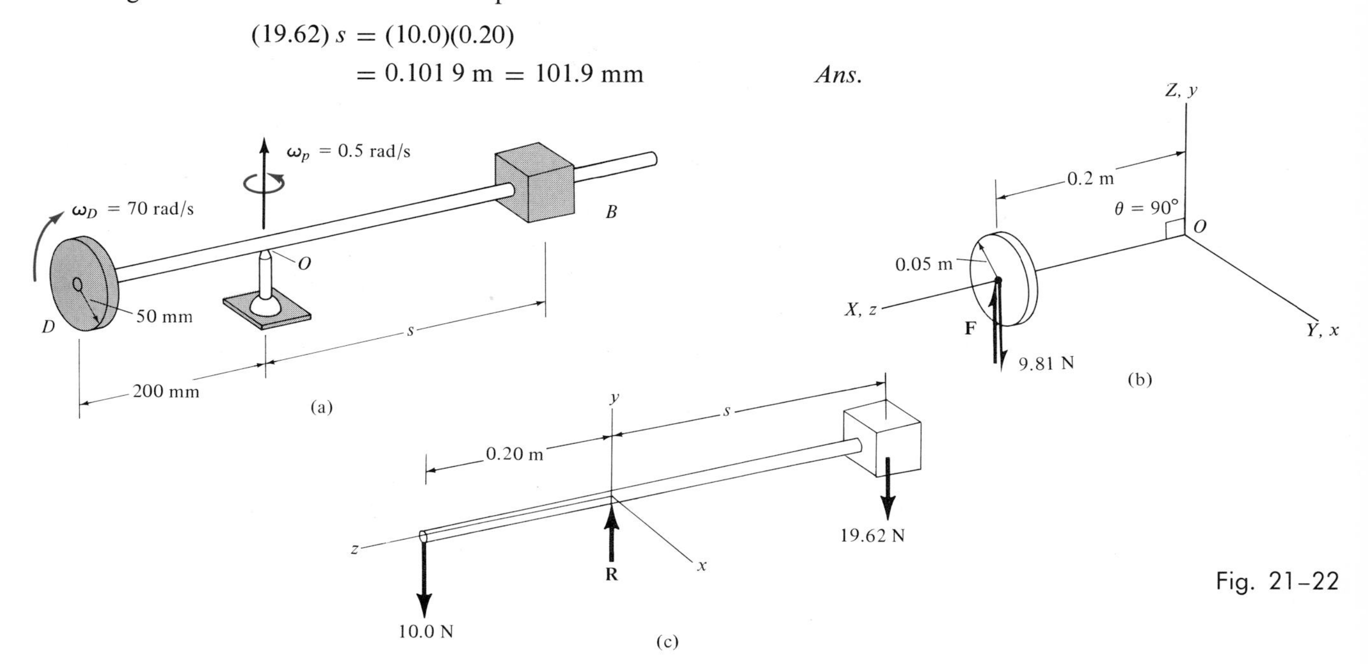

Fig. 21-22

*21–7. Torque-Free Motion

When the only external force acting on a body is caused by gravitation, the general motion of the body is referred to as *torque-free motion*. This type of motion is characteristic of planets, artificial satellites, and projectiles—provided the effects of air friction are neglected.

In order to describe the characteristics of this motion, the distribution of the body's mass will be assumed *axisymmetric*. The satellite shown in Fig. 21–23*a* is an example of such a body, where the z axis represents an axis of symmetry. The origin of the x, y, z coordinates is located at the mass center G, such that $I_{zz} = I_z$ and $I_{xx} = I_{yy} = I$ for the body. If it is assumed that gravitation is the only external force present, the summation of moments about the mass center is zero. From Eq. 21–27, this requires the angular momentum of the body to be constant, i.e.,

$$\mathbf{H}_G = \text{const} \tag{21–38}$$

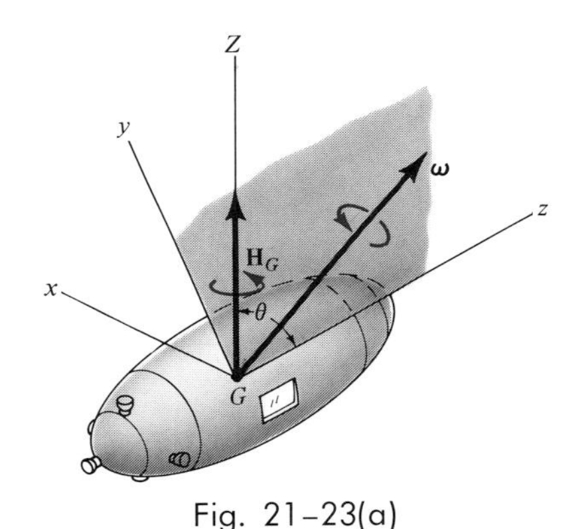

Fig. 21–23(a)

At the instant considered, it will be assumed that the inertial frame of reference is oriented such that the positive Z axis is directed along $\mathbf{H}_G$ and the y axis lies in the plane formed by the z and Z axes, Fig. 21–23*a*. The Euler angle formed between Z and z is θ, and, therefore, with this choice of axes the angular momentum may be expressed as

$$\mathbf{H}_G = H_G \sin\theta\,\mathbf{j} + H_G \cos\theta\,\mathbf{k} \tag{21–39}$$

Furthermore, using Eqs. 21–13, we have

$$\mathbf{H}_G = I\omega_x\mathbf{i} + I\omega_y\mathbf{j} + I_z\omega_z\mathbf{k} \tag{21–40}$$

where $\omega_x, \omega_y, \omega_z$ represent the x, y, z components of the body's angular velocity. Equating the respective **i, j,** and **k** components of Eqs. 21–39 and 21–40 yields

$$\omega_x = 0, \qquad \omega_y = \frac{H_G \sin\theta}{I}, \qquad \omega_z = \frac{H_G \cos\theta}{I_z} \tag{21–41}$$

or

$$\boldsymbol{\omega} = \frac{H_G \sin\theta}{I}\mathbf{j} + \frac{H_G \cos\theta}{I_z}\mathbf{k} \tag{21–42}$$

In a similar manner, equating the respective **i, j,** and **k** components of Eq. 20–13 to those of Eq. 21–42 yields

$$\dot{\theta} = 0$$

$$\dot{\phi}\sin\theta = \frac{H_G \sin\theta}{I}$$

$$\dot{\phi}\cos\theta + \dot{\psi} = \frac{H_G \cos\theta}{I_z}$$

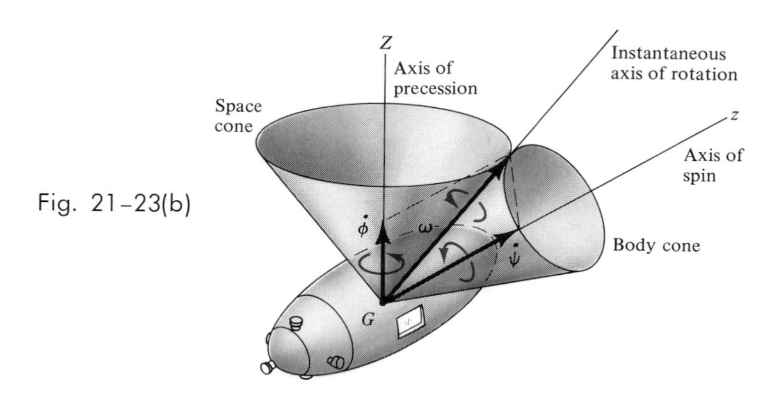

Fig. 21–23(b)

Solving, we get

$$\theta = \text{const}$$
$$\dot{\phi} = \frac{H_G}{I} \qquad (21\text{–}43)$$
$$\dot{\psi} = \frac{I - I_z}{II_z} H_G \cos\theta$$

Thus, for torque-free motion of an axisymmetrical body, the angle θ formed between the angular-momentum vector and the spin of the body remains constant. Furthermore, the angular momentum $\mathbf{H}_G$, precession $\dot{\phi}$, and spin $\dot{\psi}$ for the body remain constant at all times during the motion. Eliminating H_G from the second and third of Eqs. 21–43 yields the following relationship between the spin and precession:

$$\dot{\psi} = \frac{I - I_z}{I_z}(\dot{\phi}) \cos\theta \qquad (21\text{–}44)$$

As shown in Fig. 21–23*b*, the body precesses about the Z axis, which is fixed in direction, while it spins about the z axis. These two components of angular motion may be analyzed by using a simple cone model, introduced in Sec. 20–2. The *space cone* defining the precession is fixed from rotating, since the precession has a fixed direction, while the *body cone* rotates around the space cone's outer surface without slipping. On this basis, an attempt should be made to imagine the motion. The interior angle of each cone is chosen such that the resultant angular velocity of the body is directed along the line of contact of the two cones. This line of contact represents the instantaneous axis of rotation for the body cone, and hence the angular velocity $\boldsymbol{\omega}$ of both the body cone and the body must be directed along this line. Since the spin is a function of the moments of inertia I and I_z of the body, Eq. 21–44, the cone model in Fig. 21–23*b* is

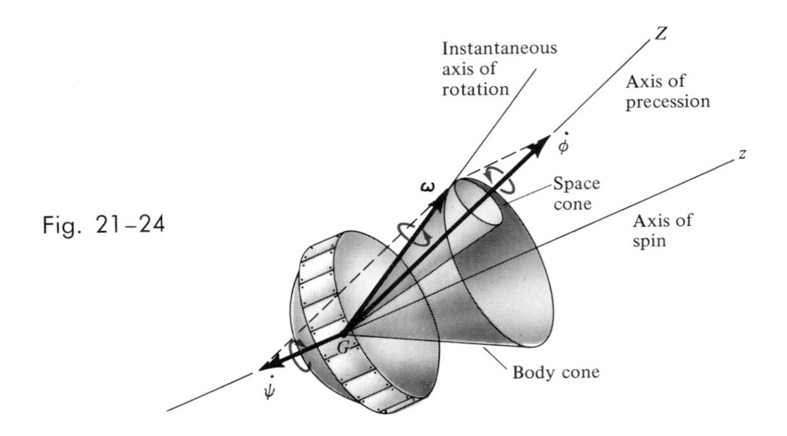

Fig. 21-24

satisfactory for describing the motion, provided $I > I_z$. Torque-free motion which meets these requirements is called *regular precession.* If $I < I_z$, the spin is negative and the precession positive. This motion is represented by the satellite motion shown in Fig. 21-24 ($I < I_z$). The cone model may again be used to represent the motion; however, to preserve the correct vector addition of spin and precession to obtain the angular velocity $\boldsymbol{\omega}$, the inside surface of the body cone must roll on the outside surface of the (fixed) space cone. This motion is referred to as *retrograde precession.*

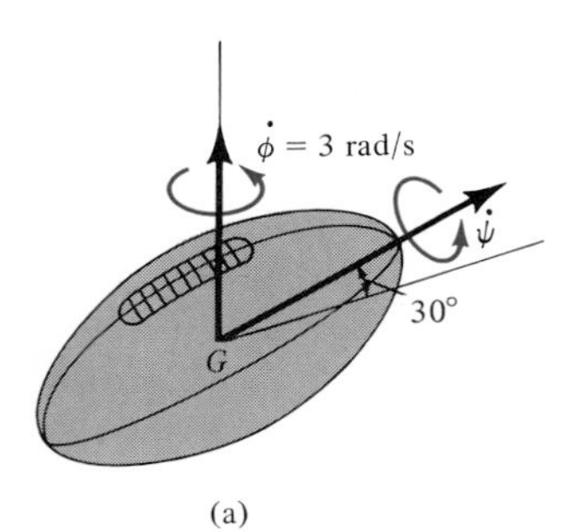

(a)

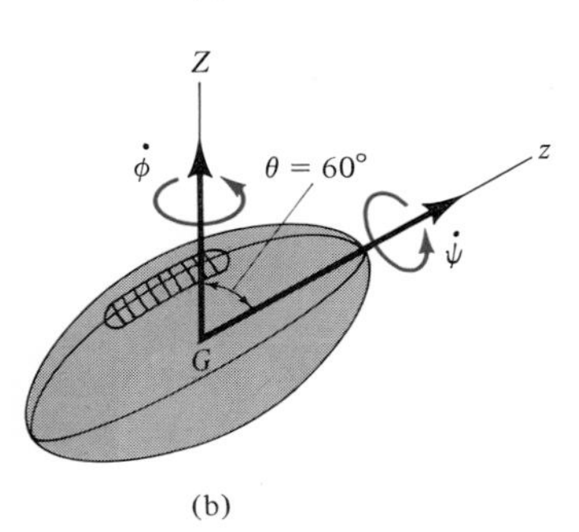

(b)

Fig. 21-25

Example 21-9

After passing a football, the motion is observed using a slow-motion projector. From the film, the spin of the football is seen to be directed 30° from the horizontal, as shown in Fig. 21-25*a*. Also, the football is seen to precess about the vertical axis at a rate of $\dot{\phi} = 3$ rad/s. If the ratio of the axial to transverse moments of inertia of the football is $\frac{1}{3}$, measured with respect to the center of gravity, determine the magnitude of the football's spin and its angular velocity. Neglect the effect of air resistance.

Solution

Since the weight of the football is the only force acting, the motion is torque-free. In the conventional sense, if the z axis is established along the axis of spin and the Z axis along the precession axis, as shown in Fig. 21-25*b*, then the angle $\theta = 60°$. Applying Eqs. 21-44 gives

$$\dot{\psi} = \frac{I - I_z}{I_z}(\dot{\phi}) \cos \theta = \frac{I - \frac{1}{3}I}{\frac{1}{3}I}(3) \cos 60°$$

$$= 3 \text{ rad/s} \qquad \textit{Ans.}$$

Using Eqs. 21-41, where $H_G = \dot{\phi} I$, Eq. 21-43, we have

$$\omega_x = 0$$

$$\omega_y = \frac{H_G \sin\theta}{I} = \frac{3I \sin 60^\circ}{I} = 2.60 \text{ rad/s}$$

$$\omega_z = \frac{H_G \cos\theta}{I_z} = \frac{3I \cos 60^\circ}{\frac{1}{3}I} = 4.50 \text{ rad/s}$$

Thus,

$$\omega = \sqrt{(\omega_x)^2 + (\omega_y)^2 + (\omega_z)^2} = \sqrt{(0)^2 + (2.60)^2 + (4.50)^2}$$
$$= 5.20 \text{ rad/s} \qquad \textit{Ans.}$$

Problems

21–49. The driving armature of a ship's engine may be approximated by the 70-kg cylinder having a radius of 125 mm. The armature is rotating with an angular velocity of $\omega_s = 300$ rad/s as the ship turns at $\omega_T = 0.2$ rad/s. Determine the vertical reactions at each of the bearings A and B due to this motion.

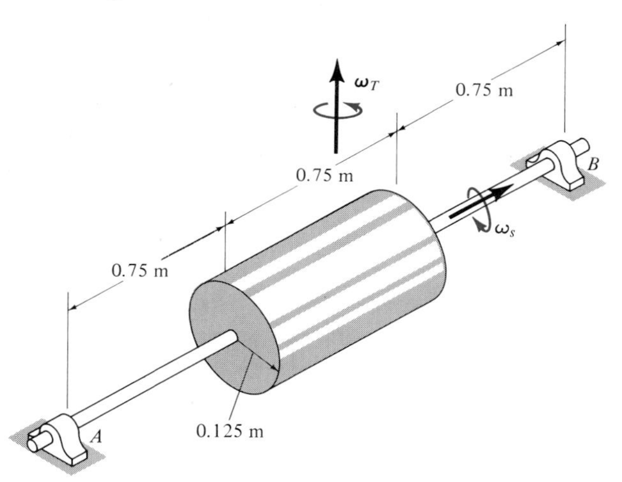

Prob. 21–49

21–50. The two disks A and C, each having a mass of $m = 6$ kg and a radius of $r = 80$ mm, are attached to the crossbar frame such that $d = 250$ mm. The vertical axis DE of the frame is subjected to an angular velocity of $\omega = 15$ rad/s, causing both disks to roll on the horizontal surface without slipping. If the shaft DE is raised slightly so that the disks leave the surface, determine the *gyroscopic bending moment* exerted at B on the frame by each of the rotating disks. If the shaft ABC were elastic, would it bend upward or downward because of this moment?

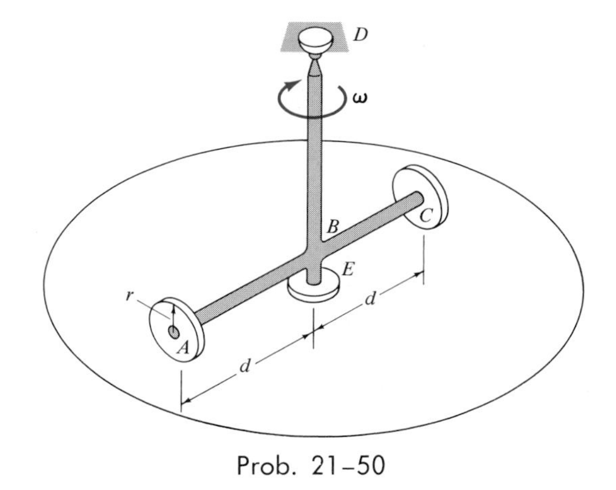

Prob. 21–50

21–50a. Solve Prob. 21–50 if each disk weighs $W = 10$ lb, and $\omega = 15$ rad/s, $r = 0.5$ ft, $d = 2$ ft.

21-51. The 20-kg disk is spinning about its center at $\omega_s = 20$ rad/s while the supporting axle is rotating at $\omega_y = 6$ rad/s. Determine the gyroscopic moment caused by the force reactions which the pin A exerts on the disk due to the motion.

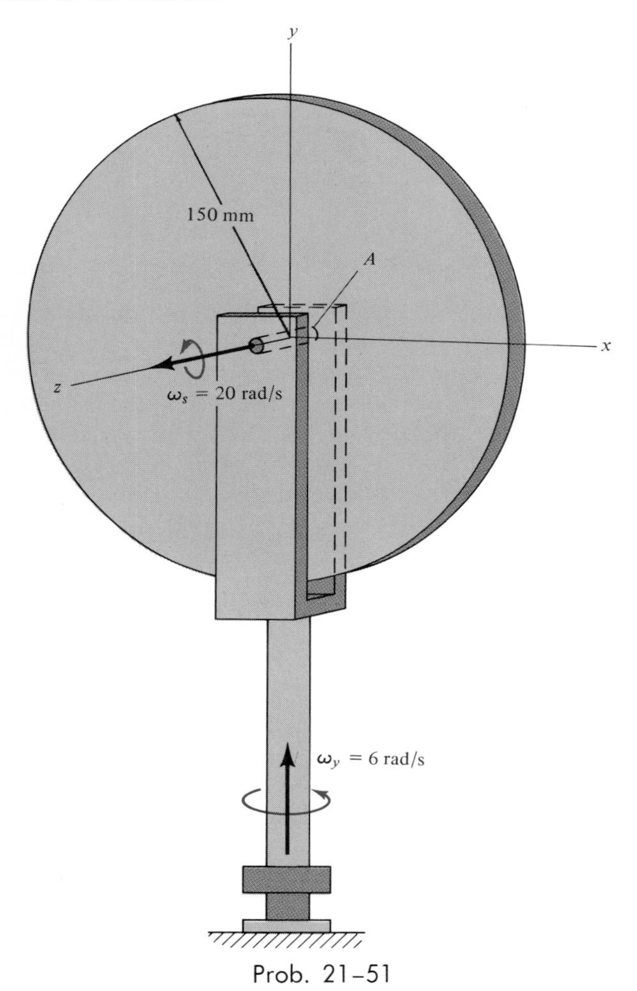

Prob. 21–51

***21-52.** The 5-kg sphere A rotates with an angular speed of $\omega_s = 160$ rad/s about the axis of the horizontal rod. If the counterbalance block B has a mass of 2 kg, determine the precession of the rod about the pivot CD. Neglect the mass of the rod.

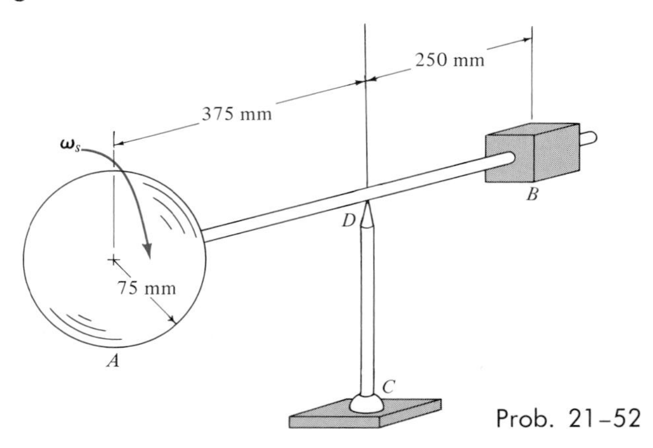

Prob. 21–52

21-53. The propeller on the single-engine airplane has a mass of 12 kg and a centroidal radius of gyration of 0.3 m computed about the axis of spin. When viewed from the front of the airplane, the propeller is turning clockwise at 200 rad/s about the spin axis. If the airplane enters a vertical curve having a radius of $\rho = 80$ m and is traveling at 200 km/h, determine the gyroscopic bending moment which the propeller exerts on the bearings of the engine when the airplane is in its lowest position, as shown.

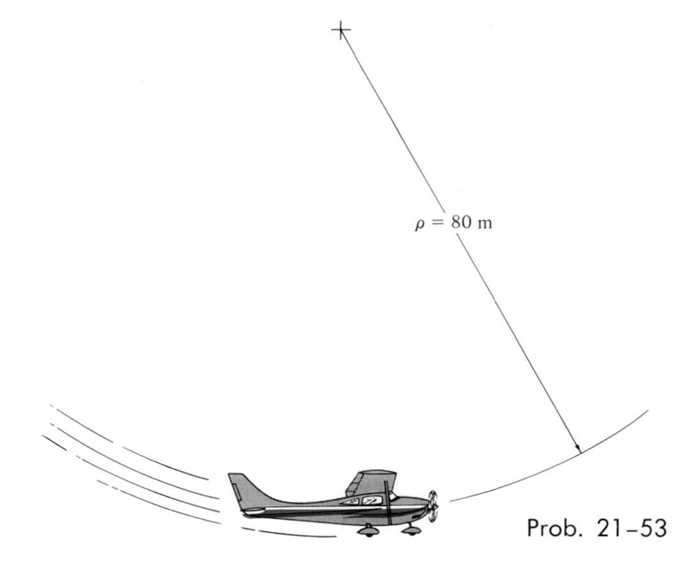

Prob. 21–53

21–54. The top has a mass of 120 g, a mass center at G, and a radius of gyration about its axis of symmetry of $k = 10$ mm. About any transverse axis acting through point O, the radius of gyration is $k_t = 15$ mm. If the top is pinned at O and the precession is $\omega_p = 0.5$ rad/s, determine the spin $\boldsymbol{\omega}_s$.

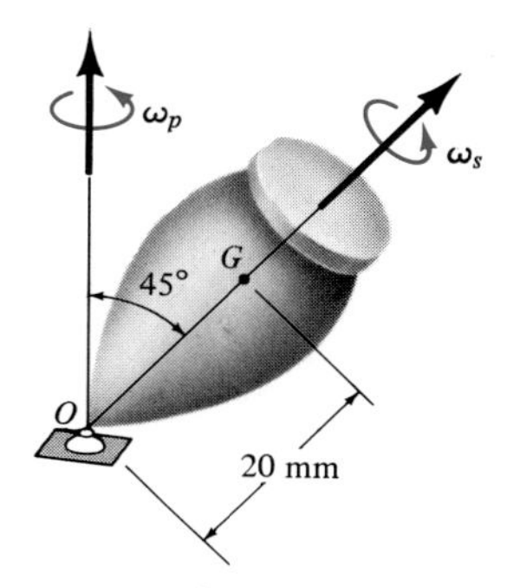

Prob. 21–54

21–55. The top consists of a disk that has a mass of $m = 6$ kg and a radius of $r = 125$ mm. The rod has a negligible mass and a length $l = 300$ mm. If the top is spinning with an angular velocity of $\omega = 300$ rad/s, determine the steady-state precessional angular velocity $\boldsymbol{\omega}_p$ of the rod when $\theta = 40°$.

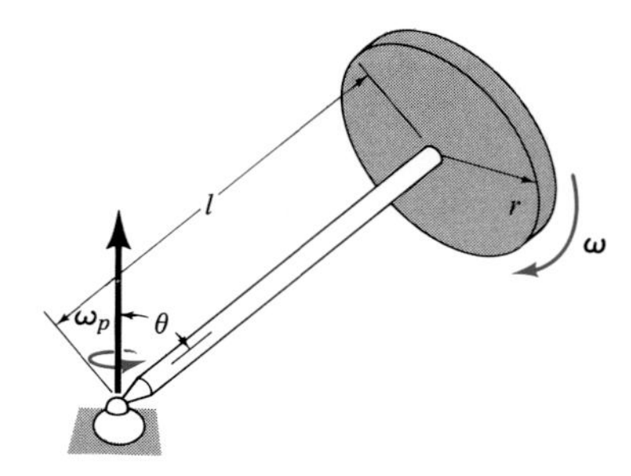

Prob. 21–55

21–55a. Solve Prob. 21–55 if the disk weighs $W_D = 3$ lb, and $r = 0.2$ ft, $l = 0.5$ ft, $\omega = 300$ rad/s, $\theta = 15°$.

***21–56.** The conical top has a mass of 0.50 kg, and the moments of inertia are $I_x = I_y = 3.0(10^{-3})$ kg · m² and $I_z = 0.9(10^{-3})$ kg · m². If it spins freely in the ball-and-socket joint at A with an angular velocity of $\omega_s =$ 650 rad/s, compute the precession of the top about the axis of the shaft AB.

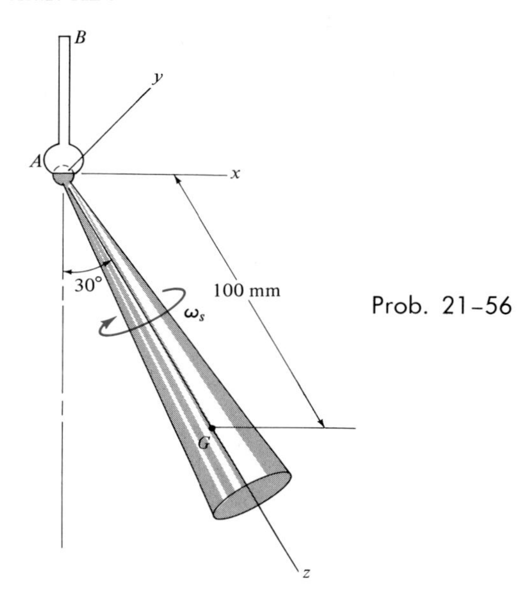

Prob. 21–56

21–57. Solve Prob. 21–55 when $\theta = 90°$.

21–58. The gyroscope consists of a uniform 500-g disk D which is attached to an axle AB that has a negligible mass. The supporting frame has a mass of 200 g and a center of gravity at G. If the disk is rotating about the axle at an angular speed of $\omega_D = 110$ rad/s, determine the constant angular velocity ω_p at which the frame precesses about the pivot point O. The frame moves in the horizontal plane.

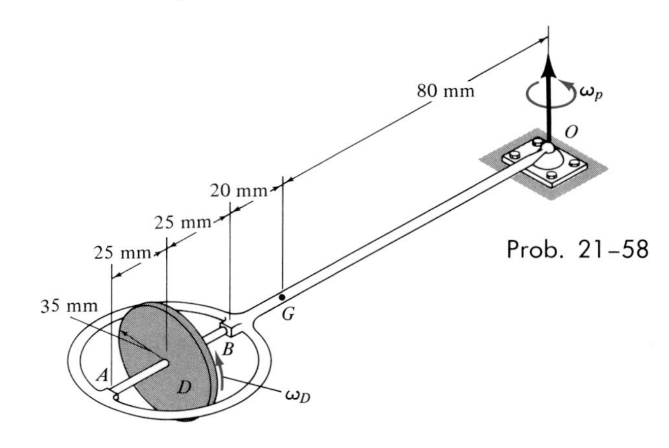

Prob. 21–58

21-59. The homogeneous cone has a mass of 6 kg and a vertex angle of 90°. If the cone rolls on the horizontal surface without slipping, determine the greatest precessional speed ω_p it can have before the tip A starts to rise from the surface.

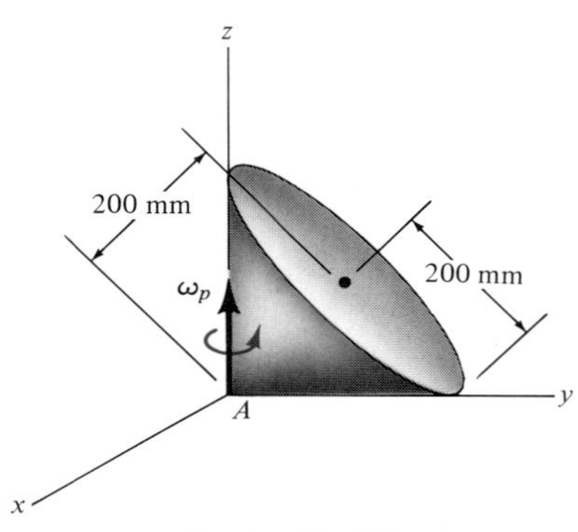

Prob. 21–59

21-61. The projectile shown is subjected to torque-free motion. The transverse and axial moments of inertia are I and I_z, respectively. If θ represents the angle between the precessional axis Z and the axis of symmetry z, and β is the angle between the angular velocity $\boldsymbol{\omega}$ and the z axis, show that β and θ are related by the equation $\tan\theta = (I/I_z)\tan\beta$.

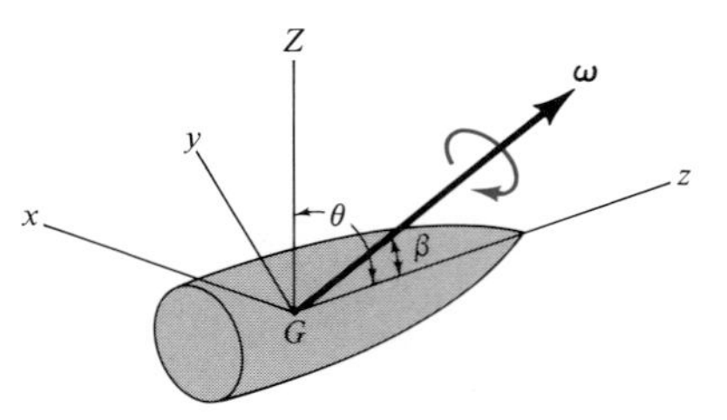

Prob. 21–61

***21-60.** The rocket has a mass of $m = 2.5$ Mg and radii of gyration of $k_z = 0.85$ m and $k_y = 2.3$ m. It is initially spinning about the z axis at $\omega_z = 0.02$ rad/s when a meteoroid M strikes it at A and creates an impulse of $\mathbf{I} = \{200\mathbf{i}\}$ kN · s. If $d = 2.5$ m, determine the axis of precession after the impact.

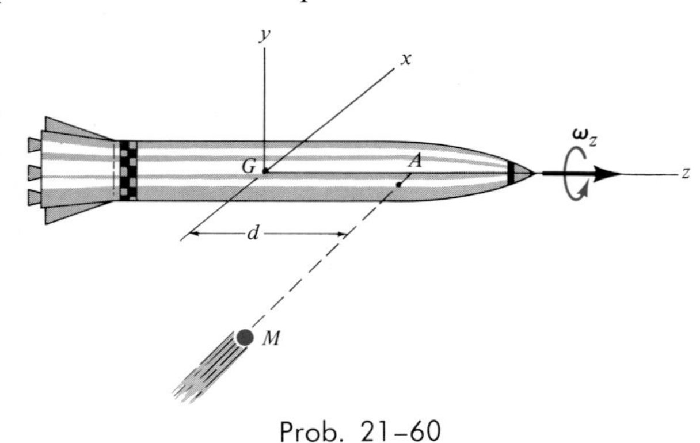

Prob. 21–60

21-62. The projectile has a mass of 0.8 kg and axial and transverse radii of gyration of $k_z = 20$ mm and $k_t = 25$ mm, respectively. If it is spinning at $\omega_s = 5$ rad/s when it leaves the barrel of a gun, determine its angular momentum. Precession occurs about the Z axis.

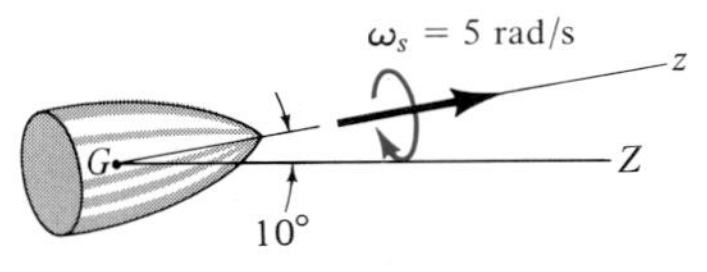

Prob. 21–62

***21-60a.** Solve Prob. 21–60 if the rocket has a weight of $W = 8000$ lb (on earth), and $k_z = 2.5$ ft, $k_y = 7$ ft, $\omega_z = 0.05$ rad/s, $\mathbf{I} = \{40\mathbf{i}\}$ lb · s, $d = 10$ ft.

21-63. While the rocket is in free flight it has a spin of 2 rad/s and precesses about an axis measured 10° from the axis of spin. If the ratio of the axial to transverse moments of inertia of the rocket is 1/15, determine the angle which the resultant angular velocity makes with the spin axis. Construct the body and space cones used to describe the motion. Is the precession regular or retrograde?

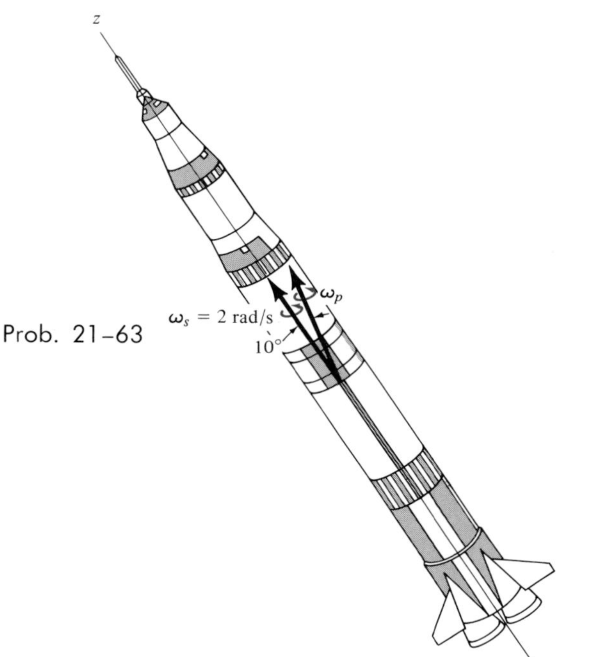

Prob. 21-63

***21-64.** The 3-kg disk is thrown in the air with a spin of $\omega_z = 10$ rad/s. If the angle θ is measured as 160°, determine the precession about the Z axis.

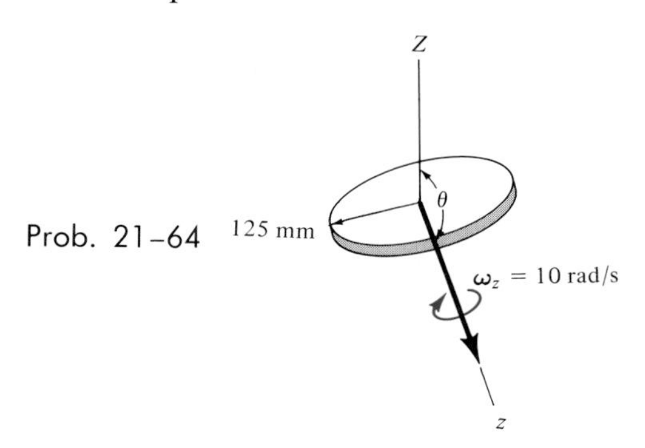

Prob. 21-64

21-65. The radius of gyration about an axis passing through the axis of symmetry of the space capsule is $k_z = 1.2$ m, and about any transverse axis passing through the center of mass G, $k_t = 1.8$ m. If the capsule has a known steady-state precession of two revolutions per hour about the z' axis, determine the rate of spin about the z axis. The capsule has a mass of $m = 1500$ kg.

Prob. 21-65

21-65a. Solve Prob. 21-65 if $k_z = 4$ ft, $k_t = 6$ ft, and the capsule has a total weight of $W = 3000$ lb on earth.

22 Vibrations

22–1. Simple Harmonic Motion

A *vibration* is the motion of a body or system of connected bodies which is randomly or uniformly repeated after a given interval of time. In engineering structures the occurrence of vibrations is widespread. Structures purposely designed to vibrate have been used for geological seismic investigations, to facilitate the packing of powdered materials such as sand or flour, and to determine the endurance or fatigue limits of machine members. Most often, however, the effects of vibrations are undesirable in engineering structures. Any vibration requires energy or power to produce it; therefore, the efficiency of machines is reduced. Furthermore, vibrations cause stress fatigue of materials, which can hasten the time of their eventual failure.

In general, there are two types of vibration, free and forced. *Free vibration* occurs when the motion is maintained by gravitational or elastic restoring forces, such as the swinging motion of a pendulum or the vibration of an elastic rod. *Forced vibration* is caused by an external periodic or intermittent force applied to the system. Both of these types of vibration may be either damped or undamped. *Undamped* vibrations can continue indefinitely because frictional effects are neglected in the analysis. Since in reality both internal and external frictional forces are present, the motion of all vibrating bodies is actually *damped.*

When the motion of a body is constrained so that it is allowed to vibrate in only one direction, it is said to have a *single degree of freedom.* A one-degree-of-freedom system requires only one coordinate to specify completely the position of the system at any time. In this book only one-degree-of-freedom systems will be discussed. The analysis of multi-degree-of-freedom systems is based on this simplified case and is thoroughly treated in textbooks devoted to vibrational theory.

The simplest type of vibrating motion having a single degree of freedom

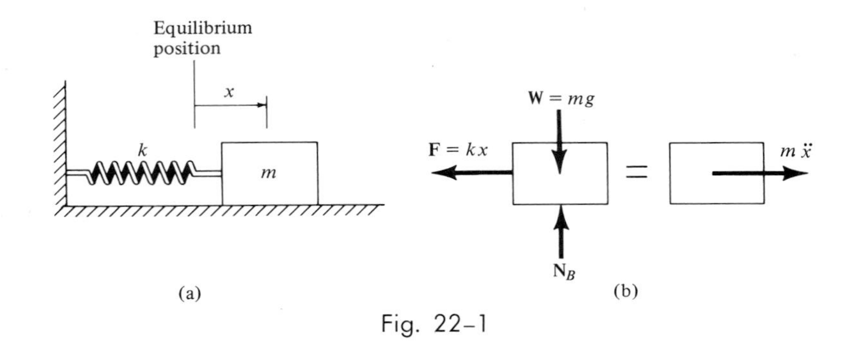

Fig. 22–1

is undamped free vibration, represented by the model shown in Fig. 22–1*a*. The block has a mass *m*, rests on a smooth surface, and is attached to a spring having a stiffness *k*. Vibrating motion is provided by displacing the block a distance *x* from its equilibrium position and allowing the spring to restore the block to its original position. As the spring pulls on the block, the block will attain a velocity such that it will proceed to move out of equilibrium when $x = 0$. Since the surface upon which the block moves is frictionless, oscillation will continue indefinitely.

The time-dependent path of motion of the block may be determined by applying the equation of motion to the block when it is in the displaced position *x*. The free-body and kinetic diagrams are shown in Fig. 22–1*b*. The elastic restoring force $F = kx$ is always directed toward the equilibrium position and it has been assumed that the acceleration **a** acts in the direction of *positive displacement*. Hence, noting that $a = d^2x/dt^2 = \ddot{x}$, we have

$$\overset{+}{\rightarrow}\Sigma F_x = ma_x; \qquad -kx = m\ddot{x}$$

Here it is seen that the acceleration $\ddot{x}$ is proportional to the displacement *x*. Motion described in this manner is called *simple harmonic motion*. Rearranging the terms into a "standard form" gives

$$\ddot{x} + p^2x = 0 \tag{22–1}$$

The constant *p* is called the *circular frequency*, expressed in rad/s, and in this case

$$p = \sqrt{\frac{k}{m}} \tag{22–2}$$

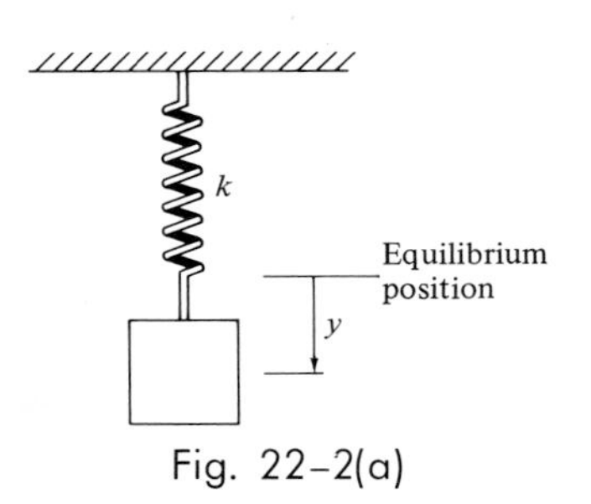

Fig. 22–2(a)

Equation 22–1 may also be obtained by considering the block to be suspended, as shown in Fig. 22–2*a*, and measuring the displacement *y* from the block's *equilibrium position*. The free-body and kinetic diagrams are shown in Fig. 22–2*b*. When the block is in equilibrium, the spring exerts an upward force of $F = W = mg$ on the block. Hence, when the block is displaced a distance y downward from this position, the magnitude of the spring force is $F = W + ky$. Applying the equation of motion gives

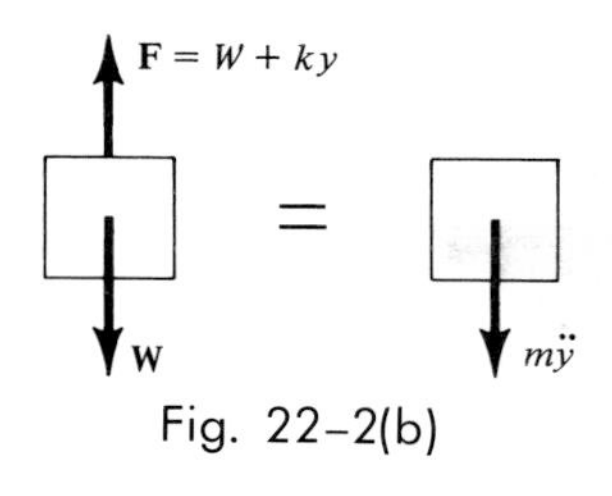

Fig. 22–2(b)

$$+\downarrow\Sigma F_y = ma_y; \qquad -W - ky + W = m\ddot{y}$$

or

$$\ddot{y} + p^2 y = 0$$

which is the same form as Eq. 22–1, where p is defined by Eq. 22–2.

Equation 22–1 is a homogeneous, second-order, linear, differential equation with constant coefficients. It can be shown, using the methods of differential equations, that the general solution of this equation is

$$x = A \sin pt + B \cos pt \tag{22–3}$$

where A and B represent two constants of integration. The block's velocity and acceleration are determined by taking successive time derivatives of Eq. 22–3, which yields

$$v = \dot{x} = Ap \cos pt - Bp \sin pt \tag{22–4}$$

$$a = \ddot{x} = -Ap^2 \sin pt - Bp^2 \cos pt \tag{22–5}$$

When Eqs. 22–3 and 22–5 are substituted into Eq. 22–1, the differential equation is indeed satisfied, and therefore Eq. 22–3 represents the true solution to Eq. 22–1.

The integration constants A and B in Eq. 22–3 are generally determined from the initial conditions of the problem. For example, suppose that the block in Fig. 22–1*a* has been displaced a distance x_1 to the right from its equilibrium position, and given an initial (positive) velocity $\mathbf{v}_1$ directed to the right. Substituting $x = x_1$ at $t = 0$ into Eq. 22–3 yields $B = x_1$. Since $v = v_1$ at $t = 0$, using Eq. 22–4 we obtain $A = v_1/p$. If these values are substituted into Eq. 22–3, the equation describing the motion becomes

$$x = \frac{v_1}{p} \sin pt + x_1 \cos pt \tag{22–6}$$

Equation 22–3 may always be expressed in terms of simple sinusoidal motion. Let

$$A = C \cos \phi \tag{22–7}$$

and

$$B = C \sin \phi \tag{22–8}$$

where C and ϕ are new constants to be determined in place of A and B. Substituting into Eq. 22–3 yields

$$x = C \cos \phi \sin pt + C \sin \phi \cos pt$$

This equation may be simplified using the trigonometric formula $\sin(\theta + \phi) = \sin\theta \cos\phi + \cos\theta \sin\phi$. Hence,

$$x = C \sin(pt + \phi) \tag{22-9}$$

If this equation is plotted on an x versus pt axis, the graph shown in Fig. 22-3 is obtained. The maximum displacement of the block from its equilibrium position is defined as the *amplitude* of vibration. From either the figure or Eq. 22-9 the amplitude is C. The *phase angle* is defined by the constant ϕ. This angle represents the amount by which the curve is displaced from the origin when $t = 0$. The constants C and ϕ are related to A and B by Eqs. 22-7 and 22-8. Squaring and adding these two equations, the amplitude becomes

$$C = \sqrt{A^2 + B^2} \tag{22-10}$$

If Eq. 22-8 is divided by Eq. 22-7, the phase angle becomes

$$\phi = \tan^{-1}\frac{B}{A} \tag{22-11}$$

Since ϕ is a constant, the sine curve, Eq. 22-9, completes one *cycle*, and hence the cyclic motion of the block is repeated in time $t = \tau$ (tau), so that $p\tau = 2\pi$ or

$$\tau = \frac{2\pi}{p} \tag{22-12}$$

This length of time is called a *period*, Fig. 22-3. From Eq. 22-2, the period may also be represented as

$$\tau = 2\pi\sqrt{\frac{m}{k}} \tag{22-13}$$

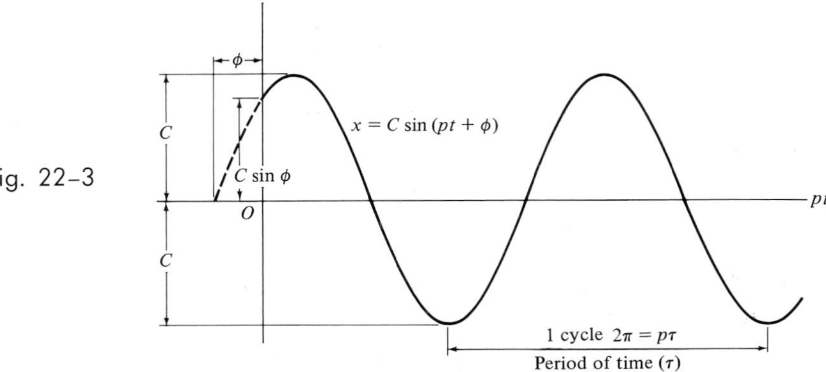

Fig. 22-3

The *frequency* f is defined as the number of cycles completed per unit of time, which is the reciprocal of the period:

$$f = \frac{1}{\tau} = \frac{p}{2\pi} \tag{22–14}$$

or

$$f = \frac{1}{2\pi}\sqrt{\frac{k}{m}} \tag{22–15}$$

The frequency is expressed in cycles/s. This ratio of units is called a *hertz* (Hz) where 1 Hz = 1 cycle/s = 2π rad/s.

22–2. Undamped Free Vibration

When a body or system of connected bodies is given an initial displacement from its equilibrium position and released, it will vibrate with a definite frequency known as the *natural frequency*. This type of vibration is called *free vibration,* since no external forces except gravitational or elastic forces act upon the body after the first displacement. Provided the *amplitude* of vibration remains *constant,* the motion is said to be *undamped.*

The undamped free vibration of a body having a single degree of freedom has the same characteristics as simple harmonic motion of the block and spring discussed in the previous section. Consequently, the body's motion is described by a differential equation of the *same form* as Eq. 22–1, i.e.,

$$\ddot{x} + p^2x = 0 \tag{22–16}$$

Hence, if the circular frequency p of the body is known, the period of vibration τ, natural frequency f, and other vibrating characteristics of the body can be established using the equations of the previous section.

PROCEDURE FOR ANALYSIS

As in the case of the block and spring, the circular frequency p of a rigid body or system of connected rigid bodies having a single degree of freedom can be determined using the following three-step procedure: *Step 1:* Draw the *free-body and kinetic diagrams* of the body when the body is displaced by a *small amount* from its equilibrium position. Locate the body with respect to its equilibrium position by using an appropriate *position coordinate* q. The vectors $m\mathbf{a}_G$ and $I_G\boldsymbol{\alpha}$, shown on the kinetic diagram, should be directed such that the body is accelerating and thereby causing an *increase* in the position coordinate.

Step 2: Apply the *equation of motion* to relate the elastic or gravitational restoring forces and couples acting on the body to the body's accelerated motion.

Step 3: Using kinematics, express the body's accelerated motion in terms of the second time derivative of the position coordinate, $\ddot{q}$. Substitute this result into the equation of motion and determine p by rearranging the terms so that the resulting equation is of the form $\ddot{q} + p^2 q = 0$.

The following examples illustrate this procedure.

Example 22–1

Determine the period of vibration for the simple pendulum shown in Fig. 22–4*a*. The bob has a mass m and is attached to a cord of length l. Neglect the size of the bob.

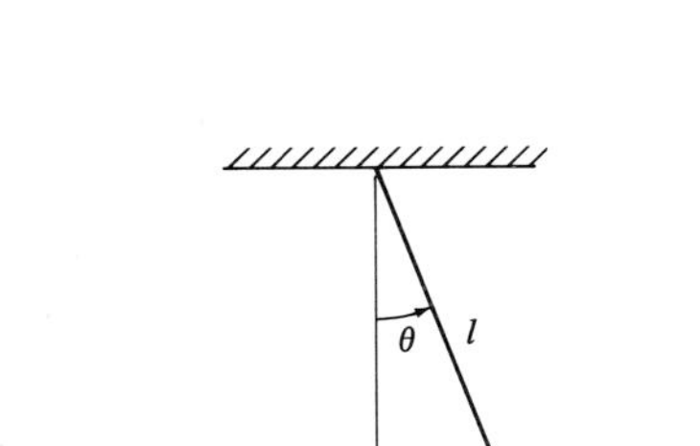

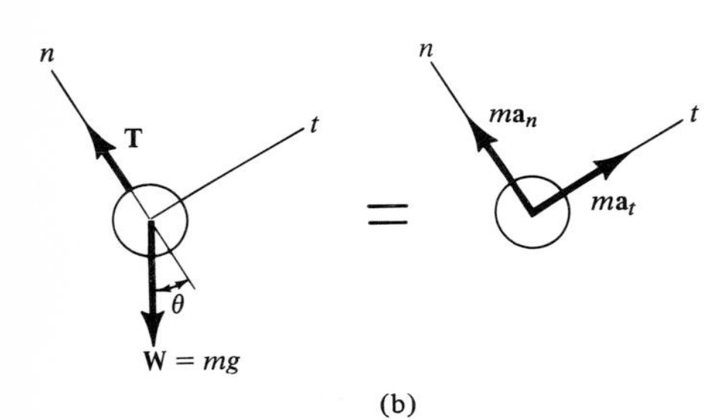

Fig. 22–4

Solution

Step 1: Motion of the system will be related to the position coordinate $(q =)\theta$. The free-body and kinetic diagrams are shown in Fig. 22–4*b*. When the bob is displaced by an angle θ at time t, the *restoring force* acting on the bob is created by the *weight component* $mg \sin\theta$. Furthermore, note that $m\mathbf{a}_t$ is shown acting in the direction of *increasing* θ.

Step 2: Applying the equation of motion in the *tangential direction,** since it involves the restoring force, yields

$$+\nearrow \Sigma F_t = ma_t; \qquad -mg \sin\theta = ma_t \tag{1}$$

Step 3: From kinematics, $a_t = d^2s/dt^2 = \ddot{s}$. Furthermore, s may be related to θ by the equation $s = l\theta$, so that $a_t = l\ddot{\theta}$. Hence, Eq. (1) reduces to the form

$$\ddot{\theta} + \frac{g}{l}\sin\theta = 0 \tag{2}$$

The solution of this equation involves the use of an elliptic integral. For *small displacements,* however, $\sin\theta \approx \theta$, in which case

$$\ddot{\theta} + \frac{g}{l}\theta = 0 \tag{3}$$

Comparing this equation with Eq. 22–16 ($\ddot{x} + p^2x = 0$), which is the "standard form" for simple harmonic motion, it is seen that $p = \sqrt{g/l}$.

*Application of the equation of motion in the n direction is not desired, since it involves the unknown cord tension **T**.

From Eq. 22-12, the period of time required for the bob to make one complete swing is therefore

$$\tau = \frac{2\pi}{p} = 2\pi\sqrt{\frac{l}{g}} \qquad \textit{Ans.}$$

This interesting result, originally discovered by Galileo Galilei through experiment, indicates that the period depends only on the length of the cord and not on the mass of the pendulum bob.

The solution of Eq. (3) is given by Eq. 22-3, where $p = \sqrt{g/l}$ and θ is substituted for x. Like the block and spring, the constants A and B in this problem may be determined if, for example, one knows the displacement and velocity of the bob at a given instant.

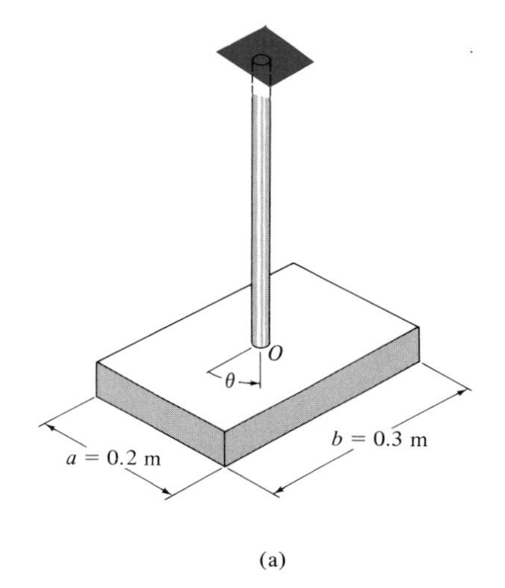

(a)

Example 22-2

The 10-kg rectangular plate shown in Fig. 22-5a is suspended at its center from a rod having a torsional stiffness of $k = 1.5\ \text{N}\cdot\text{m/rad}$. Determine the natural period of vibration of the plate when it is given a small angular displacement θ in the plane of the plate.

Solution

Step 1: The free-body and kinetic diagrams are shown in Fig. 22-5b. Since the plate is displaced in its own plane, the torsional *restoring* moment created by the rod is $M = k\theta$. This moment acts in the direction opposite to the displacement. The vector $I_O\ddot{\theta}$ acts in the direction of *positive* θ.

Step 2: Applying the equation of motion,

$$\Sigma M_O = I_O\alpha; \qquad -k\theta = I_O\ddot{\theta}$$

or

$$\ddot{\theta} + \frac{k}{I_O}\theta = 0$$

Step 3: Since this equation is in "standard form," the circular frequency is $p = \sqrt{k/I_O}$.

From Appendix D, the mass moment of inertia of the plate about an axis coincident with the rod is $I_O = \frac{1}{12}m(a^2 + b^2)$. Hence,

$$I_O = \frac{1}{12}[10][(0.2)^2 + (0.3)^2] = 0.108\ \text{kg}\cdot\text{m}^2$$

The natural period of vibration is, therefore,

$$\tau = \frac{2\pi}{p} = 2\pi\sqrt{\frac{I_O}{k}} = 2\pi\sqrt{\frac{0.108}{1.5}} = 1.69\ \text{s} \qquad \textit{Ans.}$$

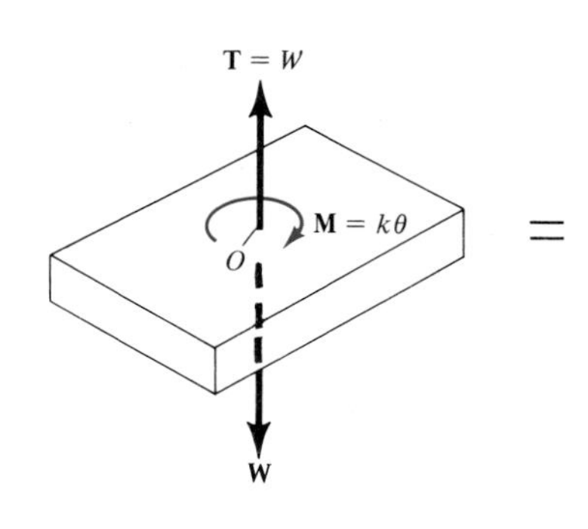

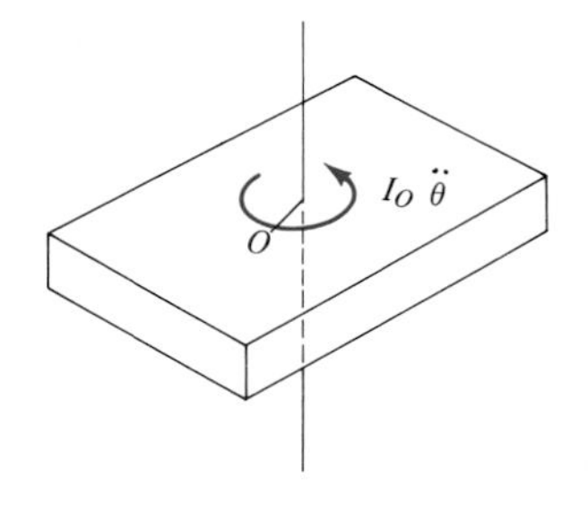

(b)

Fig. 22-5

Example 22–3

The bent rod shown in Fig. 22–6*a* has a negligible mass and supports a 5-kg collar at its end. Determine the natural period of vibration for the system.

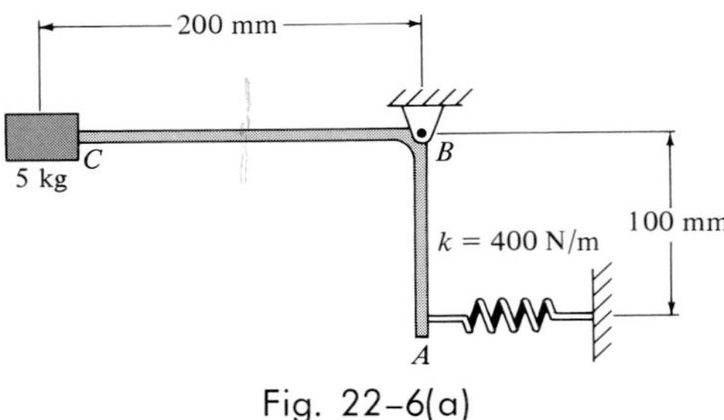

Fig. 22–6(a)

Solution

Step 1: The free-body and kinetic diagrams for the system are shown in Fig. 22–6*b*, where the rod is displaced by a small amount θ from the equilibrium position. Since the spring is subjected to an initial compression of x_{st} for statical equilibrium, for a displacement $x > x_{st}$ the spring exerts a force of $F_s = kx - kx_{st}$ on the rod. To obtain the standard form, Eq. 22–16, $(5)\mathbf{a}_y$ is assumed to act *upward*, which is in accordance with positive θ displacement.

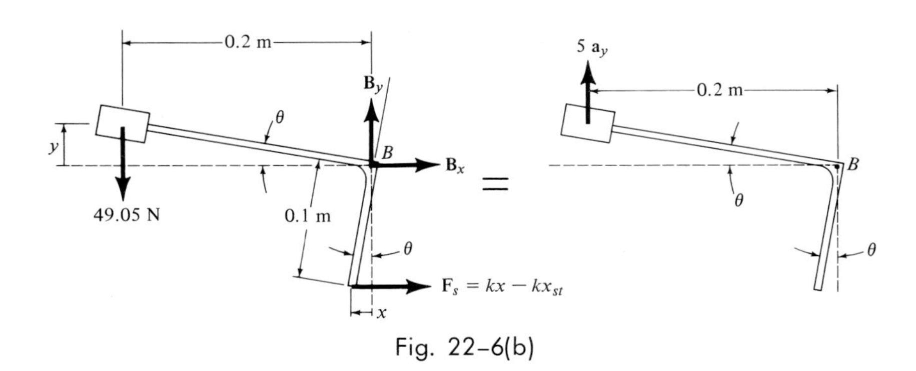

Fig. 22–6(b)

Step 2: Moments will be summed about point B to eliminate the unknown reaction at this point. Since θ is small,

$$\circlearrowleft+\Sigma M_B = \Sigma(M_k)_B; \quad kx(0.1) - kx_{st}(0.1) + 49.05(0.2) = -5a_y(0.2)$$

The second term on the left side, $-kx_{st}(0.1)$, represents the moment created by the spring force which is necessary to hold the collar in *equilibrium,* i.e., at $x = 0$. Since this moment is equal and opposite to the moment 49.05(0.2) created by the weight of the collar, these two terms cancel in the above equation, so that

$$kx(0.1) = -5a_y(0.2) \tag{1}$$

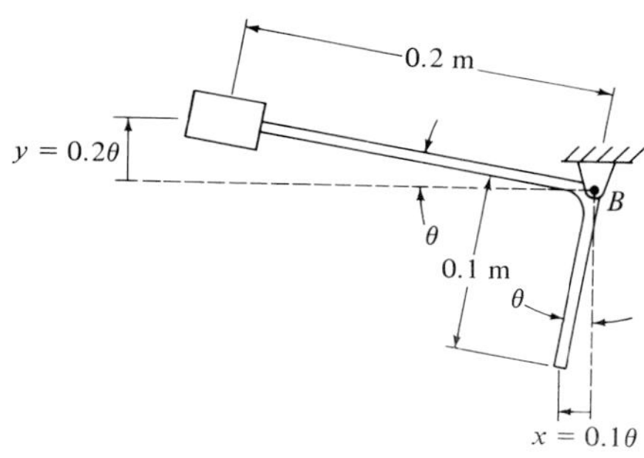

Fig. 22–6(c)

Step 3: From kinematics, the displacement of the spring and the collar may be related to the angle θ, Fig. 22–6c. Since θ is small, $x = 0.1\theta$ and $y = 0.2\theta$. Therefore, $a_y = \ddot{y} = 0.2\ddot{\theta}$. Substituting into Eq. (1) yields

$$400(0.1\theta)0.1 = -5(0.2\ddot{\theta})0.2$$

Rewriting this equation in standard form gives

$$\ddot{\theta} + 20\theta = 0$$

Comparing with $\ddot{x} + p^2x = 0$ (Eq. 22–16), we have

$$p^2 = 20, \qquad p = 4.47 \text{ rad/s}$$

The natural period of vibration is therefore,

$$\tau = \frac{2\pi}{p} = \frac{2\pi}{4.47} = 1.40 \text{ s} \qquad \textit{Ans.}$$

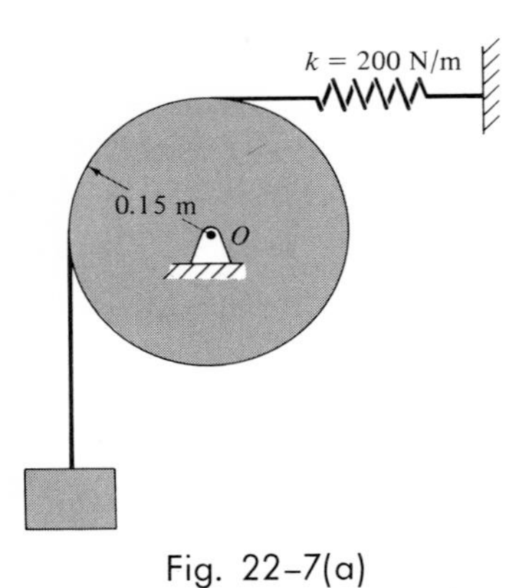

Fig. 22–7(a)

Example 22–4

A 10-kg block is suspended from a cord that passes over a 5-kg disk, as shown in Fig. 22–7*a*. The spring has a stiffness of $k = 200$ N/m. Determine the natural period of vibration for the system.

Solution

Step 1: The *system* consists of the disk which undergoes a rotation defined by the angle θ and the block which translates by an amount s. The free-body and kinetic diagrams for both the disk and the block are shown in Fig. 22–7*b*. The vector $I_O\ddot{\theta}$ acts in the direction of *positive* θ, and consequently $10\mathbf{a}_s$ acts downward in the direction of *positive* s.

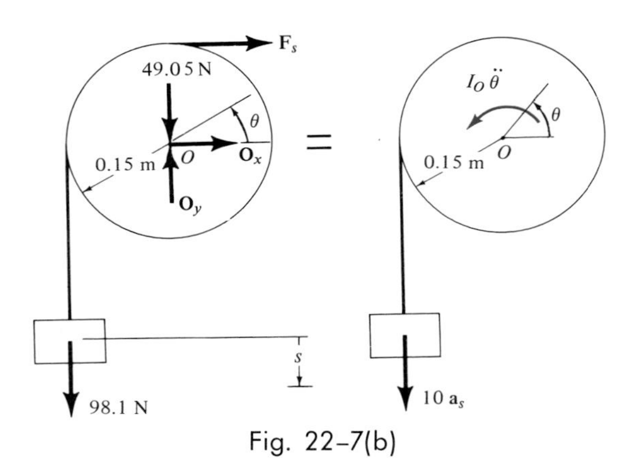

Fig. 22–7(b)

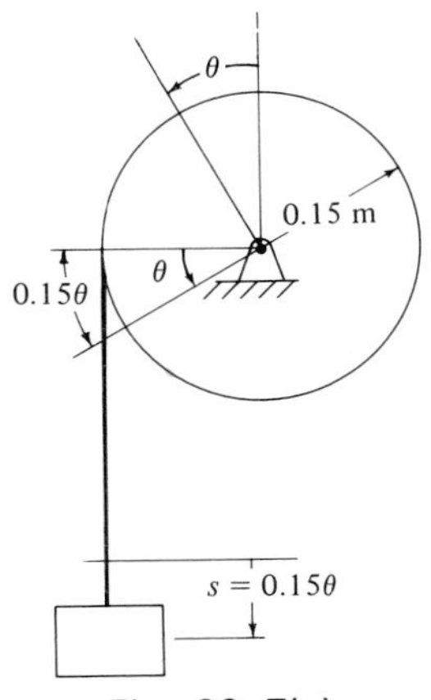

Fig. 22–7(c)

Step 2: Summing moments about point O to eliminate the reactions $\mathbf{O}_x$ and $\mathbf{O}_y$, realizing that $I_O = \frac{1}{2}mr^2$, yields

$$\circlearrowleft +\Sigma M_O = \Sigma(M_k)_O;$$
$$0.15(98.1) - F_s(0.15) = \frac{1}{2}(5)(0.15)^2\ddot{\theta} + 0.15(10)(a_s) \qquad (1)$$

Step 3: As shown on the kinematic diagram in Fig. 22–7*c*, a small positive displacement θ of the disk causes the block to lower by an amount $s = 0.15\theta$; hence, $a_s = \ddot{s} = 0.15\ddot{\theta}$. When $\theta = 0°$, the spring force required for *equilibrium* of the disk is 98.1 N, acting to the right. For a displacement θ, the spring force is $F_s = (200 \text{ N/m})(0.15\theta \text{ m}) + 98.1$ N. Substituting these results into Eq. (1) and simplifying yields

$$\ddot{\theta} + 16\theta = 0$$

Hence,

$$p^2 = 16, \qquad p = 4 \text{ rad/s}$$

Therefore, the natural period of vibration is

$$\tau = \frac{2\pi}{p} = \frac{2\pi}{4} = 1.57 \text{ s} \qquad \textit{Ans.}$$

Problems

22–1. When a 2-kg block is suspended from a spring, the spring is stretched a distance of 40 mm. Determine the natural frequency and the period of vibration for a 0.5-kg block attached to the same spring.

22–2. A spring has a stiffness of 800 N/m. If a 2-kg block is attached to the spring, pushed 50 mm above its equilibrium position, and released from rest, determine the equation which describes the block's motion. Assume that positive displacement is measured downward.

22–3. A spring is stretched 200 mm by a 15-kg block. If the block is displaced 100 mm downward from its equilibrium position and given a downward velocity of 0.75 m/s, determine the equation which describes the motion. What is the phase angle? Assume that positive displacement is measured downward.

***22–4.** If the block in Prob. 22–3 is given an upward velocity of 2 m/s when it is displaced downward a distance of 150 mm from its equilibrium position, determine the equation which describes the motion. What is the amplitude of the motion? Assume that positive displacement is measured downward.

22–5. A block having a mass of $m_b = 3$ kg is suspended from a spring having a stiffness of $k = 200$ N/m. If the block is pushed $y = 50$ mm upward from its equilibrium position and then released from rest, determine the equation which describes the motion. What is the amplitude and the natural frequency of the vibration? Assume that positive displacement is measured downward.

22–5a. Solve Prob. 22–5 if the block has a weight of $W_b = 2$ lb, and $k = 2$ lb/in., $y = 1$ in.

22–6. A 6-kg block is suspended from a spring having a stiffness of $k = 200$ N/m. If the block is given an upward velocity of 0.4 m/s when it is 75 mm above its equilibrium position, determine the equation which describes the motion and the maximum upward displacement of the block measured from the equilibrium position. Assume that positive displacement is measured downward.

22–7. A pendulum has a 0.5-m-long cord and is given a tangential velocity of 0.2 m/s toward the vertical, from a position of $\theta = 0.3$ rad from the vertical. Determine the equation which describes the angular motion.

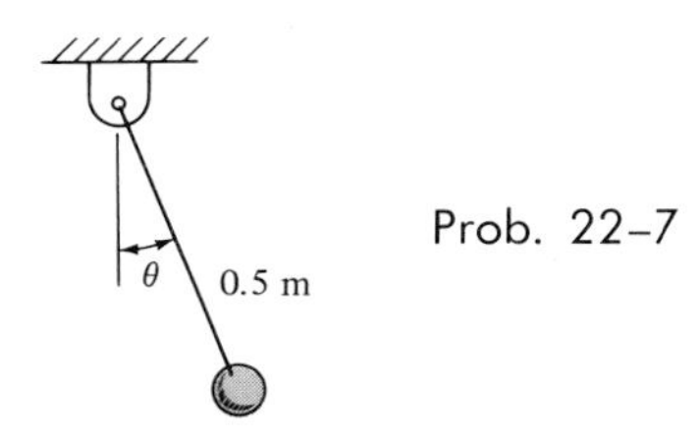

Prob. 22–7

***22–8.** Determine to the nearest degree the maximum angular displacement of the bob in Prob. 22–7 if it is initially displaced $\theta = 0.2$ rad from the vertical and given a tangential velocity of 0.15 m/s away from the vertical.

22-9. Determine the frequency of vibration for the block-and-spring mechanisms.

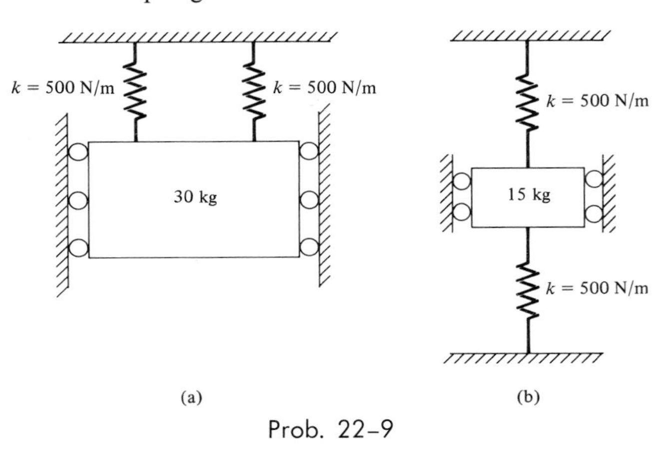

Prob. 22-9

22-10. The semicircular disk has a mass of $m = 3$ kg. Determine the period of oscillation if it is displaced a small amount and released. $r = 100$ mm.

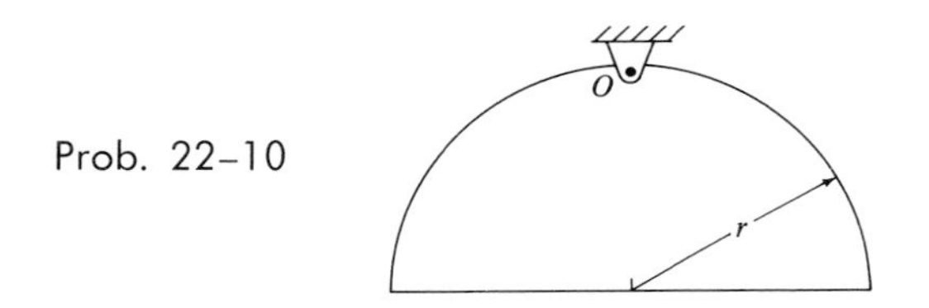

Prob. 22-10

22-10a. Solve Prob. 22-10 if the semicircular disk weighs $W = 20$ lb and $r = 1$ ft.

22-11. The thin hoop is supported by a knife edge. Determine the period of oscillation for small amplitudes of swing.

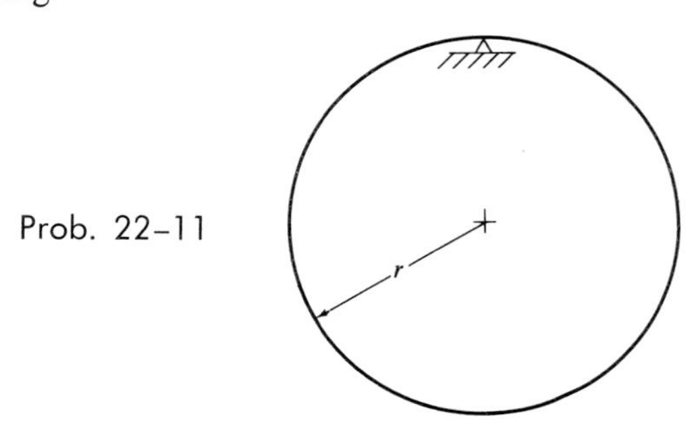

Prob. 22-11

***22-12.** The 2-kg block is fixed to the end of a rod assembly that has negligible weight. If both springs are unstretched when the assembly is in the position shown, determine the natural period of vibration for the block when it is rotated slightly about the pivot at O and released.

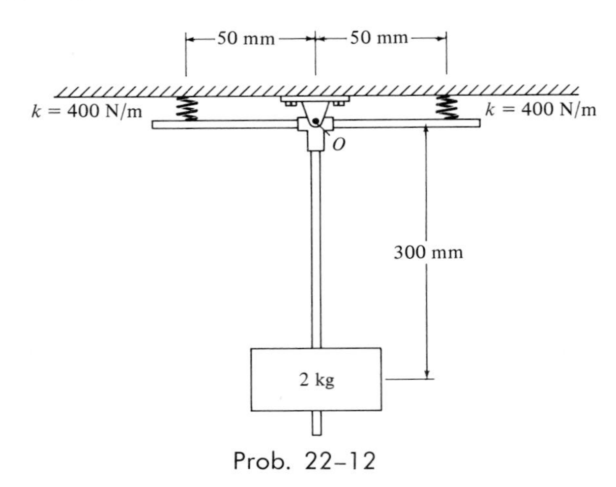

Prob. 22-12

22-13. The uniform beam has a mass of 150 kg/m and a moment of inertia of $I_O = 250\ \text{kg}\cdot\text{m}^2$ calculated about the pin at O. If the lower end is displaced a small amount and released from rest, determine the frequency of vibration. Each spring has a stiffness of $k = 500$ N/m and is unstretched when the beam is hanging from the vertical.

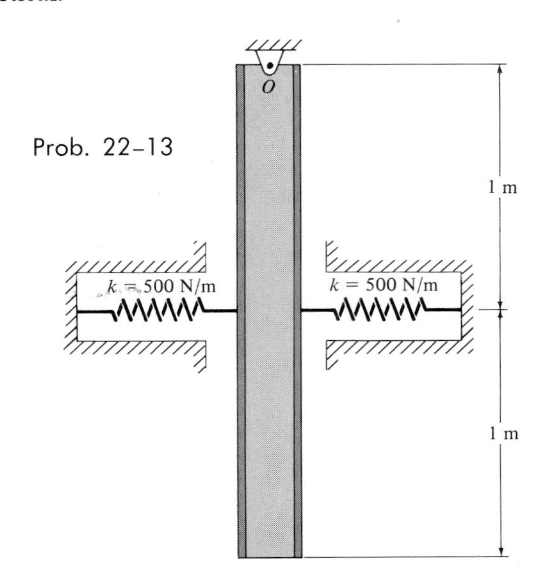

Prob. 22-13

22–14. A platform A having an unknown mass is supported by *four* springs each having the same stiffness k. When nothing is on the platform, the period of vertical vibration is measured as 3.90 s; whereas, if a 2-kg block is supported on the platform, the period of vertical vibration is 4.10 s. Compute the mass of a block placed on the (empty) platform which causes the platform to vibrate vertically with a period of 4.60 s. What is the stiffness k of each of the springs?

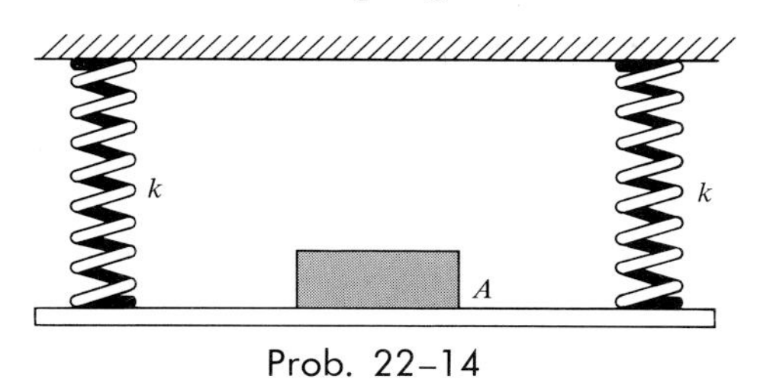

Prob. 22–14

22–15. The disk, having a mass of $m_d = 5$ kg, is pinned at its center O and supports the block A that has a mass of $m_A = 2$ kg. If the belt, which passes over the disk, is not allowed to slip at its contacting surface, compute the period of vibration of the system. $r = 150$ mm and $k = 600$ N/m.

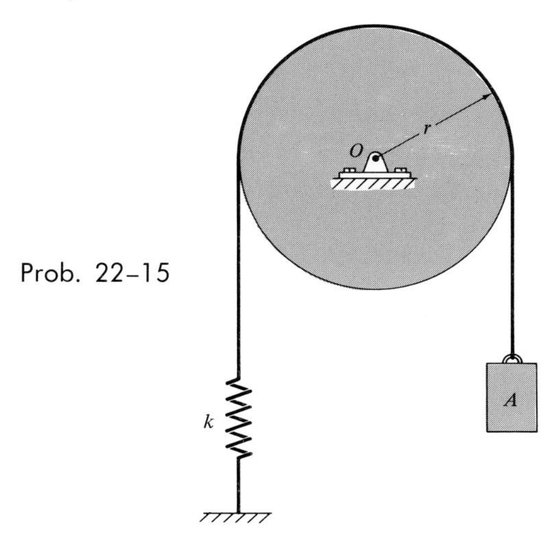

Prob. 22–15

22–15a. Solve Prob. 22–15 if the disk weighs $W_d = 20$ lb, block A weighs $W_A = 5$ lb, $r = 0.75$ ft, and $k = 120$ lb/ft.

***22–16.** The body of arbitrary shape has a mass m, mass center at G, and a radius of gyration about G of k_G. If the body is displaced by a slight amount θ from its equilibrium position and released, determine the period of oscillation.

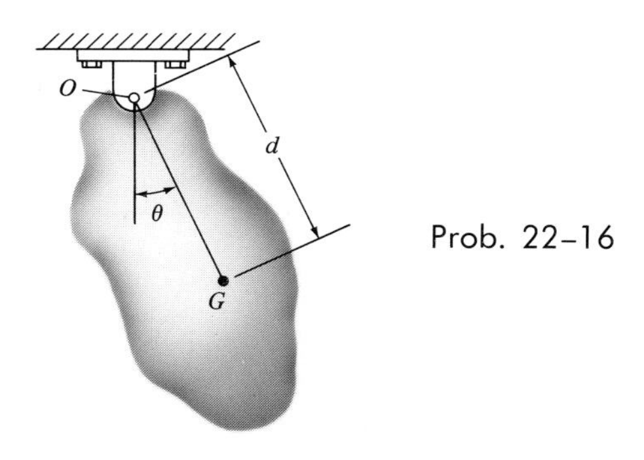

Prob. 22–16

22–17. The connecting rod is supported by a knife edge at A and the period of oscillation is measured as $\tau_A = 3.50$ s. It is then removed and rotated 180° so that it is supported by the knife edge at B. In this case the period of oscillation is measured as $\tau = 3.96$ s. Determine the location d of the center of gravity G, and compute the radius of gyration k_G. *Hint:* See Prob. 22–16.

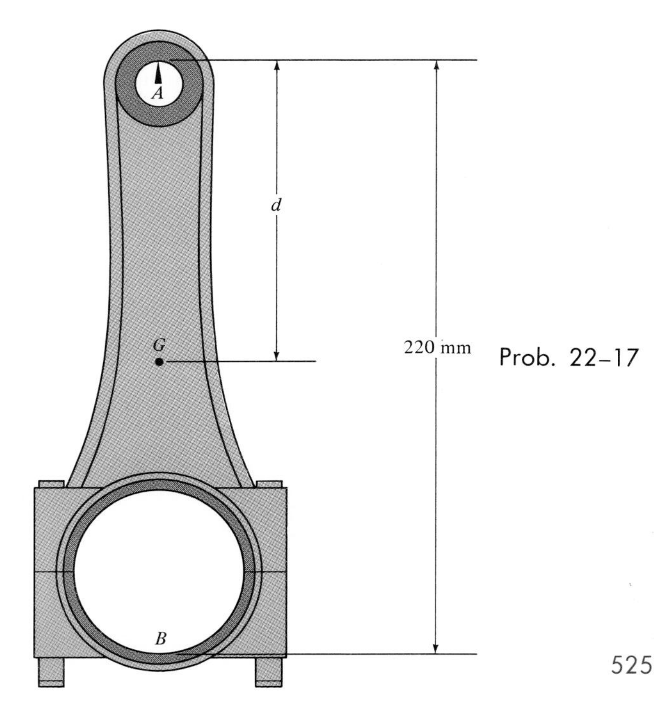

Prob. 22–17

22–18. The bell has a mass of 250 kg, a center of mass at G, and a radius of gyration about point D of $k_D = 0.4$ m. The tongue consists of a slender rod attached to the inside of the bell at C. If a 2-kg mass is attached to the end of the rod, determine the length l of the rod so that the bell will "ring silent," i.e., so that the period of vibration of the tongue is the same as that of the bell. For the calculation, neglect the small distance between C and D and neglect the mass of the rod.

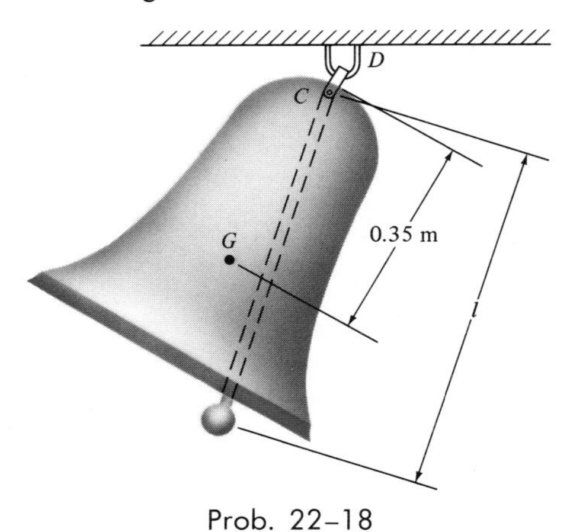

Prob. 22–18

22–19. Determine the *torsional stiffness* k of rod AB if the 2-kg thin rectangular plate has a natural period of vibration of $\tau = 0.3$ s as it oscillates around the axis of the rod. *Hint:* The torsional stiffness is defined from $M = k\theta$ and is measured in N · m/rad.

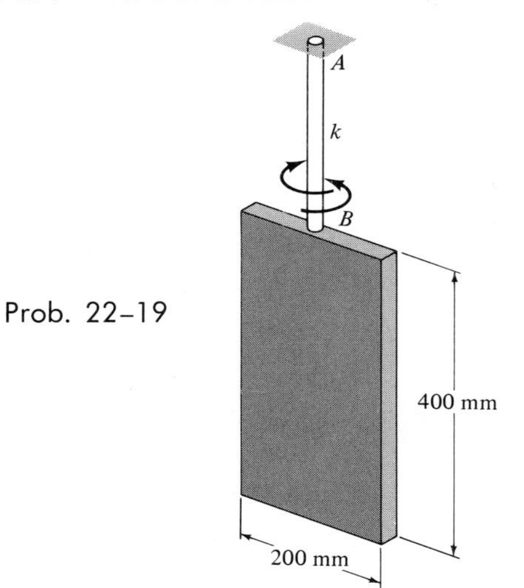

Prob. 22–19

***22–20.** The pointer on a metronome supports a slider A, having a mass of $m_A = 0.5$ kg, which is positioned at a fixed distance $d = 150$ mm from the pivot O of the pointer. A torsional spring exerts a restoring moment on the pointer having a magnitude of $M = 1.2\theta$, where θ represents the angle of swing measured in radians and M is measured in N · m. If the spring is untorqued when the pointer is in the vertical position, determine the period of vibration. Neglect the mass of the pointer.

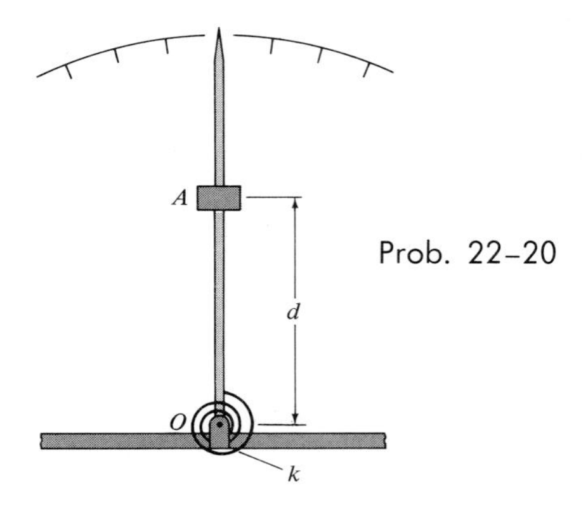

Prob. 22–20

***22–20a.** Solve Prob. 22–20 if the slider A has a weight of $W_A = 0.5$ lb; $M = 0.5\theta$, where M is measured in lb · ft and θ is in radians; and $d = 0.25$ ft.

22–21. A uniform board is supported on two wheels which rotate in opposite directions at a constant angular speed. If the coefficient of friction between the wheels and board is μ, determine the frequency of vibration of the board if it is displaced slightly, a distance x from the midpoint between the wheels, and released.

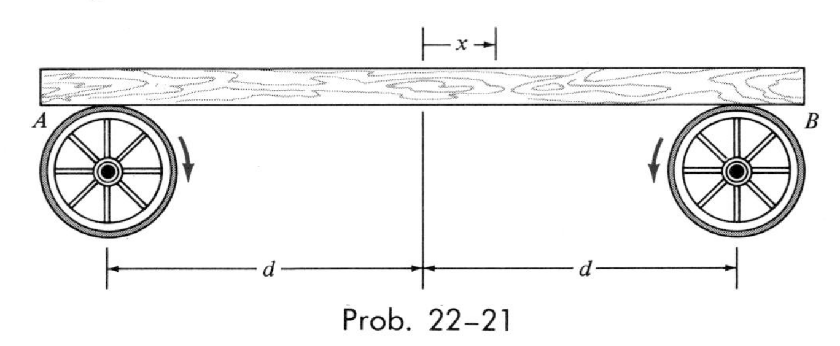

Prob. 22–21

22-22. The U-tube has a cross-sectional area of 25 mm² and is partially filled with a liquid having a density of $\rho = 700\ \text{kg/m}^3$. Determine the period of vibration of the liquid when the valve V is opened quickly to allow the liquid to oscillate.

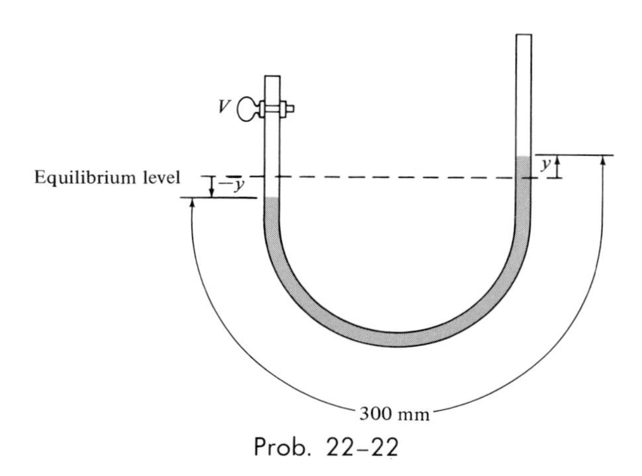

Prob. 22-22

22-23. The spool has a mass of 5 kg and a radius of gyration of $k_O = 125$ mm. If it rolls without slipping, determine the period of vibration when it is displaced slightly and released. The springs are unstretched before the spool is rotated.

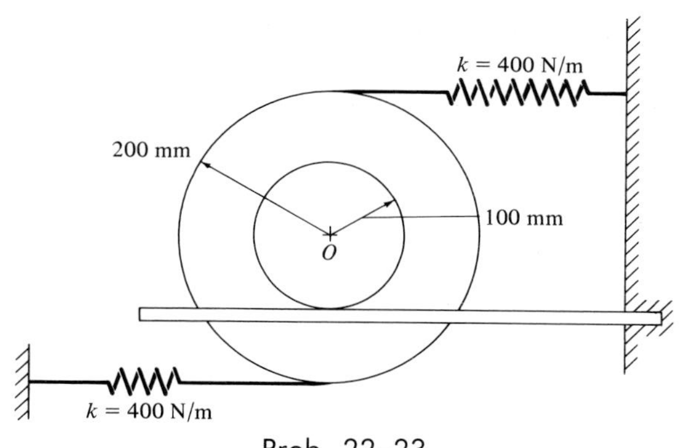

Prob. 22-23

***22-24.** If a 2-kg cylinder is displaced a small amount along the curved surface, determine the frequency at which it oscillates when it is released. The cylinder rolls without slipping.

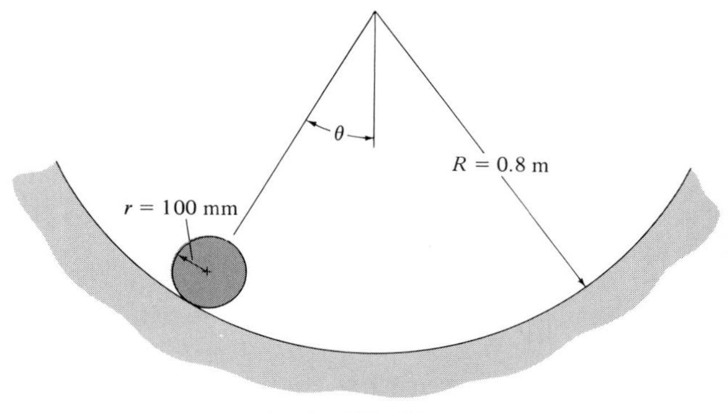

Prob. 22-24

22-25. The system shown consists of a spring with a stiffness of $k = 50$ N/m and an unstretched length of $l = 200$ mm, a bar with a negligible mass, and a small sphere A with a mass of $m = 1.5$ kg. Determine the frequency for small oscillations θ of the sphere.

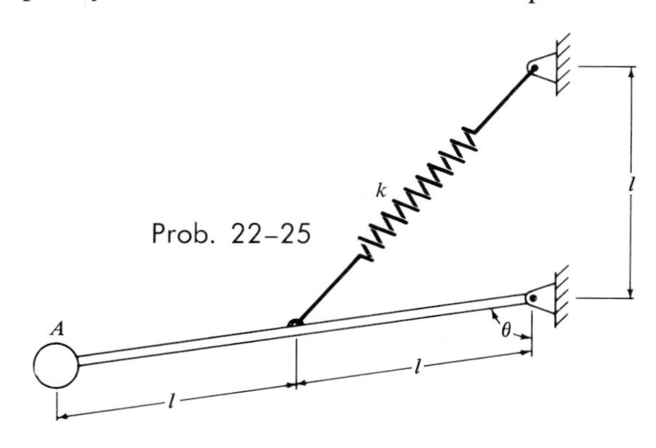

Prob. 22-25

22-25a. Solve Prob. 22-25 if $k = 25$ lb/ft, $l = 1.25$ ft, and the sphere A weighs $W = 10$ lb.

22–3. Energy Methods

The simple harmonic motion of a body, discussed in the previous section, is due only to gravitational and elastic restoring forces acting on the body. Since these types of forces are *conservative,* it is possible to use the conversion-of-energy theorem to obtain the body's natural frequency or period of vibration. To show how to do this, consider again the block and spring in Fig. 22–8. When the block is displaced an arbitrary amount x from the equilibrium position, the kinetic energy is $T = \frac{1}{2}mv^2 = \frac{1}{2}m\dot{x}^2$ and the potential energy is $V = \frac{1}{2}kx^2$. By the conservation-of-energy theorem, Eq. 14–22, it is necessary that

$$T + V = \text{const}$$

$$\tfrac{1}{2}m\dot{x}^2 + \tfrac{1}{2}kx^2 = \text{const} \qquad (22\text{–}17)$$

The differential equation describing the *accelerated motion* of the block can be obtained by *differentiating* this equation with respect to time, i.e.,

$$m\dot{x}\ddot{x} + kx\dot{x} = 0$$

$$\dot{x}(m\ddot{x} + kx) = 0$$

Since the velocity $\dot{x}$ is not *always* zero in a vibrating system,

$$\ddot{x} + p^2x = 0, \qquad p = \sqrt{k/m}$$

which is the same as Eq. 22–1.

For a system of connected bodies, determination of the natural frequency or the equation of motion by time differentiation of the energy equation is advantageous, since it is *not necessary* to dismember the system to account for reactive and connective forces which do no work. There is also an added advantage in that the circular frequency p may be obtained *directly*. For example, consider the total mechanical energy of the block and spring in Fig. 22–8, when the block is at its *maximum displacement*. In this position the mass is temporarily at rest, so that the kinetic energy is zero and the potential energy, stored in the spring, is a maximum. Therefore, Eq. 22–17 becomes $\frac{1}{2}kx_{\text{max}}^2 = \text{const}$. At the instant the block passes the equilibrium position, the kinetic energy of the block is a

Fig. 22–8

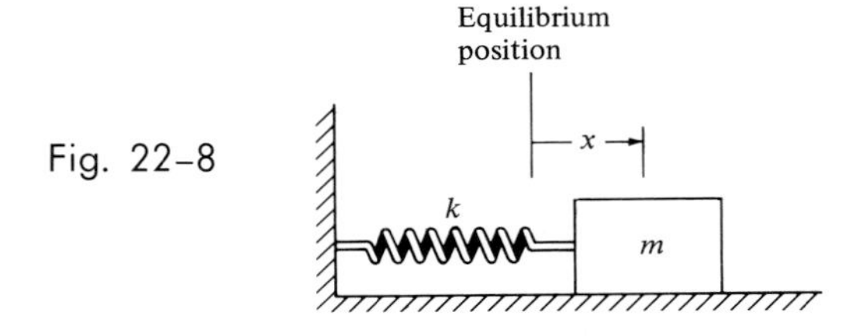

maximum and the potential energy of the spring is zero. Hence, Eq. 22–17 becomes $\frac{1}{2}m(\dot{x})^2_{max} = \text{const}$. Since the vibrating motion of the block is *harmonic,* the *solution* for the displacement and velocity may be written in the form of Eq. 22–9 and its time derivative, i.e.,

$$x = C\sin(pt + \phi), \qquad \dot{x} = Cp\cos(pt + \phi)$$

so that

$$x_{max} = C, \qquad (\dot{x})_{max} = Cp$$

Applying the conservation-of-energy theorem ($T + V = \text{const}$) yields

$$V_{max} = T_{max}; \qquad \tfrac{1}{2}kx^2_{max} = \tfrac{1}{2}m(\dot{x})^2_{max} = \text{const}$$

or

$$kC^2 = mC^2p^2$$

Solving for p yields

$$p = \sqrt{\frac{k}{m}}$$

which is identical to Eq. 22–2.

PROCEDURE FOR ANALYSIS

The following two-step procedure should be applied when solving for the circular frequency p of a body or system of connected bodies using the conservation-of-energy theorem.

Step 1: Draw a free-body diagram of the body when it is displaced by a *small amount* from its equilibrium position and define the location of the body from its equilibrium position by an appropriate position coordinate q. Formulate the energy for the body, $T + V = \text{const}$, in terms of the position coordinate.* Recall that, in general, the kinetic energy must account for both the body's translational and rotational motion, $T = \frac{1}{2}mv_G^2 + \frac{1}{2}I_G\omega^2$ (Eq. 18–1); and that the potential energy is the sum of the gravitational and elastic potential energies of the body, $V = V_g + V_e$ (Eq. 18–15).

Step 2: Take the time derivative of the energy equation and factor out the common terms. The resultant differential equation represents the equation of motion for the system. The value of p is obtained after rearranging the terms in the standard form $\ddot{q} + p^2q = 0$.

The following examples illustrate this two-step procedure.

*It is suggested that the material in Sec. 18–5 be reviewed.

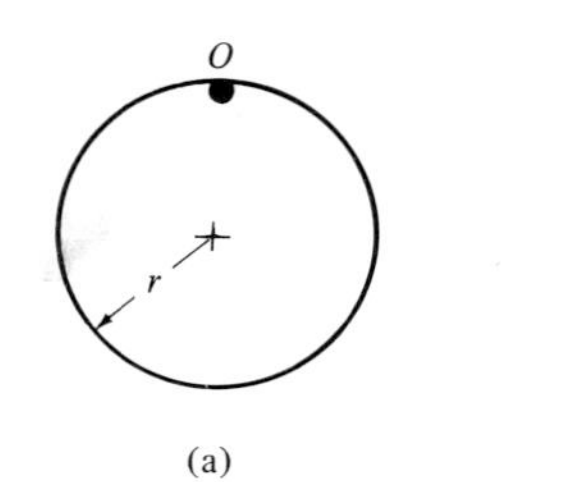

(a)

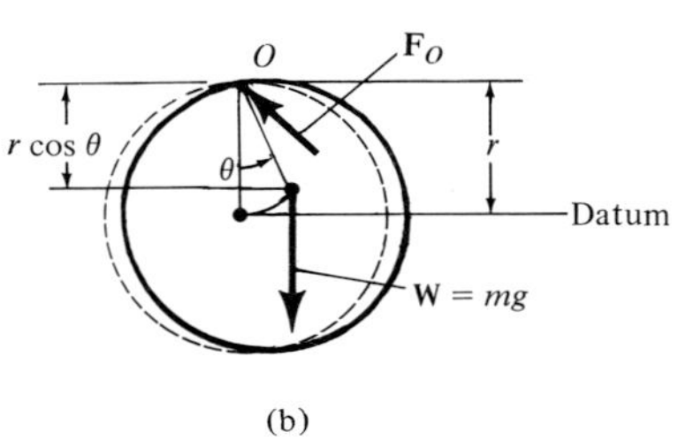

(b)
Fig. 22–9

Example 22–5

The thin hoop shown in Fig. 22–9*a* is supported by a peg at *O*. Determine the period of oscillation for small amplitudes of swing. The hoop has a mass m.

Solution

Step 1: A free-body diagram of the hoop when it is displaced a small amount $(q =)\ \theta$ from the equilibrium position is shown in Fig. 22–9*b*. Using Appendix D and the parallel-axis theorem to determine I_O, the kinetic energy can be expressed as

$$T = \tfrac{1}{2}I_O\omega^2 = \tfrac{1}{2}[mr^2 + mr^2]\dot{\theta}^2 = mr^2\dot{\theta}^2$$

If a horizontal datum is placed through the center of gravity of the hoop when $\theta = 0$, then the center of gravity moves upward $r(1 - \cos\theta)$ in the displaced position. For *small angles,* the $\cos\theta$ may be replaced by the first two terms of its series expansion, $\cos\theta = 1 - \theta^2/2 + \cdots$. Therefore, the potential energy is

$$V = mgr\left[1 - \left(1 - \frac{\theta^2}{2}\right)\right] = mgr\frac{\theta^2}{2}$$

The total energy in the system is

$$T + V = mr^2\dot{\theta}^2 + mgr\frac{\theta^2}{2}$$

Step 2: Taking the time derivative of the energy equations yields

$$mr^2 2\dot{\theta}\ddot{\theta} + mgr\theta\dot{\theta} = 0$$
$$mr\dot{\theta}(2r\ddot{\theta} + g\theta) = 0$$

Since $\dot{\theta}$ is not always equal to zero, from the terms in parentheses

$$\ddot{\theta} + \frac{g}{2r}\theta = 0$$

Hence,

$$p = \sqrt{\frac{g}{2r}}$$

so that

$$\tau = \frac{2\pi}{p} = 2\pi\sqrt{\frac{2r}{g}} \qquad \textit{Ans.}$$

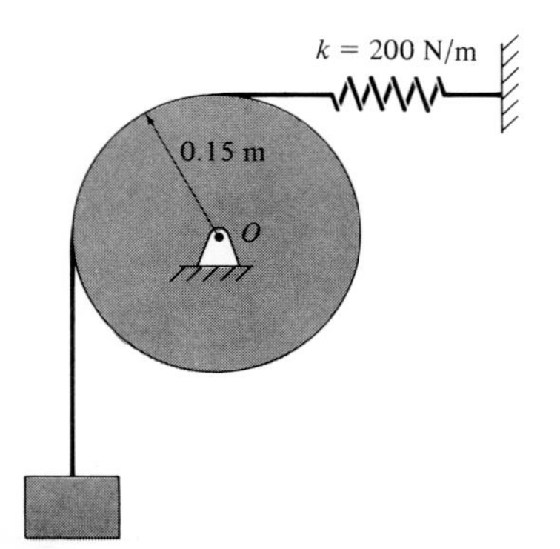

Fig. 22–10(a)

Example 22–6

A 10-kg block is suspended from a cord wrapped around a 5-kg disk, as shown in Fig. 22–10*a*. If the spring has a stiffness of $k = 200$ N/m, determine the natural period of vibration for the system.

Solution I

Step 1: The free-body diagrams of the block and disk when they are displaced by respective amounts s and θ from the equilibrium position, are shown in Fig. 22–10*b*. Since $s = 0.15\theta$, the kinetic energy of the system is

$$T = \tfrac{1}{2}m_b v_b^2 + \tfrac{1}{2}I_O\omega_d^2$$
$$= \tfrac{1}{2}(10)(0.15\dot{\theta})^2 + \tfrac{1}{2}[\tfrac{1}{2}(5)(0.15)^2](\dot{\theta})^2$$
$$= 0.141(\dot{\theta})^2$$

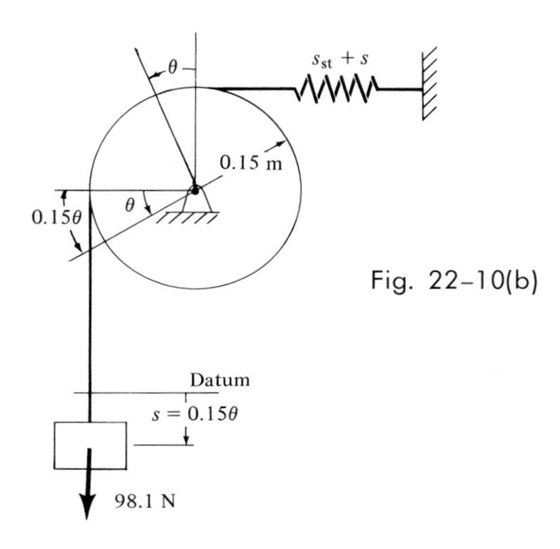

Fig. 22–10(b)

Establishing the datum at the equilibrium position of the block and realizing that the spring stretches s_{st} for equilibrium, we can write the potential energy as

$$V = \tfrac{1}{2}k(s_{st} + s)^2 - Ws$$
$$= \tfrac{1}{2}(200)(s_{st} + 0.15\theta)^2 - 98.1(0.15\theta)$$

The total energy for the system is, therefore,

$$T + V = 0.141(\dot{\theta})^2 + 100(s_{st} + 0.15\theta)^2 - 14.72\theta$$

Step 2: Taking the time derivative of the energy equation yields

$$0.282(\dot{\theta})\ddot{\theta} = 200(s_{st} + 0.15\theta)0.15\dot{\theta} - 14.72\dot{\theta} = 0$$

Since $s_{st} = 98.1/200 = 0.4905$ m, the above equation reduces to the standard form

$$\ddot{\theta} + 16\theta = 0$$

so that

$$p = \sqrt{16} = 4 \text{ rad/s}$$

Thus,

$$\tau = \frac{2\pi}{p} = \frac{2\pi}{4} = 1.57 \text{ s} \qquad \textit{Ans.}$$

Solution II
The diagrams shown in Fig. 22–10c represent the positions of the system for maximum kinetic energy and maximum potential energy, respectively.

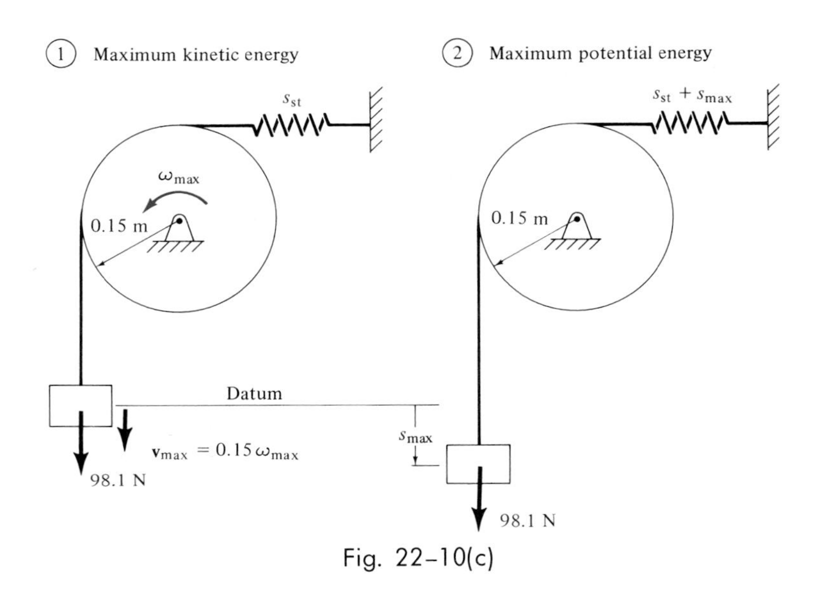

Fig. 22–10(c)

The horizontal datum is chosen at the elevation of the block when the system is in *equilibrium*. Hence, the maximum angular velocity of the disk and the maximum velocity of the block occur when the block is located at the datum. From kinematics, the velocity of the block is $v_{max} = 0.15\omega_{max}$.

The potential energy for the system is maximum when the spring is stretched and the system is temporarily at rest. At this instant the block is displaced an amount s_{max}, while the spring is elongated by an amount $s_{st} + s_{max}$, where s_{st} represents the *static deflection* of the spring, $s_{st} = 98.1/200 = 0.4905$ m. The conservation-of-energy theorem for the system may be written for the two positions shown in Fig. 22–10c, in which case

$$\{T_1\} + \{V_1\} = \{T_2\} + \{V_2\}$$

$$\{\tfrac{1}{2}m_b(v_b)_1^2 + \tfrac{1}{2}I_O(\omega_d)_1^2\} + \{\tfrac{1}{2}ks_{st}^2\} = \{\tfrac{1}{2}m_b(v_b)_2^2 + \tfrac{1}{2}I_O(\omega_d)_2^2\}$$
$$+ \{\tfrac{1}{2}k(s_{st} + s_{max})^2 - Ws_{max}\}$$

Since the system undergoes simple harmonic motion, $s_{max} = C$, $v_{max} = Cp$, and $\omega_{max} = Cp/0.15$. Therefore, substituting into the above equation yields

$$\left\{\tfrac{1}{2}(10)(Cp)^2 + \tfrac{1}{2}[\tfrac{1}{2}(5)(0.15)^2]\left(\frac{Cp}{0.15}\right)^2\right\} + \{\tfrac{1}{2}(200)(0.4905)^2\} = \{0 + 0\} + \{\tfrac{1}{2}(200)(0.4905 + C)^2 - 98.1(C)\}$$

Expanding and rearranging terms, we can write this expression as

$$6.25(Cp)^2 = 100(C)^2$$

Solving for p gives

$$p = \sqrt{\frac{100}{6.25}} = 4 \text{ rad/s}$$

Hence, the period of vibration is

$$\tau = \frac{2\pi}{p} = \frac{2\pi}{4} = 1.57 \text{ s} \qquad \textit{Ans.}$$

Problems

22–26. Solve Prob. 22–11 using energy methods.

22–27. Solve Prob. 22–13 using energy methods.

***22–28.** Solve Prob. 22–16 using energy methods.

22–29. Solve Prob. 22–24 using energy methods.

22–30. Solve Prob. 22–15 using energy methods.

22–30a. Solve Prob. 22–15a using energy methods.

22–31. Using energy methods, determine the differential equation of motion of the 2-kg block when it is displaced slightly and released. The horizontal surface is smooth and the springs are originally unstretched.

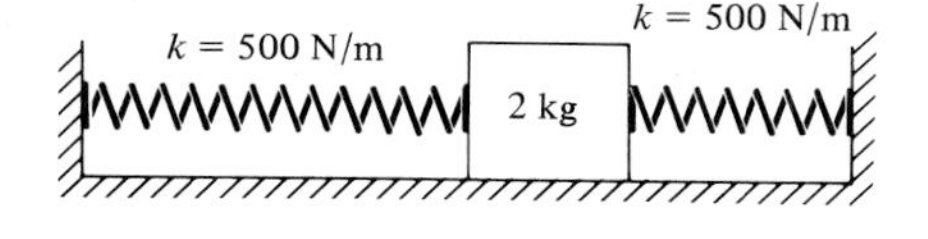

Prob. 22–31

***22–32.** If the disk has a mass of 3 kg, determine the natural frequency of vibration. The springs are originally unstretched.

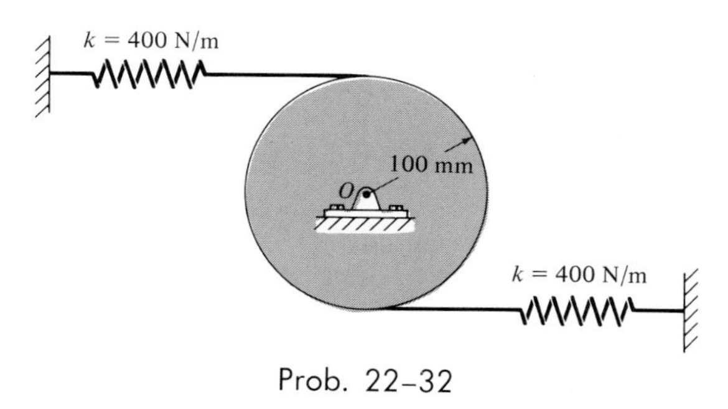

Prob. 22–32

22–33. The 2-kg disk is pin-connected at its midpoint. Determine the period of vibration of the disk if the springs have sufficient tension in them to prevent the belt from slipping on the disk. *Hint:* Assume that the initial stretch in each spring is δ_O. This term will cancel out after taking the time derivative of the energy equation.

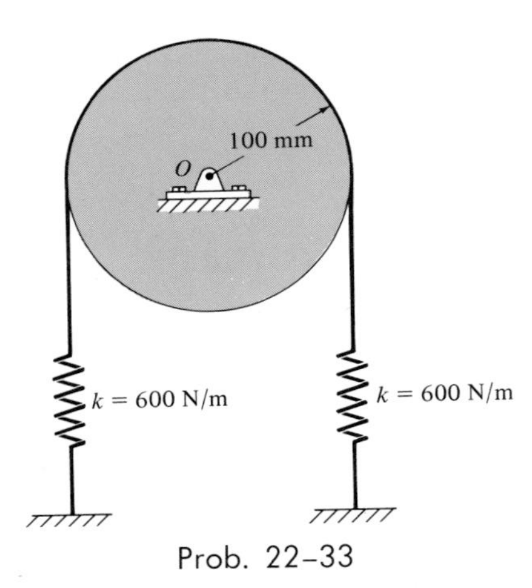

Prob. 22–33

22–34. The wheel has a mass of 50 kg. If it is suspended from a knife edge at O and displaced slightly from the equilibrium position, it is observed to have a period of vibration of 1.56 s. Determine the moment of inertia of the wheel about its mass center G.

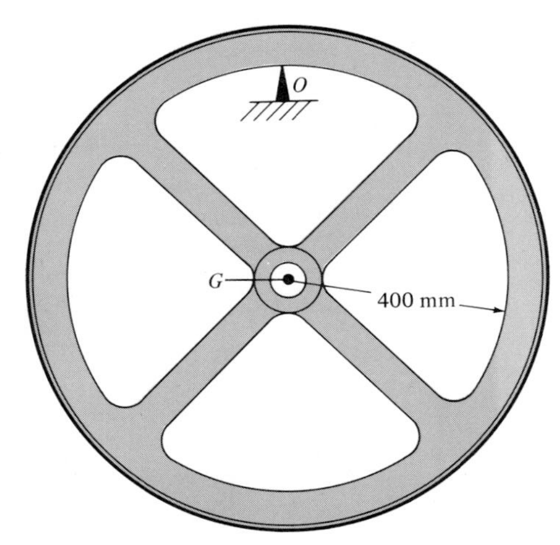

Prob. 22–34

22–35. Determine the period of vibration of the pendulum. Consider the two rods to be slender, each having a mass of $m = 6$ kg/m and length $l = 400$ mm.

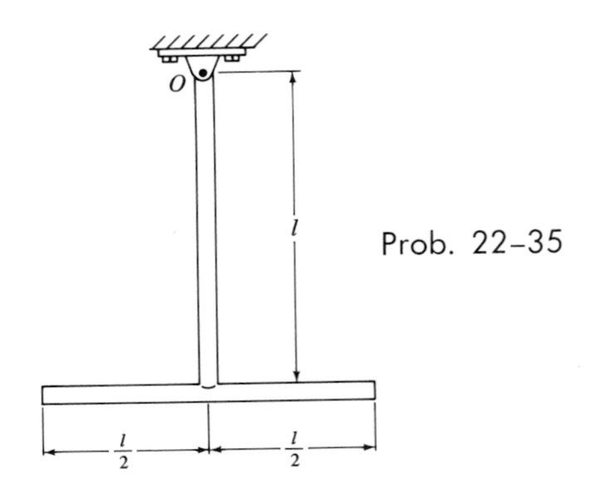

Prob. 22–35

22–35a. Solve Prob. 22–35 if the rods each have a weight of $W = 10$ lb/ft and $l = 2$ ft.

***22–36.** Determine the period of vibration of the 3-kg sphere. Neglect the mass of the rod.

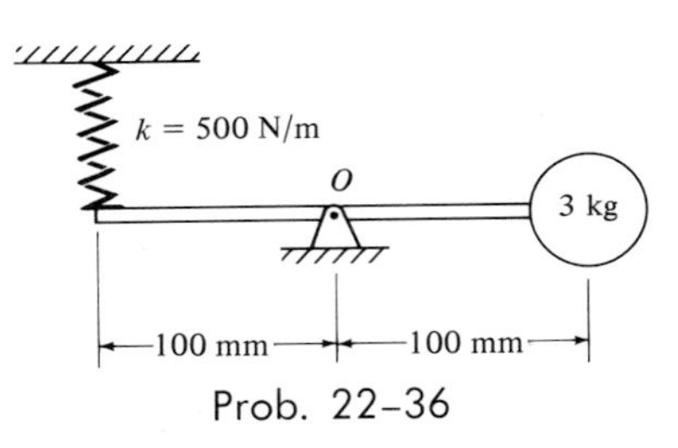

Prob. 22–36

22–37. The 5-kg sphere is attached to a rod of negligible mass. Determine the natural frequency of vibration.

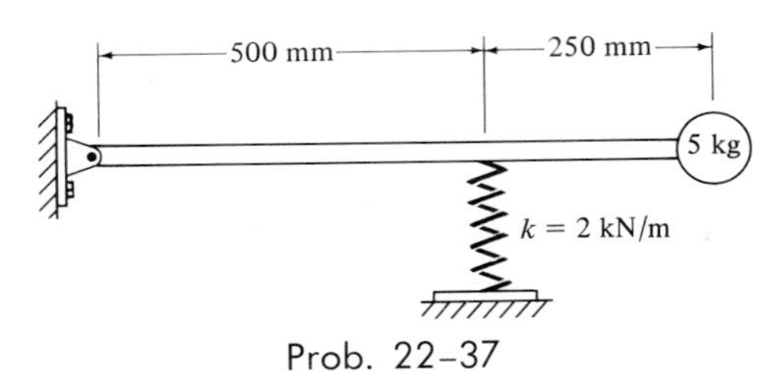

Prob. 22–37

22–38. Determine the natural frequency of vibration of the 2-kg disk. Assume that the force of friction is great enough so that the disk does not slip on the surface of the plane while it is oscillating.

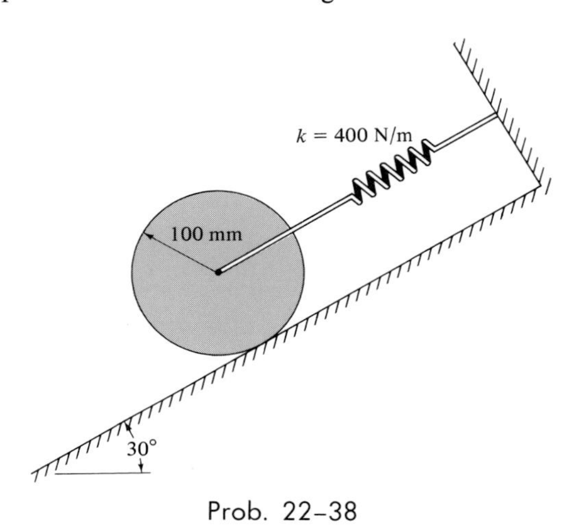

Prob. 22–38

22–39. Using energy methods, determine the differential equation of motion of the 3-kg spool. Assume that it does not slip at the surface of contact as it oscillates. The radius of gyration of the spool about its center of mass is $k_G = 125$ mm.

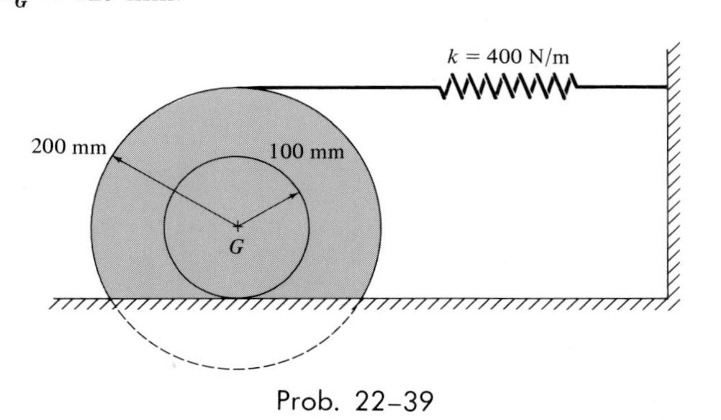

Prob. 22–39

***22–40.** The slender rod has a mass of $m = 6$ kg/m and a length of $l = 400$ mm. If it is supported in the horizontal plane by a ball-and-socket joint at A and a cable at B, determine the natural frequency of vibration when the end B is given a small horizontal displacement and then released.

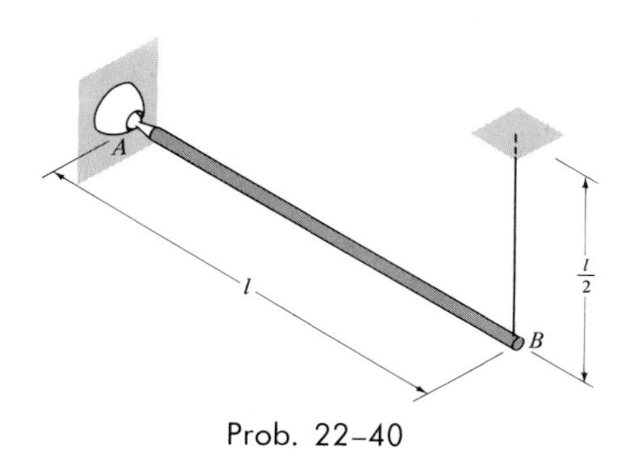

Prob. 22–40

***22–40a.** Solve Prob. 22–40 if the rod has a weight of $W = 4$ lb/ft and $l = 2$ ft.

22-4. Undamped Forced Vibration

Undamped forced vibration is considered to be one of the most important types of vibrating motion in engineering work. The principles which describe the nature of this motion may be applied to the analysis of forces which cause vibration in various types of machines and structures.

Periodic Force Excitation. The block and spring shown in Fig. 22-11*a* provide a convenient "model" which represents the vibrational characteristics of a system subjected to a periodic excitation force $F = F_0 \sin \omega t$.

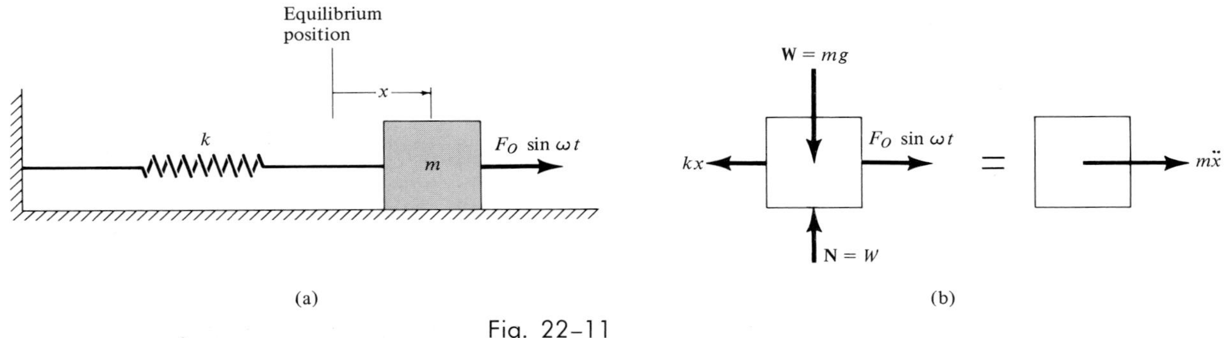

Fig. 22-11

This force has a maximum magnitude of F_0 and a *forcing frequency* ω. The free-body and kinetic diagrams for the block are shown in Fig. 22-11*b*, where x defines the displacement of the spring. Applying the equation of motion yields

$\xrightarrow{+} \Sigma F_x = ma_x;$ $\qquad F_0 \sin \omega t - kx = m\ddot{x}$

or

$$\ddot{x} + \frac{k}{m}x = \frac{F_0}{m}\sin \omega t \tag{22-18}$$

This equation is referred to as a nonhomogeneous second-order differential equation. The general solution consists of a complementary solution, x_c, *plus* a particular solution, x_p.

The *complementary solution* is determined by setting the term on the right side of Eq. 22-18 equal to zero and solving the resulting homogeneous equation which is equivalent to Eq. 22-1. The solution is defined by Eq. 22-3, i.e.,

$$x_c = A \sin pt + B \cos pt \tag{22-19}$$

where p is the circular frequency, $p = \sqrt{k/m}$ (Eq. 22-2).

Since the motion is periodic, the *particular solution* of Eq. 22–18 may be determined by assuming a solution of the form

$$x_p = C \sin \omega t \tag{22–20}$$

where C is a constant. Taking the second time derivative and substituting into Eq. 22–18 yields

$$-C\omega^2 \sin \omega t + \frac{k}{m}(C \sin \omega t) = \frac{F_0}{m} \sin \omega t$$

Factoring out $\sin \omega t$ and solving for C gives

$$C = \frac{F_0/m}{\dfrac{k}{m} - \omega^2} = \frac{F_0/k}{1 - \left(\dfrac{\omega}{p}\right)^2} \tag{22–21}$$

Substituting into Eq. 22–20, we obtain the particular solution

$$x_p = \frac{F_0/k}{1 - \left(\dfrac{\omega}{p}\right)^2} \sin \omega t \tag{22–22}$$

The *general solution* is, therefore,

$$x = x_c + x_p = A \sin pt + B \cos pt + \frac{F_0/k}{1 - \left(\dfrac{\omega}{p}\right)^2} \sin \omega t \tag{22–23}$$

Here x describes two types of vibrating motion of the block. The *complimentary solution* x_c defines the *free vibration,* which depends upon the circular frequency $p = \sqrt{k/m}$ and the constants A and B, Fig. 22–12*a*. Specific values for A and B are obtained by evaluating Eq. 22–23 at a given instant when the displacement and velocity are known. The *particular solution* x_p describes the *forced vibration* of the block caused by the applied force $F = F_0 \sin \omega t$, Fig. 22–12*b*. The resultant vibration x is shown in Fig. 22–12*c*. Since all vibrating systems are subject to *friction,* however, the free vibration, x_c, will, in time, dampen out. For this reason the free vibration is referred to as *transient,* and the forced vibration is called *steady state,* since it is the only vibration that remains, Fig. 22–12*d*.

From Eq. 22–21 it is seen that the *amplitude* of forced vibration depends upon the *frequency ratio* ω/p. If the *magnification factor* MF is defined as the ratio of the amplitude of steady-state vibration, $(x_p)_{\max}$, to the static deflection F_0/k, which is caused by the amplitude of the periodic force, F_0, then, from Eq. 22–21,

$$\text{MF} = \frac{(x_p)_{\max}}{F_0/k} = \frac{1}{1 - \left(\dfrac{\omega}{p}\right)^2} \tag{22–24}$$

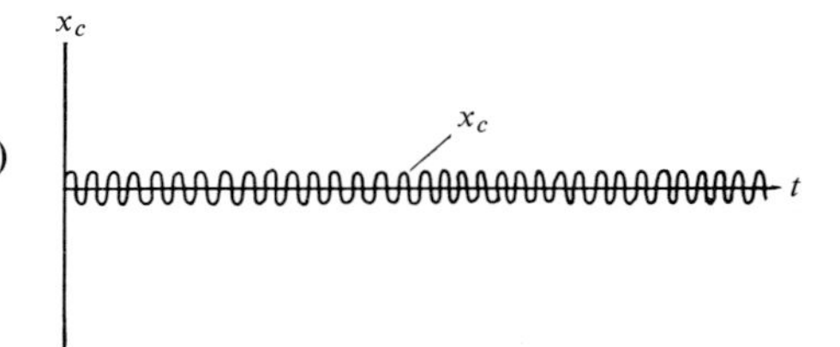

(a)

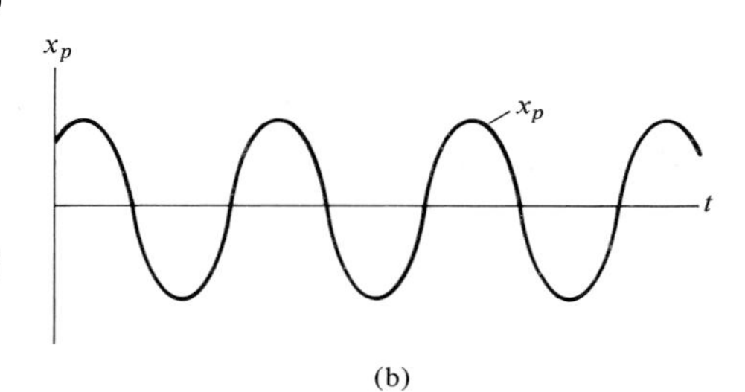

(b)

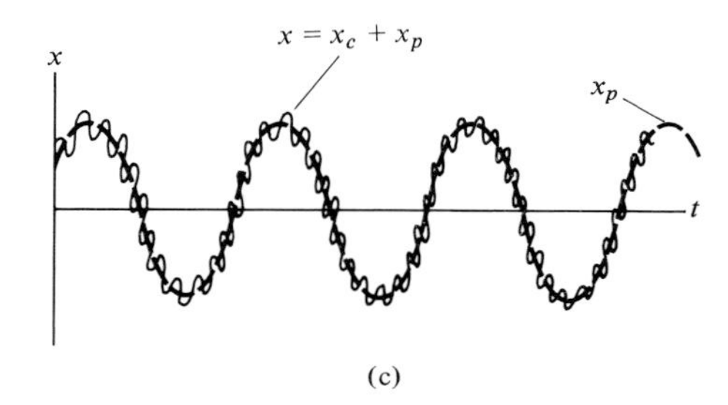

(c)

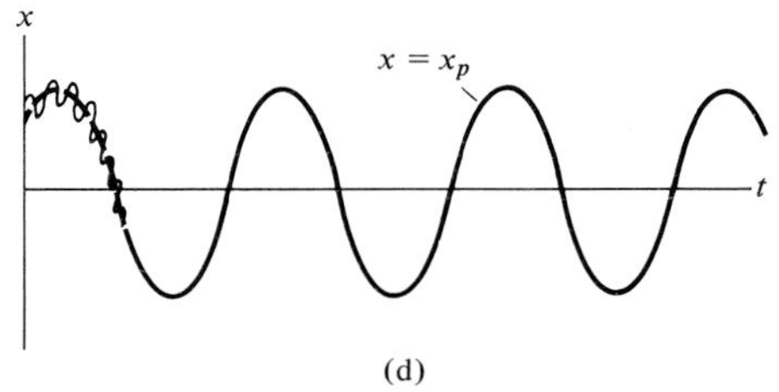

(d)

Fig. 22–12

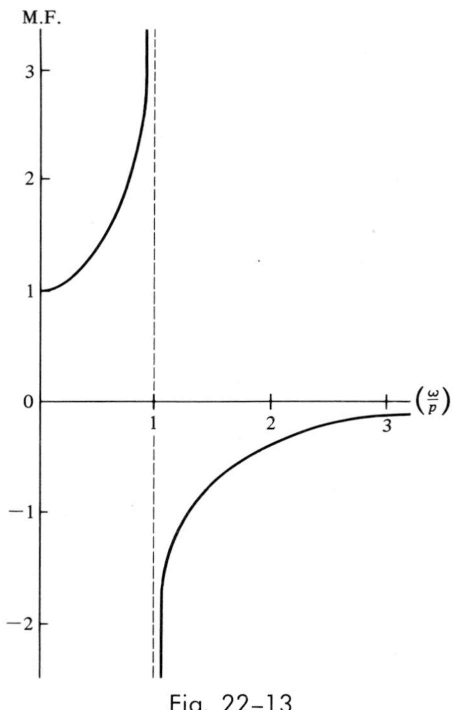

Fig. 22–13

This equation is graphed in Fig. 22–13, where it is seen that for $\omega \approx 0$, the MF ≈ 1. In this case, because of the small frequency ω, the vibration of the block will be in phase with the applied force **F.** If the force or displacement is applied with a frequency close to the natural frequency of the system, i.e., $\omega/p \approx 1$, the amplitude of vibration of the block becomes extremely large. This condition is called *resonance,* and in practice, resonating vibrations can cause tremendous stress and rapid failure of parts. When the cyclic force $F_0 \sin \omega t$ is applied at high frequencies $(\omega > p)$, the value of the MF becomes negative, indicating that the motion of the block is out of phase with the force. Under these conditions, as the block is displaced to the right, the force acts to the left, and vice versa. For extremely high frequencies $(\omega \gg p)$ the inertia of the mass prevents the block from following the force or displacement. As a result, the block remains almost stationary, and hence the MF is approximately zero.

Periodic Displacement Excitation. Forced vibrations can also arise from the periodic excitation of the foundation of a system. The model shown in Fig. 22–14*a* represents the periodic vibration of a block which is caused by

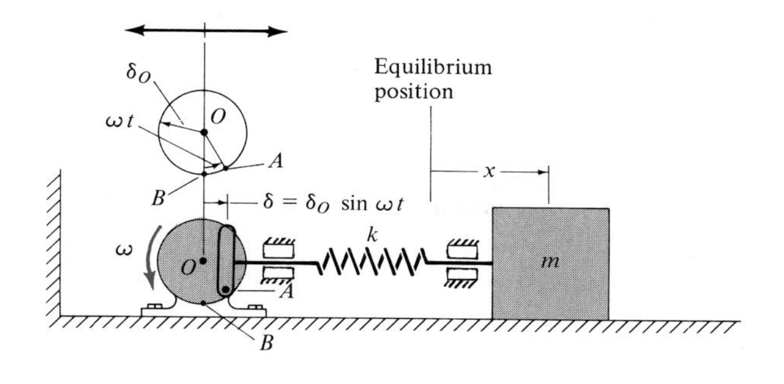

(a)

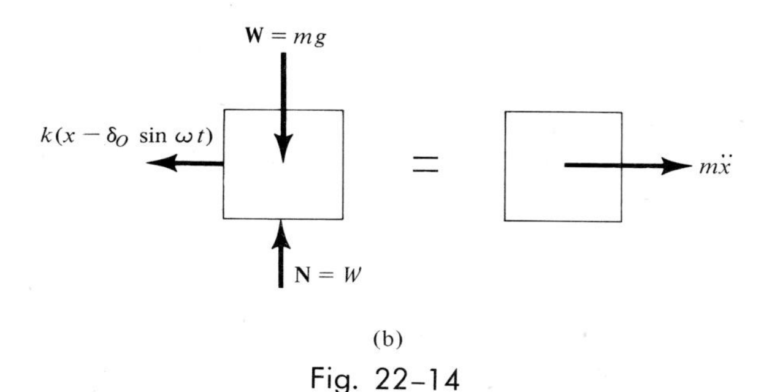

(b)

Fig. 22–14

harmonic movement $\delta = \delta_O \sin \omega t$ of the support. The free-body and kinetic diagrams for the block in this case are shown in Fig. 22–14*b*. The coordinate x is measured from the point of zero displacement of the support, i.e., when the radius vector OA coincides with OB, Fig. 22–14*a*. Therefore, general displacement of the spring is $(x - \delta_O \sin \omega t)$. Applying the equation of motion yields

$$\overset{+}{\rightarrow}\Sigma F_x = ma_x; \qquad -k(x - \delta_O \sin \omega t) = m\ddot{x}$$

or

$$\ddot{x} + \frac{k}{m}x = \frac{k\delta_O}{m}\sin \omega t \tag{22–25}$$

By comparison, this equation is identical to the form of Eq. 22–18, *provided* F_O *is replaced* by $k\delta_O$. If this substitution is made into the solutions defined by Eqs. 22–21, 22–22, and 22–23, the results are appropriate for describing the motion of the block when subjected to the support displacement $\delta = \delta_O \sin \omega t$.

Example 22–7

The instrument shown in Fig. 22–15 is attached to a platform P, which in turn is supported by *four* springs, each having a stiffness of $k = 800$ N/m. Initially the platform is at rest when the floor is subjected to a displacement of $\delta = 10 \sin (8t)$ mm, where t is measured in seconds. If the total mass of the instrument and platform is 20 kg, determine the vertical displacement x of the platform, measured from the equilibrium position, as a function of time. What would be the floor vibration to cause resonance?

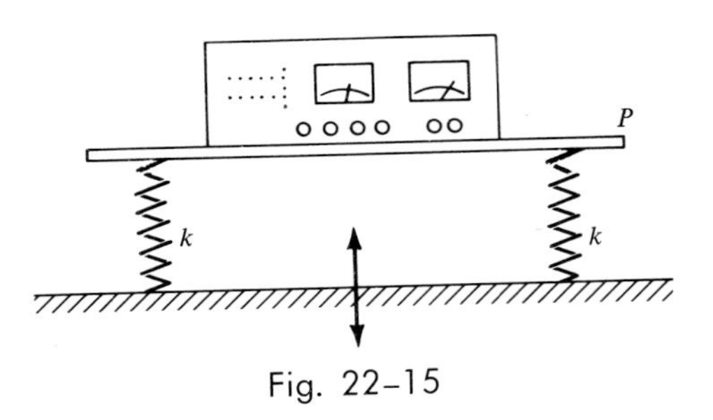

Fig. 22–15

Solution

Since the induced vibration is caused by the displacement of the supports, the motion is described by Eq. 22–23, with F_o replaced by $k\delta_o$, i.e.,

$$x = A \sin pt + B \cos pt + \frac{\delta_o}{1 - \left(\frac{\omega}{p}\right)^2} \sin \omega t \tag{1}$$

Here $\delta = \delta_o \sin \omega t = 10 \sin (8t)$ mm, so that

$$\delta_o = 10 \text{ mm}, \qquad \omega = 8 \text{ rad/s}$$

$$p = \sqrt{\frac{k}{m}} = \sqrt{\frac{4(800)}{20}} = 12.6 \text{ rad/s}$$

From Eq. 22–22, with $k\delta_o$ replacing F_o, the amplitude of vibration caused by the floor displacement is

$$(x_p)_{\max} = \frac{\delta_o}{1 - \left(\frac{\omega}{p}\right)^2} \tag{2}$$

$$= \frac{10}{1 - \left(\frac{8}{12.6}\right)^2} = 16.7 \text{ mm}$$

Hence, Eq. (1) and its time derivative become

$$x = A \sin (12.6t) + B \cos (12.6t) + 16.7 \sin (8t)$$

$$\dot{x} = A(12.6) \cos (12.6t) - B(12.6) \sin (12.6t) + 133.3 \cos (8t)$$

The constants A and B are evaluated from these equations. Since $x = \dot{x} = 0$ at $t = 0$, then

$$0 = 0 + B + 0; \qquad B = 0$$

$$0 = A(12.6) - 0 + 133.3; \qquad A = -10.6$$

The vibrating motion is therefore described by the equation

$$x = -10.6 \sin (12.6t) + 16.7 \sin (8t) \qquad \textit{Ans.}$$

Resonance will occur when the amplitude of vibration caused by the floor displacement approaches infinity. From Eq. (2), this requires that

$$\omega = p = 12.6 \text{ rad/s} \qquad \textit{Ans.}$$

Problems

22–41. If the block is subjected to the impressed force $F = F_0 \cos \omega t$, show that the differential equation of motion is $\ddot{y} + (k/m)y = (F_0/m) \cos \omega t$, where y is measured from the equilibrium position of the block. What is the general solution of this equation?

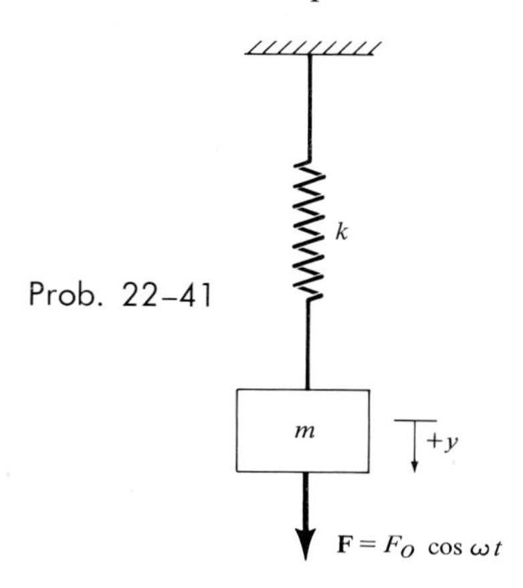

Prob. 22–41

22–42. A 4-kg block is suspended from a spring that has a stiffness of $k = 600$ N/m. The block is drawn downward 50 mm from the equilibrium position and released from rest at $t = 0$. If the support moves with an impressed displacement of $\delta = (10 \sin 4t)$ mm, where t is measured in seconds, determine the equation which describes the vertical motion of the block. Assume that positive displacement is measured downward.

22–43. A 5-kg block is suspended from a spring having a stiffness of 400 N/m. If the block is acted upon by a vertical force of $F = (5 \sin 8t)$ N, where t is measured in seconds, determine the equation which describes the motion of the block when the block is pulled down 100 mm from the equilibrium position and released from rest at $t = 0$. Assume that positive displacement is measured downward.

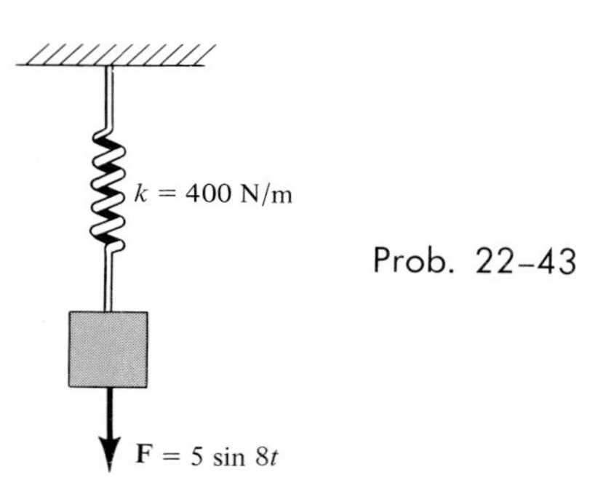

Prob. 22–43

***22–44.** The spring shown stretches 200 mm when it is attached to a 5-kg block. Determine the equation which describes the motion of the block when it is pulled 50 mm below its equilibrium position and released from rest at $t = 0$. The block is subjected to the impressed force of $F = 200 \sin 2t$, where F is measured in newtons and t in seconds. Assume that positive displacement is measured downward.

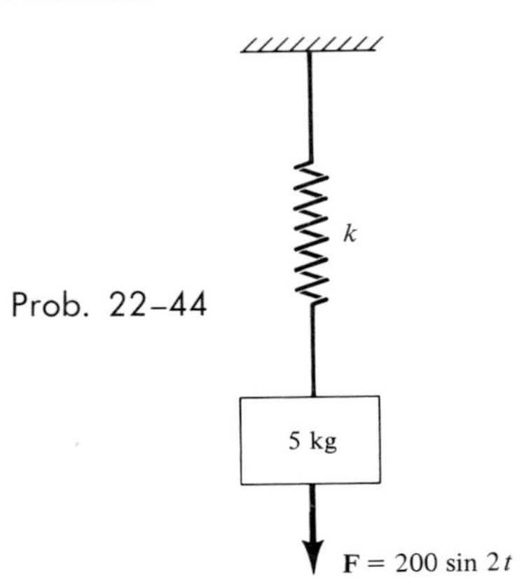

Prob. 22–44

22–45. The engine is mounted on a foundation block which is spring-supported. Describe the steady-state vibration of the system if the block and engine have a total mass of $m = 800$ kg and the engine, when running, creates an impressed force of $F = (50 \sin 2t)$ N, where t is measured in seconds. Assume that the system vibrates only in the vertical direction, with the positive displacement measured downward, and that the total stiffness of the springs can be represented as $k = 2$ kN/m.

Prob. 22–45

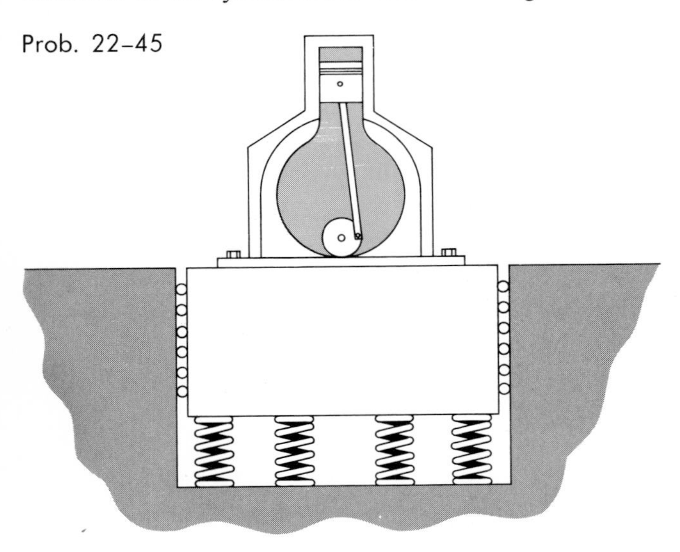

22–45a. Solve Prob. 22–45 if $W = 1500$ lb; $F = (70 \sin 4t)$ lb, where t is in seconds; and $k = 1750$ lb/ft.

22–46. Determine the rotational speed ω of the engine in Prob. 22–45 which will cause resonance.

22–47. The electric motor has a mass of 25 kg and is supported on a horizontal beam having a negligible mass. The motor turns an eccentric flywheel which is equivalent to an unbalanced 2-kg mass located 100 mm from the axis of rotation. If the static deflection of the beam is 50 mm as a result of the weight of the motor, determine the angular speed of the flywheel at which resonance will occur. *Hint:* See the first part of Example 22–8.

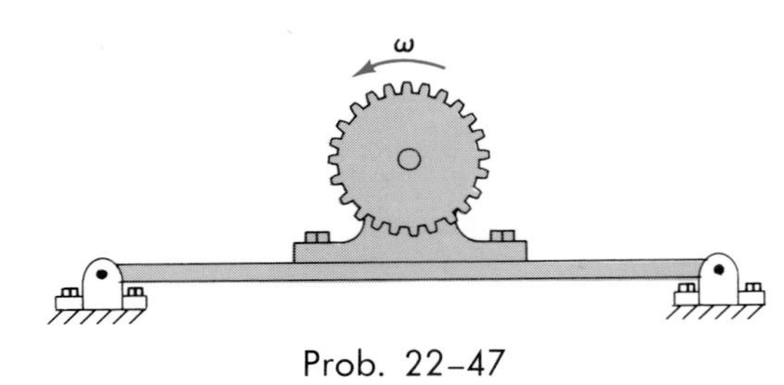

Prob. 22–47

***22–48.** What will be the amplitude of steady-state vibration of the motor in Prob. 22–47 if the angular speed of the flywheel is 16.5 rad/s? *Hint:* See the first part of Example 22–8.

22–49. What is the amplitude of steady-state vibration of the motor in Prob. 22–47 when the angular speed of the flywheel is 12 rad/s? *Hint:* See the first part of Example 22–8.

22–50. The instrument is centered uniformly on a platform P, which in turn is supported by *four* springs, each spring having a stiffness of $k = 200$ N/m. If the floor is subjected to a vibration of $\omega = 2$ Hz having a vertical displacement amplitude of $\delta_0 = 10$ mm, determine the vertical displacement amplitude of the platform and instrument. The instrument and the platform have a total mass of $m = 8$ kg.

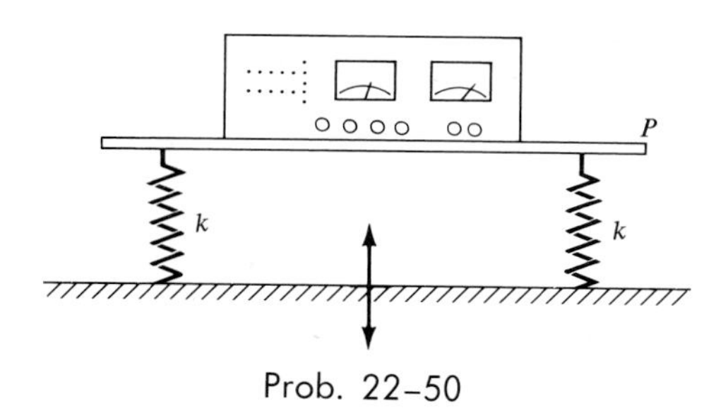

Prob. 22–50

22–50a. Solve Prob. 22–50 if $k = 100$ lb/ft, $\omega = 3$ Hz, $\delta_0 = 0.1$ ft, and $m = 20$ lb.

22–51. The light elastic rod supports a 2-kg sphere. When a 10-N vertical force is applied to the sphere, the rod deflects 10 mm. If the wall oscillates with a harmonic frequency of 2 Hz and has an amplitude of 15 mm, determine the amplitude of vibration for the sphere.

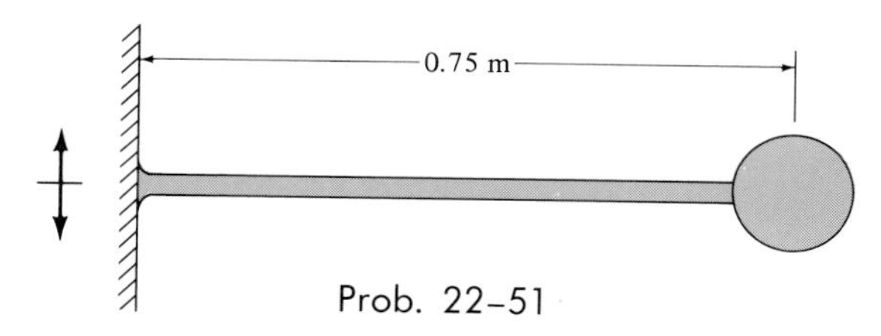

Prob. 22–51

***22–52.** The 300-kg trailer is pulled with a constant speed over the surface of a road which may be approximated by a cosine curve having an amplitude of 50 mm and wave length of 4 m. If the two springs s which support the trailer each have a stiffness of 800 N/m, determine the speed v which will cause the greatest vibration (resonance) of the trailer. Neglect the weight of the wheels.

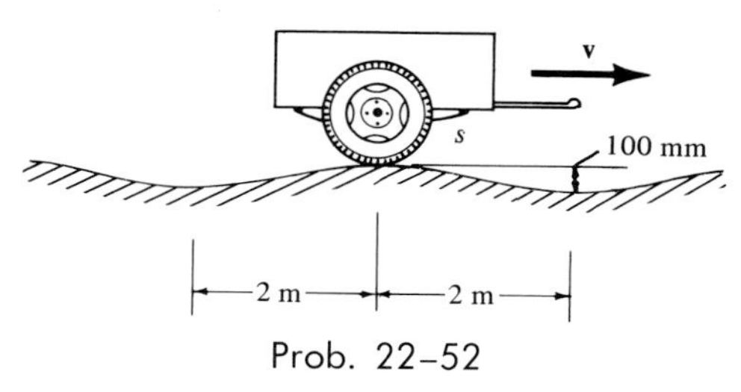

Prob. 22–52

22–53. Determine the amplitude of vibration of the trailer in Prob. 22–52 if the speed $v = 10$ km/h.

*22–5. Viscous Damped Free Vibration

The vibration analysis considered thus far has not included the effects of friction or damping in the system, and as a result, the solutions obtained are only in close agreement with the actual motion. Since all vibrations die out in time, the presence of damping forces should be included in the analysis.

In most cases damping is attributed to the resistance created by the substance such as water, oil, or air, in which the system vibrates. Provided the body moves slowly through this substance, the resistance to motion is directly proportional to the body's speed. The type of force developed under these conditions is called a *viscous damping force*. The magnitude of this force may be expressed by an equation of the form

$$F = c\dot{x} \tag{22-26}$$

where the constant c is called the *coefficient of viscous damping* and has units of $\text{N} \cdot \text{s/m}$ when the force F is measured in newtons and the velocity $\dot{x}$ is measured in m/s.

The vibrating motion of a body or system having viscous damping may be characterized by the block and spring shown in Fig. 22–16*a*. The effect of damping is provided by the *dashpot* connected to the block on the right side. Damping occurs when the piston P moves to the right or left within the enclosed cylinder. The cylinder contains a fluid, and the motion of the piston is retarded since the fluid must flow around or through a small hole in the piston. The dashpot is assumed to have a coefficient of viscous damping c.

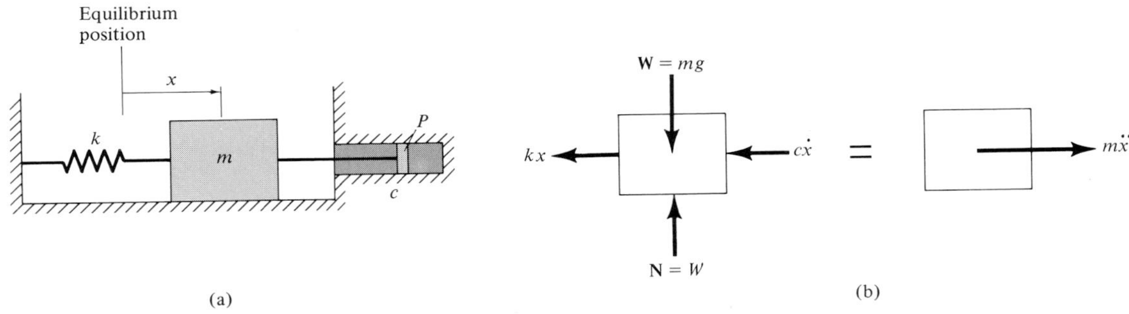

Fig. 22–16

If the block is displaced a distance x from its equilibrium position, the resulting free-body and kinetic diagrams are shown in Fig. 22–16*b*. Both the spring force kx and the damping force $c\dot{x}$ oppose the forward motion of the block, so that applying the equation of motion yields

$$\overset{+}{\rightarrow}\Sigma F_x = ma_x; \qquad -kx - c\dot{x} = m\ddot{x}$$

or

$$m\ddot{x} + c\dot{x} + kx = 0 \tag{22-27}$$

This linear, second-order, homogeneous, differential equation has solutions of the form

$$x = e^{\lambda t}$$

where e is the base of the natural logarithm and λ is a constant. The value of λ may be obtained by substituting this solution into Eq. 22–27, which yields

$$m\lambda^2 e^{\lambda t} + c\lambda e^{\lambda t} + k e^{\lambda t} = 0$$

or

$$e^{\lambda t}(m\lambda^2 + c\lambda + k) = 0$$

Since $e^{\lambda t}$ is always positive, the solution is possible provided

$$m\lambda^2 + c\lambda + k = 0$$

Hence, by the quadratic formula, the two values of λ are

$$\lambda_1 = -\frac{c}{2m} + \sqrt{\left(\frac{c}{2m}\right)^2 - \frac{k}{m}}, \quad \lambda_2 = -\frac{c}{2m} - \sqrt{\left(\frac{c}{2m}\right)^2 - \frac{k}{m}} \qquad (22\text{–}28)$$

The general solution of Eq. 22–27 is therefore a linear combination of exponentials which involves both of these roots. There are three possible combinations of λ_1 and λ_2 which must be considered for the general solution. Before discussing these combinations, however, it is first necessary to consider the definition of the *critical damping coefficient* c_c as the value of c which makes the radical in Eqs. 22–28 equal to zero, i.e.,

$$\left(\frac{c_c}{2m}\right)^2 - \frac{k}{m} = 0$$

or

$$c_c = 2m\sqrt{\frac{k}{m}} = 2mp \qquad (22\text{–}29)$$

Here the value of p is the circular frequency $p = \sqrt{k/m}$ (Eq. 22–2).

Overdamped System. When $c > c_c$, the roots λ_1 and λ_2 are both real. The general solution of Eq. 22–27 may then be written as

$$x = Ae^{\lambda_1 t} + Be^{\lambda_2 t} \qquad (22\text{–}30)$$

Motion corresponding to this solution is *nonvibrating*. The effect of damping is so strong that when the block is displaced and released, it simply creeps back to its original position without oscillating. The system is said to be *overdamped.*

Critically Damped System. If $c = c_c$, then $\lambda_1 = \lambda_2 = -c_c/2m = -p$. This situation is known as *critical damping,* since it represents a condition where c has the smallest value necessary to cause the system to be overdamped. Using the methods of differential equations, it may be shown that the solution to Eq. 22–27 for critical damping is

$$x = (A + Bt)e^{-pt} \qquad (22\text{–}31)$$

Underdamped System. Most often $c < c_c$, in which case the system is referred to as *underdamped.* In this case the roots λ_1 and λ_2 are complex numbers and it may be shown that the general solution of Eq. 22–27 can be written as

$$x = D[e^{-(c/2m)t} \sin (p_d t + \phi)] \tag{22–32}$$

where D and ϕ are constants generally determined from the initial conditions of the problem. The constant p_d is called the *damped natural frequency* of the system. It has a value of

$$p_d = \sqrt{\frac{k}{m} - \left(\frac{c}{2m}\right)^2} = p\sqrt{1 - \left(\frac{c}{c_c}\right)^2} \tag{22–33}$$

where the ratio c/c_c is called the *damping factor.*

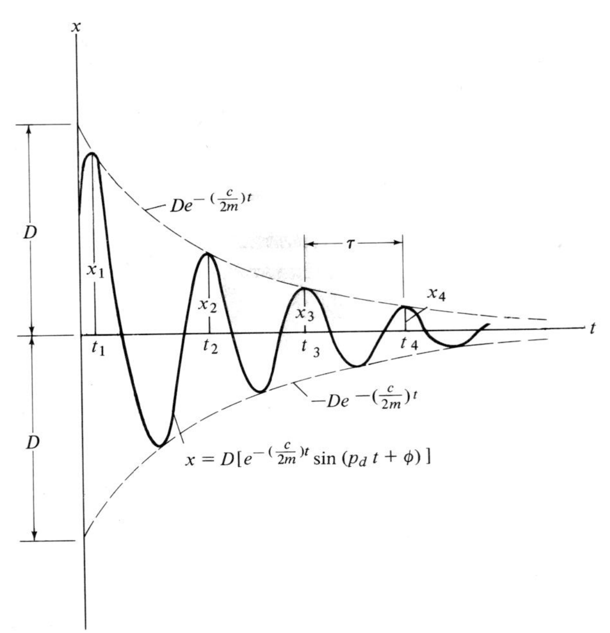

Fig. 22–17

The graph of Eq. 22–32 is shown in Fig. 22–17. The initial limit of motion, D, diminishes with each cycle of vibration, since motion is confined within the bounds of the exponential curve. Using the damped natural frequency p_d, the period of damped vibration may be written as

$$\tau_d = \frac{2\pi}{p_d} \tag{22–34}$$

Since $p_d < p$ (Eq. 22–33), the period of damped vibration τ_d will be greater than that of free vibration, $\tau = 2\pi/p$.

*22–6. Viscous Damped Forced Vibration

The most general case of single-degree-of-freedom vibrating motion occurs when the system includes the effects of forced motion and induced damping. The analysis of this particular type of vibration is of practical value when applied to systems having significant damping characteristics.

If a dashpot is attached to the block and spring shown in Fig. 22–11*a*, the differential equation which describes the motion becomes

$$m\ddot{x} + c\dot{x} + kx = F_O \sin \omega t \tag{22–35}$$

A similar equation may be written for a block and spring having a periodic support displacement, Fig. 22–14*a*, which includes the effects of damping. In that case, however, F_O is replaced by $k\delta_O$. Since Eq. 22–35 is nonhomogeneous, the general solution is the sum of a complementary solution, x_c, and a particular solution, x_p. The complementary solution is determined by setting the right side of Eq. 22–35 equal to zero and solving the homogeneous equation, which is equivalent to Eq. 22–27. The solution is therefore given by Eq. 22–30, Eq. 22–31, or Eq. 22–32, depending upon the values of λ_1 and λ_2. Because all systems contain friction, however, this solution will dampen out with time. Only the particular solution, which describes the *steady-state vibration* of the system, will remain. Since the applied forcing function is harmonic, the steady-state motion will also be harmonic. Consequently, the particular solution will be of the form

$$x_p = A' \sin \omega t + B' \cos \omega t \tag{22–36}$$

The constants A' and B' are determined by taking the necessary time derivatives and substituting them into Eq. 22–35, which after simplification yields

$$[-A'm\omega^2 - cB'\omega + kA'] \sin \omega t + [-B'm\omega^2 + cA'\omega + kB'] \cos \omega t = F_O \sin \omega t$$

Since this equation holds for all time, the constant coefficients of $\sin \omega t$ and $\cos \omega t$ may be equated, i.e.,

$$-A'm\omega^2 - cB'\omega + kA' = F_O$$
$$-B'm\omega^2 + cA'\omega + kB' = O$$

Solving for A' and B', realizing that $p^2 = k/m$, yields

$$A' = \frac{\left(\frac{F_O}{m}\right)(p^2 - \omega^2)}{(p^2 - \omega^2)^2 + \left(\frac{c\omega}{m}\right)^2}, \qquad B' = \frac{-F_O \frac{c\omega}{m^2}}{(p^2 - \omega^2)^2 + \left(\frac{c\omega}{m}\right)^2} \tag{22-37}$$

It is also possible to express Eq. 22–36 in a form similar to Eq. 22–9,

$$x_p = C' \sin(\omega t - \phi') \tag{22-38}$$

in which case the constants C' and ϕ' are

$$C' = \frac{F_O/k}{\sqrt{\left[1 - \left(\frac{\omega}{p}\right)^2\right]^2 + \left(2\frac{c}{c_c}\frac{\omega}{p}\right)^2}}$$

$$\phi' = \tan^{-1}\left(\frac{c\omega/k}{1 - \left(\frac{\omega}{p}\right)^2}\right) \tag{22-39}$$

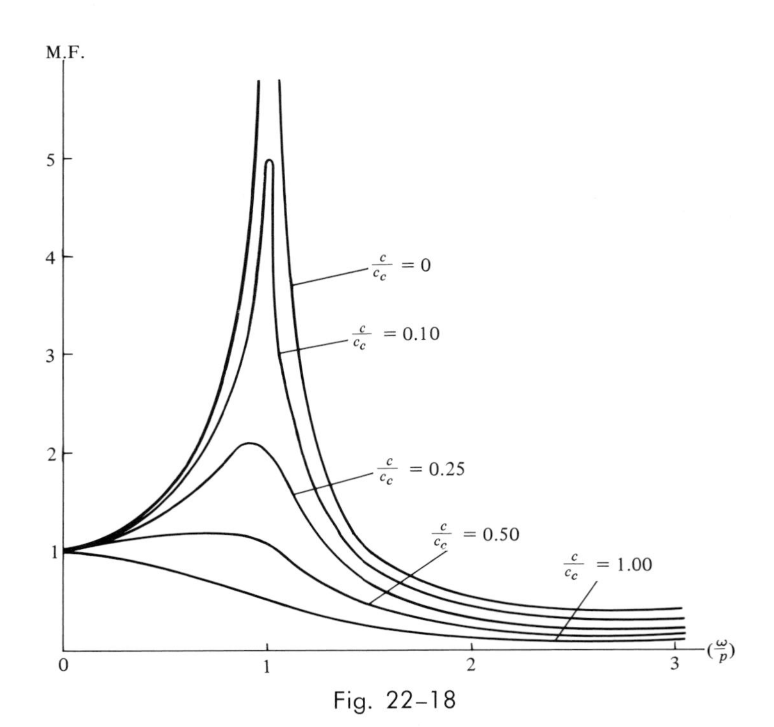

Fig. 22–18

The angle ϕ' represents the phase difference between the applied force and the resulting steady-state vibration of the damped system.

The *magnification factor* MF has been defined in Sec. 22–4 as the ratio of the amplitude of deflection caused by the forced vibration to the deflection caused by the force $\mathbf{F}_O$. From Eq. 22–38, the forced vibration has an amplitude of C'; thus,

$$\text{MF} = \frac{C'}{F_O/k} = \frac{1}{\sqrt{\left[1 - \left(\frac{\omega}{p}\right)^2\right]^2 + \left(2\frac{c}{c_c}\frac{\omega}{p}\right)^2}} \qquad (22\text{–}40)$$

The MF is plotted in Fig. 22–18 versus the frequency ratio ω/p for various values of the damping factor c/c_c. It can be seen from this graph that the magnification of the amplitude increases as the damping factor decreases. Resonance obviously occurs only when the damping is zero and the frequency ratio equals 1.

*22–7. Electrical Circuit Analogues

The characteristics of a vibrating mechanical system may be represented by an electric circuit. Consider the circuit shown in Fig. 22–19*a,* which consists of an inductor L, a resistor R, and a capacitor C. When a voltage $E(t)$ is applied, it causes a current of magnitude i to flow through the circuit. As the current flows past the inductor the voltage drop is $L(di/dt)$, when it flows across the resistor the drop is Ri, and when it arrives at the capacitor the drop is $(1/C)\int i\,dt$. Since current cannot flow past a capacitor, it is only possible to measure the charge q acting on the capacitor. The charge may, however, be related to the current by the equation $i = dq/dt$. Thus, the voltage drops which occur across the inductor, resistor, and capacitor may be written as $L\,d^2q/dt^2$, $R\,dq/dt$, and q/C, respectively. According to Kirchhoff's voltage law, the applied voltage balances the sum of the voltage drops around the circuit. Therefore,

$$L\frac{d^2q}{dt^2} + R\frac{dq}{dt} + \frac{1}{C}q = E(t) \qquad (22\text{–}41)$$

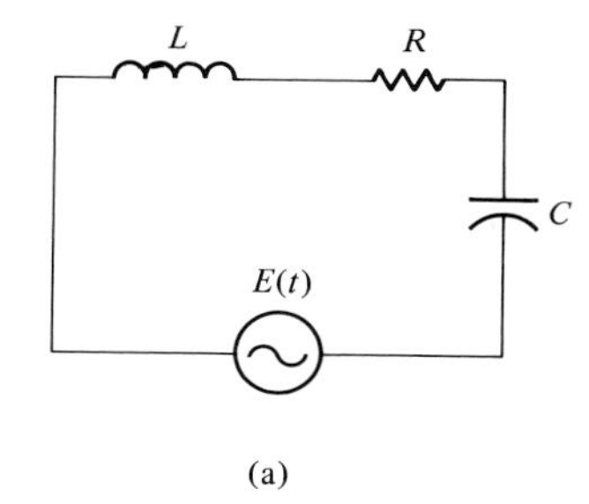

(a)

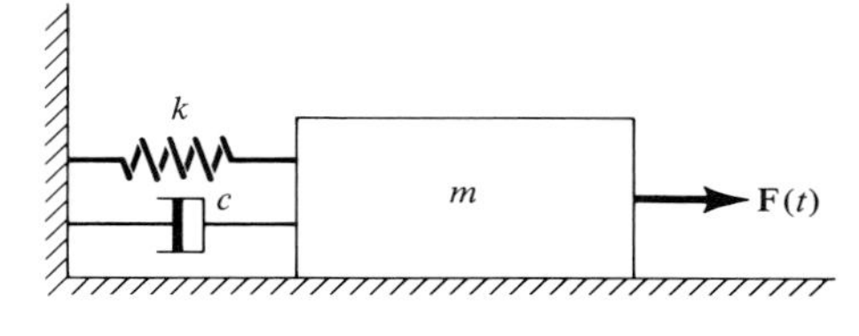

(b)

Fig. 22–19

Consider now the model of a single-degree-of-freedom mechanical system, Fig. 22–19*b,* which is subjected to both a general forcing function $F(t)$ and damping. The equation of motion for this system was established in the previous section and can be written as

$$m\frac{d^2x}{dt^2} + c\frac{dx}{dt} + kx = F(t) \qquad (22\text{–}42)$$

By comparison, it is seen that Eqs. 22–41 and 22–42 have the same form, hence mathematically the problem of analyzing an electric circuit is the same as that of analyzing a vibrating mechanical system. The analogues between the two equations are given in Table 22–1.

This analogy has important application to experimental work, for it is much easier to simulate the vibration of a complex mechanical system using an electric circuit, which can be constructed on an analogue computer, *rather* than to make an equivalent mechanical spring and dashpot model.

Table 22–1 Electrical–Mechanical Analogues

Electrical		*Mechanical*	
Electric charge	q	Displacement	x
Electric current	i	Velocity	dx/dt
Voltage	$E(t)$	Applied force	$F(t)$
Inductance	L	Mass	m
Resistance	R	Viscous damping coefficient	c
Reciprocal of capacitance	$1/C$	Spring stiffness	k

Example 22–8

The 30-kg electric motor shown in Fig. 22–20 is supported by *four* springs, each spring having a stiffness of 200 N/m. If the rotor R is unbalanced such that its effect is equivalent to a 4-kg mass located 60 mm from the axis of rotation, determine the amplitude of vibration when the rotor is turning at $\omega = 10$ rad/s. The damping factor is $c/c_c = 0.15$.

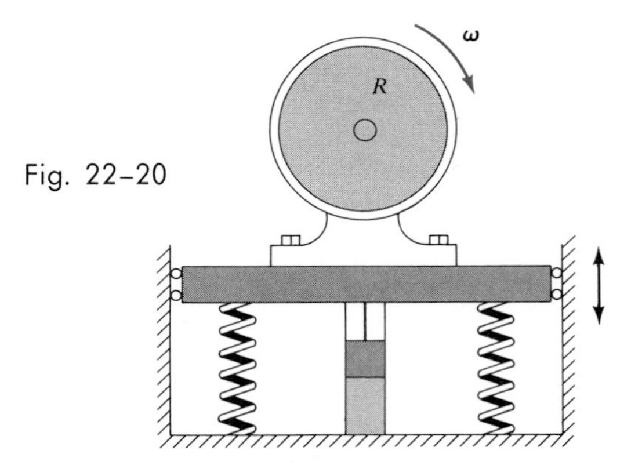

Fig. 22–20

Solution

The periodic force which causes the motor to vibrate is the centrifugal force due to the unbalanced effect of the rotor. This force has a constant magnitude of

$$F_O = ma_n = mr\omega^2 = 4\text{ kg}(0.06\text{ m})(10\text{ rad/s})^2 = 24\text{ N}$$

Oscillation in the vertical direction may be expressed in the periodic form $F = F_O \sin \omega t$, where $\omega = 10$ rad/s. Thus,

$$F = 24 \sin 10t$$

The stiffness of the entire system of four springs is $k = 4(200) = 800$ N/m. Therefore, the circular frequency of vibration is

$$p = \sqrt{\frac{k}{m}} = \sqrt{\frac{800}{30}} = 5.16\text{ rad/s}$$

Since the damping factor is known, the steady-state amplitude may be determined from the first of Eqs. 22–39, i.e.,

$$C' = \frac{F_O/k}{\sqrt{\left[1 - \left(\frac{\omega}{p}\right)^2\right]^2 + \left(2\frac{c}{c_c}\frac{\omega}{p}\right)^2}}$$

$$= \frac{24/800}{\sqrt{\left[1 - \left(\frac{10}{5.16}\right)^2\right]^2 + \left[2(0.15)\frac{10}{5.16}\right]^2}}$$

$$= 0.010\,7\text{ m} = 10.7\text{ mm} \qquad \textit{Ans.}$$

Problems

22–54. A block having a mass of 0.5 kg is suspended from a spring having a stiffness of 120 N/m. If a dashpot provides a damping force of 2.5 N when the speed of the block is 0.2 m/s, determine the period of free vibration.

22–55. A block, having a mass of $m = 2$ kg, is suspended from a spring having a stiffness of $k = 350$ N/m. The support to which the spring is attached is given a simple harmonic motion which may be expressed by $\delta = (10 \sin 2t)$ mm, where t is in seconds. If the damping factor is $c/c_c = 0.8$, determine the phase angle ϕ of forced vibration.

22–55a. Solve Prob. 22–55 if the block has a weight of $W = 5$ lb; $k = 72$ lb/ft; $\delta = 0.125 \sin 2t$, where δ is in ft and t is in seconds; and $c/c_c = 0.96$.

***22–56.** Determine the magnification factor of the block, spring and dashpot combination in Prob. 22–55.

22–57. The 10-kg block is subjected to the action of the harmonic force $F = (150 \cos 6t)$ N, where t is measured in seconds. Write the equation which describes the steady-state motion.

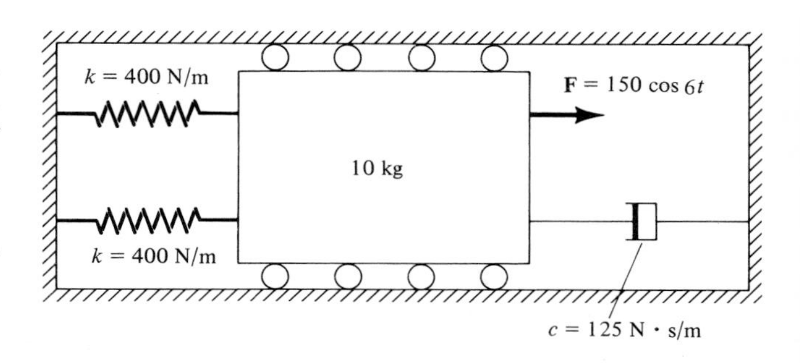

Prob. 22–57

22–58. The barrel of a cannon has a mass of 500 kg, and after firing it recoils a distance of 0.75 m. If it returns to its original position by means of a single recuperator having a damping coefficient of 2 kN · s/m, determine the required stiffness of each of the two springs fixed to the base and attached to the barrel so that the barrel recuperates without vibration.

22–59. A block, having a mass of 5 kg, is suspended from a spring that has a stiffness of $k = 900$ N/m. If it is given an upward velocity of 0.6 m/s from its equilibrium position at $t = 0$, determine the position of the block as a function of time. Assume that positive displacement of the block is downward, and that motion takes place in a medium which furnishes a damping force of $F = (50|v|)$ N, where v is measured in m/s.

***22–60.** The block, having a mass of $m = 2$ kg, is immersed in a liquid such that the damping force acting on the block has a magnitude of $F = (30|v|)$ N, where v is measured in m/s. If the block is pulled down $y_1 = 150$ mm and released from rest, describe the motion. The spring has a stiffness of $k = 400$ N/m. Assume that positive displacement is measured downward.

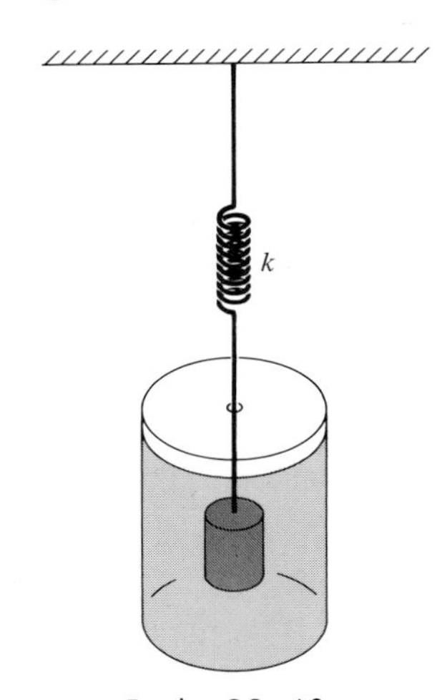

Prob. 22–60

***22–60a.** Solve Prob. 22–60 if the block has a weight of $W = 10$ lb; $F = (0.5\ |v|)$ lb, where v is measured in ft/s; $y_1 = 0.25$ ft; and $k = 48$ lb/ft.

22–61. The 4-kg circular disk is attached to three springs, each spring having a stiffness of $k = 200$ N/m. If the disk is immersed in a fluid and given a downward velocity of 0.3 m/s at the equilibrium position, determine the equation which describes the motion. Assume that positive displacement is measured downward, and that fluid resistance acting on the disk furnishes a damping force having a magnitude of $F = (60|v|)$ N, where v is measured in m/s.

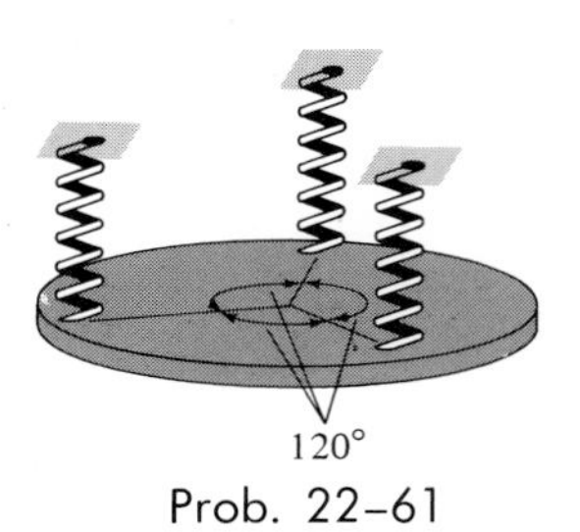

Prob. 22–61

22–62. The damping factor, c/c_c, may be determined experimentally by measuring the successive amplitudes of vibrating motion of a system. If two of these maximum displacements can be approximated by x_1 and x_2, as shown in Fig. 22–17, show that the ratio $\ln x_1/x_2 = 2\pi(c/c_c)/\sqrt{1 - (c/c_c)^2}$. The quantity $\ln x_1/x_2$ is called the *logarithmic decrement*.

22–63. The block shown in Fig. 22–16 has a mass of 20 kg and the spring has a stiffness of $k = 400$ N/m. When the block is placed out of equilibrium and released, two successive amplitudes are measured as $x_1 = 150$ mm and $x_2 = 87$ mm. Determine the coefficient of viscous damping, c. *Hint:* See Prob. 22–62.

***22–64.** The 50-kg electric motor is fastened to the mid-point of a pinned supported beam having negligible mass. It is found that the beam deflects 35 mm when the motor is not running. If the motor turns an eccentric flywheel which is equivalent to an unbalanced mass of 4 kg located 100 mm from the axis of rotation, determine the amplitude of steady-state vibration when the motor is turning at 20 rad/s. The damping factor is $c/c_c = 0.20$.

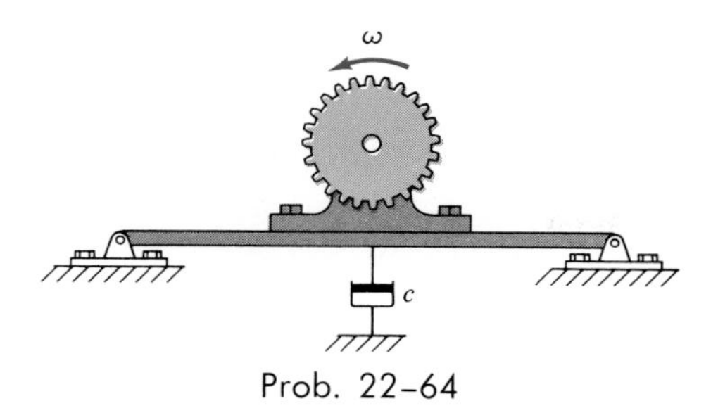

Prob. 22–64

22–65. Determine the mechanical analog for the electrical circuit. What differential equations describe the mechanical and electrical systems?

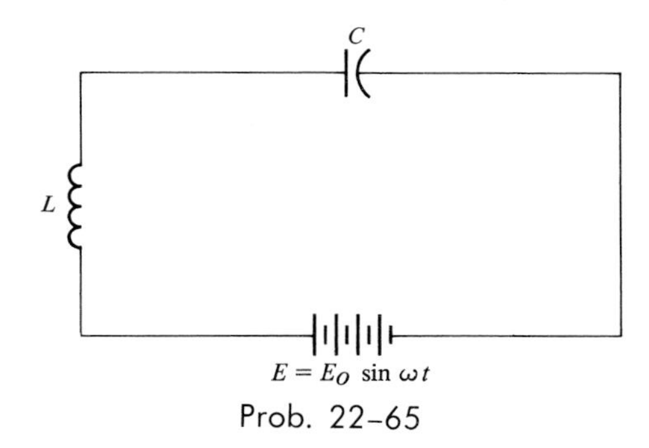

Prob. 22–65

22–66. Draw the electric circuit that is analogous to the mechanical system shown. Determine the differential equation which describes the motion of the current in the circuit.

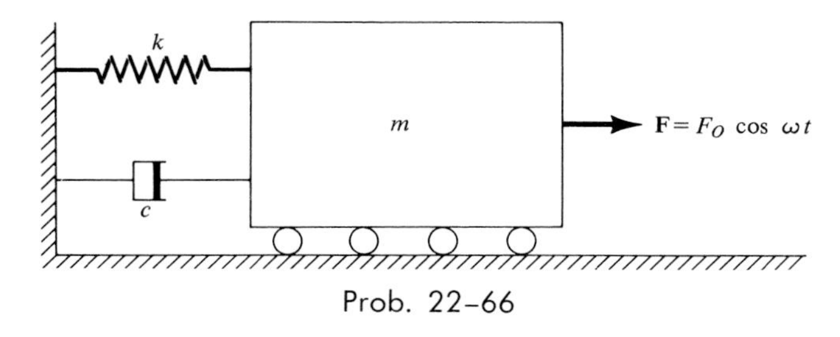

Prob. 22–66

22–67. Draw the electrical circuit that is equivalent to the mechanical system shown. What is the differential equation which describes the motion of the current in the circuit?

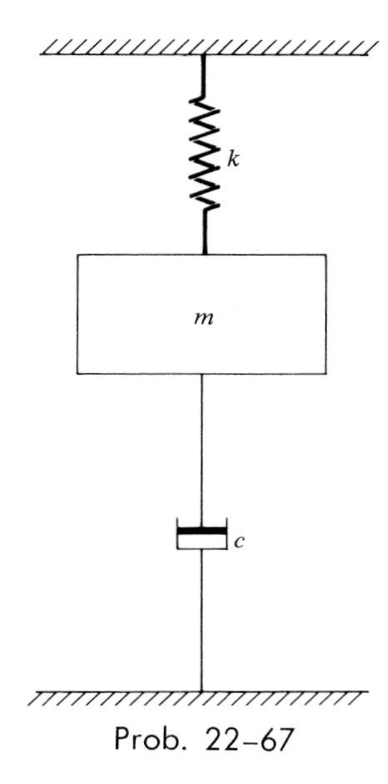

Prob. 22–67

* **22–68.** Determine the differential equation of motion for the damped vibratory system shown. What type of motion occurs?

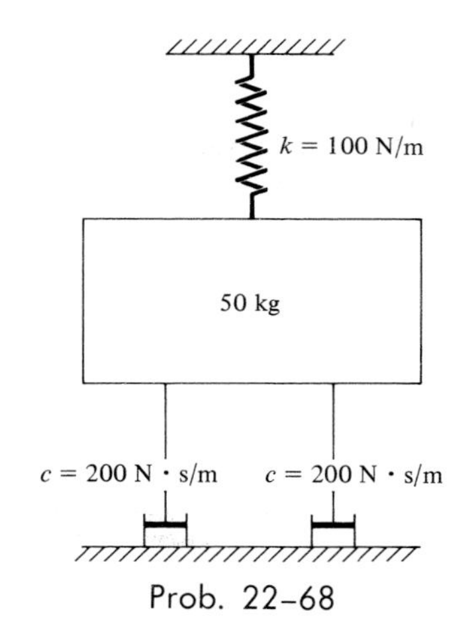

Prob. 22–68

22–69. Draw the electrical circuit that is equivalent to the mechanical system shown. What is the differential equation which describes the motion of the current in the circuit?

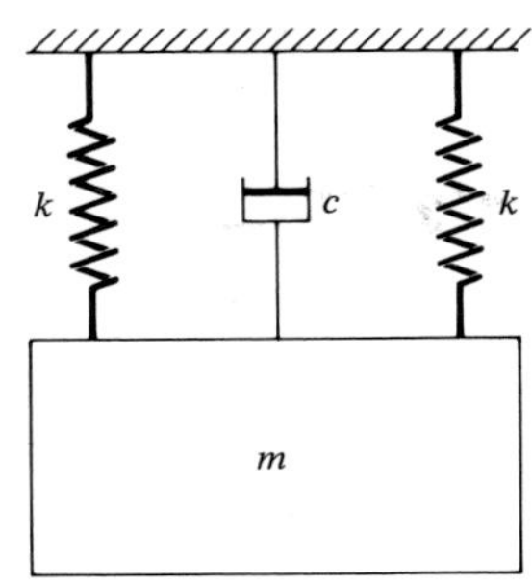

Prob. 22–69

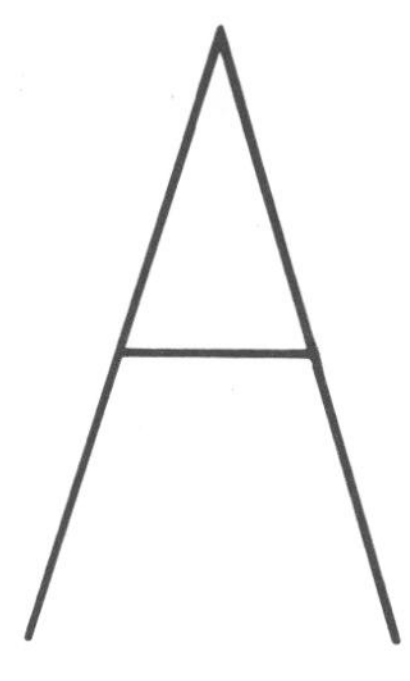

Units of Measurement

A–1. System of Units

Four fundamental quantities commonly used as a basis for measurement in mechanics are force, mass, length, and time. Since *all* four of these quantities are *related* by the equation of motion, $\mathbf{F} = m\mathbf{a}$, one *cannot* arbitrarily select units for measuring *all* these quantities; instead, the equality $\mathbf{F} = m\mathbf{a}$ is maintained if three of the four units, called *primary units,* are arbitrarily defined and the fourth unit is derived from the equation.

Absolute System of Units. A system of units which is defined on the basis of length, time, and mass is referred to as an *absolute system,* since measurements of these quantities can be made at *any location.* As shown in Table A–1, the International System of Units (SI) is absolute, since it specifies length in metres (m), time in seconds (s), and mass in kilograms (kg). The unit of force, called a newton (N), is *derived* from $\mathbf{F} = m\mathbf{a}$. Thus,

Table A–1

Type of System	*Name of System*	*Length*	*Time*	*Mass*	*Force*
Absolute	International System of Units (SI)	metre (m)	second (s)	kilogram (kg)	newton* (N) $\left(\frac{\text{kg} \cdot \text{m}}{\text{s}^2}\right)$
Gravitational	British Gravitational (FPS)	foot (ft)	second (s)	slug* $\left(\frac{\text{lb} \cdot \text{s}^2}{\text{ft}}\right)$	pound (lb)

*Derived unit.

1 newton is equal to a force required to give 1 kilogram of mass an acceleration of 1 m/s^2. ($N = kg \cdot m/s^2$)

Gravitational System. A system of units which is defined on the basis of length, time, and force is referred to as a *gravitational system*. This is because force is measured in a gravitational field, and hence its magnitude depends upon where the measurement is made. In the FPS system of units, Table A–1, length is in feet (ft), time is in seconds (s), and force is in pounds (lb). The unit of mass, called a slug, is *derived* from $\mathbf{F} = m\mathbf{a}$. Hence, 1 slug is equal to an amount of matter accelerated at 1 ft/s^2 when acted upon by a force of 1 lb ($slug = lb \cdot s^2/ft$).

A–2. The International System of Units

The International System of units, abbreviated SI for the French "Système International d'Unités," is a modern version of the metric system which received worldwide recognition at the 11th International Conference of Weights and Measures in 1960. This absolute system of units is used throughout the book, since it is intended, in time, to become a universal standard for measurement.

Base and Derived Units. Only the seven arbitrarily defined *base units* listed in Table A–2 exist in the SI system. All other units are *derived* from these. For example, as previously stated, the unit of force, the newton (N), is derived from the equation of motion [$1\ N = 1\ kg \cdot m/s^2$]. Another derived unit which is used in dynamics is the joule (J), defined as the work done by a force of 1 newton when the force is displaced 1 metre along its line of action ($1\ J = 1\ N \cdot m$).

Table A–2 Primary SI Units

Quantity	*Base Unit*	*SI Symbol*
Length	metre	m
Mass	kilogram	kg
Time	second	s
Electrical current	ampere	A
Amount of substance	mole	mol
Temperature	kelvin	K
Luminous intensity	candela	cd

Prefixes. Since units often measure quantities that can vary considerably in magnitude, prefixes, representing multiples and submultiples, often are used to modify units.* In the SI system, prefixes are increments of three digits such as those shown in Table A–3. Attaching a prefix to a unit in effect creates a new unit; thus if a multiple or submultiple unit is raised to a power, the power applies to this new unit, not just to the original unit without the multiple or submultiple. For example, $(2\ \text{kN})^2 = (2000\ \text{N})^2 = 4(10^6)\ \text{N}^2$. Also, $1\ \text{mm}^2 = 1\ (\text{mm})^2$, *not* $1\ \text{m}(\text{m}^2)$. Note that the SI system does not include the multiple deca (10) or the submultiple centi (0.01) which form part of the old metric system. Except for some volume or area measurements, the use of these prefixes is to be avoided in science and engineering.

Table A–3 Prefixes

Multiple	*Exponential Form*	*Prefix*	*SI Symbol*
1 000 000 000	10^9	giga	G
1 000 000	10^6	mega	M
1 000	10^3	kilo	k
Submultiple			
0.001	10^{-3}	milli	m
0.000 001	10^{-6}	micro	μ
0.000 000 001	10^{-9}	nano	n

Rules for Use. The following rules are given for the proper use of the various SI symbols:

1. A symbol is *never* written with a plural "s" since it may be confused with the unit for second (s).
2. Symbols are always written in lowercase letters with these two exceptions: symbols for the two largest prefixes shown in Table A–3, giga and mega, are capitalized as G and M, respectively; and symbols named after an individual are capitalized; e.g., N and J.
3. Quantities defined by several units which are multiples of one another are separated by a *dot* to avoid confusion with prefix notation, as illustrated by $\text{N} = \text{kg} \cdot \text{m/s}^2 = \text{kg} \cdot \text{m} \cdot \text{s}^{-2}$. Also, m · s (metre-second), whereas, ms (milli-second).
4. Physical constants or numbers having several digits on either side of the decimal point should be reported with a *space* between every three digits rather than with a comma, e.g., 73 569.213 427. In the case of four digits on either side of the decimal, the spacing is optional, e.g., 8537 or 8 537. Furthermore, always try to use decimals and avoid fractions; i.e., write 15.25, *not* $15\frac{1}{4}$.

*The kilogram is the only *base unit* which is defined with a prefix.

5. Compound prefixes should not be used, e.g., kms (kilo-micro-second) should be expressed as ms (milli-second). It is also best to keep numerical values between 0.1 and 1000; otherwise, a suitable prefix should be chosen. Thus, a force of 50 000 N is written as 50 kN; and so on.
6. With the exception of the base unit the kilogram, in general avoid the use of a prefix in the denominator of a composite unit. For example, do not write N/mm, but rather kN/m.
7. Although not expressed in multiples of 10, the minute, hour, and so on, are retained for practical purposes as multiples of the second. Furthermore, plane angular measurement is made using radians (rad). In this book, degrees will sometimes be used, where $360° = 2\pi$ rad. Fractions of a degree, however, should be expressed in decimal form rather than in minutes, as in 10.4°, not 10°24′.
8. When performing calculations, represent each of the numbers in terms of their *base or derived units* by converting prefixes to powers of 10. The final result can then be expressed using a *single* prefix. For example,

$$(50 \text{ kN})(60 \text{ nm}) = [50(10^3) \text{ N}][60(10^{-9}) \text{ m}]$$
$$= 3000(10^{-6}) \text{ N} \cdot \text{m} = 3 \text{ mN} \cdot \text{m}$$

When learning to use SI units, it is generally agreed that one should *not* think in terms of conversion factors between systems. Instead, it is better to think *only* in terms of SI units. A feeling for these units can only be gained through experience. Study, for example, the geometry and the loads acting on the structures and machines illustrated as problems throughout this book. As a memory aid, it might be helpful to recall that a standard flashlight battery or a small apple weighs about 1 newton. Your body is a suitable reference for small distances. For example, the millimetre scale in Fig. A–1 can be used to measure, say, the width of the tips of three or four fingers pressed together—about 50 mm; or the width of the small fingernail—about 10 mm. For most people, a stretched walking pace is about 1 m long.

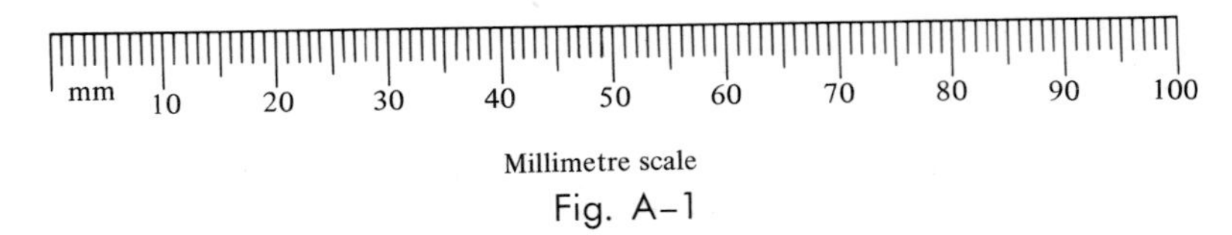

Millimetre scale

Fig. A–1

Conversion of Units. In some cases it may be necessary to convert from one system of units to another. In this regard, Table A–4 provides a list of direct conversion factors between FPS and SI units for the important physical quantities encountered in dynamics. For example, using Table A–4, 20 lb · ft = 20(1.3558) = 27.116 N · m.

Table A–4 Conversion Factors

Derived Quantity	*Unit of Measurement (FPS)*	*To Convert (FPS) to (SI), Multiply (FPS) Units by*	*Unit of Measurement (SI)*
Force	lb	4.4482	N
Mass	lb_{mass}	0.4536	kg
Moment of a force Couple	lb · ft	1.3558	N · m
Linear load intensity	lb/ft	14.594	N/m
Surface load intensity Pressure	lb/ft^2	47.8800	Pa
Length	ft	0.3048	m
Area	ft^2	0.092 90	m^2
Volume	ft^3	0.028 32	m^3
Density	lb_{mass}/ft^3	16.0187	kg/m^3
Moment of inertia	$lb \cdot ft \cdot s^2$	1.3558	$kg \cdot m^2$
Velocity	ft/s	0.3048	m/s
Acceleration	ft/s^2	0.3048	m/s^2
Work, Energy	ft · lb	1.3558	J
Power	ft · lb/s	1.3558	W

Problems

A–1. Represent each of the following with units having an appropriate prefix: (a) 6540 m, (b) 5200 kN, and (c) 0.0621 ms.

A–2. Evaluate each of the following and express with units having an appropriate prefix: (a) $(4 \text{ kN})^2$, (b) $(0.03 \text{ mm})^2$, and (c) $(200 \text{ s})^3$.

A–3. Is there a difference between m · kg and mkg? Explain.

***A–4.** Represent each of the following combinations of units in the correct SI form: (a) μMN, (b) kμm/Ms, and (c) MN/ks^2.

A–5. Represent each of the following combinations of units in the correct SI form: (a) g/ms, (b) N/nm, and (c) mm/(kg · μs).

A–6. Convert (a) 200 lb · ft to N · m, (b) 300 lb/ft^3 to Mg/m^3, (c) 6 ft/h to μm/s.

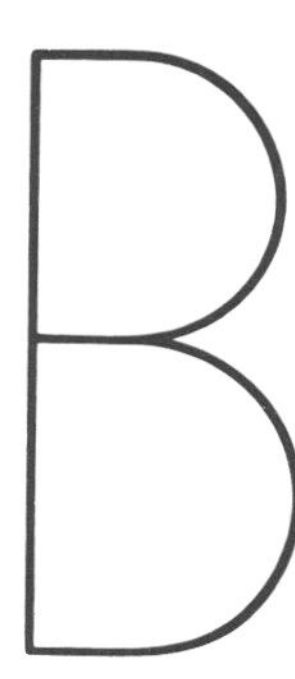

Vector Analysis

B–1. Fundamentals

The following discussion provides a brief review of the vector analysis used in statics. A more detailed treatment of these topics is given in *SI Engineering Mechanics: Statics.*

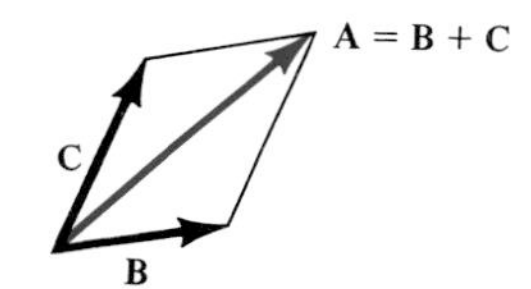

Fig. B–1

Vector. A vector, **A,** is a quantity which has a magnitude and direction, and adds according to the parallelogram law. As shown in Fig. B–1, **A** = **B** + **C,** where **A** is the *resultant vector* and **B** and **C** are *component vectors.*

Unit Vector. A unit vector, $\mathbf{u}_A$, has a magnitude of one "dimensionless" unit and acts in the same direction as **A.** It is determined by dividing **A** by its magnitude A, i.e.,

$$\mathbf{u}_A = \frac{\mathbf{A}}{A} \tag{B–1}$$

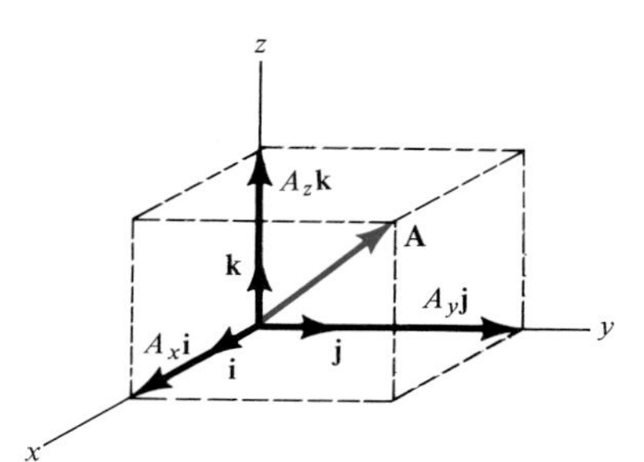

Fig. B–2

Cartesian Vector Notation. The directions of the positive x, y, z axes are defined by the Cartesian unit vectors, **i, j, k,** respectively.

As shown in Fig. B–2, vector **A** is formulated by the addition of its x, y, z components as

$$\mathbf{A} = A_x\mathbf{i} + A_y\mathbf{j} + A_z\mathbf{k} \tag{B–2}$$

The *magnitude* of **A** is determined from

$$A = \sqrt{A_x^2 + A_y^2 + A_z^2} \tag{B–3}$$

The *direction* of **A** is defined in terms of its *coordinate direction angles,* α, β, γ, measured from the *tail* of **A** to the *positive* x, y, z axes, Fig. B–3. These angles are determined from the *direction cosines* which represent

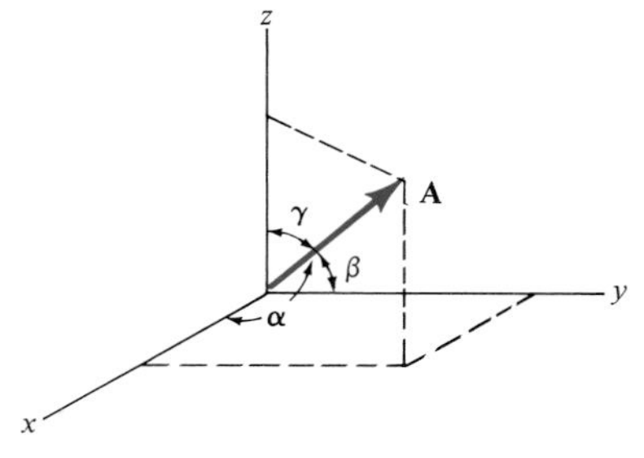

Fig. B–3

the **i, j, k** components of the unit vector $\mathbf{u}_A$; i.e., from Eqs. B–1 and B–2,

$$\mathbf{u}_A = \frac{A_x}{A}\mathbf{i} + \frac{A_y}{A}\mathbf{j} + \frac{A_z}{A}\mathbf{k} \tag{B–4}$$

so that the direction cosines are

$$\cos\alpha = \frac{A_x}{A}, \qquad \cos\beta = \frac{A_y}{A}, \qquad \cos\gamma = \frac{A_z}{A} \tag{B–5}$$

Hence $\mathbf{u}_A = \cos\alpha\,\mathbf{i} + \cos\beta\,\mathbf{j} + \cos\gamma\,\mathbf{k}$, and using Eq. B–3, it is seen that

$$\cos^2\alpha + \cos^2\beta + \cos^2\gamma = 1 \tag{B–6}$$

The Cross Product. The cross product of two vectors **A** and **B**, which yields the resultant vector **C**, is written as

$$\mathbf{C} = \mathbf{A} \times \mathbf{B} \tag{B–7}$$

and reads **C** equals **A** "cross" **B**. The *magnitude* of **C** is

$$C = AB\sin\theta \tag{B–8}$$

where θ is the angle made between the *tails* of **A** and **B** ($0° \leqslant \theta \leqslant 180°$). The *direction* of **C** is determined by the right-hand rule, whereby the fingers of the right hand are curled *from* **A** *to* **B** and the thumb points in the direction of **C**, Fig. B–4. This vector is perpendicular to the plane containing vectors **A** and **B**.

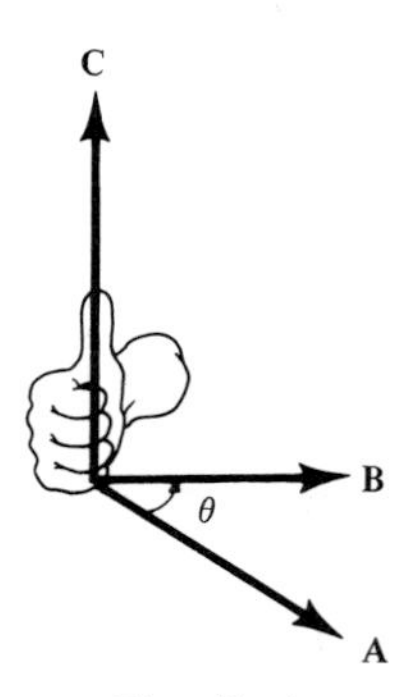

Fig. B–4

Provided **A** or **B** is not zero, then, if $\mathbf{A} \times \mathbf{B} = \mathbf{0}$, **A** is *parallel* to **B**, since $\theta = 0°$ or $180°$.

The vector cross product is *not* commutative, i.e.,

$$\mathbf{A} \times \mathbf{B} \neq \mathbf{B} \times \mathbf{A}$$

Rather,

$$\mathbf{A} \times \mathbf{B} = -\mathbf{B} \times \mathbf{A} \tag{B–9}$$

The distributive law is valid, i.e.,

$$\mathbf{A} \times (\mathbf{B} + \mathbf{D}) = \mathbf{A} \times \mathbf{B} + \mathbf{A} \times \mathbf{D} \tag{B–10}$$

And, the cross product may be multiplied by a scalar m in any manner, i.e.,

$$m(\mathbf{A} \times \mathbf{B}) = (m\mathbf{A}) \times \mathbf{B} = \mathbf{A} \times (m\mathbf{B}) = (\mathbf{A} \times \mathbf{B})m \tag{B–11}$$

If **A** and **B** are expressed in Cartesian component form, then the cross product, Eq. B–7, may be evaluated by expanding the determinant

$$\mathbf{C} = \mathbf{A} \times \mathbf{B} = \begin{vmatrix} \mathbf{i} & \mathbf{j} & \mathbf{k} \\ A_x & A_y & A_z \\ B_x & B_y & B_z \end{vmatrix} \tag{B–12}$$

which yields

$$\mathbf{C} = (A_yB_z - A_zB_y)\mathbf{i} - (A_xB_z - A_zB_x)\mathbf{j} + (A_xB_y - A_yB_x)\mathbf{k}$$

Recall that the cross product is used in statics to define the moment of a force **F** about point O, in which case

$$\mathbf{M}_O = \mathbf{r} \times \mathbf{F} \tag{B–13}$$

where **r** is a position vector directed from point O to *any point* on the line of action of force **F.**

The Dot Product. The dot product of two vectors **A** and **B,** which yields a scalar, is defined as

$$\mathbf{A} \cdot \mathbf{B} = AB \cos \theta \tag{B–14}$$

and reads **A** "dot" **B.** The angle θ is formed between the *tails* of **A** and **B** ($0° \leqslant \theta \leqslant 180°$). Provided **A** or **B** is not zero, then, if $\mathbf{A} \cdot \mathbf{B} = 0$, **A** is *perpendicular* to **B,** since $\theta = 90°$.

The dot product is commutative, i.e.,

$$\mathbf{A} \cdot \mathbf{B} = \mathbf{B} \cdot \mathbf{A} \tag{B–15}$$

The distributive law is valid, i.e.,

$$\mathbf{A} \cdot (\mathbf{B} + \mathbf{D}) = \mathbf{A} \cdot \mathbf{B} + \mathbf{A} \cdot \mathbf{D} \tag{B–16}$$

And, scalar multiplication can be performed in any manner, i.e.,

$$m(\mathbf{A} \cdot \mathbf{B}) = (m\mathbf{A}) \cdot \mathbf{B} = \mathbf{A} \cdot (m\mathbf{B}) = (\mathbf{A} \cdot \mathbf{B})m \tag{B–17}$$

If **A** and **B** are expressed in Cartesian component form, then the dot product, Eq. B–14, can be determined from

$$\mathbf{A} \cdot \mathbf{B} = A_x B_x + A_y B_y + A_z B_z \qquad \text{(B–18)}$$

The dot product may be used to determine the *angle θ formed between two vectors*. From Eq. B–14,

$$\theta = \cos^{-1}\left(\frac{\mathbf{A} \cdot \mathbf{B}}{AB}\right) \qquad \text{(B–19)}$$

It is also possible to find the *projected magnitude of a vector in a given direction* using the dot product. For example, the magnitude of the projection of vector **A** in the direction of **B**, Fig. B–5, is defined by $A \cos \theta$. From Eq. B–14, this magnitude is

$$A \cos \theta = \mathbf{A} \cdot \frac{\mathbf{B}}{B} = \mathbf{A} \cdot \mathbf{u}_B \qquad \text{(B–20)}$$

where $\mathbf{u}_B$ represents a unit vector acting in the direction of **B**, Fig. B–5.

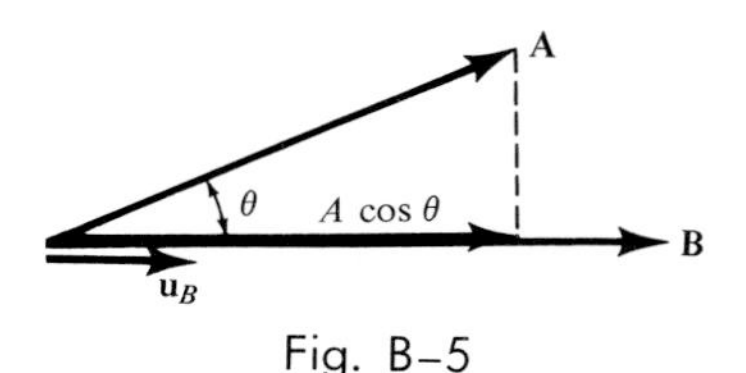

Fig. B–5

B–2. Differentiation and Integration of Vector Functions

The rules for differentiation and integration of the sums and products of scalar functions apply as well to vector functions. Consider, for example, the two vector functions $\mathbf{A}(s)$ and $\mathbf{B}(s)$. Provided these functions are smooth and continuous for all s,

$$\frac{d}{ds}(\mathbf{A} + \mathbf{B}) = \lim_{\Delta s \to 0} \frac{\Delta(\mathbf{A} + \mathbf{B})}{\Delta s}$$

$$= \lim_{\Delta s \to 0} \frac{\mathbf{A}(s + \Delta s) - \mathbf{A}(s)}{\Delta s} + \lim_{\Delta s \to 0} \frac{\mathbf{B}(s + \Delta s) - \mathbf{B}(s)}{\Delta s}$$

or

$$\frac{d}{ds}(\mathbf{A} + \mathbf{B}) = \frac{d\mathbf{A}}{ds} + \frac{d\mathbf{B}}{ds} \qquad \text{(B–21)}$$

In a similar manner, from Eq. B–21, it may be shown that

$$\int (\mathbf{A} + \mathbf{B})\, ds = \int \mathbf{A}\, ds + \int \mathbf{B}\, ds \qquad \text{(B–22)}$$

Using the fundamental definition of the derivative, the derivative of the product of a scalar function $f(s)$ and a vector function $\mathbf{A}(s)$ is written as

$$\frac{d}{ds}[f(s)\mathbf{A}(s)] = \frac{df}{ds}\mathbf{A} + f\frac{d\mathbf{A}}{ds} \tag{B–23}$$

If vector $\mathbf{A}$ is constant in both magnitude and direction for all values of s, then $\mathbf{A}(s) = \mathbf{C}$ and Eq. B–23 becomes

$$\frac{d}{ds}[f(s)\mathbf{C}] = \frac{df}{ds}\mathbf{C} = g(s)\mathbf{C} \tag{B–24}$$

When Eq. B–24 is integrated, the constant vector $\mathbf{C}$ can be factored out of the integral, which yields

$$\int \mathbf{C}g(s)\,ds = \mathbf{C}\int g(s)\,ds \tag{B–25}$$

To obtain the derivative for the cross product of two vectors $\mathbf{A}$ and $\mathbf{B}$, write

$$\frac{d}{ds}(\mathbf{A} \times \mathbf{B}) = \lim_{\Delta s \to 0} \frac{\mathbf{A}(s + \Delta s) \times \mathbf{B}(s + \Delta s) - \mathbf{A}(s) \times \mathbf{B}(s)}{\Delta s}$$

Adding and subtracting the cross-product term $\mathbf{A}(s) \times \mathbf{B}(s + \Delta s)$ with the terms in the numerator of the above equation gives

$$\frac{d}{ds}(\mathbf{A} \times \mathbf{B}) = \lim_{\Delta s \to 0} \frac{\mathbf{A}(s + \Delta s) \times \mathbf{B}(s + \Delta s) - \mathbf{A}(s) \times \mathbf{B}(s + \Delta s)}{\Delta s} + \lim_{\Delta s \to 0} \frac{\mathbf{A}(s) \times \mathbf{B}(s + \Delta s) - \mathbf{A}(s) \times \mathbf{B}(s)}{\Delta s}$$

Using the distributive property of the cross product, we have

$$\frac{d}{ds}(\mathbf{A} \times \mathbf{B}) = \lim_{\Delta s \to 0} \frac{[\mathbf{A}(s + \Delta s) - \mathbf{A}(s)] \times \mathbf{B}(s + \Delta s)}{\Delta s} + \lim_{\Delta s \to 0} \frac{\mathbf{A}(s) \times [\mathbf{B}(s + \Delta s) - \mathbf{B}(s)]}{\Delta s}$$

In the limit,

$$\frac{d}{ds}(\mathbf{A} \times \mathbf{B}) = \left(\frac{d\mathbf{A}}{ds} \times \mathbf{B}\right) + \left(\mathbf{A} \times \frac{d\mathbf{B}}{ds}\right) \tag{B–26}$$

The order of cross-product multiplication is important in this formula, since the cross-product operation is noncommutative, i.e., $\mathbf{A} \times \mathbf{B} = -\mathbf{B} \times \mathbf{A}$, Eq. B–9.

In a similar manner, it may be shown that for the dot product

$$\frac{d}{ds}(\mathbf{A} \cdot \mathbf{B}) = \frac{d\mathbf{A}}{ds} \cdot \mathbf{B} + \mathbf{A} \cdot \frac{d\mathbf{B}}{ds} \tag{B–27}$$

For a constant vector **C**, the dot- and cross-product operations involving vector integration may be written as

$$\int \mathbf{C} \cdot \mathbf{A}(s)\, ds = \mathbf{C} \cdot \int \mathbf{A}(s)\, ds \tag{B–28}$$

and

$$\int \mathbf{C} \times \mathbf{A}(s)\, ds = \mathbf{C} \times \int \mathbf{A}(s)\, ds \tag{B–29}$$

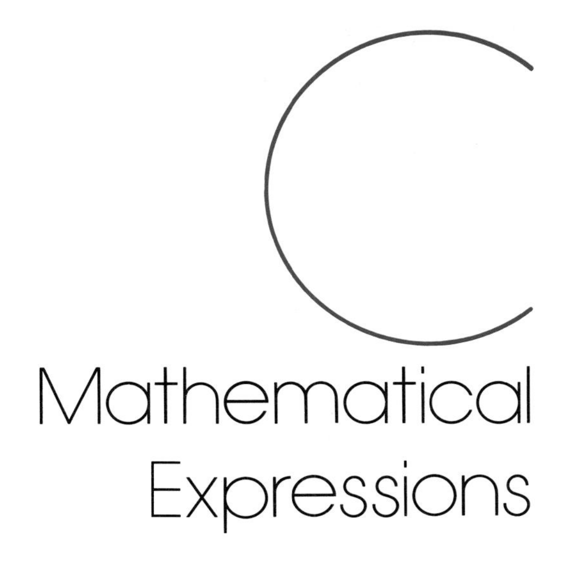

Mathematical Expressions

Quadratic Formula:

If $ax^2 + bx + c = 0$, then $x = \dfrac{-b \pm \sqrt{b^2 - 4ac}}{2a}$

Hyperbolic Functions:

$\sinh x = \dfrac{e^x - e^{-x}}{2}$, $\cosh x = \dfrac{e^x + e^{-x}}{2}$, $\tanh x = \dfrac{\cosh x}{\sinh x}$

Trigonometric Identities:

$\sin^2\theta + \cos^2\theta = 1$
$\sin(\theta \pm \phi) = \sin\theta\cos\phi \pm \cos\theta\sin\phi,$
$\sin 2\theta = 2\sin\theta\cos\theta$
$\cos(\theta \pm \phi) = \cos\theta\cos\phi \mp \sin\theta\sin\phi$
$\cos 2\theta = \cos^2\theta - \sin^2\theta,$

$\cos\theta = \pm\sqrt{\dfrac{1 + \cos 2\theta}{2}}$, $\sin\theta = \pm\sqrt{\dfrac{1 - \cos 2\theta}{2}}$

$\tan\theta = \dfrac{\sin\theta}{\cos\theta},$

$1 + \tan^2\theta = \sec^2\theta$

Power-series Expansions:

$$\sin x = x - \frac{x^3}{3!} + \frac{x^5}{5!} - \frac{x^7}{7!} + \cdots$$

$$\cos x = 1 - \frac{x^2}{2!} + \frac{x^4}{4!} - \frac{x^6}{6!} + \cdots$$

$$\sinh x = x + \frac{x^3}{3!} + \frac{x^5}{5!} + \cdots$$

$$\cosh x = 1 + \frac{x^2}{2!} + \frac{x^4}{4!} + \cdots$$

Derivatives:

$$\frac{d}{dx}(u^n) = nu^{n-1}\frac{du}{dx}$$

$$\frac{d}{dx}(uv) = u\frac{dv}{dx} + v\frac{du}{dx}$$

$$\frac{d}{dx}\left(\frac{u}{v}\right) = \frac{v\frac{du}{dx} - u\frac{dv}{dx}}{v^2}$$

$$\frac{d}{dx}(\sin u) = \cos u\frac{du}{dx}$$

$$\frac{d}{dx}(\cos u) = -\sin u\frac{du}{dx}$$

$$\frac{d}{dx}(\tan u) = \sec^2 u\frac{du}{dx}$$

$$\frac{d}{dx}(\cot u) = -\csc^2 u\frac{du}{dx}$$

$$\frac{d}{dx}(\sec u) = \tan u \sec u\frac{du}{dx}$$

$$\frac{d}{dx}(\csc u) = -\csc u \cot u\frac{du}{dx}$$

$$\frac{d}{dx}(\sinh u) = \cosh u\frac{du}{dx}$$

$$\frac{d}{dx}(\cosh u) = \sinh u\frac{du}{dx}$$

Integrals:

$$\int x^n\,dx = \frac{x^{n+1}}{n+1},\ n \neq -1$$

$$\int \frac{dx}{a+bx} = \frac{1}{b}\ln(a+bx)$$

$$\int \frac{x\,dx}{a+bx^2} = \frac{1}{2b}\ln\left(x^2+\frac{a}{b}\right)$$

$$\int \frac{dx}{(a+bx)^2} = -\frac{1}{b(a+bx)}$$

$$\int \frac{dx}{a^2-x^2} = \frac{1}{2a}\ln\frac{a+x}{a-x},\ a^2 > x^2$$

$$\int \sqrt{a+bx}\,dx = \frac{2}{3b}\sqrt{(a+bx)^3}$$

$$\int x\sqrt{a+bx}\,dx = \frac{-2(2a-3bx)\sqrt{(a+bx)^3}}{15b^2}$$

$$\int \sqrt{a^2-x^2}\,dx = \frac{1}{2}\left[x\sqrt{a^2-x^2}+a^2\sin^{-1}\frac{x}{a}\right],\ a>0$$

$$\int \sqrt{x^2\pm a^2}\,dx = \frac{1}{2}\left[x\sqrt{x^2\pm a^2}\pm a^2\ln\left(x+\sqrt{x^2\pm a^2}\right)\right]$$

$$\int x\sqrt{a^2-x^2}\,dx = -\tfrac{1}{3}\sqrt{(a^2-x^2)^3}$$

$$\int x^2\sqrt{a^2-x^2}\,dx = -\frac{x}{4}\sqrt{(a^2-x^2)^3}+\frac{a^2}{8}\left(x\sqrt{a^2-x^2}+a^2\sin^{-1}\frac{x}{a}\right),\ a>0$$

$$\int x^2\sqrt{x^2\pm a^2} = \frac{x}{4}\sqrt{(x^2\pm a^2)^3}\mp\frac{a^2}{8}x\sqrt{x^2\pm a^2}-\frac{a^4}{8}\ln\left(x+\sqrt{x^2\pm a^2}\right)$$

$$\int \sin x\,dx = -\cos x$$

$$\int \cos x\,dx = \sin x$$

$$\int e^{ax}\,dx = \frac{1}{a}e^{ax}$$

$$\int \sinh x\,dx = \cosh x$$

$$\int \cosh x\,dx = \sinh x$$

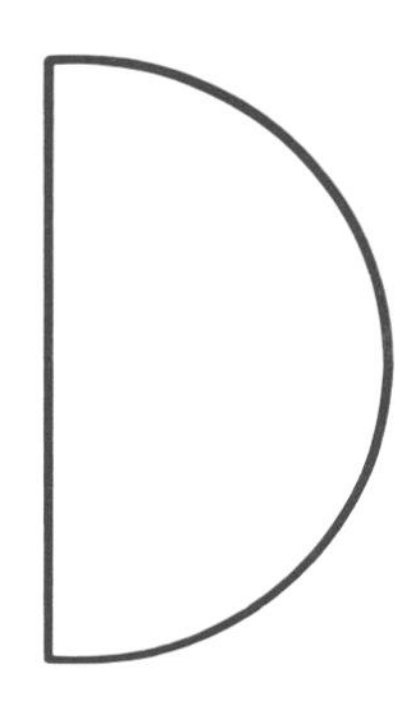

Mass Moments and Products of Inertia of Homogeneous Solids

Sphere

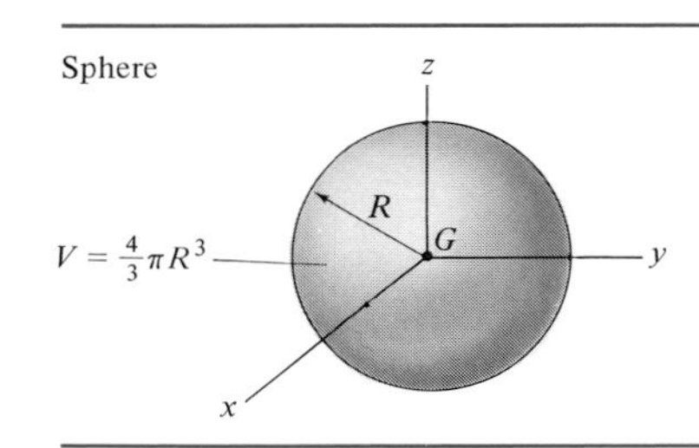

$$I_{xx} = I_{yy} = I_{zz} = \frac{2}{5}mR^2$$

$$I_{xy} = I_{xz} = I_{yz} = 0$$

Hemisphere

$V = \frac{2}{3}\pi R^3$

$\frac{3R}{8}$

$$I_{xx} = I_{yy} = 0.259mR^2$$

$$I_{zz} = \frac{2}{5}mR^2$$

$$I_{xy} = I_{xz} = I_{yz} = 0$$

Cylinder

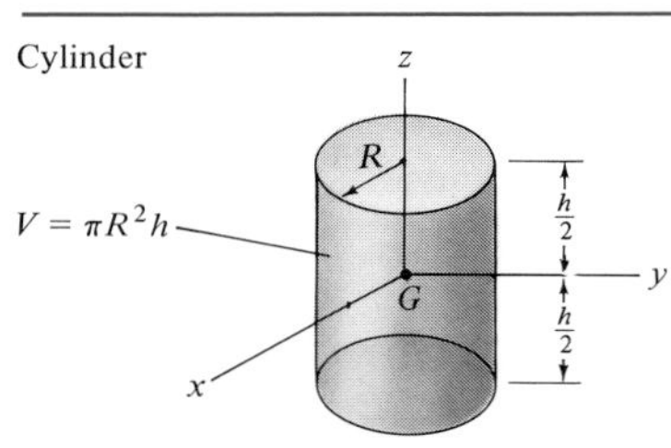

$$I_{xx} = I_{yy} = \frac{1}{12}m(3R^2 + h^2)$$

$$I_{zz} = \frac{1}{2}mR^2$$

$$I_{xy} = I_{xz} = I_{yz} = 0$$

Semicylinder

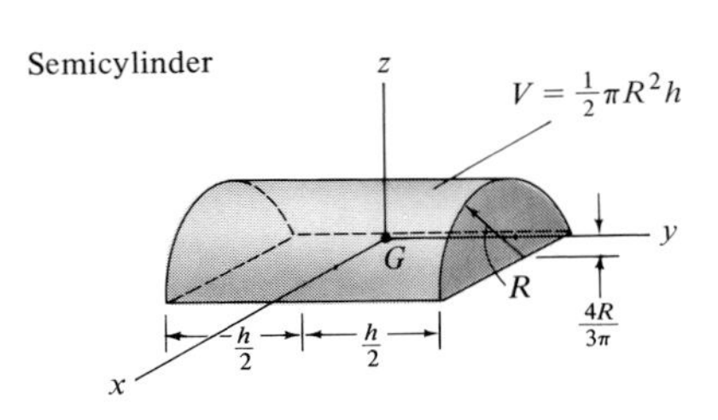

$$I_{xx} = 0.0699mR^2 + \frac{m}{12}h^2$$
$$I_{yy} = 0.320mR^2$$
$$I_{zz} = \frac{1}{12}m(3R^2 + h^2)$$
$$I_{xy} = I_{xz} = I_{yz} = 0$$

Slender rod

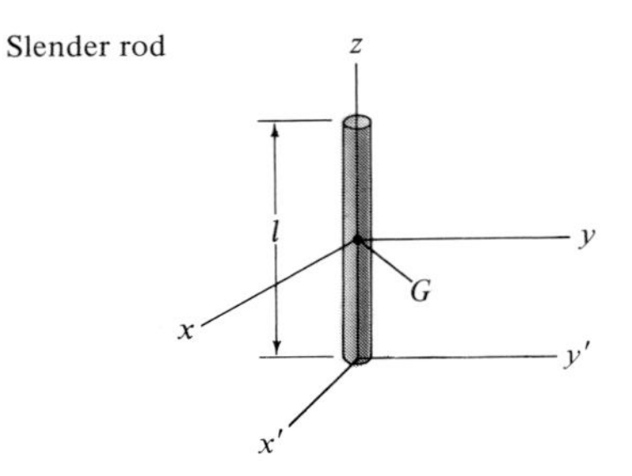

$$I_{xx} = I_{yy} = \frac{1}{12}ml^2$$
$$I_{x'x'} = I_{y'y'} = \frac{1}{3}ml^2$$
$$I_{zz} = 0$$
$$I_{xy} = I_{xz} = I_{yz} = 0$$

Thin circular disk

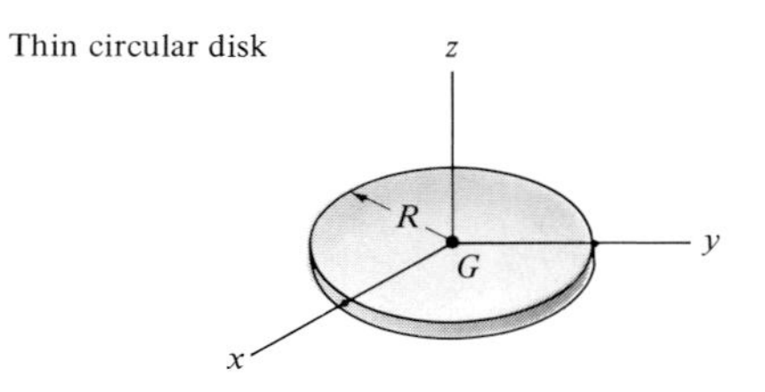

$$I_{xx} = I_{yy} = \frac{1}{4}mR^2$$
$$I_{zz} = \frac{1}{2}mR^2$$
$$I_{xy} = I_{xz} = I_{yz} = 0$$

Thin ring

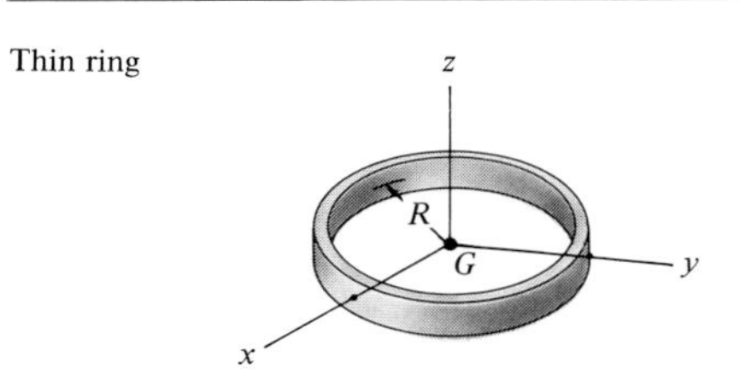

$$I_{xx} = I_{yy} = \frac{1}{2}mR^2$$
$$I_{zz} = mR^2$$
$$I_{xy} = I_{xz} = I_{yz} = 0$$

Cone

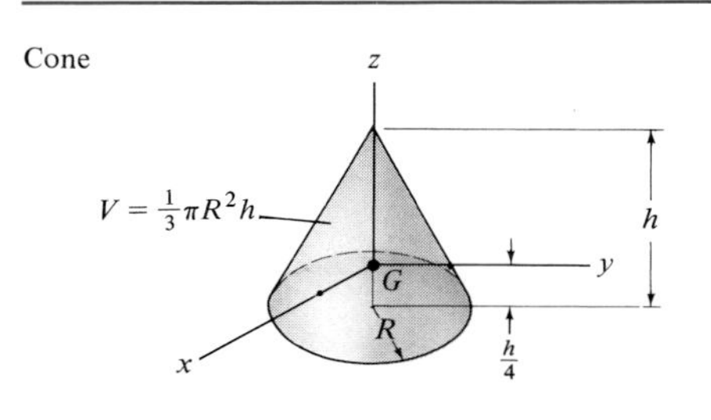

$$I_{xx} = I_{yy} = \frac{3}{80}m(4R^2 + h^2)$$
$$I_{zz} = \frac{3}{10}mR^2$$
$$I_{xy} = I_{xz} = I_{yz} = 0$$

Rectangular block

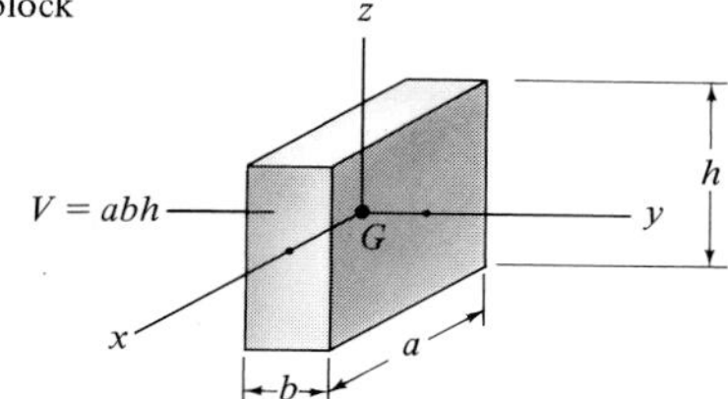

$$I_{xx} = \frac{m}{12}(b^2 + h^2)$$

$$I_{yy} = \frac{m}{12}(a^2 + h^2)$$

$$I_{zz} = \frac{m}{12}(a^2 + b^2)$$

$$I_{xy} = I_{xz} = I_{yz} = 0$$

Thin rectangular plate

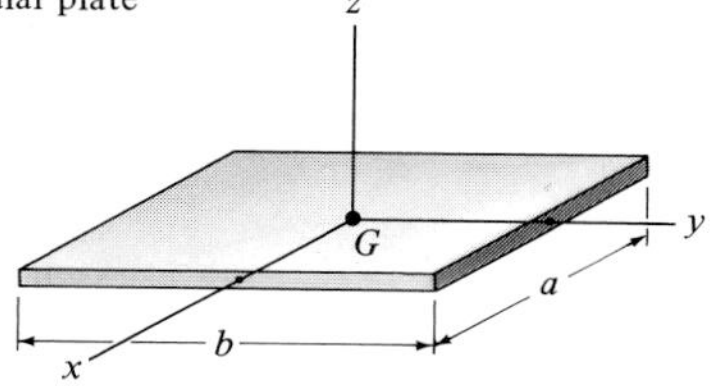

$$I_{xx} = \frac{m}{12}b^2$$

$$I_{yy} = \frac{m}{12}a^2$$

$$I_{zz} = \frac{m}{12}(a^2 + b^2)$$

$$I_{xy} = I_{xz} = I_{yz} = 0$$

Answers*

12-1. 2.17 m/s^2.

12-2. 6.94 s, 96.5 m.

12-3. 13 m/s, 76 m, 8.33 s.

12-5. 382.5 m, 189 m/s, 65 m/s^2.

12-5a. 605 ft, 361 ft/s, 146 ft/s^2.

12-6. 13 m/s $\rightarrow$, 76 m.

12-7. 240 mm, 60 mm/s, 20 mm/s^2.

12-9. (a) 26.7 s, 3792.6 mm, (b) 13.3 s.

12-10. $v_{avg} = 0.333$ m/s, $v_{sp} = 1$ m/s.

12-10a. $v_{avg} = 0.167$ ft/s, $v_{sp} = 2.167$ ft/s.

12-11. (a) 256 mm, (b) 748 mm/s, 300 mm/s^2.

12-13. -1.01 m/s^2.

12-14. 14.7 m/s, 2.57 m/s^2.

12-15. 35.7 m.

12-15a. 391.8 ft.

12-17. 1.29 m.

12-18. -59.7 mm/s^5, -1.07 m/s.

12-19. 2.63 m/s.

12-21. 11.17 km/s.

12-22. $v = R\sqrt{\dfrac{2g_o(y_o - y)}{(R + y)(R + y_o)}}$, 3.02 km/s.

12-23. (a) 45.5 m/s, (b) 100 m/s.

*Note: Answers to every fourth problem are omitted.

12-25. $t = 0$, $v = 0$, $a = 0.8$ m/s^2,
$t = 30^-$ s, $v = 24$ m/s, $a = 0.8$ m/s^2,
$t = 40$ s, $v = 24$ m/s, $a = 0$.

12-25a. $t = 0$, $v = 0$, $a = 4$ ft/s^2,
$t = 30^-$ s, $v = 120$ ft/s, $a = 4$ ft/s^2,
$t = 40$ s, $v = 120$ ft/s, $a = 0$.

12-26. (15 s, 300 m/s), (20 s, 475 m/s),
$s = 4187.5$ m.

12-27. (10 s, 50 m/s), (12.5 s, 75 m/s).

12-29. (2 s, 8 m/s), (10 s, 24 m/s), (12 s, 18 m/s)
(2 s, 8 m), (10 s, 136 m), (12 s, 178 m).

12-30. 20.0 m/s, 31.6 m/s, 40.0 m/s.

12-30a. 24.5 ft/s, 42.4 ft/s, 51.0 ft/s.

12-31. 8916.7 m, 475 m/s.

12-33. 87.5 m.

12-34. 71.7 m/s^2, 112.6 m/s^2.

12-35. 90 m/s, 34.0 s.

12-35a. 150 ft/s, 34.5 s.

12-37. 5.31 m/s^2, $\alpha = 53.0°$, $\beta = 37.1°$, $\gamma = 90°$.

12-38. 80.2 m/s^2, $x = 42.7$ m, $y = 16.0$ m, $z = 14.0$ m.

12-39. 87.4 m/s.

12-41. 15.2 m/s, 66.8°, 10 m.

12-42. 2.87 m, 19.9 m.

12-45. $v_A = 12.4$ m/s, $v_B = 11.1$ m/s.

12-45a. 35.86 ft/s, 32.1 ft/s.

12–46. 2.16 s, 16.3 m/s.

12–47. 0.19 m, 1.19 m.

12–49. $\theta_1 = 5.8°$, $\theta_2 = 79.4°$.

12–50. $\theta_1 = 66.3°$, $\theta_2 = 23.7°$, 41.9 s.

12–50a. $\theta_1 = 75.0°$, $\theta_2 = 15.0°$, 66.0 s.

12–51. 3.84 m.

12–53. 2.32 m/s^2.

12–54. 4.16 m/s, 0.865 m/s^2.

12–55. $\mathbf{v} = \{5\mathbf{u}_r + 4t\mathbf{u}_z\}$ m/s, $v = 7.81$ m/s;
$\mathbf{a} = \{4\mathbf{u}_z\}$ m/s^2, $a = 4$ m/s^2.

12–55a. $\mathbf{v} = \{10t\mathbf{u}_r - 3\mathbf{u}_z\}$ ft/s, $v = 20.2$ ft/s;
$\mathbf{a} = \{10\,\mathbf{u}_r\}$ ft/s^2, $a = 10$ ft/s^2.

12–57. $\dot{\mathbf{a}} = (\dddot{r} - 3\dot{r}\dot{\theta}^2 - 3r\dot{\theta}\ddot{\theta})\mathbf{u}_r +$
$(3\dot{r}\ddot{\theta} + r\dddot{\theta} + 3\ddot{r}\dot{\theta} - r\dot{\theta}^3)\mathbf{u}_\theta + \dddot{z}\mathbf{u}_z$.

12–58. $v_r = -500/\sqrt{1 + \theta^2}$, $v_\theta = 500\theta/\sqrt{1 + \theta^2}$,
$v_r = -353.6$ mm/s, $v_\theta = 353.6$ mm/s.

12–59. 1.125 rad/s.

12–61. 1.32 m/s.

12–62. 8.66 m/s^2.

12–63. $v = 800 \sec^2\theta$ mm/s, $a = 3200 \sec^2\theta \tan\theta$ mm/s^2.

12–65. $v = 100\sqrt{1 + \sin^2\theta}$ mm/s,
$a = 200\sqrt{1 + \cos^2\theta}$ mm/s^2.

12–65a. $v = 0.75\sqrt{1 + \sin^2\theta}$ ft/s, $a = 2.25\sqrt{1 + \cos^2\theta}$ ft/s^2.

12–66. 5.66 m/s.

12–67. 181.0 m/s^2.

12–69. 2.33 m/s^2.

12–70. 19.4 m/s.

12–70a. 69.3 ft/s.

12–71. 1.16 m/s^2.

12–73. 5.99 Mm.

12–74. 10.24 m/s^2.

12–75. 17.9 m/s^2.

12–75a. 0.907 ft/s^2.

12–77. 50 m/s^2.

12–78. 9.36 m.

12–79. 1.96 s.

12–81. $\alpha = 139.1°$, $\beta = 49.6°$, $\gamma = 95.1°$ or $\alpha = 40.9°$,
$\beta = 130.4°$, $\gamma = 84.9°$.

12–82. $\pm\mathbf{u}_b$ where $\mathbf{u}_b = -0.349\mathbf{i} - 0.0127\mathbf{j} - 0.937\mathbf{k}$.

12–83. 6 m/s.

12–85. 8 m/s.

12–85a. 20 ft/s.

12–86. 0.5 m/s.

12–87. 2 m/s.

12–89. $v_A = 11$ m/s $\rightarrow$, $v_B = 3$ m/s. $\rightarrow$

12–90. 874.8 km/h.

12–90a. 736.5 mi/h.

12–91. 1.80 m/s, 1.40 m.

12–93. 49.1 km/h.

12–94. 212.1 m/s.

12–95. 1.44 m/s, $a_B = a_{B/C} = 0.671$ m/s^2, $\varphi = 56.3°$.

12–95a. 1.60 ft/s, $a_B = a_{B/C} = 0.182$ ft/s^2, $\varphi = 51.4°$.

12–97. 102.0 km/h, 838.8 km/h^2 $^{69°}$◺.

12–98. 80.4 km/h $^{10.1°}$◺, 3263.1 km/h^2 $^{60.1°}$◺.

12–99. 18.5 m/s.

12–101. 2.4 m/s, 2.05 m/s^2.

13–1. 41.7 nN.

13–2. 75.9 N, 20.4 kg, 3.72 m/s^2.

13–3. 333.5 N, 401.54 N.

13–5. 4.38 m/s.

13–5a. 11.5 ft/s.

13–6. 3.43 m/s.

13–7. 240 N, 1.8 m.

13–9. 4.43 m/s.

13–10. 5.78 N · s/m, 32.1 N.

13–10a. 1.08 lb · s/ft, 16.14 lb.

13–11. 4.58 s, 1.25 m.

13–13. $F_D = 177.8$ kN, $F_E = 355.5$ kN, $F_t = 639.9$ kN.

13–14. 5.89 m/s^2, 11.8 N.

13–15. 13.74 kg.

13–15a. 28.2 lb.

13–17. $T_A = 54.9$ N, $T_B = 27.5$ N, $a_A = 1.96$ m/s^2, $a_B = 3.92$ m/s^2.

13–18. (a) $a_A = a_B = 1.43$ m/s^2, (b) $a_A = 5.10$ m/s^2, $a_B = 1.96$ m/s^2.

13–19. 0.815 s.

13–21. 210 N.

13–22. $F_f = -68.0$ N, $N_C = 153.1$ N.

13–23. $F_r = -39.2$ N, $F_\theta = 0$, $F_z = 392.4$ N.

13–25. 460.8 N.

13–25a. 262.0 lb.

13–26. $r = 0.198$ m.

13–27. $N_C = 1.11$ N, $F_C = 4.90$ N.

13–29. $F = m(r_1^2 v_1^2/(r_1 - v_2 t)^3)$
$\theta = (v_1/v_2)[(r_1/r) - 1]$.

13–30. 2.70 kN, 0.

13–30a. 26.7 lb, 0.

13–31. $1.992(10^{30})$ kg.

13–33. 24.4 m/s.

13–34. 12.2 m/s.

13–35. 663.6 N.

13–35a. 94.0 lb.

13–37. $T = 3.07$ N, $N_B = 0.806$ N.

13–39. 1.02 kN.

13–41. 3.84 m/s, 588.9 N.

13–42. 521.5 N, 8.32 m/s^2.

13–43. 1.15 m/s.

13–45. 65.4°.

13–45a. 76.5°.

13–46. 64.1°.

13–47. 3.84 m/s.

13–49. 0.352 m.

13–50. $\theta = 14.6°$, $\varphi = 14.0°$.

13–50a. $\theta = 15.4°$, $\varphi = 15.1°$.

13–51. 8.16 m/s.

13–53. $1.79(10^9)$ km.

13–54. 6.75 km/s.

13–55. 444.6 km/s,
775.0 Mm.

13–55a. $5.53(10^4)$ ft/s,
$4.84\,(10^8)$ mi.

13–57. 4.70 Mm, crash.

13–58. 6.21 km/s, 2.64 h, 479.0 m/s.

13–59. (a) 2.60 km/s,
(b) 6.38 km/s,
(c) $T_c = 2.19$ h
$T_e = 6.14$ h.

13–61. 10.1 km/s.

13–62. 30.8 km/s, $1/r = 5.01(10^{-13})\cos\theta + 61.08\,(10^{-13})$.

13–63. 1.93 km/s.

14–1. 4.38 m/s.

14–2. 3.43 m/s.

14–3. 4.52 m/s.

14–5. 0.933 m.

14–5a. 2.34 ft.

14–6. 13.5 kN.

14–7. 6.87 m/s.

14–9. 7.00 m/s.

14–10. 3.58 m/s.

14–10a. 4.34 ft/s.

14–11. 44.1 kN, 68 mm.

14–13. $x_B = 157.0$ mm.

14–14. 2.33 m.

14–15. 43.9 N.

14–15a. 23.0 lb.

14–17. 28.3 m/s.

14–18. 173.5 mm.

14–19. 0.153 m.

14–21. $U_{1-2} = GM_e m\left(\frac{1}{r_2} - \frac{1}{r_1}\right)$.

14–22. 83.1 km/h.

14–23. 851.7 kW.

14–25. 102.2 kW.

14–25a. 146.6 hp.

14–26. 0.46.

14–27. 16.0°.

14–29. 0.285 m/s.

14–30. 1.60 kW.

14–30a. 0.525 hp.

14–31. 2.28 kW.

14–33. 52.9 g.

14–34. 17.83 m/s.

14–35. 6.64 m/s.

14–35a. 25.4 ft/s.

14–37. 3.84 m/s, 5.89 N.

14–38. 285.4 mm.

14–39. 10.38 m/s.

14–41. 0.5 m.

14–42. 36.8 mm.

14–43. $v_B = 10.64$ m/s, $T_B = 141.5$ N.
$v_C = 9.47$ m/s, $T_C = 48.7$ N.

14–45. 6.82 m/s.

14–45a. 18.8 ft/s.

14–46. 3.34 m/s.

14–49. 7.66 m/s.

14–50. $x_A = 193.6$ mm, $x_B = 96.8$ mm.

14–50a. $x_A = 0.557$ ft, $x_B = 0.279$ ft.

15–1. 20.4 s.

15–2. 11.8 N, 11.8 m/s.

15–3. $F = 19.44$ kN, $T = 12.50$ kN.

15–5. 1615.4 m.

15–5a. 4040.5 ft.

15–6. 11.5 m/s.←

15–7. $T = 18.15$ kN.

15–9. 22.2 kN.

15–10. $(v_A)_2 = 6.54$ m/s, $(v_B)_2 = 13.08$ m/s, $T = 26.2$ N.

15–10a. $(v_A)_2 = 18.6$ ft/s, $(v_B)_2 = 37.2$ ft/s, $T = 6.92$ lb.

15–11. 15 kN · s both cases.

15–13. 164.9 N.

15–14. 24.8 m/s.

15–15. 0.575 s.

15–15a. 0.686 s.

15–17. 2.06 s.

15–18. 0.791 m/s.←

15–19. 0.5 m/s, 16.9 kJ.

15–21. 5.84 m/s, 350.0 N · s.

15–22. 733.4 m/s.

15–23. (a) 1.75 m/s, (b) 2.11 m/s.

15–25. 1.86 m/s, 780.5 N.

15–25a. 4.55 ft/s, 132.2 lb.

15–26. 7.06 m.

15–27. 0.242 m.

15–29. 1.19 m/s.

15–30. 8.62 m/s.

15–30a. 27.5 ft/s.

15–31. $(v_A)_2 = 0.4$ m/s, $(v_B)_2 = 1.6$ m/s.

15–33. 0.901.

15–34. 0.657 m.

15–35. 0.181 m.

15–35a. 3.23 ft.

15–37. 0.258 m.

15–38. 0.304 m, 0.240 m.

15–41. 33.8 km/h.

15–42. 1.50 m/s.

15–43. $(v_A)_2 = 6.90$ m/s, $(v_B)_2 = 75.6$ m/s.

15–45. 4.50 Mg · m^2/s.

15-45a. 1.31(10^6) slug · ft²/s.

15-46. $\mathbf{H}_O = \{12.5\mathbf{k}\}$ kg · m²/s.

15-47. $\mathbf{H}_P = \{-9.6\mathbf{i} + 2\mathbf{j} + 12.5\mathbf{k}\}$ kg · m²/s.

15-49. 9.5 m/s.

15-50. 4.38 m/s.

15-50a. 19.04 ft/s.

15-51. 12.42 km/s, 12.55 Mm.

15-53. 3.83 m/s, 208.9 mm.

15-54. $v_2 = v_1/(1 - 8\pi(a/l))$.

15-55. 3.6 m/s, 2.69 J.

15-55a. 13.12 ft/s, 1.94 ft · lb.

15-58. 45.0 N.

15-59. 15.8 N.

15-61. 97.2 N.

15-62. 858.1 N.

15-63. 388.4 N.

15-65. $P = 3.12$ N, $F = 6.24$ N.

15-65a. $P = 2.72$ lb, $F = 8.17$ lb.

15-66. 0.808 m/s².

15-67. $A_x = 1767.1$ N←, $A_y = 2564.8$ N↓, $M = 883.6$ N · m.↶

15-69. 115.1 mm/s².

15-70. 12.5 kN.

15-70a. 5823.0 lb.

15-71. 81.4 mm/s².

15-73. 15.75 kN.

15-74. 401.3 m/s.

15-75. 559.2 m/s.

15-75a. 522.8 ft/s.

15-77. $dm/dt = m_o\left(\dfrac{a_o + g}{v_{e/r}}\right)e^{[(a_o + g)/v_{e/r}]t}$.

15-78. $(7.85t + 0.32)$ N.

15-79. $(9.81t + 0.5)$ N.

16-1. $\theta_2 = 3.32$ rev., $t = 1.67$ s.

16-2. (a) 181.5 rev., (b) 16.5 rev.

16-3. 6.17 m/s.

16-5. 100 s.

16-5a. 37.5 s.

16-6. $r_A = r_B = 31.8$ mm, 1.91 canisters marked per minute.

16-7. $\omega = 28.6$ rad/s, $\theta = 24.1$ rad, $v_p = 7.16$ m/s, $a_p = 205.06$ m/s².

16-9. 60 mm/s.↑

16-10. 125 mm/s↑, 37.5 mm/s².↓

16-10a. 0.45 ft/s↑, 0.15 ft/s².↓

16-11. 183.8 mm/s↑, 5.62 mm/s².↓

16-13. 0.860 m/s.

16-14. $\alpha_B = \omega_A^2 t(r_A^2 + r_B^2)/2\pi r_B^3$.

16-15. 2.5 m/s, 13.1 m/s².

16-15a. 6.0 ft/s, 37.4 ft/s².

16-17. $v_{CD} = -l\,\omega \sin\theta$, $a_{CD} = -l\,\omega^2 \cos\theta - l\,\alpha \sin\theta$.

16-18. $v = l\,\omega \csc^2\theta$.

16-19. $v_p = -2v_A \cot\theta$.

16-21. $\omega = 19.25$ rad/s. $\alpha = 204.3$ rad/s².

16-22. $\omega = v/2r$.

16-23. 13.3 rad/s, 1 rev.

16-25. 27.1 m/s.

16-25a. 72.4 ft/s.

16-26. $v = -r_1\,\omega \sin\theta - r_1^2\,\omega \sin 2\theta / (2\sqrt{(r_1 + r_2)^2 - (r_1 \sin\theta)^2})$.

16-27. $\omega_A = 2(\sqrt{2}\cos\theta - 1)/(3 - 2\sqrt{2}\cos\theta)$

16-29. 1.025 m/s↓, 3.54 rad/s.↻

16-30. $v_B = 5.20$ m/s ↘30°, $v_C = 6$ m/s.→

16-30a. $v_B = 6.93$ ft/s ↘30°, $v_C = 8$ ft/s.→

16-31. 120 mm/s→, 1.20 rad/s.↶

16-33. 100 mm/s→, 1 rad/s.↻

16-34. 300 mm/s→, 3 rad/s.↻

16-35. 8 m/s. ←

16-35a. 9.6 ft/s. ←

16-37. 61.3 m/s. 48.5°↘

16–38. $\omega_A = 4$ rad/s ↻, $\omega_B = 1$ rad/s ↺, $v_D = 60$ mm/s.↑

16–39. 14 rad/s.↻

16–41. 1.02 m/s.→

16–42. $\omega_{BC} = 2.17$ rad/s↻, $\omega_{AB} = 2.25$ rad/s.↻

16–43. 12 rad/s.↻

16–45. 24 m/s.→

16–45a. 8 ft/s.→

16–46. 312.5 mm/s. 8.13°

16–47. 120 mm/s.→

16–49. $\omega_A = 4$ rad/s ↻, $\omega_B = 1$ rad/s ↺, $v_D = 60$ mm/s.↑

16–50. 8 m/s. ←

16–50a. 9.6 ft/s. ←

16–51. 14 rad/s.↻

16–53. 2.17 rad/s↻, 2.25 rad/s. ↻

16–54. 4.47 m/s. 63.4°

16–55. 1.1 m/s. 15°

16–55a. 3.75 ft/s. 15°

16–57. $v_A = 353.6$ mm/s 45°, $v_B = 750$ mm/s.→

16–58. $v_E = 4.76$ m/s. 40.9°

16–59. 48 rad/s.↺

16–61. 7.06 m/s.↓

16–62. 25.4 rad/s²↺, 5.21 m/s².↓

16–63. $a_A = 8$ m/s²↑, $a_B = 20.59$ m/s². 29.1°

16–65. $\alpha_{AB} = 119.6$ rad/s²↻, $\alpha_{BC} = 105.4$ rad/s².↺

16–65a. $\alpha_{AB} = 36.2$ rad/s²↻, $\alpha_{BC} = 27.8$ rad/s².↺

16–66. $a_B = 8.0$ m/s²↑, $a_C = 50.63$ m/s². 9.09°

16–67. 160 rad/s²↻, 1.0 m/s².→

16–69. 40 rad/s²↺, 2.0 m/s².↓

16–70. 46.9 mm/s²→, 0.938 rad/s².↺

16–70a. 0.680 ft/s²→, 2.27 rad/s².↺

16–71. $a_A = 155.8$ mm/s²←, $\alpha_{AB} = 0$.

16–73. $\omega_{DC} = 0.410$ rad/s↺, $\alpha_{DC} = 1.04$ rad/s².↺

16–74. $\omega_{AC} = 0$, $\omega_F = 40$ rad/s↻, $\alpha_{AC} = 404.0$ rad/s²↻, $\alpha_F = 538$ rad/s².↻

16–75. 1.25 rad/s↻, 17.1 rad/s².↺

16–75a. 2 rad/s↻, 43.7 rad/s².↺

16–77. $\mathbf{v}_B = \{0.6\mathbf{i} + 2.4\mathbf{j}\}$ m/s, $\mathbf{a}_B = \{-14.25\mathbf{i} + 8.4\mathbf{j}\}$ m/s².

16–78. $\mathbf{a}_B = \{-0.4\mathbf{i} - 12\mathbf{j}\}$ m/s².

16–79. $\mathbf{a}_B = \{0.85\mathbf{i} - 12.4\mathbf{j}\}$ m/s².

16–81. $\mathbf{v}_A = \{-2.63\mathbf{i} - 6\mathbf{j}\}$ m/s, $\mathbf{a}_A = \{12\mathbf{i} - 9.19\mathbf{j}\}$ m/s².

16–82. $\mathbf{v}_A = \{-2.63\mathbf{i} - 6\mathbf{j}\}$ m/s, $\mathbf{a}_A = \{12.375\mathbf{i} - 9.19\mathbf{j}\}$ m/s².

16–83. 0.667 rad/s↺, 3.08 rad/s².↻

16–85. 2.67 rad/s↺, 20 rad/s².↺

16–85a. 2 rad/s↺, 13.5 rad/s².↺

16–86. 0.375 rad/s↻, 2.43 rad/s².↺

16–87. $\omega_A = 0$, $\alpha_A = 14.43$ rad/s².↺

17–1. $(1/3)\, ml^2$.

17–2. $(1/12)\, m(a^2 + h^2)$.

17–3. $(2/5)\, mr^2$.

17–5. 57.7 mm.

17–5a. 1.15 ft.

17–6. $2\, mb^2/5$.

17–7. $(2/5)\, mr^2$.

17–9. $m[R^2 + (3/4)a^2]$.

17–10. 1.86 kg · m².

17–10a. 1.69 slug · ft².

17–11. 2.25 kg · m².

17–13. 7.963 kg · m².

17–14. 0.0826 kg · m², 0.393 m.

17–15. 1.67 kg · m².

17–15a. 31.08 slug · ft².

17–17. 4861.6 N.

17–18. (a) 14.2 s, (b) 21.9 s.

17–19. $a_G = 16.35$ m/s²←, $v = 110.74$ m/s.→

17–21. 11.38 m/s, 33.4°.

17–22. $F_B = 780$ N, $N_A = 516.4$ N, $N_B = 1396.5$ N.

17–23. 10.4°.

17–25. 1075.5 N.

17-25a. 263.1 lb.

17-26. 903.4 N.

17-27. $s = 1.0$ km, $N_A = 3948.6$ N, $N_B = 24.6$ N.

17-29. 68.0 m.

17-30. (a) $C_y = 7.22$ N, $B_y = 7.22$ N,
(b) $C_y = B_y = 14.72$ N.

17-30a. (a) $C_y = B_y = 0.706$ lb, (b) $C_y = B_y = 5$ lb.

17-31. $A_x = 10.75$ kN, $A_y = 19.35$ kN.

17-33. $A_x = 10.32$ kN, $A_y = 19.35$ kN.

17-34. $N_C = 735.75$ N, $F_C = 183.7$ N.

17-35. 4.76 rad/s^2.

17-35a. 10.12 rad/s^2.

17-37. 0.902 m.

17-38. 7.84 m/s.

17-39. $T_A = 85.7$ N.

17-41. $O_x = 0$, $O_y = 27.81$ N.

17-43. 0.482 m, $O_x = 0$.

17-45. 14.2 rad/s^2.

17-45a. 22.85 rad/s^2.

17-46. $\alpha = 3.87$ rad/s^2↻, $A_x = 23.2$ N, $A_y = 2058.2$ N.

17-47. 1.93 rad/s, 45°.

17-49. 10.4 s.

17-50. $O_x = 160.0$ N, $O_y = 78.48$ N, $\alpha = 210.0$ rad/s^2.

17-50a. $O_x = 5$ lb, $O_y = 10$ lb, $\alpha = 38.6$ rad/s^2.↻

17-51. 10.4 s.

17-53. 0.714 s.

17-54. 0.664 s.

17-55. 14.72 rad/s^2, 0, 4.905 m/s^2.

17-55a. 32.2 rad/s^2, 0, 16.1 ft/s^2.

17-57. 0.060 kg · m^2.

17-58. 24.8 rad/s.↺

17-59. $T = 26.67\,(0.09 - x^2)$ N.

17-62. 195.2 mm/s.

17-63. 2.66 rad/s^2.

17-65. 7.17 rad/s^2.

17-65a. 4.32 rad/s^2.

17-66. 14.9 rad/s^2.

17-67. 34.0°.

17-69. 41.2 m/s^2.

17-70. 1.204 m.

17-70a. 3.57 ft.

17-71. $\alpha_A = \alpha_B = 43.6$ rad/s^2, $T = 15.7$ N.

17-73. $v_o = 1.47$ m/s, $\omega = 8.32$ rad/s, yes.

17-74. 575.1 N.

17-75. 0.665 rad/s^2.

17-75a. 0.346 rad/s^2.

17-76. 16.4 m/s.

17-77. 0, 0.147 rad/s^2, 0.0897 m/s^2.

18-1. 6.60 rad/s.

18-2. 17.16 rad/s.

18-5. 85.1 mm.

18-5a. 0.155 ft.

18-6. 15.1 rad/s.

18-7. 1.34 rev, $A_x = 111.2$ N, $A_y = 167.6$ N.

18-9. 5.06 rad/s.

18-10. 0.990 m/s.→

18-10a. 6.95 ft/s.→

18-11. 5.32 m/s.↑

18-13. 3.68 m/s.

18-14. 4.71 m/s.

18-15. 70.5°.

18-15a. 70.5°.

18-17. 6.60 rad/s.

18-18. 17.16 rad/s.

18-19. 4.71 m/s.

18-21. 3.70 rad/s.

18-22. 1.80 m.

18-23. 2.58 m/s.←

18–25. 14.9 rad/s.

18–25a. 10.01 rad/s.

18–26. 7.58 rad/s.

18–27. 9.87 rad/s.

18–29. 3.83 rad/s.

18–30. 3.48 m/s.

18–30a. 12.90 ft/s.

18–31. 3.72 m/s.

18–33. 52.0°.

19–1. 3.26 kg · m²/s, 4.8 kg · m²/s.

19–2. 0.580 rad/s.

19–3. 44.1 N, 16.2 m.

19–5. 9.42 m/s.

19–5a. 25.1 ft/s.

19–9. (a) 21.4 rad/s, (b) 20.8 rad/s, (c) 21.1 rad/s.

19–10. 18 m/s.

19–10a. 18.73 ft/s.

19–11. 15 rad/s.

19–13. 1.59 m/s.

19–14. 126.7 rad/s.

19–15. 1.784 N · mm.

19–15a. 0.0343 ft · lb.

19–17. 61.2 rad/s.

19–18. 2.0 rad/s.

19–19. 1.12 m/s.

19–21. 3.45 rad/s.

19–22. 0.169 rad/s.

19–23. 0.024 rad/s.

19–25. 3.75 rad/s.

19–25a. 0.5 rad/s.

19–26. 0.0246 rad/s.

19–27. 76.4 N · m.

19–29. 16.7 mm.

19–30. 75 mm.

19–30a. 0.375 ft.

19–31. $h = (2/3)\ l$.

19–34. 22.4°.

19–35. 90.4 mm/s.

19–35a. 0.452 ft/s.

19–37. 1.96 m/s.

20–1. $\boldsymbol{\omega} = \{35.35\mathbf{j} + 37.35\mathbf{k}\}$ rad/s, $\dot{\boldsymbol{\omega}} = \{-70.7\mathbf{i}\}$ rad/s².

20–2. (a) $\dot{\boldsymbol{\omega}} = \omega_t\, \omega_s \mathbf{j}$, (b) $\dot{\boldsymbol{\omega}} = -\omega_s \omega_t \mathbf{k}$.

20–3. $\boldsymbol{\omega} = \omega_x \mathbf{i} + \omega_y \mathbf{j} + \omega_s \mathbf{k}$, $\boldsymbol{\alpha} = \omega_y \omega_s \mathbf{i} - \omega_x \omega_s \mathbf{j}$.

20–5. $\mathbf{v}_A = \{0.9\mathbf{i} + 2.53\mathbf{j} - 1.90\mathbf{k}\}$ m/s, $\mathbf{a}_A = \{-0.316\mathbf{i} + 0.144\mathbf{j} - 0.054\mathbf{k}\}$ m/s².

20–5a. $\mathbf{v}_A = \{20\mathbf{i} + 76.4\mathbf{j} - 43.6\mathbf{k}\}$ ft/s, $\mathbf{a}_A = \{-70.9\mathbf{i} + 28\mathbf{j} - 8\mathbf{k}\}$ ft/s².

20–6. $\omega_s = 40$ rad/s, $\omega_p = 5$ rad/s.

20–7. $\mathbf{v}_A = \{0.08\mathbf{i} - 0.346\mathbf{j} + 0.2\mathbf{k}\}$ m/s, $\mathbf{a}_A = \{-0.138\mathbf{i} - 0.116\mathbf{j} - 0.173\mathbf{k}\}$ m/s².

20–9. $\boldsymbol{\omega} = \{3.26\mathbf{i} + 9.55\mathbf{j}\}$ rad/s; $\omega = 10.09$ rad/s; $\boldsymbol{\alpha} = \{-14.70\mathbf{k}\}$ rad/s².

20–10. $\boldsymbol{\omega}_P = \{-25\mathbf{j}\}$ rad/s; $\boldsymbol{\alpha} = \{-3750\mathbf{i}\}$ rad/s².

20–10a. $\boldsymbol{\omega}_P = \{-40\mathbf{j}\}$ rad/s; $\boldsymbol{\alpha} = \{-4800\mathbf{i}\}$ rad/s².

20–11. $\omega = 8.25$ rad/s, $\dot{\omega} = 25.3$ rad/s².

20–13. $\alpha = 77.0°$, $\beta = 97.4°$, $\gamma = 15°$, No, finite rotations are not vectors.

20–15. $\mathbf{v}_B = \{1.0\mathbf{k}\}$ m/s.

20–15a. $\mathbf{v}_B = \{2.67\mathbf{k}\}$ ft/s.

20–17. $\mathbf{v}_B = \{0.75\mathbf{i}\}$ m/s.

20–18. $\boldsymbol{\omega} = \{0.204\mathbf{i} - 0.612\mathbf{j} + 1.361\mathbf{k}\}$ rad/s, $\mathbf{v}_B = \{-0.333\mathbf{j}\}$ m/s.

20–19. $\boldsymbol{\omega} = \{0.769\mathbf{i} - 2.31\mathbf{j} + 0.513\mathbf{k}\}$ rad/s, $\mathbf{v}_B = \{-0.333\mathbf{j}\}$ m/s.

20–21. $\mathbf{v}_P = \{-0.707\mathbf{i} + 0.707\mathbf{j} + 0.424\mathbf{k}\}$ m/s, $\mathbf{a}_P = \{-3.535\mathbf{i} - 4.808\mathbf{j} + 4.242\mathbf{k}\}$ m/s².

20–22. $\mathbf{v}_A = \{-2.77\mathbf{i} + 0.80\mathbf{j} - 1.39\mathbf{k}\}$ m/s, $\mathbf{a}_A = \{-8.48\mathbf{i} - 13.26\mathbf{j} - 2.64\mathbf{k}\}$ m/s².

20–23. $\mathbf{v}_B = \{-121.92\mathbf{i} - 30\mathbf{j} + 51.96\mathbf{k}\}$ m/s, $\mathbf{a}_B = \{319.4\mathbf{i} - 887.4\mathbf{j} - 90\mathbf{k}\}$ m/s².

20-25. $\mathbf{v}_A = \{-3.90\mathbf{i} - 1.125\mathbf{j} + 1.95\mathbf{k}\}$ m/s, $\mathbf{a}_A = \{4.15\mathbf{i} - 17.63\mathbf{j} + 3.51\mathbf{k}\}$ m/s^2.

20-25a. $\mathbf{v}_A = \{-5.20\mathbf{i} - 1.125\mathbf{j} + 1.95\mathbf{k}\}$ ft/s, $\mathbf{a}_A = \{3.20\mathbf{i} - 13.36\mathbf{j} + 1.76\mathbf{k}\}$ ft/s^2.

20-26. $\mathbf{v}_B = \{10.67\mathbf{i} - 3\mathbf{j} + 5.20\mathbf{k}\}$ m/s, $\mathbf{a}_B = \{31.28\mathbf{i} + 9.72\mathbf{j} - 46.80\mathbf{k}\}$ m/s^2.

20-27. $\mathbf{v}_G = \{-12\mathbf{i} + 16.5\mathbf{j} + 2\mathbf{k}\}$ m/s, $\mathbf{a}_G = \{-99\mathbf{i} - 53\mathbf{j} + 75\mathbf{k}\}$ m/s^2.

20-29. $\mathbf{v}_A = \{20.78\mathbf{i} + 30\mathbf{j} - 12\mathbf{k}\}$ m/s, $\mathbf{a}_A = \{8.54\mathbf{i} - 111.17\mathbf{j} - 26.23\mathbf{k}\}$ m/s^2.

20-30. $\mathbf{v}_P = \{-10.5\mathbf{i} + 150\mathbf{j} + 10\mathbf{k}\}$ m/s, $\mathbf{a}_P = \{-450\mathbf{i} - 15.75\mathbf{j} - 29\,998\mathbf{k}\}$ m/s^2.

20-30a. $\mathbf{v}_P = \{-16\mathbf{i} + 375\mathbf{j} + 4\mathbf{k}\}$ ft/s, $\mathbf{a}_P = \{-600\mathbf{i} - 12.8\mathbf{j} - 56\,247\mathbf{k}\}$ ft/s^2.

20-31. $\mathbf{v}_G = \{-3.47\mathbf{i} - 3.2\mathbf{j} + 8.54\mathbf{k}\}$ m/s.

20-33. $\mathbf{v}_C = \{1.5\mathbf{i} - 1.4\mathbf{j} - 1.2\mathbf{k}\}$ m/s, $\mathbf{a}_C = \{29.76\mathbf{i} + 10.9\mathbf{j} - 18.05\mathbf{k}\}$ m/s^2.

20-34. $\mathbf{v}_C = \{43.72\mathbf{i} - 46.12\mathbf{j} - 16.49\mathbf{k}\}$ m/s, $\mathbf{a}_C = \{63.42\mathbf{i} - 96.76\mathbf{j} + 355.90\mathbf{k}\}$ m/s^2.

21-1. $\frac{m}{6}ah$.

21-2. $\frac{m}{12}a^2$.

21-3. $\frac{mah}{5}$.

21-5. 0.0193 kg · m^2.

21-5a. 0.349 slug · ft^2.

21-6. 0.0350 kg · m^2.

21-7. 0.0678 kg · m^2.

21-9. 3.54 (10^{-3}) kg · m^2.

21-10. 28.14 (10^{-3}) kg · m^2.

21-10a. 0.1035 slug · ft^2.

21-11. 0.644 kg · m^2.

21-14. $\mathbf{H} = \{-6.363\,(10^{-4})\mathbf{i} + 2.64\,(10^{-4})\mathbf{j} + 0.135\mathbf{k}\}$ kg · m^2/s.

21-15. 15.87 J.

21-15a. 1.05 ft · lb.

21-17. 0.271 J.

21-18. 0.105 kg · m^2/s.

21-19. 26.77 rad/s.

21-21. 38.2 mm.

21-22. 12.13 rad/s.

21-23. $\mathbf{u}_A = 0.196\mathbf{i} - 0.981\mathbf{k}$, $(\mathbf{H}_o)_2 = (\mathbf{H}_o)_1 = \{0.106\mathbf{i} - 0.106\mathbf{k}\}$ kg · m^2/s.

21-25. 4.17 m/s.

21-25a. 15.2 ft/s.

21-26. 30.02 rad/s.

21-27. 30.98 rad/s.

21-29. $\Sigma M_x = (d/dt)(I_x\omega_x - I_{xy}\omega_y - I_{xz}\omega_z) - \Omega_z(I_y\omega_y - I_{yz}\omega_z - I_{yx}\omega_x) + \Omega_y(I_z\omega_z - I_{zx}\omega_x - I_{zy}\omega_y)$, etc.

21-30. $\Sigma M_x = I_x\dot{\omega}_x - I_{xy}\dot{\omega}_y - I_{xz}\dot{\omega}_z - \Omega_z(I_y\omega_y - I_{yz}\omega_z - I_{yx}\omega_x) + \Omega_y(I_z\omega_z - I_{zx}\omega_x - I_{zy}\omega_y)$, etc.

21-33. $A_x = B_x = 0$, $A_z = 57.25$ N, $B_z = 45.8$ N.

21-34. $F_A = 55.07$ N, $F_B = 43.04$ N.

21-35. $B_z = 5.66$ N, $A_x = B_x = 2.83$ N, $A_y = B_y = 4.25$ N.

21-35a. $B_z = 4.25$ lb, $A_x = B_x = 2.13$ lb, $A_y = B_y = 3.19$ lb.

21-37. 9.36 N · m.

21-38. $B_x = 90$ N, $A_y = B_y = 0$, $A_z = B_z = 19.62$ N.

21-39. 990.5 N.

21-41. $B_x = 0$, $A_y = 4.84$ N, $B_y = 11.17$ N, $A_z = B_z = 39.24$ N.

21-42. $B_x = 7.2$ N, $A_y = 4.84$ N, $B_y = 11.17$ N, $A_z = 40.29$ N, $B_z = 38.19$ N.

21-43. $T = 141.4$ N, $F_A = 281.1$ N.

21-45. $T_B = 65.2$ N, $M_y = M_z = 0$, $A_x = 0$, $A_y = 108$ N, $A_z = 76.97$ N.

21-45a. $T_B = 47.1$ lb, $M_y = M_z = 0$, $A_x = 0$, $A_y = 93.2$ lb, $A_z = 57.1$ lb.

21-46. $\theta = 31.4°$, $M_y = M_z = 0$, $A_x = 0$, $A_y = 5.92$ N, $A_z = 111.3$ N.

21-47. $\theta_D = 139.1°$, $m_D = 0.661$ kg, $\theta_F = 40.9°$, $m_F = 1.323$ kg.

21-49. $A_y = 357.9$ N, $B_y = 328.8$ N.

21-50. 13.6 N · m; downward.

21-50a. 34.9 lb · ft; downward.

21–51. 27 N · m.

21–53. 150 N · m.

21–54. 3924.4 rad/s.

21–55. 34.29 rad/s (high precession), 1.30 rad/s (low precession).

21–55a. 22.84 rad/s (high precession), 3.04 rad/s (low precession).

21–57. 1.26 rad/s.

21–58. 22.9 rad/s.

21–59. 6.83 rad/s.

21–62. $4.51\ (10^{-3})\ \text{kg} \cdot \text{m}^2/\text{s}$.

21–63. 0.673°, regular precession.

21–65. $\dot{\psi} = 2.35$ rev/h.

21–65a. $\dot{\psi} = 2.35$ rev/h.

22–1. 4.98 Hz, 0.201 s.

22–2. $x = -0.05 \cos (20t)$.

22–3. $x = 0.107 \sin (7t) + 0.100 \cos (7t)$, $\phi = 43.1°$.

22–5. $x = -0.05 \cos (8.165t)$, 0.05 m, 1.30 Hz.

22–5a. $x = -0.0833 \cos (19.7t)$, 1 in., 3.13 Hz.

22–6. $x = -0.0693 \sin (5.77t) - 0.075 \cos (5.77t)$, 0.102 m.

22–7. $\theta = 0.0903 \sin (4.43t) + 0.300 \cos (4.43t)$.

22–9. (a) 0.919 Hz, (b) 1.30 Hz.

22–10. 0.675 s.

22–10a. 0.924 s.

22–11. $\tau = 2\pi \sqrt{\dfrac{2r}{g}}$.

22–13. 0.632 Hz.

22–14. 12.34 N/m, 7.44 kg.

22–15. 0.544 s.

22–15a. 0.391 s.

22–17. $d = 124.4$ mm, $k_G = 602.7$ mm.

22–18. 0.457 m.

22–19. 2.92 N · m/rad.

22–21. $f = \dfrac{1}{2\pi} \sqrt{\dfrac{d}{\mu g}}$.

22–22. 0.777 s.

22–23. 0.356 s.

22–25. 0.718 Hz.

22–25a. 0.267 Hz.

22–26. $\tau = 2\pi \sqrt{\dfrac{2r}{g}}$.

22–27. 0.632 Hz.

22–29. 0.486 Hz.

22–30. 0.544 s.

22–30a. 0.391 s.

22–31. $\ddot{x} + 500x = 0$.

22–33. 0.181 s.

22–34. $4.09\ \text{kg} \cdot \text{m}^2$.

22–35. 1.23 s.

22–35a. 1.52 s.

22–37. 2.12 Hz.

22–38. 1.84 Hz.

22–39. $\ddot{\theta} + 468.3\theta = 0$.

22–41. $y = A \sin pt + B \cos pt + \dfrac{F_o/m}{\left(\dfrac{k}{m} - \omega^2\right)} \cos \omega t$.

22–42. $y = (-3.66 \sin 12.25t + 50 \cos 12.25t + 11.2 \sin 4t)$ mm.

22–43. $y = (-55.9 \sin 8.94t + 100 \cos 8.94t + 62.5 \sin 8t)$ mm.

22–45. $x_p = (41.7 \sin 2t)$ mm.

22–45a. $x_p = (0.0697 \sin 4t)$ ft.

22–46. 1.58 rad/s.

22–47. 14.01 rad/s.

22–49. 22.0 mm.

22–50. 17.3 mm.

22–50a. 0.223 ft.

22–51. 21.9 mm.

22–53. 19.5 mm.

22–54. 0.687 s.

22–55. 13.9°.

22–55a. 10.2°.

22–57. $x = 0.172 \cos (6t - 59.6°)$ m.

22–58. 1 kN/m.

22–59. $x = -48.2e^{-5t} \sin (12.45t)$ mm.

22–61. $x = 31e^{-7.5t} \sin (9.68t)$ mm.

22–63. 15.5 N · s/m.

22–65. $m\ddot{x} + kx = F_o \sin \omega t,\ L\ddot{q} + \frac{1}{c}q = E_o \sin \omega t.$

22–66. $L\ddot{q} + R\dot{q} + \frac{1}{c}q = E_o \cos \omega t.$

22–67. $L\ddot{q} + R\dot{q} + \frac{1}{c}q = 0.$

22–69. $L\ddot{q} + R\dot{q} + \frac{2}{c}q = 0.$

A–1. 6.54 km, 5.20 MN, 62.1 μs.

A–2. 16 MN^2, 0.90 nm^2, 8 Ms^3.

A–3. Yes, m · kg (metre) (kilogram), 1 m kg = 1 g.

A–5. kg/s, GN/m, km/(kg · s).

A–6. 271.2 N · m, 4.81 Mg/m^3, 508.0 μm/s.

Index